AF818050

MATHEMATICS IN INDUSTRY 30

Series Editors
Hans Georg Bock, IWR, Heidelberg University, Heidelberg, Germany
Frank de Hoog, CSIRO, Canberra, Australia
Avner Friedman, Ohio State University, Columbus, OH, USA
Arvind Gupta, University of British Columbia, Vancouver, BC, Canada
André Nachbin, IMPA, Rio de Janeiro, RJ, Brazil
Tohru Ozawa, Waseda University, Tokyo, Japan
William R. Pulleyblank, United States Military Academy, West Point, NY, USA
Torgeir Rusten, Det Norske Veritas, Hoevik, Norway
Fadil Santosa, University of Minnesota, Minneapolis, MN, USA
Jin Keun Seo, Yonsei University, Seoul, Korea
Anna-Karin Tornberg, Royal Institute of Technology (KTH), Stockholm, Sweden

THE EUROPEAN CONSORTIUM
FOR MATHEMATICS IN INDUSTRY

SUBSERIES

Managing Editor
Michael Günther, University of Wuppertal, Wuppertal, Germany

Series Editors
Luis L. Bonilla, University Carlos III Madrid, Escuela, Leganes, Spain
Otmar Scherzer, University of Vienna, Vienna, Austria
Wil Schilders, Eindhoven University of Technology, Eindhoven, The Netherlands

Mathematics in Industry focuses on the research and educational aspects of mathematics used in industry and other business enterprises. Books for *Mathematics in Industry* are in the following categories: research monographs, problem-oriented multi-author collections, textbooks with a problem-oriented approach, conference proceedings. Relevance to the actual practical use of mathematics in industry is the distinguishing feature of the books in the *Mathematics in Industry* series.

More information about this series at http://www.springer.com/series/4650

István Faragó • Ferenc Izsák • Péter L. Simon
Editors

Progress in Industrial Mathematics at ECMI 2018

Springer

ECMI
EUROPEAN CONSORTIUM FOR MATHEMATICS IN INDUSTRY

Editors
István Faragó
Department of Applied Analysis and Computational Mathematics &
ELTE-MTA Numnet Research Group
Eötvös Loránd University
Budapest, Hungary

Department of Differential Equations
Mathematical Institute
Budapest University of Technology and Economics
Budapest, Hungary

Ferenc Izsák
Department of Applied Analysis and Computational Mathematics &
ELTE-MTA Numnet Research Group
Eötvös Loránd University
Budapest, Hungary

Péter L. Simon
Department of Applied Analysis and Computational Mathematics &
ELTE-MTA Numnet Research Group
Eötvös Loránd University
Budapest, Hungary

ISSN 1612-3956 ISSN 2198-3283 (electronic)
Mathematics in Industry
The European Consortium for Mathematics in Industry
ISBN 978-3-030-27549-5 ISBN 978-3-030-27550-1 (eBook)
https://doi.org/10.1007/978-3-030-27550-1

Mathematics Subject Classification (2010): 49-XX, 60-XX, 65-XX, 68-XX, 35-XX, 76-XX, 78-XX, 80-XX, 91-XX, 93-XX

© Springer Nature Switzerland AG 2019
This work is subject to copyright. All rights are reserved by the Publisher, whether the whole or part of the material is concerned, specifically the rights of translation, reprinting, reuse of illustrations, recitation, broadcasting, reproduction on microfilms or in any other physical way, and transmission or information storage and retrieval, electronic adaptation, computer software, or by similar or dissimilar methodology now known or hereafter developed.
The use of general descriptive names, registered names, trademarks, service marks, etc. in this publication does not imply, even in the absence of a specific statement, that such names are exempt from the relevant protective laws and regulations and therefore free for general use.
The publisher, the authors, and the editors are safe to assume that the advice and information in this book are believed to be true and accurate at the date of publication. Neither the publisher nor the authors or the editors give a warranty, expressed or implied, with respect to the material contained herein or for any errors or omissions that may have been made. The publisher remains neutral with regard to jurisdictional claims in published maps and institutional affiliations.

This Springer imprint is published by the registered company Springer Nature Switzerland AG.
The registered company address is: Gewerbestrasse 11, 6330 Cham, Switzerland

Preface

The 20th European Conference on Mathematics for Industry, ECMI 2018, was held in Budapest from 18th to 22nd June 2018. Hungary is well-known for its outstanding achievements in pure mathematics, but much less known for its contributions to applied mathematics, in spite of the works of outstanding scientists like Gyula Farkas, Theodore von Kármán, John von Neumann, or Rudolf E. Kalman. Therefore, it was the privilege of the Hungarian mathematics community to have the opportunity to reinforce the contacts with the major European network promoting industrial mathematics, by bringing together more than 350 researchers for intellectual interaction for 5 days.

BUDAPEST 2018

The European Consortium for Mathematics in Industry (ECMI) organized its first international conference in Oberwolfach, in 1983, followed by a series of conferences, a persistent objective of which has been to galvanize interaction between academy and industry, leading to innovations in both fields. The 20th Conference, ECMI 2018, inspired multidisciplinary research along these lines further leading to the formulation of real-life challenges, where mathematical technologies provided significant new insights. Following the traditions of ECMI, the conference focused on various fields of industrial and applied mathematics, such as Applied Physics, Biology and Medicine, Cybersecurity, Data Science, Economy, Finance and Insurance, Energy, Production Systems, Social Challenges, Vehicles and Transportation. These themes nicely fit to current distinguished national research programs in Hungary, in particular programs on Autonomous Vehicles, Digital Factories, Brain Research, or Precision Agriculture supported by the EU and the National Research, Development and Innovation Office. The conference was jointly organized by the János Bolyai Mathematical Society, the

Institute of Mathematics at Eötvös Loránd University, and the Institute for Computer Science and Control of the Hungarian Academy of Sciences (MTA SZTAKI). The newly appointed Minister of Innovation and Technology, László Palkovics, was kind enough to patronize our conference. The statistics of the conference were more than satisfactory. In addition to the nine plenary talks, given by world class researchers, we had 50 minisymposia, and 45 contributed talks and poster presentations, running in 7 parallel sessions. Altogether there were more than 350 participants, from around 40 countries. More than 50 participants were students.

The Scientific Committee was set up as follows:

- László Monostori, MTA SZTAKI, Budapest and Fraunhofer Project Center at SZTAKI, Chair
- Dietmar Hömberg, Weierstrass Institute and Technische Universität, Berlin, co-chair
- Adérito Araújo, President of ECMI, University of Coimbra
- Helen Byrne, University of Oxford
- Raimondas Čiegis, Vilnius Gediminas Technical University

Group photo of the participants of the 20th ECMI Conference, Budapest

- István Faragó, Eötvös Loránd University, Budapest
- Zoltán Horváth, Széchenyi István University, Győr
- Sergey Lupuleac, Saint Petersburg State Polytechnic University
- Alessandra Micheletti, Universitá di Milano
- Claudia Nunes, University of Lisbon
- Ronny Ramlau, Johannes Kepler University, Linz
- Angela Stevens, University of Münster

The plenary talks covered several major areas of applied and industrial mathematics, such as network theory, numerical methods of PDEs, mathematics of tomography, mechanical models, traffic management, control theory, cancer research, and environmental modelling. The plenary speakers were:

- Paola Goatin, INRIA Sophia Antipolis – Team ACUMES, France
- Stefan Kurz, TU Darmstadt and Robert Bosch GmbH, Germany
- Knut-Andreas Lie, SINTEF Digital, Mathematics and Cybernetics, Oslo, Norway
- László Lovász, Hungarian Academy of Sciences, Hungary
- Christophe Prud'homme, University of Strasbourg and Cemosis, France
- Samuli Siltanen, University of Helsinki, Finland
- Gábor Stépán, Budapest University of Technology and Economics, Hungary
- Andrew Stuart, CalTech, USA
- Anna Marciniak-Czochra, University of Heidelberg, Germany, delivering the Alan Tayler Memorial Lecture

The plenary talk given by László Lovász, President of the Hungarian Academy of Sciences, has been recorded, processed, and made available by the eLearning Department of MTA SZTAKI at the address: http://www.bolyai.hu/ECMI2018_video.html.

According to the tradition of ECMI conferences, the winner of the Anile prize, honoring Professor Angelo Marcello Anile (1948–2007) of the University of Catania, was announced at the opening ceremony of the conference. The prize is given to a young researcher for an excellent PhD thesis in industrial mathematics. The Anile prize, in 2018, was awarded to Peter Gangl, Johannes Kepler Universität Linz.

Anile prize is awarded to Peter Gangl

The Hansjörg Wacker Memorial Prize established in memory of ECMI founding member Hansjörg Wacker (1939–1991), who was Professor at the Johannes Kepler University, Linz, is awarded for the best mathematical dissertation at the Masters level on an industrial project. The Hansjörg Wacker Memorial Prize, in 2018, was awarded to Edvin Åblad, Chalmers University.

The conference venue was the Danubius Hotel Hélia. As part of the social program, an ECMI reception was held on Tuesday evening to create an opportunity for ECMI members to meet each other. The conference gala dinner was held in a Danube river cruise on Europa boat. During this event, Hilary Ockendon and István Faragó, who initiated the ECMI membership of Hungary, were elected to be Honorary Members of ECMI.

Hansjörg Wacker Memorial Prize is awarded to Edvin Åblad

The organizers express their deepest gratitude to everybody involved in the success of this meeting, the plenary speakers, the members of the Scientific Committee, the organizers of the minisymposia, the contributing authors, and all the participants of the conference.

It is our pleasure to acknowledge the financial support of Graphisoft, Secudit, Morgan Stanley, the Hungarian Academy of Sciences, and the EPIC Center of

MTA SZTAKI, providing the financial basis for the participation of many young researchers.

On behalf of the organizers

Péter L. Simon, István Faragó, László Gerencsér and Ferenc Izsák

Budapest, Hungary István Faragó
Budapest, Hungary Ferenc Izsák
Budapest, Hungary Péter L. Simon

Organization

Scientific Committee

- László Monostori, chair, MTA SZTAKI, Budapest and Fraunhofer Project Center at SZTAKI
- Dietmar Hömberg, co-chair, President of ECMI, Weierstrass Institute and Technische Universität Berlin
- Adérito Araújo, University of Coimbra
- Helen Byrne, University of Oxford
- Raimondas Čiegis, Vilnus Gediminas Technical University
- István Faragó, Eötvös Loránd University, Budapest
- Zoltán Horváth, Széchenyi István University, Győr
- Sergey Lupuleac, Saint Petersburg State Polytechnical University
- Alessandra Micheletti, Universitá di Milano
- Cláudia Nunes, University of Lisbon
- Ronny Ramlau, Johannes Kepler University, Linz
- Angela Stevens, University of Münster

Organizing Committee

- Péter L. Simon, chair, János Bolyai Mathematical Society and Eötvös Loránd University, Budapest
- Aurél Galántai, Óbudai University, Budapest
- László Gerencsér, MTA SZTAKI, Budapest
- András Hajdu, Univerity of Debrecen
- Tibor Illés, Budapest University of Technology and Economics
- Ferenc Izsák, secretary, Eötvös Loránd University, Budapest
- Dezső Miklós, Alfréd Rényi Institute of Mathematics, Budapest

- Gábor Molnár-Sáska, János Bolyai Mathematical Society and Morgan Stanley, Budapest
- Gergely Röst, University of Szeged
- András Zempléni, Eötvös Loránd University, Budapest

Contents

Part I
Modeling of Industrial Processes

Imbalance Determination for Wind Turbines

Jenny Niebsch and Ronny Ramlau

Abstract In the growing field of clean energy extraction from wind the topic of rotor imbalances of wind turbines is of vital importance for the operation, safety and lifetime consumption of the turbines. The vibrations induced by imbalances lead to damages of important components, high repair expenses, and reduced output. The state of the art procedure to identify rotor imbalance is an expensive on-site procedure. We replace that procedure by a method that only uses the vibrations of the turbine during operation for the imbalance determination. To this end, a mathematical model of the turbine in the shape of an operator or matrix A was constructed that maps the imbalance p to the resulting vibrations u. Thus the problem of reconstructing an unknown imbalance from measured vibration data forms an inverse ill-posed problem that requires regularization techniques for its stable solution. We developed such a method, first for the case that the vibration data are collected during an operation with constant rotational speed. Later the situation of operation with variable speed was investigated and more sophisticated algorithms were developed for that case.

1 Introduction

The rotor of a wind turbine consists of a hub and usually three rather large blades that are supposed to have the same mass distribution. Due to tolerances in the production process the mass distribution can vary and the rotor is imbalanced. Mass imbalances can also occur when water penetrates the blades or ice is accumulated. A mass imbalance acts like an eccentric additional mass. During the rotation this mass induces centrifugal forces that lead to the vibration of the turbine in radial direction

J. Niebsch (✉)
RICAM, Linz, Austria
e-mail: jenny.niebsch@oeaw.ac.at

R. Ramlau
Industrial Mathematics Institute, Johannes Kepler University, Linz, Austria
e-mail: ronny.ramlau@jku.at

© Springer Nature Switzerland AG 2019

I. Faragó et al. (eds.), *Progress in Industrial Mathematics at ECMI 2018*, Mathematics in Industry 30, https://doi.org/10.1007/978-3-030-27550-1_1

with the same frequency as the rotational frequency. Those unwanted vibrations put a load onto the tower and the drive train that increases the abrasion and decreases the lifespan of the turbine. By determining the imbalance, i.e., the additional mass and its position, a counterweight of the same mass can be placed opposite this position. The rotor is balanced or at least the vibrations can be reduced to a tolerable level. A second possible cause of vibrations are aerodynamic imbalances. They arise mainly from deviations in the pitch angles of the three blades or because of changes in the blades profile due to abrasion. The vibrations induced by aerodynamic imbalances have radial, axial (direction of the drive train) and torsion components. They are detected by a visual examination of the blades, and optical methods are employed to find pitch angle deviations [1, 4, 5].

The determination of both kinds of imbalances requires a team on-site. First aerodynamic imbalances are removed. Afterwards a data based model is used to determine the mass imbalance. To generate that model vibration data have to be collected during two runs of the turbine with a fixed frequency, with and without a test weight. This procedure is very time consuming and expensive.

The goal of several projects of the authors was to replace the expensive existing methods by a methods based on a mathematical model that allows for the computation of the imbalance(s) off-line using only vibration data collected by a Condition Monitoring System (CMS).

The problem was solved in a first step for mass imbalances only with vibrational data collected during the operation with a fixed frequency. In a second project, aerodynamic imbalances from pitch angle deviation were included and reconstructed simultaneously with the mass imbalances. A third project investigated the reconstruction of mass imbalances in case the vibration data were collected during operation with variable rotational speed.

After stating the mathematical model for this problem we will present the three approaches followed by the description of the results we achieved.

2 Mathematical Model

The relation of a dynamical load $p(x,t)$ induced by an imbalance (mass and/or aerodynamic) and the displacement $u(x,t)$ that is the consequence of that load can be described by a partial differential equation (PDE) that is derived by the physical laws and based on some simplifying assumptions. We have followed [3] to approximate the PDE or rather an equivalent energy formulation by a system of ordinary differential equations (ODE) using a Finite-Element approach, see Fig. 1. The resulting ODE system described by the operator L is given as

$$(L\mathbf{u})(t) = \mathbf{M}\mathbf{u}''(t) + \mathbf{S}\mathbf{u}(t) = \mathbf{p}(t), \tag{1}$$

where $\mathbf{M}$ and $\mathbf{S}$ denote the mass and stiffness matrix of the turbine, $\mathbf{u}$ and $\mathbf{p}$ are the vectors of displacement and load, resp., at each degree of freedom (dof) of the

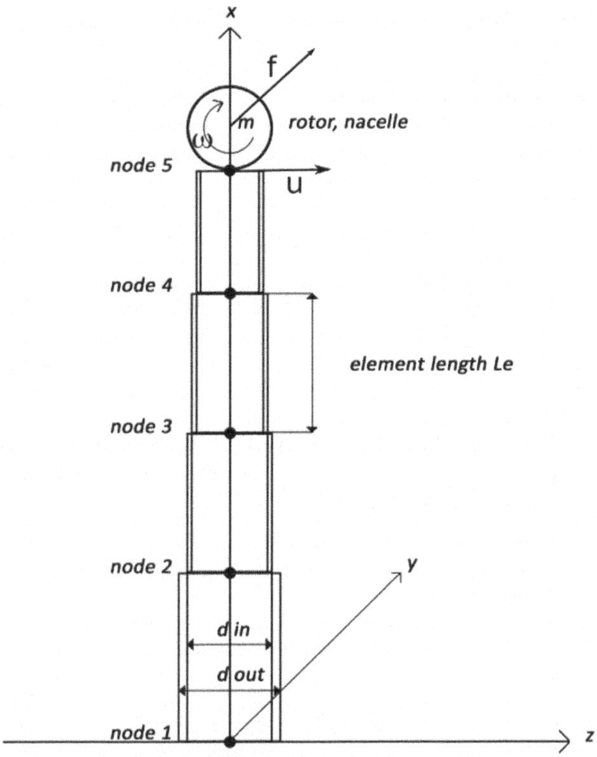

Fig. 1 Model of a wind turbine: the tower is divided into 4 element with 5 nodes, the nacelle including the rotor is treated as a point mass at node 5

model. Unfortunately, Eq. (1) can not be used directly to compute $\mathbf{p}$ from measured displacement data $\mathbf{u}$. Since the data are usually corrupted by noise and the operator L as a differential operator can not be stably evaluated, other methods to find $\mathbf{p}$ and with it the imbalance have to be employed.

In our approaches we transform Eq. (1) into an operator equation of the form

$$A\mathbf{p}(t) = \mathbf{u}(t) \quad \text{or} \quad A\mathbf{x} = \mathbf{u}, \tag{2}$$

where $\mathbf{x}$ represents the imbalance we want do reconstruct. The presentation of the operator A depends on the load we consider. It is a matrix for a load from a mass imbalance and constant rotational frequency, see Sect. 3. For a load from mass and aerodynamic imbalances, A becomes a nonlinear operator, see in Sect. 4. In case of a mass imbalance and variable rotational speed, A is described by tensor products of matrices and an integral operator, see in Sect. 5.

For given vibrational displacement data $\mathbf{u}$ we have to find the load $\mathbf{p}$ and the imbalance within. This is an inverse and usually ill-posed problem which can be stably solved using regularization techniques, e.g., Tikhonov regularization, truncated singular value decomposition (TSVD), or iterative methods, [2].

3 Mass Imbalances with Constant Frequency Data

A mass imbalance is an additional mass Δm that is eccentric from the rotor center by a radius r and angular position φ to a zero mark, usually one blade. The imbalance $(\Delta m \cdot r, \varphi)$ is given in the units kg and degree. It can be described in complex terms as

$$f = \Delta m r e^{i\varphi}. \tag{3}$$

The load that is induced when f rotates with a constant angular velocity ω_0 is given by

$$p(t) = f e^{i\omega_0 t} \tag{4}$$

for each dof of our model. Thus $\mathbf{f}$ and $\mathbf{p}$ are the vectors collecting imbalances, or loads, resp., at all dof. Since we have only one rotor that can be imbalanced, the vector $\mathbf{p}(t)$ has only one nonzero entry.

The resulting vibration is of the same frequency as the load, thus the ansatz $\mathbf{u(t)} = \mathbf{u_0} e^{i\omega_0 t}$ is inserted into Eq. (1). That leads to the algebraic system

$$\left(\mathbf{M} + \omega_0^{-2}\mathbf{S}\right)^{-1} \mathbf{f} = \mathbf{u_0}. \tag{5}$$

The dimension can be reduced by restricting the vectors $\mathbf{f}$ and $\mathbf{u_0}$ to its nonzero entries. The solution of the inverse problem in this case is easily computed, c.f. [9].

4 Mass and Aerodynamic Imbalances

Aerodynamic imbalances also induce displacements in radial direction z (see Fig. 1) but also in axial direction y and torsions around the x-axis. The forces F and moments M from pitch angle deviations are derived via a nonlinear blade element momentum method, [6], and added to the forces induced by mass imbalance. In the combined case the load vector is split into a sine and cosine part with the forces and moments from both imbalances as coefficients:

$$\begin{aligned} \mathbf{p}(t) &= \mathbf{p}_c \cos(\omega_0 t) + \mathbf{p}_s \sin(\omega_0 t) \qquad \text{with} \\ \mathbf{p}_c &:= \left(0, \cdots, 0, F_z^c, M_x^c, 0, M_z^c\right)^T \\ \mathbf{p}_s &:= \left(0, \cdots, 0, F_z^s, M_x^s, 0, M_z^s\right)^T. \end{aligned} \tag{6}$$

We can measure the amplitude c and phase shift γ of the vibration with frequency ω at each sensor position. We have three sensors for the measurement of the displacement in y-direction, the displacement in z-direction and torsion β_x around the x-axis. The data vector $\mathbf{y}$ arranged in sine and cosine parts has the form $\mathbf{y} = \left(c_y \cos(\gamma_y), c_z \cos(\gamma_z), c_{\beta_x} \cos(\gamma_{\beta_x}), c_y \sin(\gamma_y), c_z \sin(\gamma_z), c_{\beta_x} \sin(\gamma_{\beta_x})\right)^T$ for all three sensors. The operator A from Eq. (2) can be written as

$$\mathbf{y} = \mathbf{A}(\mathbf{x}) = \begin{pmatrix} \mathbf{B_r p}_{c,r}(\mathbf{x}) \\ \mathbf{B_r p}_{c,s}(\mathbf{x}) \end{pmatrix} \tag{7}$$

where

$$\mathbf{p}_{c,r}(\mathbf{x}) := \begin{pmatrix} F_z^c \\ M_x^c, \\ M_z^c \end{pmatrix}, \quad \mathbf{p}_{s,r}(\mathbf{x}) := \begin{pmatrix} F_z^s \\ M_x^s, \\ M_z^s \end{pmatrix},$$

and $\mathbf{B}_r$ is the matrix $(-\omega_0^2\mathbf{M}+\mathbf{S})^{-1}$ restricted to the positions of the nonzero entries of the load and the measurements. $\mathbf{x} = (\theta_1, \theta_2, \theta_3, \Re(f), \Im(f))$ is the vector that contains the 3 pitch angles and real and imaginary part of the mass imbalance. For the solution of the nonlinear inverse problem the Tikhonov functional is minimized by a steepest decent algorithm using the Frechet derivative of $\mathbf{A}$. For the details we refer to [7].

5 Mass Imbalances with Variable Frequency Data

In case the vibration data are measured during operation with variable rotational speed, i.e., $\omega(t)$, the load at the ith dof from a mass imbalance f_i, c.f. (3), has the form

$$p_i(t) = f_i[\omega^2(t) - i\omega'(t)]e^{i\phi(t)} \tag{8}$$

with $\phi'(t) = \omega(t)$. Hence the approach from Sect. 3 is not applicable. In [8] the representation of A in (2) in terms of tensor products was derived. With the definition of the Volterra integral operator

$$(Kp_i)(t) = \int_0^t (t-\theta)p_i(\theta)d\theta \tag{9}$$

and the tensor products

$$\mathscr{K} = \mathbf{M}^{-1} \otimes K \qquad \mathscr{A} = (\mathbf{M}^{-1}\mathbf{S}) \otimes (-K) \tag{10}$$

we have

$$((\mathbf{I} - \mathscr{A})^{-1}\mathscr{K}\mathbf{p})(t) = \mathbf{u}(t). \tag{11}$$

After adapting this operator further to the special situation of wind turbines a direct connection between the searched for imbalance **f** and the data measured at the sensor positions $\mathbf{u}_s$ can be made assuming that $\omega(t)$, which is known only from discrete measurements, can be approximated by a C^1-function.

The regularization is done using a TSVD of the operator, c.f. [8].

6 Results

The algorithm reconstructing mass imbalances from fixed frequency data was tested successfully in various applications and with data from real wind turbines, cf. Fig. 2. Here, a 350 kg imbalance at blade B ($\varphi = 330°$) was reconstructed from noisy vibration data. The reconstructed imbalance (red) has an absolute value of 331 kg, the angle is about 332°. The reconstruction error is in the range of the data noise level of about 5%. The yellow arrows indicate the balancing weights that need to be placed at blade A an C in order to compensate the imbalance of B. The algorithm was recently included in the CMS software from our project partner Bachmann Monitoring GmbH, Germany.

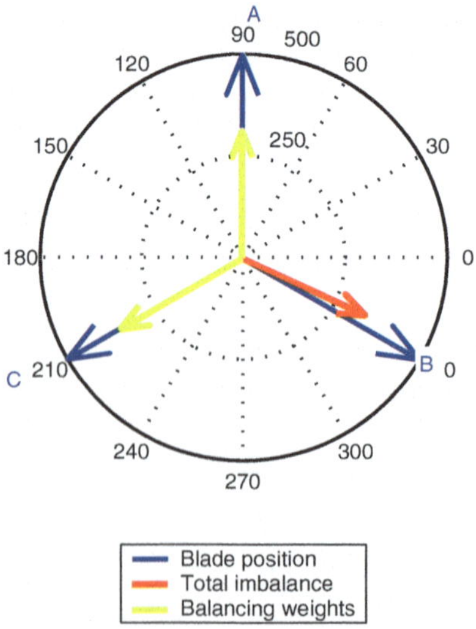

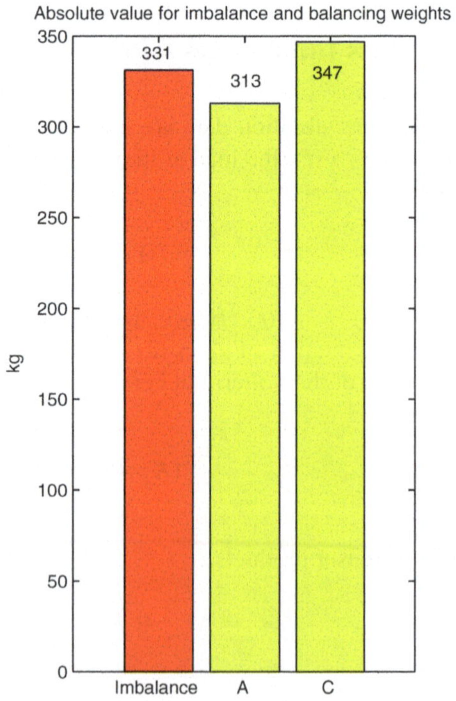

Fig. 2 Reconstruction of a 350 kg imbalance at blade B ($\varphi = 330°$), with 5% data error (red) and computed balancing weights at the other two blades (yellow)

The simultaneous reconstruction of mass imbalances and pitch angle deviation was successfully applied to artificial data. The test for real turbine data and therefore the practical application was prevented by the fact that the computation of the forces and moments from pitch angle deviation requires profile and airfoil data from the turbine under consideration. Those data turned out to be considered as highly confidential and were not made available for us.

The most recent algorithm for reconstructing imbalances from variable frequency data worked well for test cases with artificial data. Here, tests with real data and different measurement conditions are the task of future work.

Acknowledgement The presented work was funded by FFG 818098, FFG 841288.

References

1. Caselitz, P., Giebhardt, J.: Rotor condition monitoring for improved operational safety of offshore wind energy converters. ASME J. Solar Energy Eng. **127**, 253–261 (2005)
2. Engl, H.W., Hanke, M., Neubauer, A.: Regularization of Inverse Problems. Springer, Netherlands (2000)
3. Gasch, R., Knothe, K.: Strukturdynamik, vol. 2. Springer, Berlin (1989)
4. Hameed, Z., Hong, Y.S., Cho, Y.M., Ahn, S.H., Song, C-K.: Condition monitoring and fault detection of wind turbines and related algorithms: a review. Renew. Sust. Energy Rev. **13**, 1–39 (2009)
5. Hansen, M.: Aerodynamics of Wind Turbines. Earthscan, London (2008)
6. Ingram, G.: Wind Turbine Blade Analysis using the Blade Element Momentum Method. Note on the BEM Method. Durham University, Durham (2005)
7. Niebsch, J., Ramlau, R.: Simultaneous estimation of mass and aerodynamic rotor imbalances for wind turbines. J. Math. Ind. **4**, 12 (2014). https://doi.org/10.1186/2190-5983-4-12
8. Niebsch, J., Ramlau, R., Soodhalter, K.: Solution of coupled differential equations arising from imbalance problems. ETNA **46**, 89–106 (2017)
9. Ramlau, R., Niebsch, J.: Imbalance estimation without test masses for wind turbines. ASME J. Solar Energy Eng. **131**(1), 011010 (2009)

Sparse Representations for Uncertainty Quantification of a Coupled Field-Circuit Problem

Roland Pulch and Sebastian Schöps

Abstract We consider a model of an electric circuit, where differential algebraic equations for a circuit part are coupled to partial differential equations for an electromagnetic field part. An uncertainty quantification is performed by changing physical parameters into random variables. A random quantity of interest is expanded into the (generalised) polynomial chaos using orthogonal basis polynomials. We investigate the determination of sparse representations, where just a few basis polynomials are required for a sufficiently accurate approximation. Furthermore, we apply model order reduction with proper orthogonal decomposition to obtain a low-dimensional representation in an alternative basis.

1 Introduction

In science and engineering, complex applications require an advanced modelling by multiphysics systems or coupled systems. We examine a coupled field-circuit problem of an electric circuit, where differential algebraic equations (DAEs) for circuit components are combined with partial differential equations (PDEs) for electromagnetic components, see [8].

Uncertainty quantification (UQ) investigates the impact of variations in input parameters on a quantity of interest (QoI). Often parameters are remodelled into random variables. The random QoI can be expanded in the (generalised) polynomial chaos, where orthogonal basis polynomials are involved, see [12]. Sparse representations aim for a reduced set of basis functions with a given accuracy of approximation. Many methods for sparse representations have been derived and studied, see [2, 3, 5] and the references therein. Alternatively, methods of

R. Pulch (✉)
Universität Greifswald, Institute of Mathematics and Computer Science, Greifswald, Germany
e-mail: pulchr@uni-greifswald.de; roland.pulch@uni-greifswald.de

S. Schöps
Technische Universität Darmstadt, Centre for Computational Engineering, Darmstadt, Germany
e-mail: schoeps@temf.tu-darmstadt.de

© Springer Nature Switzerland AG 2019
I. Faragó et al. (eds.), *Progress in Industrial Mathematics at ECMI 2018*,
Mathematics in Industry 30, https://doi.org/10.1007/978-3-030-27550-1_2

model order reduction (MOR) yield low-dimensional (dense) approximations of the random QoI, see [6, 7].

We apply this UQ concept to the coupled field-circuit problem [11]. On the one hand, sparse representations are determined by neglecting basis functions with small coefficients. On the other hand, MOR using proper orthogonal decomposition (POD) identifies a low-dimensional approximation in an alternative basis. Our aim is to obtain approximations with as few basis functions as possible, while still maintaining some accuracy. Computational effort during the offline phase, i.e., evaluations of the multiphysics systems, is not saved by the proposed methods. However, the online evaluation cost of the polynomials can be reduced in the first approach.

2 Coupled Field-Circuit Problem

We investigate the rectifier circuit depicted in Fig. 1. The model consists of a circuit part and a field part. Modified nodal analysis (MNA) [4] produces a system of DAEs

$$\begin{aligned} \mathbf{A}_C \tfrac{\mathrm{d}}{\mathrm{d}t}\mathbf{q}_C(\mathbf{u},t) + \mathbf{A}_R\mathbf{r}(\mathbf{u},t) + \mathbf{A}_L\mathbf{j}_L + \mathbf{A}_V\mathbf{j}_V + \mathbf{A}_M\mathbf{j}_M + \mathbf{A}_D\mathbf{j}_D + \mathbf{A}_I\mathbf{i}(t) &= \mathbf{0}, \\ \tfrac{\mathrm{d}}{\mathrm{d}t}\boldsymbol{\phi}_L(\mathbf{j}_L,t) - \mathbf{A}_L^\top\mathbf{u} &= \mathbf{0}, \\ \mathbf{A}_V^\top\mathbf{u} - \mathbf{v}(t) &= \mathbf{0}, \end{aligned} \tag{1}$$

with incidence matrices $\mathbf{A}_\star$, node voltages $\mathbf{u}(t)$, branch currents $\mathbf{j}_L(t), \mathbf{j}_V(t)$, sources $\mathbf{i}(t)$, $\mathbf{v}(t)$ and constitutive relations $\mathbf{q}_C(\cdot,t)$, $\mathbf{r}_C(\cdot,t)$, $\boldsymbol{\phi}_L(\cdot,t)$. Initial values are considered in the time interval $t \in [t_0, t_{\mathrm{end}}]$. We apply Shockley's model

$$j_{D,k} = I_{S,k}\left(\exp\left(\mathbf{A}_{D,k}^\top\mathbf{u}/U_{\mathrm{TH},k}\right) - 1\right), \qquad k = 1,2,3,4 \tag{2}$$

for the four diodes with parameters $I_{S,k}$, $U_{\mathrm{TH},k}$, where $\mathbf{A}_{D,k}$ denotes the kth column of $\mathbf{A}_D$. We involve a refined model for the transformer (dashed box in Fig. 1) given

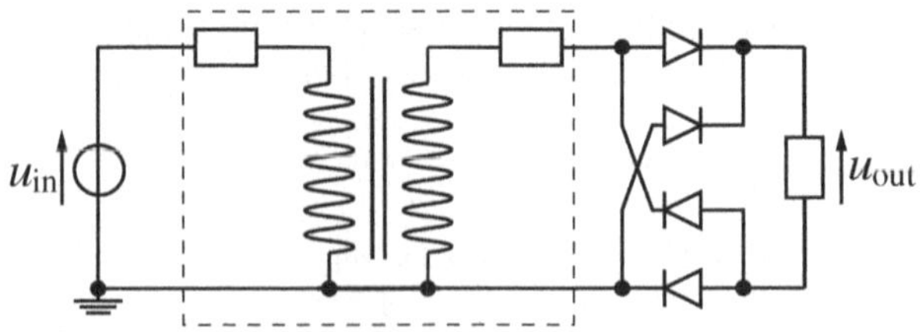

Fig. 1 Diagram of rectifier circuit. A PDE model is used for the components in dashed box

by the two-dimensional (2D) magnetostatic approximation of Maxwell's equations

$$\begin{aligned} \nabla \cdot (\boldsymbol{\nu}(\|\nabla A(t,\mathbf{x})\|,\mathbf{x})\ \nabla A(t,\mathbf{x})) &= \boldsymbol{\chi}(\mathbf{x})^\top \mathbf{j}_M(t) \qquad \text{for } \mathbf{x} \in \Lambda \\ \frac{\mathrm{d}}{\mathrm{d}t} \int_\Lambda \boldsymbol{\chi}(\mathbf{x})\ A(t,\mathbf{x})\ \mathrm{d}\mathbf{x} &= \mathbf{A}_M^\top \mathbf{u}(t) \end{aligned} \tag{3}$$

on the spatial domain $\Lambda \subset \mathbb{R}^2$. The magnetic vector potential $A : [t_0, t_{\text{end}}] \times \Lambda \to \mathbb{R}$ is unknown. The lumped currents and voltages are distributed and integrated by the winding function $\boldsymbol{\chi} : \Lambda \to \mathbb{R}^2$, see [9]. The reluctivity is $\boldsymbol{\nu} : \mathbb{R} \times \Lambda \to \mathbb{R}^{2\times 2}$. In the iron core region, it reads as $\boldsymbol{\nu}(B, \mathbf{x}) = \nu(B)\mathbf{I}_2$ (identity matrix $\mathbf{I}_2$) using Brauer's model

$$\nu(B) = \kappa_1 \exp\left(\kappa_2 B^2\right) + \kappa_3 \tag{4}$$

with the magnetic field $B = \|\nabla A\|$ and the parameters $\kappa_1, \kappa_2, \kappa_3$. A finite element method yields a nonlinear system of algebraic equations. More details on this coupled problem can be found in [8]. Now we define the output voltage as QoI.

3 Stochastic Model

We consider uncertainties in $q = 11$ parameters: the parameters of Shockley's model (2) for each diode separately (8 parameters) and the three parameters of Brauer's model (4). We describe the uncertainties by independent uniform probability distributions with 20% variation around each mean value. The random variables are $\mathbf{p} : \Omega \to \Pi$ with event space Ω and parameter domain $\Pi \subset \mathbb{R}^q$. The joint probability density function is constant on the cuboid Π. Let $y : [t_0, t_{\text{end}}] \times \Pi \to \mathbb{R}$ be the random output voltage (QoI) of the coupled problem.

The expected value of a function $f : \Pi \to \mathbb{R}$ reads as

$$\mathbb{E}[f] = \frac{1}{\text{volume}(\Pi)} \int_\Pi f(\mathbf{p})\ \mathrm{d}\mathbf{p}. \tag{5}$$

The expected value (5) implies an inner product $< f, g >= \mathbb{E}[fg]$ for two square-integrable functions. The accompanying norm is $\|f\|_{L^2} = \sqrt{< f, f >}$. We define the basis polynomials $(\Phi_i)_{i\in\mathbb{N}}$ with $\Phi_i : \Pi \to \mathbb{R}$ by $\Phi_i(\mathbf{p}) = \phi_{i_1}(p_1)\phi_{i_2}(p_2)\cdots\phi_{i_q}(p_q)$, where ϕ_ℓ denotes the (normalised) Legendre polynomial of degree ℓ. There is a one-to-one mapping between the indices i and the multi-indices $i_1, \ldots, i_q$. It follows that $(\Phi_i)_{i\in\mathbb{N}}$ represents a complete orthonormal system satisfying $< \Phi_i, \Phi_j >= \delta_{ij}$.

We assume that the random process $y(t,\cdot)$ is square-integrable for each t. Consequently, the (generalised) polynomial chaos expansion

$$y(t,\mathbf{p}) = \sum_{i=1}^{\infty} w_i(t)\Phi_i(\mathbf{p}) \tag{6}$$

exists pointwise for each t. The coefficient functions are given by the inner products $w_i(t) =< y(t,\cdot), \Phi_i(\cdot) >$. The infinite series (6) is truncated to a finite sum

$$\tilde{y}^{(I)}(t,\mathbf{p}) = \sum_{i\in I} \tilde{w}_i(t)\Phi_i(\mathbf{p}) \tag{7}$$

with a finite index set $I \subset \mathbb{N}$ and approximations $\tilde{w}_i$ of the coefficients. Typically, all polynomials up to a total degree d are included in an index set I^d. The number of basis polynomials becomes $\left|I^d\right| = \frac{(d+q)!}{d!q!}$.

Stochastic collocation techniques yield approximations of the unknown coefficient functions, see [12]. A quadrature rule is determined by nodes $\{\mathbf{p}^{(1)},\ldots,\mathbf{p}^{(s)}\} \subset \Pi$ and weights $\{\gamma_1,\ldots,\gamma_s\} \subset \mathbb{R}$. The approximations become

$$\tilde{w}_i(t) = \sum_{j=1}^{s} \gamma_j \, y(t,\mathbf{p}^{(j)}) \, \Phi_i(\mathbf{p}^{(j)}) \tag{8}$$

for $i = 1,\ldots,m$ and w.l.o.g. $I^d = \{1,\ldots,m\}$. Thus the coupled problem (1),(3) has to be solved s-times for different realisations of the parameters.

4 Sparse Approximation

The aim is to find an index set $J \subset I^d$ with $|J| \ll |I^d|$ for fixed total degree d, while the error is still below some threshold. The total error $y - \tilde{y}^{(J)}$ consists of three parts: (1) the truncation error ($\mathbb{N} \to I^d$), (2) the error of the numerical method ($w_i \to \tilde{w}_i$), and (3) the additional sparsification error ($I^d \to J$). We assume that the errors (1) and (2) are sufficiently small and focus on the error (3).

The relative L^2-error of the sparsification reads as

$$E(t;J) = \frac{\left\|\tilde{y}^{(I^d)}(t,\cdot) - \tilde{y}^{(J)}(t,\cdot)\right\|_{L^2}}{\left\|\tilde{y}^{(I^d)}(t,\cdot)\right\|_{L^2}} = \left(\frac{\sum_{i\in I^d\setminus J} \tilde{w}_i(t)^2}{\sum_{i\in I^d} \tilde{w}_i(t)^2}\right)^{\frac{1}{2}} \tag{9}$$

for each t including the coefficients (8). Given an error tolerance $\varepsilon > 0$, we obtain an optimal index set

$$J_t = \operatorname{argmin}\left\{ \left|J'\right| \;:\; J' \subseteq I^d \;\text{ and }\; E(t, J') < \varepsilon \right\} \tag{10}$$

with respect to the error (9) for each time point. A global index set is given by

$$\hat{J} = \bigcup_{t \in [t_0, t_{\text{end}}]} J_t, \tag{11}$$

which is sufficiently accurate with respect to the tolerance ε for all times. More details can be found in [6].

5 Model Order Reduction

Alternatively, we use an MOR with proper orthogonal decomposition (POD), see [1], to determine a low-dimensional approximation of the polynomial surrogate. Let $I^d = \{1, \ldots, m\}$ and $\tilde{\mathbf{w}} = (\tilde{w}_1, \ldots, \tilde{w}_m)^\top$. A transient simulation of the coupled problem yields the coefficients (8) by the stochastic collocation technique. We collect snapshots $\tilde{\mathbf{w}}(t_0), \tilde{\mathbf{w}}(t_1), \ldots, \tilde{\mathbf{w}}(t_{k-1})$ for discrete time points in a matrix $\mathbf{W} \in \mathbb{R}^{m \times k}$. A singular value decomposition yields

$$\mathbf{W} = \mathbf{U}\mathbf{S}\mathbf{V}^\top \qquad \text{with} \qquad \mathbf{S} = \operatorname{diag}(\sigma_1, \sigma_2, \ldots, \sigma_{\min\{m,k\}})$$

including the singular values $\sigma_1 \geq \sigma_2 \geq \cdots$ and orthogonal matrices $\mathbf{U} \in \mathbb{R}^{m \times m}$, $\mathbf{V} \in \mathbb{R}^{k \times k}$. Let $\mathbf{u}_1, \ldots, \mathbf{u}_m$ be the columns of the matrix $\mathbf{U}$. We arrange the orthogonal projection matrix $\mathbf{P}_r = (\mathbf{u}_1 \cdots \mathbf{u}_r) \in \mathbb{R}^{m \times r}$ ($\mathbf{P}_r^\top \mathbf{P}_r = \mathbf{I}_r$) for each dimension $r \leq \min\{m, k\}$. Given $\tilde{\mathbf{w}}(t) \in \mathbb{R}^m$, the best approximation with respect to the reduced basis reads as $\bar{\mathbf{w}}_r(t) = \mathbf{P}_r^\top \tilde{\mathbf{w}}(t)$ for any t. Vice versa, we obtain the approximation $\tilde{\mathbf{w}}(t) \approx \mathbf{P}_r \bar{\mathbf{w}}_r(t)$ for given $\bar{\mathbf{w}}_r(t)$ and any t. We define the (dense) low-dimensional approximation, cf. (7),

$$\tilde{y}^{(I^d)}(t, \mathbf{p}) \approx \sum_{i=1}^{m} \left[\sum_{j=1}^{r} u_{ij} \bar{w}_j(t) \right] \Phi_i(\mathbf{p}) = \sum_{j=1}^{r} \bar{w}_j(t) \underbrace{\left[\sum_{i=1}^{m} u_{ij} \Phi_i(\mathbf{p}) \right]}_{=:\Psi_j(\mathbf{p})} \tag{12}$$

with the new orthonormal basis polynomials $\{\Psi_1, \ldots, \Psi_r\}$ and associated coefficients $\bar{w}_1, \ldots, \bar{w}_r$. An error estimate for an approximation of this kind is given in [7].

6 Numerical Results

In the coupled problem (1), (3), we supply a harmonic oscillation with period $T = 0.02$ as input voltage. We use the Stroud-5 quadrature rule with $s = 243$ nodes, which is exact for all polynomials up to total degree 5, see [10]. In each node, we perform a monolithic time integration in $[0, 2T]$ by the implicit Euler method. The step size $\Delta t = 10^{-4}$ is used in time, whereas a smaller step size does hardly change the numerical results. This time integration yields $k = 401$ snapshots in equidistant time points. We choose the polynomial degree $d = 3$, i.e., $|I^3| = m = 364$ due to $q = 11$ random parameters.

Figure 2 (left) shows the maximum of the coefficients (8) in time. All coefficients of degree three are (at least) three orders of magnitudes smaller than the coefficient of degree zero. This property suggests that the relative truncation error is below 0.1%. Furthermore, Fig. 2 (right) demonstrates a fast decay, which indicates some potential for a sparse approximation as described in Sect. 4. For given error tolerances $\varepsilon \in [10^{-4}, 10^{-1}]$, we determine the cardinalities $\max_{t \in [0,2T]} |J_t|$ (pointwise) and $|\hat{J}|$ (union) with the index sets (10) and (11), respectively. Figure 3 (left) illustrates the cardinalities in dependence on the error tolerances. The potential for a global sparse approximation using (11) is bad, because more than 80 basis polynomials are required.

Alternatively, we apply the POD technique from Sect. 5. We choose the reduced dimensions $r = 1, \ldots, 20$ and obtain the approximations (12). Figure 3 (right) depicts the maximum in time of the relative L^2-errors for these approximations. Now we achieve an efficient low-dimensional representation, where less than 20 basis polynomials ($r \ll m$) yield a small error.

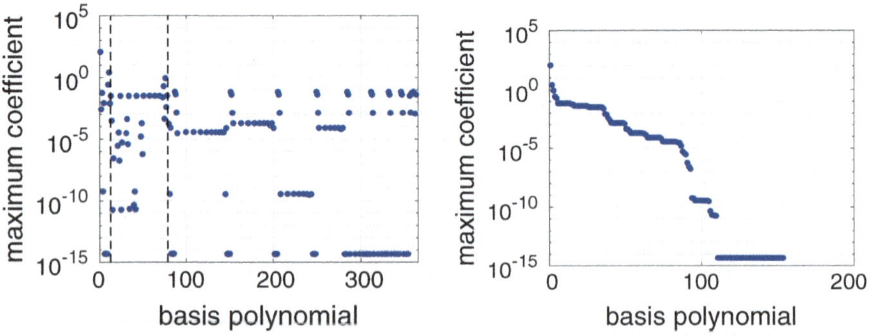

Fig. 2 Maximum of $\{\tilde{w}_i(t) \;:\; t \in [0, 2T]\}$ for $i = 1, \ldots, 364$ with coefficients (8), left: dashed lines separate the coefficients of degree zero/one, two and three, right: coefficients in descending order

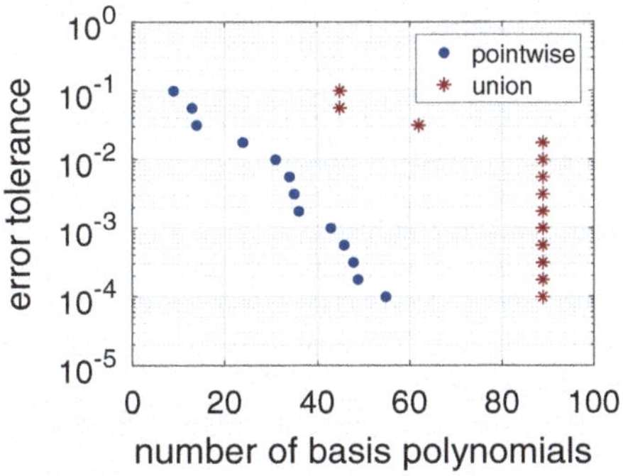

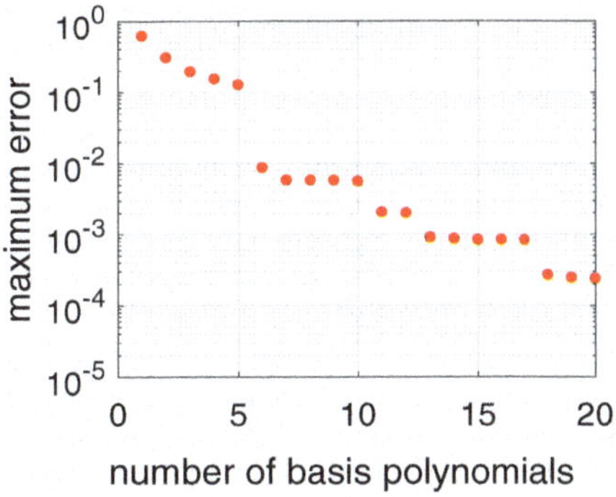

Fig. 3 Relation between number of basis polynomials and error tolerance or maximum relative L^2-errors, left: sparsification by neglecting basis polynomials, right: basis selection using POD

7 Conclusions

We performed an UQ of a coupled DAE-PDE system modelling a field-circuit problem. Sparse approximations of the random QoI were identified using its orthogonal polynomial expansion. In this test example, an appropriate sparse representation could not be achieved on the global time interval by simply neglecting basis polynomials. Alternatively, we obtained an efficient low-dimensional approximation by changing to another orthogonal polynomial basis, which was identified via an MOR.

Acknowledgements The research of the second author is supported by the Excellence Initiative of the German Federal and State Governments and by the Graduate School of Computational Engineering at Technische Universität Darmstadt.

References

1. Antoulas, A.: Approximation of Large-Scale Dynamical Systems. SIAM Publications, Philadelphia (2005)
2. Blatman, G., Sudret, B.: Adaptive sparse polynomial chaos expansion based on least angle regression. J. Comput. Phys. **230**(6), 2345–2367 (2011)
3. Doostan, A., Owhadi, H.: A non-adapted sparse approximation of PDEs with stochastic inputs. J. Comput. Phys. **230**(8), 3015–3034 (2011)
4. Ho, C.W., Ruehli, A., Brennan, P.: The modified nodal approach to network analysis. IEEE Trans. Circ. Syst. **22**(6), 504–509 (1975)
5. Jakeman, J.D., Narayan, A., Zhou, T.: A generalized sampling and preconditioning scheme for sparse approximation of polynomial chaos expansions. SIAM J. Sci. Comput. **39**(3), A1114–A1144 (2017)
6. Pulch, R.: Model order reduction and low-dimensional representations for random linear dynamical systems. Math. Comput. Simulat. **144**, 1–20 (2018)

7. Pulch, R.: Model order reduction for random nonlinear dynamical systems and low-dimensional representations for their quantities of interest. Math. Comput. Simulat. **166**, 76–92 (2019)
8. Schöps, S.: Multiscale Modeling and Multirate Time-Integration of Field/Circuit Coupled Problems. VDI Verlag. Fortschritt-Berichte VDI, Reihe 21, Nr. 398 (2011)
9. Schöps, S., De Gersem, H., Weiland, T.: Winding functions in transient magnetoquasistatic field-circuit coupled simulations. COMPEL **29**(2), 2063–2083 (2013)
10. Stroud, A.J.: Approximate Calculation of Multiple Integrals. Prentice-Hall, Englewood Cliffs (1971)
11. Tsukerman, I.A., Konrad, A., Meunier, G., Sabonnadiere, J.C.: Coupled field-circuit problems: trends and accomplishments. IEEE Trans. Magn. **29**(2), 1701–1704 (1993)
12. Xiu, D.: Numerical Methods for Stochastic Computations: a Spectral Method Approach. Princeton University Press, Princeton (2010)

On a Dry Spinning Model Using Two-Phase Flow

Manuel Wieland, Walter Arne, Robert Feßler, Nicole Marheineke, and Raimund Wegener

Abstract On the basis of a mixture model ansatz we propose a three-dimensional two-phase flow fiber model for dry spinning processes, which are characterized by solvent evaporation and fiber-air interaction. Employing dimensional reduction this model is embedded into an efficient numerical framework, such that simulations of industrial spinning setups become feasible.

1 Introduction

In dry spinning processes multiple hot jets consisting of polymer and solvent are extruded from nozzles into a spinning chamber, where they form out and are drawn down by a take up roller. In the spinning chamber solvent evaporates out of the jets due to a hot airstream, which leads to solidification of the spun fibers. The presence of fiber-air interactions in industrial spinning setups creates the need of a fully two-way coupled simulation of the dry spun fibers with the surrounding airflow. Since the three-dimensional multiphase/-scale problem is in general not computable by direct numerical simulations due to its complexity we derived a dimensionally reduced fiber model in [7] using assumptions on slenderness, radial symmetry and linearization which allowed the embedding into an efficient numerical solution framework. In this paper we focus on a detailed deduction of the underlying three-dimensional fiber model that is based on a mixture model ansatz [2] and we highlight the performance of the numerical framework for an industry-related spinning setup.

M. Wieland (✉) · N. Marheineke
Universität Trier, Lehrstuhl Modellierung und Numerik, Trier, Germany
e-mail: manuel.wieland@itwm.fraunhofer.de; wieland@uni-trier.de; marheineke@uni-trier.de

W. Arne · R. Feßler · R. Wegener
Fraunhofer ITWM, Kaiserslautern, Germany
e-mail: walter.arne@itwm.fraunhofer.de; robert.fessler@itwm.fraunhofer.de; raimund.wegener@itwm.fraunhofer.de

© Springer Nature Switzerland AG 2019

I. Faragó et al. (eds.), *Progress in Industrial Mathematics at ECMI 2018*,
Mathematics in Industry 30, https://doi.org/10.1007/978-3-030-27550-1_3

2 Three-Dimensional Dry Spinning Model

We develop a stationary three-dimensional dry spinning model for a viscous uni-axial fiber. Let $\Omega \subset \mathbb{R}^3$ be the a priori unknown fiber domain whose boundary $\partial\Omega = \Gamma_{\text{in}} \cup \Gamma_{\text{fr}} \cup \Gamma_{\text{out}}$ consists of the fixed inlet at the nozzle Γ_{in}, the free lateral fiber surface Γ_{fr} and the outlet Γ_{out}. In Ω we consider balance laws for mass, momentum and energy for the two phases i, $i \in \{p, d\}$—polymer and diluent. Thereafter we employ the mixture model ansatz [2] to reduce the balance laws for momentum and energy for the single phases to balance laws for the mixture. The quantities for the single phases are indicated by the respective index i, $i \in \{p, d\}$.

Phase Balances Let ρ_i and $\mathbf{v_i}$, $i \in \{p, d\}$, be the partial densities and velocities for polymer and diluent phases in the mixture. The stationary mass balances for the two phases read

$$\nabla \cdot (\rho_i \mathbf{v_i}) = 0, \qquad i \in \{p, d\}. \tag{1a}$$

The stationary momentum balances for the polymer and diluent phases are

$$\nabla \cdot (\rho_i \mathbf{v_i} \otimes \mathbf{v_i}) = \nabla \cdot \boldsymbol{\Sigma}_i{}^{\mathrm{T}} + \mathbf{f_i}, \qquad i \in \{p, d\}, \tag{1b}$$

with the respective stress tensors $\boldsymbol{\Sigma}_i$. The fields $\mathbf{f_i}$ denote the body force densities acting on phase i. Neglecting effects of inner friction and convective terms due to pressure fluctuations as well as energy transfer caused by body forces, the stationary energy balances for the polymer and diluent phases are modeled as

$$\nabla \cdot (\rho h_i \mathbf{v_i}) = \nabla \cdot (C_i \nabla T), \qquad i \in \{p, d\}, \tag{1c}$$

where h_i are the partial enthalpies of polymer and diluent in the mixture, ρ denotes the mixture density and the right hand sides represent the energy transport by heat conduction at mixture temperature T und thermal conductivities C_i.

Mixture Model Ansatz The idea of the mixture model is to consider only one linear momentum equation as sum of the phase balances (1b) and only one energy balance equation as sum of (1c). The mixture density ρ is given by $\rho = \rho_p + \rho_d$. Similarly, also the mixture stress tensor $\boldsymbol{\Sigma}$, total body force $\mathbf{f}$, mixture enthalpy h as well as the mixture thermal conductivity C are the sums of the quantities of the single phases. For the mixture we assume ideality, i.e., the enthalpy of mixing is zero and the volume does not change under mixing. This leads to the relations

$$h = \frac{\rho_p}{\rho} h_p^0 + \frac{\rho_d}{\rho} h_d^0, \qquad 1 = \frac{\rho_p}{\rho_p^0} + \frac{\rho_d}{\rho_d^0},$$

where h_i^0, ρ_i^0 denote the enthalpies and material densities of pure polymer and solvent. For the stress tensor $\boldsymbol{\Sigma}$ we assume incompressibility and a Newtonian fluid with dynamic mixture viscosity μ, i.e., $\boldsymbol{\Sigma} = -p\mathbf{I} + \mu(\nabla\mathbf{v} + (\nabla\mathbf{v})^{\mathrm{T}})$ with mixture pressure p, mixture velocity $\mathbf{v}$ and $\mathbf{I} \in \mathbb{R}^{3,3}$ denoting the unit matrix. The definition of the mixture velocity $\mathbf{v}$ requires a different treatment: Since the consideration of only one linear momentum balance does not close our model, we have to employ constitutive relations for the differences between the phase velocities and the mixture velocity. In our dry spinning scenario we consider the polymer phase as dominating phase and the diluent phase as secondary phase. Therefore, we fix the polymer velocity as mixture velocity, i.e., $\mathbf{v} = \mathbf{v_p}$. Then, only one constitutive relation for the difference between the mixture velocity and the diluent velocity $\mathbf{v_{pd}} = \mathbf{v} - \mathbf{v_d}$ is needed. We use Fick's law in a version, which is linear with respect to the diluent mass fraction ρ_d/ρ, namely $\rho_d\mathbf{v_{pd}} = \rho D\nabla\,(\rho_d/\rho)$, with D denoting the diffusion coefficient of the diluent in the polymer. This formulation of Fick's law is appropriate to obtain an efficiently evaluable linear advection-diffusion equation for the polymer mass fraction in the dimensionally reduced fiber model (cf. Sect. 3). Employing Fick's law the mass balances for polymer and diluent (1a) become

$$\nabla\cdot(\rho_p\mathbf{v}) = 0, \qquad \nabla\cdot(\rho_d\mathbf{v}) - \nabla\cdot\left(\rho D\nabla\left(\frac{\rho_d}{\rho}\right)\right) = 0. \tag{2a}$$

Moreover, summing up the momentum phase balances (1b) and neglecting diffusive parts in the stresses yields the mixture momentum balance

$$\nabla\cdot(\rho\mathbf{v}\otimes\mathbf{v}) - \nabla\cdot\left(\rho D\left(\mathbf{v}\otimes\nabla\left(\frac{\rho_d}{\rho}\right) + \nabla\left(\frac{\rho_d}{\rho}\right)\otimes\mathbf{v}\right)\right) = \nabla\cdot\boldsymbol{\Sigma}^{\mathrm{T}} + \mathbf{f}. \tag{2b}$$

Analogously using Fick's law and the phase balances (1c) we obtain the total energy balance for the mixture

$$\nabla\cdot(\rho h\mathbf{v}) - \nabla\cdot\left(h_d^0\rho D\nabla\left(\frac{\rho_d}{\rho}\right)\right) = \nabla\cdot(C\nabla T). \tag{2c}$$

Model Closing The free boundary value problem (BVP) formed by (2a), (2b), (2c) is closed by appropriate boundary conditions (bc):

Kinematic/dynamic bc, Γ_{fr}: $\quad \mathbf{v}\cdot\boldsymbol{\nu} = 0, \qquad \boldsymbol{\Sigma}\cdot\boldsymbol{\nu} = \mathbf{f}_\star,$

Mass/heat flux bc, Γ_{fr}: $\quad -\rho D\nabla\,(\rho_d/\rho)\cdot\boldsymbol{\nu} = j,$

$$-C\nabla T\cdot\boldsymbol{\nu} = \alpha(T - T_\star) + j(\delta - h_d^0),$$

Inlet bc, Γ_{in}: $\quad \mathbf{v} = \mathbf{v}_0, \qquad \rho_d = \rho_{d,0}, \qquad T = T_0,$

Outlet bc, Γ_{out}: $\quad \mathbf{v} = \mathbf{v}_1.$

Apart from body forces $\mathbf{f}$ due to gravity, surface forces $\mathbf{f}_\star$ due to the surrounding airflow are considered. The geometry is specified via the kinematic boundary condition on Γ_{fr} with unit outer normal vector $\boldsymbol{\nu}$. At the free fiber surface Γ_{fr} the diluent density jumps due to the solvent evaporation. We model the diluent mass flux j by the difference of the diluent density in the air at the fiber surface ς and away from the fiber $\rho_{d,\star}$ with convective mass transfer coefficient β, i.e., $j = \beta(\varsigma - \rho_{d,\star})$. For the numerical treatment we introduce the transfer coefficient $\gamma = \varrho\beta$ with $\varrho = \varsigma\rho/\rho_d$, i.e., $j = \gamma\left(\rho_d/\rho - \rho_{d,\star}/\varrho\right)$. The temperature is continuous at the fiber surface, whereas the heat flux also has a jump due to solvent evaporation with evaporation enthalpy δ of the diluent. The heat flux is described—analogously to the mass flux—by the difference of the temperature at the fiber surface and away from the fiber $T_\star$ with heat transfer coefficient α.

3 Dimensionally Reduced Model Equations

The direct numerical simulation of the three-dimensional fiber model from Sect. 2—especially considering a two-way coupling with airflow computations—is computationally extremely demanding and thus in general not possible. Hence we propose a dimensionally reduced, efficiently evaluable fiber model, under the assumptions of slenderness and radial symmetry [7]: We employ one-dimensional equations for fiber velocity u and stress σ resulting from averaging the Newtonian stress tensor $\boldsymbol{\Sigma}$ and the momentum balance (2b) over circular fiber cross-sections. We combine these one-dimensional equations with two-dimensional advection-diffusion equations for the polymer mass fraction $c = \rho_p/\rho = 1 - \rho_d/\rho$ and temperature T revealing the radial effects that are essential due to evaporation [1, 4, 5]. These two-dimensional equations are obtained from (2a), (2c) by radial symmetry and linearization around the cross-sectional averaged polymer mass fraction $\bar{c}$ and temperature $\bar{T}$, see [7] for details. We make the system dimensionless using the initial fiber radius R_0, the initial fiber speed u_0, the initial fiber temperature T_0, the fiber length L as well as initial mixture density ρ_0 and initial specific heat capacity q_0:

System 1 (One-Two-Dimensional BVP) *One-dimensional equations, $z \in (0, 1)$:*

$$\partial_z u = \frac{1}{3R^2}\sigma, \qquad \partial_z \sigma = \frac{\mathrm{Re}}{3}\frac{c_0}{\bar{c}}\frac{1}{R^2}\sigma - \frac{\mathrm{Re}}{\mathrm{Fr}^2}\frac{c_0}{\bar{c}}\frac{1}{u} - \mathrm{ReM}\frac{1}{R} f_{air}, \tag{3a}$$

with boundary conditions at inlet $z = 0$ and outlet $z = 1$: $u(0) = 1, \quad u(1) = \mathrm{Dr},$
Two-dimensional equations, $(r, z) \in (0, 1)^2$:

$$u\partial_z c - \frac{1}{\epsilon \mathrm{Pe_D}}\frac{\bar{c}}{R^2 r}\partial_r(r\partial_r c) = 0, \qquad \rho q u \partial_z T - \frac{1}{\epsilon \mathrm{Pe_C}}\frac{1}{R^2 r}\partial_r(r\partial_r T) = 0, \tag{3b}$$

with boundary conditions at inlet $z = 0$*, fiber surface* $r = 1$ *and symmetry boundary* $r = 0$*:*

$$c\big|_{z=0} = c_0, \qquad \partial_r c\big|_{r=0} = 0, \qquad \frac{1}{\text{Pe}_\text{D}}\frac{\rho}{R}\partial_r c\big|_{r=1} = j\big|_{r=1},$$

$$T\big|_{z=0} = 1, \qquad \partial_r T\big|_{r=0} = 0, \qquad -\frac{1}{\text{Pe}_\text{C}}\frac{1}{R}\partial_r T\big|_{r=1} = \big(\alpha(T - T_\star) + j(\delta - h_d^0)\big)\big|_{r=1},$$

Constitutive laws and geometric relation:

$$\rho^{-1} = c\,(\rho_p^0)^{-1} + (1-c)\,(\rho_d^0)^{-1}, \qquad q = cq_p^0 + (1-c)q_d^0,$$

$$R = \sqrt{\frac{c_0}{\bar{c}\rho u}}, \qquad q_d^0 = \partial_T h_d^0.$$

The averaging of the polymer mass balance (2a) results in an explicit expression for the unknown fiber radius R, i.e., we face no free BVP anymore. The temperature derivatives of h_p^0, h_d^0 and h are in particular the specific heat capacities q_p^0, q_d^0 and q for constant pressure. The characteristic dimensionless parameters are Reynolds number $\text{Re} = \rho_0 u_0 L/\mu$, Froude number $\text{Fr} = u_0/\sqrt{gL}$, length ratio $\epsilon = R_0/L$ and draw ratio $\text{Dr} = u_1/u_0$ with take up speed u_1 as well as Peclet numbers $\text{Pe}_\text{D} = u_0 R_0/D$, $\text{Pe}_\text{C} = \rho_0 q_0 u_0 R_0/C$, respectively. The further dimensionless air-drag associated parameter M is a scalar field due to its dependence on airflow quantities, see [3]—also for an appropriate air-drag model.

4 Numerical Framework, Simulation Results and Discussion

The numerical computation of the fiber solution of System 1 is performed in MATLAB. For the ordinary differential equation (3a) we use the routine `bvp4c.m` providing a collocation method with a Runge–Kutta scheme of fourth order. On top we build a suitable continuation strategy (homotopy method) as presented in [7]. The solution of the advection-diffusion equation (3b) is implicitly given in terms of Green's functions and leads for the surface values to Volterra integral equations of second kind with singular kernel, which we can solve very efficiently by the product integration method, see [6, 7]. The coupling of the one- and two-dimensional fiber equations is then done iteratively.

To demonstrate the performance of our proposed fiber model we consider a spinning setup due to gravity. In Fig. 1 the numerical solution for a typical setup is shown: the polymer mass fraction grows from the nozzle to the fiber end due to solvent evaporation. The polymer mass fraction at the fiber boundary ($r = 1$) rises faster than the averaged mass fraction indicating an inhomogeneous polymer distribution over the fiber cross-sections. The fiber temperature behavior is dominated by evaporation effects: the averaged as well as the surface fiber temperature drop down

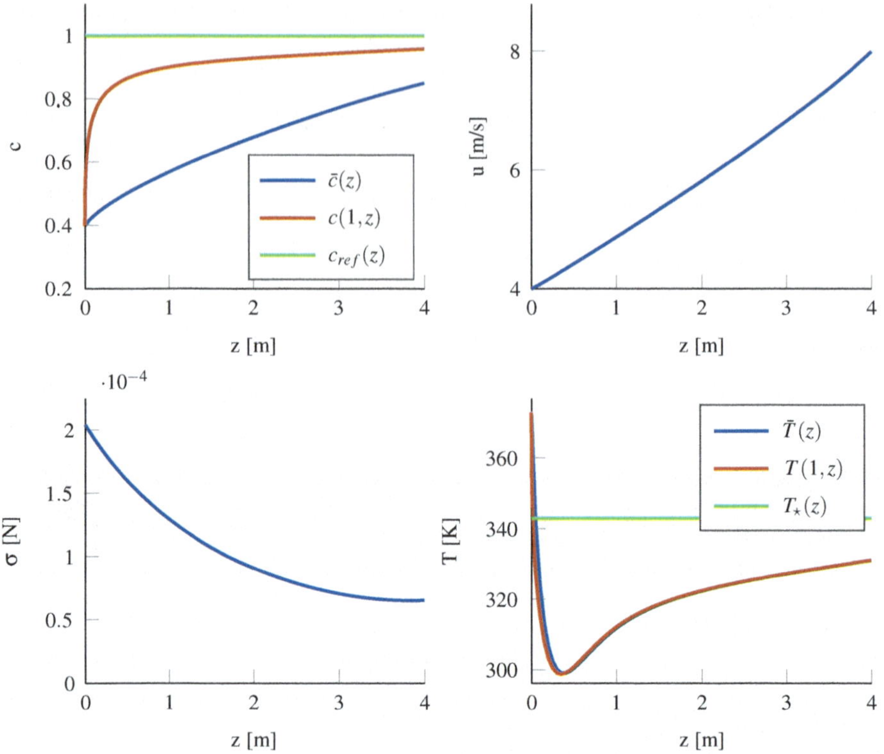

Fig. 1 Fiber solution for the process parameters: $L = 4\,\text{m}$, $R_0 = 7.5 \cdot 10^{-5}\,\text{m}$, $u_0 = 4\,\text{m/s}$, $u_1 = 8\,\text{m/s}$, $T_0 = 373.15\,\text{K}$, $c_0 = 0.4$, $T_\star = 343.15\,\text{K}$, $\rho_{d,\star} = 0\,\text{kg/m}^3$ and $f_{air} = 0\,\text{N/m}$. The remaining parameters are chosen constant in the region of a typical industrial spinning setup yielding (Re, Fr, ϵ, Pe_D, Pe_C, Dr) = (3.03, $6.39 \cdot 10^{-1}$, $1.88 \cdot 10^{-5}$, $3 \cdot 10^5$, $3.38 \cdot 10^3$, 2)

directly at the nozzle and approach the surrounding temperature $T_\star$ further down in the spinning chamber.

The numerical simulation is performed on an Intel Core i7-6700 CPU (4 cores, 8 threads) with 16 GBytes of RAM and takes only around 61 s. In contrast a simulation of the respective three-dimensional problem (cf. Sect. 2) takes several hours. This computational speed makes simulations of multiple fibers in a two-way coupling with airflow simulations under the use of complex rheological models for transition coefficients, dynamic viscosity and other physical parameters feasible. For industrially relevant simulation results we refer to [7].

Acknowledgement This work has been supported by German DFG, project 251706852, MA 4526/2-1, WE 2003/4-1.

References

1. Gou, Z., McHugh, A.J.: Two-dimensional modeling of dry spinning of polymer fibers. J. Non-Newton. Fluid **118**(2–3), 121–136 (2004)
2. Manninen, M., Taivassalo, V.: On the mixture model for multiphase flow. VTT Publications **288**, 1–67 (1996)
3. Marheineke, N., Wegener, R.: Modeling and application of a stochastic drag for fibers in turbulent flows. Int. J. Multiphase Flow **37**, 136–148 (2011)
4. Ohzawa, Y., Nagano, Y.: Studies on dry spinning. I. Fundamental equations. J. Appl. Polym. Sci. **13**, 257–283 (1969)
5. Ohzawa, Y., Nagano, Y.: Studies on dry spinning. II. Numerical solutions for some polymer–solvent systems based on the assumption that drying is controlled by boundary-layer mass transfer. J. Appl. Polym. Sci. **14**(7), 1879–1899 (1970)
6. Wieland, M., Arne, W., Feßler, R., Marheineke, N., Wegener, R.: Product integration method for simulation of radial effects in dry spinning processes. PAMM **18**(1), e201800055 (2018)
7. Wieland, M., Arne, W., Feßler, R., Marheineke, N., Wegener, R.: An efficient numerical framework for fiber spinning scenarios with evaporation effects in airflows. J. Comput. Phys. **384**, 326–348 (2019)

Modeling Bimaterial 3D Printing Using Galvanometer Mirror Scanners

Daniel Bandeira and Marta Pascoal

Abstract Three-dimensional printing is a process for building new parts with a specified shape. Despite its increasing popularity, printers capable of working with more than one material are yet unavailable. In this work we model the design and the operation of an apparatus for printing with two materials, namely printing a component which includes a previously constructed inner structure. The structure that supports the second material brings difficulties, resulting from the possible "shaded" areas on the printing surface. The problem is addressed assuming the installation of galvanometer mirror scanners as additional light sources on the walls of the printer, and it is modeled in two steps: finding the least number of emitters to use, so that the whole part can be constructed, as well as their position; and assigning them with each cell of the part to be reached. The first step is formulated as a set covering problem. The second is formulated as a linear integer problem and aiming at optimizing two objectives: the number of emitters activated per layer and the quality of the printed part. Methods for solving the problems are described and tested.

1 Introduction

Three-dimensional printing, or 3D printing, is an additive process for rapid free form manufacturing, where the final object (known as part) is created by the addition of successive thin layers of material. Each layer corresponds to a cross-section of the part to be constructed, and the printer draws each layer as if it was a 2D printing [3, 7]. Printings are made of a single material, which can vary from resin to

---

D. Bandeira
Department of Mathematics, University of Coimbra, Coimbra, Portugal

M. Pascoal (✉)
CMUC, Department of Mathematics, University of Coimbra, Coimbra, Portugal

Institute for Systems Engineering and Computers – Coimbra, Coimbra, Portugal
e-mail: marta@mat.uc.pt

© Springer Nature Switzerland AG 2019
I. Faragó et al. (eds.), *Progress in Industrial Mathematics at ECMI 2018*,
Mathematics in Industry 30, https://doi.org/10.1007/978-3-030-27550-1_4

ceramics or metal (among others). One of the technologies used for 3D printing is stereolithography (SL). In this case, each layer is added using liquid resin exposed to a laser light, usually fixed at the top of the printer and able to reach the printing platform. Only the zone of the resin that is reached by the laser beam is cured. Then the platform that supports the model moves to get ready for printing the next layer. This type of 3D printing is fairly standard nowadays, and has become quite popular due to its ability to produce new parts quickly and at a low cost.

The present work focuses on a process analogous to SL, but with the goal of printing an object in which the resin covers a previously constructed 3D grid structure of a different material, like metal. This type of components has application to custom orthotics, intelligent components, complex or fragile parts where over-injection or other options are not feasible or not economically sustainable. In this case the metal structure may block the laser light, thus preventing the cure of shaded areas. This work addresses the question of placing additional galvanometer mirror scanners on the walls of the printer to overcome this issue. Their positions depend on the part to print and are fixed from the beginning of the printing process. However, the laser beam reflected by each galvanometer scanner can be oriented with the goal of reaching the shaded areas. Hereafter galvanometer scanners and laser are sometimes simply referred to as emitters.

As explained before, the part is divided into layers, each one evenly partitioned in squares, called voxels. Assuming that both the voxels to cure and the possible locations for the emitters are known, the problem is modeled in two parts:

- Emitters location problem (ELP): The goal of which is to find the emitters' position that minimizes the number of emitters required to complete the printing.
- Emitters assignment problem (EAP): Using the solution of the ELP, it is then necessary to determine the voxels of each layer that each emitter should reach.

The rest of the text is organized as follows. In Sects. 2 and 3 integer linear optimization models are presented for these two problems. The formulations are empirically tested for a case study in Sect. 4. Finally, concluding remarks are discussed.

2 Emitters Location Problem

The goal of the ELP is to find the minimum number of emitters that allows to print a given part, as well as their location. To do this, let us consider that there are m voxels that the laser light needs to cure and n possible positions for the laser emitters. The emitters coverage matrix is defined as $A = [a_{ij}]_{i=1,\dots,m;\, j=1,\dots,n}$, such that

$$a_{ij} = \begin{cases} 1 \text{ if the emitter at position } j \text{ can reach the voxel } i \\ 0 \text{ otherwise} \end{cases}, \quad i = 1, \dots, m, \ j = 1, \dots, n.$$

The matrix A can be calculated by geometric arguments, as shown in [1]. Let also x_j be binary decision variables, such that

$$x_j = \begin{cases} 1 \text{ if the emitter at position } j \text{ is installed} \\ 0 \text{ otherwise} \end{cases}, \quad j = 1, \ldots, n.$$

The objective function of the ELP is the total number of emitters to use, which is to be minimized. This is given by

$$\sum_{j=1}^{n} x_j.$$

We say that the emitter j covers the voxel i, or that i is covered by j, if it is able to reach it by means of a laser beam, for any $j = 1, \ldots, n$, $i = 1, \ldots, m$. At least one emitter needs to cover each voxel, in order to cure the material. Therefore, a solution for the ELP is feasible if any voxel i, can be reached by at least one emitter, that is, if

$$\sum_{j=1}^{n} a_{ij} x_j \geq 1, \quad i = 1, \ldots, m.$$

Thus, the ELP can be formulated as the set covering problem below,

$$\begin{aligned} \text{minimize} \quad & \sum_{j=1}^{n} x_j \\ \text{subject to} \quad & Ax \geq 1 \\ & x \in \{0, 1\}^n \end{aligned} \tag{1}$$

The optimal value of problem (1) is the number of emitters required to ensure the complete printing of the part. Its optimal solution provides the positions where the emitters should be installed, which corresponds to the indices j such that $x_j = 1$, $j = 1, \ldots, n$. The set covering problem is a classical combinatorial optimization problem which has been shown to be NP-complete [4, 8], therefore, exact methods may be limited to solve it as the size of the problem grows.

3 Emitters Assignment Problem

Assume now that n_2 emitters have been installed in the positions determined by the ELP. The goal of the EAP is then to select the emitter to assign to each voxel. Also consider that p layers of the part need to be printed, each one with m_k voxels to cure, $k = 1, \ldots, p$.

Let k be a layer to print, $k = 1, \dots, p$, x_j be variables similar to those used in the ELP, and y_{ij} be binary variables defined by

$$y_{ij} = \begin{cases} 1 \text{ if the emitter } j \text{ is activated to cure voxel } i \\ 0 \text{ otherwise} \end{cases}, \; i = 1, \dots, m_k, \; j = 1, \dots, n_2.$$

Two aspects are taken into account for defining the objective functions, the operability of the system and the quality of the printing. The first is expressed by the number of emitters used on each layer of the part, and the second as the distortion of the laser light when it reaches the layer, denoted by z_1 and z_2, respectively. With this respect it should be noted that the light beam has the shape of a circle at its origin, but the circle is distorted as an ellipse when reaching the layer, every time its incidence angle is not exactly 90°. Therefore, similarly to the ELP, the first criteria, to minimize, can be expressed by

$$z_1(x, y) = \sum_{j=1}^{n_2} x_j.$$

The second depends on the emitter that reaches each voxel, y_{ij}, and the beam's angle of incidence, $\theta_{ij} \in [0, \frac{\pi}{2}]$, $i = 1, \dots, m_k$, $j = 1, \dots, n_2$, which can be calculated by a procedure similar to the emitters coverage matrix [1]—see Fig. 3a. Thus, minimizing the distortion of the laser light corresponds to maximizing the function

$$z_2(x, y) = \sum_{i=1}^{m_k} \sum_{j=1}^{n_2} \theta_{ij} y_{ij}.$$

The choice of the emitter used to reach a particular voxel must take two aspects into account: the uniqueness of this solution and its viability. Assignment constraints can be used to ensure the first one

$$\sum_{j=1}^{n_2} y_{ij} = 1, \; i = 1, \dots, m_k, \tag{2}$$

whereas the second depends on the constraints

$$\sum_{j=1}^{n_2} a_{ij} y_{ij} = 1, \; i = 1, \dots, m_k, \tag{3}$$

where $A = [a_{ij}]_{i=1,\dots,m_k; j=1,\dots,n_2}$ is the submatrix of the emitters coverage matrix, restricted to the set of voxels to cure at layer k and the emitters installed in the printer. The other aspect to consider is the emitters that are activated to print the

k-th layer. For the emitters used in each layer, the covering conditions introduced in Sect. 2 can be used,

$$\sum_{j=1}^{n_2} a_{ij} x_j \geq 1, \ i = 1, \ldots, m_k. \tag{4}$$

The variables y_{ij} and x_j are related, because when the emitter j is activated to reach a voxel i, it is activated for the entire layer. This corresponds to imposing the constraints

$$y_{ij} \leq x_j, \ i = 1, \ldots, m_k, \ j = 1, \ldots, n_2,$$

and, by aggregating these conditions, we can derive the equivalent constraints

$$\sum_{i=1}^{m_k} y_{ij} \leq m_k x_j, \ j = 1, \ldots, n_2. \tag{5}$$

Finally, it should be noted that when (3) and (5) hold, the constraints (4) are satisfied as well. Combining all the information, we obtain the following biobjective linear integer problem,

$$\begin{aligned}
\text{minimize} \quad & z_1(x, y) \\
\text{maximize} \quad & z_2(x, y) \\
\text{subject to} \quad & (2), (3), (5) \\
& y_{ij} \in \{0, 1\}, \ i = 1, \ldots, m_k, \ j = 1, \ldots, n_2 \qquad \text{(6a)} \\
& x_j \in \{0, 1\}, \ j = 1, \ldots, n_2. \qquad \text{(6b)}
\end{aligned}$$

In general, the objective functions z_1 and z_2 are conflicting, which means that there is no feasible solution that optimizes both simultaneously. Solving the problem considering only z_1 or z_2 provides an idea of how much the two objective functions may range, but as an alternative to the usual concept of optimality in single-objective problems, in these cases we usually seek for efficient solutions. A solution is said to be efficient if there is no other which is strictly better than the first with respect to the two objective functions simultaneously. The approaches for finding the efficient solutions of biobjective integer problems can be classified into a priori, interactive or a posteriori, depending on whether how one efficient solution is chosen among the all set. In the first case the decision maker (DM) knows how the relative importance of the two objective functions, which can be aggregated accordingly before one solution is found. In the second case, partial efficient solutions are shown to the DM, who then guides the search for an acceptable solution. The last case consists of computing all the efficient solutions and then let the DM express the preferences with respect to that whole set. A single solution must be chosen before printing a given part, however it is not clear in advance how the two objective functions are related, thus an a posteriori approach is more indicated for the EAP. Several methods have been proposed to find the efficient solutions of biobjective integer problems like the EAP. For instance, two-phase methods or the ϵ-constrained method, among others [5, 6]. This topic is studied in [2].

4 Computational Experiments

The formulations presented above were tested for a case study consisting of constructing the cube with 5 faces shown in Fig. 1a. The thickness of the metal grid is considered equal to the thickness of the resin layers, that is, 1. This is also the value used as the width of the voxels. The parameters for printing the cube are:

- The length of each segment of the metal grid, l_M.
- The thickness of the resin added on each side of the metal grid, $l_P = 1$.
- The number of divisions of the metal grid, which is assumed to be uniform, n_M.
- The distance between the cube to print and the side walls of the printer, where emitters can be installed, h, which depends on the size of the part, but ensuring that the printing platform is of size 1250×1250 units.
- The height of the printing area, fixed to $h_V = 1250$ units.

The used length unit corresponds to 0.2 mm, the length of the side of the voxels. Each layer contains $n_V \times n_V$ voxels. The remaining characteristics of the problems solved are summarized in Fig. 1b. The linear problems were solved using CPLEX 12.7, whereas MATLAB R2016b was used for the remaining calculations. The presented run times are mean values obtained for 30 repetitions on an Intel® i7-6700 Quadcore of 3.4 GHz, with 8 Mb of cache and 16 Gb of RAM.

For the ELP it was assumed that a laser is already fixed at the center of the top of the printer. Additionally, 80 possible locations are considered for other emitters on the side walls. The solutions of problem (1) given by CPLEX are presented in Table 1. According to the results, between 2 and 4 emitters besides the top one are

(a)

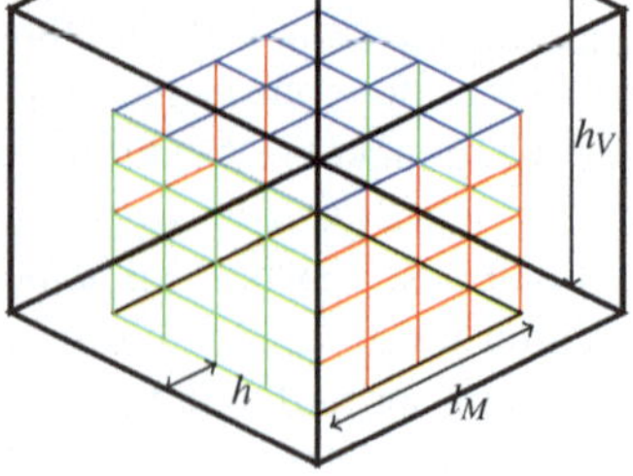

(b)

Test	n_V	n_M	h
T1	200	5	525
T2	200	10	525
T3	200	20	525
T4	300	5	475
T5	300	10	475
T6	300	20	475
T7	500	5	375
T8	500	10	375
T9	500	20	375

Fig. 1 Case study. (**a**) Printing area and object to print. (**b**) Test parameters

Table 1 ELP solutions and run times

Test	Emitters' positions	Time (s)
T1	(1, 1000, 250) and (1250, 1, 250)	5.03
T2	(1, 1000, 1000), (1, 251, 250) and (1250, 1250, 250)	19.13
T3	(1250, 1, 1000), (1, 1, 250), (1, 1250, 250) and (1250, 1250, 250)	140.03
T4	(1, 1, 750) and (1250, 1250, 250)	9.89
T5	(1, 750, 1000), (1, 251, 500) and (1250, 1000, 500)	34.98
T6	(1, 1, 500), (1250, 1250, 500), (1, 750, 250) and (1250, 501, 250)	1323.36
T7	(1, 1, 750) and (1250, 1250, 500)	27.52
T8	(1, 1, 500), (1250, 1250, 500) and (1250, 1, 250)	40.70
T9	(1, 1, 500), (1250, 1250, 500) and (1250, 1, 250)	144.05

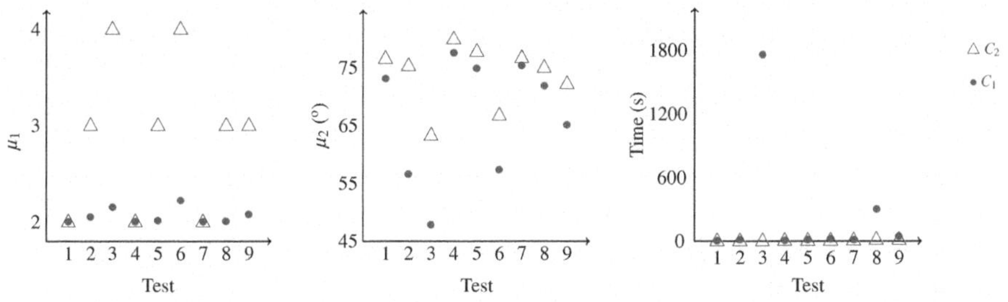

Fig. 2 EAP solutions and run times

required for completing the printing. Although most problems were solved in less than 3 min, one of them required almost half an hour, which reflects the hardness of the problem.

Using the solutions in Table 1, the EAP was considered when optimizing one objective function at a time. The approach that optimizes z_i is represented below by C_i, $i = 1, 2$. Figure 2 shows the mean results for the number of fixed emitters required for printing each layer, μ_1, the mean value of θ, μ_2, and the run times regarding printing the whole part for each method. In terms of solutions the approach C_1 always finds a way to print the part using 2 or 3 emitters per layer besides the top one, while this only happens with C_2 when a broader grid is considered. For the remaining cases applying C_2 implies using 3 or 4 emitters. The average angles of incidence of the beam over the voxels are between 60° and 80° when using approach C_2 and between 45° and 80° when using C_1. The results are worse, i.e., the angle of incidence is smaller, when the grid is denser. The approach C_1 is more sensitive to this change than C_2. As explained next, small angles of incidence may lead to a distorted final part. In general in average the two approaches run fast, a few seconds, and in approximately the same CPU time. However, the tests T3 and T8 were harder to solve using the approach C_1 than using C_2, around 30 and 5 min, respectively.

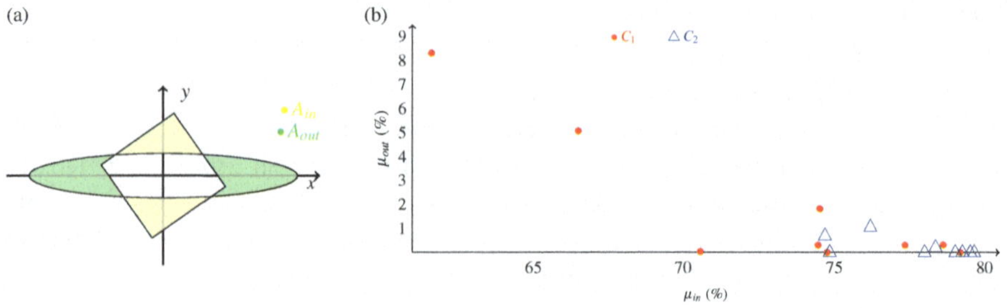

Fig. 3 Quality of the printed part. (**a**) Laser distortion on a voxel. (**b**) Distortion areas

A measure of the quality of the produced part should also be taken into account. As mentioned earlier, in general, the laser beam reaches the printing surface as an ellipse because of angle θ. We have considered that the centers of the laser and of the voxel are aligned, thus, two situations may affect the quality of the part: a region beyond the target voxel may be cured, leading to an outer area A_{out}, and part of the target voxel may lack the cure, leading to an inner area A_{in}. Both are illustrated in Fig. 3a.

The mean values of A_{in} and A_{out} were calculated for the same case study. Standard lasers for stereolithography have a radius of 0.05 mm, so, taking into account the considered unit of measurement, the laser has radius 0.25 units. Figure 3b shows the mean value of the percentage relative to the voxel area of A_{in} and A_{out}, respectively μ_{in} and μ_{out}. In all cases an area of voxels is left to cure and for some of them there is also an area reached outside the voxels. The mean value of A_{in} was above 60% for all tests. The main reasons are the assumption that the laser reaches only the center of voxels and considering voxels whose sides are twice the diameter of the laser beam. Working with smaller voxels would result in a reduction of this area, but would increase the values of A_{out}. The area A_{out} is almost null for most of the cases. The instance with the highest values of A_{out} corresponds to approach C_1 when applied to test T3.

The polymer at a given voxel may be affected by a beam pointing at neighbor voxels. Likewise, only the outer area of voxels in the border is relevant for the quality of the part. Therefore, the expressions of A_{in} and A_{out} are only estimate measures for the printing quality. Additionally, current 3D printing processes include a post-printing finishing phase where all part is exposed to UV light to cure any liquid resin left. This can reduce the theoretical values of A_{in} and allows to improve the produced part.

5 Conclusions

This work addressed the bimaterial 3D printing problem based on the installation of galvanometer scanners on the walls of a printer. The problem is treated as locating the emitters and assigning them with the voxels of a given part, which were formulated and tested for a case study. The software CPLEX was able to find exact solutions for the considered instances. Nevertheless, these are computationally hard problems, thus heuristics should be designed to prevent cases for which this does not happen or when no commercial specialized software is available. Moreover, the unexpectedly cured area of the extreme solutions of the EAP was relatively small, while the uncured area of the part seems fairly high. In practice this latter issue can be addressed with a post-printing finishing phase, a standard 3D printing procedure. Our model can also restrict the emitter positions having in mind to reduce the laser distortion, although that may compromise the full printing of the part. Finally, the presented approach can still be used to print parts with more than two materials, by handling the product of the first print as the inner structure of the next one.

Acknowledgements This work was developed within the Project PT2020-POCI-SII and DT 17963: NEXT.Parts, Next-Generation of Advanced Hybrid Parts, from the program Portugal 2020, through COMPETE 2020—POCI. The work of MP was partially supported by the Portuguese Foundation for Science and Technology under project grants UID/MAT/00324/2019 and UID/MULTI/00308/2019.

References

1. Bandeira, D., Pascoal, M., Mateus, A., Reis Silva, M.: Multi-material 3d printing using stereolithography: an optimization approach. To appear in Advanced Materials Research. http://www.mat.uc.pt/~marta/NextG/BandeiraEtAl17.pdf
2. Bandeira, D., Pascoal, M., Santos, B.: Bimaterial 3D printing using galvanometer scanners. To appear in Optimization and Engineering (2019). https://doi.org/10.1007/s11081-019-09433-6
3. Burns, M.: Automated Fabrication: Improving Productivity in Manufacturing. Prentice-Hall, Englewood Cliffs (1993)
4. Caprara, A., Toth, P., Fischetti, M.: Algorithms for the set covering problem. Ann. Oper. Res. **98**, 353–371 (2000)
5. Ehrgott, M.: Multicriteria Optimization. Springer, Berlin (2006)
6. Ehrgott, M., Gandibleux, X.: Multiobjective Combinatorial Optimization—Theory, Methodology, and Applications, pp. 369–444. Springer, Boston (2002)
7. Gibson, I., Rosen, D., Stucker, B.: Additive Manufacturing Technologies. Springer, New York (2015)
8. Karp, R.M.: Reducibility among combinatorial problems. In: Complexity of Computer Computations, pp. 85–103. Springer, Berlin (1972)

Thermal Transport Equations and Boundary Conditions at the Nanoscale

Marc Calvo-Schwarzwälder, Matthew G. Hennessy, Pol Torres, Timothy G. Myers, and F. Xavier Alvarez

Abstract The Guyer–Krumhansl equation is an extension to the classical Fourier law that is particularly appealing from a theoretical point of view because it provides a link between kinetic and continuum models and is based on well-defined physical parameters. Here we show how, subjected to a specific boundary condition analogous to the slip conditions for fluids, the Guyer–Krumhansl equation yields promising results in predicting the effective thermal conductivity of nanowires with circular and rectangular cross-sections.

1 Introduction

The classical equations fail to describe heat transfer at small length scales [2] and, given the increasing number of miniaturised electrical components, it is therefore crucial to find a valid description of heat flow for these situations. In particular, for applications involving heat removal, there is a general concern about predicting correctly the thermal conductivity. For instance, it has been observed that the thermal conductivity of Si nanowires with a diameter of 37 nm decreases by approximately 87% with respect to the bulk value [8]. This decrease can be related to the emergence of new thermal transport regimes which are not captured by Fourier's law.

M. Calvo-Schwarzwälder (✉) · M. G. Hennessy · T. G. Myers
Centre de Recerca Matemàtica, Bellaterra, Barcelona, Spain
e-mail: mcalvo@crm.cat; mhennessy@crm.cat; tmyers@crm.cat

P. Torres · F. X. Alvarez
Departament de Física, Universitat Autònoma de Barcelona, Bellaterra, Barcelona, Spain
e-mail: pol.torres@uab.cat; xavier.alvarez@uab.cat

© Springer Nature Switzerland AG 2019
I. Faragó et al. (eds.), *Progress in Industrial Mathematics at ECMI 2018*,
Mathematics in Industry 30, https://doi.org/10.1007/978-3-030-27550-1_5

The Guyer–Krumhansl (G-K) equation [5, 6] is a popular extension to Fourier's law which includes memory and non-local effects in a simple way. In this paper we attempt to capture the size dependence of the thermal conductivity by the hydrodynamic model, based on the G-K equation, combined with suitable boundary conditions suggested by the similarities of the governing equations with those used in fluid dynamics.

2 Mathematical Formulation

We consider a nanowire (NW) of length L^*, suspended in vacuum, subject to a heat flux $\mathbf{Q}^*$ by fixing the temperature at the left and right end at different temperatures $T_0^* > T_1^*$, where the * notation refers to dimensional quantities. Under steady state assumption and neglecting radiation at the surface of the NW, heat flow is governed by

$$\mathbf{Q}^* - \ell^{*2}\nabla^2\mathbf{Q}^* = -k^*\nabla T^*, \qquad \nabla \cdot \mathbf{Q}^* = 0, \tag{1}$$

which represent the hydrodynamic equation (which is a vectorial equation) and conservation of energy. The quantities $k^*(T^*)$ and $\ell^*(T^*)$ are the bulk thermal conductivity and a non-local length related to the phonon mean free path. In the original equations derived by Guyer and Krumhansl, ℓ^* is exactly the mean free path, but other definitions of this parameter can be found in the literature. Here we use a non-local length and a thermal conductivity computed via the kinetic collective model (KCM), which has shown excellent results in predicting the (bulk) thermal conductivity in different media [11].

The boundary conditions for (1) will be specified for each of the considered cross-sections. In general, the components of the heat flux normal to the surface of the NW will be assumed to be zero, whereas we will impose slip conditions with a slip length ℓ_s^* for the components which are parallel to the surface. Following other authors [1, 9, 10, 13], the slip length will be assumed to be proportional to the non-local length ℓ^*, i.e., we assume $\ell_s^* = C\ell^*$. In previous studies, this parameter was allowed to depend on the temperature. Here we introduce a novelty by assuming that it might also depend on the Knudsen number, which is related to ℓ^*. Finally, the left and right ends of the NW will be assumed to be at constant temperatures T_0^* and T_1^* respectively.

Assuming that the longitudinal axis is described by the variable z^*, the effective thermal conductivity (ETC) will be computed via

$$k_{\text{eff}}^* = \frac{Q^*}{-\partial T^*/\partial z^*} \tag{2}$$

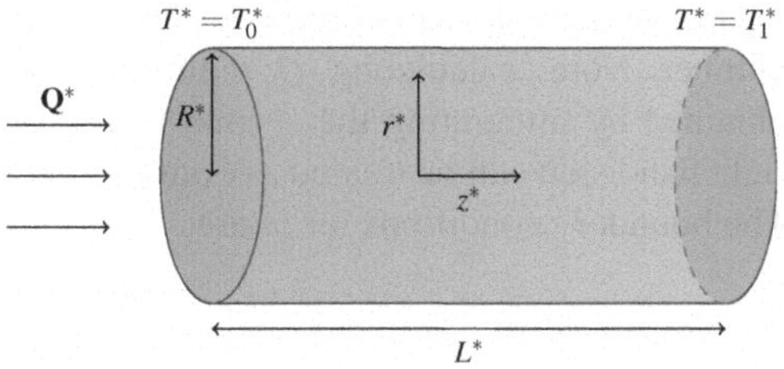

Fig. 1 Heat flux through a circular nanowire of length L^* and radius R^* induced by a temperature difference $\Delta T = T_0^* - T_1^*$

where Q^* is the total heat flux per unit area across a cross-section S^*,

$$Q^* = |S^*|^{-1} \int_{S^*} \mathbf{Q}^* \cdot d\mathbf{S}^*. \tag{3}$$

2.1 Circular Nanowires

We consider a NW with a circular cross-section of radius R^*; see Fig. 1. In addition, we assume $R^* \ll L^*$ or, alternatively, $R^*/L^* = \epsilon \ll 1$. Since the NW exhibits rotational symmetry, we neglect angular dependencies and thus we express the heat flux in terms of its axial and radial components, $\mathbf{Q}^* = v^*\mathbf{r} + w^*\mathbf{z}$. The boundary conditions for v^* and w^* are

$$v^*|_{r^*=0} = \left.\frac{\partial w^*}{\partial r^*}\right|_{r^*=0} = v^*|_{r^*=R^*} = \left[w^* + C\ell^* \frac{\partial w^*}{\partial r^*}\right]_{r^*=R^*} = 0. \tag{4}$$

The boundary conditions at $r^* = 0$ follow from the axial symmetry of the NW, whereas on the surface of the NW we have imposed zero flux and slip conditions for v^* and w^* respectively. The parameter C encodes the information about phonon-boundary scattering and roughness. In the literature it is essentially treated as a fitting parameter [9, 13]. Here we take $C = \exp(-R^*/\ell^*)$ to model the transition from diffusive ($R^* \gg \ell^*$, $C \to 0$) to ballistic ($R^* \ll \ell^*$, $C \to 1$) transport. Further discussion on this parameter can be found in Refs. [3, 4, 9, 10, 13].

The problem is non-dimensionalised via the new variables $z = z^*/L^*$, $r = r^*/R^*$, $v = v^*/(v_0^*)$, $w = w^*/w_0^*$, $T = (T^* - T_1^*)/\Delta T$ and $k = k^*/k_0^*$, where $\Delta T = T_0^* - T_1^*$, k_0^* is a reference value for the bulk thermal conductivity and w_0^*, v_0^* are (unknown) typical values of the flux components. Balancing terms in the energy equation in (1) requires $v_0^* = \epsilon w_0^*$, i.e. that heat is mainly transported in the axial direction. For w_0^* we choose the typical scale $w_0^* = k_0^* \Delta T/L^*$. In the new variables the total heat flux per unit area becomes $Q = 2\int_0^1 wr dr$ and the governing equations take the form

$$\frac{1}{r}\frac{\partial}{\partial r}(rv) + \frac{\partial w}{\partial z} = 0, \qquad w - \frac{\mathrm{Kn}^2}{r}\frac{\partial}{\partial r}\left(r\frac{\partial w}{\partial r}\right) = -k\frac{\partial T}{\partial z}, \qquad \frac{\partial T}{\partial r} = 0, \tag{5}$$

where we have neglected terms of order ϵ and Kn= ℓ^*/R^* is called the Knudsen number. Note, calculating Q requires only knowledge about w, which can be obtained by integrating the second equation in (5). Furthermore, the term on the right hand side can be treated as constant in the variable r due to the third equation. The boundary conditions for w are

$$\left.\frac{\partial w}{\partial r}\right|_{r=0} = \left[w + \mathrm{Kn}C\frac{\partial w}{\partial r}\right]_{r=1} = 0. \tag{6}$$

The problem can be solved analytically in terms of Bessel functions of the first kind, giving [3]

$$w = -k\left(1 - \frac{I_0(r/\mathrm{Kn})}{I_0(1/\mathrm{Kn}) + CI_1(1/\mathrm{Kn})}\right)\frac{\partial T}{\partial z}, \tag{7}$$

$$Q = -k\left(1 - \frac{2\mathrm{Kn}I_1(1/\mathrm{Kn})}{I_0(1/\mathrm{Kn}) + CI_1(1/\mathrm{Kn})}\right)\frac{\partial T}{\partial z}, \tag{8}$$

and hence we find

$$k_{\mathrm{eff}}/k = 1 - \frac{2\mathrm{Kn}I_1(1/\mathrm{Kn})}{I_0(1/\mathrm{Kn}) + CI_1(1/\mathrm{Kn})}. \tag{9}$$

Using Taylor expansions we can obtain simple expressions in the limits of large and small Knudsen numbers,

$$k_{\mathrm{eff}}/k \approx 1 - 2\mathrm{Kn}, \qquad \text{for Kn} \ll 1, \tag{10a}$$

$$k_{\mathrm{eff}}/k \approx \frac{1}{2\mathrm{Kn}}, \qquad \text{for Kn} \gg 1, \tag{10b}$$

In particular, in the limit $R^* \ll \ell^*$ we find that $k^* \propto 1/R^*$, which agrees with experimental results [8].

2.2 Rectangular Nanowires

We now assume that the NW has a rectangular cross-section of height $2H^*$ and width $2W^*$, as depicted in Fig. 2. Without loss of generality, we can assume $H^* \leq W^* \ll L^*$ or, introducing the aspect ratios $\phi = H^*/W^*$ and $\epsilon = H^*/L^*$, $\epsilon \ll \phi \leq 1$. Upon splitting the flux in its Cartesian components, $\mathbf{Q}^* = u^*\mathbf{x} + v^*\mathbf{y} + w^*\mathbf{z}$,

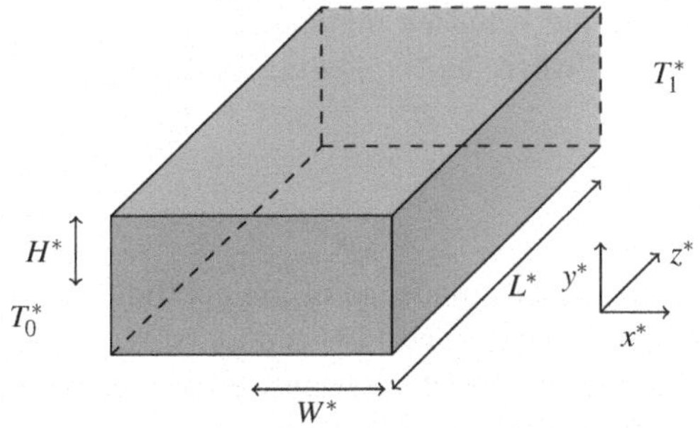

Fig. 2 A slab of length L^* and with a rectangular cross-section of dimensions $2W^* \times 2H^*$ is held at different temperatures $T_0^* > T_1^*$ at the left and right ends respectively, which induces a heat flux $\mathbf{Q}^*$

we can write the boundary conditions as

$$u^*|_{x^*=0} = \left.\frac{\partial v^*}{\partial x^*}\right|_{x^*=0} = \left.\frac{\partial w^*}{\partial x^*}\right|_{x^*=0} = \left.\frac{\partial u^*}{\partial y^*}\right|_{y^*=0} = v^*|_{y^*=0} = \left.\frac{\partial w^*}{\partial y^*}\right|_{y^*=0} = 0, \tag{11a}$$

$$u^*|_{x^*=W^*} = \left[v^* + C\ell^*\frac{\partial v^*}{\partial x^*}\right]_{x^*=W^*} = \left[w^* + C\ell^*\frac{\partial w^*}{\partial x^*}\right]_{x^*=W^*} = 0, \tag{11b}$$

$$\left[u^* + C\ell^*\frac{\partial u^*}{\partial y^*}\right]_{y^*=H^*} = v^*|_{y^*=H^*} = \left[w^* + C\ell^*\frac{\partial w^*}{\partial y^*}\right]_{y^*=H^*} = 0, \tag{11c}$$

where C may now vary along the surface of the NW to account for the behaviour of phonons near corners [14]. By construction, the total heat flux per unit area is

$$Q^* = \frac{1}{4H^*W^*}\int_{-H^*}^{H^*}\int_{-W^*}^{W^*} w^* dx^* dy^*, \tag{12}$$

hence calculating Q^* requires only knowledge about w^*.

Using a similar strategy as in the circular case, defining the dimensionless quantities $x = x^*/W^*$, $y = y^*/H^*$, $z = z^*/L^*$, $T = (T^* - T_0^*)/\Delta T$, $k = k^*/k_0^*$, $u = u^*/(\phi^{-1}\epsilon w_0^*)$, $v = v^*/(\epsilon w_0^*)$, and $w = w^*/w_0^*$, we find $\partial T/\partial x = \partial T/\partial y = 0$ at leading order and hence the problem of computing the ETC can be reduced to solving

$$w - \mathrm{Kn}^2\left(\phi^2\frac{\partial^2 w}{\partial x^2} + \frac{\partial^2 w}{\partial y^2}\right) = -k\frac{\partial T}{\partial z}, \tag{13}$$

subject to

$$\left.\frac{\partial w}{\partial x}\right|_{x=0} = \left.\frac{\partial w}{\partial y}\right|_{y=0} = \left[w + \phi\mathrm{Kn}C\frac{\partial w}{\partial x}\right]_{x=1} = \left[w + \mathrm{Kn}C\frac{\partial w}{\partial y}\right]_{y=1} = 0, \tag{14}$$

where the Knudsen number is now defined as $\mathrm{Kn} = \ell^*/H^*$. Upon writing $w = w_F \cdot \hat{w}$, where $w_F = -k\partial w/\partial z$, the ETC is then computed via

$$k_{\mathrm{eff}}/k = \int_0^1 \int_0^1 \hat{w} dx dy. \tag{15}$$

Due to the dependence of the coefficient C on the variables x and y, the problem defined by (13) and (14) cannot be solved analytically in general. However, under the assumption that C is uniform on the boundary, a separable solution for (13) can be constructed, giving

$$k_{\mathrm{eff}}/k = \sum_{n,m\geq 0} \frac{c(\eta_n)c(\mu_m)}{1 + \mathrm{Kn}^2\left(\phi^2\eta_n^2 + \mu_m^2\right)}, \qquad c(t) = \frac{2\sin^2(t)}{t(t + \cos(t)\sin(t))}, \tag{16}$$

where η_n and μ_m satisfy

$$\cot(\eta_n) = \phi C \mathrm{Kn} \eta_n, \qquad \cot(\mu_m) = C \mathrm{Kn} \mu_m. \tag{17}$$

In the case of a non-uniform slip coefficient $C(x, y)$, we can give analytical expressions for the ETC in the limits of small and large Knudsen numbers [4],

$$k_{\mathrm{eff}}/k \approx 1 - \mathrm{Kn}\left[\int_0^1 \frac{dx}{1 + C(x, 1)} + \phi \int_0^1 \frac{dy}{1 + C(1, y)}\right], \quad \text{for } \mathrm{Kn} \ll 1, \tag{18a}$$

$$k_{\mathrm{eff}}/k \approx \frac{1}{\mathrm{Kn}}\left[\int_0^1 C^{-1}(x, 1)dx + \phi \int_0^1 C^{-1}(1, y)dy\right], \quad \text{for } \mathrm{Kn} \gg 1. \tag{18b}$$

3 Results

In Fig. 3 we compare our analytical expressions against experimental data provided by Refs. [7, 8]. The data for $\ell^*(T^*)$ and $k^*(T^*)$ is obtained from an open based on the kinetic collective model [11].

In the circular case, our choice of C leads to excellent results except for temperatures below ~150 K for the case of $R^* = 28$ nm. In the rectangular case, taking a constant slip coefficient $C = 4$, which agrees with previous works [11, 12], gives excellent results in the low and high temperature limits. The intermediate regime is well captured by a exponential expression $C(T^*) = 4\exp(-T^*/T_c^*)$ with a fitting parameter T_c^*. In the case of silicon, the fitting parameter was found to be $T_c^* \approx 9$ K.

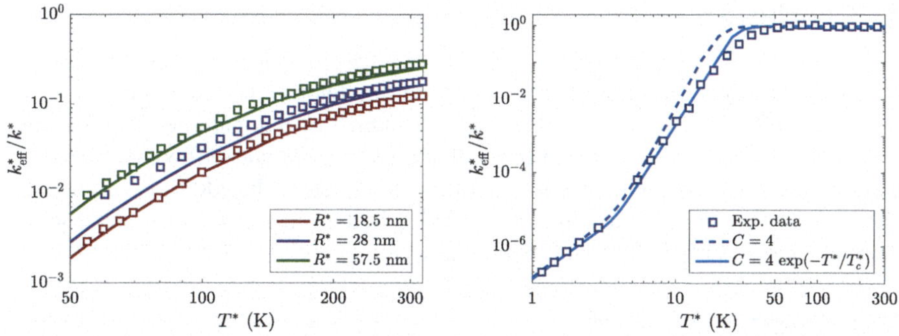

Fig. 3 Left: Validation of the analytical expression (9) with $C(T^*) = \exp(-R^*/\ell^*(T^*))$, represented by solid lines, against experimental data for Si [8], represented by squares. Different colors represent nanowires with different radii. Right: Validation of the analytical expression (16) for different forms of the slip coefficient, against experimental data for Si [7], represented by squares. For the temperature-dependent coefficient we have used $C = 4\exp(-T^*/T_c^*)$ with $T_c^* = 9$ K. The dimensions of the slab correspond to $\phi \approx 0.641$

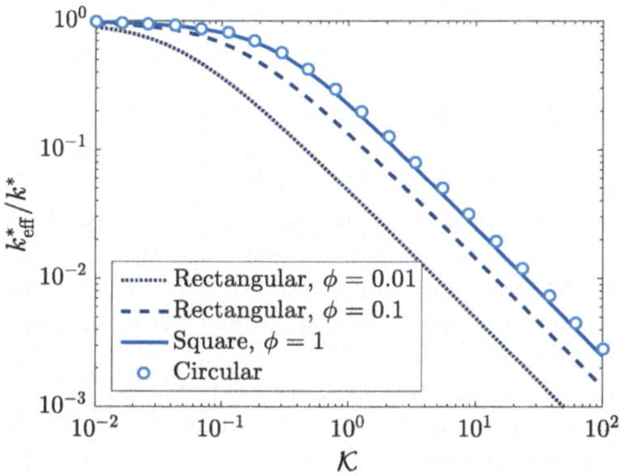

Fig. 4 Comparison of the predicted ETC for different cross-sections in terms of the effective Knudsen number $\mathscr{K} = \ell^*/\sqrt{A^*}$, where A^* is the area of a cross-section, independently of the shape considered

Being able to predict the thermal conductivity for a NW with either a circular or rectangular cross-section enables us to decide which geometry is the most efficient heat transporter. For this we introduce, for a fixed cross-sectional area A^*, the effective Knudsen number $\mathscr{K} = \ell^*/\sqrt{A^*}$. In the circular case we have Kn$= \sqrt{\pi}\mathscr{K}$, whereas in the rectangular case Kn$= 2\phi^{-1/2}\mathscr{K}$. In Fig. 4 we can see that the optimal case corresponds to the circular and square ($\phi = 1$) cross-sections, whilst in the thin film limit ($\phi \to 0$) we obtain a less efficient conductor.

4 Conclusions

By choosing a specific form of the slip-length, the hydrodynamic model is able to predict the ETC in nanowires with a reduced number of free parameters, providing excellent results. Furthermore, in the ballistic ($\mathrm{Kn} \gg 1$) and diffusive ($\mathrm{Kn} \ll 1$) regimes the ETC is obtained without fitting parameters. In addition, among the different cross-sections, the circular nanowire is the most efficient transporter of thermal energy, with the thin film being the worst.

Acknowledgements M. C. acknowledges that the research leading to these results has received funding from "La Caixa" Foundation. M. H. has received funding from the European Union's Horizon 2020 research and innovation programme under the Marie Skłodowska-Curie grant agreement No. 707658. T. M. acknowledges the support of a Ministerio de Ciencia e Innovacion grant MTM2017-82317-P. P. T. and F. X. A. acknowledge the financial support of the Spanish Ministry of Economy and Competitiveness under Grant Consolider nanoTHERM CSD2010-00044, TEC2015-67462-C2-2-R (MINECO/FEDER), TEC2015-67462-C2-1-R (MINECO/FEDER). The authors have been partially funded by the CERCA Programme of the Generalitat de Catalunya.

References

1. Alvarez, F.X., Jou, D., Sellitto, A.: Phonon hydrodynamics and phonon-boundary scattering in nanosystems. J. Appl. Phys. **105**(1), 014317 (2009)
2. Cahill, D.G., Ford, W.K., Goodson, K.E., Mahan, G.D., Majumdar, A., Maris, H.J., Merlin, R., Phillpot, S.R.: Nanoscale thermal transport. J. Appl. Phys. **93**(2), 793–818 (2003)
3. Calvo-Schwarzwälder, M., Hennessy, M.G., Torres, P. Myers, T.G., Alvarez, F.X.: A slip-based model for the size-dependent effective thermal conductivity of nanowires. Int. Commun. Heat Mass Transfer **91**, 57–63 (2018)
4. Calvo-Schwarzwälder, M., Hennessy, M.G., Torres, P. Myers, T.G., Alvarez, F.X.: Effective thermal conductivity of rectangular nanowires based on phonon hydrodynamics. Int. J. Heat Mass Transfer **126**, 1120–1128 (2018)
5. Guyer, R.A., Krumhansl, J.A.: Solution of the linearized phonon Boltzmann equation. Phys. Rev. **148**(2), 766 (1966)
6. Guyer, R.A., Krumhansl, J.A.: Thermal conductivity, second sound, and phonon hydrodynamic phenomena in nonmetallic crystals. Phys. Rev. **148**(2), 778 (1966)
7. Inyushkin, A.V., Taldenkov, A.N., Gibin, A.M., Gusev, A.V., Pohl, H.-J.: On the isotope effect in thermal conductivity of silicon. Phys. Stat. Sol. (C) **1**(11), 2995–2998 (2004)
8. Li, D., Wu, Y. Kim, P., Shi, L. Yang, P., Majumdar, A.: Thermal conductivity of individual silicon nanowires. Appl. Phys. Lett. **83**(14), 2934–2936 (2003)
9. Sellitto, A., Alvarez, F.X., Jou, D.: Temperature dependence of boundary conditions in phonon hydrodynamics of smooth and rough nanowires. J. Appl. Phys. **107**(11), 114312 (2010)
10. Sellitto, A., Alvarez, F.X., Jou, D.: Geometrical dependence of thermal conductivity in elliptical and rectangular nanowires. Int. J. Heat Mass Transfer **55**(11), 3114–3120 (2012)
11. Torres, P., Torelló, A., Bafaluy, J., Camacho, J., Cartoixà, X., Alvarez, F.X.: First principles kinetic-collective thermal conductivity of semiconductors. Phys. Rev. B **95**(4), 165407 (2017)
12. Zhang, Z. M.: Nano/Microscale Heat Transfer. McGraw Hill, New York (2017)
13. Zhu, C.-Y., YOu, W., Li, Z.-Y.: Nonlocal effects and slip heat flow in nanolayers. Sci. Rep. **7**, 9568 (2017)
14. Ziman, J.M.: Electrons and Phonons: The Theory of Transport Phenomena in Solids. Oxford University Press, Oxford (1960)

On the Lifespan of Lithium-Ion Batteries for Second-Life Applications

Daniel Müller and Kai Peter Birke

Abstract The second-life concept adds monetary value to disused automotive batteries. In turn, this could lead to a higher market share of electric transportation by reducing the total costs for the consumer. However, the ageing of batteries limits their total lifetime and the non-linear ageing behaviour at later stages can diminish the benefit of second-life application. With a model-based ageing study, we show that the lifetime can be doubled by introducing a two-stage anode porosity.

1 Introduction

Like many other electrochemical cells, lithium-ion batteries exhibit a reduction in performance during usage and storage. This degradation over the batteries' lifetime is commonly called ageing and is the result of unwanted side-reactions or mechanical and structural changes in different parts of the cell. For some applications, especially when high energy- or power-density is required, this degradation necessitates the replacement of the battery pack. Recycling consequently terminates the conventional life-cycle of a battery. Yet, under certain conditions, reapplication of those batteries can still provide economic and ecological benefit.

1.1 Second-Life

The approach of reusing batteries in a secondary application is called second-life. With increasing electrification in the transport sector, the amount of replaced batteries from electric vehicles is going to rise significantly. Therefore, a growing economic and ecological benefit is ascribed to the second-life concept. Traction

D. Müller (✉) · K. P. Birke
Electrical Energy Storage Systems, Institute for Photovoltaics, University of Stuttgart, Stuttgart, Germany
e-mail: daniel.mueller@ipv.uni-stuttgart.de; peter.birke@ipv.uni-stuttgart.de

© Springer Nature Switzerland AG 2019
I. Faragó et al. (eds.), *Progress in Industrial Mathematics at ECMI 2018*, Mathematics in Industry 30, https://doi.org/10.1007/978-3-030-27550-1_6

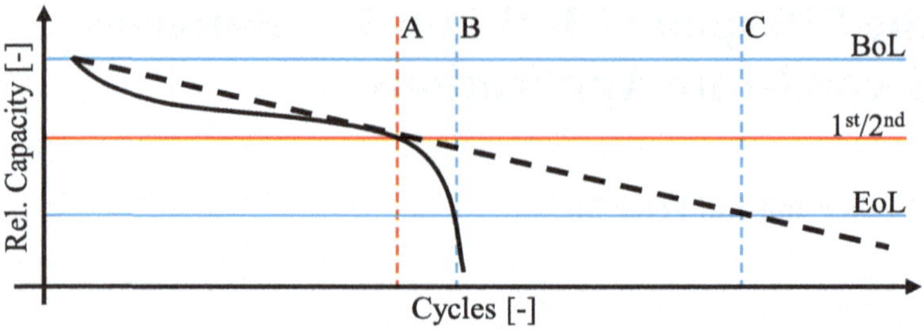

Fig. 1 Schematic capacity retention curve for idealised linear ageing (dashed black line) and non-linear ageing (solid black line). The red markers show the assumed transition between primary and secondary application with a criterion based on the remaining capacity. For non-linear ageing, the sudden-death effect reduces the second-life (A to B) significantly, while for idealised linear ageing the usable time (A to C) is much higher

batteries are an excellent candidate for second-life application. High requirements from the automotive sector lead to an early replacement and therefore the possibility for less demanding second-life application with favorable operating conditions, for example as stationary storage systems.

The point in time of transition between primary and second-life application is defined by an End-of-Life (EoL) criterion for the intended usage. For mobile applications, it is often assumed to be 80% of the battery's initial capacity. This transition criterion is crucial for second-life applications since at around the same value for residual capacity or State-of-Health (SoH) a change in ageing behaviour has been observed. Figure 1 shows two ageing curves, a solid black line for non-linear and a dashed black line for idealised linear ageing. The solid red line depicts the transition criterion based on the remaining capacity, while the dashed red line shows the resulting cycle number at the time of transition. For both ageing curves, the Beginning-of-Life (BoL) state, the transition criterion between applications and the terminal EoL criterion are the same. With those assumptions, the idealised linear ageing offers a decent period for second-life application (A to C) while for the actual non-linear ageing this period is significantly reduced (A to B).

1.2 Sudden-Death

While ignoring the initial cycles for formation, the ageing behaviour of lithium-ion batteries can in a simplified way be separated into two phases. A phase of linear ageing behaviour from the start until an inflexion point in the capacity retention, where the ageing becomes non-linear. This change in degradation per cycle is called sudden-death and has been documented at a 1.5-fold increase of inner resistance and around 80% of residual capacity [1]. The solid line in Fig. 1 shows a sudden-death (around point A). A cause for the sudden-death is a change in the predominant ageing mechanism. During the linear stage, this is the formation of solid electrolyte interface (SEI), but ageing-induced lithium plating is responsible for the non-linearity. Commonly, plating of metallic lithium is related to charging at low temperatures or with high C-rates, but this heterogeneous ageing-induced plating also occurs at mild operating conditions. Additionally, once started it amplifies itself. Literature sources specify pore clogging due to SEI growth in the anode, causing a kinetic hindrance for the ion transport in the electrode, as the main reason for this failure mode [2–4]. Besides the layer growth, sudden-death can be induced mechanically by local compression [5] or caused by loss of active material [6].

In this model-based analysis, we investigate the effect of graded porosity in graphite anodes with regard to second-life application and the sudden-death ageing effect. Traditionally, batteries have a constant porosity across the thickness of each electrode and the separator. To compensate for the increased layer growth at the separator side of the anode, we introduce a two-stage porosity for the anode or in other words an anode consisting out of two layers with different porosities.

2 Model Description

The employed model is based on the electrochemical pseudo 2 dimensional (P2D) approach from Doyle et al. A detailed description and deduction of these underlying equations can be found in the corresponding literature [7–11]. Those governing equations are based on charge and species conservation in the solid and electrolyte phase, respectively. These partial differential equations were solved in a commercial implementation of the finite element method (FEM). The Butler-Volmer kinetics expression as well as appropriate boundary conditions provide the relation between the liquid electrolyte phase and the solid phase of the electrode particles.

Modelling of the ageing mechanism is realised similar to the model presented by Yang et al. [12]. We only consider lithium plating and SEI formation, both assumed to be irreversible, at the anode and both are implemented as competing side-reactions with the intercalation. This results in the local current density being the sum of the current densities of intercalation, lithium plating and SEI formation. For the lithium intercalation, the Butler-Volmer equation, with cathodic and anodic part, and an exchange current density depending on rate constants, transfer coefficients as well as liquid and solid lithium concentrations are used for calculation. Following the work of Safari et al. [13] and Darling et al. [14] about SEI formation, we use cathodic Tafel equations for the SEI formation as well as for the lithium plating. Exchange current densities for the side reactions are calculated in a simplified manner and used for adjusting the ageing progress.

Both side reactions consume lithium ions and deposit their products on the anode particles. The deposition leads to a growing layer, which increases the resistance, influencing the surface overpotential and the potential drop across the layer. At the same time, it reduces the porosity of the anode, which directly relates to a hindrance of ion transport in the electrolyte.

Cell parameters belong to an experimental high-energy cell with thick electrodes and are provided by a German automotive supplier. The simulation uses a constant current/constant voltage (CC/CV) strategy for charging. Discharge is CC only. C/2 current rates are applied for the whole simulation. Between the single phases, 10-min relaxation periods are included.

3 Graded Porosity in Graphite Anodes

In this section, we compare simulated cycling results of two cell designs, a conventional design and a two-stage anode porosity design. Apart from the anode porosity, the cell parameters are identical. As an explanation, the expressions porosity and electrolyte volume fraction are interchangeable in this model approach. In the conventional design, the electrolyte volume fraction ϵ_l of the anode amounts to $\epsilon_l = 0.26$, with an anode thickness of d_A. The graded anode basically consists of two layers, a layer with lower electrolyte volume fraction of $\epsilon_l = 0.23$ facing the current collector and a second layer with higher electrolyte volume fraction of $\epsilon_l = 0.29$ at the separator side. Both layers are equally thick with $d_L = d_A/2$, this results in the same averaged starting porosity as the conventional cell and therefore the same nominal capacity of both designs. Even though further optimisation is possible, we chose a two-stage porosity over a more sophisticated porosity profile having an acceptable additional effort during electrode production in mind.

Figure 2 shows the normalised discharge capacity Q_{DC} over cycle number n for both mentioned cell designs. The reference design, blue line, exhibits the expected ageing behaviour. A decreasing ageing rate, a reduction of capacity loss per cycle, until around 260 cycles. At a remaining capacity of approximately $Q_{DC} = 0.75$ the rate increases, there is an inflexion point in the capacity retention curve, and the battery experiences a sudden-death. After 250 cycles, the electrolyte volume fraction in the anode at the anode/separator boundary has decreased below $\epsilon_l = 0.04$, this can be assumed to equal congested pores. During a couple of additional cycles, the remaining capacity of this cell fades quickly.

For the first 150–200 cycles, the behaviour of the cell with two-stage porosity is basically identical. The orange curve also shows a kink after around 240 cycles, but it is considerably less severe than for the traditional design. This first visible change in ageing rate happens when the electrolyte volume fraction also falls below $\epsilon_l = 0.04$ at the border of the low porosity layer in the middle of the electrode. This means, ion transport into the rear part of the electrode is severely limited. Here, the sudden-death event occurs after more than 500 cycles, compared to 260 for the conventional cell. At this point, the remaining capacity falls below $Q_{DC} = 0.5$ and the electrolyte volume fraction reaches an even lower value of $\epsilon_l = 0.02$ at the anode/separator boundary.

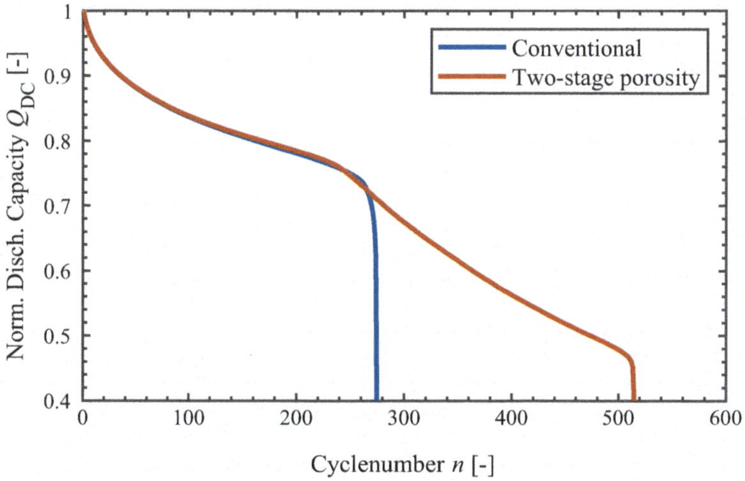

Fig. 2 Normalised discharge capacity over the number of cycles. The blue line shows the ageing rate of a conventional anode with constant initial porosity. A two-stage porosity with a lower porosity layer at the current collector and a higher porosity layer at the separator is simulated for the orange line

4 Conclusion

With this ageing simulation, we show the potentially disastrous influence of the non-linear ageing behaviour, the sudden-death effect, on the second-life concept. By utilising the two-stage porosity in the negative electrode, we can increase the cycle count until the sudden-death twofold. Applying the before mentioned transition criterion from primary application to second-life of $Q_{DC} = 0.8$ and an EoL at $Q_{DC} = 0.6$, the number of second-life cycles increases almost by a factor of 2 when the two-stage anode porosity is used. Shifting the transition or EoL criterion to lower capacities will increase the two-stage porosity designs advantage even further.

References

1. Ecker, M., Nieto, N., Käbitz, S., Schmalstieg, J., Blanke, H., Warnecke, A., Sauer, D.U.: Calendar and cycle life study of $Li(NiMnCo)O_2$-based 18650 lithium-ion batteries. J. Power Sources **248**, 839–851 (2014)
2. Broussely, M., Biensan, P., Bonhomme, F., Blanchard, P., Herreyre, S., Nechev, K., Staniewicz, R.J.: Main aging mechanisms in Li ion batteries. J. Power Sources **146**, 90–96 (2005)
3. Burns, J.C., Kassam, A., Sinha, N.N., Downie, L.E., Solnickova, L., Way, B.M., Dahn, J.R.: Predicting and extending the lifetime of Li-Ion batteries. J. Electrochem. Soc. **160**, A1451–A1456 (2013)
4. Schuster, S.F., Bach, T., Fleder, E., Müller, J., Brand, M., Sextl, G., Jossen, A.: Nonlinear aging characteristics of lithium-ion cells under different operational conditions. J. Energy Storage **1**, 44–53 (2015)

5. Bach, T.C., Schuster, S.F., Fleder, E., Müller, J., Brand, M.J., Lorrmann, H., Jossen, A., Sextl, G.: Nonlinear aging of cylindrical lithium-ion cells linked to heterogeneous compression. J. Energy Storage **5**, 212–223 (2016)
6. Dubarry, M., Truchot, C., Liaw, B.Y., Gering, K., Sazhin, S., Jamison, D., Michelbacher, C.: Evaluation of commercial lithium-ion cells based on composite positive electrode for plug-in hybrid electric vehicle applications. Part II. Degradation mechanism under 2 C cycle aging. J. Power Sources **196**, 10336–10343 (2011)
7. Doyle, M., Fuller, T.F., Newman, J.: Modeling of galvanostatic charge and discharge of the lithium/polymer/insertion cell. J. Electrochem. Soc. **140**, 1526–1533 (1993)
8. Fuller, T.F., Doyle, M., Newman, J.: Simulation and optimization of the dual lithium ion insertion cell. J. Electrochem. Soc. **141**, 1–10 (1993)
9. Fuller, T.F., Doyle, M., Newman, J.: Relaxation phenomena in lithium-ion-insertion cells. J. Electrochem. Soc. **141**, 982–990 (1994)
10. Doyle, M., Newman, J.: Modeling the performance of rechargeable lithium-based cells: design correlations for limiting cases. J. Power Sources **54**, 46–51 (1995)
11. Newman, J., Thomas-Alyea, K. E.: Electrochemical Systems. Wiley, Hoboken (2004)
12. Yang, X.-G., Leng, Y., Zhang, G., Ge, S., Wang, C.-Y.: Modeling of lithium plating induced aging of lithium-ion batteries: transition from linear to nonlinear aging. J. Power Sources **360**, 28–40 (2017)
13. Safari, M., Morcrette, M., Teyssot, A., Delacourt, C.: Multimodal physics-based aging model for life prediction of li-ion batteries. J. Electrochem. Soc. **156**, 145–153 (2009)
14. Darling, R., Newman, J.: Modeling side reactions in composite $LiMn_2O_4$ electrodes. J. Electrochem. Soc. **145**, 990–998 (1998)

Modeling and Simulation Approaches for the Production of Functional Parts in Micro Scale

Andreas Luttmann, Mischa Jahn, and Alfred Schmidt

Abstract In this paper, simulation approaches for the partial melting of metallic workpieces in micro scale are presented. The underlying model considers heat transport in the whole workpiece, the solid-liquid phase transition assuming a sharp interface and the fluid flow in the liquid part including surface tension effects. Depending on whether the solid-liquid interface is handled either by an interface-tracking or an interface-capturing approach, two different numerical schemes based on an ALE finite element method are presented. A crucial aspect for both methods is the geometrical evolution of the solid-liquid-gas triple junction due to the non-material movement of the solid-liquid interface. Yielding mutual advantages and disadvantages, both methods can be used in alternation in a combined approach. Numerical results are shown for melting the tip of a thin steel wire by a laser beam.

1 Introduction

In modern production engineering processes, miniaturization is of growing interest. Due to the challenges arising when transferring cold forming processes from macro to micro range, such as size effects [6], the Collaborative Research Center micro cold forming was founded in 2007. Its central concern is the supply of methods and processes for a systematic design of reliable micro cold forming processes. For this purpose, numerical simulations are a powerful tool to support process development, since they can offer additional insight and make experiments partially obsolete. In the case of complex process chains, it is convenient to simulate only one or several process steps and to formulate a model for the interaction between all steps. For the process chain and simulation approach described in the following, this has been done in [5] using cause-effect networks.

A. Luttmann (✉) · M. Jahn · A. Schmidt
Zentrum für Technomathematik, Universität Bremen, Bremen, Germany
e-mail: andreasl@math.uni-bremen.de; mischa@math.uni-bremen.de; schmidt@math.uni-bremen.de

© Springer Nature Switzerland AG 2019
I. Faragó et al. (eds.), *Progress in Industrial Mathematics at ECMI 2018*, Mathematics in Industry 30, https://doi.org/10.1007/978-3-030-27550-1_7

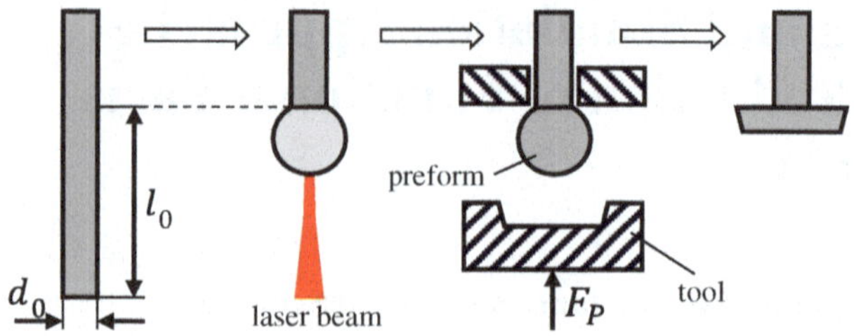

Fig. 1 Two-level cold forming process: In the master forming step, a preform is generated by melting the end of the wire by a coaxial laser beam. In the following cold forming step, a closed die is used to calibrate the preform

As an example, we consider a thermal upsetting process for a thin steel wire with diameter $d_0 < 1$ mm as sketched in Fig. 1. In the master forming step, the metallic wire is partially molten by a coaxial laser beam within a shielding gas atmosphere. Due to the fact that surface tension exceeds gravitational force (shape-balance effect) in micro range [6], the melt forms a nearly perfect sphere which solidifies after the laser is switched off. The generated preform is then calibrated in a subsequent forming step in a closed die. By using the two-level cold forming process upset ratios $s := \frac{l_0}{d_0} \gg 200$ can be achieved while a conventional multi-level cold forming process is limited by the value $s = 2.1$ and decreases if d_0 does.

In the following, modeling and simulation of mass and heat transport for the master forming step are considered. Two different approaches are presented using either an interface-tracking or interface-capturing approach for the solid-liquid interface. After comparing both the advantages and disadvantages with special attention to the solid-liquid-gas triple junction, a combined approach is proposed and numerical results for the master forming step are shown.

2 Model

Mass and heat transport are modeled within continuum mechanics by coupling conservation equations for mass, momentum and energy. Assuming a sharp solid-liquid interface $\Gamma_{ls}(t)$ let $\Omega(t) := \Omega_l(t) \cup \Omega_s(t) \cup \Gamma_{ls}(t) \subset \mathbb{R}^3$, $t \in [t_0, t_N]$, denote the time dependent physical domain as sketched in Fig. 2. On $\Gamma_{\{s,l\}}(t)$ the outer normal to $\Omega_{\{s,l\}}(t)$ is given by $\mathbf{n}_{\{s,l\}}(t)$ and $\mathbf{n}_{ls}(t)$ is the outer normal to $\Omega_l(t)$ on $\Gamma_{ls}(t)$. The geometrical evolution in time can be fully described by the normal velocities of $\Gamma_s(t)$, $\Gamma_l(t)$ and $\Gamma_{ls}(t)$ which we denote by $v_{\mathbf{n},s}$, $v_{\mathbf{n},l}$ and $v_{\mathbf{n},ls}$.

The material movement of particles is characterized by the material velocity $\mathbf{u}(t)$. In the liquid part, fluid dynamics are modeled by the incompressible Navier–Stokes equations including surface tension effects and a kinematic boundary condition for

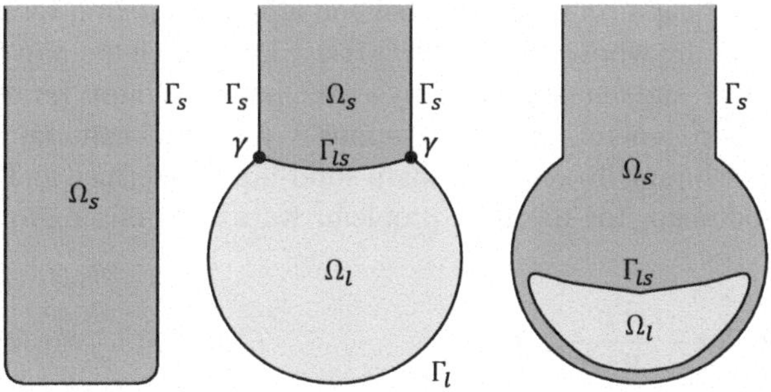

Fig. 2 $2d$ geometrical setting for different topology of $\Gamma_{ls}(t)$: initial state (left), melting (middle) and solidification (right). At the outer boundary, $\Gamma_{\{s,l\}}(t) := \partial\Omega_{\{s,l\}}(t) \cap \partial\Omega(t)$ denotes the solid resp. free capillary boundary and $\gamma(t) := \bar{\Gamma}_s(t) \cap \bar{\Gamma}_l(t) \cap \bar{\Gamma}_{ls}(t)$ the solid-liquid-gas triple junction

the free capillary boundary. In the solid part, we assume no material movement and a no-slip condition at the solid-liquid interface:

$$\partial_t \mathbf{u} + \mathbf{u} \cdot \nabla \mathbf{u} - \frac{1}{Re} \Delta \mathbf{u} + \nabla p = f_u(T), \qquad \nabla \cdot \mathbf{u} = 0 \qquad \text{in } \Omega_l(t), \tag{1}$$

$$\sigma \cdot \mathbf{n}_l = -\frac{1}{We} \mathscr{K} \mathbf{n}_l, \qquad \mathbf{u} \cdot \mathbf{n}_l = v_{\mathbf{n},l} \quad \text{on } \Gamma_l(t), \tag{2}$$

$$\mathbf{u} = 0 \qquad \text{on } \Omega_s(t) \cup \Gamma_s(t) \cup \Gamma_{ls}(t), \qquad v_{\mathbf{n},s} = 0 \qquad \text{on } \Gamma_s(t). \tag{3}$$

Here, p is the pressure, T is the temperature, $f_u(T)$ accounts for buoyancy in Boussinesq-Approximation, σ is the stress tensor, $\mathscr{K}$ is the sum of principal curvatures and Re and We are the Reynolds and Weber number.

In the interface-tracking approach, energy conservation is modeled by the heat equation in each subdomain, a Stefan condition at the solid-liquid interface, which accounts for latent heat and its non-material geometrical evolution, and boundary fluxes:

$$\partial_t T + \mathbf{u} \cdot \nabla T - \frac{q_{\{s,l\}}}{RePr} \Delta T = 0 \qquad \text{in } \Omega_{\{s,l\}}(t), \tag{4}$$

$$-Ste \left[\frac{q_{\{s,l\}}}{RePr} \nabla T \cdot \mathbf{n}_{ls} \right] = v_{\mathbf{n},ls} \qquad \text{on } \Gamma_{ls}(t), \tag{5}$$

$$\frac{q_{\{s,l\}}}{RePr} \nabla T \cdot \mathbf{n}_{\{s,l\}} = g(T) \qquad \text{on } \Gamma_{\{s,l\}}(t). \tag{6}$$

Here, $q_{\{s,l\}}$ are subdomain dependent constants with respect to material parameters, Pr and Ste are the Prandtl and Stefan number, and $g(T)$ accounts for boundary fluxes due to the laser beam, cooling by shielding gas and thermal radiation.

In the interface-capturing approach, an enthalpy formulation for the two-phase Stefan problem in the whole domain $\Omega(t)$ is used, in which the non-linear relation between enthalpy e and temperature T by a function β accounts for the solid-liquid phase transition. In general, this yields a mushy region of material that is neither entirely solid nor liquid. To ensure a sharp solid-liquid interface and the definition of a liquid subdomain for the flow problem, we assign this region to the solid subdomain:

$$\partial_t e + \mathbf{u} \cdot \nabla e - \frac{q}{Re\,Pr} \Delta T = 0, \qquad T = \beta(e) \qquad \text{in } \Omega(t), \tag{7}$$

$$\Omega_l(t) = \{x \in \Omega(t) \colon T(x,t) > T_m\}, \qquad \Gamma_{ls}(t) = \partial\Omega_l(t) \cap \Omega(t). \tag{8}$$

Here, q reflects spatial differences in material parameters, T_m is the melting temperature and boundary heat fluxes are the same as in Eq. (6).

A problem in both modeling approaches is the geometrical evolution of the solid-liquid-gas triple junction $\gamma(t)$. This can be directly seen for the interface-tracking approach, because due to the non-material evolution of the solid-liquid interface, the normal velocities $v_{\mathbf{n},s}$, $v_{\mathbf{n},l}$ and $v_{\mathbf{n},ls}$ of the adjacent boundaries cannot be fulfilled at once. This incompatibility will be left open for the model and we will address it in the numerical method in the next section.

3 Numerical Method

The numerical solution is based on an ALE finite element method for the Navier–Stokes equations with a free capillary surface using the FORTRAN-Code "NAVIER" [1] in a 2d rotational symmetric version. Coupling the flow problem with energy conservation using the interface-tracking approach is substantially based on [2]. For the interface-capturing approach, the finite element method [3] has been adapted to solve Eq. (7). In either case, the subproblems for fluid flow, energy and geometrical evolution are decoupled from each other. For a detailed description of all methods and results see [4].

We now address the discrete evolution of the triple junction $\gamma(t)$ as sketched in Fig. 3 and how the overdetermination due to $v_{\mathbf{n},s}$, $v_{\mathbf{n},l}$ and $v_{\mathbf{n},ls}$ is resolved. For convenience, we will not add additional indices to indicate discrete quantities and use the same notation as before. In the interface-tracking approach, we simply omit one condition and set the velocity $\mathbf{v}$ of $\gamma(t)$ according to

$$\mathbf{v} \cdot \mathbf{n}_{ls} = v_{\mathbf{n},ls} \qquad \text{and} \qquad \begin{cases} \mathbf{v} \cdot \mathbf{n}_s = v_{\mathbf{n},s} & \text{for } v_{\mathbf{n},ls} \geq 0 \quad \text{(melting)} \\ \mathbf{v} \cdot \mathbf{n}_l = v_{\mathbf{n},l} & \phantom{\text{for }} v_{\mathbf{n},ls} < 0 \text{ (solidification)} \end{cases}, \tag{9}$$

such that the evolution of $\gamma(t)$ is always tangential to $\partial\Omega$. The violation of the third condition can thereby be interpreted as an additional approximation error.

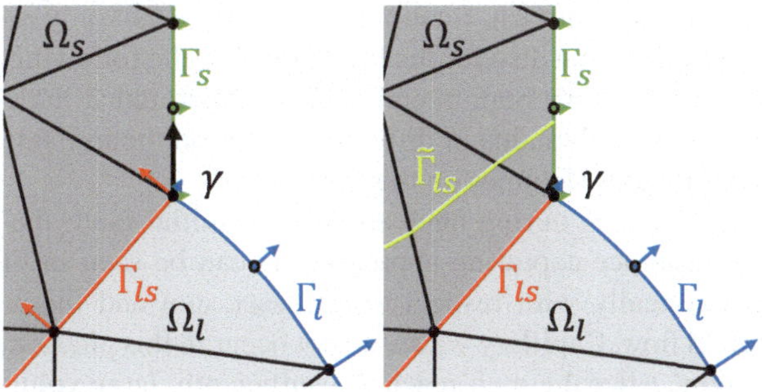

Fig. 3 Discrete boundary evolution around triple junction: The evolution is determined by the velocities in each vertex and edge midpoint. For the interface-tracking (left) as well as the interface-capturing (right) approach, the normal velocities $v_{\mathbf{n},\{s,l,ls\}}$ are indicated in each point. Additionally, the resulting velocity $\mathbf{v}$ of $\gamma(t)$ is shown (thick)

In the interface-capturing approach, the evolution of the solid-liquid phase boundary is not linked to the edges. Instead, Eq. (8) is adapted to the triangulation $\mathscr{S}(t)$ by

$$\Omega_l(t) = \cup\{S \in \mathscr{S}(t)\colon T|_S > T_m\}, \qquad \Gamma_{ls}(t) = \partial\Omega_l(t) \cap \mathring{\Omega}(t), \tag{10}$$

such that a triangle S can change its phase state within one time step. Due to the singular evolution, we have either no overdetermination at $\gamma(t)$ due to $v_{\mathbf{n},s} = v_{\mathbf{n},l} = v_{\mathbf{n},ls} = 0$, or a singular jump of $\gamma(t)$ if a neighboring triangle at the boundary changes its phase. If this jump introduces a kink to $\Gamma_l(t)$, which typically occurs during melting, an artificial capillary wave due to an imbalance between surface tension and inner forces originates at the kink.

Comparing both approaches, these capillary waves only occur in the interface-capturing approach. On the other hand, the interface-tracking approach is not able to handle topology changes of $\Gamma_{ls}(t)$ (see Fig. 2). The remedy is to switch between both approaches during runtime and use the best approach for each situation. Switching between both approaches is straightforward as it requires either calculating e from T by approximating β^{-1} when switching to the interface-capturing approach or a remeshing procedure that integrates $\tilde{\Gamma}_{ls}(t) := \partial\{x \in \Omega(t)\colon T(x,t) > T_m\} \cap \mathring{\Omega}(t)$ as $\Gamma_{ls}(t)$ onto edges of the new mesh when switching to the interface-tracking approach.

4 Numerical Results

Numerical results are shown for a steel wire with diameter $d_0 = 0.4\,\mathrm{mm}$ that is heated for 100 ms with a laser power of $P = 130\,\mathrm{W}$. Dimensionless quantities in this case are $Re \approx 0.57$, $Pr \approx 0.13$, $We \approx 3.38 \cdot 10^{-6}$ and $Ste \approx 4.21$. Due to the

2d rotational symmetric approach, rotational symmetry is not only assumed for the geometrical configuration but also for the fluid flow. For the following visualization, the 2d geometry of the wire's bottom end is shown mirrored at the symmetry axis to show the mesh and subdomains in different shades on the left side and the fluid flow as well as temperature isolines on the right side.

Since topology changes during nucleation of an initial melt, the simulation is started using the interface-capturing approach. As can be seen in Fig. 4, the melt starts growing spherically with respect to the laser spot and thermal circulation determines the fluid flow. Capillary waves do not occur in this phase, since the lower boundary is straight. After the melt reaches a sufficiently large volume, remeshing is performed to switch to the interface-tracking approach for the melting phase.

Shape changes due to surface tension occur as soon as the triple junction $\gamma(t)$ reaches the radial boundary. As shown in Fig. 5, the melt starts forming a semi-sphere before growing further into an upwards moving full sphere. In this phase the fluid flow gets about 100 times larger compared to thermal circulation seen before. Due to overheating, melting continues for another 107 ms after laser switch-off.

Switching back to the interface-capturing approach, solidification starts beginning from the solid material and thermal circulation becomes predominant again. Since a nearly perfect sphere has already been formed at the wire's bottom end, the geometry barely changes and artificial capillary waves do not occur. Later on,

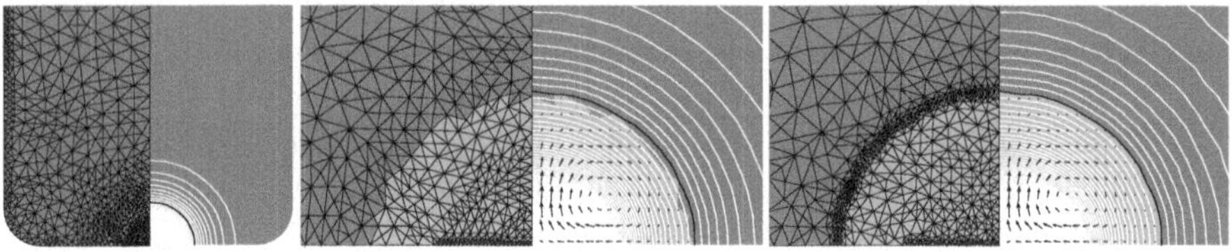

Fig. 4 End of nucleation phase: Switch from interface-capturing (zoomed out left, middle) to interface-tracking approach (right) at $t = 0.32$ ms

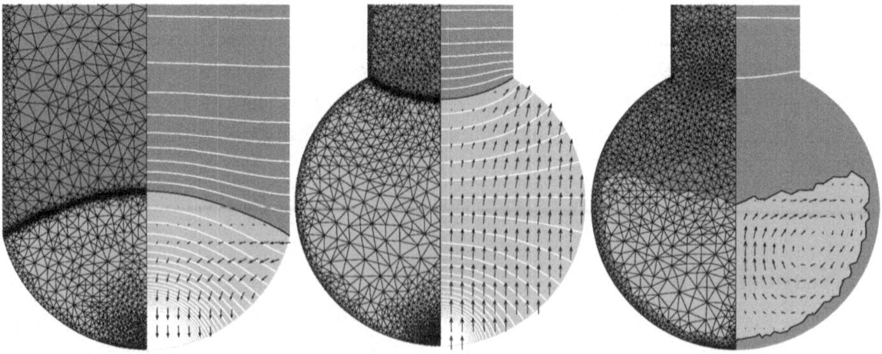

Fig. 5 Melting and solidification: Formation of a semi-sphere at $t = 6.5$ ms (left), upwards moving full sphere at $t = 100$ ms (middle) and solidification from all sides at $t = 400$ ms (right)

topology is changing due to simultaneous solidification originating from the free capillary boundary. Finally, solidification ends at $t = 605$ ms and a preform with an upset ratio of $s \approx 7.3$ and a nearly perfect spherical shape has been achieved.

5 Conclusion

Neither the interface-tracking nor the interface-capturing approach is capable of simulating the whole master forming step in a satisfying manner. Since drawbacks of both approaches occur in different time periods, a combined approach based on alternation between both methods has been proposed. By doing so, the whole process can be simulated in a satisfying manner without suffering from any drawback of either method.

Acknowledgements The authors gratefully acknowledge the financial support by the DFG (German Research Foundation) for the subproject A3 within the Collaborative Research Center SFB 747 Mikrokaltumformen—Prozesse, Charakterisierung, Optimierung.

References

1. Bänsch, E.: Finite element discretization of the Navier–Stokes equations with a free capillary surface. Numer. Math. **88**(2), 203–235 (2001)
2. Bänsch, E., Paul, J., Schmidt, A.: An ALE finite element method for a coupled Stefan problem and Navier–Stokes equations with free capillary surface. Int. J. Numer. Methods Fluids **71**(10), 1282–1296 (2013)
3. Elliot, C.M.: On the Finite element approximation of an elliptic variational inequality arising from an implicit time discretization of the stefan problem. IMA J. Numer. Anal. **1**(1), 115–125 (1981)
4. Luttmann, A.: Modellierung und Simulation von Prozessen mit fest-flüssig Phasenübergang und freiem Kapillarrand. Dissertation. Universität Bremen (2018)
5. Rippel, D., Schattmann, C., Jahn, M., Lütjen, M., Schmidt, A.: Application of cause-effect-networks for the process planning in laser rod end melting. In: MATEC Web of Conferences, vol. 190, p. 15005 (2018)
6. Vollertsen, F.: Categories of size effects. Prod. Eng. **2**(4), 377–383 (2008)

Polynomial Chaos Approach to Describe the Propagation of Uncertainties Through Gas Networks

Stephan Gerster, Michael Herty, Michael Chertkov, Marc Vuffray, and Anatoly Zlotnik

Abstract The ability of gas-fired power plants to ramp quickly is used to balance fluctuations in the power grid caused by renewable energy sources, which in turn leads to time-varying gas consumption and fluctuations in the gas network. Since gas system operators assume nearly constant gas consumption, there is a need to assess the risk of these stochastic fluctuations, which occur on shorter time scales than the planning horizon. We present a mathematical formulation for these stochastic fluctuations as a generalization of isothermal Euler equations. Furthermore, we discuss control policies to damp fluctuations in the network.

1 Introduction

The low cost of natural gas has driven an expansion of gas-fired power plants. In parallel, there has been an expansion of renewable power generation such as wind or solar power. Due to their limited controllability, other grid resources must respond to counteract fluctuations. Different types of controlling the intermittent resources are under consideration. But the use of fast-responding traditional generation, like gas-fired power stations, is the current state-of-practice [14]. Whereas gas systems operators have traditionally met demands that evolve slowly in a well-known pattern, the control of intermittent generation causes unknown fluctuations on a shorter time scale [6, 7].

The meaningfulness of mathematical models depends on the scale of phenomena of interest. Algebraic models may be sufficient to describe average states in a gas network [3, 15]. If there is interest in dynamics on shorter time scales, isothermal Euler equations, which form a 2×2 system of hyperbolic balance laws, provide a

S. Gerster (✉) · M. Herty
IGPM, RWTH Aachen University, Aachen, Germany
e-mail: gerster@igpm.rwth-aachen.de; herty@igpm.rwth-aachen.de

M. Chertkov · M. Vuffray · A. Zlotnik
CNLS, Los Alamos National Laboratory, Los Alamos, NM, USA
e-mail: chertkov@lanl.gov; vuffray@lanl.gov; azlotnik@lanl.gov

© Springer Nature Switzerland AG 2019

I. Faragó et al. (eds.), *Progress in Industrial Mathematics at ECMI 2018*,
Mathematics in Industry 30, https://doi.org/10.1007/978-3-030-27550-1_8

suitable model [2, 19]. A mathematical theory for hyperbolic systems on networks has been developed for the Euler equations [9] and for the p-system [8].

Extensions to stochastic influences are only partial. Sampling-based methods use numerical quadrature to compute the statistics of interest, which leads to high computational cost [1]. Therefore, a splitting into a stationary and a fluctuating component is proposed in [5–7].

At the expense of higher computational cost, we describe uncertainties in nonlinear systems, where the gas dynamics may deviate significantly from steady states. We borrow the idea to represent stochastic processes by orthogonal polynomials from [4, 10, 12, 16–18]. This approach is known as stochastic Galerkin formulation with a generalized polynomial chaos (gPC) expansion. The stochastic input is represented as a truncated gPC series expansion. It is substituted into the governing equations and projected by a stochastic Galerkin method to obtain deterministic evolution equations for the gPC modes. To the best of our knowledge, there is currently no hyperbolic stochastic Galerkin formulation for fluid dynamic equations on *networks*. We propose to handle boundary conditions in a sample-based way, while the stochastic Galerkin formulation is used within the pipe. This flexible ansatz allows to consider various coupling conditions.

2 Stochastic Gas Flow on Networks

For simplicity, we study a single node, where L ingoing and R outgoing pipes meet. The stochastic density of the gas $\rho(t,x;\xi)$ and the mass flux $q(t,x;\xi)$, defined on $(t,x) \in [0,T] \times [0,x_{\text{end}}]$, are parameterized by a possibly multidimensional random variable ξ and are described by the isothermal Euler equations

$$\frac{\partial}{\partial t}\begin{pmatrix} \rho^{(j)}(t,x;\xi) \\ q^{(j)}(t,x;\xi) \end{pmatrix} + \frac{\partial}{\partial x}\begin{pmatrix} q^{(j)}(t,x;\xi) \\ \frac{q^{(j)}(t,x;\xi)^2}{\rho^{(j)}(t,x;\xi)} + a^2 \rho^{(j)}(t,x;\xi) \end{pmatrix} = -\frac{f}{2D}\begin{pmatrix} 0 \\ \frac{q^{(j)}(t,x;\xi)|q^{(j)}(t,x;\xi)|}{\rho^{(j)}(t,x;\xi)} \end{pmatrix}. \tag{1}$$

The parameters are the speed of sound $a > 0$, the friction factor $f > 0$ and the diameter $D > 0$. We write $y := (\rho, q)^{\text{T}}$ as abbreviation and consider *deterministic initial values* $y^{(j)}(0,x;\xi) = y_0^{(j)}(x)$ only. *Stochastic coupling conditions* for ingoing pipes $y_\ell^{(j)}(t;\xi) := y^{(j)}(t, x_{\text{end}};\xi)$ with $j = 1,\ldots,L$ and outgoing pipes $y_r^{(j)}(t;\xi) := y^{(j)}(t,0;\xi)$ with $j = L+1,\ldots,L+R$, which have the same spatial and material properties, are implicitly given by $L+R$ algebraic equations

$$\mathcal{C} \; : \; \mathbb{R}^{2(L+R)} \to \mathbb{R}^{L+R}, \quad \mathcal{C}\Big[y_\ell^{(1)},\ldots,y_\ell^{(L)},\, y_r^{(L+1)},\ldots,y_r^{(L+R)}\Big](t;\xi) = 0. \tag{2}$$

To set remaining degrees of freedom, we assume the source term being negligibly small near and inside pipe intersections. For subsonic flows there is one negative and one positive characteristic velocity. Lax curves, which are tangent to the eigenvectors of the Jacobian of the flux function, are denoted as $\mathcal{L}_{\ell,r}(\cdot;\, \mathbf{y}) : \mathbb{R}^+ \to \mathbb{R}^2,\ \theta \mapsto \mathcal{L}_{\ell,r}(\theta;\, \mathbf{y})$ and connect the old traces within the pipe with the new traces at the edge. These curves are found e.g. in [8]. The new traces are determined by

$$\mathcal{C}\Big[\mathcal{L}_\ell(\theta^{(1)};\, y_\ell^{(1)}),\ldots,\mathcal{L}_\ell(\theta^{(L)};\, y_\ell^{(L)}),\ \mathcal{L}_r(\theta^{(L+1)};\, y_r^{(L+1)}),\ldots,\mathcal{L}_r(\theta^{(L+R)};\, y_r^{(L+R)})\Big](t;\xi) = 0.$$

We summarize coupling conditions for two pipes, namely one ingoing (in) and one outgoing (out).

1. Compressors: There is conservation of mass, i.e. $q^{(\mathrm{out})}(t, 0;\, \xi) = q^{(\mathrm{in})}(t, x_{\mathrm{end}};\, \xi)$, and the pressure law can be described by fixing the following:

$$\text{Boost Ratio} \qquad \rho^{(\mathrm{out})}(t,0;\,\xi) = \mathrm{BR}(t)\rho^{(\mathrm{in})}(t, x_{\mathrm{end}};\,\xi) \tag{BR}$$

$$\text{Output Pressure} \qquad a^2\rho^{(\mathrm{out})}(t,0;\,\xi) = \mathrm{OP}(t) \tag{OP}$$

2. Gas-fired power plant: The pressure is preserved, i.e. $\rho^{(\mathrm{out})}(t, 0;\, \xi) = \rho^{(\mathrm{in})}(t, x_{\mathrm{end}};\, \xi)$. A prescribed withdrawal $w(t)$ determines the remaining boundary conditions by

$$q^{(\mathrm{out})}(t,0;\,\xi) = q^{(\mathrm{in})}(t, x_{\mathrm{end}};\,\xi) - w(t). \tag{GP}$$

A junction with one ingoing ($j = 1$) and two outgoing pipes ($j = 2, 3$) is modelled by postulating equality of pressure and conservation of mass, i.e.

$$\rho^{(1)}(t, x_{\mathrm{end}};\,\xi) = \rho^{(2)}(t,0;\,\xi) = \rho^{(3)}(t,0;\,\xi) \ \text{ and } \ q^{(1)}(t, x_{\mathrm{end}};\,\xi) = q^{(2)}(t,0;\,\xi) + q^{(3)}(t,0;\,\xi). \tag{JU}$$

These conditions form the boundary control that regulates in-, outflows and compressor power.

3 Stochastic Galerkin Formulation

The functional dependence on the stochastic input is described a priori as a series expansion

$$\mathcal{G}_K[y](t,x;\,\xi) := \sum_{k=0}^{K} \hat{y}_k(t,x)\phi_k(\xi), \qquad \hat{y}_k(t,x) := \frac{\mathbb{E}\big[y(t,x;\,\xi)\phi_k(\xi)\big]}{\mathbb{E}\big[\phi_k^2(\xi)\big]}. \tag{PC}$$

An orthogonal gPC basis is chosen such that $\mathbb{E}\big[\phi_i(\xi)\phi_j(\xi)\big] = 0$ for all $i \neq j$. Using a multiindex notation $\boldsymbol{k} := (k_1, \ldots, k_N)$ we may extend definition (PC) to the multidimensional case as

$$\mathcal{G}_K[y](t,x;\xi) := \sum_{\|\boldsymbol{k}\|_1 \le K_{\mathrm{PC}}} \hat{y}_{\boldsymbol{k}}(t,x)\phi_{\boldsymbol{k}}(\xi) \quad \text{with} \quad \phi_{\boldsymbol{k}}(\xi) := \phi_{k_1}(\xi_1) \cdot \ldots \cdot \phi_{k_N}(\xi_N). \tag{mPC}$$

Expansion (mPC) is of the form (PC) with $K = (N + K_{\mathrm{PC}})!(N!K_{\mathrm{PC}}!)^{-1} - 1$. Hence, the computational complexity grows fast, although it is a sparse basis [12]. A stochastic Galerkin formulation of the balance law (1) is deduced in [11] by inserting expansion (PC) into Eq. (1). The formulations of the nonlinear terms have no explicit expression. We mention here only that the gPC modes $\hat{\rho} := (\hat{\rho}_0, \ldots, \hat{\rho}_K)^{\mathrm{T}}$ and $\hat{q} := (\hat{q}_0, \ldots, \hat{q}_K)^{\mathrm{T}}$ are described by a hyperbolic balance law of the form

$$\frac{\partial}{\partial t}\begin{pmatrix}\hat{\rho}\\ \hat{q}\end{pmatrix} + \frac{\partial}{\partial x}\begin{pmatrix}\hat{q}\\ \widehat{\frac{q^2}{\rho}}(\hat{\rho},\hat{q}) + a^2\hat{\rho}\end{pmatrix} = -\frac{f}{2D}\begin{pmatrix}\mathbb{O}\\ \widehat{\frac{q|q|}{\rho}}(\hat{\rho},\hat{q})\end{pmatrix} \tag{3}$$

with $\mathbb{O} := (0, \ldots, 0)^{\mathrm{T}}$ and we refer the interested reader to [11]. For a fixed realisation ξ the old traces at the edges are $\mathcal{G}_K\big[y^{(j)}\big](t,0;\xi)$ and $\mathcal{G}_K\big[y^{(j)}\big](t,x_{\mathrm{end}};\xi)$. The new traces are determined by the coupling conditions (2). Their gPC modes are

$$\hat{y}_k^{(j)}(t,x_{\mathrm{end}}) = \frac{\mathbb{E}\Big[y_\ell^{(j)}(t;\xi)\phi_k(\xi)\Big]}{\mathbb{E}\big[\phi_k^2(\xi)\big]} \quad \text{and} \quad \hat{y}_k^{(j)}(t,0) = \frac{\mathbb{E}\Big[y_r^{(j)}(t;\xi)\phi_k(\xi)\Big]}{\mathbb{E}\big[\phi_k^2(\xi)\big]} \quad \text{for} \quad k = 0,\ldots,K. \tag{4}$$

4 Numerical Results

All pipes have equal length $x_{\mathrm{end}} = 100\,\mathrm{km}$. The parameters are $a = 376\,\mathrm{m/s}$, $f = 0.01$ and $D = 1$ m. At the beginning of each simulation the network is in a deterministic steady state [13]. Therefore, we can apply the dimensional transformations

$$\bar{t} := \frac{at}{x_{\mathrm{end}}}, \quad \bar{x} := \frac{x}{x_{\mathrm{end}}}, \quad \bar{\rho}^{(j)}(\bar{t},\bar{x};\xi) := \frac{\rho^{(j)}(t,x;\xi)}{\rho_0^{(j)}(0)}, \quad \bar{q}^{(j)}(\bar{t},\bar{x};\xi) := \frac{q^{(j)}(t,x;\xi)}{a\rho_0^{(j)}(0)}.$$

We denote as $\bar{\mathcal{C}}$ the scaled boundary conditions and solve the scaled system

$$\frac{\partial}{\partial \bar t}\begin{pmatrix}\bar\rho^{(j)}(\bar t,\bar x;\xi)\\ \bar q^{(j)}(\bar t,\bar x;\xi)\end{pmatrix}+\frac{\partial}{\partial \bar x}\begin{pmatrix}\bar q^{(j)}(\bar t,\bar x;\xi)\\ \frac{\bar q^{(j)}(\bar t,\bar x;\xi)^2}{\bar\rho^{(j)}(\bar t,\bar x;\xi)}+\bar\rho^{(j)}(\bar t,\bar x;\xi)\end{pmatrix} = -\frac{x_{\text{end}} f}{2D}\begin{pmatrix}0\\ \frac{\bar q^{(j)}(\bar t,\bar x;\xi)|\bar q^{(j)}(\bar t,\bar x;\xi)|}{\bar\rho^{(j)}(\bar t,\bar x;\xi)}\end{pmatrix},$$

$$\bar{\mathcal{C}}\Big[\bar y_\ell^{(1)},\dots,\bar y_\ell^{(L)},\bar y_r^{(L+1)},\dots,\bar y_r^{(L+R)}\Big](\bar t;\xi)=0.$$

The hyperbolic system (3) is solved numerically with the local Lax–Friedrichs flux in space and Heun's method in time. The integrals in Eq. (4) are determined by Gaussian quadrature.

First, we consider one ingoing (left) and one outgoing pipe (right). The mass flux at the right edge is uniformly between 30 and 50 kg/s perturbed. We simulate a gas-fired power plant with a fixed withdrawal $w(t) = 10$ kg/s such that $q^{(1)}(0, x) = 50$ kg/s and $q^{(2)}(0, x) = 40$ kg/s. The steady state is determined for the pressure $a^2\rho_0^{(1)}(0) = 10$ bar. Figure 1 shows the means (green) and the 1.0-confidence regions (grey) for the pressures and mass fluxes. We observe a propagation of uncertainties, which arise from the right boundary, into the domain. The withdrawal does not cause additional fluctuations.

For a compressor, the steady state and boost ratio are chosen such that the input pressures at the left edges of each pipe are $a^2\rho_0^{(1)}(0) = a^2\rho_0^{(2)}(0) = 10$ bar. The mass flux $q^{(1)}(0, x) = q^{(2)}(0, x) = 40$ kg/s is constant. Figure 2 compares the compressor policy (BR), using a fixed boost ratio, with the policy (OP), where the output pressure is fixed. A deterministic output pressure makes uncertainties in the pressure in the outgoing pipe decrease. However, perturbations of the pressure in the ingoing pipe increase. Furthermore, fluctuations in the mass fluxes are higher.

Next, we consider a network with one ingoing (pipe 1) and two outgoing pipes (pipe 2 and pipe 3). Initially, the network is in a steady state with $a^2\rho_0^{(1)}(0) = 10$ bar and $q^{(1)}(0, x) = 50$ kg/s, $q^{(2)}(0, x) = q^{(3)}(0, x) = 25$ kg/s. The first source of uncertainty is the right edge of pipe 2. There, the mass flux $q^{(2)}(t, x_{\text{end}}; \xi)$ is uniformly between 15 and 35 kg/s perturbed. Independently, the right edge of pipe 3

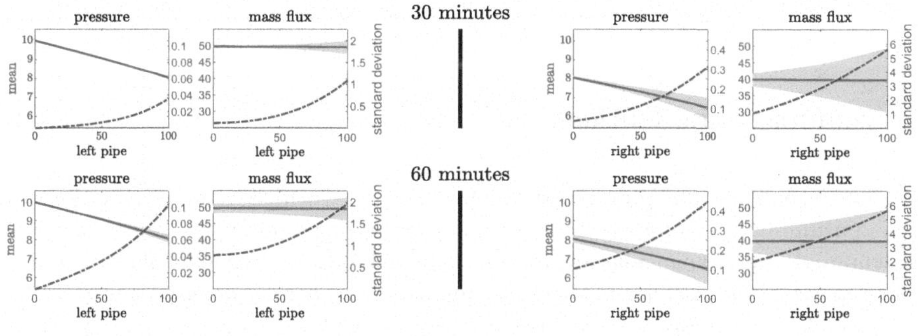

Fig. 1 Fixed withdrawal of gas; units for pressure in bar, mass flux in kg/s

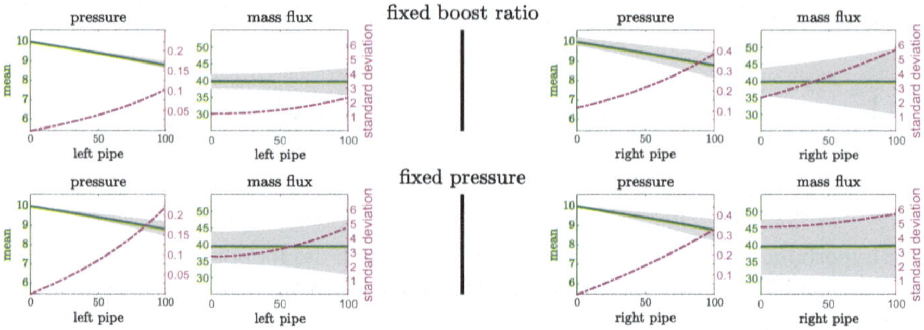

Fig. 2 Two pipes coupled by a compressor; units for pressure in bar, mass flux in kg/s; $t = 60$ min

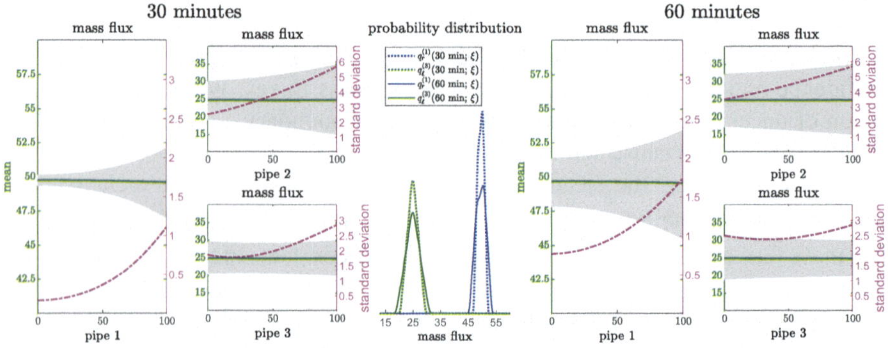

Fig. 3 Junction with one ingoing and two outgoing pipes satisfying the coupling conditions (JU); units for pressure in bar, mass flux in kg/s

is uniformly between 20 and 30 kg/s smaller perturbed. Figure 3 shows how larger perturbations from pipe 2 propagate into pipes 1 and 3. Note the non-monotonic increase in the standard deviation in pipe 3 after 60 min, although the width of the confidence region is monotone decreasing. This is because the perturbations at the left side are peak-shaped, whereas those at the right side are uniformly distributed.

5 Summary and Outlook

We have analyzed the propagation of uncertainties, which arise from unknown gas consumption, through systems of gas pipes. Uncertainties are represented as polynomial chaos expansions and their propagation is described by isothermal Euler equations. Control policies to damp fluctuations have been compared numerically. A combined simulation of gas-grid systems, as well as regulation policies that guarantee pressure and mass fluxes in a prescribed confidence region are a matter of further research. The influence of nonlinear terms compared to linearized models should be studied.

Acknowledgements This work is supported by DFG HE5386/14,15, BMBF 05M18PAA, DFG-GRK 2326, Student Guest Program of Los Alamos National Laboratory.

References

1. Abgrall, R., Congedo, P.M., Geraci, G., Iaccarino, G.: An adaptive multiresolution semi-intrusive scheme for UQ in compressible fluid problems. Int. J. Numer. Methods Fluids **78**, 595–637 (2015)
2. Banda, M.K., Herty, M., Klar, A.: Coupling conditions for gas networks governed by the isothermal Euler equations. Netw. Heterogen. Media **1**, 295–314 (2006)
3. Brouwer, J., Gasser, I., Herty, M.: Gas pipeline models revisited: Model hierarchies, non-isothermal models, and simulations of networks. Multiscale Model. Simul. **9**, 601–623 (2011)
4. Cameron, R.H., Martin, W.T.: The orthogonal development of non-linear functionals in series of Fourier-Hermite functionals. Ann. Math. **48**(2), 385–392 (1947)
5. Chertkov, M., Korotkevich, A.: Adiabatic approach for natural gas pipeline computations. In: IEEE 56th Annual Conference on Decision and Control, pp. 5634–5639. IEEE, Piscataway (2017)
6. Chertkov, M., Backhaus, S., Lebedev, V.: Cascading of fluctuations in interdependent energy infrastructures: gas-grid coupling. Appl. Energy **160**, 541–551 (2015)
7. Chertkov, M., Fisher, M., Backhaus, S., Bent, R., Misra, S.: Pressure fluctuations in natural gas networks caused by gas-electric coupling. In: 48th Hawaii International Conference on System Sciences, pp. 2738–2747. IEEE, Piscataway (2015)
8. Colombo, R.M., Garavello, M.: A well-posed Riemann problem for the p-system at a junction. Netw. Heterogen. Media **1**, 495–511 (2006)
9. Colombo, R.M., Garavello, M.: Euler system for compressible fluids at a junction. J. Hyperbolic Differ. Equ. **5**, 547–568 (2008)
10. Després, B., Poëtte, G., Lucor, D.: Uncertainty quantification for systems of conservation laws. J. Comput. Phys. **228**, 2443–2467 (2009)
11. Gerster, S., Herty, M., Sikstel, A.: Hyperbolic stochastic Galerkin formulation for the p-system. J. Comput. Phys. **395**, 186–204 (2019)
12. Ghanem, R.G., Spanos, P.D.: Stochastic Finite Elements: A Spectral Approach. Springer, New York (1991)
13. Gugat, M., Herty, M.: Existence of classical solutions and feedback stabilization for the flow in gas networks. ESAIM Control Optim. Calc. Var. **17**, 28–51 (2011)
14. Moniz, E., Meggs, A., et al. The Future of Natural Gas. An Interdisciplinary MIT study. MIT Energy Initiative, Cambridge (2011)
15. Osiadacz, A.: Nonlinear programming applied to the optimum control of a gas compressor station. Int. J. Numer. Methods Eng. **15**, 1287–1301 (1980)
16. Pettersson, P., Iaccarino, G., Nordström, J.: A stochastic Galerkin method for the Euler equations with Roe variable transformation. J. Comput. Phys. **257**, 481–500 (2014)
17. Wiener, N.: The homogeneous chaos. Am. J. Math. **60**(4), 897–936 (1938)
18. Xiu, D., Karniadakis, G.E.: The Wiener-Askey polynomial chaos for stochastic differential equations. SIAM J. Sci. Comput. **24**, 619–644 (2002)
19. Zlotnik, A., Roald, L., Backhaus, S., Chertkov, M., Andersson, G.: Control policies for operational coordination of electric power and natural gas transmission systems. In: 2016 American Control Conference (ACC), pp. 7478–7483. IEEE, Piscataway (2016)

Mathematical Modelling of Nanocrystal Growth

Claudia Fanelli, Timothy G. Myers, and Vincent Cregan

Abstract We will describe a model for the process of synthesizing nanoparticles of a specific size from a liquid solution. Initially, we will consider a single particle model that accounts for monomer diffusion in solution around the particle and kinetic reactions at the particle surface. For the far-field bulk concentration, a mass conservation expression is used. Based on a small dimensionless parameter, we propose a pseudo-steady state approximation to the model. The model is then extended to a system of N particles. Numerical solutions for the time-dependent average particle radius compared against experimental data are shown to have excellent agreement.

1 Introduction

Nanoparticles are units of matter with dimensions between 1 and 100 nanometers (nm) that have gained a lot of interest during recent decades, due to their wide variety of applications in biomedicine, environmental-related problems, electronics and catalysis [5]. They have unique chemical, physical, mechanical, and optical properties. Gold nanoparticles provide an excellent example: at the nanoscale, the motion of the gold's electrons is confined and, because of that, they react differently with light compared at a larger scale. The result is that gold nanoparticles are not yellow as we expect, but can appear purple or red. Moreover, adjusting their size, gold nanoparticles can be tuned according to the purpose: for example, they can

C. Fanelli (✉) · T. G. Myers
Centre de Recerca Matemàtica/BGSMath, Barcelona, Spain

Universitat Politècnica de Catalunya, Barcelona, Spain
e-mail: cfanelli@crm.cat; tmyers@crm.cat

V. Cregan
MACSI/PMTC, University of Limerick, Limerick, Ireland
e-mail: vincent.cregan@ul.ie

© Springer Nature Switzerland AG 2019

I. Faragó et al. (eds.), *Progress in Industrial Mathematics at ECMI 2018*,
Mathematics in Industry 30, https://doi.org/10.1007/978-3-030-27550-1_9

selectively accumulate in tumors in order to identify diseased cells and to target laser destruction of the tumor avoiding healthy cells.

It is clear that many properties of nanoparticles are size dependent. Hence, the ability to create nanoparticles of a specific size is crucial. In order to do this, we need a clear understanding of the growth process.

Using the precipitation method (i.e. the creation of a solid from a solution) monodisperse spherical nanoparticles can be generated. The standard approach is to apply the classical La Mer and Dinegar synthesis strategy where nucleation and growth are separated [4]. Moreover, the growth involves two different stages: the *focusing period*, where the mean radius of the particles increases rapidly, and the *defocusing period*, where the growth gets slower and the size distribution broadens. The first phase leads to the desired result of monodisperse nanoparticles. In the second phase we can observe a phenomenon called *Ostwald ripening*, a process by which larger particles grow at the expense of the smaller ones, which dissolve due to their much higher solubility. This process produces monomer, which is subsequently used to support growth of the larger particles. However, this simultaneous growth and dissolution leads to the unwanted defocusing of the particle size distribution (PSD). Recently, it has been shown that the PSD can be refocused by changing the reaction kinetics [1]. The mathematical challenge is to model the process of synthesizing nanoparticles of the required size from a liquid solution.

2 Mathematical Model

Growth occurs due to the diffusion of monomer molecules from the bulk to the surface of the nanoparticles of radius r. The monomer concentration $C(r, t)$ follows the diffusion equation described by

$$\frac{\partial C}{\partial t} = \frac{D}{r^2}\frac{\partial}{\partial r}\left(r^2 \frac{\partial C}{\partial r}\right), \qquad r_p < r < r_p + \delta \tag{1}$$

where D is the diffusion coefficient and r is the distance from the center of the particle with radius r_p. We consider a diffusion layer of width δ around the particle, where the concentration adjusts from C_i, which is the monomer concentration adjacent to the surface, to C_b, the far–field concentration. Growth requires $C_b > C_i$ and it is assumed $\delta \gg r_p$. The concentration C is subject to

$$C(r_p, t) = C_i, \qquad C(r_p + \delta, t) = C_b(t), \qquad C(r, 0) = C_0, \tag{2}$$

where the value C_0 represents the initial well-mixed and uniform concentration of the monomer solution. Note that in practice C_i is difficult to measure, hence it is a standard to work in terms of the particle solubility C_s, which is defined by the

Ostwald–Freundlich equation. In this case the boundary condition may be written,

$$C(r_p, t) = C_s + \frac{D}{k}\frac{\partial C}{\partial r}\bigg|_{r=r_p} = C_\infty \exp\left(\frac{\alpha}{r_p}\right) + \frac{D}{k}\frac{\partial C}{\partial r}\bigg|_{r=r_p} \tag{3}$$

where C_∞ is the solubility of the bulk material. The parameter α is the capillary length and it is usually of the order of 1–6 nm. In the LSW theory it is assumed that $\alpha \ll r$ in order to omit higher terms of the exponential expansion. However, when $\alpha \approx r$, at the start of the growth process, it can play a dominant role. The expression for the time-dependent bulk concentration C_b is obtained via mass conservation of the monomer atoms in the particle and surrounding solution. The volume of solute per particle is $1/N_0$, where N_0 is the population density and noting that the molar volume $V_M = M_p/\rho_p$, we obtain

$$C_b(t) \approx C_0 - \frac{4\pi}{3}\frac{\rho_p}{M_p}N_0 r_p^3 = C_0 - \frac{4\pi N_0}{3V_M}r_p^3. \tag{4}$$

The particle radius r_p is also an unknown function of time that can be determined by balancing mass added to the crystal with the incoming monomer flux, hence

$$\frac{\mathrm{d}r_p}{\mathrm{d}t} = V_M D \frac{\partial C}{\partial r}\bigg|_{r=r_p}, \qquad r_p(0) = r_{p,0}, \tag{5}$$

where $r_{p,0}$ the initial particle radius.

3 Pseudo-Steady State

The time scales suggest that diffusion is a much faster process than growth, hence the concentration has sufficient time to equilibrate to its steady-state value as the growth slowly proceeds. As the 'constants' of integration may be time-dependent and come from applying the boundary conditions where the radius depends on time, the evolution is better described as a pseudo–steady state. Neglecting the time derivative in (1) and defining $C_s = C_{eq}$ to be constant, we obtain

$$C = C_b + \frac{kr_p^2}{D(r_p+\delta) + k\delta r_p}\left[C_b - C_{eq}\right]\left(1 - \frac{\delta + r_p}{r}\right). \tag{6}$$

The assumption $\delta \gg r_p$ will reduce (6) to

$$C = C_b + \frac{kr_p^2}{(D + kr_p)\delta}\left[C_b - C_{eq}\right]\left(1 - \frac{\delta}{r}\right), \tag{7}$$

hence

$$\frac{dr_p}{dt} = \frac{Dk}{D + kr_p}\left[a^3 - b^3 r_p^3\right] \tag{8}$$

where $a^3 = V_M(C_0 - C_{eq})$ and $b^3 = 4\pi N_0/3$. The model is invalid for small times because, provided the fluid is initially well-mixed, the width of the adjustment zone where the concentration increases from the particle edge to the bulk value δ is such that $\delta(0) = 0 < r_p$. Moreover, various authors have assumed that the process is driven solely by diffusion or surface kinetics, which leads to a slightly simpler solution form. However, we can note that for purely diffusion driven growth the value of the concentration in the solute adjacent to the particle must be exactly equal to the equilibrium concentration of the particle throughout the process, and the surface kinetics limit requires that the concentration is constant in space throughout the process. From this point of view it seems clear that the reductions are physically unrealistic.

When the pseudo-steady assumptions hold, setting $\delta \gg r_p$ and taking $C_{eq} = C_\infty e^{\alpha/r_p}$ to be constant leads to an implicit solution for $t(r)$ of the form

$$t - t_0 = \frac{1}{6a^2bk}\left[\left\{\ln\frac{a^2 + abr_p + b^2r_p^2}{(a - br_p)^2} - \ln\frac{a^2 + abr_{p0} + b^2r_{p0}^2}{(a - br_{p0})^2}\right\} + 2\sqrt{3}\left\{\arctan\left(\frac{a + 2br_p}{\sqrt{3}a}\right) - \arctan\left(\frac{a + 2br_{p0}}{\sqrt{3}a}\right)\right\}\right].$$

The arctan term is always negligible compared to the log term and if it is dropped from the model the errors will be of the order 0.1%. Removing this term we find that both diffusion and kinetic driven processes are accurately approximated by a solution of the form

$$r_p = \frac{r_m}{2}\frac{\left[1 + 2f(r_{p0})\exp\left(\frac{t-t_0}{G}\right) - \sqrt{-3 + 12f(r_{p0})\exp\left(\frac{t-t_0}{G}\right)}\right]}{\left[-1 + f(r_{p0})\exp\left(\frac{t-t_0}{G}\right)\right]}. \tag{9}$$

A feature made clear from the parameter $G = (ak + bD)/(6a^2b^2kD)$ is that ak and bD are interchangeable: it does not matter if we define them the opposite way round, the result is the same. Physically this means that *the model cannot distinguish between diffusion or reaction driven growth*. An important consequence of this analysis is the observation that there are at most two independent controlling parameters for the growth model, namely G, r_m. If the maximum radius, r_m, is measured then there is only a single controlling parameter, whereas previous authors have used up to eight. With just the single fitting parameter, we may obtain more accurate results, as shown in Fig. 1. We compare the results from Eq. (9) with the models analysed in four different papers. In all cases, Eq. (9) provides a better approximation. For further details see Myers and Fanelli [6].

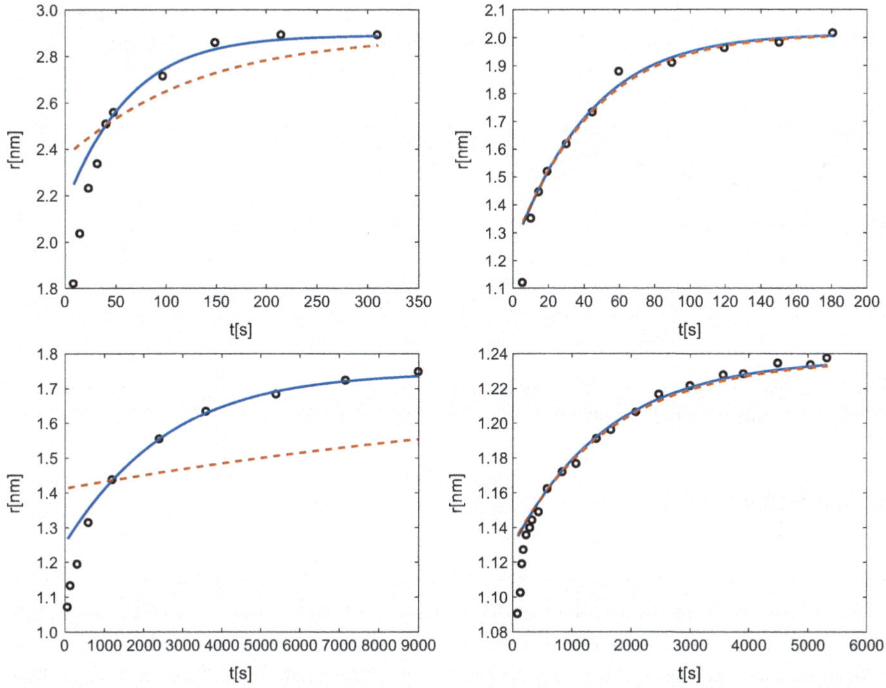

Fig. 1 Comparison of the model (blue solid lines) with four different models (red dashed lines) and their experimental values (black dots): Chuang et al. [3] (top left), Bullen et al. [2] (top right), Pan et al. [7] (bottom left) and Su et al. [9] (bottom right)

4 The *N* Particles System

We now consider a system of N particles that follow a normal standard distribution. The bulk material is assumed well-mixed and the particles are separated at large distances compared to their radii such that there are no interparticle diffusional interactions, therefore we may consider the same equations obtained before for each particle in the system. Note that now the mass balance has to take into account that we have the contribution of N particles. Thus, we call r_i the ith particle radius, $r_{i,0}$ its initial value and $C_{s,i}$ its solubility, and we write the ordinary differential equation (8) for each particle. Finally, the equation for the mass conservation becomes

$$\frac{N_p}{N_0} M_p C_0 = M_p C_b(t) \left[\frac{N_p}{N_0} - \frac{4\pi}{3} \sum_{i=1}^{N_p} {r_i}^3 \right] + \frac{4\pi \rho_p}{3} \sum_{i=1}^{N_p} {r_i}^3. \tag{10}$$

In Fig. 2 the model is compared with the experimental study of Peng et al. [8] for the growth kinetics of cadmium selenide (CdSe) nanoparticles in a nonaqueous solution.

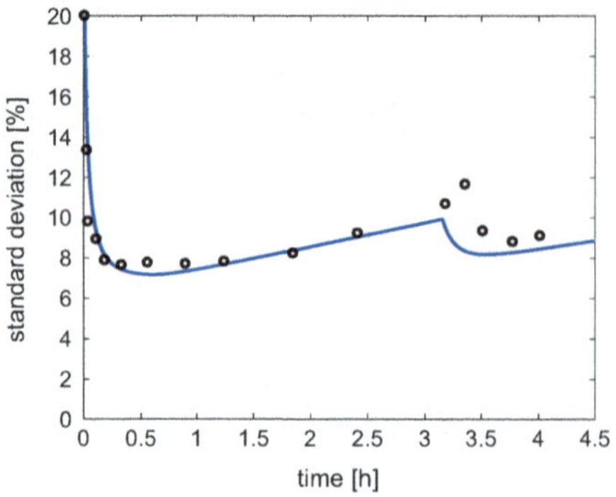

Fig. 2 Standard deviation and mean radius evolution of a system of 1000 CdSe nanoparticles (solid line) compared to experimental values of Peng et al. [8] (dots)

5 Conclusions

We presented a model for the growth of a single nanoparticle, which was extended to a system of N particles. The analysis of the model leads to several important conclusions. First of all, it was shown that the standard pseudo-steady state model is invalid for early times, leading to incorrect values for the diffusion and surface kinetic coefficient when the fitting analysis includes all the experimental data. It is also shown that the model cannot distinguish between diffusion or surface reaction driven growth and the simplifications made following this criteria are physically unrealistic. Moreover, an explicit formula for the variation of the radius as a function of time, depending on just two unknown non-dimensional parameters was presented. This makes data fitting a much simpler process.

References

1. Bastús, N.G., Comenge, J., Puntes, V.: Kinetically controlled seeded growth synthesis of citrate-stabilized gold nanoparticles of up to 200 nm: size focusing versus Ostwald ripening. Langmuir **27**(17), 11098–11105 (2011)
2. Bullen, C.R., Mulvaney, P.: Nucleation and growth kinetics of CdSe nanocrystals in octadecene. Nano Lett. **4**(12), 2303–2307 (2004)
3. Chuang, X., Hongxun, H., Wei, C. and Jingkang, W.: Crystallization kinetics of CdSe nanocrystals synthesized via the TOPTOPOHDA route. J. Cryst. Growth **310**(15), 3504–3507 (2008)
4. La Mer, V.K., Dinegar, R.: Theory, production and mechanism of formation of monodispersed hydrosols. J. Am. Chem. Soc. **72**(11), 4847–4854 (1950)
5. Mantzaris, N.V.: Liquid-phase synthesis of nanoparticles: particle size distribution dynamics and control. Chem. Eng. Sci. **60**(17), 4749–4770 (2005)
6. Myers, T.G., Fanelli, C.: On the incorrect use and interpretation of the model for colloidal, spherical crystal growth. J. Colloid Interface Sci. **536**, 98–104 (2019)

7. Pan, B., He, R., Gao, F., Cui, D., Zhang, Y.: Study on growth kinetics of CdSe nanocrystals in oleic acid/dodecylamine. J. Cryst. Growth **286**(2), 318–323 (2006)
8. Peng, X., Wickham, J., Alivisatos, A.P.: Kinetics of II–VI and III–V colloidal semiconductor nanocrystal growth:"focusing" of size distributions. J. Am. Chem. Soc. **120**(21), 5343–5344 (1998)
9. Su, H., Dixon, J.D., Wang, A.Y., Low, J., Xu, J., Wang, J.: Study on growth kinetics of CdSe nanocrystals with a new model. Nanoscale Res. Lett. **5**(5), 823–828 (2010)

A Mean-Field Evacuation Simulation

Claudia Totzeck

Abstract We discuss a mean-field simulation of an evacuation scenario. We model the crowd which needs to be evacuated using a probability measure μ. The controls are represented by external assistants formulated by ordinary differential equations. The task of evacuation is written as optimal control problem. Under the assumption that μ has an L^2-density, we state the corresponding first order optimality condition using a Lagrangian approach in the L^2-topology. Based on this we solve the problem with an instantaneous control algorithm. Simulation results of an evacuation scenario underline the feasibility of the approach and show the behaviour that is expected to fit the requirements posed by the cost functional.

1 Introduction

The modelling and simulation of large particle systems were investigated by many research groups in the last decades. Applications range from school of fish to flocks of birds, herd of sheep and crowds of pedestrians [2, 3, 7]. One frequent modelling assumption is the three-phase model of interaction. It postulates that individuals have repulsive influence on each other, if their distance is very small. On the other hand, if the individual is alone and finds others far away, he or she tends to move towards the crowd. And in case the individuals are surrounded by others with distance in the range of their comfort zone, there is neither repulsive nor attractive interaction [6]. This kind of interactions can be modelled with the help of potentials proposed by Cucker and Smale [8] or D'Orsogna et al [9]. Due to the curse of dimensions it is common practise to approximate the large particle system using a mean-field approximation. A first numerical comparison of the particle model and the mean-field model can be found in [1], which shows that the behaviour of the kinetic and the particle model agrees for a large number of individuals.

C. Totzeck (✉)
TU Kaiserslautern, Kaiserslautern, Germany
e-mail: totzeck@mathematik.uni-kl.de

© Springer Nature Switzerland AG 2019

I. Faragó et al. (eds.), *Progress in Industrial Mathematics at ECMI 2018*,
Mathematics in Industry 30, https://doi.org/10.1007/978-3-030-27550-1_10

A natural extension of the models is the optimisation of the crowd behaviour using external agents as controllers. Due to the high dimensions of the state models, we need tailored optimisation techniques for the simulation. In the following we use an instantaneous control approach which updates the controllers based on the current states. Compared to an optimal control approach, this has the advantage that we do not need to store the full information of the forward solve to update the controllers. The instantaneous control approach was successfully applied to control traffic flow [10] and the Navier–Stokes equations [11]. In the following we model an evacuation scenario using a probability measure μ to model the crowd and assistants described by a ODE system. The velocities of the assistants are the controls. We formulate the task as optimal control problem and derive its first order optimality system under the assumption that μ has a L^2-density. This allows us to use a standard Lagrangian approach in the L^2-framework which has the advantage that the adjoint equations can be implemented similarly to the state equation. We acknowledge that the natural topology corresponding to μ would be the Wasserstein metric, but then the adjoint would be vector-valued and thus inappropriate for numerical investigations. The relations of first order optimality systems in L^2 and Wasserstein sense mentioned here will be discussed in all details in [5].

The article is organized as follows: in Sect. 2 we describe the state model. Then we discuss the objective functional that models the task of leading the crowd to an assembly point E_{des} and state the optimal control problem. In Sect. 3 we state the first order optimality conditions for the control problem using the L^2-topology. This serves as basis for the instantaneous control algorithm which we employ for the numerical results shown in Sect. 4.

2 State Model and Control Problem

The state model is given by a partial differential equation coupled to ordinary differential equations (ODE) system for the assistants. Let d denote the dimension of the state and velocity space and $[0, T]$ the time interval of interest. The probability measure modelling the crowd is denoted by $\mu \in \mathscr{C}([0, T], \mathscr{P}_2(\mathbb{R}^d \times \mathbb{R}^d))$, $\mu : t \mapsto \mu(t, x, v)$, the positions of the M assistants are collected in the vector $g := (g_i)_{i=1,\dots,M} \in \mathscr{C}([0, T], \mathbb{R}^{dM})$. We propose the dynamic

$$\partial_t \mu + \nabla_x \cdot (v \mu) = \nabla_v \cdot ((K_1 * \mu + K_2 - \alpha v) \mu), \qquad \mu(0, x, v) = \mu_0(x, v), \tag{1a}$$

$$\frac{d}{dt} g = u, \qquad g(0) = g_0. \tag{1b}$$

The interactions are defined through K_1 and K_2 which we assume to be gradients of potentials P_1 and P_2. Indeed, for the numerics we choose the Morse potentials [9]

$$P_i(x, y) = R_i \exp(-|x - y|/r_i) + A_i \exp(-|x - y|/a_i), \quad i \in \{1, 2\} \tag{2}$$

$$K_1(x, y) = \nabla_x P_1(x, y), \qquad K_2(x) = \sum_{k=1}^{M} \nabla_x P_2(x, g_k(t)). \tag{3}$$

The repulsion and attraction forces are adjustable via the strength variables R_i, A_i and the distance variables r_i, a_i for $i = 1, 2$, where the index 1 corresponds to the interaction of the crowd and the index 2 to the interaction of the crowd and assistants. The velocities of the assistants collected in u are the controls of the problem. We propose the following cost functional to model the evacuation

$$J(\mu, u) = \int_0^T \frac{\sigma_1}{4T} |\mathbb{V}(\mu(t))|^2 + \frac{\sigma_2}{2T} |\mathbb{E}(\mu(t)) - E_{\text{des}}|^2 + \frac{\sigma_3}{2MT} \|u(t)\|^2 dt, \tag{4}$$

with variance and center of mass given by

$$\mathbb{V}(\mu(t)) := \int_{\mathbb{R}^d \times \mathbb{R}^d} |x|^2 \, d\mu(t, x, v), \qquad \mathbb{E}(\mu(t)) := \int_{\mathbb{R}^d \times \mathbb{R}^d} x \, d\mu(t, x, v),$$

respectively. Hence, the first term forces the controls to keep the crowd close together and the second term intends to lead the crowd to the predefined assembly point E_{des}. The third term is the usual penalty term. On the one hand it minimizes the energy used by the controllers, on the other hand it regularizes the cost functional and helps in the proof of the existence of the minimizer.

We assume to have a maximal velocity $u_{\max}$ for the assistants, which is represented by the admissible set

$$U_{\text{ad}} = \{u \in L^2 \colon \|u\| \leq u_{\max}\}.$$

The control problem is summarized as

Problem 1 Find $(\mu, u) \in \mathscr{C}([0, T], \mathscr{P}_2(\mathbb{R}^d \times \mathbb{R}^d)) \times U_{\text{ad}}$ such that

$$\min_{(\mu,u)} J(\mu, u) \quad \text{subject to} \quad (1). \tag{CP}$$

3 First Order Optimality Conditions

In the following we briefly summarize the derivation of the optimization algorithm. Details can be found in [13], where the strategy is applied to a different application. In the following we assume to have enough regularity for all formal computations. In particular, we assume that μ has a L^2-density which allows to use the L^2-topology for the Lagrangian approach to compute the optimality system. Indeed, using the standard Lagrangian approach, we formally derive the first order optimality system. We denote the state $y = (\mu, g, y_0)$ with $y_0 = (\mu_0, g_0)$ and corresponding the state space by $\mathscr{Y}$. The adjoint states $p = (\varphi, \xi, \eta)$ are assumed to belong to the space $\mathscr{Z}$ with dual $\mathscr{Z}^*$. Then the Lagrangian corresponding to (CP) is given by

$$\mathscr{L}(y, p, u) = J(\mu, u) + \langle e(y, u), p \rangle_{\mathscr{Z}^*, \mathscr{Z}},$$

where $e\colon \mathscr{Y} \times U \to \mathscr{Z}^*$ is the state mapping given by

$$\begin{aligned}\langle e(y, u), (\varphi, \xi) \rangle_{\mathscr{Z}^*, \mathscr{Z}} = &- \int_0^T \int_{\mathbb{R}^d \times \mathbb{R}^d} (\partial_t \varphi + v \cdot \nabla_x \varphi + S(\mu) \cdot \nabla_v \varphi) \mu d(x, v) dt \\ &+ \int_0^T (\frac{d}{dt} g - u) \cdot \xi dt + \int_{\mathbb{R}^d \times \mathbb{R}^d} \varphi(T, x, v) \mu(T, x, v) - \varphi(0, x, v) \mu(0, x, v) d(x, v) \\ &- \int_{\mathbb{R}^d \times \mathbb{R}^d} (\mu(0, x, v) - \mu_0) \eta_\mu d(x, v) + (g(0) - g_0) \cdot \eta_g\end{aligned}$$

with $\eta = (\eta_\mu, \eta_g)$ being the multipliers for the initial conditions. Here and in the following we use the abbreviation

$$S(\mu(t, x, v)) := K_1 * \mu + K_2 - \alpha v.$$

Formally, solving $d\mathscr{L} \overset{!}{=} 0$ leads to the following first order optimality condition

Proposition 1 *The optimality condition for* (CP) *reads*

$$\int_0^T (\frac{\sigma_3}{TM} u^* - \xi) \cdot (u - u^*) dt \geq 0 \quad \text{for all} \quad u \in U_{ad}, \tag{5a}$$

where the adjoint $p = (\varphi, \xi, \eta) \in \mathscr{Z}$ *satisfies*

$$\partial_t \varphi + v \cdot \nabla_x \varphi = -S(\mu) \cdot \nabla_v \varphi + d_\mu S(\mu)[\varphi] - d_\mu J, \qquad \frac{d}{dt} \xi = -d_g S(\mu)[\varphi], \tag{5b}$$

$$d_\mu J(t, x, v) = \frac{\sigma_1}{T} \mathbb{V}(\mu(t, x, v)) |x - \mathbb{E}(\mu(t, x, v))|^2 + \frac{\sigma_2}{T} (\mathbb{E}(\mu(t, x, v)) - E_{des}) \cdot x$$

supplemented with the terminal conditions $\varphi(T, x, v) = 0$ *for all* (x, v) *and* $\xi(T) = 0$.

With the help of system (5) we can derive a projected gradient descent method to compute numerical results. In order to reduce the memory consumption, we do not consider the optimal control problem stated above, but rather the corresponding instantaneous control problem. The latter is obtained when restricting the above computations to a shorter time-interval and proceeding iteratively until the final time T is reached. In fact, we compute the forward solution and its adjoint on one time step before updating the controls for the next time step. See [4] for more details.

4 Simulation

The following simulation is computed on the time interval $[0, 10]$ with $M = 5$ assistants. The time step is $dt = 0.013$. The assistants are assumed to have a maximal velocity of $u_{\max} = 0.1$. The parameters characterising the interaction of the crowd are $A_1 = 0.02$, $R_1 = 0.05$, $a_1 = 1$ and $r_1 = 0.2$. The friction parameter is denoted by $\alpha = 0.1$. The interaction parameters of the crowd and the assistants are $A_2 = 0.02$, $R_2 = 0.01$, $a_2 = 0.5$ and $r_2 = 0.2$. The destination or assembly point is marked in green at the position $(-0.2, -0.2)$. The cost functional parameters are $\sigma_1 = 0.1$, $\sigma_2 = 100$ and $\sigma_3 = 1e^{-6}$. These values indicate that the position of the center of mass of the crowd is more important for the evacuation than the variance of the crowd. The discretization stepsize in the space domain is $hx = hy = 0.027$ and in the velocity domain it holds $hvx = hvy = 0.0014$.

The implementations of the forward and adjoint systems are based on a Strang splitting scheme [12]. We apply a semi-Lagrangian solver in the space domain and a semi-implicit finite-volume scheme in the velocity domain. All computations are based on a fixed grid. The semi-Lagrangian solver transports the information along characteristic curves. To obtain these curves we solve ODEs using a second order Runge–Kutta scheme. To assure that the terminal point of each transport step is a grid point, we do a polynomial reconstruction based on cubic Bezier curves. In the velocity space we employ a second order finite volume scheme where the advection is approximated with the help of a Lax–Wendroff flux. A van-Leer limiter is used to intercept oscillations caused by non-smooth solutions.

The optimisation algorithm computes the state solution and the corresponding adjoint solution on one time step. Based on this information the gradient is computed and the controls are updated. Due to the maximal velocity constraint of the assistants, we need to project the controls to the feasible set. For more information on the implementation and the instantaneous control approximation see [13].

The numerical results are shown in Fig. 1. We see snapshots of the evacuation simulation at different points in time. The trajectories of the assistants are shown in red. The current position of each assistant is represented by a red triangle. The assembly point is highlighted by the green marker. We see that the probability measure representing the crowd, is following the assistants appropriately. One of the agents stays behind to attract the upper right part of the crowd. At the final time $T = 10$ the crowd is gathered around the assembly point E_{des} as desired. These results show that the proposed algorithm is appropriate to model a simple evacuation scenario using controllable assistants.

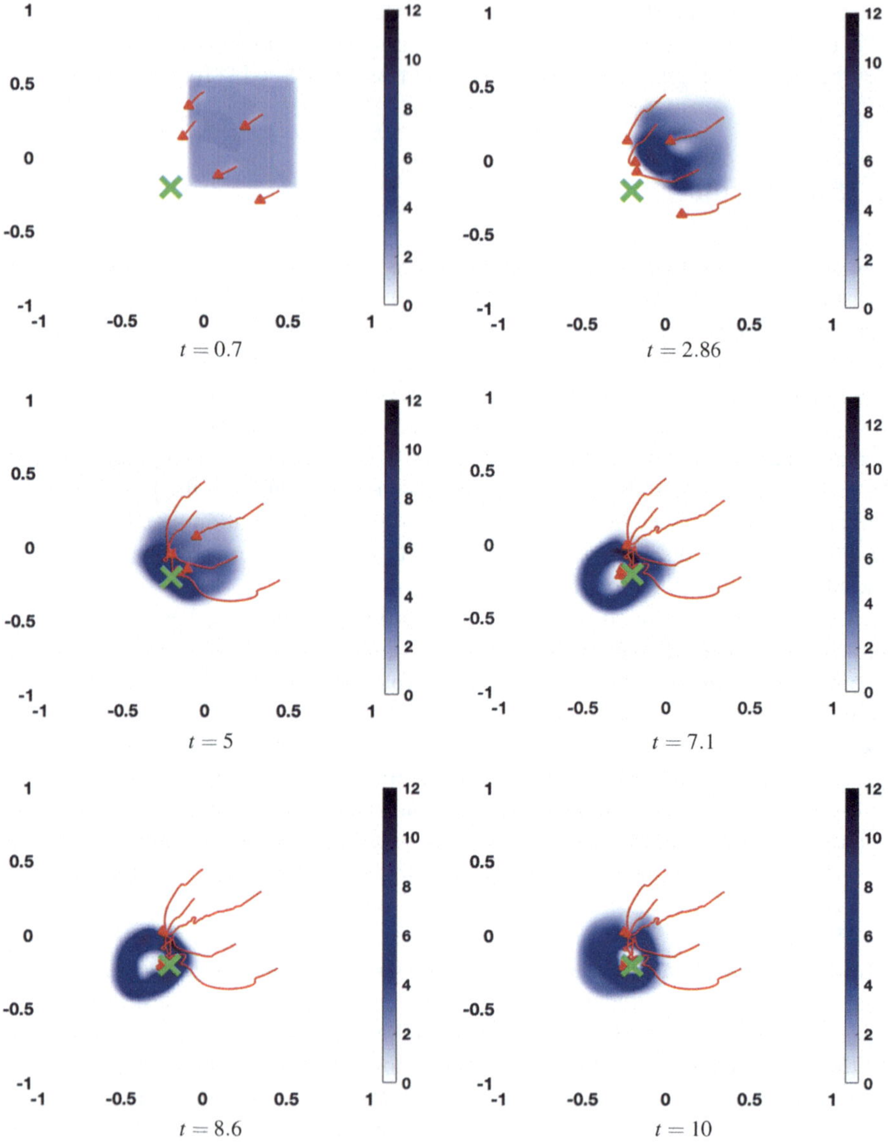

Fig. 1 Simulation of the evacuation with 5 assistants

References

1. Albi, G., Pareschi, L.: Appl. Math. Lett. **26**, 397–401 (2013)
2. Albi, G., Bongini, M., Cristiani, E., Kalise, D.: SIAM J. Appl. Math. **76**(4), 1683–1710 (2016)
3. Burger, M., Di Francesco, M., Markowich, P., Wolfram, M.-T.: discrete and continuous dynamical systems B. Am. Inst. Math. Sci. **19**(5), 1311–1333 (2014)
4. Burger, M., Pinnau, R., Roth, A., Totzeck, C., Tse, O.: Instantaneous control of interacting particle systems in the mean-field limit. arXiv:1903.12407
5. Burger, M., Pinnau, R.,Totzeck, C., Tse, O.: Mean-field optimal control and optimality conditions in the space of probability measures. arXiv:1902.05339
6. Carrillo, J.A., Fornasier, M., Toscani, G., Vecil, F.: Mathematical modeling of collective behavior in socio-economic and life sciences. In: Naldi, G., Pareschi, L., Toscani, G. (eds.) Modeling and Simulation in Science, Engineering and Technology. Birkhäuser, Boston (2010)
7. Carrillo, J.A., Choi, Y.-P., Hauray, M.: In: Munteean, A., Toschi, F. (eds.): Collective Dynamics from Bacteria to Crowds. CISM. Springer, Vienna (2014)
8. Cucker, F., Smale, S.: Emergent behavior in flocks. IEEE Trans. Auto. Control **52**, 852–862 (2007)
9. D'Orsogna, M.R., Chuang, Y.L., Bertozzi, A.L.,Chayes, L.S.: Self-propelled particles with soft-core interactions: patterns, stability, and collapse. Phys. Rev. Lett. **96**, 104302 (2006)
10. Herty, M., Kirchner, C., Klar, A.: Instantaneous control for traffic flow. Math. Methods Appl. Sci. **30**, 153–169 (2007)
11. Hinze M.: Instantaneous closed loop control of the Navier–Stokes system. SIAM J. Control Optim. **44**, 564–583 (2005)
12. Strang, G.: On the construction and comparison of difference schemes. SIAM J. Num. Anal. **5**, 506–517 (1968)
13. Totzeck, C.: Asymptotic Analysis of Optimal Control Problems and Global Optimization. Doktorarbeit TU Kaiserslautern Verlag Dr. Hut, München (2017)

A Poroelasticity Model Using a Network-Inspired Porosity-Permeability Relation

Menel Rahrah, Fred J. Vermolen, Luis A. Lopez-Peña, and Bernard J. Meulenbroek

Abstract Compressing a porous material or injecting fluid into a porous material can induce changes in the pore space, leading to a change in porosity and permeability. In a continuum scale PDE model, such as Biot's theory of linear poroelasticity, the Kozeny–Carman equation is commonly used to determine the permeability of the porous medium from the porosity. The Kozeny–Carman relation assumes that there will be flow through the porous medium at a certain location as long as the porosity is larger than zero at this location. In contrast, from discrete network models it is known that percolation thresholds larger than zero exist, indicating that the fluid will stop flowing if the average porosity becomes smaller than a certain value dictated by these thresholds. In this study, the difference between the Kozeny–Carman equation and the equation based on the percolation theory, is investigated.

1 Introduction

Having a good estimation of permeability is of a pivotal importance for the description of different physical processes. However, mainly due to the complexity of the connected pore space, it has been very difficult to formulate satisfactory theoretical models for the relation between the porosity and the permeability. One of the most largely used methods remains the Kozeny–Carman approach. In this study, we briefly introduce a new approach for the permeability that is derived on a micro-scale network model. We refer to this approach as the network-inspired relation. The Kozeny–Carman relation assumes that the pore space is fully connected, therefore, flow through the porous medium is possible as long as the average porosity is larger than zero. In contrast, the new network-inspired approach states that the permeability is positive only if the porosity is larger than a specific percolation

M. Rahrah (✉) · F. J. Vermolen · L. A. Lopez-Peña · B. J. Meulenbroek
Delft Institute of Applied Mathematics, Delft, The Netherlands
e-mail: M.Rahrah@tudelft.nl; F.J.Vermolen@tudelft.nl; L.A.LopezPena@tudelft.nl; B.J.Meulenbroek@tudelft.nl

© Springer Nature Switzerland AG 2019

I. Faragó et al. (eds.), *Progress in Industrial Mathematics at ECMI 2018*, Mathematics in Industry 30, https://doi.org/10.1007/978-3-030-27550-1_11

threshold, that depends on the topology of the network. As application, we consider the flow of an incompressible fluid through a poroelastic porous medium.

2 Governing Equations

The model provided by Biot's theory of linear poroelasticity with single-phase flow [1] is used in this study to determine the local displacement of the grains of a porous medium and the fluid flow through the pores, assuming that the deformations are very small. We assume that the fluid-saturated porous medium has a linearly elastic solid matrix and is saturated by an incompressible Newtonian fluid. Let $\Omega \subset \mathbb{R}^3$ denote the computational domain with boundary Γ, and $\mathbf{x} = (x, y, z) \in \Omega$. Furthermore, t denotes time, belonging to a half-open time interval $I = (0, T]$, with $T > 0$. The initial boundary value problem of an incompressible fluid flow in a deformable porous medium in the two-field $(\mathbf{u}/p)$ formulation, where $\mathbf{u}$ and p are the unknown functions, is stated as follows [4]:

$$\text{equilibrium equations:} \quad -\nabla \cdot \boldsymbol{\sigma}' + (\nabla p + \rho g \mathbf{e}_z) = \mathbf{0} \quad \text{on } \Omega \times I; \tag{1a}$$

$$\text{continuity equation:} \quad \frac{\partial}{\partial t}(\nabla \cdot \mathbf{u}) + \nabla \cdot \mathbf{v} = 0 \quad \text{on } \Omega \times I, \tag{1b}$$

where $\boldsymbol{\sigma}'$ and $\mathbf{v}$ are defined by the following equations

$$\text{Biot's constitutive equations:} \quad \boldsymbol{\sigma}' = \lambda(\nabla \cdot \mathbf{u})\mathbf{I} + \mu(\nabla \mathbf{u} + \nabla \mathbf{u}^T); \tag{2}$$

$$\text{Darcy's law:} \quad \mathbf{v} = -\frac{\kappa}{\eta}(\nabla p + \rho g \mathbf{e}_z). \tag{3}$$

Here, $\boldsymbol{\sigma}'$ denotes the effective stress tensor, p the pore pressure, ρ the fluid density, g the gravitational acceleration, $\mathbf{u}$ the displacement vector, $\mathbf{v}$ Darcy's velocity, λ and μ the Lamé coefficients; κ the permeability of the porous medium and η the fluid viscosity. In addition, appropriate boundary and initial conditions are specified in Sect. 3.

2.1 The Porosity-Permeability Relations

In this study, we consider the spatial dependency of the porosity and the permeability of the porous medium. The porosity θ is computed from the displacement vector using the porosity-dilatation relation (see [2, 3])

$$\theta(\mathbf{x}, t) = 1 - \frac{1 - \theta_0}{\exp(\nabla \cdot \mathbf{u})}, \tag{4}$$

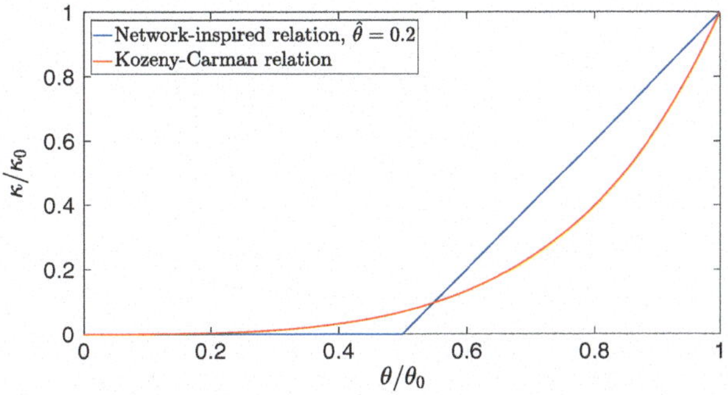

Fig. 1 The normalised permeability κ/κ_0 as a function of the normalised porosity θ/θ_0

with θ_0 the initial porosity. Subsequently, the permeability can be determined using the Kozeny–Carman equation [5]

$$\kappa(\mathbf{x}, t) = \frac{d_s^2}{180} \frac{\theta(\mathbf{x}, t)^3}{(1 - \theta(\mathbf{x}, t))^2}, \tag{5}$$

where d_s is the mean grain size of the soil. The Kozeny–Carman relation assumes that the permeability becomes zero if and only if the porosity also becomes zero. A new approach for the relation between the porosity and the permeability is inspired by the fluid flow through the edges (channels) of a network. In a network with a random topology, the network-inspired porosity-permeability relation states:

$$\kappa(\mathbf{x}, t) = \begin{cases} 0 & \theta \leq \hat{\theta} \\ \frac{\theta - \hat{\theta}}{\theta_0 - \hat{\theta}} \kappa_0 & \theta > \hat{\theta} \end{cases}, \tag{6}$$

where κ_0 is the initial permeability computed using the Kozeny–Carman relation and $\hat{\theta}$ the percolation threshold, which represents the minimal porosity needed to have connection via voids or channels from one end to the other. This percolation threshold depends on the topology of the network. The permeability obtained using both relations is depicted in Fig. 1, as function of the porosity.

3 Problem Formulation

The following numerical experiment is designed to study the different relations for the porosity and the permeability. As shown in Fig. 2, the infiltration of a fluid into a porous medium is studied. In addition, a vertical load is applied on a part of the top edge of the domain, in order to create a region with a high density of the grains which will emphasise the difference between the porosity-permeability relations. We assume that the flow pattern is axisymmetric. Therefore, we determine the solution for a fixed azimuth. Hence, the computational domain Ω is an L-shaped two-dimensional surface with cylindrical coordinates $\mathbf{r} = (r, z)$ and boundary Γ.

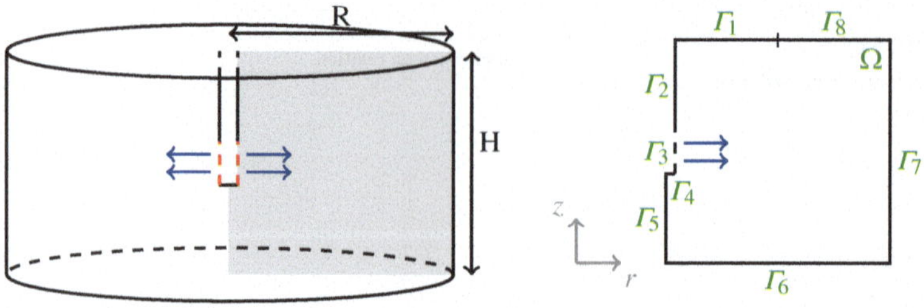

Fig. 2 Sketch of the setup for the aquifer problem: (left) physical problem and (right) numerical discretisation. Taking advantage of the symmetry of geometry and boundary conditions, only the grey region is discretised

The fluid is injected into the soil through a filter placed on boundary segment Γ_3, using a pump pressure. The vertical load is applied on boundary segment Γ_8. Furthermore, the injection tube is fitted with a casing (boundary segments Γ_2 and Γ_4) and a perforated section (boundary segment Γ_3) to prevent loose material from entering and potentially clogging the injection tube. More precisely, the boundary conditions for this problem are given as follows:

$$\frac{\kappa}{\eta}(\nabla p + \rho g \mathbf{e}_z) \cdot \mathbf{n} = 0 \quad \text{on} \quad \mathbf{r} \in \Gamma \setminus \Gamma_3 \cup \Gamma_7; \tag{7a}$$

$$p = \rho g(H - z) + p_{pump} \quad \text{on} \quad \mathbf{r} \in \Gamma_3; \tag{7b}$$

$$p = \rho g(H - z) \quad \text{on} \quad \mathbf{r} \in \Gamma_7; \tag{7c}$$

$$\boldsymbol{\sigma}'\mathbf{n} = \mathbf{0} \quad \text{on} \quad \mathbf{r} \in \Gamma_1 \cup \Gamma_7; \tag{7d}$$

$$\mathbf{u} \cdot \mathbf{n} \leq 0 \quad \text{on} \quad \mathbf{r} \in \Gamma_2 \cup \Gamma_3 \cup \Gamma_4; \tag{7e}$$

$$(\boldsymbol{\sigma}'\mathbf{n}) \cdot \mathbf{t} = 0 \quad \text{on} \quad \mathbf{r} \in \Gamma \setminus \Gamma_1 \cup \Gamma_7 \cup \Gamma_8; \tag{7f}$$

$$\mathbf{u} \cdot \mathbf{n} = 0 \quad \text{on} \quad \mathbf{r} \in \Gamma_5 \cup \Gamma_6; \tag{7g}$$

$$\boldsymbol{\sigma}'\mathbf{n} = (0, -\sigma_0')^T \quad \text{on} \quad \mathbf{r} \in \Gamma_8, \tag{7h}$$

where $\mathbf{t}$ and $\mathbf{n}$ are the unit tangent and the outward normal vectors. Further, p_{pump} is a prescribed pump pressure and σ_0' is the intensity of a uniform vertical load. Note that the boundary conditions on boundary segment Γ_5 appear as a result of symmetry. The initial condition is $\mathbf{u}(\mathbf{r}, 0) = 0$ for $\mathbf{r} \in \Omega$.

4 Numerical Results

To solve the discretised problem of (1), the Galerkin finite element method with triangular Taylor-Hood elements [2], is adopted. Quadratic basis functions are used for the approximation of the displacements, while the pressure field is approximated

Table 1 An overview of the values of the model parameters

Property	Symbol	Value	Unit
Young's modulus	E	$35 \cdot 10^6$	Pa
Poisson's ratio	ν	0.3	–
Fluid viscosity	η	$1.307 \cdot 10^{-3}$	$\text{Pa} \cdot \text{s}$
Fluid density	ρ	1000	kg/m^3
Gravitational acceleration	g	9.81	m/s^2
Initial porosity	θ_0	0.4	–
Mean grain size	d_s	$314 \cdot 10^{-6}$	m
Pump pressure	p_{pump}	10^5	Pa
Uniform load	σ_0'	10^7	N/m^2

by continuous piecewise linear functions. In addition, the backward Euler method is applied for the time integration. The computational domain is an L-shaped surface with radius $R = 1.0\,\text{m}$, height $H = 2.0\,\text{m}$, filter radius $R_f = 10.0\,\text{cm}$ and filter length $L_f = 1.0\,\text{m}$. The filter is placed between $z = 0.5$ and $z = 1.5$, while the vertical load is applied between $r = 0.5$ and $r = 1.0$. The domain is discretised using a regular triangular grid, with $\Delta r = \Delta z = 0.05$. The time step size is chosen to be $\tau = 0.5$. Furthermore, values for some model parameters have been chosen (see Table 1).

The Lamé coefficients λ and μ are related to Young's modulus E and Poisson's ratio ν by: $\lambda = \frac{\nu E}{(1+\nu)(1-2\nu)}$ and $\mu = \frac{E}{2(1+\nu)}$. The impact of the porosity-permeability relations on the fluid flow is defined in this study as the impact on the time average of the volumetric flow rate $\overline{Q}$ at a distance $R - R_f$ from the injection filter. We compute the volume flow rate using the velocity field as described by Darcy's law (3). The velocity field is obtained from the gradient of the pressure ∇p by applying the finite element method with piecewise linear approximation. For the Kozeny–Carman relation as well as for the network-inspired relation, $\overline{Q}$ is depicted in Fig. 3 as a function of the percolation threshold. In this figure, $\overline{Q}$ is normalised by dividing on the time average of the volumetric flow rate obtained by the Kozeny–Carman

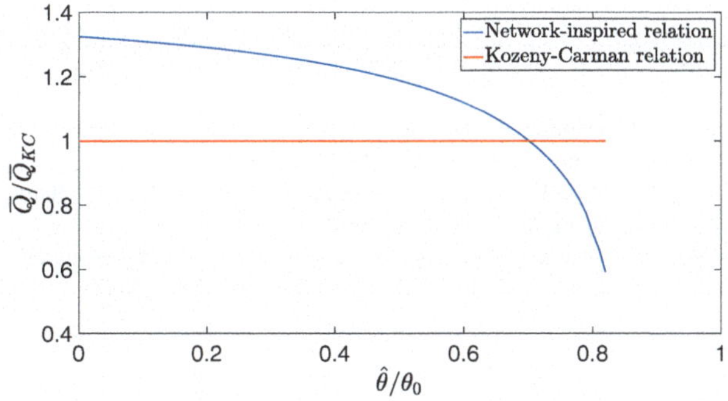

Fig. 3 The time average of the volumetric flow rate $\overline{Q}$ as a function of the percolation threshold $\hat{\theta}$

relation $\overline{Q}_{KC}$. As expected from Fig. 1, for low percolation thresholds the network-inspired relation results in higher flow rates than the Kozeny–Carman relation. In addition, the flow rate changes significantly as a function of the percolation threshold. Hence, the fluid flow depends on the topology of the connected pore space.

5 Conclusions

In this study, a three-dimensional poroelasticity problem is designed in order to analyse the applicability of the microscopic network-inspired porosity-permeability relation on the macro-scale. Furthermore, the results obtained with this relation are compared with the Kozeny–Carman relation, which is often used for this type of physical problems. To determine the displacements of the grains that are needed to compute the porosity, Biot's model for poroelasticity is used. Since the topology of macro-scale porous media is not known, computations are performed with different values of the percolation threshold. The numerical results indicate that for low percolation thresholds the network-inspired relation results in higher flow rates than the Kozeny–Carman relation, as expected from Fig. 1. In addition, it is shown that the flow rate changes significantly as a function of the percolation threshold which means that the water flow depends on the topology of the connected pore space.

Acknowledgements The work of M. Rahrah was supported by the Netherlands Organisation for Scientific Research NWO (project number 13263). The work of L.A. Lopez-Peña was supported by the Mexican Institute of Petroleum IMP (through the PCTRES grant).

References

1. Biot, M.A.: General theory of three-dimensional consolidation. J. Appl. Phys. **12**, 155–164 (1941)
2. Rahrah, M., Vermolen, F.: Monte Carlo assessment of the impact of oscillatory and pulsating boundary conditions on the flow through porous media. Transp. Porous Med. **123**, 125–146 (2018)
3. Tsai T.-L., Chang K.-C., Huang, L.-H.: Body force effect on consolidation of porous elastic media due to pumping. J. Chin. Inst. Eng. **29**, 75–82 (2006)
4. Wang, H.F.: Theory of Linear Poroelasticity with Applications to Geomechanics and Hydrogeology. Princeton University Press, Princeton (2000)
5. Wang S.-J., Hsu K.-C.: Dynamics of deformation and water flow in heterogeneous porous media and its impact on soil properties. Hydrol. Process. **23**, 3569–3582 (2009)

Motion of a Spherical Particle Attached to the Interface Between Two Viscous Fluids

Galina Lyutskanova-Zhekova and Krassimir Danov

Abstract The motion of small particles, attached to fluid interfaces, is important for the production of 2D-ordered micro- and nano-layers, which are applied for the production of solar panels, CCDs, and bio-memory chips. The problem was solved semi-analytically for water/air interface and three-phase contact angles $\alpha \leq 90°$, using the Mehler–Fox transformation (Zabarankin, Proc R Soc A 463:2329–2349, 2007). We propose a numerical method, based on the gauge formulation of the Stokes equations for two viscous fluids, for calculating the velocity field, pressure, and drag force coefficient. The method is applicable for all values of α and fluid viscosities. The weak singularity of the solutions at the three-phase contact line is studied and the respective phase diagram is calculated. The isolation of the type of singularity helps us to construct an efficient second-order numerical scheme, based on the ADI approach. The problem is solved numerically for different particle positions at the interface and ratios of the fluid viscosities.

1 Introduction

The 2D layers of micro- and nano-particles, attached to interfaces, are related to the production of antireflective surface coverages in solar panels, CCD, and bio-memory chips. The quality of these layers depend on the values of the contact angle, α, and the mobility of particles at interfaces. For small particles, α is measured from the translational motion of individual particles, attached to fluid–fluid interfaces [1].

G. Lyutskanova-Zhekova (✉)
Institute of Mathematics and Informatics, Bulgarian Academy of Sciences, Sofia, Bulgaria

Faculty of Mathematics and Informatics, Sofia University, Sofia, Bulgaria
e-mail: g.zhekova@math.bas.bg; g.zhekova@fmi.uni-sofia.bg

K. Danov
Faculty of Chemistry and Pharmacy, Sofia University, Sofia, Bulgaria
e-mail: kd@lcpe.uni-sofia.bg

© Springer Nature Switzerland AG 2019

I. Faragó et al. (eds.), *Progress in Industrial Mathematics at ECMI 2018*,
Mathematics in Industry 30, https://doi.org/10.1007/978-3-030-27550-1_12

The principles of the numerical solution of the respective Stokes problem and the drag force coefficient are studied in [2]. If one of the fluid phases is air, then the problem has a semi-analytical solution in terms of the Mehler–Fox integral transformation [6], which is valid only for particles more immersed in the fluid phase ($\alpha \leq 90°$). Analytical approximations for the drag and diffusion coefficient of a spherical particle, attached to flat interface between two immiscible fluids, are constructed for the case of a vanishing viscosity ratio between the fluid phases [3].

The aim of the present study is to develop an effective numerical method for calculating the velocity field, pressure, and drag coefficient in the case of two fluid phases with arbitrary viscosities and three-phase contact angles $0 < \alpha < 180°$.

2 Formulation of the Problem

A small spherical particle of radius R is attached to the interface between two infinite incompressible viscous Newtonian fluids (Fig. 1). For small capillary numbers, the perturbations of the dividing surface due to the particle motion are so small that the surface is flat. Thus, the three-phase contact line is a circumference of radius $r_c = R \sin \alpha$, where α is the central angle (Fig. 1). Its center is chosen to be the origin of Cartesian coordinate system with axis of revolution Oz and unit basis vectors $\mathbf{e}_x$, $\mathbf{e}_y$, and $\mathbf{e}_z$.

The particle translates parallel to the interface along the y-axis with known constant velocity V. The fluid motion is so slow that the inertia terms in the Navier–Stokes equations can be neglected. Thus, we can describe the sought-out local velocities $\mathbf{v}_k$ for both phases as solutions of the dimensionless Stokes equations, $\nabla \cdot \mathbf{v}_k = 0$ and $\nabla p_k = \nabla^2 \mathbf{v}_k$ $(k = 1, 2)$, where p_k is the pressure on the both sides of the interface, ∇ is the spatial gradient and subscripts "1" and "2" denote the

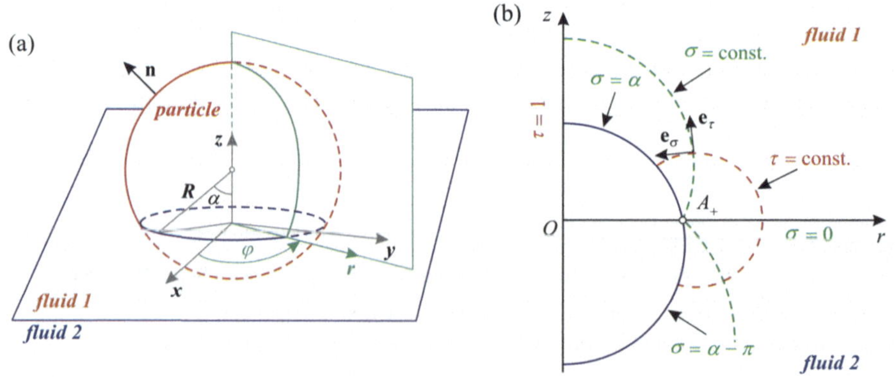

Fig. 1 (**a**) Sketch of a spherical particle attached to the plane interface between two fluids. (**b**) Toroidal coordinate system of revolution with coordinates τ and σ

upper and lower phases, respectively. For simplicity of notations, all dimensionless geometrical parameters are scaled with r_c, the velocity vectors—with V, and the pressures—with $\eta_k V/r_c$, where η_k is the dynamic viscosity. In order to close the system, we apply the following boundary conditions (BCs) for the Stokes problem: the no-slip BCs at the particle surface S_p for both fluid phases $\mathbf{v}_k = \mathbf{e}_y$ $(k = 1, 2)$; the kinematic BCs at the interface $\mathbf{v}_1 = \mathbf{v}_2$ and $\mathbf{v}_1 \cdot \mathbf{e}_z = \mathbf{v}_2 \cdot \mathbf{e}_z = 0$; the dynamic BC at the interface $\mu_1(\partial\mathbf{v}_1/\partial z) \times \mathbf{e}_z = \mu_2(\partial\mathbf{v}_2/\partial z) \times \mathbf{e}_z$, where the dimensionless viscous coefficients are given by $\mu_k := \eta_k/(\eta_1+\eta_2)$ $(k = 1, 2)$. Finally, the physical values of $\mathbf{v}_k$ and p_k vanish at large distances from the particle.

The original form of the Stokes equations is not convenient for a computer modeling because it consists of a system of elliptic differential equations for $\mathbf{v}_k$ and unknown functions p_k. In fact p_k are calculated, using the continuity equation $\nabla \cdot \mathbf{v}_k = 0$. Thus, we use the gauge formulation [5], which introduces vector, $\mathbf{w}_k$, and scalar, ξ_k, potentials by using the following definitions: $\nabla^2\mathbf{w}_k = 0$ and $\nabla\xi_k = \mathbf{w}_k - \mathbf{v}_k$ $(k = 1, 2)$. The scalar potentials are defined with respect to a constant, so that we define $\xi_k \to 0$ at infinity. The substitution of $\mathbf{w}_k$ and ξ_k into the continuity equation leads to the Possion equation, $\nabla^2\xi_k = \nabla \cdot \mathbf{w}_k$, and that into the momentum balance equation—to the formula $p_k = -\nabla^2\xi_k$. Thus, the Stokes problem is reduced to a well-defined system of elliptic partial differential equations (PDEs). In such a way, the number of degrees of freedom increases and, thus, we specify the following additional boundary condition $\mu_1(\partial\xi_1/\partial z) = \mu_2(\partial\xi_2/\partial z)$ at $z = 0$.

The BCs for the vector potentials in cylindrical coordinates (r, φ, z) follow directly from the BCs for the velocities and in particular: the no-slip BCs have the form $\mathbf{w}_k - \nabla\xi_k = \mathbf{e}_y$ at S_p; the kinematic BCs are $w_{1r} = w_{2r}$, $w_{1\varphi} = w_{2\varphi}$, $w_{1z} - (\partial\xi_1/\partial z) = w_{2z} - (\partial\xi_2/\partial z) = 0$ at $z = 0$; the dynamic BCs have the form $\mu_1(\partial w_{1r}/\partial z) = \mu_2(\partial w_{2r}/\partial z)$, $\mu_1(\partial w_{1\varphi}/\partial z) = \mu_2(\partial w_{2\varphi}/\partial z)$.

In cylindrical coordinates (r, φ, z) (Fig. 1), the Fourier expansion of the solution with respect to the polar angle, φ, contains only the first Fourier mode [2, 6]. The components of the vector and scalar potentials can be presented as $w_{kr} = a_{kr} \sin\varphi$, $w_{k\varphi} = a_{k\varphi} \cos\varphi$, $w_{kz} = a_{kz} \sin\varphi$, $\xi_k = b_k \sin\varphi$. The amplitudes $(a_{kr}, a_{k\varphi}, a_{kz}, b_k)$ depend on the radial, r, and vertical, z, coordinates and the 3D problem is reduced to 2D system of eight PDEs in cylindrical coordinates. The system is closed with respective BCs for the amplitudes of the first Fourier mode. From numerical viewpoint, it is convenient to uncouple the considered PDE system, using new functions: $a_{kr} = 2(u_{k0} + u_{k2})$, $a_{k\varphi} = 2(u_{k0} - u_{k2})$, $a_{kz} = 2u_{k1}$, $b_k = b_{k1} + u_{k0}r + u_{k2}r + u_{k1}z$ $(k = 1, 2)$. Thus, we considerably simplify the problem to the following homogeneous system of PDEs $(k = 1, 2)$: $L_0[u_{k0}] = 0$, $L_1[u_{k1}] = 0$, $L_2[u_{k2}] = 0$, $L_1[b_{k1}] = 0$, where the dimensionless Laplace operators L_j have the following form:

$$L_j[u] = \frac{1}{r}\frac{\partial}{\partial r}\left(r\frac{\partial u}{\partial r}\right) + \frac{\partial^2 u}{\partial z^2} - \frac{j^2 u}{r^2}\ (j = 0, 1, 2). \tag{1}$$

To construct an efficient numerical scheme, the complex domains are transformed into rectangles, introducing modified toroidal coordinates τ and σ (Fig. 1): $rh = 1 - \tau^2$ and $zh = 2\tau \sin\sigma$, where $h = 1 + \tau^2 - 2\tau\cos\sigma$ is the normalized metric coefficient. The position of the fluid–fluid interface and those of the upper and lower particle surfaces are $\sigma = 0$, $\sigma = \sigma_1 = \alpha$ and $\sigma = \sigma_2 = \alpha - \pi$, respectively (Fig. 1). At the axis of revolution one has $\tau = 1$ and the three-phase contact line corresponds to the pole, A_+, where $\tau = 0$. The expressions for the Laplace operators, L_j ($j = 0, 1, 2$), in toroidal coordinates are

$$L_j[u] = \frac{h^3}{4\tau(1-\tau^2)}\frac{\partial}{\partial\tau}\left[\frac{\tau(1-\tau^2)}{h}\frac{\partial u}{\partial\tau}\right] + \frac{h^3}{4\tau^2}\frac{\partial}{\partial\sigma}\left(\frac{1}{h}\frac{\partial u}{\partial\sigma}\right) - \frac{j^2h^2}{(1-\tau^2)^2}u. \quad (2)$$

The functions in the system of PDEs are dependent on each other because of the BCs. For the first Fourier mode in terms of the new functions, u_{kj} and b_{k1} ($k = 1, 2$ and $j = 0, 1, 2$), we derive as follows: (1) for the no-slip BCs at the upper and lower particle surfaces ($k = 1, 2$ and $\sigma = \sigma_k$) :

$$2\left[(1+\tau^2)\cos\sigma - 2\tau\right]u_{k2} + \left[(1-\tau^2)\sin\sigma\right]u_{k1} = 0, \quad (3)$$

$$u_{k2} + \frac{(1-\tau^2)\sin\sigma}{8\tau}\frac{\partial}{\partial\sigma}\left[b_{k1} + \frac{1-\tau^2}{h}(u_{k0}+u_{k2}) + \frac{2\tau\sin\sigma}{h}u_{k1}\right] = 0, \quad (4)$$

$$u_{k0} - u_{k2} = \frac{1}{2}, \; b_{k1} + \frac{1-\tau^2}{h}(u_{k0}+u_{k2}) + \frac{2\tau\sin\sigma}{h}u_{k1} = 0; \quad (5)$$

(2) for the scalar potential and the kinematic BCs at the fluid–fluid interface ($\sigma = 0$):

$$u_{10} = u_{20}, \; u_{12} = u_{22}, \; b_{11} = b_{21}, \quad (6)$$

$$u_{k1} - \frac{(1-\tau)^2}{2\tau}\frac{\partial b_{k1}}{\partial\sigma} - \frac{1-\tau^2}{2\tau}\frac{\partial}{\partial\sigma}(u_{k0}+u_{k2}) = 0 \; (k = 1, 2); \quad (7)$$

(3) for the scalar potential and the dynamic BCs at the fluid–fluid interface ($\sigma = 0$):

$$\mu_1\frac{\partial b_{11}}{\partial\sigma} = \mu_2\frac{\partial b_{21}}{\partial\sigma}, \; \mu_1\frac{\partial u_{10}}{\partial\sigma} = \mu_2\frac{\partial u_{20}}{\partial\sigma}, \; \mu_1\frac{\partial u_{12}}{\partial\sigma} = \mu_2\frac{\partial u_{22}}{\partial\sigma}. \quad (8)$$

At the axis of revolution ($\tau = 1$), the natural BCs for symmetry of the solutions are used. The values of all functions vanish at infinity. At the three-phase contact line ($\tau = 0$), all functions are constants, which do not depend on σ.

3 Singularity Diagrams

The semi-analytical results for water–air interface and $\alpha \le 90°$ [6] show that the pressure solutions are regular at the three-phase contact line. In the general case, functions p_k can have a weak singularity at $\tau = 0$, while $\mathbf{v}_k$ are regular functions at $\tau = 0$. The boundary between singular and regular solutions for the pressure at $\tau = 0$ corresponds to the weaker possible logarithmic singularity, $p_k = (A_k + B_k\sigma) \ln \tau \sin \varphi + \cdots$, where A_k and B_k are unknown constants. We substitute this asymptotic expansion with respect to τ ($\tau \to 0$) into the Stokes equations and solve the obtained leading order problem for the velocity functions. Subsequently, the general representations are substituted into the BCs. This leads to a homogeneous linear system for A_k and B_k, which has a nontrivial solution when its determinant is equal to zero, i.e.

$$\mu_1 \frac{\sin(2\alpha) - 2\alpha \cos(2\alpha)}{\sin \alpha - \alpha \cos \alpha} = \mu_2 \frac{\sin(2\alpha) - 2(\alpha - \pi) \cos(2\alpha)}{\sin \alpha - (\alpha - \pi) \cos \alpha}. \tag{9}$$

The solutions of (9) are shown in Fig. 2 (note that by definition $\mu_2 = 1 - \mu_1$). For $0.2 < \mu_1 < 0.8$, the values of pressure at the contact line are constants for all angles α. The pressure has no singularity for all values of the viscosities when the contact angle obeys the inequality $\alpha_b < \alpha < \pi - \alpha_b$, where $\alpha_b = 51.2733°$ is the smallest positive root of the equation $\sin(2\alpha_b) = 2(\alpha_b - \pi) \cos(2\alpha_b)$.

The pressure function has a stronger than logarithmic singularity, i.e. $p_k = [A_k \cos(\lambda\sigma) + B_k \sin(\lambda\sigma)]\, \tau^\lambda \sin \varphi + \cdots$, inside the regions, shown in Fig. 2b, where from a physical viewpoint $-0.5 < \lambda < 0$. Following an analogous procedure for the solution of the leading order problem, we arrive to the respective transcendental equation for the singularity parameter, λ. Figure 2 shows the dependence of λ on the central angle, α, and viscosity ratio, μ_1, for $\alpha \le \alpha_b$. Because of the symmetry, the picture is analogous replacing μ_1 with μ_2 and α with $\pi - \alpha$. One sees that with

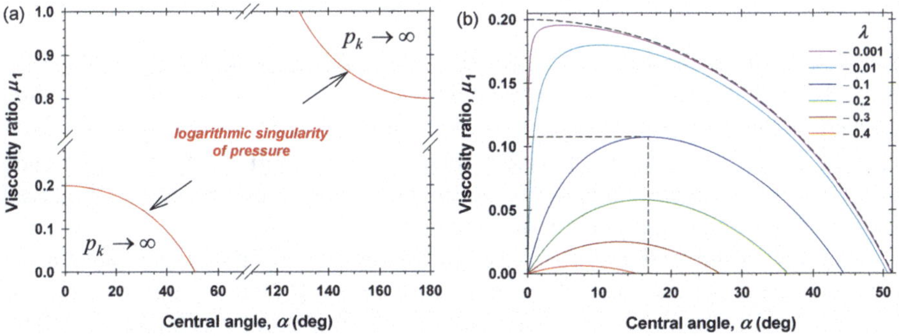

Fig. 2 (**a**) Diagram of the weak singularity of functions p_k. (**b**) Lines with fixed values of the singularity parameter, λ, in the region $\alpha < \alpha_b$

the decrease of viscosity ratio μ_1 the pressure has stronger singularity. For example, the dashed line in Fig. 2 shows that if $\mu_1 = 0.1075$, then the strongest singularity of $\lambda = -0.1$ takes place at central angle equal to 16.8°. In all cases $\lambda > -0.5$, so the singularity is weak and the integral from the pressure over the particle surface converges (the drag force is finite).

4 Numerical Method and Numerical Results

In order to solve the problem, we introduce numerical time t and seek the stationary solution of the parabolic problem:

$$\frac{\partial \mathbf{f}_k}{\partial t} = T[\mathbf{f}_k] + S[\mathbf{f}_k] \ (k = 1, 2), \tag{10}$$

with appropriate BCs imposed, where $\mathbf{f}_k = (u_{k0}, u_{k1}, u_{k2}, b_{k1})$ is the vector of the solutions at phase k and $S[\cdot]$ and $T[\cdot]$ are operators that act at σ- and τ-direction, respectively:

$$S[\mathbf{f}_k] = h\frac{\partial}{\partial \sigma}\left(\frac{1}{h}\frac{\partial \mathbf{f}_k}{\partial \sigma}\right), \quad T[\mathbf{f}_k] = (T_0[u_{k0}], T_1[u_{k1}], T_2[u_{k2}], T_1[b_{k1}]), \tag{11}$$

$$T_j[u] = \frac{h\tau}{1-\tau^2}\frac{\partial}{\partial \tau}\left[\frac{\tau(1-\tau^2)}{h}\frac{\partial u}{\partial \tau}\right] - \frac{4\tau^2 j^2}{(1-\tau^2)^2}u. \tag{12}$$

Let us introduce a rectangular mesh with time step size δ_t and space step sizes δ_τ and δ_σ in τ and σ-direction, respectively. On this grid, the solution at the moment t is denoted by $\mathbf{f}_k^{(0)}$ and this at the moment $t+2\delta_t$—by $\mathbf{f}_k^{(2)}$. Using the Crank–Nicolson method, we reduce the problem to the following algebraic system:

$$(\mathbf{U} - \delta_t \mathbf{T})(\mathbf{U} - \delta_t \mathbf{S})[\mathbf{f}_k^{(2)} - \mathbf{f}_k^{(0)}] = 2\delta_t \mathbf{T}[\mathbf{f}_k^{(0)}] + 2\delta_t \mathbf{S}[\mathbf{f}_k^{(0)}], \tag{13}$$

where $\mathbf{U}$ is the unit operator and the BCs are approximated, using second-order finite differences [4]. In order to solve the latter, we use the alternative direction implicit method—at first, we solve the problem in the τ-direction $(\mathbf{U} - \delta_t \mathbf{T})[\mathbf{f}_k^*] = 2\delta_t \mathbf{T}[\mathbf{f}_k^{(0)}] + 2\delta_t \mathbf{S}[\mathbf{f}_k^{(0)}]$ and then in the σ-direction $(\mathbf{U} - \delta_t \mathbf{S})[\mathbf{f}_k^{(2)} - \mathbf{f}_k^{(0)}] = \mathbf{f}_k^*$. In such a way, we reduce the problem to two linear algebraic systems, which can be solved using the direct elimination numerical method.

In order to validate the results, we compare the values of the drag coefficient, f, computed using our method to the semi-analytic results [6] for $\alpha \le 90°$ and fluid–air interface (see Table 1). The computations are performed for $\delta_\sigma = 0.017$, $\delta_\tau = 0.05$ and different time steps δ_t. The relative error is less than 1% and the CPU time is less than 10 s in all studied cases. As it can be expected, the decrease of the grid size decreases the relative error of f. For example, if $\alpha = 15°$ and the rectangular

Table 1 Comparison between the calculated and exact values of the drag force coefficient, f

α (deg)	δ_t	CPU time (s)	f (calculated)	f (exact)	Rel. error (%)
15	0.10	2.534	1.4306	1.4374	0.473
30	0.15	7.332	1.4013	1.3392	0.612
60	0.45	7.504	1.2522	1.2509	0.104
75	0.60	5.242	1.1473	1.1370	0.906
90	0.60	8.798	0.9916	1.0000	0.840

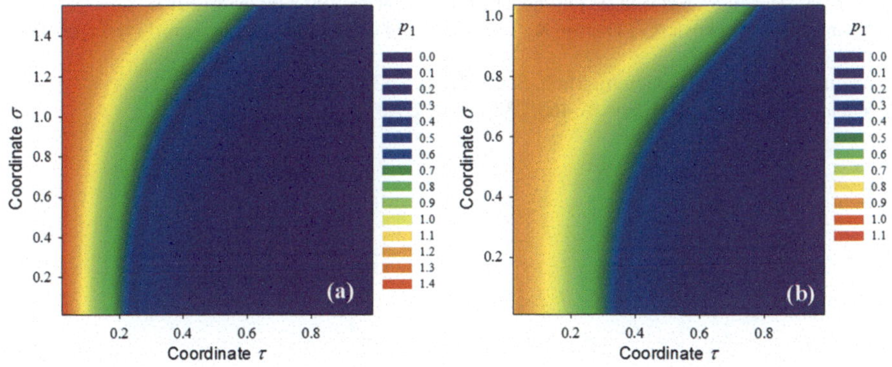

Fig. 3 Pressure distribution for air/fluid interfaces: (**a**) $\alpha = 90°$. (**b**) $\alpha = 60°$

domain is divided by 20×20 then the relative error is 0.16%, while for 30×30—it decreases to 0.017%. The respective CPU time for calculations increases from 4.3 to 11 s—it triples by increasing the number of space-discretisation steps by factor of 2.25. Analogous trends hold true for all values of the contact angle.

Figure 3 shows the pressure distribution for air/water interface and two values of the three-phase contact angle. It is well illustrated that the pressure maximum for $\alpha = 90°$ is at the contact line, while that for $\alpha = 60°$—it is shifted along the particle surface inside the fluid phase. Using the proposed numerical method, it is possible to perform a systematic study for wide ranges of physical parameters.

5 Conclusion

The hydrodynamic problem for the translation of a spherical particle, attached to a fluid/fluid interface, is simplified using the gauge formulation. The introduction of appropriate functions and toroidal coordinates reduces the 3D Stokes equations to a 2D system of eight homogeneous PDEs. The system is coupled because of the complex BCs. The developed efficient ADI type second-order numerical scheme gives possibility for fast and precise calculations of all physical parameters (velocity vector and pressure fields, and drag force coefficient).

Acknowledgement The work of Galina Lyutskanova–Zhekova has been partially supported by the Sofia University "St. Kl. Ohridski" under contract No. 80-10-139/25.04.2018.

References

1. Dani, A., Keiser, G., Yeganeh, M.S., Maldarelli, C.: Langmuir **31**, 13290–13302 (2015)
2. Danov, K., Dimova, R., Pouligny, B.: Phys. Fluids **12**, 2711–2722 (2000)
3. Dorr, A., Hardt, S., Masoud, H., Stone, H.: J. Fluid Mech. **790**, 607–618 (2016)
4. Lyutskanova-Zhekova, G., Danov, K.: Lecture Notes Computer Science, vol. 11189, pp. 433–440. Springer (2019)
5. Weinan, E., Liu, J.-G.: Comm. Math. Sci. **1**(2), 317–332 (2003)
6. Zabarankin, M.: Proc. R. Soc. A **463**, 2329–2349 (2007)

Refinement of Surfaces of Industrial Objects

Gábor Renner and György Gyurecz

Abstract Refinement of industrial (e.g. car-body) surfaces is performed by evaluation of the shape and distribution of reflection lines or highlight lines. In the paper, we propose a method to semi-automatically evaluate and improve the quality of the highlight line structures. The correspondence between the shape of the highlight lines and the surface parameters is highly complicated and strongly nonlinear. In the paper, a genetic process is proposed for the computation of the parameters (control points) of the surfaces, that corresponds to the corrected highlight line structure.

1 Introduction

Car-body surfaces are high quality (Class A) surfaces. For the evaluation of their quality, sensitive methods were developed; these are mainly the evaluation of the shape and distribution of reflection lines or highlight lines. A highlight line structure is a series of highlight lines, they represent visually the reflection and shape error characteristics of the surface. Highlight lines are calculated as the surface imprint of a linear light source array, placed above the surface. The structure of the highlight lines is evaluated by their pattern and the individual shape of the highlight lines. High-quality surfaces can be characterized by uniform or smoothly changing highlight line patterns and smooth highlight lines.

A method for displaying highlight lines was developed by Beier and Chen [1]. Methods for designing and correcting surfaces using highlight lines were first published by Klass [2] and later Kaufmann and Klass [3]. The relation between highlight lines and the defining parameters of the surfaces can be expressed by a highly non-linear equation system, which is too time consuming to solve, and the

G. Renner (✉)
Computer and Automation Research Institute, Budapest, Hungary
e-mail: renner@vision.sztaki.hu

G. Gyurecz
Óbuda University, Budapest, Hungary

© Springer Nature Switzerland AG 2019

I. Faragó et al. (eds.), *Progress in Industrial Mathematics at ECMI 2018*,
Mathematics in Industry 30, https://doi.org/10.1007/978-3-030-27550-1_13

result is not always satisfactory. The method developed by Zhang and Cheng [4] introduces a great number of simplifications to obtain a linear system of equation to modify surface parameters through highlight lines.

In the paper, we propose a method to evaluate and improve the quality of the highlight line structures. The evaluation is carried out in two steps. First, distance and angle functions are computed to quantify the error in the highlight line structure. Then the highlight points are corrected, and based on these points the corrected highlight line segments are constructed. For the computation of the parameters of the surfaces that correspond to the corrected highlight line structure a genetic process was developed. The genetic operators and parameters of the genetic process are adjusted to the specific technical problem of surface refinement by highlight lines. We discuss the genetic representation and the fitness function of the genetic process.

2 Highlight Lines

Highlight lines are calculated as a set of discrete highlight points. These are points on the surface where the corresponding surface normal and the light source line intersect each other. Let **L**(λ) = **A**+**B**λ a line representing a light source, where **A** is a point on **L**(λ), and **B** is a vector defining the direction of the line (Fig. 1).

The shape of a surface **S**(u,v) is defined by an array of control points $\mathbf{P}_{i,j}$ in Bézier, B-spline or NURBS representations. The signed perpendicular distance *d*(u,v) between the normal **N**(u,v) at a surface point **S**(u,v) and the linear light source is:

$$d(u, v) = \frac{|[B \times N(u, v)] \cdot [A - S(u, v)]|}{\|B \times N(u, v)\|} \tag{1}$$

For a surface point on the highlight line, d(u,v) = 0 holds, which must be solved for the (u,v) parameter of the surface.

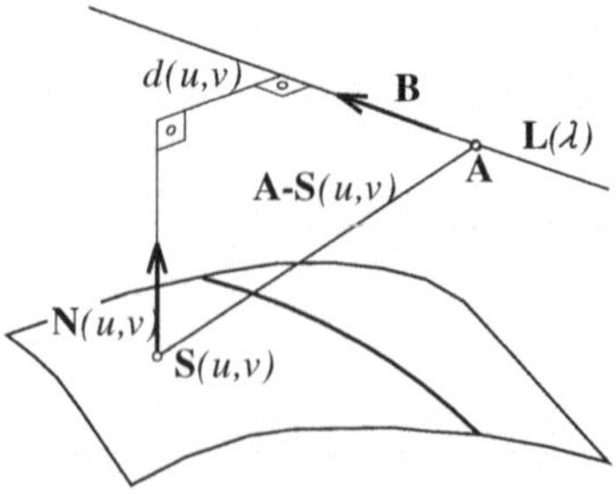

Fig. 1 Distance interpreted between surface normal and the light source

The identification of the defected highlight curve segments $\mathbf{C}_i$, i = 0...N, is performed interactively, and their endpoints $\mathbf{A}_i$ and $\mathbf{B}_i$ and tangents $\mathbf{T}_i{}^1$ and $\mathbf{T}_i{}^2$ at the endpoints are searched.

3 Refinement of the Highlight Line Pattern

The pattern of the defected highlight curve segments is evaluated on sequences s_j, j = 0...M of highlight points $\mathbf{E}_{0,0},\ldots\mathbf{E}_{i,j}\ldots\mathbf{E}_{N,M}$ spanning over the defective segment in crosswise direction. The sequences include correct highlight curve points $\mathbf{E}_{0,j}$, $\mathbf{E}_{1,j}$ and $\mathbf{E}_{N-1,j}$, $\mathbf{E}_{N,j}$ at the ends; they ensure continuity of corrected highlight segments with the adjoining unaffected region. We evaluate the error in the structure of the highlight pattern by d_j distance and $\propto_j$ angle functions defined on s_j, sequences. The distance function represents the inequalities of the structure in crosswise direction; the angle function characterizes the structural error along the highlight curves.

Point sequences start with points $\mathbf{E}_{0,0}\ldots\mathbf{E}_{0,M}$ parametrically equally spaced on highlight curve C_0. They are determined by the location of the furthest endpoints $\mathbf{A}_\mathbf{i}$ and $\mathbf{B}_i$ and the M number of sequences. The subsequent points are calculated based on the shortest perpendicular distance between subsequent highlight curves.

Let $\mathbf{E}'_{i,j} = \mathbf{C}_i(t)$ a point on the current, and $\mathbf{E}'_{i-1,j} = \mathbf{C}_{i-1}(t)$ a point on the previous highlight curve (Fig. 2). Point $\mathbf{E}'_{i,j}$ is in the perpendicular direction if

$$\mathbf{H}'_{i,j} = \mathbf{E}'_{i,j} - \mathbf{E}'_{i-1,j}$$

$$\boldsymbol{H}'_{i,j} \cdot \boldsymbol{T}'_{i,j} = 0$$

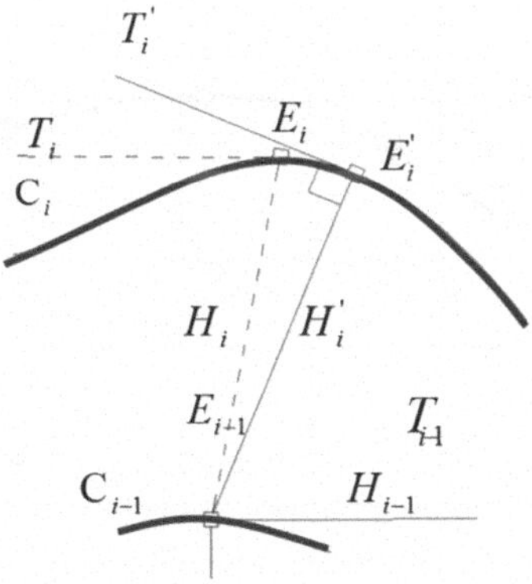

Fig. 2 Calculation of the evaluation points

where

$$\mathbf{T'}_{i,j} = \mathbf{C'}_i(t) \text{ at } \mathbf{H'}_{i,j}$$

The location of evaluation point $\mathbf{E}_{i,j}$ is in the surrounding of $\mathbf{E'}_{i,j}$ where:

$$\mathbf{T}_{i,j} \cdot \mathbf{T}_{i-1,j} = 1$$

$$\mathbf{T}_{i-1,j} = \mathbf{C'}_i(t) \text{ at } \mathbf{H}_{i-1,j}$$

The distance error function is defined by the distances $d_{i,j}$ between the consecutive sequence elements:

$$d_{i,j} = \left\| \boldsymbol{E}_{i+1,j} \right\| - \left\| \boldsymbol{E}_{i,j} \right\| \tag{2}$$

The angle error function is defined by angles $\alpha_{i,j}$ between the consecutive vectors $\boldsymbol{H}_i$:

$$\alpha_{i,j} = arccos\left(\frac{\boldsymbol{H}_{i+1} \cdot \boldsymbol{H}_i}{\left\| \boldsymbol{H}_{i+1} \right\| \cdot \left\| \boldsymbol{H}_i \right\|} \right) \tag{3}$$

In Fig. 3, an example of distance function is presented. For the angle function a similar figure can be obtained. The sequence i = 0..N of the functions correspond to the defective highlight curves. The rapid and irregular changes represent the defects in the highlight line structure. The function values at i = −2, −1, N+1, N+2 correspond to the adjoining correct highlight curves.

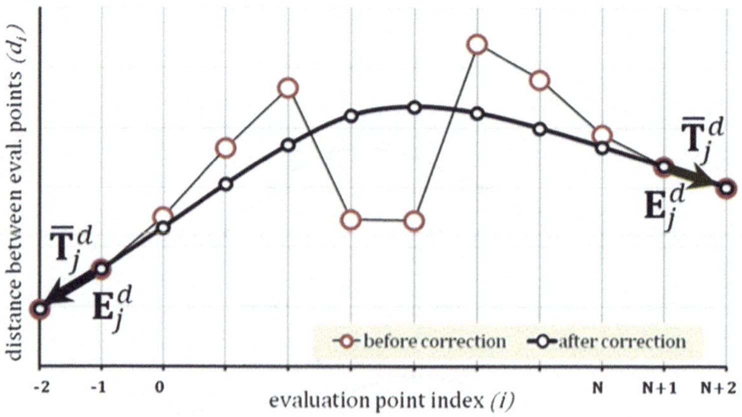

Fig. 3 The distance evaluation function before and after correction

4 Calculation of Corrected Highlight Curves

A correct highlight line pattern can be characterized by smooth evaluation functions, without oscillations shown in Fig. 3, and maintains continuity with the adjoining correct pattern. For the correction of highlight curves, smooth evaluation functions are calculated by least square approximation method. Continuity with the correct highlight line structure of the adjoining region is ensured by constraints on endpoints and end tangents $\mathbf{T}_i{}^1$ and $\mathbf{T}_i{}^2$. The tangents are calculated as: $\mathbf{T}_j{}^1 = \mathbf{E}_{0,j} - \mathbf{E}_{1,j}$ and $\mathbf{T}_j{}^2 = \mathbf{E}_{N-1,j} - \mathbf{E}_{N,j}$.

Based on the new evaluation functions, points of the new highlight curves are obtained. The points $\boldsymbol{R}_{i,j}$ of the new highlight curves are calculated starting from point $\boldsymbol{E}_{i,j|i=1}$ and moving in the direction $\alpha_{\mathrm{i,j}}$ with the distance $\mathrm{d_{i,j}}$ defined by the angle and distance functions. The new $\mathbf{C}_i$ highlight curves are computed as cubic B-Splines, constructed from the new $\mathbf{R}_{(\mathbf{i,j})}$ points by constrained least squares curve fitting.

5 Surface Correction by Genetic Algorithm

Genetic algorithms apply the mechanism of evolution in finding optimal solution to complex non-linear problems (Goldberg [5]). The components of a GA process, including the coding (chromosomes) genetic operators (crossover, mutation), fitness function depend on the specific technical problem to be solved. Their selection must be analyzed and tested carefully. Our goal is to adjust the parameters of surfaces that produces the desired highlight lines by using a GA. Although GA does not provide unique mathematical solution to the problem, it is able to arrive at a nearly optimal solution. In the paper, we discuss two basic items of the genetic process; genetic representation and fitness function. More details on the genetic process can be found in [6], where the setting of other genetic parameters and also their influence on the genetic process is discussed.

Surface modifications are performed by modifying the control points $\mathbf{P}_{i,j}$. Control points, that have influence on the surface region to be corrected are included in the genetic representation. A gene g_γ consist of control point modification

$$g_\gamma = \Delta\mathbf{P}_{i,j}\ (x, y, z,) \tag{4}$$

where x, y and z are Cartesian co-ordinates of $\Delta\mathbf{P}_{i,j}$, while γ is the identifier of genes within a chromosome of the surface.

Fitness function consists of two components: accuracy and shape similarity. Accuracy is based on distance, while shape similarity on angle difference of tangent vectors between corresponding points of actual and desired highlight curves.

Denote h_i^{des} the desired, and h_i^{cre} the highlight line, created during the genetic search and $d_i\ (t_k)$ the deviation between corresponding highlight points at different parameters t of highlight lines. Then, the distance error is

$$f_{dist} = \sum_{i=1}^{l} \left(\sum_{k=1}^{n_i} \left(d_i\ (t_k) - \frac{1}{n_i} \cdot \sum_{k=1}^{n_i} d_i\ (t_k) \right)^2 \cdot \frac{1}{n_i} \right), \tag{5}$$

where $d_i\ (t_k) = \left| h_i^{cre}\ (t_k) - h_i^{des}\ (t_k) \right|$ while n_i denotes the number of examined highlight points. Variable l indicates the number of highlight lines. Angle difference f_{ang} is calculated in the same manner, except the deviation is composed as follows:

$$d_i\ (t_k) = \arccos \left(\frac{h_i^{des}\ (t_k) \cdot h_i^{cre}\ (t_k)}{\left| h_i^{des}\ (t_k) \right| \ \left| h_i^{cre}\ (t_k) \right|} \right) \tag{6}$$

Distance error component promotes the creation of accurate highlight lines, but their shape is often poor. The angle error component behaves in the opposite way: it promotes producing highlight lines with good shape similarity, but on the expenses of their accuracy. We eliminated the disadvantages of fitness components by letting the distance dominate in the beginning of the search and make the angle dominate at the end.

6 Results

The proposed surface refinement method was tested on several industrial surfaces of different complexity, size and error domains. The application of our method starts with the evaluation and correction of the highlight line structure of the surface as discussed above. The surface correction is performed by a genetic search with a fitness function reflecting accuracy and shape similarity. Thus, fitness provides a combined measure of the deviation between erroneous and corrected highlight line structures. The change of fitness through generations is shown in Fig. 4 in case of the test example. GA runs until the user defined stop criteria (e.g., the residual error) is fulfilled.

Figure 5 demonstrates results visually by the example of a car-body panel. Left picture shows the surface with the original erroneous highlight lines, the right picture shows the surface with the corrected highlight lines. The defected region is indicated by white circle. The quality of the surface compared to the original one can be evaluated by the reduction of the fitness to 6% of its starting value, and visually by comparing the highlight line patterns.

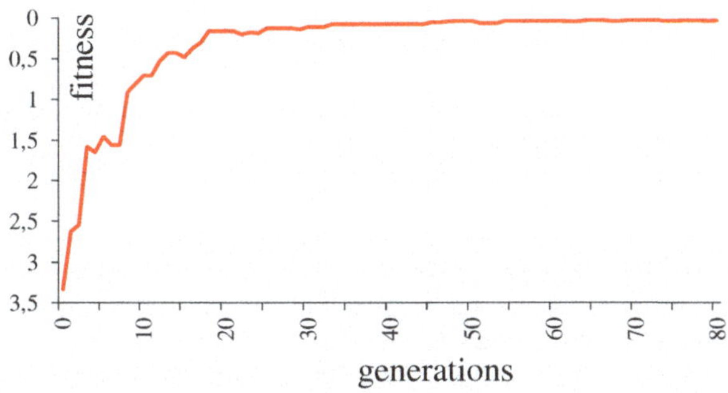

Fig. 4 The change of fitness through generations

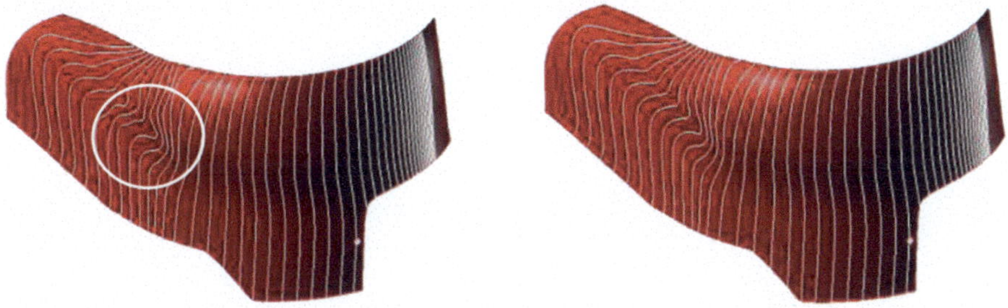

Fig. 5 Car-body surfaces; defective and corrected surfaces

7 Conclusion

Evaluation of the quality and refinement of industrial surfaces based on their highlight line structure is presented. The defective surface area is selected interactively, the evaluation and correction of highlight lines is automated. Defining parameters (control points) of the surfaces, that corresponds to the corrected highlight line structure are computed through a genetic algorithm, without computing highly nonlinear correlation between control points and highlight lines. Best performing genetic operators, fitness function, strategies and parameters of the genetic process were determined. Although genetic algorithms are not able to find a unique optimal solution, they are suitable to search for a nearly optimal solution in a very complex nonlinear search space, and thus, to solve very complex technical problems.

The method is applicable to surfaces of any kind of CAD representations, and wide range of errors in the highlight line structure and consequently in the surface. The applicability of the method was proved on several car-body and other industrial surfaces.

References

1. Beier, K.P., Chen, Y.: Highlight-line algorithm for real time surface quality assessment. Comput. Aided Des. **26**(4), 268–277 (1994)
2. Klass, R.: Correction of local surface irregularities using reflection lines. Comput. Aided Des. **12**(2), 73–78 (1980)
3. Kaufmann, E., Klass, R.: Smoothing surfaces using reflection lines for families of splines. Comput. Aided Des. **20**(6), 312–316 (1988)
4. Zhang, C., Cheng, F.: Removing local Irregularities of NURBS surfaces by modifying highlight lines. Comput. Aided Des. **30**(12), 923–930 (1998)
5. Goldberg, D.E.: Genetic algorithms for search, optimization, and machine learning. Addison-Wesley Professional, Boston (1989)
6. Gyurecz, G.Y., Renner, G.: Correcting fine structure of surfaces by genetic algorithm. Acta Polytech. Hung. **8**, 181–190 (2011)

Extended Gaussian Approximation for Modeling the Quantum Dynamics of Localized Particles

Omar Morandi

Abstract We derive a quantum model that provides some corrections to the classical motion of nearly localized particles. Our method is based on the assumption that the particle wave function is strongly localized and represented by a Gaussian shape. As an application of our method, we describe the motion of a particle in a 2D non harmonic potential.

1 Introduction

During the last decade, various approaches have been proposed in order to describe the quantum dynamics of nearly localized particles. New models that extend the concept of classical trajectory to the quantum mechanical context have been proposed [1–11]. One of the major advantages of developing methods in which the particle dynamics shows analogies with the classical Newtonian dynamics, is the possibility to interpret the quantum motion in term of a classical corrected transport.

In this paper, we derive the evolution equation for a localized quantum particle. We develop a model that preserves the classical description of particle motion in terms of trajectories. We assume that the wave function is described by a Gaussian-like wave packet. In order to apply our method to a general situation, in our ansatz we insert a set of time dependent parameters that modulate the Gaussian shape. Finally, the particle motion is expressed by a system of nonlinear differential equations.

The Gaussian beam approximation is a popular method used to describe nearly localized particles [12]. Similarly to the Gaussian beam approximation, our expansion procedure is based on the projection of the solution over a set of functions which are modulated by a Gaussian whose width changes in time according to the quantum evolution equation.

O. Morandi (✉)
University of Florence, Florence, Italy
e-mail: omar.morandi@unifi.it

© Springer Nature Switzerland AG 2019
I. Faragó et al. (eds.), *Progress in Industrial Mathematics at ECMI 2018*,
Mathematics in Industry 30, https://doi.org/10.1007/978-3-030-27550-1_14

Our approach is designed to describe the motion of heavy particles which are typically well localized. We will discuss the application of our method to the 2D motion of a particle in the presence of a confining non harmonic potential.

In this paper we extend the results obtained in Ref. [13], where the 1D case with some extension to the 2D problem, was considered.

2 Description of the Model: Extended Gaussian Ansatz

We consider a particle defined by the wave function $\psi(\mathbf{r}) \in L^2(\mathbb{R}^d)$, where d is the dimension of the space. The particle evolution is given by the time dependent Schrödinger equation

$$i\frac{\partial \psi}{\partial t} = \left(-\frac{1}{2}\Delta_{\mathbf{r}} + U(\mathbf{r})\right)\psi \, . \tag{1}$$

In order to simplify the notation, we set the Planck constant and the particle mass to one. U is the external potential that is assumed to be known. Our model consists on the following ansatz. We represent the particle wave function in polar coordinates and we shift the spatial coordinate by a time dependent vector $\mathbf{s} \in \mathbb{R}^d$. We expand both the phase and the modulus of the wave function on the basis set of the harmonic oscillator centred in $\mathbf{s}$. We obtain

$$\psi(\mathbf{r}) = \sqrt{\sum_{\{n\}=0}^{\infty} a_{\{n\}}(t) h^{\sigma}_{\{n\}}(\mathbf{r}-\mathbf{s})} e^{-\frac{1}{2}\sum_{n=1}^{d} r_i^2 \sigma_i + i\sum_{\{n\}=0}^{\infty} \chi_{\{n\}}(t) h^{\sigma}_{\{n\}}(\mathbf{r}-\mathbf{s})} \tag{2}$$

We have introduced the compact notation $\{n\}$ to indicate the sequence of d integers $\{n\} \doteq (n_1, n_2, \ldots, n_d)$ and

$$h^{\sigma}_{\{n\}}(\mathbf{r}) \doteq \prod_{i=1}^{d} h^{\sigma_i}_{n_i}(r_i) \, . \tag{3}$$

The functions $h^{\sigma_i}_{n_i}(r_i)$ are the normalized Hermite functions. For the details concerning the definition of $h^{\sigma_i}_{n_i}(r_i)$ we refer to [13, Eq. (19)]. In the previous equation, $\chi_{\{n\}}$ and $a_{\{n\}}$ together with $\mathbf{s} \in \mathbb{R}^d$ and σ_i with $i = 1, \ldots, d$ represent the new unknowns of the problem. In particular, the parameter $\mathbf{s} \in \mathbb{R}^d$ represents the mean of the particle position and σ_i provides the width of the Gaussian packed.

The main interest of our approach is to derive the evolution equation of the parameters. By using the ansatz of Eq. (2) in Eq. (1) after some algebra it is possible to derive the set of evolution equations for all the parameters in a closed form. We give here the final form of the evolution equations for the phase $\chi_{\{n\}}$, the modulus

$a_{\{n\}}$, the Gaussian width $\boldsymbol{\sigma}$ and the centre of the expansion $\mathbf{s}$ of the particle wave function:

$$\begin{aligned}\frac{\mathrm{d}\chi_{\{n\}}}{\mathrm{d}t} =& \left(\prod_j^d \sigma_j^{1/4}\right) \sum_{\{r\},\{s\}} \left(\frac{R_{\{r\}}R_{\{s\}}}{4} - \chi_{\{r\}}\chi_{\{s\}}\right) \sum_{i=1}^d \sigma_i \sqrt{r_i s_i}\, \mathbb{A}^1_{n_i,r_i-1,s_i-1} \prod_{j\neq i} \mathbb{A}^1_{n_j,r_j,s_j} \\ &- \sum_{i=1}^d \frac{\sigma_i n_i R_{\{n\}}}{2} + \frac{\pi^{d/4}}{4} \sum_{i=1}^d \sigma_i^{3/4} \left(\sqrt{2}\delta_{n_i,2} - \delta_{n_i,0}\right) \left(\prod_{j\neq i}^d \sigma_j^{-1/4} \delta_{n_j,0}\right) \\ &+ 2\sum_{i=1}^d \chi_{\{n;n_i\to n_i+1\}} S_i \sigma_i \sqrt{2(n_i+1)} + \sum_i M_i \left(\frac{2n_i+1}{2}\chi_{\{n\}} + \sqrt{(n_i+2)(n_i+1)}\,\chi_{\{n;n_i\to n_i+2\}}\right) \\ &+ \int_{\mathbb{R}^d} U(\mathbf{s}-\mathbf{r}) h^{\sigma}_{\{n\}} e^{-\sum_{n=1}^d r_i^2 \sigma_i}\, \mathrm{d}\mathbf{r} \end{aligned} \tag{4}$$

$$\begin{aligned}\frac{\mathrm{d}a_{\{n\}}}{\mathrm{d}t} =& - \sum_{i=1}^d \frac{M_i}{\sqrt{2\sigma_i}} \left[a_{\{n\}} \frac{2n_i+1}{2} + a_{\{n;n_i\to n_i-2\}} \sqrt{n_i(n_i-1)}\right] - 2\sum_{i=1}^d a_{\{n;n_i\to n_i-1\}} \sigma_i S_i \sqrt{n_i} \\ &+ 2\left(\prod_j^d \sigma_j^{1/4}\right) \sum_{\{r\},\{s\}} a_{\{r\}}\chi_{\{s\}} \sum_{i=1}^d \sigma_i \sqrt{n_i s_i}\, \mathbb{A}^1_{n_i-1,r_i,s_i-1} \prod_{j\neq i} \mathbb{A}^1_{n_j,r_j,s_j} \end{aligned} \tag{5}$$

$$\frac{\mathrm{d}\sigma_i}{\mathrm{d}t} = -\, 2M_i\sigma_i \tag{6}$$

$$\frac{\mathrm{d}s_i}{\mathrm{d}t} = \sqrt{2\sigma_i}\, S_i \;, \tag{7}$$

where $M_i = 2\sigma_i \sum_{\{m\}} a_{\{m\}} \left(\chi_{\{m\}} m_i + \chi_{\{m;m_i\to m_i+2\}} \sqrt{(m_i+1)(m_i+2)}\right)$ and $S_i = \sum_{\{m\}} a_{\{m\}} \chi_{\{m;m_i\to m_i+1\}} \sqrt{m_i+1}$. In particular, our set of equations contains the additional parameters R_n which can be expressed in terms of the variables $a_{\{n\}}$. The details concerning this point are given in Ref. [13]. Moreover, δ denotes the Kronecker's delta and the matrix $\mathbb{A}$ is defined as

$$\mathbb{A}^{\sigma}_{n,r,s} = \pi^{1/4} \int_{\mathbb{R}} h^{\sigma}_n(x) h^{\sigma}_r(x) h^{\sigma}_s(x) e^{-x^2\sigma}\, \mathrm{d}x \;.$$

Explicit form of $\mathbb{A}$ can be found in Ref. [13]. Finally, for the indexes we have introduced the following notation: $\{r; r_i \to a\} \doteq (r_1, \ldots, r_{i-1}, a, r_{i+1}, \ldots, r_d)$ represents the set of indexes in which the i-th term is substituted by a.

3 Numerical Simulations: 2D Case

In order to illustrate our method we perform a numerical test case. We solve the evolution equations in the case of a two-dimensional system. We consider the following double well potential

$$U(x, y) = -\frac{\omega_x}{2}x^2 + V_3 x^3 + V_4 x^4 + V_5 xy + V_6 x^2 y + \frac{\omega_y}{2} y^2 \;. \tag{8}$$

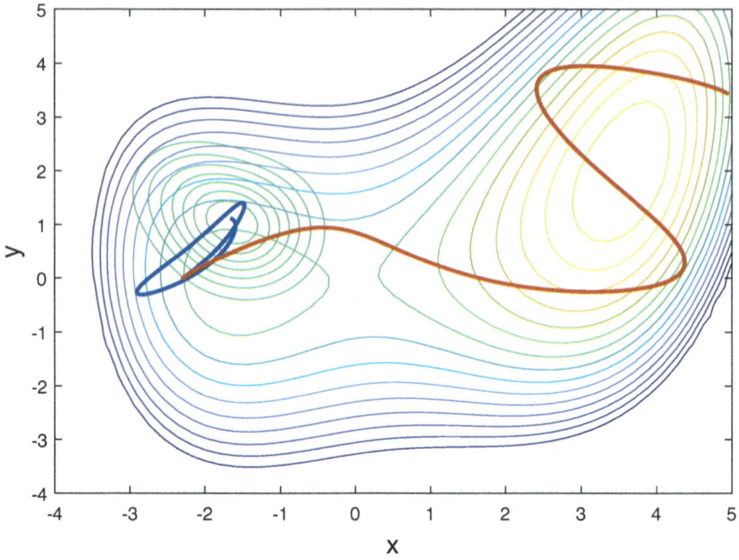

Fig. 1 Evolution of a initially localized Gaussian pulse inside the potential profile (8) (Coloured lines represent the contour plot of U). The panel refers to the time $t = 3$. The contour plot of the solution is depicted by blue lines, the trajectory of the centre of the wave function and the trajectory of the classical motion are depicted by, respectively, a light blue and a red line

We take the following values of the parameters $\omega_x = 1$, $\omega_y = 1$, $V_3 = 0.1$, $V_4 = -0.05$, $V_5 = 0.2$, $V_6 = 0.1$. As initial condition, we have considered a Gaussian beam localized around the left minima of the potential profile. The initial momentum $\mathbf{p} = (1, 1)$. The result of the simulation for the time $t = 3$ is depicted in Fig. 1. In our simulation, we have solved the system of Eqs. (4)–(7) by fixing a cut off on the indexes $\{n\}$. We evaluate the following parameters: a_{n_1,n_2}, χ_{n_1,n_2} with $0 \leq n_1 \leq 3$, $0 \leq n_2 \leq 3$. We plot by continuous blue curves the contour of the solution. In order to follow the evolution of the particle, we have depicted the trajectory of the mean particle position by a light blue continuous line. In order to appreciate the difference between the classical and the quantum corrected dynamics, we have depicted by red continuous curves the classical trajectories obtained by solving the Newton equation. Our simulation shows that the quantum solution differs significantly from the classical one. Taking into account the spreading of the wave function around the mean particle position becomes crucial in order to capture the correct behaviour of the particle motion.

4 Conclusions

We have derived a quantum model for particles characterized by wave functions with Gaussian shape. The oscillations of the particle wave function around the mean particle position are reproduced by Hermite polynomials. The particle motion is described by a set of time dependent parameters. We have applied our method to investigate the motion of a nearly localized particle in a 2D structure.

Acknowledgement This work was supported by National Group of Mathematical Physics (GNFM-INdAM).

References

1. Maddox, J.B., Bittner, E.R.: J. Chem. Phys. **115**, 6309 (2001)
2. Wang, L., Zhang, Q., Xu, F., Cui, X.-D., Zheng, Y.: Int. J. Quantum Chem. **115**, 208 (2015)
3. Horowitz, J.M.: Phys. Rev. E **85**, 031110 (2012)
4. Poirier, B.: Trajectory-based derivation of classical and quantum mechanics. In: Hughes, K.H., Parlant G. (eds.) Quantum Trajectories, CCP6, Daresbury Laboratory (2011).
5. Singer, K.: Mol. Phys. **85**, 701 (1995)
6. Bronstein, Y., Depondt, P., Finocchi, F., Saitta, A.M.: Phys. Rev. B **89**, 214101 (2014)
7. Morandi, O.: J. Phys. A: Math. Theor. **43**, 365302 (2010)
8. Morandi, O.: J. Math. Phys. **53**, 063302 (2012)
9. Sellier, J.M., Nedjalkov, M., Dimova, I.: Phys. Rep. **577**, 1 (2015)
10. Muscato, O., Wagner, W.: SIAM J. Sci. Comput. **38**(3), 1483 (2016)
11. Ceriotti, M., Bussi, G., Parrinello, M.: Phys. Rev. Lett. **103**, 030603 (2009)
12. Jin, S., Wei, D., Yin, D.: J. Comput. Appl. Math. **265**, 199 (2014)
13. Morandi, O.: J. Phys. A: Math. Theor. **51**, 255301 (2018)

Reduced 3D Model of a Passive Admixture Transport in Shallow, Elongated and Weakly Curved Natural Water-Stream

Konstantin Nadolin and Igor Zhilyaev

Abstract The main goal of the study is a validation of a simplified 3D mathematical model for passive admixture spreading in shallow flows. The tested model is oriented to the hydrological and ecological problems, and it can be applied to natural streams like rivers and channels. The earlier proposed model of the '*elongated, shallow and weakly curved stream*' (Nadolin, Mat Model 21(2):4–28, 2009) takes into account the structure of a stream-bed for evaluation of flow velocity in every point of domain. This is a model advantage, which allows calculation of the admixture spreading in a channel with varying width and depth more accurately than by using in-depth averaged models. For example, we can observe the opposite flow in a near-surface zone, which may be caused e.g. by the wind. The results of numerical experiments show that this reduced 3D model adequately describes the admixture spreading processes in natural streams with acceptable accuracy.

1 Introduction

Mathematical models of various types are used for evaluation of the hydrological characteristics of streams and for simulation of the admixture spreading [3, 4, 9]. The most accurate are three-dimensional models, which are based on full equations of turbulent motion. However, the high accuracy of these simulations cannot be obtained in practice because the data of the real hydrological measurements are not precise enough and no initial and boundary conditions for 3D partial differential equations are available. In addition, the complexity and computational costs of

K. Nadolin (✉)
Institute for Mathematics, Mechanics and Computer Science, Southern Federal University, Rostov-on-Don, Russia
e-mail: kanadolin@sfedu.ru

I. Zhilyaev
Institute of Polymer Engineering, FHNW University of Applied Sciences and Arts Northwestern Switzerland, Windisch, Switzerland
e-mail: Igor.Zhilyaev@fhnw.ch

© Springer Nature Switzerland AG 2019
I. Faragó et al. (eds.), *Progress in Industrial Mathematics at ECMI 2018*,
Mathematics in Industry 30, https://doi.org/10.1007/978-3-030-27550-1_15

numerical experiments with 3D mathematical models are increased due to the geometry of the model domain, which is extremely elongated along the flow direction. Natural water flows have significant difference in size of they length, width, and depth. The ratio between the average depth and width for the typical lowland river varies from 1:10 to 1:200.

Note, that systematical analysis of the admixture spreading in the long stream of viscous fluid was initiated by G.I.Taylor [10, 11] and R.Aris [1]. The term "Taylor dispersion" is now widely used in literature as the name for this problem [5].

The main aim of this work is to validate the simplified mathematical model for spreading process in natural streams.

In [6], the equations with reduced dimensions for channel flow hydrodynamics and mass transfer is proposed. The hydrodynamical system of the reduced 3D mathematical model was studied in [7].

This article focuses on testing of a reduced mathematical model of a shallow and elongated stream first proposed in [6]. The model is verified by comparing the data of hydrological experiment, published in [2] and the results obtained on the base of the reduced model. The numerical experiments were performed by finite-element software COMSOL©[8].

2 Problem Statement

Let us consider a relatively slow stream in a non-deformable rigid bed $z = h(x, y)$. The channel flow is shallow, elongated, and weakly curved. In a mathematical sense, the *'shallow and elongated'* assumption means that the stream bed geometry has the ratio $D : W : L \approx \epsilon \ll 1$. Here D is the average depth, W is the average width and L is the length of the section of the stream under consideration; $\epsilon \ll 1$—a value that is used as a small parameter. *'Weakly curved'* means that $\partial h \partial y \sim \epsilon$ and $\partial h \partial x \sim \epsilon^2$.

Let us introduce Cartesian coordinates such that the plane (xy) is located on the flow surface and z-axis is directed toward the bottom. We assume that the x-axis is directed along the flow, and the y-axis is perpendicular to x and directed from the left to the right bank. The origin lies in the inlet section at equal distances from the banks. According to [6] the equations of the 3D reduced mathematical model for the passive admixture transport in *shallow, elongated and weakly curved stream* in dimensionless variables can be written as:

$$\frac{\partial c}{\partial t} + u\frac{\partial c}{\partial x} + v\frac{\partial c}{\partial y} + w\frac{\partial c}{\partial z} = \frac{\partial}{\partial z}\left(d\frac{\partial c}{\partial z}\right) - \lambda c \tag{1}$$

$$c\big|_{t=0} = c^0, \qquad \left.\frac{\partial c}{\partial x}\right|_{x=0} = \pi_0, \qquad \left.\frac{\partial c}{\partial z}\right|_{z=h} = \left.\frac{\partial c}{\partial z}\right|_{z=\xi} = 0 \tag{2}$$

$$\frac{\partial}{\partial z}\left(\nu\frac{\partial u}{\partial z}\right) = -ReGI, \qquad u\big|_{z=h} = 0, \qquad \left.\frac{\partial u}{\partial z}\right|_{z=\xi} = 0 \tag{3}$$

$$\frac{\partial p}{\partial z} = G, \qquad p\Big|_{z=\xi} = 0 \tag{4}$$

$$\frac{\partial}{\partial z}\left(\nu \frac{\partial v}{\partial z}\right) = Re \frac{\partial p}{\partial y}, \qquad v\Big|_{z=h} = 0, \qquad \frac{\partial v}{\partial z}\Bigg|_{z=\xi} = 0 \tag{5}$$

$$\frac{\partial w}{\partial z} = -\left(\frac{\partial u}{\partial x} + \frac{\partial v}{\partial y}\right), \qquad w\Big|_{z=h} = 0 \tag{6}$$

$$\frac{\partial \xi}{\partial t} + u\Big|_{z=\xi} \frac{\partial \xi}{\partial x} + v\Big|_{z=\xi} \frac{\partial \xi}{\partial y} - w\Big|_{z=\xi} = 0 \tag{7}$$

Here c is the concentration of admixture; u, v and w are the components of a velocity vector along the longitudinal (x), transversal (y) and vertical (z) directions, respectively. The known function $h(x, y)$ describes the shape of the stream-bed and the unknown function $\xi(t, x, y)$ describes a slightly deformable free surface of the flow. The known functions $c^0(x, y, z)$ and $\pi_0(t, y, z)$ set the initial distribution of concentration and its inflow through the inlet, respectively.

Equations (1)–(7) contain a set of parameters: d—the dimensionless coefficient of the turbulent diffusion in z- direction; λ—the decay factor for the admixture; ν—the normalized viscosity, which allows taking into account changes in the viscosity of the turbulent flow in accordance with the Boussinesq turbulence hypothesis; Re is the Reynolds number; G is the gravity parameter and I is the slope of the flow.

For more details about derivation of Eqs. (1)–(6) see [6].

3 Solution of the Hydrodynamics System

Equations (1) and (2) form the concentration system, and Eqs. (3)–(7) form the hydrodynamics system of the model for the shallow and elongated stream. These subsystems are consistent according to the precision of the approximation [6].

The hydrodynamic subsystem does not depend on the concentration subsystem and its solution can be explicitly written in the form of integrals

$$p = G(z - \xi), \qquad u = ReGI\,(J_2 - \xi J_1) \tag{8}$$

$$v = ReG \frac{\partial \xi}{\partial y}\,(J_2 - \xi J_1) \tag{9}$$

$$w = ReG\left(I \frac{\partial}{\partial x}\,(J_4 - \xi J_3) + \frac{\partial}{\partial y}\left((J_4 - \xi J_3)\,\frac{\partial \xi}{\partial y}\right)\right) \tag{10}$$

Here we introduced the notations

$$J_1 = \int_z^{h(x,y)} \frac{d\tau}{\nu}, \quad J_2 = \int_z^{h(x,y)} \frac{\tau d\tau}{\nu}, \quad J_3 = \int_z^{h(x,y)} J_1 d\tau, \quad J_4 = \int_z^{h(x,y)} J_2 d\tau \tag{11}$$

The pressure and velocity components in (8) are expressed in terms of the free surface function ξ, which is determined from the kinematic boundary condition (7).

Combination of (8) and (7) allows performing the kinematic boundary condition (7) in the following form

$$\frac{\partial \xi}{\partial t} = ReG\Big[I\left(\frac{\partial}{\partial x}(J_4 - \xi J_3) - (J_2 - \xi J_1)\frac{\partial \xi}{\partial x}\right)$$

$$+ (J_4 - \xi J_3)\frac{\partial^2 \xi}{\partial y^2} + \frac{\partial \xi}{\partial y}\left(\frac{\partial}{\partial y}(J_4 - \xi J_3) - (J_2 - \xi J_1)\frac{\partial \xi}{\partial y}\right)\Big] \quad (12)$$

where functions (11) and their derivatives are calculated within $z = \xi$ (i.e. on the free surface).

Equation (12) was solved with the finite-element software COMSOL© [8]. For detailed description of these numerical experiments, see [7].

4 Testing the Model

To verify the proposed model, we used the data which was published in [2], where the transfer of an admixture in the Severn River was studied. The section of river under study flows through the territory of Wales (Great Britain) between the settlements of Llanidloes and Caersws. In article [2], the observations of British hydrologists, who studied the distribution of tracer—coloring matter, were published. The diffusion coefficient of this substance is 10^{-6} cm/s.

The goal of that experiment was to collect and publish data of diffusive transfer of admixture for testing mathematical models proposed by various authors. The concentration of admixture was monitored in a section of a river about 14 km long by 6 observation stations located downstream. British authors describe in detail the geometry of the river-bed and the flow velocity in the considered section of the river, as well as other hydrological characteristics of the water stream obtained as a result of measurements that lasted more than 10 h.

The width of the channel on the considered area was measured at 86 points and varies from 13 to 48 m with an average value of 20 m. The depth of the flow was measured in each of the 86 sections with an interval of 1 m. (The average depth was 0.6 m.)

Considering that the average distance between measuring stations is 2 km, the approximate value of the parameter ϵ is 0.01, which satisfies the requirements of the mathematical model (1)–(7).

Thus, the British authors provided data required for the mathematical modeling and performing computational experiments to calculate the mass transfer of passive admixture in a natural water flow using proposed reduced 3D mathematical model (1)–(7).

Figure 1 shows the reconstructed flow region. The reconstruction of the river-bed geometry was made on the base of data presented in [2] for the section of the river Severn between stations A and F.

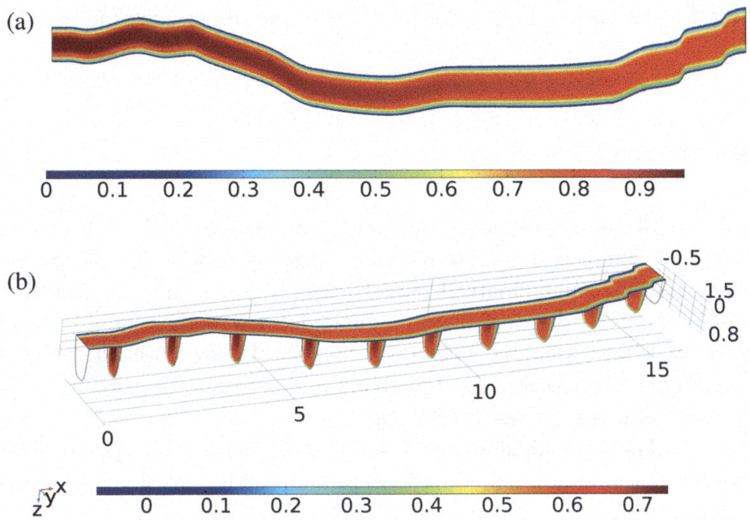

Fig. 1 Reconstructed stream-bed function $h(x, y)$ and the velocity field: (**a**)—the horizontal plane view (depth difference is colored according to presented scale); (**b**)—a set of segments with cross-sections and colored velocity field

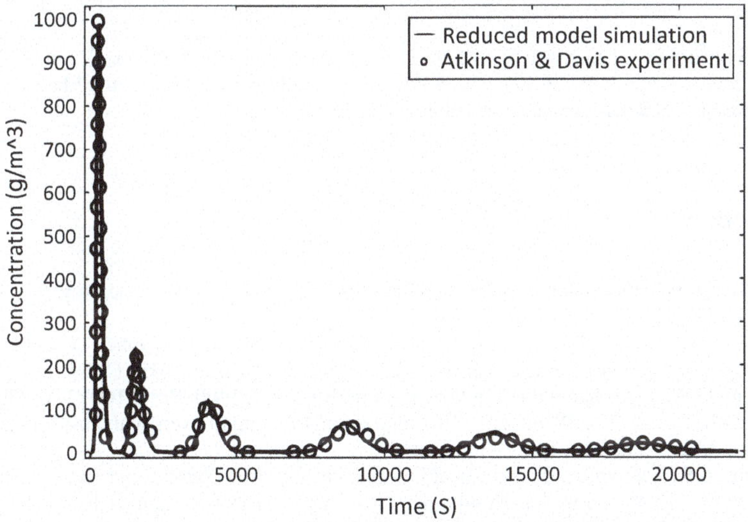

Fig. 2 The concentration at times when admixture passes each of the six measurement stations (A–F)

Figure 2 shows the values of concentration at different times. The solid line corresponds to the concentration of a substance calculated using a reduced 3D model of a long, shallow and slightly curved flow (1)–(7). Circles on the graph depict the results of measurements of the concentration of a substance at times when the admixture flows through the cross-sections of the stream near the observational stations.

5 Conclusions

The simulation of the passive admixture spreading in channel flows based on complete 3D hydrodynamic and mass transfer equations system is very computationally expensive. Therefore, mathematical models, which give a simplified but adequate description of the process, could be implemented. Such models should consider the key features of natural streams. The equations of a shallow, elongated and weakly curved stream (1)–(7) describe the flow dynamics as a three-dimensional, however, they are much simpler than the full 3D equations.

The proposed mathematical model of a long shallow and weakly curved flow can be applied only for simulation of slow flows, which can be described by steady-state model equations.

The results of the numerical simulation that are given in this article show that the proposed reduced 3D model of a long shallow flow adequately describes its hydrodynamics and mass transfer of the passive admixture. It can be used to simulate the spreading of pollutants in such streams.

Acknowledgements The research is partly supported by a grant from the Vladimir Potanin Charity Fund (project GPK-96/17). The results of the study will be used in the Master Program "Computational Mechanics and Information Technologies".

References

1. Aris, R.: On the dispersion of solute in a fluid flowing through a tube. Proc. Roy. Soc. London Ser. A. **235**, 67–77 (1956)
2. Davis P., Atkinson T., Wigley T.: Longitudinal dispersion in natural channels: 1. Experimental results from the river severn, U.K. hydrol. Earth Syst. Sci. **4**(3), 345–353 (1999)
3. Knight, D.W.: River hydraulics—a view from midstream. J. Hydr. Res. **51**(1), 2–18 (2013)
4. Luk, G.K.Y., Lau, Y.L. Watt, W.E.: Two-dimensional mixing in rivers with unsteady pollutant source. J. Env. Eng. **116**, 125–143 (1990)
5. Monin, A.S., Yaglom, A.M.: Statistical Fluid Mechanics. MIT Press, Cambridge (1979)
6. Nadolin, K.A.: An approach to simulating passive mass transport in channel flows. Mat. Model. **21**(2), 4–28 (2009)
7. Nadolin, K.A., Zhilyaev, I.V.: A reduced 3D hydrodynamic model of a shallow, long, and weakly curved stream. Water Res. **44**(2), 237–245 (2017)
8. Pryor R.W.: Multiphysics Modeling Using COMSOL: A First Principle Approach. Jones & Bartlett Publisers, Sudbury, MA (2011)
9. Stansby, P.K.: Coastal hydrodynamics—present and future. J. Hydr. Res. **51**(4), 341–350 (2013)
10. Taylor, G.I.: Dispersion of soluble matter in solvent flowing slowly through a tube. Proc. Roy. Soc. London Ser. A. **219**, 186–203 (1953)
11. Taylor, G.I.: The dispersion of matter in turbulent flow through a pipe. Proc. Roy. Soc. London Ser. A. **223**, 446–468 (1954)

European Gas Prices Dynamics: EEX Ad-Hoc Study

Yaroslava Khrushch, Susann Rudolf, Aleksandra Detkova, and Ivan P. Yamshchikov

Abstract This paper regards the dynamics of gas spot prices on one of European energy exchanges—EEX. A detailed description of the price dynamics is provided alongside with several multi-factor models for daily gas prices. An original approach to the development of such multi-factor daily price models is proposed. Specifically, daily price models taking into account non-integer power of time variable tend perform relatively well on the horizon of several weeks despite the heteroskedasticity of the daily prices.

1 Introduction

One of the most important determinant for gas prices is crude oil price [1, 13]. There are several approaches to oil price forecasting that use regressive analysis [4, 5, 10], generalized autoregressive conditional heteroskedasticity (GARCH)-based methods [8, 9, 11], applying support vector machines [14] or artificial neural networks [15]. However, there is no consensus on the best approach to such problem. Meanwhile market participants and policy makers constantly need to make price forecasts. This paper tries to look at gas prices as is, without pinning them to the prices of oil, and see what information can be extracted from the data per se. In this paper we focus on the gas prices data from the EEX exchange (European Energy Exchange AG). We describe the dataset that we have obtained from EEX and compare different models that could be used to forecast these prices.

Y. Khrushch · S. Rudolf
University of Applied Sciences Zittau/Görlitz, Zittau, Germany

A. Detkova (✉)
Department of Economics, Leipzig University, Leipzig, Germany
e-mail: sonara@bk.ru

I. P. Yamshchikov
Max Planck Institute for Mathematics in the Sciences, Leipzig, Germany

© Springer Nature Switzerland AG 2019
I. Faragó et al. (eds.), *Progress in Industrial Mathematics at ECMI 2018*, Mathematics in Industry 30, https://doi.org/10.1007/978-3-030-27550-1_16

2 Data Description

There are three main types of gas products that can be traded at the EEX gas spot market: contracts with constant base volume to be delivered within 24 h on the following day (GND or Gas-Next-Day); contracts with constant base volume to be delivered within 48 h on the following weekend (GWE); contracts with variable volume to be delivered within a flexible amount of delivery hours on the current trading day (GWID).

The following data fields are available in the EEX files: Commodity ID, Daily Reference Price per Market Area, Trading Date, Delivery Date, Market Area (which is the place of delivery), Delivery Hours, Open Price (which is the first bid/ask price on a Trading Date), Max Price and Min Price, referring to the maximum and minimum price of a contract, Last Price (which is the last bid/ask price on a Trading Date), Daily Volume and Number of Trades. The dataset is split according to the commodity IDs (GND, GWE and GWID) and the place of delivery that corresponds to one of the virtual trading hubs or market areas: NCG stands for NetConnect Germany GmbH & Co. KG and GPL for GASPOOL Balancing Services GmbH mainly refer to the gas transport facilities within Germany, whereas the TTF stands for Gastransport Services B.V. to Dutch Title Transfer Facility and is situated in the Netherlands.

3 Regression Model

Gas price behaviour is rather specific. It has periods of rapid growth and fast declines. These jumps and falls could be associated with a number of reasons starting from geopolitical events and finishing with technological break-throughs as well as purely supply and demand market factors. Technological factors' influence results in long lasting trends that might be incorporated into the model on a time horizon of several months that is also a typical time-frame for an industrial forecast. At the same time news about scientific, technological or political shocks can affect the markets sporadically but only for a short period of time. Can one predict gas prices assuming that the influence of non-market factors would be sporadic and non-lasting? How precise can one be in the forecasts that would use historical gas data exclusively? This paper addresses these questions.

The time-series of daily gas-prices provided by EEX is stationary under augmented Dickey-Fuller test for unit roots. This means that we can apply regression analysis in order to see, which factors might influence the dynamics of gas prices. These would naturally be weather (that could be paralleled with monthly seasonality), day of the week, price on the day before (or a longer time-lag of the price) etc. One would like to know which of these factors is more significant when we talk about price dynamics. We assume that the values of the studied time-series y_t exist for all observation time points treating the series as a repetitive pattern of

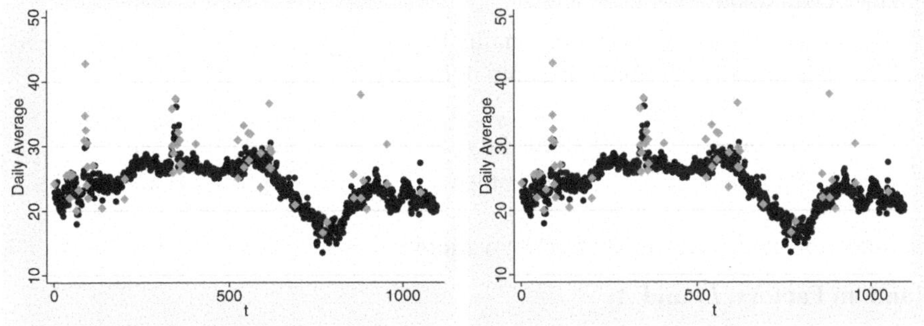

Fig. 1 Dynamics of prices for different regions with marked outliers according to Irvin criterium, NetConnect Germany GmbH & Co. KG (left), GASPOOL Balancing Services GmbH (right)

length 5 corresponding to 5 days trading days of the week. We hope that this bias is eventually incorporated within the seasonal component. We also introduce a more intricate dependence on time that we further discuss in detail. Following standard technique to the commodity pricing [2] outliers should be excluded from the time series according to the Irvin criterion [6]. In Fig. 1 time series with highlighted outliers are presented.

A regressand of the model is a spot price Pr_t, where the spot price is understood as a Daily-Average price, however all further reasoning could be applied to a different type of the price that one would like to forecast. Naturally, price time series depends on several factors that can be classified into three groups: trend factors T_t, cyclic factors C_t and random factors L_t and A_t.

Trend Modelling, T_t

In [12] the trend of the electricity prices is approximated with a linear function, however as we can see in Fig. 1 linear approximation is not a good fit for the case. The change in the price does not have a strong dependence on time, but is highly seasonal and has a number of visually random drops and jumps. This makes it reasonable to use a polynomial trend for modelling such data. On the interval $t \in [1, 1291]$ *growth* of a polynomial $P_n(t), n \in N$ is relatively high $(y'' = n(n-1)P_{n-2}(t),\ \ n \in N,\ \ n > 2)$. If we want to model the dynamics of the gas prices with the help of the polynomials $P_n(t)$, the appropriate coefficients should be of the order $a \cdot 10^{-6}$. At the same time we can look on a nonlinear regression using the terms like $t^{1/p},\ \ p \in N$, see in [7]. This allows to build the following model

$$T_p = \sum_{i=1}^{p} \alpha_i t^{1/i} + \alpha_0, \quad T_p'' = \sum_{i=1}^{p} \alpha_i \frac{(i-1)}{i^2} \cdot \frac{1}{t^{2-1/i}}.$$

Iterating through different types of polynomials we have found the case with $p = 5$ to be the most suitable with this particular task.

Cyclic Component, C_t

Cyclic component that describe seasonality is $C_t = \sum_{j=1}^{11} b_j t_j + \sum_{k=1}^{6} a_k d_k + a_7 w$, where t_j is a dummy variable for month defined as

$$t_1 = \begin{cases} 1, \text{ for June;} \\ 0, \text{ for other,} \end{cases} \quad t_2 = \begin{cases} 1, \text{ for July;} \\ 0, \text{ for other,} \end{cases} \quad \cdots \quad t_{11} = \begin{cases} 1, \text{ for April} \\ 0, \text{ for other months.} \end{cases}$$

Analogously d_k is a weekly dummy variable.

Random Factors, L_t and A_t

The random component is standardly described as $L_t = \frac{1}{\sigma\sqrt{2\pi t}} e^{\frac{(t-\mu)^2}{2\sigma^2 t}}$. Here we also need to pay a separate attention to the outliers that we have mentioned above. One can model anomaly jumps and drops seen in Fig. 1 with two variables S_1 and S_2. Since we can see a clear change of trend we need to introduce the third variable S_3 using the method introduced in [3]. The abnormal behaviour therefore is described as

$$A_t = \sum_{j=1}^{3} c_j S_j + S_3 \left(\sum_{p=1}^{5} d_{p+1} t^{1/p} \right) + L_t \left((f_1 + \upsilon f_4) S_1 + (f_2 + \upsilon f_5) S_2 + \upsilon f_3 \right), \tag{1}$$

where, $\upsilon = sin(\frac{2\pi t}{365.25})$ and S_1, S_2 and S_3 are defined as

$$S_1 = \begin{cases} 1, \text{ if } Pr_t > 30; \\ 0, \text{ other,} \end{cases} \quad S_2 = \begin{cases} 1, \text{ if } Pr_t < 20; \\ 0, \text{ other,} \end{cases} \quad S_3 = \begin{cases} 1, \text{ if } t < t^*; \\ 0, \text{ other,} \end{cases}$$

where t^* is a moment of a trend-change, see [7].

The spot price Pr_t is described as $Pr_t = T_t + C_t + L_t + A_t$. Excluding the outliers and determining the model coefficients with the method of least squares one can build a forecast. Table 1 shows the estimated quality for the basic models for all time periods as well as estimated coefficients used in these models.

Models NCG (y^b), GPL (y^d), TTF (y^f) take into account the prices on the previous day (y_{t-1}) and on the day before yesterday (y_{t-2}). Models NCG (y^a), NCG (y^c), GPL (y^e) have negative coefficient associated with variable w, meaning that the 1st day of the week tends to have a higher price than the coming days of the week. According to TTF (y^e) Monday, Tuesday and Wednesday have higher prices than Thursday and Friday. Whereas in TTF (y^f), that takes into account the price yesterday (y_{t-1}) and the day before yesterday (y_{t-2}), only Monday and Wednesday have significant coefficients, naturally. All the models perform well enough in terms of R^2 and root-mean-square error (RMSE). Let us now compare two different time-periods. Period II would correspond to a short-time price forecast and Period III to the long-time one. The results are demonstrated in Table 1. Comparison of these models for different time-intervals that presented in Table 2 shows that they all

Table 1 The coefficients and performing results of the models estimated for 3 periods, starting date is from the 30th of May 2011

	Period till 31.03.2015						Period till 18.04.2015		Period till 31.01.2015	
Coeff.	NCG (y^a)	NCG (y^b)	GPL (y^c)	GPL (y^d)	TTF (y^e)	TTF (y^f)	NCG (y^g)	NCG (y^h)	NCG (y^i)	NCG(y^j)
t	0.11	0.04	0.23	0.03	0.14	0.06	0.129	0.058	0.078	–
$t^{1/5}$	–	–	−201.83	–	–	–	–	–	–	
$t^{1/3}$	38.29	14.24	218.95	10.03	53.95	25.23	50.079	23.867	35.615	7.703
$t^{1/2}$	−15.76	−5.81	−50.44	−3.80	−20.92	−9.65	−19.199	−8.932	−12.987	−1.976
s_1	4.02	3.05	3.52	2.74	8.09	5.21	–	–	–	–
s_2	−3.05	−2.22	−2.53	−1.79	−2.89	−1.99	–	–	–	–
s_3	–	–	104.98	−4.17	28.09	16.59	25.471	16.251	25.604	16.730
s_3t	−0.10	−0.03	0.14	0.01	−0.17	−0.08	−0.122	−0.056	−0.072	–
$s_3t^{1/5}$	–	–	−48.44	–	–	–	–	–	–	–
$s_3t^{1/3}$	−38.28	−12.93	–	–	−61.15	−29.62	−51.709	4.523	−37.507	−9.746
$s_3t^{1/2}$	15.58	5.20	–	–	23.90	11.36	19.668	1.842	13.571	2.765
$sint$	−28.24	−11.95	–	–	–	–	–	–	–	–
L_t	–	–	−0.92	–	670.67	–	–	–	–	–
s_1L_t	–	–	1.67	0.54	–	–	6.432	4.768	6.559	4.934
s_2L_t	–	–	–	–	962.27	–	−3.046	−2.417	−3.092	−2.489
sinS2	–	–	–	–	–	–	1.073	0.487	1.282	0.828
w	−0.07	−0.10	−0.08	−0.09	–	–	−0.69	−0.093	−0.074	−0.095
t_1	−1.02	−0.54	−0.34	–	−1.48	−0.81	−1.02	−0.748	−1.04	−0.702
t_2	−1.00	−0.36	−0.01	–	−1.38	−0.68	−1.00	−0.575	0.928	−0.417

(continued)

Table 1 (continued)

	Period till 31.03.2015						Period till 18.04.2015		Period till 31.01.2015	
t_3	−0.78	−0.24	0.68	–	−0.87	−0.43	−0.78	−0.455	0.759	−0.368
t_4	0.54	0.29	1.87	0.33	−0.47	0.16	0.448	0.241	0.503	0.281
t_5	0.95	0.19	2.34	0.28	1.20	0.48	1.035	0.411	1.079	0.446
t_6	1.59	0.39	3.44	0.50	2.30	1.07	1.882	0.895	1.933	0.952
t_7	1.20	0.13	2.72	–	2.13	0.95	1.653	0.704	1.721	0.787
t_8	0.89	0.33	2.14	–	1.13	0.47	0.858	0.382	0.929	0.431
t_9	1.37	0.31	2.18	–	1.23	0.41	1.014	0.281	0.838	0.33
t_{10}	0.61	−0.07	1.53	–	0.58	0.04	0.191	−0.198	−0.030	−0.412
t_{11}	0.20	−0.14	0.30	–	0.06	−0.21	0.16	−0.16	−0.003	−0.274
d_1	–	–	–	–	0.33	0.29	–	–	–	–
d_2	–	–	–	–	0.33	–	–	–	–	–
d_3	–	–	–	–	0.43	0.22	–	–	–	–
d_4	–	–	–	–	0.30	–	–	–	–	–
d_5	–	–	–	–	0.26	–	–	–	–	–
y_{t-1}	–	0.27	–	0.33	–	0.31	–	0.159	–	−0.18
y_{t-2}	–	0.28	–	0.23	–	0.15	–	0.193	–	0.222
y_{t-3}	–	–	–	0.12	–	–	–	0.065	–	–
R^2	0.997	0.997	0.997	0.997	0.997	0.998	0.9972	0.9975	0.9972	0.9975
F_{prob}	17,733.63	19,220.29	1311.94	12,743.33	18,149.13	22,190.47				
$RMSE$	1.401	1.316	1.352	1.386	1.229	1.179	1.291	1.237	1.298	1.247
$E(\epsilon_i)$	8.22e−10	−1.27e−09	0.0002	0.0002	−0.00004	−0.00004	−0.00002	−0.000019	−9.05e−06	0.00007

Table 2 The forecasts in comparison with actual prices

	RMSE	Std. dev.	Min	Max		RMSE	Std. dev.	Min	Max
Actual prices NCG Period I	–	1.145	19.146	25.381	Actual prices NCG Period II	–	3.477	11.407	50.937
NCG y^a	1.204	0.157	21.739	22.334	NCG (y^g)	1.291	3.233	17.042	33.587
NCG y^b	1.495	0.225	22.025	22.793	NCG (y^h)	1.237	3.258	16.745	34.361
Actual prices TTF Period I	–	1.256	20.052	26.088	Actual prices NCG Period III	–	3.547	11.407	50.937
TTF y^e	1.149	0.161	20.762	21.284	NCG (y^i)	1.298	3.305	16.967	33.586
TTF y^f	1.149	0.167	20.823	21.455	NCG(y^j)	1.247	3.327	16.644	34.289

perform comparatively well. Variables s_1 and s_2 were introduced in Eq. (1) in order to describe unexpected jumps or drops of the prices. If one wants to take further factors into consideration (e.g. daily temperatures or some political context that might influence the prices of gas) it is the significance of these variables that is to change. Table 2 compares the forecasts with actual prices on different test periods.

4 Conclusion

This paper provides a simple yet flexible approach for forecasting gas prices on EEX market exchange. Gas prices have sophisticated time dynamics and are hard to forecast without additional data analysis (i.e. media publications, affecting the industry, political decision and historic context). However, using several basic techniques such as regression analysis one can estimate the price dynamics to some extent. First of all, for a regressive model focused on gas prices one rarely needs to include prices with time lag of more than 2–3 days. Second, despite intuitive notions substitution of time variable with a power of it (in this particular case t^r where $1/5 \leqslant r \leqslant 1$) could give better results for the forecast in terms of the RMSE on the test period. There is a number of rules of thumb that produce comparable forecasts, yet a more rigorous investigation of the connection between different time scales on energy market are out of the scope of this particular work that mostly deals with the empiric results. Third, one should apply some sort of filtering of outliers to build a comprehensive mid-term model not even mentioning a long-term one. In this particular paper Irwin criterion is shown to be effective.

Acknowledgements The research was partially carried out while this author Ivan P. Yamshchikov was in UAS Zittau, Germany. This research is supported by the European Union in the FP7-PEOPLE-2012-ITN Program under Grant Agreement Number 304617 (FP7 Marie Curie Action, Project Multi-ITN STRIKE—Novel Methods in Computational Finance). Short reference for contract: PITN-GA-2012-304617 STRIKE.

References

1. Asche, F., Misund, B., Sikveland, M.: The relationship between spot and contract gas prices in Europe. Energy Econ. **38**, 212–217 (2013)
2. Emery, G.W., Liu, Q.W.: An analysis of the relationship between electricity and natural gas futures prices. J. Futur. Mark. **22**(2), 95–122 (2002)
3. Gujarati, D.N.: Basic Econometrics. Tata McGraw-Hill Education, New York, NY (2009)
4. Hamilton, J.D.: Oil and the macroeconomy since World War II. J. Polit. Econ. **91**(2), 228–248 (1983)
5. Hamilton, J.D.: Understanding Crude Oil Prices (No. w14492). National Bureau of Economic Research, Cambridge (2008)
6. Irvin J.O.: On a criterion for the rejection of outlying observation. Biometrika **17**, 238–250 (1925)

7. Kimenta, J.: Elements of Econometrics. University of Michigan Press, Ann Arbor (1986)
8. Mohammadi, H., Su, L.: International evidence on crude oil price dynamics: applications of ARIMA-GARCH models. Energy Econ. **32**(5), 1001–1008 (2010)
9. Morana, C.: A semiparametric approach to short-term oil price forecasting. Energy Econ. **23**(3), 325–338 (2001)
10. Narayan, P.K., Gupta, R.: Has oil price predicted stock returns for over a century? Energy Econ. **48**, 18–23 (2015)
11. Narayan, P.K., Liu, R.: A unit root model for trending time-series energy variables. Energy Econ. **50**, 391–402 (2015)
12. Schmidt, T.: Modelling energy markets with extreme spikes. In: Mathematical Control Theory and Finance, pp. 359–375. Springer, Berlin (2008)
13. Villar, J.A., Joutz, F.L.: The relationship between crude oil and natural gas prices. Energy Information Administration, Office of Oil and Gas, 1–43 (2006)
14. Xie, W., Yu, L., Xu, S., Wang, S.: A new method for crude oil price forecasting based on support vector machines. In: International Conference on Computational Science, pp. 444–451. Springer, Berlin (2006)
15. Yu, L., Wang, S., Lai, K.K.: Forecasting crude oil price with an EMD-based neural network ensemble learning paradigm. Energy Econ. **30**(5), 2623–2635 (2008)

Modelling Time-of-Flight Transient Currents with Time-Fractional Diffusion Equations

M. Luísa Morgado and Luís F. Morgado

Abstract In this work we explore the use of tempered fractional derivatives in the modelling of transient currents in disordered materials. We particularly focus on the numerical approximation of the involved problems. As it is known, the solutions of fractional differential equations usually exhibit singularities in the origin in time, and therefore, a decreasing of the convergence order of standard numerical schemes may be expected. In order to overcome this, we propose a finite difference scheme on a time graded mesh, in which the grading exponent can be properly chosen, taking into account the singularity type. Numerical results are presented and discussed.

1 Introduction

Since the 1960s there is an increasing interest and research on organic semiconductors, due to their particular characteristics (transparency, flexibility, low cost), as a material for the fabrication of optoelectronic devices, such as organic solar cells, light emitting diodes and light emitting transistors. The charge carrier mobility μ of these materials is one of the main properties of interest and the Time-of-Flight (TOF) technique is one of the preferred methods to estimate it. In the TOF experiment, a transient current $I(t)$ through a thin layer of material sandwiched between two parallel electrodes is obtained, as a result of the motion of excess charge carriers generated by a laser or voltage pulse, under the influence of an externally applied electric field E directed normally to the electrodes. These

M. L. Morgado (✉)
Center for Computational and Stochastic Mathematics (CEMAT), Lisbon and Departamento de Matemática, Universidade de Trás-os-Montes e Alto Douro, UTAD, Vila Real, Portugal
e-mail: luisam@utad.pt

L. F. Morgado
Instituto de Telecomunicações, Lisboa, and Departamento de Física, Universidade de Trás-os-Montes e Alto Douro, UTAD, Vila Real, Portugal
e-mail: lmorgado@utad.pt

© Springer Nature Switzerland AG 2019
I. Faragó et al. (eds.), *Progress in Industrial Mathematics at ECMI 2018*,
Mathematics in Industry 30, https://doi.org/10.1007/978-3-030-27550-1_17

transient currents usually exhibit an anomalous dispersive character [5] with two regions with power-law behavior, separated by the transit time t_{tr} : $\sim t^{-1+\alpha}$, if $t < t_{tr}$ and $\sim t^{-1-\alpha}$, if $t > t_{tr}$ with $0 < \alpha < 1$. An estimate for μ is calculated from t_{tr}, the instant when the two power-law curves intersect. Such behavior is attributed to the trapping of carriers, in localized states distributed in the mobility gap, for times τ, or waiting times, determined by a relaxation function $\Psi(\tau)$ with an asymptotic time dependence of the form $\Psi(\tau) \sim \tau^{-\alpha}$.

Diffusion-advection equations have been widely used to describe the evolution of carrier density in the materials, but it is known that in the case of disordered materials, integer order models do not describe accurately the process [8] and that is the reason why in the latest decades the use of fractional time derivatives have been proposed to improve those models. For example, in [3] the following model was considered:

$$D_t^\alpha y(x,t) = -v\frac{\partial y(x,t)}{\partial x} + D\frac{\partial^2 y(x,t)}{\partial x^2} + f(x,t),\ t \in (0,T],\ x \in (0,L), \tag{1}$$

with initial condition

$$y(x,0) = g(x), \quad x \in (0,L), \tag{2}$$

and boundary conditions

$$y(0,t) = \phi_0(t), \ \ y(L,t) = \phi_L(t), \quad t \in (0,T], \tag{3}$$

where $0 < \alpha < 1$ and the fractional derivative is of the Caputo type which, for the considered values of α, is given by [2]:

$$D_t^\alpha y(t) = \frac{1}{\Gamma(1-\alpha)}\int_0^t (t-s)^{-\alpha} y'(s)\, ds. \tag{4}$$

It was assumed that $v > 0$ is the average fluid velocity, $D > 0$ is the diffusion coefficient and g, f, ϕ_0 and ϕ_L are continuous functions in their respective domains.

Here we consider the general model (which obviously reduces to (1)–(3) in the case where $\lambda = 0$ with $v(x) \equiv v > 0$ and $D(x) \equiv D > 0$):

$$\mathbb{D}_t^{\alpha,\lambda} u(x,t) = \frac{\partial}{\partial x}\left(-v(x)u(x,t) + D(x)\frac{\partial u(x,t)}{\partial x}\right) + f(x,t),\ \text{in}(0,T] \times (0,L), \tag{5}$$

$$u(x,0) = g(x), \quad x \in (0,L), \tag{6}$$

$$u(0,t) = \phi_0(t), \ \ u(L,t) = \phi_L(t), \quad t \in (0,T], \tag{7}$$

where $\mathbb{D}_t^{\alpha,\lambda} u(x,t)$ is the tempered Caputo derivative with respect to the variable t of the function $u(x,t)$ [1]:

$$\mathbb{D}_t^{\alpha,\lambda}(u(t)) = e^{-\lambda t} D_t^{\alpha}\left(e^{\lambda t}u(t)\right) \tag{8}$$
$$= \frac{e^{-\lambda t}}{\Gamma(1-\alpha)}\int_0^t \frac{1}{(t-s)^{\alpha}} \frac{d\left(e^{\lambda s}u(s)\right)}{ds} ds, \quad 0<\alpha<1,\ \lambda \geq 0.$$

Note that if in the equation above, we consider $\lambda = 0$, the definition of the usual Caputo derivative (4) is recovered.

The total measured current $I(t)$, produced by the extraction of carriers from the space between the electrodes, placed at $x = 0$ and $x = L$, is given [4] by the space average of the current density $j(x,t)$, and since

$$j(x',t) = -\frac{d}{dt}\int_0^{x'} qu(x,t)dx, \tag{9}$$

where q is the carrier electrical charge, we obtain

$$\frac{I(t)}{q} = -\frac{d}{dt}\int_0^{L}(L-x)u(x,t)dx. \tag{10}$$

2 Numerical Method and Results

In order to approximate the solution of (5)–(7), we first take (8) into account, and note that (5) can be written as

$$D_t^{\alpha}\left(e^{\lambda t}u(x,t)\right) = -\frac{\partial\left(e^{\lambda t}v(x)u(x,t)\right)}{\partial x} + \frac{\partial}{\partial x}\left(D(x)\frac{\partial\left(e^{\lambda t}u(x,t)\right)}{\partial x}\right) + e^{\lambda t}f(x,t).$$

Therefore, if we consider the function $y(x,t) = e^{\lambda t}u(x,t)$, and we determine the solution $y(x,t)$ of problem:

$$D_t^{\alpha}y(x,t) = -\frac{\partial\left(v(x)y(x,t)\right)}{\partial x} + \frac{\partial}{\partial x}\left(D(x)\frac{\partial y(x,t)}{\partial x}\right) + e^{\lambda t}f(x,t), \tag{11}$$

$$y(x,0) = g(x), \quad x \in (0,L), \tag{12}$$

$$y(0,t) = e^{\lambda t}\phi_0(t), \quad y(L,t) = e^{\lambda t}\phi_L(t), \quad t \in (0,T], \tag{13}$$

then the solution of (5)–(7) is obtained through $u(x,t) = e^{-\lambda t}y(x,t)$. Therefore, we first approximate the solution of problem (11)–(13), by adapting the numerical scheme proposed in [3]. We now briefly describe the numerical approach.

We consider a uniform mesh in the interval $[0, L]$, defined by the grid-points $x_i = ih$, $i = 0, 1, \ldots, K$, where $h = \frac{L}{K}$, and we use the following second order finite difference approximations:

$$\frac{\partial \left(v(x)y(x,t)\right)}{\partial x}\Big|_{x=x_i} \approx \frac{v(x_{i+1})y(x_{i+1},t) - v(x_{i-1})y(x_{i-1},t)}{2h}, \tag{14}$$

$$\begin{aligned} \frac{\partial}{\partial x}\left(D(x)\frac{\partial y(x,t)}{\partial x}\right)\Big|_{x=x_i} \approx \frac{1}{h^2}\Bigg(D\left(x_i + \frac{h}{2}\right) y(x_{i+1},t) - \\ - \left(D\left(x_i + \frac{h}{2}\right) + D\left(x_i - \frac{h}{2}\right)\right) y(x_i,t) + \\ + D\left(x_i - \frac{h}{2}\right) y(x_{i-1},t)\Bigg), \quad i = 1, \ldots, K-1. \end{aligned} \tag{15}$$

For the numerical approximation of the Caputo derivative of order α on the interval $[0, T]$, we will use the non-uniform mesh

$$t_i = \left(\frac{i}{n}\right)^r T, \tag{16}$$

where $r \geq 1$ is the so-called grading exponent. The length of each one of the intervals defined with this partition is:

$$\tau_i = t_{i+1} - t_i = \frac{(i+1)^r - i^r}{n^r} T, \quad i = 0, 1, \ldots, n-1.$$

Note that if $r = 1$ we obtain a uniform mesh. We then use the following approximation for the Caputo derivative (see [3]):

$$D^\alpha y(t_k) \approx \frac{1}{\Gamma(2-\alpha)} \sum_{j=0}^{k-1} \tau_j^{-\alpha} a_{j,k} \left(y(t_{j+1}) - y(t_j)\right) = \tilde{D}^\alpha y_k, \tag{17}$$

where

$$a_{j,k} = \left(\frac{k^r - j^r}{(j+1)^r - j^r}\right)^{1-\alpha} - \left(\frac{k^r - (j+1)^r}{(j+1)^r - j^r}\right)^{1-\alpha}, \quad j = 0, 1, \ldots, k-1, \;\; k = 1, \ldots, n. \tag{18}$$

Concerning the order of the approximation we have (see [7]):

$$\left| D_t^\alpha y(t_k) - \tilde{D}^\alpha y_k \right| \leq C k^{-\min\{2-\alpha, r\alpha\}},$$

which gives us an information about the proper choice of the grading exponent (see [7]).

Using (17), we obtain:

$$D_t^\alpha y(x_i, t_l) \approx \frac{1}{\Gamma(2-\alpha)} \sum_{j=0}^{l-1} \tau_j^{-\alpha} a_{j,l} \left(y(x_i, t_{j+1}) - y(x_i, t_j) \right), \tag{19}$$

$i = 1, \ldots, K-1$, $l = 1, \ldots, n$, where the coefficients $a_{j,l}$ are defined in (18). Denoting by $Y_i^l \approx y(x_i, t_l)$, $f_i^l = f(x_i, t_l)$, $D\left(x_i \pm \frac{h}{2}\right) = D_{i\pm\frac{1}{2}}$, and substituting (14), (15) and (19) in (11), we obtain the following implicit numerical scheme:

$$\frac{1}{\Gamma(2-\alpha)} \sum_{j=0}^{l-1} \tau_j^{-\alpha} a_{j,l} \left(Y_i^{j+1} - Y_i^j \right) = \frac{D_{i+\frac{1}{2}} Y_{i+1}^l - \left(D_{i+\frac{1}{2}} + D_{i-\frac{1}{2}} \right) Y_i^l + D_{i-\frac{1}{2}} Y_{i-1}^l}{h^2}$$

$$- \frac{v_{i+1} Y_{i+1}^l - v_{i-1} Y_{i-1}^l}{2h} + e^{\lambda t_l} f_i^l, \quad i = 1, \ldots, K-1,\ l = 1, \ldots, n, \tag{20}$$

where, according to the initial and boundary conditions (12) and (13), we have

$$Y_i^0 = g(x_i), \quad i = 1, \ldots, K-1,$$
$$Y_0^l = e^{\lambda t_l} \phi_0(t_l), \quad Y_K^l = e^{\lambda t_l} \phi_L(t_l), \quad l = 1, \ldots, n.$$

After having determined the unknowns Y_i^l, $i = 1, \ldots, K-1$, $l = 1, \ldots, n$, the solution of (5)–(7) at the mesh-points will be given by $u(x_i, t_l) \approx U_i^l = e^{-\lambda t_l} Y_i^l$.

Figures 1, 2, and 3 present some numerical simulations for transient currents, considering a narrow (gaussian profile) of photogenerated carriers, at the position $x = 0.2$ of a layer of thickness $L = 1$. Both advection and diffusion coefficients are constants: $v = 0.01$ and $D = 0$, for Fig. 1 and $v = 0.01$ and $D = 0.01$, for Fig. 2.

In Fig. 3 we show the current behavior when the carrier velocity has two different values along the the material layer, $v = 0.001$ in the first half and $v = 0.05$

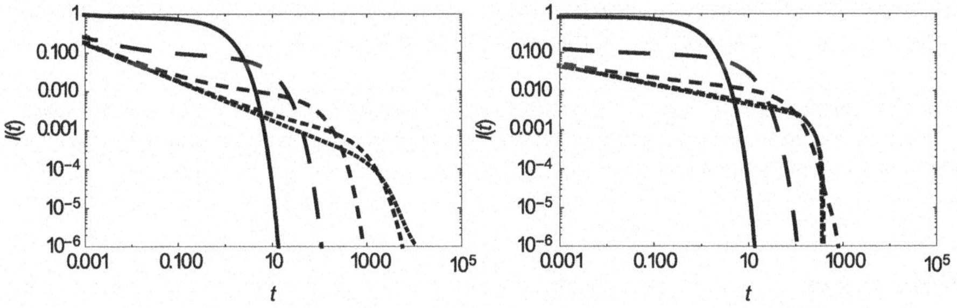

Fig. 1 Transient current for $\alpha = 0.5$ (left) and $\alpha = 0.75$ (right), with $\lambda = 0$ (tiny dash), $\lambda = 0.001$ (small dash), $\lambda = 0.01$ (medium dash), $\lambda = 0.1$ (large dash) and $\lambda = 1$ (solid); $v = 0.01$ and $D = 0$

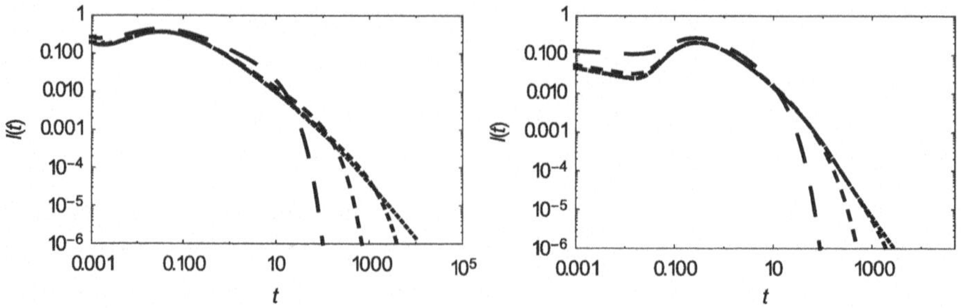

Fig. 2 As in Fig. 1, with $\lambda = 0$ (tiny dash), $\lambda = 0.001$ (small dash), $\lambda = 0.01$ (medium dash) and $\lambda = 0.1$ (large dash); $\upsilon = 0.01$ and $D = 0.01$

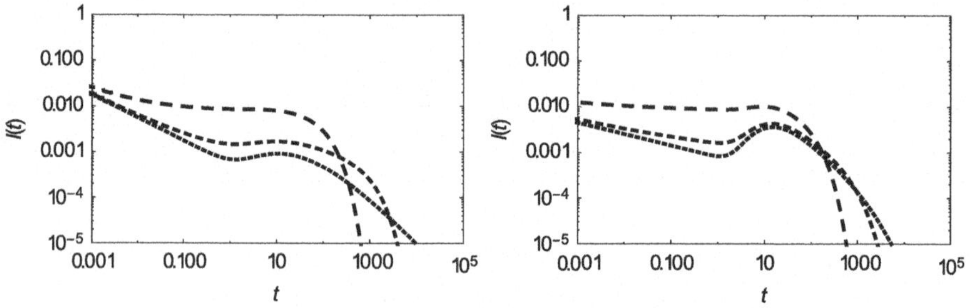

Fig. 3 As in Fig. 1, for two layers material with $\lambda = 0$ (tiny dash), $\lambda = 0.001$ (small dash) and $\lambda = 0.01$ (medium dash); $\upsilon = 0.001$ in the first layer and $\upsilon = 0.05$ in the second; $D = 0$ in both layers

in the second half, in a diffusion-less situation, $D = 0$. These numerical results are in agreement with the analytical results related to the TOF experiments in [6]. Since in most of the cases, the analytical solution is not known, numerical methods are necessary and the one presented here is able to model many situations. In a forthcoming paper, we will prove that this numerical scheme is stable and convergent, namely with convergence orders of $(2 - \alpha)$ and 2, time and space, respectively, with a proper choice of the grading exponent which takes into account some regularity assumptions on the solution of such problems.

Acknowledgements The authors acknowledge the Fundação para a Ciência e a Tecnologia (Portuguese Foundation for Science and Technology) through projects UID/Multi/04621/2013 and PEst-OE/EEI/LA0008/2013, respectively.

References

1. Baeumer, B., Meerschaert, M.M.: Tempered stable Lévy motion and transient super-diffusion. J. Comput. Appl. Math. **233**, 2438–2448 (2010)
2. Diethelm, K.: The Analysis of Fractional Differential Equations: An Application-Oriented Exposition Using Differential Operators of Caputo Type. Springer-Verlag, Berlin (2010)

3. Morgado, L.F., Morgado, M.L.: Numerical modelling transient current in the time-of-flight experiment with time-fractional advection-diffusion equations. J. Math. Chem. **53**, 958–973 (2015)
4. Philippa, B.W., White, R.D., Robson, R.E.: Analytic solution of the fractional advection-diffusion equation for the time-of-flight experiment in a finite geometry. Phys. Rev. E **84**, 041138 (2011)
5. Scher, H., Montroll, E.: Anomalous transit-time dispersion in amorphous solids. Phys. Rev. B **12**(6), 2455–2477 (1975)
6. Sibatov, R.T., Morozova, E.: Tempered Fractional Model of Transient Current in Organic Semiconductor Layers, in Theory and Applications of Non-Integer Order Systems, pp. 287–295. Springer International Publishing, Berlin (2017)
7. Stynes, M., O'Riordan, E., Gracia, J.L.: Error analysis of a finite difference method on graded meshes for a time-fractional diffusion equation. SIAM J. Numer. Anal., **55**, 1057–1079 (2017)
8. Uchaikin,V.V., Sibatov, R.T.: Fractional Kinetics in Solids: Anomalous Charge Transport in Semiconductors, Dielectrics and Nanosystems. World Scientific, Singapore (2012)

Reverse Logistics Modelling of Assets Acquisition in a Liquefied Petroleum Gas Company

Cristina Lopes, Aldina Correia, Eliana Costa e Silva, Magda Monteiro, and Rui Borges Lopes

Abstract In the business of liquefied petroleum gas (LPG), the LPG cylinder is the main asset and a correct planning of its needs is critical. This work addresses a challenge, proposed at an European Study Group with Industry by a Portuguese energy sector company, where the objective was to define an assets acquisition plan, i.e., to determine the amount of LPG cylinders to acquire, and when to acquire them, in order to optimize the investment. The used approach to find the solution of this problem can be divided in three phases. First, it is necessary to forecast demand, sales and the return of LPG bottles. Subsequently, this data can be used in a model for inventory management. Classical inventory models, such as the Wilson model, determine the Economic Order Quantity (EOQ) as the batch size that minimizes the total cost of stock management. A drawback of this approach is that it does not take into account reverse logistics, which in this challenge (i.e. the return of cylinders) plays a crucial role. At last, because it is necessary to consider the return rate of LPG bottles, reverse logistic models and closed loop supply chain models are explored.

1 Problem Description

This work addresses an industrial challenge that consisted in planning the acquisition of liquefied petroleum gas (LPG) cylinders. The challenge was proposed at an European Study Group with Industry, by a Portuguese company of the energy sector (named ALPHA for confidentiality reasons) that started its activity in 2006

C. Lopes (✉)
LEMA, CEOS.PP, ISCAP—Polytechnic of Porto, Porto, Portugal
e-mail: cristinalopes@iscap.ipp.pt

A. Correia · E. Costa e Silva
CIICESI, ESTG—Polytechnic of Porto, Porto, Portugal
e-mail: aic@estg.ipp.pt; eos@estg.ipp.pt

M. Monteiro · R. B. Lopes
CIDMA, University of Aveiro, Aveiro, Portugal

© Springer Nature Switzerland AG 2019
I. Faragó et al. (eds.), *Progress in Industrial Mathematics at ECMI 2018*,
Mathematics in Industry 30, https://doi.org/10.1007/978-3-030-27550-1_18

focusing in the production and distribution of biofuel and, since then, has extended its business areas to other fuels and energy. In this company, the LPG cylinder business started in 2012, and since then it has experienced a continuous growth. ALPHA currently commercializes propane gas in two types of cylinders: type A with capacity 9 kg, and type B with capacity 45 kg.

In Portugal, companies selling LPG cylinders are also responsible for collecting the empty cylinders, regardless of the company from which the previous cylinders were bought (direct replacement policy) [7]. The empty bottles returned to the company can be reinserted in the system, filled again and sold to the clients. As the acquisition of new bottles is expensive, reusing is the key. Cylinders are assets owned by the companies: each competitor can only refill its own cylinders. Companies experiencing growth have to purchase additional cylinders to meet demand. The cylinder is the main asset and a correct planning of its needs is critical.

The industrial challenge was to find a model to forecast the demand and return rate of each type of cylinder, and to define an assets acquisition plan, i.e., to determine *when to order* to the external supplier new LPG bottles (Order Point) and *how many* should be bought (batch size), in order to optimize the investment.

2 Literature Review

Classical inventory models, such as Wilson's deterministic model [4, 9], determine the Economic Order Quantity (EOQ) as the batch size that minimizes the total cost of stock management. The total cost is the sum of three components:

- C_A Acquisition Costs (price of acquiring the assets)
- C_S Setup costs (fixed cost for every order, transportation, collect)
- C_H Holding costs (insurances, taxes, rent, electricity, salary, opportunity costs)

Once the forecast of demand, sales and return of LPG cylinders is determined, an EOQ model can be used for inventory management [2].

The EOQ model is an attempt to estimate the best order quantity by balancing the conflicting costs of holding stock and of placing replenishment orders. The effect of order quantity on stock-holding costs is that, the larger the order quantity for a given item, the longer will be the average time in stock and the greater will be the storage costs. On the other hand, the placing of a large number of small-quantity orders produces a low average stock, but a much higher cost in terms of the number of orders that need to be placed and the associated administrative and delivery costs.

Another classical approach is the Continuous Review Policy (s,Q), which considers probabilistic demand.

A drawback of these approaches is that they do not take into account reverse logistics, which in this challenge (i.e. the return of cylinders) plays a crucial role. The plan should take in account the empty bottles that are returned to the company, which can be either reused or disposed of. Therefore we started by applying to the data two inventory models with reverse flows found in literature, and then developed a deterministic model and stochastic model tailored for this case study.

2.1 Inventory Models in Literature with Reverse Flows

Richter [6] extended the EOQ model to allow the incorporation of used products, which were repaired and incorporated in production. It assumes a stationary demand in a model with two shops, where the first shop is producing new products and repairing products used by the second shop.

Also considering deterministic demand and reverse logistics is the model proposed by Teunter [8], differing in allowing to consider varying disposal rates and disaggregating holding costs. In this model (Fig. 1, *M* manufacturing batches and *R* recovery batches succeed each other.

We implemented in an Excel file, for the company to use, all the formulas from Teunter model for computing the total cost per unit of time (case M = 1), the optimal batch size for manufacturing Q_m and for recovery Q_r and the number of recovery batches.

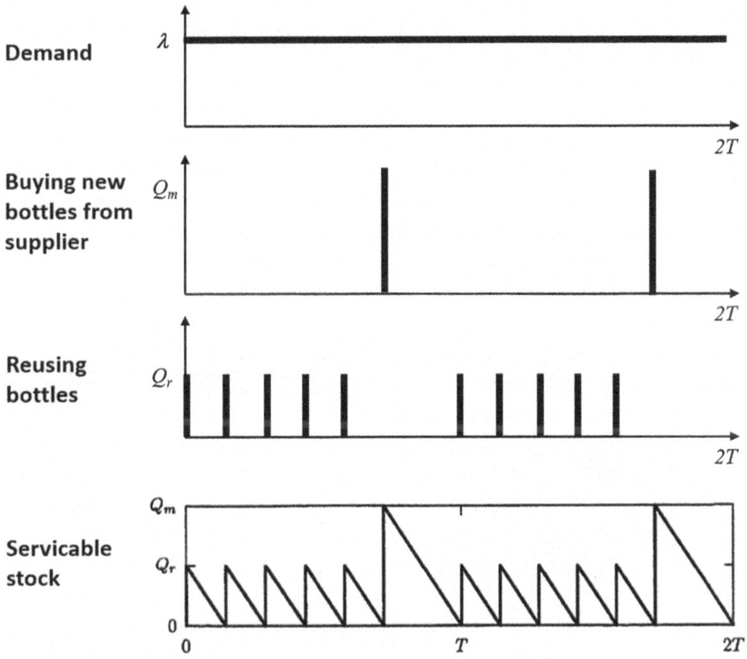

Fig. 1 Inventory stock model according to Teunter model [8] for case M = 1, R = 5

Other developments on the EOQ model are by Alivoni et al. [1]. They propose a stochastic model where production or purchase of new items integrates product reuse, in order to identify the need of placing a production/purchasing order to avoid stock-out situations.

3 Inventory Models Developed for the Company Based on Continuous Replenishment

The models presented before do not contemplate all the specifications required in this case study. In our case, the returned items arrive continuously, not in discrete moments, and can have three different destinations, as depicted in Fig. 2. Most of the returned LPG bottles (98%) only need cleaning, and some of them (about 2%) need requalification. At the moment, because this business is relatively new for the company, there is no LPG bottles that need to be disposed of, but in the future this situation can also occur. The costs and time for each of these processes are different.

The previous model considered that both the acquired and returned bottles arrive at discrete moments periodically in time, but actually that only happens with the acquired bottles. The returned bottles arrive continuously to the warehouse, and are continuously cleaned and requalified and filled (with rate $u + d$), as depicted in Fig. 3. Therefore, a continuous replenishment model could be adapted to this case study. In this setting, two cases can happen:

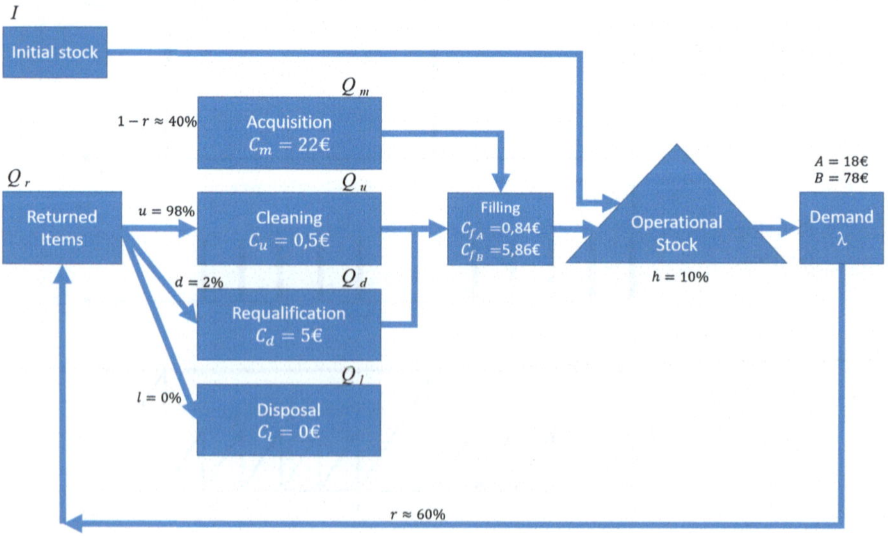

Fig. 2 Reverse flows and Inventory stock costs in our case study

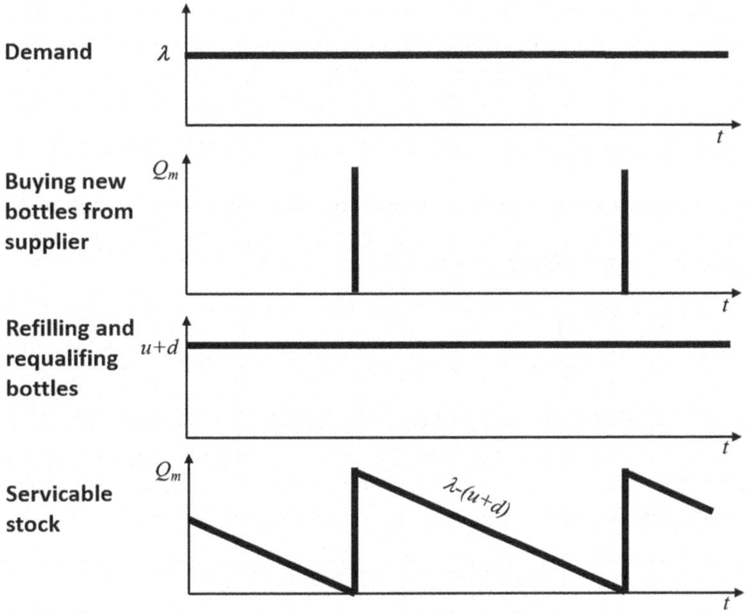

Fig. 3 Deterministic Inventory stock model D developed for our case study

Case $\lambda > u+d$: If demand exceeds the incoming flow, sometimes we have to buy new cylinders from supplier. We derived a Deterministic Model D with continuous returns for this case.

Case $\lambda \leq u + d$: If the returned bottles are enough to respond to the demand, buying new bottles from the supplier is unnecessary. To address this case we present the Deterministic Model R without purchases.

3.1 Model D: Deterministic Continuous Returns

We developed a deterministic model D, based on EOQ [9], which considers deterministic continuous constant demand, deterministic discrete replenishment from supplier, and deterministic continuous constant replenishment from returned bottles, for the case when returns are not enough to respond to the demand and hence the company has to buy new bottles from the supplier (Fig. 3).

As in the classical EOQ formula, in this model the total costs considered are the sum of the acquisition costs C_A, setup costs C_S and holding Costs C_H. The acquisition costs in Eq. (1) consider the cases where new bottles are acquired from the supplier with a cost C_m, the bottles are reused with just a cleaning cost C_u, or the case where the returned bottles have to be requalified with a cost C_d. In this three cases, a constant filling cost is also included. In the future, a disposal cost C_l

could also be considered. At the moment, because this business is relatively new for the company, there is no LPG bottles that need to be disposed of. Hence, the rate of bottles returned and disposed of (l) is zero. The acquisition costs are:

$$C_A = C_m(1-r)(\lambda - I) + C_u u(\lambda - I) + C_d(r-u)(\lambda - I) \tag{1}$$

where λ is the constant demand rate (units/units of time), I is the initial stock r is the return rate, u is the rate of bottles returned and cleaned and $d = r - u$ is the rate of bottles returned and requalified. The setup costs are:

$$C_S = \frac{K_m(\lambda - I)(1-r)}{Q_m} + \frac{K_u(\lambda - I)u}{Q_u} + \frac{K_d(\lambda - I)(r-u)}{Q_d} \tag{2}$$

where K_m is the production fixed setup costs, K_u is the reuse fixed setup costs, K_d is the requalification fixed setup costs, Q_m is the batch size for buying new bottles, Q_u is the batch size for reusing bottles, and Q_d is the batch size for requalifying bottles. The holding costs are:

$$C_H = h_m\frac{(1-r)Q_m}{2} + h_u\frac{uQ_u}{2} + h_d\frac{(r-u)Q_d}{2} + h_i\frac{I}{2} \tag{3}$$

where h_m is the holding cost per new item bought per year, h_u is the holding cost per reused item per year, h_d is the holding cost per requalified item per year, and h_i is the holding cost per existent item in stock per year.

By deriving the total costs, it is possible to obtain the expression for the optimal quantities Q_m^* (batch size for buying new bottles), Q_u^* (batch size for reuse) and Q_d^* (batch size for requalification) that minimize the total costs.

$$Q_m^* = \sqrt{\frac{2K_m(\lambda - I)}{h_m}} \qquad Q_u^* = \sqrt{\frac{2K_u(\lambda - I)}{h_u}} \qquad Q_d^* = \sqrt{\frac{2K_d(\lambda - I)}{h_d}} \tag{4}$$

From the triangle in Fig. 4 we can find the Order Point: $\frac{OP}{l} = \lambda - (u+d) \Leftrightarrow OP = (\lambda - (u+d))\,l$

This model was also implemented in an Excel file for the company to use.

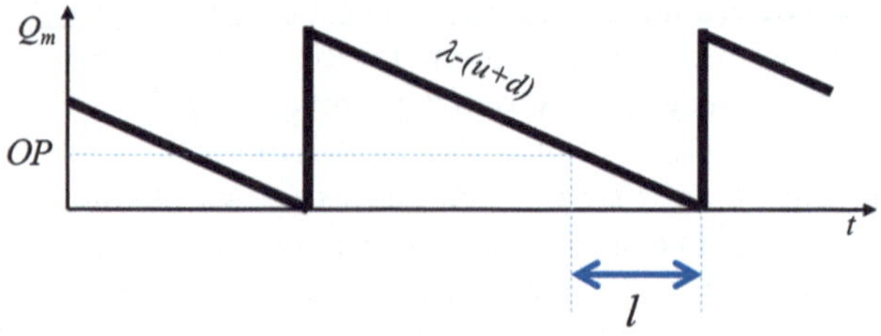

Fig. 4 Stock quantity across time, lead time, and order point in model D

3.2 Model R: Deterministic Without Purchases

Our Deterministic Model R without purchases considers deterministic continuous constant demand, unnecessary replenishment from supplier, and deterministic continuous constant replenishment from returned bottles. In this setting, there is a period T_1 where there is simultaneously continuous replenishment of bottles (with rate $u + d$) and demand being satisfied (with rate λ), and a period T_2 where replenishment is interrupted and there is only demand being satisfied.

Therefore, from the slopes of the main triangles in Fig. 5, we have:

$$T_1 = \frac{M}{u + d - \lambda} \qquad T_2 = \frac{M}{\lambda} \qquad M = Q - \lambda \cdot T_1 = Q\left(1 - \frac{\lambda}{u + d}\right) \tag{5}$$

where M is the maximum stock level, and the batch size corresponds to the total production during period T_1, i.e., $Q = (u + d)T_1$. The total costs are given by:

$$TC(Q) = C_u u(D - I) + C_d d(D - I) + +(K_u + K_d)\frac{D}{Q} + +C_h \frac{Q}{2}\left(1 - \frac{\lambda}{u + d}\right) \tag{6}$$

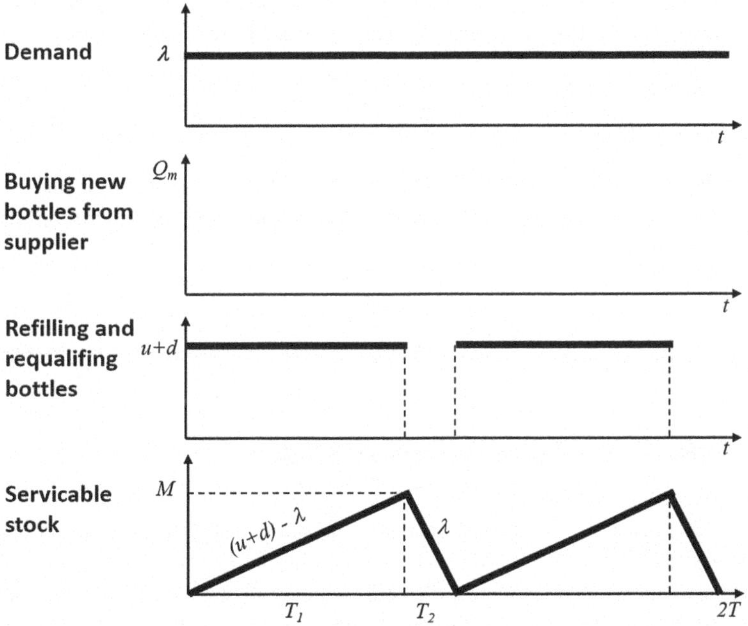

Fig. 5 Deterministic inventory stock model R for continuous returns without purchases

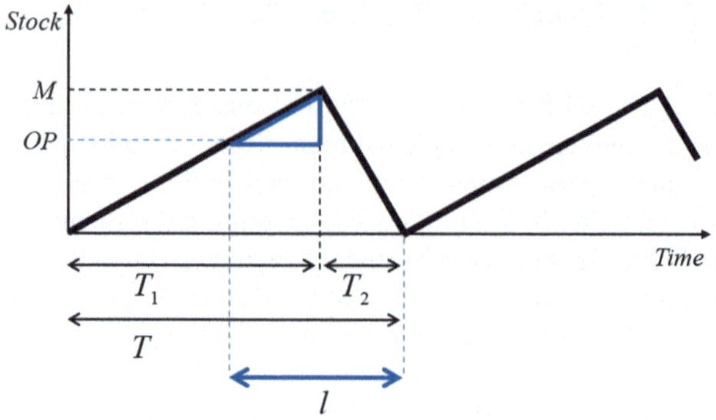

Fig. 6 Stock quantity across time, lead time, and order point in model R

being D the demand for the planning horizon (year) and λ the daily demand. Deriving the total costs, the optimal quantity $Q*$ that minimizes the total cost is:

$$Q^* = \sqrt{\frac{2(K_u + K_d)D}{C_h}}\sqrt{\frac{u+d}{u+d-\lambda}} \tag{7}$$

If the lead time l is longer than the period of demand ($l > T_2$) then from the slope in the blue triangle in Fig. 6 we can derive the Order Point (formula (8)).

$$\frac{M - OP}{l - T_2} = u + d - \lambda \Leftrightarrow OP = M - (u + d - \lambda)(l - T_2) \tag{8}$$

Replacing M and T_2 using Eq. (5), we can obtain the order point OP as a function that depends only on the quantity of bottles Q, the demand rate and reutilization rate, and lead times:

$$OP = Q\left(1 + \frac{u+d}{\lambda}\right) + l\,(\lambda - (u + d)) \tag{9}$$

3.3 Model S: Stochastic Inventory Model

At first we assumed a deterministic constant demand and return rate, but in fact it is not constant nor deterministic. It shows seasonality and trend. To correctly plan the acquisition of new cylinders from the supplier, we proceeded to forecast not only the demand, but also the reverse logistic flows.

Forecasting of demand and returns was made using exponential smoothing and moving averages to compute seasonal coefficients and forecast demand and returns. Multiple regression models and Artificial neural networks were also used to

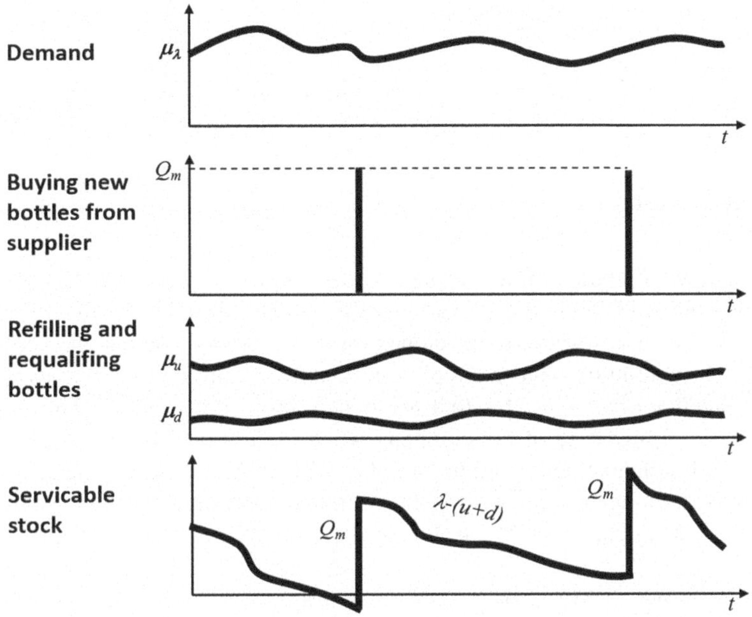

Fig. 7 Stochastic inventory stock model S

forecast [5]. Afterwords, we used a weighted linear combination of the probability density functions as in [3] for the final forecast. The forecasted mean and RMSE was used as input values for the stochastic inventory models developed for the case study.

For this, we present a stochastic inventory model S, based on the continuous review policy (s,Q), which considers continuous stochastic demand, discrete replenishment from supplier, continuous stochastic replenishment from returned bottles, and constant lead times, as depicted in Fig. 7.

Assuming demand during lead time is $dl \sim N(\mu_{dl}, \sigma_{dl})$, then the Order Point is:

$$OP = \mu_{dl} + z_\alpha \sigma_{dl} \tag{10}$$

where $z_\alpha = \Phi^{-1}(1-\alpha)$ is the safety factor for a given Level of Service $1-\alpha$. The demand is replaced by a difference of normal random variables $\lambda - (u+d)$ where $\lambda \sim N(\mu_\lambda, \sigma_\lambda)$, $u \sim N(\mu_u, \sigma_u)$ and $d \sim N(\mu_d, \sigma_d)$. Assuming independence, we have:

$$\mu_{\lambda-(u+d)} = \mu_\lambda - \mu_u - \mu_d \qquad \sigma_{\lambda-(u+d)} = \sqrt{\sigma_\lambda^2 + \sigma_u^2 + \sigma_d^2} \tag{11}$$

Finally, the order point is given by

$$OP = (\mu_\lambda - \mu_u - \mu_d)l + Z_\alpha \sqrt{l(\sigma_\lambda^2 + \sigma_u^2 + \sigma_d^2)} \tag{12}$$

4 Conclusions

A Portuguese company in the energy sector posed a challenge to define the acquisition plan of LPG bottles. To answer this industrial challenge, three inventory models with reverse flows were developed for the company. The model D considers deterministic continuous constant replenishment from returned LPG bottles and also discrete batches of new bottles that are bought from the supplier. The model R considers the future case of the company when the returned bottles cover for the demand, and replenishment from the supplier will be unnecessary. Finally, model S was developed to approach the angle that demand and returns are not constant but continuous and stochastic, with discrete replenishment from the supplier.

Acknowledgements COST Action TD1409, Mathematics for Industry Network (MI-NET), COST-European Cooperation in Science and Technology; CIDMA-Center for Research and Development in Mathematics and Applications; FCT-Portuguese Foundation for Science and Technology, project UID/MAT/04106/2013. We would like to thank Ana Sapata from University of Evora, and Claudio Henriques, Fabio Henriques e Mariana Pinto from University of Aveiro for their contributions during the European Study Group.

References

1. Alinovi, A., Bottani, E., Montanari, R.: Reverse logistics: a stochastic EOQ-based inventory control model for mixed manufacturing/remanufacturing systems with return policies. Int. J. Prod. Res. **50**(5), 1243–1264 (2012)
2. Ballou, R.H.: Business Logistics Management, 4th edn. Prentice Hall, Upper Saddle River (2006)
3. Cassettari, L., Bendato, I., Mosca, M., Mosca, R.: A new stochastic multi source approach to improve the accuracy of the sales forecasts. Foresight **19**(1), 48–64 (2017)
4. Harris, F.W.: Operations Cost, Factory Management Series. Shaw, Chicago (1915)
5. Lopes, I.C., Costa e Silva, E., Correia, A., Monteiro, M., Borges Lopes, R.: Combining data analysis methods for forecasting liquefied petroleum gas cylinders demand. V Workshop on Computational Data Analysis and Numerical Methods, ESTG, Instituto Politécnico do Porto, Portugal, (2018)
6. Richter, K.: The extended EOQ repair and waste disposal model. Int. J. Prod. Econ. **45**(1–3), 443–447 (1996)
7. Sousa, J.: Background of Portuguese domestic energy consumption at european level. In: IT4Energy International Workshop on Information Technology for Energy Applications (2012)
8. Teunter, R.H.: Economic ordering quantities for recoverable item inventory systems. Nav. Res. Logist. **48**(6), 484–495 (2001)
9. Wilson, R.H.: A scientific routine for stock control. Harv. Bus. Rev. **13**, 116–28 (1934)

Part II
Biological and Medical Models and Applications

Cyclic Structure Induced by Load Fluctuations in Adaptive Transportation Networks

Erik Andreas Martens and Konstantin Klemm

Abstract Transport networks are crucial to the functioning of natural systems and technological infrastructures. For flow networks in many scenarios, such as rivers or blood vessels, acyclic networks (i.e., trees) are optimal structures when assuming time-independent in- and outflow. Dropping this assumption, fluctuations of net flow at source and/or sink nodes may render the pure tree solutions unstable even under a simple local adaptation rule for conductances. Here, we consider tree-like networks under the influence of spatially heterogeneous distribution of fluctuations, where the root of the tree is supplied by a constant source and the leaves at the bottom are equipped with sinks with fluctuating loads. We find that the network divides into two regions characterized by tree-like motifs and stable cycles. The cycles emerge through transcritical bifurcations at a critical amplitude of fluctuation. For a simple network structure, depending on parameters defining the local adaptation, cycles first appear close to the leaves (or root) and then appear closer towards the root (or the leaves). The interaction between topology and dynamics gives rise to complex feedback mechanisms with many open questions in the theory of network dynamics. A general understanding of the dynamics in adaptive transport networks is essential in the study of mammalian vasculature, and adaptive transport networks may find technological applications in self-organizing piping systems.

1 Introduction

Network modeling deals with the rules for establishing and removing connections between the entities that make up a system. For instance, social networks display a much larger amount of triangles than expected under entirely random wiring.

E. A. Martens (✉)
Kgs. Lyngby, Denmark
e-mail: eama@dtu.dk

K. Klemm
IFISC (CSIC-UIB), Campus Universitat de les Illes Balears, Palma de Mallorca, Spain
e-mail: klemm@ifisc.uib-csic.es

© Springer Nature Switzerland AG 2019

I. Faragó et al. (eds.), *Progress in Industrial Mathematics at ECMI 2018*,
Mathematics in Industry 30, https://doi.org/10.1007/978-3-030-27550-1_19

A possible explanation is that nodes (i.e., persons) are more likely to introduce their already existing friends to each other [1]. Similarly, for biological networks, a simple network growth rule of node copying and random perturbation of edges mimics genome duplication and thereby reproduces statistical features of protein interaction networks [2, 3].

More recent models of adaptive networks involve bidirectional dependence between a dynamics in the networked system and the dynamic modification of its link structure [4–6]. Here we study this dependence specifically for the case of a network for transport and distribution, motivated by the vascular (blood circulatory) system in higher animals. This system fulfills the task of transport from one central source (heart/lung) to spatially distributed sinks. Assuming a constant in-flow from the source and a constant outflow into sinks, the optimal distribution system in terms of energy consumption is a tree [7, 8], i.e., a cycle-free connected network. When load at the sinks fluctuates, however, networks involving cycles become optimal as shown by Corson [9] and Katifori with colleagues [10].

Here we combine this insight with local adaptation [11, 12] rather than global optimization. Indeed, such models are relevant in vascular physiology, where arterioles adapt their diameter and wall thickness on time scales from seconds over days to months in dependence on local flow variables including pressure and flow shear [13, 14]. The conductances of the flow network self-organize towards balanced pressure fluctuations. We observe that cycles form only when the amplitude of load fluctuations exceeds a threshold. With the source placed at the top and all sinks in the bottom layer of a hierarchical network, as illustrated in Fig. 1, we show that cycle formation is localized: depending on a parameter of the adaptation rule, there is a transition between cycle formation close to the source or cycle formation close to the sinks.

2 Model

V denotes the set of nodes of the network with $N = |V| < \infty$ and $A \subseteq N \times N$ the set of edges. The edges are bidirectional, so $(i, j) \in A$ implies $(j, i) \in A$. Each node is assigned a pressure p_i. The edge flow is $f_{ij} > 0$ from node i to j. Furthermore we assume that the network is resistive and linear, i.e., it is Ohmian with $f_{ij} = C_{ij}(p_i - p_j)$, where an edge carries the property of a conductance between nodes i and j with $C_{ij} = C_{ji} > 0$ only if $(i, j) \in A$; otherwise $C_{ij} = C_{ji} = 0$.

Here, we study tree-like networks of height H as illustrated in Fig. 1, with cross-edges on every branching level, $l = 0, \ldots, H$. Cross-edges lead to cyclic structure. A cycle (red triangle) is a connected subnetwork of m nodes such that each node has exactly two neighbours. We focus on two types of networks: the *simply augmented tree* (SAT), where cross-edges only form triangular submotifs (i.e., excluding dotted cross-edges) and *fully augmented tree* (FAT) where all displayed cross-edges are allowed.

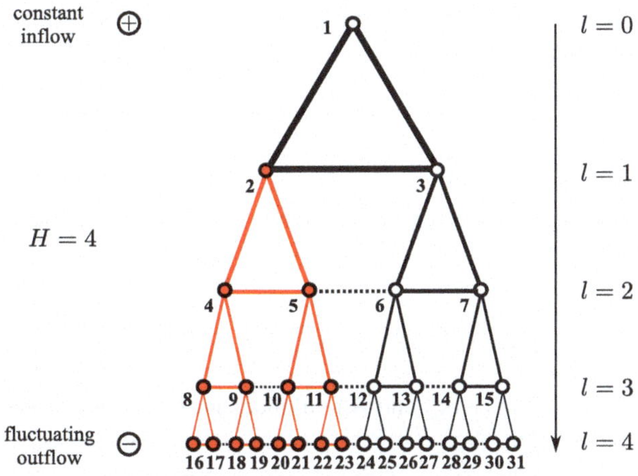

Fig. 1 'Simple Augmented Trees' (SAT) are tree-like structures with *cross-edges* (full horizontal edges) that connect only nodes in each minimal subtree (e.g., edges highlighted in red), thus forming a *'tree of triangles'*; the 'Fully Augmented Tree' (FAT) connects all nodes within one tree level l by a path. The branching level in the cut tree is denoted by l. Cross-edges introduce cycles to the network. The root of the tree (top) and the fluctuating sinks in the leaves of the tree (bottom) drive the flow. Cross-edges emerge depending on the strength of the fluctuation in the leaves. For SATs, the dynamics in a triangular submotif at level l (red edges) will only depend on downstream fluctuations (red nodes)

To model *sources and sinks* in the network, we include non-zero nodal flows h_i. A proper subset S of the node set V is chosen as the set of sink nodes. The set S is time-independent and typically comprises the most peripheral nodes, where capillaries connect to the vein network. With a tree structure underlying the network, S is chosen as the set of leaves of the tree. Focusing on the networks based on symmetric trees of height H (cf. Fig. 1), we have the $|S| = n = 2^H$ leaves as sink nodes. For each sink node $i \in S$, the nodal flow $h_i(t)$ is non-positive at all times $t \in \mathbb{R}$. A single node in $V \setminus S$ is chosen as the source node and indexed as node 1 for simplicity. For the networks based on symmetric trees, the source node is the root of the tree. The source node has a positive nodal flow $h_1(t) = 1$ for all $t \in \mathbb{R}$. For all other nodes $j \in V \setminus (S \cup \{1\})$, we set $h_j(t) = 0$ for all $t \in \mathbb{R}$. Mass balance requires that $\sum_{k \in V} h_k(t) = 0$ for all $t \in \mathbb{R}$.

Assuming that the accumulation rate of fluid at any node is nearly instantaneous, or that vessels are inelastic, the nodal accumulation rate becomes negligible [12, 15], and we may express mass balance by invoking Kirchhoff's first law,

$$\sum_j C_{ij}(p_i - p_j) = h_i, \tag{1}$$

which is re-written in vector/matrix notation by defining the nodal flow $\mathbf{h} := (h_i)_{i\in V}$ and the Kirchhoff matrix $\mathbf{K} = (K_{ij})_{i,j\in V}$ with $K_{ij} := (\delta_{ij} \sum_{j'} C_{ij'}) - C_{ij}$,

$$\mathbf{K} \cdot \mathbf{p} = \mathbf{h} \tag{2}$$

f which is solved for $\mathbf{p} := (p_i)_{i\in V}$.

To impose *adaptive dynamics to the network*, we postulate the generic ad-hoc law for the conductances [12]:

$$\frac{d}{dt} C_{ij} = \alpha_1 C_{ij} |p_j - p_i|^\gamma - \alpha_2 C_{ij}. \tag{3}$$

Thus, the first term on the right hand side induces growth proportional to the power dissipated along the edge, thus mitigating rising pressure differences by increasing the conductance along the edge. The network adapts towards minimizing power consumption. The last term prevents unlimited growth of the conductances.

Rescaling variables with $C'_{ij} := h_1^{-1}(\alpha_2/\alpha_1)^{1/\gamma} C_{ij}$ and $p'_i := (\alpha_1/\alpha_2)^{1/\gamma} p_i$, $h' := h/h_1$ (so that $h'_1 = 1$), $t' := \alpha_2 t$, the resulting dimensionless model reads

$$\frac{d}{dt'} C'_{ij}(t) = C'_{ij}(t)[|p'_j(t) - p'_i(t)|^\gamma - 1], \tag{4}$$

$$\mathbf{K}'(t) \cdot \mathbf{p}'(t) = \mathbf{h}'(t). \tag{5}$$

where we drop the primes and omit the argument t from now on. Note that the solvability condition of the Kirchhoff equation [12], $\sum_j h'_j(t) = 0$, follows from Fredholm's alternative and has the physical interpretation of mass conservation.

We consider sinks with varying load, compliant with $\sum_{k\in V} h_k(t) = 0$. For each time t, there is a sink node $s \in S$ so that the nodal flow $\mathbf{h}(t)$ is the vector $\mathbf{g}^{(s)}$ with components

$$g_i^{(s)} = \begin{cases} +1 & \text{if } i \text{ is the source (root) node.} \\ -\frac{1}{n} - \frac{a}{\sqrt{2}} =: h_- & \text{if } i = s \\ -\frac{1}{n} + \frac{1}{n-1}\frac{a}{\sqrt{2}} =: h_+ & \text{if } i \in S \setminus \{s\} \\ 0 & \text{otherwise.} \end{cases} \tag{6}$$

This reflects the situation where at each point t in time, one of the sinks has higher load (h_-) than the others (having load h_+). Independent of time, the source (root node) has an inflow of $+1$. All other nodes have neither in- nor outflow. The driving amplitude a is a parameter of the model, obeying $a \in [0, a_{\max}]$ with $a_{\max} = \sqrt{2}(|S| - 1)/|S|$. This ensures $h_- \le h_+ \le 0$ meaning sink nodes actually behave as sinks at all times. In the extreme case $a = a_{\max}$, only the sink with the higher load is on ($h_- = -1$) while all others are off ($h_+ = 0$). This case reproduces the single moving sink as employed by Katifori et al. [10].

Following [12], we assume that (1) all sinks are in the high load state for the same fraction of time and (2) the variation of sink load occurs on a time scale faster than the adaptation of the conductances. Thus, Eq. (4) are effectively equivalent to an averaged form as follows,

$$\frac{d}{dt}C_{ij}(t) = C_{ij}(t)[\langle|p_j(t) - p_i(t)|^\gamma\rangle - 1], \tag{7}$$

where time t corresponds to the slow time and the average $\langle.\rangle$ is taken uniformly over the $n = |S|$ assignments of high load sinks,

$$\langle|p_j(t) - p_i(t)|^\gamma\rangle := n^{-1}\sum_{s\in S}|p_j^{(s)}(t) - p_i^{(s)}(t)|^\gamma \tag{8}$$

and

$$\mathbf{K}(t)\cdot\mathbf{p}^{(s)}(t) = \mathbf{g}^{(s)}\,. \tag{9}$$

For simplicity, we drop the averaging brackets $\langle\cdot\rangle$ from now on.

3 Analysis

3.1 *The Case of Exponent $\gamma = 2$*

Our previous analysis [12] of the model concentrated on the case $\gamma = 2$. With this choice, the growth of an edge with conductance C_{ij} in (3) is proportional to the power (dissipated energy per time) over this same edge. For the height $H = 1$, the simply/fully augmented tree becomes a triangle of nodes $V = \{1, 2, 3\}$ as shown in Fig. 2. The two sink nodes, indexed 2 and 3 are connected to each other with a conductance $C_- := C_{23}$, and each of them also to the source node, indexed 1, with the symmetric conductance $C_\wedge := C_{12} = C_{13} =$ of the cut-edges. We are interested in (slow time) stationary solutions of the model driven with load fluctuations of amplitude a as a single parameter. The stationary conductance C_- of the cross-edge connecting the two sinks is of particular interest. The solutions undergo a transcritical bifurcation at parameter value $a_c = 1/\sqrt{6} \approx 0.408$. For sub-critical fluctuation amplitude, $a < a_c$, the uniquely stable solution branch has $C_- = 0$ for the cross edge and $C_\wedge > 0$ for the cut edge which grows monotonically with a. For super-critical amplitude, $a > a_c$, the stable branch has a positive conductance for the cross-edge, $C_- > 0$, which grows almost linearly with increasing amplitude a, while the cut-edge stays exactly constant. Thus, the cross-edge short circuits fluctuations so that the two cut-edges may stay constant.

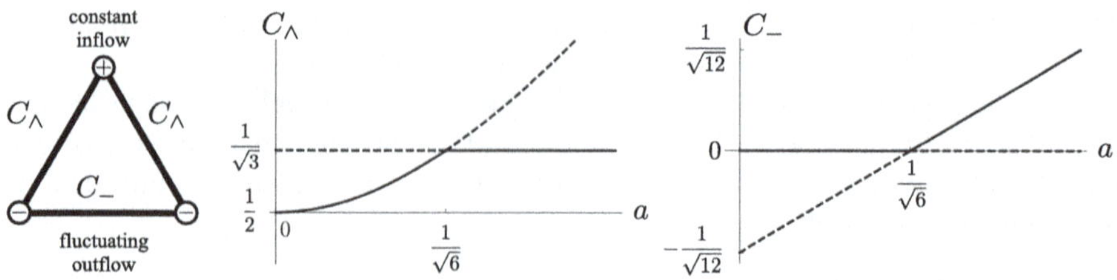

Fig. 2 Solution branches for the triangular graph with $H = 1$ undergo a sub-critical bifurcation where the conductance of the cross-edge, C_-, becomes non-zero

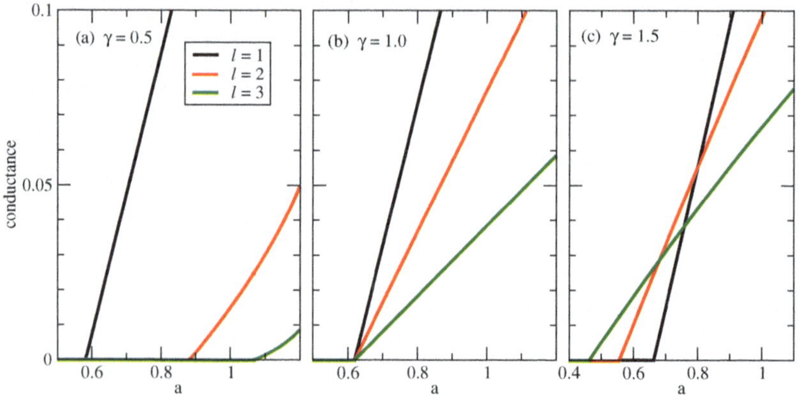

Fig. 3 Conductances of cross-edges in simply augmented trees of height $H = 3$ as a function of sink fluctuation parameter. The three panels distinguish exponent values (**a**) $\gamma = 0.5$, (**b**) $\gamma = 1.0$ and (c) $\gamma = 1.5$ which affects the transition order for different levels l

In simply augmented trees with more layers ($H > 1$, see Fig. 1), all cross-edge conductances undergo transcritical bifurcations from zero to non-zero as well [12, 16]. The parameter value $a_c^{(l,H)}$ at the transition depends both on the level l of the cross-edge and the system height H. In a given system with height H, as the fluctuation amplitude a increases, cross-edges at the sink nodes first undergo a transcritical transition from zero to positive conductance. As the amplitude a is increased further, cross-edges at the next level become non-zero, thus following a strict ordering $a_c^{(l+1,H)} < a_c^{(l,H)}$ for all $l \in \{1, \ldots, H-1\}$. The resulting ordering for $\gamma = 2$ is qualitatively similar to the case illustrated in Fig. 3(c) where $gamma = 1.5$. The strict ordering in terms of level l may be directly linked to the topology of the SAT, which (including its cross edges) forms a tree of triangular submotifs [16].

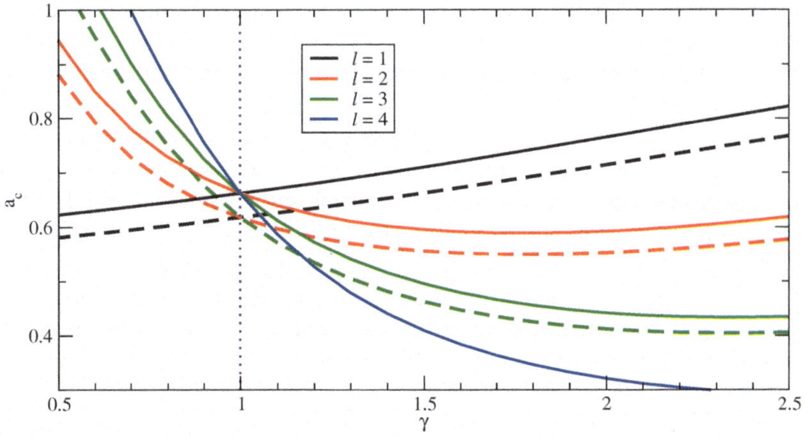

Fig. 4 Critical fluctuation amplitude values $a_c^{(l,H)}$ for cross-edges in simply augmented trees of height $H = 3$ (dashed lines) and $H = 4$ (solid lines), as a function of exponent γ

3.2 *Effect of Varying Exponent γ*

Figure 3 shows the influence of γ on the transcritical bifurcations in the simply augmented tree of height $H = 3$. For $\gamma = 1.5$, the transitions occur in the same ordering as known for $\gamma = 2$, i.e. from sink node level (here $l = 3$) towards source node level. For $\gamma = 0.5$, the order of transitions is reversed, while $\gamma = 1$ has all transitions at the same parameter value.

Figure 4 shows the γ-dependence of critical values of parameter a for cross-edges at different levels l in systems of heights $H = 3$ and $H = 4$. This confirms, here numerically for the provided cases, the change of behaviour at $\gamma = 1$. For $\gamma < 1$, the onset of cross-edge conductances happens first close to the source and moves downward in the system. For $\gamma > 1$ this order is reversed.

In fully augmented trees, the ordering of the critical amplitudes, $a_c^{(l,H)}$, follow a more complex pattern and is subject to further study [16].

4 Conclusion and Discussion

We have studied conditions for the presence or absence of cycle forming edges in a model of vascular networks under load fluctuations. Variation of the exponent γ in the local pressure dependence of conductance adaptation changes the order by which cycles arise in the simply augmented trees (Fig. 1), either first close to the sink or close to the root. For network structures with less symmetry, preliminary analysis finds a more complex ordering sequence. This reflects that, generally, the interaction between topology and dynamics gives rise to complex feedback mechanisms posing open questions in the theory of network dynamics. Future work on this model ought to include a comparison to empirical vascular networks. Both the network structures themselves and measurements on the flow through branches of the network are becoming available [17–19].

In view of data and for better alignment with real structures, the tree-like networks of this model may be augmented further. In a first step, one may include further edges inside a level as indicated by the dashed lines in Fig. 1. For this fully augmented tree, preliminary analysis [12] has shown that cross-edges become conductive in a pattern similar to that of the simply augmented tree at exponent $\gamma = 2$. Further results will be reported elsewhere [16]. A complete quantitative understanding of the transitions in this system would bring us closer to a general theory for the emergence of cycles in transport networks in the presence of fluctuations.

A further interesting step would be to abandon imposed network structures altogether and cast the adaptation dynamics into real two- or three-dimensional space. Having both conductance c and pressure p as scalar fields, adaptation of conductances can be described [16] by an equation as

$$\partial_t c(x,t) = c(x,t)[(\nabla p(x,t))^2 - 1] \tag{10}$$

as a proposal for the real-space analog of the adaptation rule in Eq. (4).

Beyond modeling the self-organization of transport networks in nature, this branch of research has bearings also in technical applications. *Programmable materials* is a branch of technology to generate complex objects by self-assembly of their suitably programmed constituents [20]. Recent ideas and advances point in the direction of evolutionary materials that are capable of self-repair and adaptation to changing environmental conditions [21]. A pertinent example are urban water-supply systems where pipes self-adapt the flow capacity in response to local demand fluctuations in a city with evolving population density [20].

Acknowledgements EAM and KK acknowledge travel funding from Action CA15109, European Cooperation for Statistics of Network Data Science (COSTNET). KK acknowledges funding from MINECO through the Ramón y Cajal program and through project SPASIMM, FIS2016-80067-P (AEI/FEDER, EU). We thank J. C. Brings Jacobsen for helpful discussions on circulatory physiology and E. Katifori on adaptive networks.

References

1. Davidsen, J., Ebel, H., Bornholdt, S: Emergence of a small world from local interactions: modeling acquaintance networks. Phys. Rev. Lett. **88**, 128701 (2002)
2. Solé, R.V., Pastor-Satorras, R., Smith, E., Kepler, T.B.: A model of large-scale proteome evolution. Adv. Complex Syst. **5**(1), 43–54 (2002)
3. Ispolatov, I., Krapivsky, P.L., Yuryev, A.: Duplication-divergence model of protein interaction network. Phys. Rev. E **71**, 061911 (2005)

4. Gross, T., Blasius, B.: Adaptive coevolutionary networks: a review. J. R. Soc., Interface **5**(20), 259–71 (2008).
5. Herrera, J.L., Cosenza, M.G., Tucci, K., González-Avella, J.C.: General coevolution of topology and dynamics in networks. Europhys. Lett. **95**(5), 58006 (2011)
6. Porter, M.A., Gleeson, J.P.: Dynamical Systems on Networks. Frontiers in Applied Dynamical Systems: Reviews and Tutorials, vol. 4 (2016)
7. Kantorovich, L.V.: On the translocation of masses. In: Doklady Akademii Nauk SSSR, vol. 37, pp. 199–201 (1942)
8. Villani, C.: Topics in Optimal Transportation. Graduate Studies in Mathematics, vol. 58. American Mathematical Society, Providence (2003)
9. Corson, F.: Fluctuations and redundancy in optimal transport networks. Phys. Rev. Lett. **104**(4), 048703 (2010)
10. Katifori, E., Szőllősi, G.J., Magnasco, M.O.: Damage and fluctuations induce loops in optimal transport networks. Phys. Rev. Lett. **104**(4), 048704 (2010)
11. Hu. D., Cai, D.: Adaptation and optimization of biological transport networks. Phys. Rev. Lett. **111**, 138701 (2013)
12. Martens, E.A., Klemm, K.: Transitions from trees to cycles in adaptive flow networks. Front. Phys. **5**(62), 1–10 (2017)
13. Jacobsen, J.C.B., Mulvany, M.J., Holstein-Rathlou, N.H.: A mechanism for arteriolar remodeling based on maintenance of smooth muscle cell activation. Am. J. Physiol. Regul. Integr. Comp. Physiol. **294**, R1379–R1389 (2008)
14. Jacobsen, J.C.B., Hornbech, M.S., Holstein-Rathlou, N.H.: A tissue in the tissue: models of microvascular plasticity. Eur. J. Pharm. Sci. **36**(1), 51–61 (2009)
15. Reichold, J., Stampanoni, M., Keller, A.L., Buck, A., Jenny, P., Weber, B.:Vascular graph model to simulate the cerebral blood flow in realistic vascular networks. J. Cereb. Blood Flow Metab. **29**(8), 1429–1443 (2009)
16. Martens, E.A., Klemm, K.: Noise-induced bifurcations in a model of vascular networks. In preparation (2019)
17. Blinder, P., Tsai, P.S., Kaufhold, P.S., Knutsen, P.M., Suhl, H., Kleinfeld, D.: The cortical angiome: an interconnected vascular network with noncolumnar patterns of blood flow. Nat. Neurosci. **16**(7), 889–97 (2013)
18. Poelma, C.: Exploring the potential of blood flow network data. Meccanica **52**(3), 489–502 (2017)
19. Alim, K.: Fluid flows shaping organism morphology. Philos. Trans. R. Soc., B **373**(1747), 1–5 (2018)
20. Campbell, T.A., Tibbits, S., Garrett, B.: The programmable world. Sci. Am. **311**(5), 60–65 (2014)
21. Papadopoulou, A., Laucks, J., Tibbits, S.: From self-assembly to evolutionary structures. Archit. Des. **87**(4), 28–37 (2017)

Time-Reversal Methods in Acousto-Elastodynamics

Franck Assous and Moshe Lin

Abstract The aim of the article is to solve an inverse problem in order to determine the presence and some properties of an elastic "inclusion" (an unknown object, characterized by elastic properties discriminant from the surrounding medium) from partial observations of acoustic waves, scattered by the inclusion. The method will require developing techniques based on Time Reversal methods. A finite element method based on acousto-elastodynamics equations will be derived and used to solve the inverse problem. Our approach will be applied to configurations modeling breast cancer detection, using simulated ultrasound waves.

1 Introduction

Time reversal (TR) is a subject of very active research for over two decades. Many international teams are currently working on the subject from theoretical, physical and numerical points of view. It was originally experimentally developed by Fink in 1992 in acoustics and showed very interesting features [8]. The principle is to take advantage of the reversibility of wave propagation phenomena, for example in acoustics, elastic or electromagnetism in an unknown medium, to back-propagate signals to the sources that emitted them. The initial experiment, was to refocus, very precisely, a recorded signal after passing through a barrier consisting of randomly distributed metal rods. Since then, numerous applications of this physical principle have been designed, for instance [13] and references therein. The first mathematical analysis can be found in [3] for a homogeneous medium and in [4, 7] for a random medium. In this article, we are basically concerned with equations of acousto-elastodynamics. As the application we have in mind are concerned with ultrasound-based elasticity imaging methods, we consider a coupled fluid/solid model. For the sake of simplicity, we consider a "layered" medium and we want to

F. Assous · M. Lin (✉)
Ariel University, Ariel, Israel
e-mail: moshelin1@walla.co.il

© Springer Nature Switzerland AG 2019

I. Faragó et al. (eds.), *Progress in Industrial Mathematics at ECMI 2018*, Mathematics in Industry 30, https://doi.org/10.1007/978-3-030-27550-1_20

determine the presence of an "inclusion" in the elastic part, from recorded acoustic waves scattered by this inclusion. However, the method does not require a priori knowledge of the physical properties of the inclusion.

2 Forward Problem

We first formulate the mathematical forward problem. We consider a two-dimensional fluid-solid domain Ω made of two parts, an acoustic one Ω_f and an elastic one Ω_s. For simplicity, we will assume that Ω is a rectangle. The acoustic part of the domain Ω_f corresponds to a homogeneous fluid, characterized by its density ρ_f and its Lamé parameter λ_f. We denote by $\partial\Omega_f$ the boundary of Ω_f and $\mathbf{n}$ is the outward unit normal to the boundary. Introduce the pressure $p(\mathbf{x}, t)$ on a time t, $\mathbf{x} = (x_1, x_2) \in \Omega_f$, and $f(\mathbf{x}, t)$ is a given source, for instance a Ricker function, the acoustic wave equation in Ω_f is written

$$\frac{1}{\lambda_f}\frac{\partial^2 p}{\partial t^2} - \operatorname{div}\left(\frac{1}{\rho_f}\nabla p\right) = f\,, \tag{1}$$

together with initial homogeneous conditions. We assume that the boundary $\partial\Omega_f$ can be split into $\partial\Omega_f = \Gamma_f \cup \Gamma_I$, where Γ_I denotes the interface between the fluid and solid part, assumed, for simplicity, to be horizontal. We supplement the system with absorbing boundary conditions [6] on $\partial\Omega_f$. Denoting by $V_p = \sqrt{\frac{\lambda}{\rho}}$ the wave velocity in the fluid, the absorbing boundary conditions on Γ_f are written

$$\frac{\partial p}{\partial t} = -V_p\,\nabla p\cdot\mathbf{n} \quad \text{on } \Gamma_f\,. \tag{2}$$

On the part Γ_I, we add an interface condition for the pressure $p(\mathbf{x}, t)$, that will be presented below, see (5). We then introduce the governing equations of linear elastodynamics for Ω_s, the solid part of the domain, characterized by the density ρ_s, and the Lamé parameters λ_s and μ_s. We assume that the boundary $\partial\Omega_s$ can be split into $\partial\Omega_s = \Gamma_s \cup \Gamma_I$. Denoting by $\mathbf{u}(\mathbf{x}, t) = (u_1(x_1, x_2, t), u_2(x_1, x_2, t))$ the velocity[1] on a time t, at a point $\mathbf{x} = (x_1, x_2) \in \Omega_s$, we have

$$\rho_s\frac{\partial^2\mathbf{u}}{\partial t^2} - \nabla\cdot(\mu_s\nabla\mathbf{u}) - \nabla((\lambda_s + \mu_s)\nabla\cdot\mathbf{u}) = 0\,, \tag{3}$$

This equation is supplemented with homogeneous initial conditions and absorbing boundary conditions on Γ_s, as proposed in [10]. For this purpose, we introduce the

[1] $\mathbf{u}(\mathbf{x}, t)$ is the velocity, that is the time derivative of the displacement. This formulation allows us to derive a pressure-velocity fluid-solid formulation, which will make easier the derivation of the variational formulation, see below.

matrix A

- $A = \begin{pmatrix} -\sqrt{\rho_s(\lambda_s + 2\mu_s)} & 0 \\ 0 & -\sqrt{\rho_s\mu_s} \end{pmatrix}$ for horizontal boundary edges,
- $A = \begin{pmatrix} -\sqrt{\rho_s\mu_s} & 0 \\ 0 & -\sqrt{\rho_s(\lambda_s + 2\mu_s)} \end{pmatrix}$ for vertical boundary edges,

and $(\tau_{ij}(\mathbf{u}))_{1\le i,j\le 2}$ the classical stress tensor $\tau_{ij}(\mathbf{u}) = \lambda_s \mathrm{div}\,\mathbf{u}\,\delta_{ij} + \mu_s(\dfrac{\partial u_i}{\partial x_j} + \dfrac{\partial u_j}{\partial x_i})$. The conditions are written, using the Einstein summation convention,

$$A_{ij}\frac{\partial u_j}{\partial t} = \tau_{ij}\, n_j, \qquad i = 1, 2\,. \tag{4}$$

Finally, we introduce the transmission conditions at the fluid-solid interface Γ_I:

$$\frac{1}{\rho_f}\frac{\partial p}{\partial x_2} = -\frac{\partial u_2}{\partial t} \tag{5}$$

$$\frac{\partial u_1}{\partial x_2} + \frac{\partial u_2}{\partial x_1} = 0, \qquad \frac{\partial p}{\partial t} = \lambda_s \frac{\partial u_1}{\partial x_1} + (\lambda_s + 2\mu_s)\frac{\partial u_2}{\partial x_2} \tag{6}$$

These conditions express the continuity of the normal component (5) and of the stress tensor (10). They will appear naturally in the pressure-velocity variational formulation, that will be basis of the finite element method.

3 Time Reversed Problem

In a second step, we formulate the time reversed acousto-elastic problem. Examples of time reversal techniques (numerical or experimental) can be found (among others!) in [1, 2, 8, 9, 12]. We first introduce the time-reversed wave equation for the acoustic part of the domain Ω_f. We denoted by $p^R(\mathbf{x}, t')$ the time-reversed pressure, defined by $p^R(\mathbf{x}, t') = p(\mathbf{x}, T_f - t)$, $\mathbf{x} \in \Omega_f$, where T_f denotes the final time. Since the wave equation involves only second order time derivatives, this definition ensures that the reverse field $p^R(\mathbf{x}, t')$ is a solution to the wave equation

$$\frac{1}{\lambda_f}\frac{\partial^2 p^R}{\partial t'^2} - \mathrm{div}\,(\frac{1}{\rho_f}\nabla p^R) = 0\,, \tag{7}$$

together with (TR) initial conditions and (TR) absorbing boundary conditions on Γ_f, analogous to (2). In addition, on the boundary Γ_{SRA}, modeling a source-receivers array (SRA) where the forward signal is recorded (see Fig. 1), we set $p^R(t') = p(T_f - t)$ which is the (recorded) source of the TR.

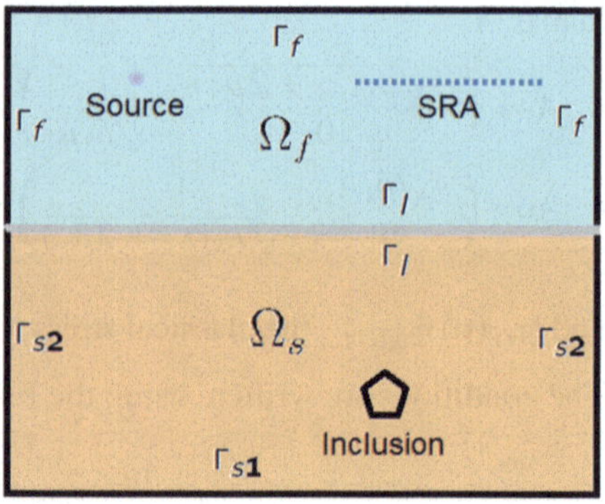

Fig. 1 Example of fluid-solid medium

Similarly, we also introduce the elastic time-reversed problem associated to Eq. (4). We denote by $\mathbf{u}^R(\mathbf{x}, t') = (u_1^R(x_1, x_2, t'), u_2^R(x_1, x_2, t'))$ the time-reversed velocity solution to linear elastodynamics, that solves

$$\rho_s \frac{\partial^2 \mathbf{u}^R}{\partial t'^2} - \nabla \cdot (\mu_s \nabla \mathbf{u}^R) - \nabla((\lambda_s + \mu_s)\nabla \cdot \mathbf{u}^R) = 0 \,, \tag{8}$$

together with (TR) initial conditions and (TR) absorbing boundary conditions on Γ_s, that have analogous expressions as in (4). Finally, we derive the time-reversed continuity transmission conditions at the interface Γ_I

$$\frac{1}{\rho_f} \frac{\partial p^R}{\partial x_2} = -\frac{\partial u_2^R}{\partial t'} \tag{9}$$

$$\frac{\partial u_1^R}{\partial x_2} + \frac{\partial u_2^R}{\partial x_1} = 0, \qquad \frac{\partial p^R}{\partial t'} = \lambda_s \frac{\partial u_1^R}{\partial x_1} + (\lambda_s + 2\mu_s) \frac{\partial u_2^R}{\partial x_2} \tag{10}$$

In order to create synthetic data, the forward and reverse formulations are approximated by the FreeFem++ package [11] which implements a finite element method in space. In this study we use a P^2 finite element method. The advancement in time is performed by using a second order in time central finite difference scheme, so that it is time reversible also on the numerical level.

4 Numerical Results

In this section, we describe numerical results obtained for a scatter identification problem, in the case of two scatters located in the elastic part. The principle of the numerical process is as follows: an incident wave is generated by a point source such that after a time T_f the total field is negligible. On the boundary Γ_{SRA} located in the fluid part, the forward signal is recorded. Then, we perform numerically a time-reversed computation, by back propagating the recorded scattered data from the SRA. However, we do not assume we know the physical properties or the number of the inclusions, nor their locations. Hence, the recorded data are back propagated in the medium without the inclusions. Finally, we intend to image the unknown scatterers in the medium—responsible of the diffraction of the incident wave—by using correlation method between the forward $\mathbf{u}^I$ and the reversed wave $\mathbf{u}^s_R$ in the same spirit as those involved for instance in time reverse migration [5]. As a first attempt, we have considered the following RTM (Reverse Time Migration) criterion:

$$RTM(\mathbf{x}) = \int_0^{T_f} \mathbf{u}^s_R(T_f - t, \mathbf{x}) \times \mathbf{u}^I(t, \mathbf{x}) dt\,, \tag{11}$$

To illustrate our purpose, we consider a two layered medium, made of fluid part (top) and of a elastic one (bottom), the elastic part sketching a breast tissue geometry and is a heterogeneous medium, as it contains a skin layer (see Fig. 2). The SRA is an horizontal line as sketched on Fig. 2. For the fluid part, we choose $\rho = 1000\,\mathrm{kg/m^3}$ and $\lambda = 2.25\,\mathrm{GPa}$, for the solid part, the same value of ρ with $\lambda = 1.83\,\mathrm{GPa}$ and $\mu = 18.33\,\mathrm{kPa}$, and for the skin (inside the solid part), $\rho = 1150\ \mathrm{kg/m^3}$, $\lambda = 6.66\,\mathrm{GPa}$ and $\mu = 66.66\,\mathrm{kPa}$. There are two elliptical inclusions with different size, shape, and elastic properties. The first one represents a benign tumor with $\rho = 1000\,\mathrm{kg/m^3}$, $\lambda = 2.16\,\mathrm{GPa}$ and $\mu = 21.66\,\mathrm{kPa}$, and the second one a malignant tumor, with the same ρ, $\lambda = 2.99\,\mathrm{GPa}$ and $\mu = 30\,\mathrm{kPa}$. Note that both inclusions are penetrable, which means that the reflection of the incident wave highlighting the inclusion is quite weak. Finally, the source used to generate the acoustic wave in the fluid part is a Ricker function of the form $f(\mathbf{x}, t) = (1 - 2\pi^2(\nu_0 t - 1)^2)e^{-\pi^2(\nu_0 t - 1)^2}$, with a central frequency $\nu_0 = 100\,\mathrm{kHz}$ and a corresponding wavelength equal to $\lambda_W = 12\,\mathrm{mm}$.

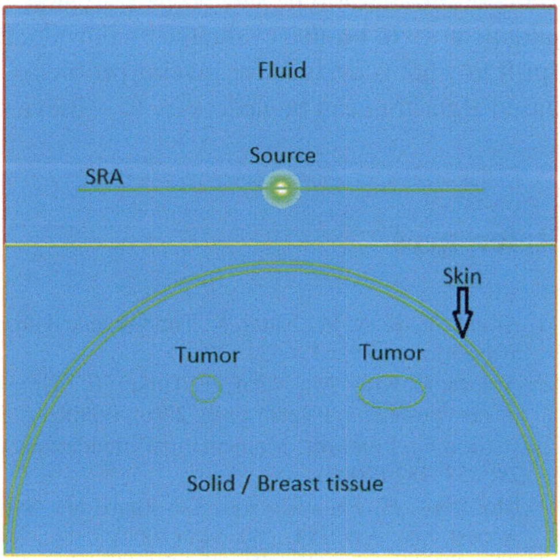

Fig. 2 Example of breast tissue with skin medium

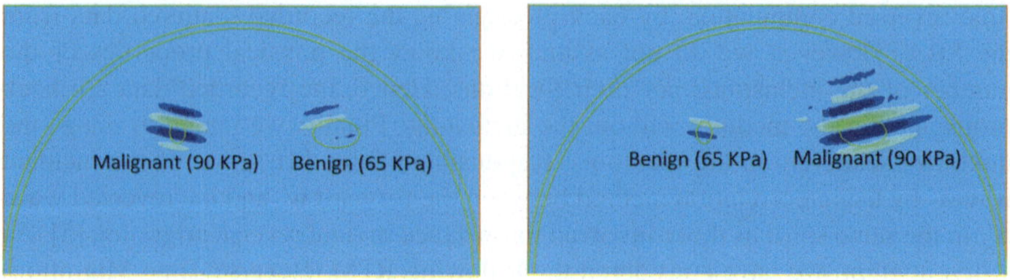

Fig. 3 RTM of Y component in case of a medium that mimics breast tissue with a skin part

Hence, the scatterers are illuminated by an incident acoustic field, that is first transmitted to the elastic medium through the interface Γ_I, and then scattered by the inclusions, before to be recorded by the SRA. The SRA being located in the fluid part, they are able to record only a scalar quantity (the pressure $p(\mathbf{x}, t)$), and not a vector velocity $\mathbf{u}(\mathbf{x}, t)$. However, as shown on images below, where the correlation image between the forward and the reversed wave is depicted (only in the elastic layer), one is able to determine the existence and location of the malignant tumor and the result is consistent also when we switch between the tumors elastic values (Fig. 3).

5 Conclusion

We proposed a time-reversal approach for acousto-elastic equations. Numerical results have been presented to illustrate the feasibility of the algorithm in a heterogeneous fluid-solid medium (breast tissue with skin), using only partial information, that is pressure recorded data in the fluid part. In a future work, more general configurations will be investigated. The quality of the obtained elasticity parameters will be also evaluated by introducing different cost functions, in the same spirit as what is derived for inverse problems. As usual in this context, optimization based algorithm can be necessary to achieve this part.

References

1. Assous, F., Kray, M., Nataf, F.: Time-reversed absorbing conditions in the partial aperture case. Wave Motion **49**, 617–631 (2012)
2. Assous, F., Kray, M., Nataf, F., Turkel, E.: Time reversed absorbing condition: application to inverse problems. Inverse Prob. **27**(6), 065003 (2011)
3. Bardos, C., Fink, M.: Mathematical foundations of the time reversal mirror. Asymptot. Anal. **29**, 157–182 (2002)
4. Blomgren, P., Papanicolaou, G., Zhao, H.: Super-resolution in time-reversal acoustics. J. Acoust. Soc. Am. **111**, 230–248 (2002)

5. Claerbout, J.F.: Imaging the Earth's Interior. Blackwell, Oxford (1985)
6. Clayton, R., Engquist, B.: Absorbing boundary conditions for acoustic and elastic wave equations. Bull. Seismol. Soc. Am. **67**, 1529–1540 (1977)
7. Clouet, J.-F., Fouque, J.-P.: A time-reversal method for an acoustical pulse propagating in randomly layered media. Wave Motion **25**, 361–368 (1997)
8. Fink, M., Wu, F., Cassereau, D., Mallart, R.: Imaging through inhomogeneous media using time reversal mirrors. Ultrason. Imaging **13–2**, 179–199 (1991)
9. Givoli, D., Turkel, E.: Time reversal with partial information for wave refocusing and scatterer identification. Comput. Methods Appl. Mech. Eng. **213–216**, 223–242 (2012)
10. Halpern, L.: Etudes des conditions aux limites absorbantes pour des schémas numériques relatifs a des équations hyperboliques linéaires. Ph.D Thesis, Paris VI University (1980)
11. Hecht, F.: New development in FreeFem++. J. Numer. Math. **20**(3–4), 251–265 (2012)
12. Kosmas, P., Rappaport, C.M.: Time reversal with the fdtd method for microwave breast cancer detection. IEEE Trans. Microwave Theory Tech. **53** (7), 2317–2323 (2005)
13. Larmat, C., Montagner, J.-P., Fink, M., Capdeville, Y., Tourin, A., Clévédé, E.: Time-reversal imaging of seismic sources and application to the great sumatra earthquake. Geophys. Res. Lett. **33**, 1–4 (2006)

An Agent Based Modeling of Spatially Inhomogeneous Host-Vector Disease Transmission

Isti Rodiah, Wolfgang Bock, and Torben Fattler

Abstract In this article, we consider a microscopic model for host-vector disease transmission based on configuration space analysis. Using Vlasov scaling we obtain the corresponding mesoscopic (kinetic) equations, describing the density of susceptible and infected compartments in space. The resulting system of equations can be seen as a generalization to a spatial host-vector disease model.

1 Introduction

Kinetic models in disease spread are often a starting point of theoretical studies in epidemiology. In disease spread dynamics, it is in many cases known how the infection dynamics evolve on the level of particle interaction. The modeling of spatial disease spread from a microscopic agent-to-agent model has already been done e.g. in a cancer model in [5] and in [1, 2] for a direct contact disease transmission. PDE models for a spatial host-vector disease dynamics have been proposed see e.g. [7, 9], however, to the authors' knowledge, there has not been shown yet, that these PDE models are well-defined Vlasov scaling limits arising from a particle system model on the agent-to-agent-interaction, hence the microscopic level.

Disease transmission represents the contact between host and vector in host-vector diseases. A series of different models for vector-borne diseases such as Dengue fever including stochastic and deterministic models have been proposed, see e.g. [10] and references therein. The models described above do not provide any

I. Rodiah (✉) · W. Bock
Department of Mathematics, Technomathematics Group, TU Kaiserslautern, Kaiserslautern, Germany
e-mail: rodiah@mathematik.uni-kl.de; bock@mathematik.uni-kl.de

T. Fattler
Department of Mathematics, Functional Analysis Group, TU Kaiserslautern, Kaiserslautern, Germany
e-mail: fattler@mathematik.uni-kl.de

© Springer Nature Switzerland AG 2019
I. Faragó et al. (eds.), *Progress in Industrial Mathematics at ECMI 2018*, Mathematics in Industry 30, https://doi.org/10.1007/978-3-030-27550-1_21

information about the spatial spread of a disease. In the SIR (Susceptible-Infected-Recovered) model case, an advection-diffusion equation has been identified as the limiting equation in a long term scaling limit, see [4]. Another approach in incorporating spatial information for the SIR model may also be found in [11] and recently [3]. On the macroscopic level, the models are very flexible for describing the different aspects of disease dynamics. To consider both microscopical modeling and spatial resolution, we describe the disease dynamics by means of an interacting particle system with suitable interaction potentials. Fundamental in this area is dynamics of the so-called marked configuration spaces [6]. These techniques together with a proper scaling of the microscopic system, the so-called Vlasov scaling, have recently been used to model the dynamics of cancer cells [5].

2 Microscopic Model

In [8], the host-vector disease transmission is modeled via marked configuration spaces. The *configuration space* Γ *over* $\mathbb{R}^2$ is defined by

$$\Gamma := \Gamma_{\mathbb{R}^2} := \left\{\gamma \subset \mathbb{R}^2 \,\middle|\, \#(\gamma \cap K) < \infty \text{ for all } K \subset \mathbb{R}^2 \text{ compact}\right\},$$

where $\#A$ denotes the cardinality of a set A. Given four copies of the space Γ, denoted by Γ^S, Γ^I, Γ^U, and Γ^V, let

$$\Gamma^4 := \left\{\vec{\gamma} := (\gamma^S, \gamma^I, \gamma^U, \gamma^V) \in \Gamma^S \times \Gamma^I \times \Gamma^U \times \Gamma^V \,\middle|\, \gamma^i \cap \gamma^j = \varnothing,\ i \neq j\right\}.$$

The model hence consists of four compartments gives as susceptible hosts (S), infected hosts (I), susceptible vectors (U), and infected vectors (V). We set up the model in the evolution of the aforementioned four-component system in the state space Γ^4. The dynamics of host and vector are described by SIS and UV model.

For a specification of the infection rates, we use a potential depending on individual to individual distance

$$[0, \infty) \ni r \mapsto \phi_R(r) := \phi(r) \in [0, \infty),$$

with $R \in (0, \infty)$. One example of the potential is shown in Fig. 1. We set $\beta_h \in [0, 1]$ to be the risk of infection for a susceptible host to be in direct contact with an infected vector and $\beta_v \in [0, 1]$ to be the risk of infection for a susceptible vector to be in direct contact with an infected host. For fixed $x \in \gamma^S$, the infection rate for a single susceptible host located at x in the surrounding $\gamma^V \in \Gamma^V$ is given by $c_h(x, \gamma^V)$. For fixed $\tilde{x} \in \gamma^U$, the infection rate for a single susceptible vector

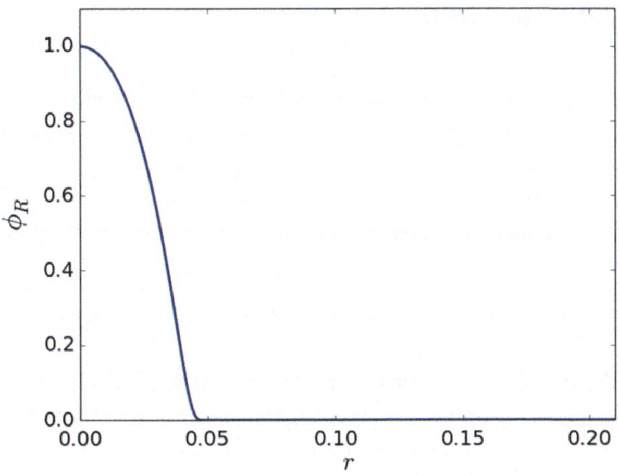

Fig. 1 The potential with $R = 0.05$

located at $\tilde{x}$ in the surrounding $\gamma^I \in \Gamma^I$ is given by $c_v(\tilde{x}, \gamma^I)$. The infection rates are the following formula:

$$c_h(x, \gamma^V) = \beta_h \sum_{\tilde{y} \in \gamma^V} \phi(|x - \tilde{y}|) \qquad \text{and} \qquad c_v(\tilde{x}, \gamma^I) = \beta_v \sum_{y \in \gamma^I} \phi(|\tilde{x} - y|)\,.$$

In host-vector disease transmission, we define a couple of generators L_h and L_v, that are the generators for host and vector, respectively. The disease dynamics are given by

$$\begin{aligned}(L_h F)(\vec{\gamma}) := &\sum_{x \in \gamma^S} c_h\left(x, \gamma^V\right)\left(F\left(\gamma^S \setminus \{x\}, \gamma^I \cup \{x\}, \gamma^U, \gamma^V\right) - F(\vec{\gamma})\right) \\ &+ \sum_{y \in \gamma^I} \alpha_h \left(F\left(\gamma^S \cup \{y\}, \gamma^I \setminus \{y\}, \gamma^U, \gamma^V\right) - F(\vec{\gamma})\right) \end{aligned} \tag{1}$$

and

$$(L_v F)(\vec{\gamma}) := \sum_{\tilde{x} \in \gamma^U} c_v\left(\tilde{x}, \gamma^I\right)\left(F\left(\gamma^S, \gamma^I, \gamma^U \setminus \{\tilde{x}\}, \gamma^V \cup \{\tilde{x}\}\right) - F(\vec{\gamma})\right), \tag{2}$$

where the function $c_h(x, \gamma^V)$ is the infection rate of host, $\alpha_h \in [0, 1]$ is the constant recovery rate of host, and the function $c_v(\tilde{x}, \gamma^I)$ is the infection rate of vector.

3 Numerical Simulation

In this section, we give a brief introduction into the numerical method which is used for simulation. The spread of the disease is modeled via a flip according to an infection rate, given via the Markov generator in Sect. 2. Since the infected individuals influence the infection rate at a certain point in the area, the computation of these rates is the main task. Briefly, the procedure of numerical implementation is as follows:

(1) Generate the state of individuals and distribute the individuals uniformly in space.
(2) Calculate the transition rate or probability for each individual.
(3) Generate random variable, then compare it with the transition rate. If the random variable is smaller than the transition rate, the state of the individual is changed.

We consider the area $[0, 1] \times [0, 1] \subset \mathbb{R}^2$ with $\beta_h = 0.1$, $\beta_v = 0.2$, and $\alpha_h = 0.14$. Figure 2 shows a spatial distribution of hosts and vectors evolving in time. In this particular case, we have the initial number of infectious hosts $I(0) = 20$, susceptible hosts $S(0) = 2480$, infectious vectors $V(0) = 0$, and susceptible vectors $U(0) = 400$. Susceptible hosts are depicted as black spots, infected host as red spots, susceptible vectors as blue spots and infected vectors as green spots.

3.1 *Infection and Recovery*

3.1.1 Comparison of Particle and Deterministic SISUV Model

In the particle model, we consider specific infection rates depending on the surrounding. However just considering the number of incidents should lead to an ODE system for a high number of particles. On the other hand, the standard ODE model assumes a uniform distribution of all particles from the beginning and neglects all behavior coming from spatial effects. Here, we compare the SISUV ODE system with the particle system for spatial uniformly distributed

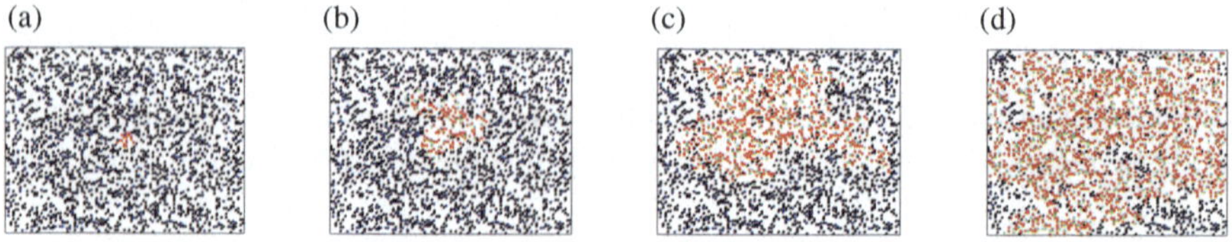

Fig. 2 SISUV model of 2,500 host and 400 vectors included infection and recovery. (**a**) $t = 0$. (**b**) $t = 2$. (**c**) $t = 5$. (**d**) $t = 10$

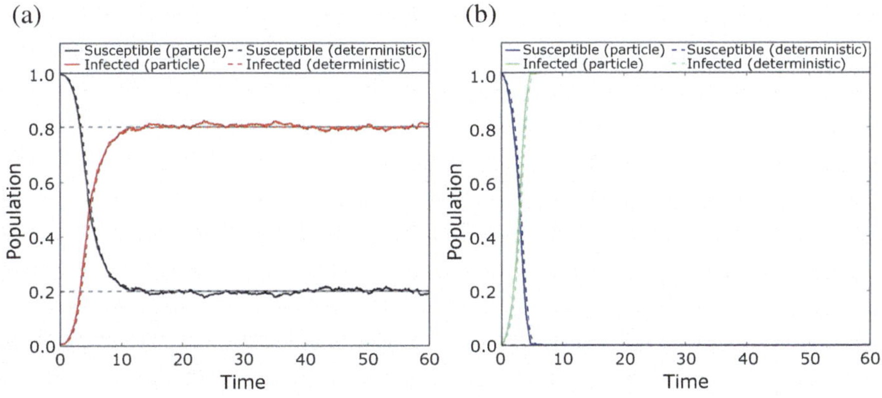

Fig. 3 The deterministic model and the particle model of 2500 host and 400 vectors with $\phi = 1$. (**a**) Host. (**b**) Vector

particles for $\phi = 1$ and $\phi = \phi_R$ from the previous section (the infection is localized).

We choose $\beta_h = 0.001$, $\beta_v = 0.002$, and $\alpha_h = 0.1$. First, we consider a constant potential of infection $\phi = 1$, i.e. a susceptible host (vector) interacts with all infected vectors (hosts) via the same rate of infection. Figure 3 shows that the particle model is in good agreement to the classical SISUV model given by the ODE system

$$\frac{d}{dt}S(t) = -\beta_h\, S(t)\, V(t) + \alpha_h\, I(t), \qquad \frac{d}{dt}I(t) = \beta_h\, S(t)\, V(t) - \alpha_h\, I(t),$$

$$\frac{d}{dt}U(t) = -\beta_v\, U(t)\, I(t), \qquad \frac{d}{dt}V(t) = \beta_v\, U(t)\, I(t),$$

where in this case $\beta_h = 0.001$, $\beta_v = 0.002$, and $\alpha_h = 0.1$.

Then, we consider the particle model with a potential of infection ϕ_R i.e., infections are possible just if susceptible host (vector) and infectious vectors (hosts) are sufficiently close to each other. Figure 4 shows that the dynamics of the particle model is "slower" than the classical SISUV model. This is due to the fact that individuals just interact locally in the particle model, while they are assumed to interact globally in the classical SISUV model. In both cases, the different simulations in Figs. 3 and 4 show the same qualitative behavior, although having different infection radius leads to different infection rates.

3.1.2 Comparison of Particle Model and Kinetic Equation

We compare the kinetic equation with averaged runs of the particle simulation. For this purpose, we choose $\beta_h = 0.1$, $\beta_v = 0.2$, $\alpha_h = 0.14$, $N = 10{,}000$, and

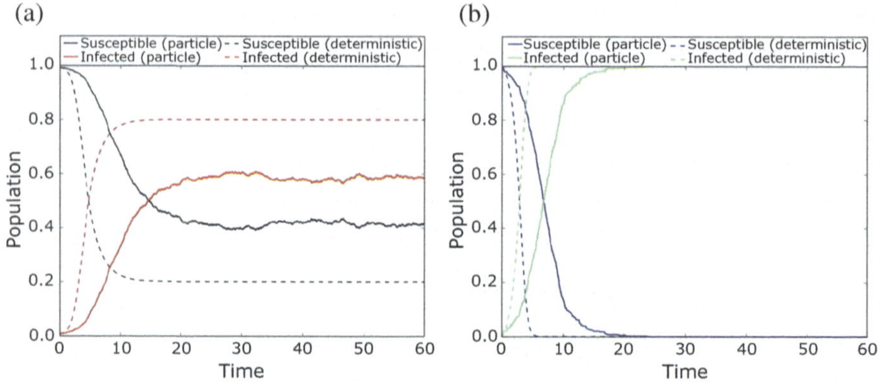

Fig. 4 The deterministic model and the particle model of 2500 host and 400 vectors with ϕ_R. (**a**) Host. (**b**) Vector

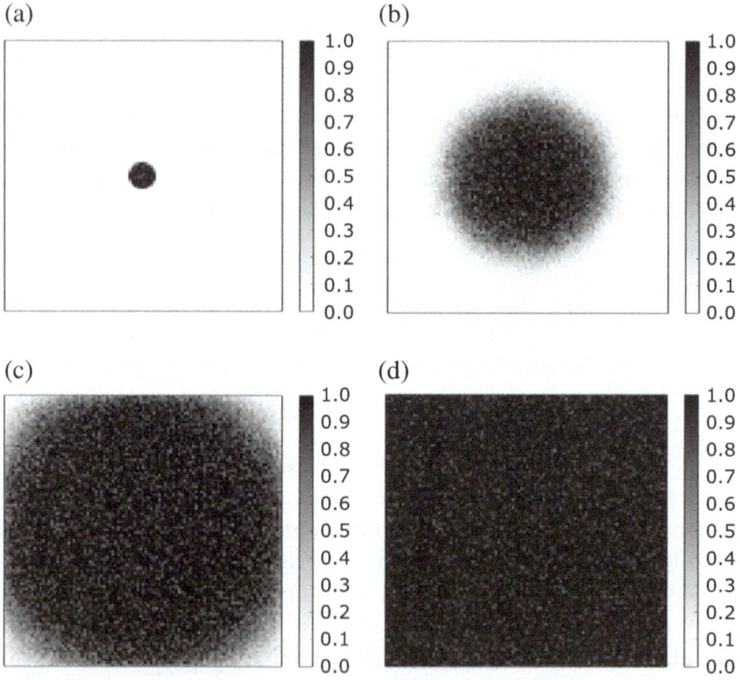

Fig. 5 Average of a hundred runs of the particle model of 10,000 hosts and 900 vectors for the infected state of host. (**a**) $t = 0$. (**b**) $t = 0.5$. (**c**) $t = 1$. (**d**) $t = 2$

$M = 900$. For the kinetic system, we consider an equidistant spatial distribution of particles. For the simulation, we use an initially random uniform spatial distribution of particles. Figure 5 shows the spatial distribution of averaged runs of the particle

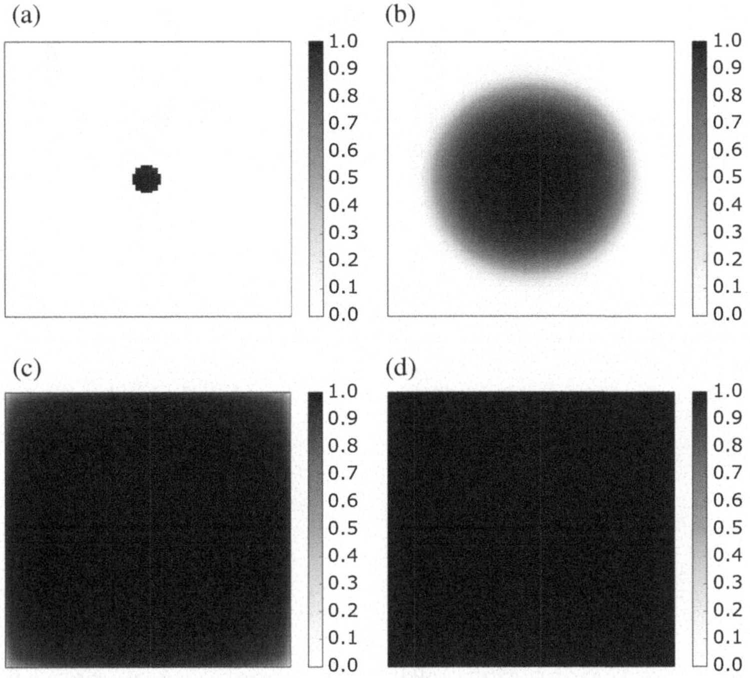

Fig. 6 Numerical solution of the kinetic equation for the infected state of host. (**a**) $t = 0$. (**b**) $t = 0.5$. (**c**) $t = 1$. (**d**) $t = 2$

model. We partition the domain $[0, 1] \times [0, 1]$ in 10,000 sub-domains. The kinetic equation is solved via a standard finite differences method with $\Delta x = \frac{1}{100}$ and $\Delta t = 0.01$. Figure 6 shows the spatial solution of the kinetic equations. Figure 7 shows the difference between the average of the particle model and the kinetic equation in each sub-domains. The comparison between the dynamics in the kinetic and the particle approximation is shown in Fig. 8.

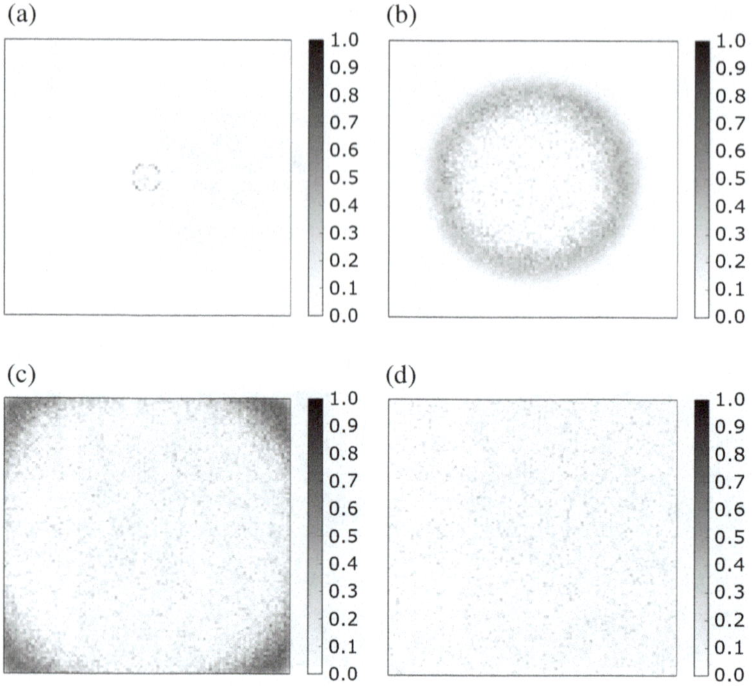

Fig. 7 Difference between the average of the particle model and the kinetic system for the infected state of host in each sub-domians. (**a**) $t = 0$. (**b**) $t = 0.5$ (**c**) $t = 1$. (**d**) $t = 2$

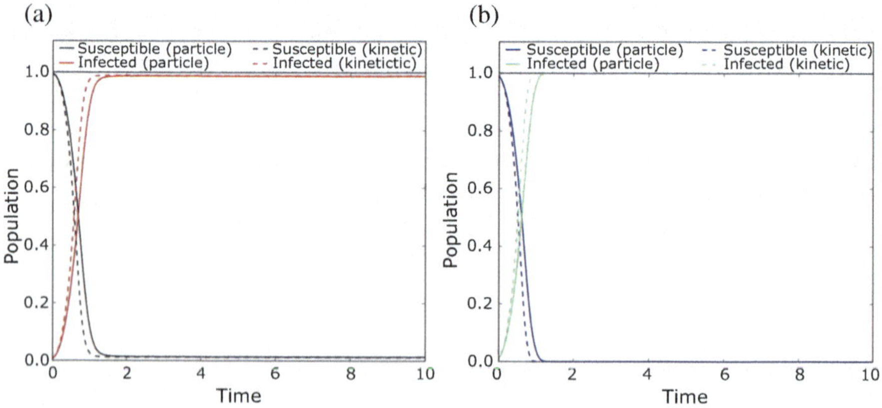

Fig. 8 Average of a hundred runs of the particle model and the kinetic system. (**a**) Host. (**b**) Vector

Acknowledgement We thank the reviewer for his comments which significantly improved the work.

References

1. Bock, W., Fattler, T., Rodiah, I., Tse, O.: An analytic method for agent based modeling of disease spreads. In: Proceedings of the 8th Jagna International Workshop (2017)
2. Bock, W., Fattler, T., Rodiah, I., Tse, O.: Numerical evaluation of agent based modeling of disease spreads. In: Proceedings of the 8th Jagna International Workshop (2017)
3. Bock, W., Jayathunga, Y.: Math. Methods Appl. Sci. **41**, 3232–3245 (2018)
4. Chalub, F., Souza, M.O.: Math. Comput. Modelling **53**, 1568–1574 (2011)
5. Finkelshtein, D.L., Friesen, M., Hatzikirou, H., Kondratiev, Y.G., Krüger, T., Kutoviy, O.: Interdisciplinary Studies of Complex Systems **7**, 5–85 (2015)
6. Finkelshtein, D.L., Kondratiev, Y.G., Oliveira, M.J.: Rep. Math. Phys. **71**, 123–148 (2013)
7. Rao, V.S.H., Durvasula, R.: Dynamic Models of Infectious Diseases, vol. 1. Vector-Borne Diseases. Springer, New York (2013)
8. Rodiah, I.: An agent based modeling disease transmission in the framework configuration spaces. Ph.D. Thesis (2018)
9. Rocha, F., Aguiar, M., Souza, M., Stollenwerk, N.: Proceedings 12th International Conference on Computational and Mathematical Methods in Science and Engineering, vol. 3, pp. 1047–1062 (2012)
10. Rocha, F., Mateus, L., Skwara, U., Aguiar, M., Stollenwerk, N.: Int. J. Comput. Math. **93**, 1405–1422 (2016)
11. Schmidtchen, M., Tse, O., Wackerle, S.: A multiscale approach for spatially inhomogeneous disease dynamics. arXiv:1602.05927 (2016)

Modelling Dengue with the SIR Model

Peter Heidrich and Thomas Götz

Abstract Severe dengue outbreaks and their consequences point out the need for prognosis and control methods which can be derived by epidemiological mathematical models. In this article we develop a model to describe observed data on hospitalized dengue cases in Colombo (Sri Lanka) and Jakarta (Indonesia). Usually, the disease is epidemiologically modelled with the *SIRUV* model consisting of the susceptible (S), infected (I) and recovered humans (R) and the uninfected (U) and infected (V) female mosquitos. Because we do not have any information about the mosquito population we reduce the model to a *SIR* model which depends on a time-dependent transmission rate $\beta(t)$ and fit it to the received data sets. To solve this, optimal control theory constructed on Pontryagin's maximum (minimum) principle is applied in order to reach the solution with numerical optimization methods. The results serve as a basis for different simulations.

1 Introduction

Severe dengue outbreaks and their consequences point out the need for prognosis and control methods which can be derived by epidemiological mathematical models. Dengue is classified as a fast emerging viral disease which occurs in over 100 tropical and subtropical endemic countries every year—especially in South East Asia, Latin America and the Western Pacific. The dengue virus is categorized in four distinct serotypes (DEN 1–4). Once infected with the virus a severe flu-like infection or in some cases a severe dengue (dengue haemorrhagic fever) may occur. In severe course of the disease dengue fever can lead to death. The disease is a mosquito-borne viral infection which is transmitted by vectors like the *Aedes aegypti*. The female mosquito absorbs the virus while feeding on the blood of an infected human. When the infected mosquito bites an uninfected human the virus can be transmitted.

P. Heidrich (✉) · T. Götz
Mathematical Institute, University Koblenz-Landau, Koblenz, Germany
e-mail: heidrich@uni-koblenz.de; goetz@uni-koblenz.de

© Springer Nature Switzerland AG 2019

I. Faragó et al. (eds.), *Progress in Industrial Mathematics at ECMI 2018*,
Mathematics in Industry 30, https://doi.org/10.1007/978-3-030-27550-1_22

Thus, the human functions as a carrier and multiplier of the virus. A transmission is followed by an incubation time of 4–10 days. Once infected, the virus is located 2–7 days in the blood. Meanwhile the patient shows the symptoms and can transmit the virus in a period of maximum 12 days to an uninfected mosquito. The recovery from the infection caused by one serotype of the virus provides lifelong immunity against this specific serotype. However, a subsequent infection with another serotype increases the risk of a severe dengue. The transmission of the disease depends on the living conditions for the vectors which are influenced by regional rainfall, temperature, humidity and the degree of urbanization. The *World Health Organisation (WHO)* hypothesizes that approximately 50–100 million infections occur every year whereby latest estimates are at 390 million infected humans of which only approximately one fourth is hospitalized or registered [6]. By private communication we received data sets of dengue cases in *Colombo (Sri Lanka)* and *Jakarta (Indonesia)* from the local *Departments of Mathematics* [1, 4]. Usually, the disease is modelled with the *SIRUV* model consisting of the *susceptible (S), infected (I)* and *recovered (R) humans* and the *uninfected (U)* and *infected (V) female mosquitos*. Because we do not have any information about the mosquito population we reduce the model to a *SIR* model applying the findings of Rocha et al. [2]

$$\begin{aligned}\frac{dS}{dt} &= \mu(N-S) - \frac{\beta(t)}{N}SI\\ \frac{dI}{dt} &= \frac{\beta(t)}{N}SI - (\alpha+\mu)I\\ \frac{dR}{dt} &= \alpha I - \mu R.\end{aligned}$$

The system is reduced from five to three ordinary differential equations (ODEs) and depends on a time dependent *transmission rate* $\beta(t)$. In order to fit the parameters of the model to the received data sets we implement an objective function

$$J(u) = \int_0^T \left(I(t) - I^d(t)\right)^2 dt + \frac{\|u\|^2}{N^2}$$

which shall be minimized with respect to u. The results serve as a basis for two numerical simulations concerning the behaviour of the dengue outbreaks.

2 Data Analysis

The available data consists of the weekly hospitalized dengue cases in the *Colombo City District* and the *Special Capital Region of Jakarta*. To reduce the noise in the data we smoothen it with a moving average. Each data point d_i is replaced by

$\bar{d}_i = \frac{1}{4}\sum_{k=0}^{3} d_{i-k}$ for all $i \geq 3$. In both cases a periodical behaviour with varying intensities concerning the peaks can be recognized. In Colombo we observe half-yearly repeating outbreaks in the midyear and at the turn of the year, the dengue outbreaks in Jakarta appear yearly in the first quarter. The results of the fast Fourier transform (FFT) underpin these observations since significant high values at two frequencies per year in Colombo and one frequency per year in Jakarta can be noticed.

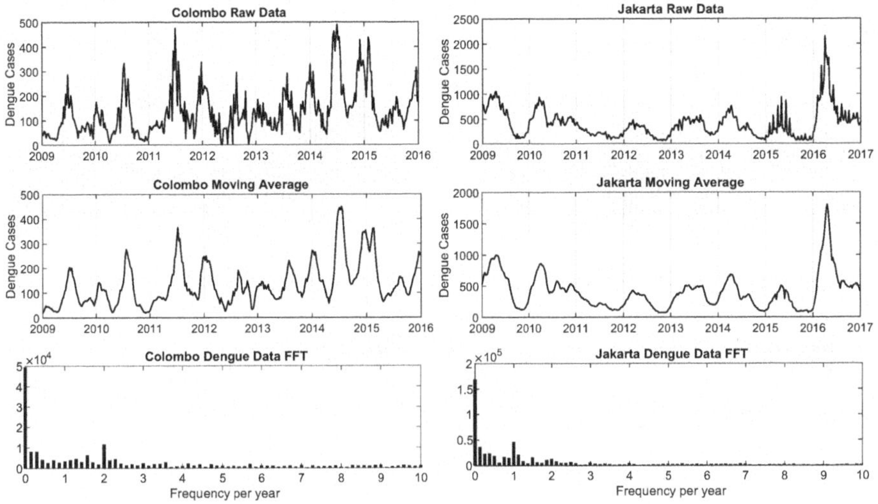

It is assumed that this behaviour relates to the weather conditions especially the precipitation, because the vectors of the disease need small amounts of standing water to lay their eggs in. We apply the FFT on the appropriate rainfall data sets and recognize that their periodical behaviour fit to the dengue data.

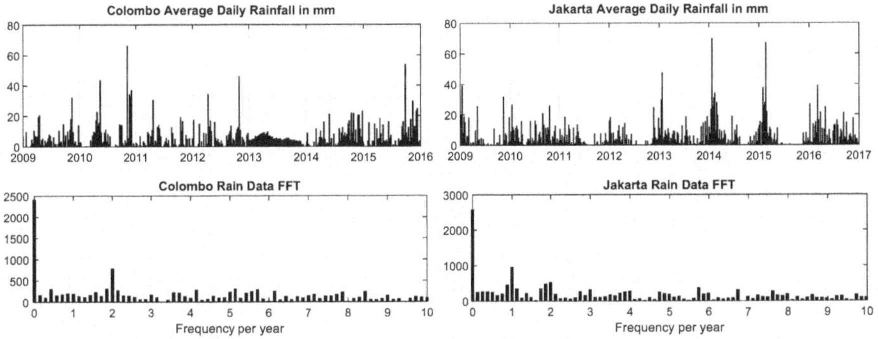

To substantiate the relation between rainfall and dengue data we use a cross-correlation and finally receive significant high values at time lags between 6 to 10 weeks. Consequently, this means that after an intensive rain period it takes approximately 2 months until the dengue cases significantly rise in the cases of Colombo and Jakarta. The clusters between precipiation and dengue data additionally show that if the average daily rainfall is stronger than approximately 15 to 20 mm a day, less dengue data points appear. Thus, we assume that in periods of very strong rainfall the eggs of the mosquitos are destroyed or washed away so that the reproduction of the vectors is restricted. In the following this border will be called *cut-off*.

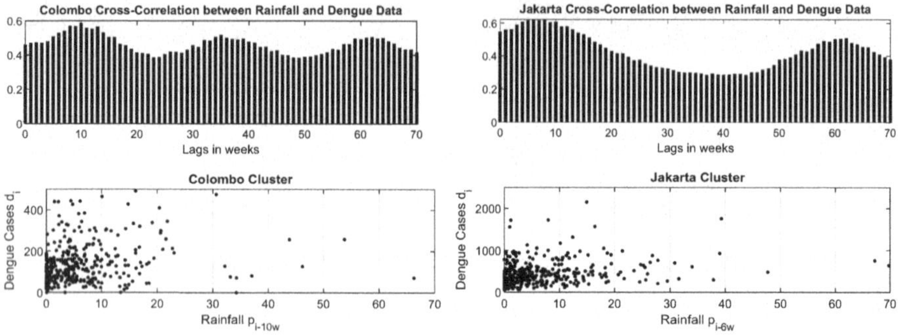

3 The SIR Model

The present *SIR* Model includes the three usual groups of *susceptible (S), infected (I)* and *recovered (R) individuals*:

$$\frac{dS}{dt} = \mu(N - S) - \frac{\beta(t)}{N} SI \tag{1}$$

$$\frac{dI}{dt} = \frac{\beta(t)}{N} SI - (\alpha + \mu) I$$

$$\frac{dR}{dt} = \alpha I - \mu R$$

$$N = S_0 + I_0 + R_0$$

$$0 \leq S_0, I_0, R_0.$$

The *total population* N is assumed to be constant because of the short time period. Consequently, the *birth* and *death rate* are equal and named with μ. The transition from infected to recoverd individuals depends on the *recovery rate* α. We omit the explicit mosquito dynamics of *uninfected* (U) and *infected* (V) vectors and use a time-dependent *transmission rate* $\beta(t)$ instead.

Simulation 1	Simulation 2
$\beta(t) = \beta_0 + \beta_1 \cos(\omega t)$	$\beta(t) = \beta_0 + \beta_1 \int_{t-\frac{\tau_2}{52}}^{t-\frac{\tau_1}{52}} p_c(\xi)d\xi \cdot \sin\left(\omega\left(t + \frac{\phi}{52}\right)\right)$

Here β_0 stands for the *average transmission rate* and β_1 for the *degree of periodical variation*. In simulation 2 a *phase-shift* ϕ is additionally included and β_1 is multiplicated with an integral of the precipitation function p_c. It is defined by

$$p_c(\xi) = \begin{cases} p(\xi), & p(\xi) < c \\ 0, & p(\xi) \geq c. \end{cases} \tag{2}$$

The continuously differentiable function $p(\xi)$ includes the rainfall data points p_i and c represents the *cut-off*. The interval $[t - \frac{\tau_2}{52}, t - \frac{\tau_1}{52}]$ is set around the time lag between precipitation and dengue data. In the case of Colombo the time lag is 10 weeks, therefore $[t - \frac{12}{52}, t - \frac{8}{52}]$ is a possible choice. To fit the model to the dengue data we solve the optimization problem

$$\min_u J(u) = \min_u \int_0^T \left(\gamma I(t) - I^d(t)\right)^2 dt + \frac{\|u\|^2}{N^2} \tag{3}$$

subject to (1). Because it is assumed that only a fraction of infected individuals are hospitalized we establish γ as *hospitalization rate*. The continuous function $I^d(t)$ includes the dengue data points d_i and u consists of the parameters that shall be fitted.

	Fitted parameters	Fixed parameters
Simulation 1	$u = (\beta_0, \beta_1, S_0, I_0, R_0)'$	$N, \mu, \alpha, \omega, \gamma$
Simulation 2	$u = (\beta_0, \beta_1, c, \tau_2, \phi, \gamma, S_0, I_0, R_0)'$	$N, \mu, \alpha, \omega, \tau_1$

The integral in $J(u)$ is based on a L^2 norm so that its minimization corresponds to a least squares method. Additionally we add a *regularization term* $\frac{\|u\|^2}{N^2}$. Its size is much smaller than the size of the integral therefore $\int_0^T \left(\gamma I(t) - I^d(t)\right)^2 dt$ dominates the minimization algorithm which is decisive for the biological context. The addition with this convex and radially unbounded regularization term has an analytical background because otherwise some parameters would disappear in the *gradient* and consequently the corresponding columns and rows in the *Hessian matrix* would be equal to zero. Thus, it would be difficult to calculate and categorize *critical points*. In a way this corresponds to a *Tikhonov regularization* [5]. The division by the size of the total population N is caused by the fact that the transmission rate $\beta(t)$ is divided by N in the SIR model and the investigation of the initial conditions S_0, I_0 and R_0 in relation to N is useful. In order to optimize (3) with *Pontryagin's maximum (minimum) principle* we introduce a *Lagrange function*

$$\mathcal{L}(u, x, \lambda) = \int_0^T \left(\gamma I(t) - I^d(t)\right)^2 dt + \frac{\|u\|^2}{N^2} + \int_0^T \left\langle \lambda(t), g(u, x(t), t) - \frac{dx(t)}{dt} \right\rangle dt$$

where $\lambda = (\lambda_S, \lambda_I, \lambda_R)'$ includes the *adjoint functions*, $x = (S, I, R)'$ consists of the *state variables*, $g = (g_S, g_I, g_R)'$ symbolizes the right terms of the ODEs in (1) and $\langle \cdot, \cdot \rangle$ stands for the scalar product. The *necessary optimality condition* for a minimum (u^*, x^*, λ^*) is fullfilled if $\nabla \mathcal{L}(u^*, x^*, \lambda^*) = 0$ holds. Solving $\frac{\partial \mathcal{L}}{\partial x_i} = 0$ via Gâteaux derivative delivers the adjoint ODEs

$$\frac{d\lambda_S}{dt} = \left(\mu + \frac{\beta(t)}{N} I\right) \lambda_S - \frac{\beta(t)}{N} I \lambda_I$$

$$\frac{d\lambda_I}{dt} = \frac{\beta(t)}{N} S \lambda_S + \left((\alpha + \mu) - \frac{\beta(t)}{N} S\right) \lambda_I - \alpha \lambda_R - 2\gamma \left(\gamma I - I^d\right)$$

$$\frac{d\lambda_R}{dt} = \mu \lambda_R$$

$$0 = \lambda_S(T), \lambda_I(T), \lambda_R(T)$$

and $\frac{\partial \mathcal{L}}{\partial \lambda_i} = 0$ leads to the ODEs in (1). In simulation 2 the gradient of $\mathcal{L}$ respect to u is given by

$$\frac{\partial \mathcal{L}}{\partial u_i} = u_i \frac{2}{N^2} + \frac{1}{N} \int_0^T \frac{\partial \beta(t)}{\partial u_i} \left(\lambda_I(t) - \lambda_S(t)\right) S(t) I(t) dt \qquad i \in \{1, \ldots, 5\}$$

$$\frac{\partial \mathcal{L}}{\partial u_6} = \gamma \frac{2}{N^2} + 2 \int_0^T I(t) \left(\gamma I(t) - I^d(t)\right) dt$$

$$\frac{\partial \mathcal{L}}{\partial u_7} = S_0 \frac{4}{N^2} + R_0 \frac{2}{N^2} - \frac{2}{N} + \lambda_S(0) - \lambda_I(0)$$

$$\frac{\partial \mathcal{L}}{\partial u_9} = R_0 \frac{4}{N^2} + S_0 \frac{2}{N^2} - \frac{2}{N} + \lambda_R(0) - \lambda_I(0).$$

u_8 is calculated by the substitution $I_0 = N - S_0 - R_0$. The *conjugate gradient method* combined with the *forward-backward sweep method* is applied to solve the optimization problem numerically until $\|J(u_{i+1}) - J(u_i)\| < 10^{-9}$ holds. [3]

4 Results

In both simulations a time-scale t in years is applied. The values of the fixed parameters N, μ and α are extracted from statistics of the *WHO* [6]. The timing of the peaks fits to the behaviour in the data sets especially in simulation 2 because of the phase shift ϕ. In Jakarta the model maps the relation between the yearly peaks whereby the inclusion of the rain data allows a more accurate dynamical behaviour. In Colombo the half-yearly varying oscillation proves more difficult to be reproduced though, the adding of the precipitation again improves the dynamics of the model. Comparing the absolute values of the fitted parameters in both locations we determine that similar results are achieved.

Jakarta	β_0	β_1	c	τ_2	γ	ϕ	S_0	I_0	R_0	N	μ	α	τ_1	ω
Sim. 1	38,6	6,0	/	/	1	/	$6,6 \cdot 10^6$	$6,0 \cdot 10^2$	$3,4 \cdot 10^6$	10^7	1/69	26	/	2π
Sim. 2	51,6	14,7	17,0	9,0	0,45	9,2	$4,8 \cdot 10^6$	$1,2 \cdot 10^3$	$5,2 \cdot 10^6$	10^7	1/69	26	4	2π

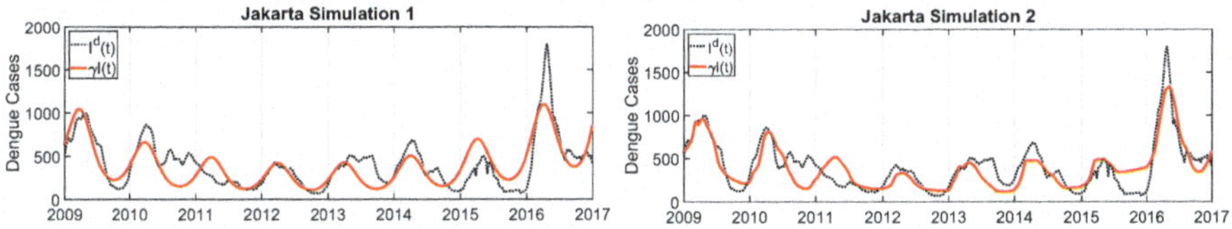

Colombo	β_0	β_1	c	τ_2	τ	ϕ	S_0	I_0	R_0	N	μ	α	τ_1	ω
Sim. 1	26,7	4,6	/	/	1	/	$1,3 \cdot 10^6$	$5,0 \cdot 10^1$	$1,5 \cdot 10^4$	$1,3 \cdot 10^6$	1/75	26	/	4π
Sim. 2	37,2	−10,0	15,0	14,00	0,44	−4,0	$9,1 \cdot 10^5$	$2,1 \cdot 10^2$	$3,9 \cdot 10^5$	$1,3 \cdot 10^6$	1/75	26	8	4π

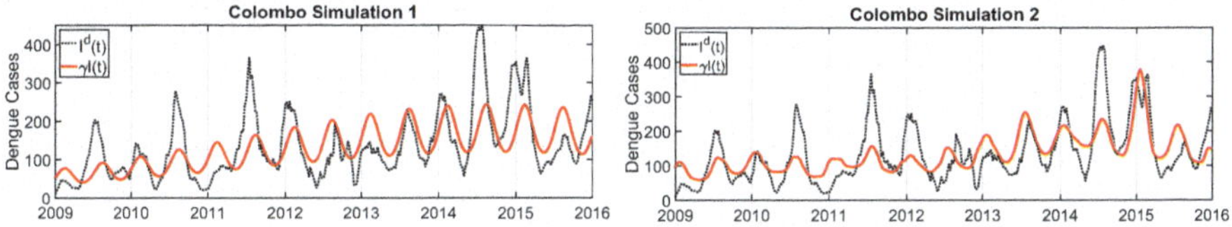

Acknowledgement Peter Heidrich wants to thank the research fund of the University Koblenz-Landau for supporting the participation in the ECMI 2018 conference.

References

1. Aldila, D.: Private Communication, Department of Mathematics. University of Indonesia, Jakarta (2017)
2. Aguiar, M., Rocha, F., Souza, M., Stollenwerk, N.: Time-scale separation and center manifold analysis describing vector-borne disease dynamics. Int. J. Comput. Math. **90**, 2105–2125 (2013)
3. Lenhart, S., Workman, J.T.: Optimal Contol Applied to Biological Models. CRC Press, Boca Raton (2007)

4. Perrera, S.: Private Communication, Department of Mathematics. University of Colombo, Sri Lanka (2016)
5. Tikhonov, A.N., Goncharsky, A., Stepanov, V.V., Yagola, A.G.: Numerical Methods for the Solution of Ill-Posed Problems. Springer, Amsterdam (2013)
6. World Health Organization. http://who.int (2018). Accessed 31 Aug 2018

Mathematical Modeling for Laser-Induced Thermotherapy in Liver Tissue

Norbert Siedow and Christian Leithäuser

Abstract Laser-induced thermotherapy (LITT) plays an important role in oncology to treat human liver tumors. LITT is a minimally invasive method causing tumor destruction due to heat ablation and coagulative effects of the tissue. Tumor tissue is much more sensitive to heat than normal healthy tissue. The big advantage of the LITT compared to other minimally invasive procedures, such as microwave ablation or radiofrequency therapy, is that the treatment primarily takes place under MRI control, such that patients are exposed to a small radiation dose. The present paper describes the mathematical modeling of laser-induced thermotherapy and shows simulation results for porcine liver.

1 Introduction

The aim of thermal ablation methods is to destroy cancer tissue by generating cytotoxic temperature for a short time without damaging vital tissue. These methods are minimally invasive and used for treating for example lunge, liver, and prostate cancer, when surgical resection is not possible or too dangerous for the patient. Tumor tissue is much more sensitive to heat than normal healthy tissue. Most proteins denature at 40–42 °C. Irreversible coagulation necrosis occurs in the temperature range of 60–100 °C. Temperatures above 150 °C result in vaporization and carbonation. This leads to an undesirable reduced thermal conductivity, and heat can not penetrate further into the tissue.

The most popular thermal ablation methods are the radiofrequency ablation technique, the LITT, and the microwave ablation.

The principle of LITT is based on the local introduction of energy via an optical fiber directly into the cancerous tissue. The laser fiber is located in a water-cooled, MR-compatible applicator. The introduction of the applicator into the tumor is done

N. Siedow (✉) · C. Leithäuser
Faunhofer-Institute for Industrial Mathematics, Kaiserslautern, Germany
e-mail: norbert.siedow@itwm.fraunhofer.de; christian.leithaeuser@itwm.fraunhofer.de

© Springer Nature Switzerland AG 2019
I. Faragó et al. (eds.), *Progress in Industrial Mathematics at ECMI 2018*,
Mathematics in Industry 30, https://doi.org/10.1007/978-3-030-27550-1_23

under the CT, while the actual treatment takes place under MRI control. Thus, the patient is exposed to only a small dose of radiation during LITT. An additional advantage is the possibility to use MR-Thermometry for active image control. MR-Thermometry methods are based on MR measured parameters depending on temperature like the longitudinal relaxation time (T_1), the diffusion coefficient (D), or the proton resonance frequency (PRF) of tissue water. The linear temperature dependence of the proton resonance frequency and its near-independence with respect to tissue type make the PRF-based methods the preferred choice for many application. For a more deeper understanding to MR-Thermometry we refer to the review paper [1].

In the following we discuss the mathematical modeling of the LITT, and compare simulation results with temperature maps from MR-Thermometry.

2 Mathematical Modeling

Let $\Omega \subset \mathbb{R}^3$ denote the geometry of the liver, which is obtained from MRI through segmentation. The boundary Γ of Ω consists of the radiating part of the adjacent applicator Γ_{rad}, which is not part of the liver, the cooled part of the applicator Γ_{cool}, and the surface of the liver Γ_{amb} (see Fig. 1). The mathematical model is described by a system of partial differential equation for the heat transfer inside the liver, the radiative transfer from the applicator into the liver tissue, and a damage function. [2–4]

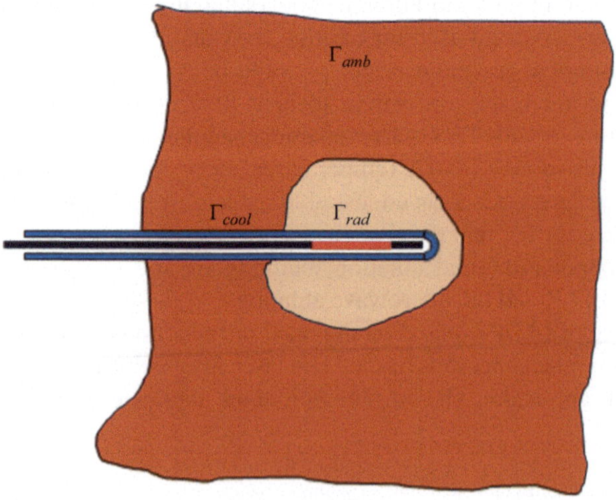

Fig. 1 Water cooled applicator with radiating laser fiber

2.1 The Temperature

The heat transfer in the liver is modeled by the well-known bio-heat equation

$$c_p \rho \frac{\partial T}{\partial t}(x,t) = \nabla \cdot (k_h \nabla T(x,t)) + \xi_b (T_b - T(x,t)) + Q_{rad},$$
$$T(x,0) = T_{init}(x), \tag{1}$$

where $T(x,t)$ denotes the temperature depending on the three-dimensional position x and the time t. c_p is the thermal conductivity, ρ the density, k_h the thermal conductivity, and ξ_b the perfusion rate due to the blood flux. T_b denote the blood temperature and Q_{rad} the energy source term due to the irradiation of the laser fiber. The initial temperature is assumed to be known $T_0(x)$.

For the heat transfer between liver and applicator and with the surroundings Robin type boundary conditions are used.

$$k_h \frac{\partial T}{\partial n} = \alpha_{cool}(T_{cool} - T), \ x \in \Gamma_{rad} \cup \Gamma_{cool},$$
$$k_h \frac{\partial T}{\partial n} = \alpha_{amb}(T_{amb} - T), \ x \in \Gamma_{amb}. \tag{2}$$

Here n is the outer normal vector, α_{cool} and α_{amb} the heat transfer coefficients with the cooling part of the applicator and the surroundings of the liver, respectively. T_{cool} denotes the cooling temperature and T_{amb} the ambient temperature.

The source term in (1) is given by

$$Q_{rad} = \frac{\mu_a}{4\pi} \int_{S^2} I(s,x) ds = \mu_a \phi(x), \tag{3}$$

where μ_a is the absorption coefficient and $\phi(x)$ the radiative energy defined as the integral of the radiative intensity $I(s,x)$ over all directions s of the whole sphere S^2.

2.2 The Radiative Transfer

The irradiation of laser light is described by the radiative transfer equation

$$s \cdot \nabla I(s,x) + (\mu_a + \mu_s) I(s,x) = \frac{\mu_s}{4\pi} \int_{S^2} P(s \cdot s') I(s', x) ds', \tag{4}$$

with the absorption and scattering coefficients, μ_a and μ_s, the scattering phase function $P(s \cdot s')$ given by the Henyey-Greenstein term

$$P(s \cdot s') = \frac{1 - g^2}{(1 + g^2 - 2g(s \cdot s'))^{3/2}}. \tag{5}$$

g is the so-called anisotropy factor. $g = 0$ describes the isotropic and $g = 1$ the anisotropic scattering. The boundary condition is given by

$$I(s,x) = F,\ x \in \Gamma_{rad}, \quad I(s,x) = 0,\ x \notin \Gamma_{rad}, \tag{6}$$

where F is an energy density defined later.

Because of the high dimension of the radiative transfer equation (4) we use the so-called P_1-approximation to approximate (4). Introducing the ansatz

$$I(s,x) = \phi(x) + 3s \cdot q(x),$$

where $q(x) = \frac{1}{4\pi} \int_{S^2} I(s,x) s ds$ is radiative flux vector, one obtains the much simpler three-dimensional diffusion equation

$$-\nabla \cdot (D \nabla \phi(x)) + \mu_a \phi(x) = 0, \quad D = \frac{1}{3(\mu_a + (1-g)\mu_s)}. \tag{7}$$

To approximate the boundary condition (6) we use the Marshak's procedure described for instance in [5]. We obtain Robin type boundary conditions

$$D \frac{\partial \phi}{\partial n}(x) = \frac{q_{app}}{A_{\Gamma_{rad}}},\ x \in \Gamma_{rad}, \quad D \frac{\partial \phi}{\partial n}(x) + b\phi(x) = 0,\ x \notin \Gamma_{rad}, \tag{8}$$

where q_{app} is the energy delivered by the laser and $A_{\Gamma_{rad}}$ the surface area of the radiating part of the fiber. The parameter $b = 0.5$ for $x \in \Gamma_{amb}$ and $b = 0$ for $x \in \Gamma_{cool}$. From the numerical point of view (7) and (8) is much easier to compute than the original system (4) and (6).

2.3 The Damage Function

The damage of the liver/cancer tissue will be described by the so-called damage function. It is common (see [2, 3]) to use the Arrhenius law

$$w(x,t) = \int_0^t A e^{-E_a/RT(x,\tau)} d\tau, \tag{9}$$

with so-call frequency factor A, activating energy E_a, and ideal gas constant R to describe the change of material properties due to coagulation. The subscript n

stands for native tissue and c for the properties of the coagulated tissue. For the optical parameters we obtain

$$\begin{aligned} \mu_a &= \mu_{an} + (1 - e^{-w})(\mu_{ac} - \mu_{an}), \\ \mu_s &= \mu_{sn} + (1 - e^{-w})(\mu_{sc} - \mu_{sn}), \\ g &= g_n + (1 - e^{-w})(g_c - g_n). \end{aligned} \tag{10}$$

3 Results

The mathematical model described above was used to simulate the heating of pig porcine. The liver geometry and applicator position were obtained from segmented MR-images. The computational geometry was generated using Open Cascade (Open Cascade SAS, Guyancourt, France) and the mesh using the software code Gmesh. The differential equations, including boundary conditions, were discretised and solved using GetDP (P. Dular and C. Geuzian, University of Liege). More details can be found in the co-work [6]. The used heat and optical parameters are listed in [4]. The time-depending simulation results were compared with data from MR-Thermometry (see Fig. 2) and thermocouples placed at different positions around the laser-applicator. The simulation as well as the MR-Thermometry are in good agreement with measured data. Looking at Fig. 2 one can see the typical shape of the temperature distribution around the radiating part of the applicator.

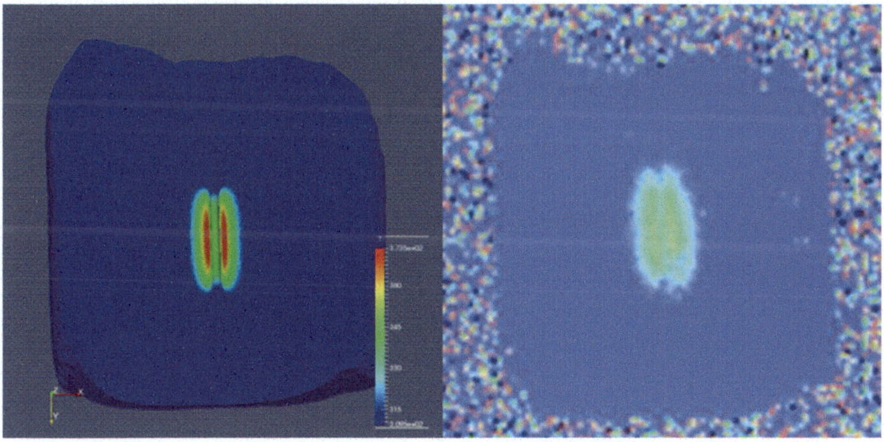

Fig. 2 Temperature simulated (left) and taken from MR-Thermometry (right)

4 Conclusions

LITT is a minimal-invasive method in the field of interventional oncology used for treating liver cancer. Mathematical modeling and computer simulation are important features for treatment planning and imaging the necrosis of the tissue. The numerical simulation is in good agreement with the MR-Thermometry and temperature measurements for porcine liver. For future work blood perfusion has to be taken into account. The blood flux of vessels and tissue has a cooling effect, which is very important for treating humans and depending on the physiology of the patient. To model these effects more research is needed.

Acknowledgements The authors acknowledge the financial support by the Federal Ministry of Education and Research of Germany in the framework of the project *proMT: Prognostische modellbasierte online MR-Thermometrie bei minimalinvasiver Thermoablation zur Behandlung von Lebertumoren* (Förderkennzeichen: 05M16AMA).

References

1. de Senneville, B.D., Quesson, B., Moonen, C.: Magnetic resonance temperature imaging. Int. J. Hyperth. Taylor Francis, **21**(6), 515–531 (2005)
2. Mohammed, Y., Verhey, J.F.: A finite method model to simulate laser interstitial thermos therapy in anatomical inhomogeneous regions. Biomed. Eng. Online **4**, 2 (2005)
3. Fasano, A., Hömberg, D., Naumov, D.: On a mathematical model for laser-induced thermotherapy. App. Math. Model. **34**, 12 (2010)
4. Hübner, F., Leithäuser, C., Bazrafshan, B., Siedow, N., Vogl, T.J.: Validation of a mathematical model for laser-induced thermotherapy in liver tissue. Lasers Med. Sci. **32**, 6 (2017)
5. Modest, M.F.: Radiative Heat Transfer. Academic Press, San Diego (2003)
6. Leithäuser, C., Hübner, F., Bazrafshan, B., Siedow, N., Vogl, T.J.: Experimental validation of a mathematical model for laser-induced thermotherapy. In: European Consortium for Mathematics in Industry. Springer, Berlin (2018)

Fiber-Based Modeling of Muscles in the Musculoskeletal System

Michael H. Gfrerer and Bernd Simeon

Abstract The aim of this contribution is to present a fiber-based modeling approach for the dynamic behavior of muscles within the musculoskeletal system. We represent the skeletal system as a rigid multi-body system which is actuated by muscles. We model each muscle as an one-dimensional cable with variable cross section undergoing large deformation and strains. In order to avoid penetration of the muscles and the skeleton, contact is considered. We use our framework to conduct a dynamic forward simulation of a simple upper limb model.

1 Introduction

The simulation of the musculoskeletal system is a common field of interest. In such simulations, the skeletal system is typically represented by a rigid multi-body system. Concerning the modeling of the muscles, in the simplest case, the muscle paths are assumed to be the straight line between the insertion points. In order to account for the physiology, the muscle paths can be enhanced by so-called 'via' or wrapping' points [1, 3, 5, 9]. Another possibility is to solve the shortest distance problem taking the constrains by the bones and other structures into account [6, 10]. In such line of action models, the muscular forces are often calculated by Hill-type muscle models [13]. Those lumped parameter models are computationally cheap but may lack realism.

More detailed models are based on nonlinear continuum mechanics. However, they have the drawback of an increased computational cost. Nevertheless, they are able to represent three-dimensional geometry and multi-scale architectures [2, 8], Furthermore, they allow for the inclusion of multi-physic effects [4, 11].

M. H. Gfrerer (✉) · B. Simeon
Felix-Klein-Zentrum für Mathematik, University of Kaiserslautern, Kaiserslautern, Germany
e-mail: gfrerer@mathematik.uni-kl.de; simeon@mathematik.uni-kl.de

© Springer Nature Switzerland AG 2019

I. Faragó et al. (eds.), *Progress in Industrial Mathematics at ECMI 2018*,
Mathematics in Industry 30, https://doi.org/10.1007/978-3-030-27550-1_24

In the present paper, we model a muscle as a three-dimensional continuum located around an one-dimensional curve in space. Due to this geometry setting we derive an one-dimensional cable model incorporating large deformations and thickness change. The total stress tensor is additively decomposed into a passive, an active and a prestress contribution. For the passive response an incompressible Kelvin-Voigt material law for finite strains is used. The active stress contribution is modeled by the relations given in [8]. We consider the coupling of the rigid-body system and the muscles at the insertion points. Furthermore, contact on the lateral surface of the muscles and the rigid bodies is incorporated.

2 Fiber-Based Model of the Muscle-Tendon Complex

We model the homogenized muscle-tendon complex by means of continuum mechanics. Thus, the motion is governed by the balance of momentum

$$\mathrm{Div}(\mathbf{FS}) + \mathbf{b}_0 = \rho_0 \ddot{\mathbf{u}}, \tag{1}$$

where $\mathbf{F}$ is the deformation gradient, $\mathbf{S}$ the second Piola-Kirchhoff stress tensor, $\mathbf{b}_0$ the volume force, ρ_0 the mass density, and $\mathbf{u}$ is the displacement. In the following, $\mathbf{C} = \mathbf{F}^\top \mathbf{F}$ denotes the right Cauchy-Green deformation tensor. We assume that the second Piola-Kirchhoff stress tensor can be additively decomposed into a passive, an active, and a prestress part, $\mathbf{S} = \mathbf{S}^p + \mathbf{S}^a + \mathbf{S}^{pre}$. The passive contribution is determined by an incompressible hyper-elastic material response

$$\mathbf{S}^p = 2\gamma^M \frac{\partial W^M}{\partial \mathbf{C}} + 2(1-\gamma^M)\frac{\partial W^T}{\partial \mathbf{C}} + p\,\mathbf{C}^{-1}, \tag{2}$$

where we have introduced the function γ^M in order to distinguish between muscle (M) and tendon (T) material. Furthermore, p is the undetermined volumetric response. In the present contribution, we use a Kelvin-Voigt material

$$\frac{\partial W^i}{\partial \mathbf{C}} = 2\mu^i \left(\mathbf{I} - \frac{1}{3}\mathbf{C}^{-1}\mathrm{tr}\mathbf{C}\right) + \frac{\eta^i}{2}\mathbf{C}^{-1}\dot{\mathbf{C}}\,\mathbf{C}^{-1}, \tag{3}$$

where μ^i and η^i with $i = M, T$ are the respective shear modulus and viscosities. In order to derive a cable model, we follow the kinematic assumptions in [7]. The initial configuration is parametrized as

$$\mathbf{X}^{3D}(\theta^1, \theta^2, \theta^3) = \mathbf{X}(\theta^1) + \phi_1(\theta^2, \theta^3)\,\mathbf{B}(\theta^1) + \phi_2(\theta^2, \theta^3)\,\mathbf{N}(\theta^1) \tag{4}$$

where $\mathbf{X}(\theta^1)$ is the centerline curve. The circular cross section is described by $\phi_1(\theta^2, \theta^3) = \theta^2 \cos(\theta^3)$, $\phi_2(\theta^2, \theta^3) = \theta^2 \sin(\theta^3)$ and

$$\mathbf{B}(\theta^1) = \frac{\mathbf{X}_{,\theta^1} \times \mathbf{X}_{,\theta^1\theta^1}}{||\mathbf{X}_{,\theta^1} \times \mathbf{X}_{,\theta^1\theta^1}||}, \quad \mathbf{N}(\theta^1) = \frac{\mathbf{X}_{,\theta^1} \times \mathbf{B}}{||\mathbf{X}_{,\theta^1} \times \mathbf{B}||}. \tag{5}$$

The current configuration is parametrized by

$$\mathbf{x}^{3D}(\theta^1, \theta^2, \theta^3) = \mathbf{x}(\theta^1) + \left(\phi_1(\theta^2, \theta^3)\, \mathbf{b}(\theta^1) + \phi_2(\theta^2, \theta^3)\, \mathbf{n}(\theta^1)\right) \Lambda(\theta^1). \tag{6}$$

The vectors $\mathbf{n}$ and $\mathbf{b}$ are defined analogously to (5). The function Λ accounts for thickness changes of the cross section during the deformation. Enforcing the incompressibility constraint on the cross section level allows us to compute the thickness change

$$\Lambda = \sqrt{\frac{||\mathbf{A}_1||}{||\mathbf{a}_1||}}, \tag{7}$$

where $\mathbf{A}_1 = \mathbf{X}_{,\theta^1}$ and $\mathbf{a}_1 = \mathbf{x}_{,\theta^1}$ are the tangent vectors to the initial and the current centerline, respectively. We assume that the active stress is generated such that it acts only along the direction of the centerline

$$\mathbf{S}^a = \gamma^M S^a \mathbf{A}_1 \otimes \mathbf{A}_1 \quad \text{with} \quad S^a = \alpha(t)\, \frac{S_{max}}{\lambda^2} \begin{cases} \exp^{-\left|\frac{\lambda/\lambda_{opt}-1}{\Delta W_{asc}}\right|^{\nu_{asc}}}, & \lambda < \lambda_{opt} \\ \exp^{-\left|\frac{\lambda/\lambda_{opt}-1}{\Delta W_{desc}}\right|^{\nu_{desc}}}, & \lambda > \lambda_{opt} \end{cases}. \tag{8}$$

For the prestress contribution we have $\mathbf{S}^{pre} = \frac{\sigma_0}{||\mathbf{a}_1||^2} \mathbf{A}_1 \otimes \mathbf{A}_1$, where σ_0 is an input parameter. By neglecting bending and shear stresses we have for the passive contribution

$$\mathbf{S}^p \approx \mathbf{S}^{p,11} \mathbf{A}_1 \otimes \mathbf{A}_1 + \left[\sum_{i=M,T} \left(2\mu^i - \frac{2\mu^i \operatorname{tr}\mathbf{C}}{3\Lambda^2} + \eta^i \frac{\mathbf{a}^1 \cdot \dot{\mathbf{a}}_1}{\Lambda^2} \right) + \frac{p}{\Lambda^2} \right] \mathbf{A}_2 \otimes \mathbf{A}_2 + \mathbf{S}^{p,33} \mathbf{A}_3 \otimes \mathbf{A}_3. \tag{9}$$

The enforcement of vanishing stress in the thickness direction allows us to statically condensate the volumetric response

$$p = -2\mu^i \left(\Lambda^2 - \frac{\operatorname{tr}\mathbf{C}}{3} \right) + \eta^i\, \mathbf{a}^1 \cdot \dot{\mathbf{a}}_1. \tag{10}$$

Thus, the passive stress contribution is given by

$$S^p \approx \left(2\mu\left(\frac{1}{||\mathbf{A}_1||^2} - \frac{||\mathbf{A}_1||}{||\mathbf{a}_1||^3}\right) + 2\eta\frac{\mathbf{a}^1\cdot\dot{\mathbf{a}}_1}{||\mathbf{a}_1||^2}\right)\mathbf{A}_1\otimes\mathbf{A}_1. \tag{11}$$

In total, the second Piola-Kirchhoff stress tensor is given by

$$\mathbf{S}=\left[\frac{\sigma_0}{||\mathbf{a}_1||^2}+\gamma^M S^a+\sum_{i=M,T}\left(2\mu^i\left(\frac{1}{||\mathbf{A}_1||^2} - \frac{||\mathbf{A}_1||}{||\mathbf{a}_1||^3}\right) + 2\eta^i\frac{\mathbf{a}^1\cdot\dot{\mathbf{a}}_1}{||\mathbf{a}_1||^2}\right)\right]\mathbf{A}_1\otimes\mathbf{A}_1. \tag{12}$$

The cable model has been coupled to a rigid body system consisting of cylinders. The coupling conditions at an end point x^c of one cable are

$$\mathbf{u}^{RB}(x^c) = \mathbf{u}^F(x^c), \quad \mathbf{F}^{RB}(x^c) + \mathbf{F}^F(x^c) = 0. \tag{13}$$

The first equation in (13) ensures the compatibility of the deformation. In the numerical model it is treated as a constraint on the end positions of the cables. Therefore, they are incorporated like inhomogeneous Dirichlet boundary conditions. The second equation in (13) is the second Newton law. Since the force acting at the insertion point of the cable model is given by

$$\mathbf{F}^F = A_0 S^{11}\mathbf{g}_1\,||\mathbf{G}_1||. \tag{14}$$

we can treat the force as an external load on the rigid body system. Furthermore, we consider frictionless contact on the lateral surfaces of the cables and the rigid bodies. Here, we consider the thickness-change (which is given by Λ) of the muscles during deformation.

3 Numerical Results

We have implemented a finite element method based on the weak form of (1) in Matlab. The unknown displacement field $\mathbf{u}$ is discretized with respect to space by cubic Hermite splines. The well known equations of motion for a rigid body system can be found for example in [12]. For the time integration we have used a backward Euler schema with one Newton step per time step (linear-implicit Euler). Contact has been realized by the penalty method.

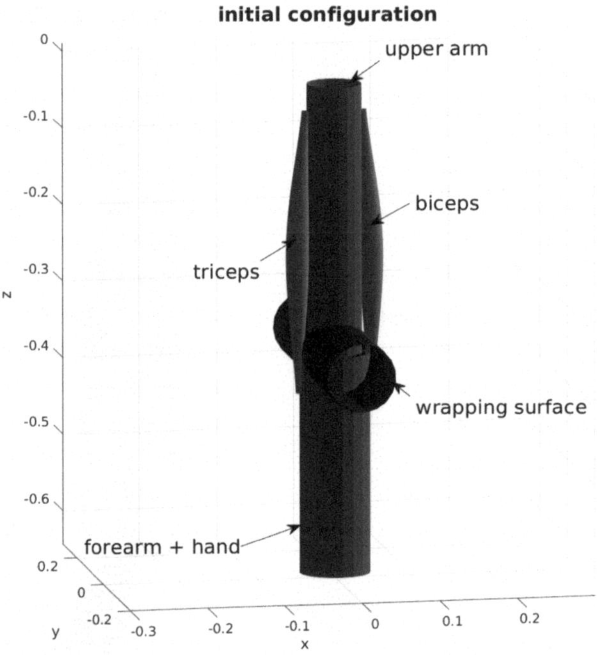

Fig. 1 Initial configuration

Table 1 Model parameters

ρ_0	900 kg/m^3	S_{max}	3×10^5 N/m^2	Length upper arm	0.351 m
μ^M	10^5 N/m^2	λ_{opt}	1.3 [−]	Length forearm	0.287 m
μ^T	10^6 N/m^2	ΔW_{asc}	0.3 [−]	Radius upper arm	0.035 m
η^M	10^4 N/m^2	ν_{asc}	4 [−]	Radius forearm	0.045 m
η^T	10^4 Ns/m^2	ΔW_{desc}	0.1 [−]	Mass upper arm	1.9241 kg
σ_0	30 N/m^2	ν_{desc}	4 [−]	Mass upper arm	1.502 kg

We consider a strongly simplified model of the upper limb, consisting of two rigid bodies and two muscles (see Fig. 1). The upper arm is fixed and the elbow joint is assumed to have only one rotational degree of freedom. An additional wrapping surface forces the triceps to bend around the elbow. The radius of both muscles is assumed to be the same and is given by the muscle ratio are

$$r_0 = 5 \times 10^{-3}(1 + 2\,f(\xi))\text{m}, \tag{15}$$

where $f(\xi) = 16(\xi^2 - 2\xi^3 + \xi^4)$ and ξ is the dimensionless position in the muscle. The muscle ratio follows the spatial distribution $\gamma^M = f(\xi)$. The remaining model parameters are given in Table 1.

Fig. 2 State at $t = 1$

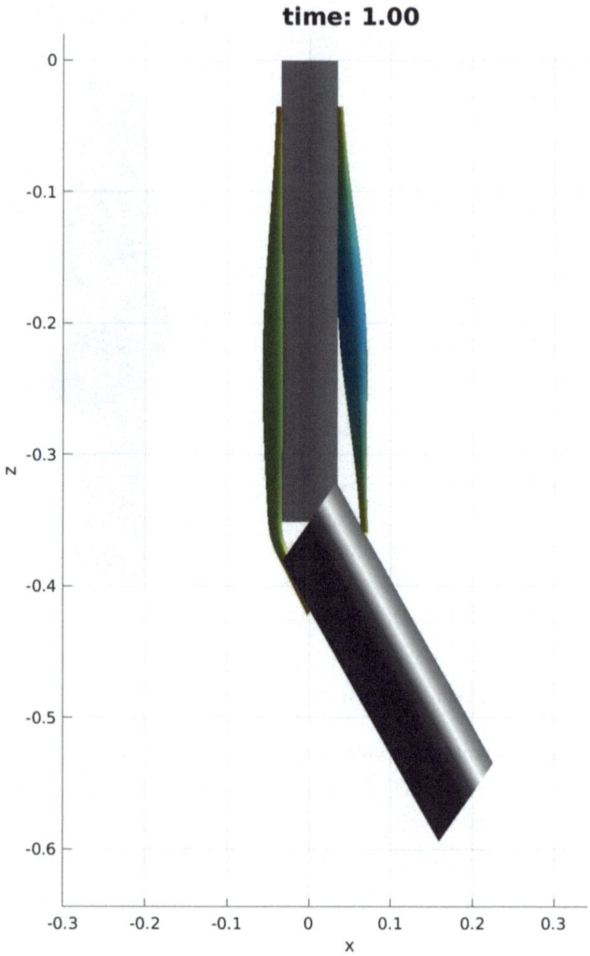

Within the dynamic forward simulation the activation level of the biceps is increased until the simulation time $t = 1$ and hold constant at $\alpha = 0.13$ onwards. The triceps reacts only passively. The state of the system is depicted at $t = 1$ and $t = 12$ in Figs. 2 and 3. Due to the activation the forearm is lifted up and the biceps develops a belly due to contraction. The rotation of the forearm is plotted over time in Fig. 4. The force between triceps and the upper arm over time is depicted in Fig. 5.

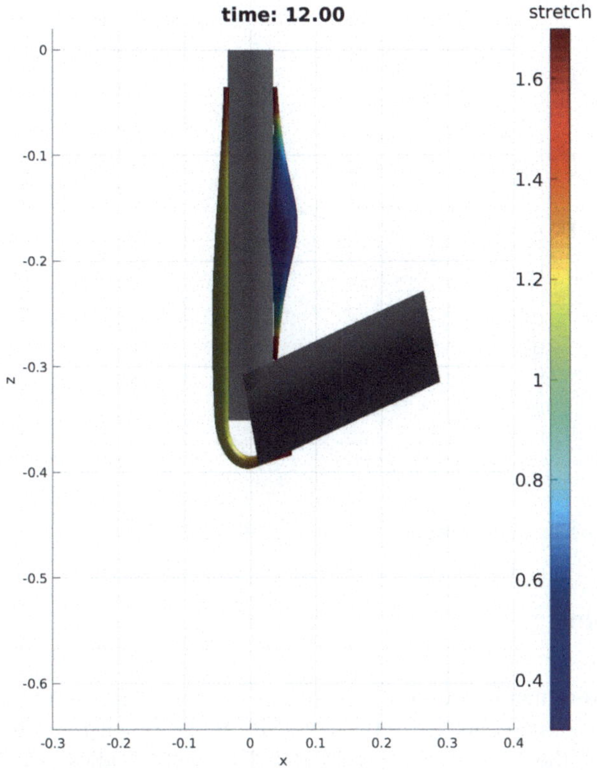

Fig. 3 State at $t = 12$ (simulation end)

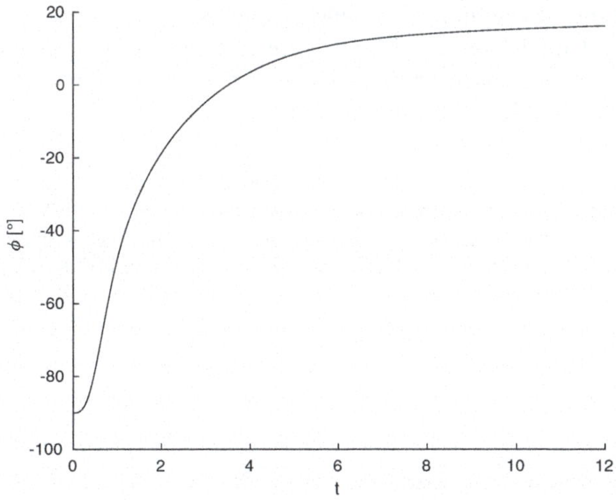

Fig. 4 Rotation of forearm

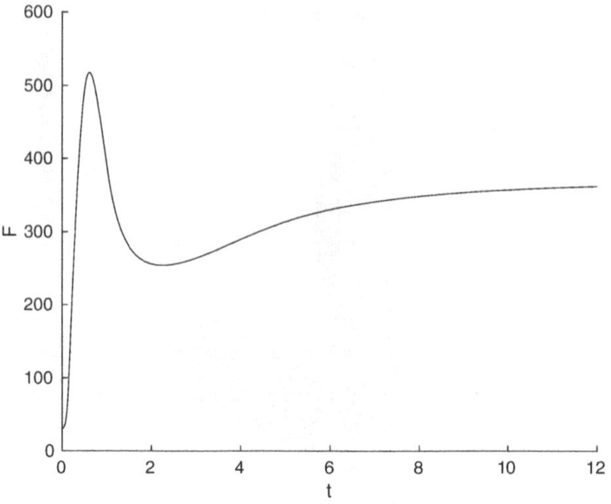

Fig. 5 Force of triceps at the upper arm

4 Conclusion

A new forward-dynamic musculoskeletal system simulation framework has been presented. All components are represented by three-dimensional bodies. Due to the stiffness of the bones they are modeled as rigid bodies, whereas muscles are modeled by one-dimensional cables derived from continuum mechanics. The advantage of this approach is the relatively low computational cost compared to models accounting for full three-dimensional kinematics, without introducing too much assumptions like in lumped parameter models.

In the present model the rigid bodies are restricted to be cylinders. In future work we plan to incorporate triangulated bone surfaces into our model. Furthermore, we would like to couple the present model with an electrochemical model on the microscopic level.

Acknowledgements The authors thank Joachim Linn from Fraunhofer ITWM Kaiserslautern for fruitful discussions on the topic. The authors acknowledge the financial support by the Federal Ministry of Education and Research of Germany in the framework of DYMARA (project number 05M16UKD).

References

1. Arnold, A.S., Salinas, S., Hakawa, D.J., Delp, S.L.: Accuracy of muscle moment arms estimated from MRI-based musculoskeletal models of the lower extremity. Comput. Aided Surg. **5**(2), 108–119 (2000)

2. Blemker, S.S., Delp, S.L.: Three-dimensional representation of complex muscle architectures and geometries. Ann. Biomed. Eng. **33**(5), 661–673 (2005)
3. Delp, S.L., Loan, J.P., Hoy, M.G., Zajac, F.E., Topp, E.L., Rosen, J.M.: An interactive graphics-based model of the lower extremity to study orthopaedic surgical procedures. IEEE Trans. Biomed. Eng. **37**(8), 757–767 (1990)
4. Heidlauf, T., Klotz, T., Rode, C., Altan, E., Bleiler, C., Siebert, T., Röhrle, O.: A multi-scale continuum model of skeletal muscle mechanics predicting force enhancement based on actin–titin interaction. Biomech. Model. Mechanobiol. **15**(6), 1423–1437 (2016)
5. Hwang, J., Knapik, G.G., Dufour, J.S., Aurand, A., Best, T.M., Khan, S.N., Mendel, E., Marras, W.S.: A biologically-assisted curved muscle model of the lumbar spine: model structure. Clin. Biomech. **37**, 53–59 (2016)
6. Maas, R.: Biomechanics and optimal control simulations of the human upper extremity. Ph.D. Thesis, Friedrich-Alexander-Universität Erlangen-Nürnberg (2014)
7. Raknes, S., Deng, X., Bazilevs, Y., Benson, D., Mathisen, K., Kvamsdal, T.: Isogeometric rotation-free bending-stabilized cables: statics, dynamics, bending strips and coupling with shells. Comput. Methods Appl. Mech. Eng. **263**, 127–143 (2013)
8. Röhrle, O., Sprenger, M., Schmitt, S.: A two-muscle, continuum-mechanical forward simulation of the upper limb. Biomech. Model. Mechanobiol. **16**(3), 743–762 (2017)
9. Rupp, T., Ehlers, W., Karajan, N., Günther, M., Schmitt, S.: A forward dynamics simulation of human lumbar spine flexion predicting the load sharing of intervertebral discs, ligaments, and muscles. Biomech. Model. Mechanobiol. **14**(5), 1081–1105 (2015)
10. Scholz, A., Sherman, M., Stavness, I., Delp, S., Kecskeméthy, A.: A fast multi-obstacle muscle wrapping method using natural geodesic variations. Multibody Syst. Dyn. **36**(2), 195–219 (2016)
11. Shorten, P.R., OCallaghan, P., Davidson, J.B., Soboleva, T.K.: A mathematical model of fatigue in skeletal muscle force contraction. J. Muscle Res. Cell Motil. **28**(6), 293–313 (2007)
12. Simeon, B.: Computational Flexible Multibody Dynamics. Springer, Berlin (2013)
13. Zajac, F.E.: Muscle and tendon properties models scaling and application to biomechanics and motor. Crit. Rev. Biomed. Eng. **17**(4), 359–411 (1989)

Improving Thermal Ablation of Liver Tumors

Matthias Andres and René Pinnau

Abstract Laser-induced interstitial thermotherapy (LITT) is a medical treatment which attempts to destroy liver tumors by thermal ablation. A realistic real-time simulation shall support the practitioner online in planning the therapy. The heat transfer inside the liver can be described by a PDE system consisting of the so-called bio-heat equation and a radiative transfer model. We model the heat loss due to blood perfusion by a simple sink term with spatially varying coefficient accounting for the presence of vessels. Using PDE-constrained optimization we demonstrate how to fit this parameter in order to minimize the deviation between the predicted and measured temperature.

1 Overview

In this work we consider a mathematical model for the laser-induced interstitial thermotherapy (LITT), which is a medical treatment to destroy liver tumors. To this a specific applicator is inserted into the liver. It has a laser emitting part at its tip and is cooled by a water flow (see Fig. 1). The mathematical description is based on the so-called bio-heat equation, which is well-studied in literature and appears to be a valid model: In [5] the mathematical simulation for an ex vivo setup was validated by a real experiment with a porcine-liver and showed promising results.

Nevertheless, the model depends on various parameters, which are only partly known from literature and often ambiguous. In this work, we solve this problem by formulating the identification of unknown parameters as an inverse problem. The medical treatment is monitored by magnetic resonance imaging which provides ground-truth data for the temperature later in the real application. This information

M. Andres (✉) · R. Pinnau
Technische Universität Kaiserslautern, Department of Mathematics, Kaiserslautern, Germany
e-mail: andres@mathematik.uni-kl.de; pinnau@mathematik.uni-kl.de

© Springer Nature Switzerland AG 2019
I. Faragó et al. (eds.), *Progress in Industrial Mathematics at ECMI 2018*,
Mathematics in Industry 30, https://doi.org/10.1007/978-3-030-27550-1_25

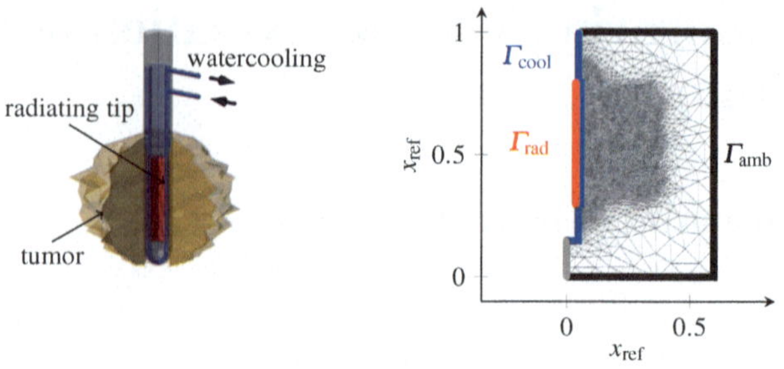

Fig. 1 Illustration of the laser applicator and the scaled computational domain with 4709 nodes

will enter the inverse problem through a cost functional and is the basis for the parameter fitting.

An inverse problem in the context of LITT has already been studied in [9]. The authors developed an optimization strategy for identifying certain parameters of the Arrhenius law, which models the coagulation of tissue due to protein denaturation caused by the increased temperature.

Here, we focus on the effect of blood perfusion. In [8] the authors modeled the temperature decrease due to blood flow in a homogenized way as a linear sink term in the bio-heat equation with a constant coefficient. The effect of a single thick vessel was studied in [6], including the knowledge of the vessel location and modeling the blood flow in terms of the Navier Stokes equation.

Our attempt is between the simple homogenized and the very complex blood flow model. We add a sink term to the bio-heat equation similar to [8], but we allow the prefactor to vary in space and time, depending on the vessel structure and the coagulation of the tissue. This allows to model local effects due to single blood vessels as well as the influence of capillary vessels. Furthermore it does not add too much complexity to the model such that this problem would become unrealistic to handle in a parameter identification context.

Having a more precise model for the blood perfusion would add great value to the simulation of LITT, especially to make a step from ex vivo to in vivo simulations. Furthermore, the position of thick vessels relative to the tumor region varies for each treatment and needs to be identified beforehand. This kind of parameter cannot be given by literature but needs to be identified for each treatment and patient individually in an automatized way.

In Sect. 2 we review the mathematical model, which was validated in [5] by experimental data for slightly different boundary conditions. In Sect. 3 we introduce a cost functional modeling the deviation of the predicted temperature data from the measured data depending on the location of vessels and compute the corresponding gradient. In Sect. 4 we demonstrate a gradient-descent method for solving the inverse problem. The last section shows possible directions for future work.

2 Mathematical Model

The process of radiative heat transfer inside the liver is described by a coupling of the so-called bio-heat equation,

$$\begin{aligned}
\alpha\partial_t T - \nabla\cdot(\kappa\nabla T) &= \hat{\mu}\mu_a\phi + \xi\,(T_b - T) && \text{on } (0,1)\times\Omega,\\
\kappa\nabla T\cdot n &= \alpha_{\text{amb/cool}}\,(T_{\text{amb/cool}} - T) && \text{on } (0,1)\times\Gamma_{\text{ambient/cooling}},\\
T(0,\cdot) &= T_0 && \text{on } \Omega,
\end{aligned} \tag{1}$$

with the SP_1 approximation (e.g., [7]) of the radiative transfer equation,

$$\begin{aligned}
-\nabla\cdot(D\nabla\phi) + \mu_a\phi &= 0 && \text{on } (0,1)\times\Omega,\\
\frac{\varepsilon}{2}\phi + D\nabla\phi\cdot n &= \varepsilon\delta\frac{q_{\text{laser}}}{x_{\text{ref}}^2|\Gamma_{\text{rad}}|}\cdot\chi_{\Gamma_{\text{rad}}} && \text{on } (0,1)\times\Gamma,
\end{aligned} \tag{2}$$

where T describes the temperature (scaled by 1 K), ϕ describes the irradiance (scaled by 1 W/m^2) and $\Gamma_{\text{rad}} \subset \mathbb{R}^3$ is the part of the applicator that emits laser light. This model is taken from [5] with slightly modified boundary conditions. A nonlinear coupling is introduced by the process of coagulation due to the increased temperature of the tissue, which enters the system through the function γ as

$$\begin{aligned}
\gamma(t,x) &= \exp\left(-t_{\text{ref}}\int_0^t A\exp\left(-\frac{E_a}{RT_{\text{ref}}T(s,x)}\right)\mathrm{d}s\right),\\
D &= \frac{\varepsilon^2}{3(\mu_a + (1-g)\mu_s)},\\
p &= p_n + (1-\gamma)\cdot(p_c - p_n),
\end{aligned}$$

for $p \in \{\mu_a, \mu_s, g, \xi_{\text{in}}, \xi_{\text{out}}\}$, where the subindices n and c stand for the respective values for native and coagulated tissue. In this work, other than, e.g., [5, 8], we consider a scaled heterogeneous blood perfusion rate ξ in the bio-heat equation, which affects the modeled heat loss due to blood flow. We model ξ as

$$\xi = \xi_{\text{out}} + u\cdot(\xi_{\text{in}} - \xi_{\text{out}}), \tag{3}$$

where $u : \Omega \to [0,1]$ can be seen as an indicator function for blood vessels. Having this interpretation in mind the values ξ_{in} and ξ_{out} model the heat loss due to blood

Table 1 Overview on occurring variables

$x_{\text{ref}} = 0.05$ m	$t_{\text{ref}} = t^*_{\text{end}} = 60$ s	$T_{\text{ref}} = 1$ K	$\phi_{\text{ref}} = 1\ \text{W}/\text{m}^2$
$\mu_{\text{ref}} = 1560\ {}^1/\text{m}$	$\varepsilon = {}^1/x_{\text{ref}}\mu_{\text{ref}}$	$\kappa = 1$	$\alpha = 345.94$
$\mu_{\text{a,n}} = 0.03205$	$\mu_{\text{a,c}} = 0.03846$	$\mu_{\text{s,n}} = 5.128$	$\mu_{\text{s,c}} = 19.23$
$g_{\text{n}} = 0.97$	$g_{\text{c}} = 0.95$	$E_{\text{a}} = 6.3\,\text{e}5\ \text{J}/\text{mol}$	$A = \exp(226.7847)\ {}^1/\text{s}$
$R = 8.31\ \text{J}/\text{molK}$	$\alpha_{\text{cool}} = 26.04$	$\alpha_{\text{amb}} = 0$	$\hat{\mu} = 8.125$
$\delta = 1.72\ \text{m}^2/\text{W}$	$q_{\text{laser}} = 28.9$ W	$\xi_{\text{in,n}} = 312.3$	$\xi_{\text{in,c}} = 312.3$
$\xi_{\text{out,n}} = 0$	$\xi_{\text{out,c}} = 0$	$T_0 = 310.15$	$T_{\text{b}} = 310.15$
$T_{\text{amb}} = 310.15$	$T_{\text{cool}} = 293.15$		

The values are based on the parameters in [5], whereas the blood perfusion rates are taken from [3]

perfusion inside and outside of thick blood vessels, respectively. An overview of the occurring variables is given in Table 1.

3 Gradient-Based Optimization

We formulate the task of identifying the location of blood vessels as an inverse problem, where we consider the function $u\colon \Omega \to [0, 1]$ in Eq. (3) as control input for the following optimization problem:

$$u_{\text{opt}} = \operatorname*{argmin}_{u:\Omega\to[0,1]} J(T, \phi, u)$$

$$\text{s.t. } (T, \phi) \text{ fulfill Eqs. (1), (2) for corresponding } u,$$

with cost functional

$$J(T, \phi, u) = \frac{\lambda_1}{2J_0} \int_\Omega (T(1, x) - T_{\text{d}}(x))^2 \, \text{d}x + \frac{\lambda_2}{J_0} \int_\Omega u \, \text{d}x.$$

The L^1 penalty for the control term is motivated, e.g., by Casas et al. [2]. The desired temperature T_{d} corresponds to the measured temperature at the final physical time t^*_{end}, and the value J_0 normalizes the cost functional to one for the initial value of u in the optimization algorithm in Sect. 4. Following [4], we derive for the reduced cost functional $f(u) = J(T(u), \phi(u), u)$, based on a formal Lagrange principle, the Riesz representation of the derivative w.r.t. u:

$$f'(u) = \frac{\lambda_2}{J_0} + \int_0^1 (\xi_{\text{in}} - \xi_{\text{out}}) (T_{\text{b}} - T) \varphi \, \text{d}t,$$

where φ is part of the adjoint state of the PDE constraint and results from solving the adjoint equation

$$
\begin{aligned}
-\alpha \partial_t \varphi - \nabla \cdot (\kappa \nabla \varphi) + \xi \varphi + f_{\mu_a} - f_\xi + f_D &= 0 \quad \text{on } (0,1) \times \Omega, \\
-\nabla \cdot (D \nabla \psi) + \mu_a \left(\psi - \hat{\mu} \varphi\right) &= 0 \quad \text{on } (0,1) \times \Omega, \\
\alpha_{\text{amb/cool}} \varphi + \kappa \nabla \varphi \cdot n &= 0 \quad \text{on } (0,1) \times \Gamma_{\text{amb/cool}}, \\
\frac{\varepsilon}{2} \psi + D \nabla \psi \cdot n &= 0 \quad \text{on } (0,1) \times \Gamma, \\
\varphi(1,\cdot) - \frac{\lambda_1}{\alpha J_0} \left(T(1,\cdot) - T_d\right) &= 0 \quad \text{on } \Omega,
\end{aligned}
\tag{4}
$$

with the auxiliary functions

$$
\begin{aligned}
f_{\mu_a} &= \int_t^1 \left(\mu_{a,c} - \mu_{a,n}\right) \phi \left(\psi - \hat{\mu} \varphi\right) \gamma \, ds \cdot F(t), \\
f_\xi &= \int_t^1 (\xi_c - \xi_n) \cdot (T_b - T) \varphi \gamma \, ds \cdot F(t), \\
f_D &= \int_t^1 \begin{bmatrix} \frac{\partial D}{\partial \mu_a} & \frac{\partial D}{\partial \mu_s} & \frac{\partial D}{\partial g} \end{bmatrix} \cdot \begin{bmatrix} \mu_{a,c} - \mu_{a,n} \\ \mu_{s,c} - \mu_{s,n} \\ g_c - g_n \end{bmatrix} \left(\nabla \psi^T \cdot \nabla \phi\right) \gamma \, ds \cdot F(t), \\
F(t) &= t_{\text{ref}} A \exp\left(-\frac{E_a}{R T_{\text{ref}} T(t)}\right) \cdot \frac{E_a}{R T_{\text{ref}} T(t)^2}.
\end{aligned}
\tag{5}
$$

4 Numerical Experiment

In this section we consider a model problem demonstrating the identification of blood vessels. First we consider the true location of the vessels u_{true} (see Fig. 2) and compute the true temperature distribution for a physical simulation time of $t^*_{\text{end}} = 60$ s, where the temperature at the last timestep is taken as T_d. Based on the gradient information of the cost functional from Sect. 3 with $\lambda_1 = 1$, we apply a projected gradient-descent algorithm with Armijo linesearch rule (e.g., [4]), which we initialize with $u = 0$. In order to solve the occurring PDEs we first discretize in time with a physical timestep of 5 s, and solve the resulting stationary PDEs using the finite element method with the help of FEniCS (see [1]). We assumed rotational invariance and thus considered the corresponding problem in two space dimensions.

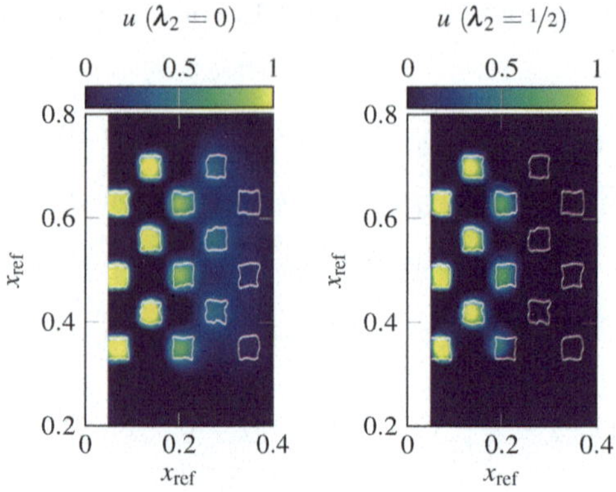

Fig. 2 Result of the gradient-descent algorithm after 19 iterations for $\lambda_2 = 0$ and $\lambda_2 = 1/2$, respectively. The contour lines indicate the function u_{true} used to compute the desired temperature T_{d}. We see that vessels close to the applicator can be identified. The temperature differences in the region more distant to the applicator seem to be too small for estimating the underlying vessel structure for $t^*_{\text{end}} = 60\,\text{s}$

5 Future Work

We discussed how the simple model of blood perfusion in the bio-heat equation can be improved by using heterogeneous coefficients which model the different influence of capillary and thick vessels to the temperature distribution. Based on a model problem we demonstrated how the identification of the unknown perfusion coefficient can be attempted via PDE-constrained optimization.

In order to make this approach applicable to the real medical treatment, there are several issues which need to be investigated. As the scaled blood perfusion rate ξ has to be identified during the therapy, this requires a fast solution of the inverse problem in real time. Furthermore the numerical treatment of the term F in Eq. (5) becomes challenging for larger simulation times.

As the data to be fitted is given in the real application by MR temperature values, it is necessary to study the effect of noisy data in the cost functional on the parameter identification, where especially a proper choice of the L^1 penalty term in the cost functional needs to be discussed further.

Acknowledgements The authors are grateful for the support of the German Federal Ministry of Education and Research (BMBF) grant no. 05M16UKE.

References

1. Alnæs, M.S., Blechta, J., Hake, J., Johansson A., Kehlet, B., Logg, A., Richardson C., Ring J., Rognes, M. E., Wells, G. N.: The FEniCS Project Version 1.5. Arch. Numer. Softw. **3**(100), 9–23 (2015)

2. Casas, E., Ryll, C., Tröltzsch, F.: Sparse optimal control of the Schlögl and FitzHugh–Nagumo systems. Comput. Meth. Appl. Mat. **13**(4), 415–442 (2013)
3. Foundation for Research on Information Technologies in Society (IT'IS): Tissue Properties. https://itis.swiss/virtual-population/tissue-properties/database/heat-transfer-rate/. Cited 26 Sept 2018
4. Hinze, M., Pinnau, R., Ulbrich, M., Ulbrich, S.: Optimization with PDE Constraints, vol. 23. Springer, Berlin (2008)
5. Hübner, F., Leithäuser, C., Bazrafshan, B., Siedow, N., Vogl, T.J.: Validation of a mathematical model for laser-induced thermotherapy in liver tissue. Lasers Med. Sci. **32**(6), 1399–1409 (2017)
6. Mohammed, Y., Verhey, J.F.: A finite element method model to simulate laser interstitial thermo therapy in anatomical inhomogeneous regions. Biomed. Eng. Online **4**(1), 2 (2005)
7. Pinnau, R.: Analysis of optimal boundary control for radiative heat transfer modeled by the $SP\_\{n\}$-system. Commun. Math. Sci. **5**(4), 951–969 (2007)
8. Puccini, S., Bär, N., Bublat, M., Kahn, T., Busse, H.: Simulations of thermal tissue coagulation and their value for the planning and monitoring of laser-induced interstitial thermotherapy (LITT). Magn. Reson. Med. **49**(2), 351–362 (2003)
9. Tse, O., Pinnau, R., Siedow, N.: Identification of temperature-dependent parameters in laser-interstitial thermo therapy. Math. Models Methods Appl. Sci. **22**(9), 1250019 (2012)

Efficient Therapy-Planning via Model Reduction for Laser-Induced Thermotherapy

Kevin Tolle and Nicole Marheineke

Abstract Laser-induced thermotherapy is a local, minimally invasive treatment for liver tumors, which uses laser radiation to destroy targeted tissue. Many factors, such as the placement of the applicator(s), the length of the treatment and the amount of radiation introduced, affect the success of the treatment. In this work, we focus on controlling the amount of laser power applied during the treatment. This results in a PDE-constrained optimal control problem. Because such problems are computationally expensive to solve directly, a space-mapping approach is used. The coarse model used in the space-mapping method is derived through a novel linearization of the constraining equations and subsequently reduced using proper orthogonal decomposition. An example problem shows the viability of this approach.

1 Introduction

An important aspect in laser-induced thermotherapy (LITT) is ensuring the complete ablation of the tumor, while preserving as much of the surrounding healthy tissue as possible. Although many different aspects of the treatment process are relevant to this goal, this work focuses solely on a novel approach for finding the ideal amount of laser power during the treatment via an optimal control problem.

The space-mapping approach, which uses a hierarchy of models in order to solve optimization problems, builds the core of this work. The nonlinear mathematical model validated in [7] serves as an accurate but expensive fine model. By applying a unique linearization and a standard model order reduction technique, a coarse model is derived. These models together build the hierarchy used by the space-mapping approach, whose performance is demonstrated in an academic example.

K. Tolle (✉) · N. Marheineke
Trier University, Trier, Germany
e-mail: tolle@uni-trier.de; marheineke@uni-trier.de

© Springer Nature Switzerland AG 2019
I. Faragó et al. (eds.), *Progress in Industrial Mathematics at ECMI 2018*,
Mathematics in Industry 30, https://doi.org/10.1007/978-3-030-27550-1_26

2 Hierarchical Modeling and Optimization

The space-mapping (SM) approach seeks to align the optimization over a fine model to a coarse model, where the coarse model shares the same physics as its fine counterpart, while remaining computationally cheaper to evaluate [2, 3]. Although most often found in engineering design, the viability of the approach for transport problems was investigated in [8].

Given a desired response y^d and a fine response $f(u_f)$ for $u_f \in U_f$, the fine control

$$u_f^* = \arg\min_{u_f} \frac{1}{2} \|f(u_f) - y^d\|^2$$

is assumed to be too expensive to compute. On the other hand, the coarse response $c(u_c)$ for $u_c \in U_c$ delivers

$$u_c^* = \arg\min_{u_c} \frac{1}{2} \|c(u_c) - y^d\|^2,$$

which can be easily computed but may lie outside the desired accuracy. The key element in the space-mapping approach is the mapping function $p : U_f \to U_c$, which is defined through

$$p(u_f) = \arg\min_{u_c} \frac{1}{2} \|c(u_c) - f(u_f)\|^2.$$

Assuming that $f(u_f^*) \approx y^d$ and/or $f(u_f^*) \approx c(u_c^*)$, the following relationship

$$p(u_f^*) = \arg\min_{u_c} \frac{1}{2} \|c(u_c) - f(u_f^*)\|^2 \approx \arg\min_{u_c} \frac{1}{2} \|c(u_c) - y^d\|^2 = u_c^* \quad (1)$$

holds. The aggressive space-mapping (ASM) method approximates (1) by solving $F(u_f^p) \stackrel{\text{def}}{=} p(u_f^p) - u_c^* = 0$ iteratively via a quasi-Newton iteration with a Broyden-type approximation of the Jacobian for u_f^p, which in turn approximates the fine control.

Fine Model Let $\Omega \subset \mathbb{R}^3$ be the domain of interest. The boundary Γ of Ω consists of the portion along the applicator Γ_{appl} and the ambient portion Γ_{amb}. The applicator, which is actively cooled, has a portion $\Gamma_{\text{rad}} \subset \Gamma_{\text{appl}}$, which radiates laser light into the surrounding tissue. The nonlinear system describing the temperature

T, radiation φ and tissue damage ζ is given for $x \in \Omega$ and $t \in [0, t_f]$ by

$$\begin{aligned} \rho\, c_p\, \partial_t T - \nabla \cdot (k\, \nabla T) &= \xi_b\, (T_b - T) + \mu_a(\zeta)\, \varphi, \qquad T(x,0) = T_0(x), \\ -\nabla \cdot (D(\zeta)\, \nabla \varphi) &= -\mu_a(\zeta)\, \varphi, \\ \partial_t \zeta &= -A\, e^{-\frac{E_a}{R\,T}}\, \zeta, \qquad \zeta(x,0) = \zeta_0(x) \end{aligned} \tag{2a}$$

with the following boundary conditions:

$$k\, \nabla T \cdot \mathbf{n} = \begin{cases} \alpha_{\text{cool}}\, (T_{\text{cool}} - T) & \text{on } \Gamma_{\text{appl}}, \\ \alpha_{\text{amb}}\, (T_{\text{amb}} - T) & \text{on } \Gamma_{\text{amb}} \end{cases} \tag{2b}$$

and

$$D(\zeta)\, \nabla \varphi \cdot \mathbf{n} = \begin{cases} |\Gamma_{\text{rad}}|^{-1}\, q_{\text{eff}} & \text{on } \Gamma_{\text{rad}}, \\ 0 & \text{on } \Gamma_{\text{appl}} \backslash \Gamma_{\text{rad}}, \\ -\frac{1}{2}\, \varphi & \text{on } \Gamma_{\text{amb}}, \end{cases} \tag{2c}$$

where the effective laser power q_{eff} is assumed to be proportional to the actual laser power q_{appl} according to the relationship $q_{\text{eff}}(t) = (1 - \beta_q)\, q_{\text{appl}}(t)$ for all $t \in [0, t_f]$. The absorption coefficient β_q describes the amount of power directly absorbed by the coolant. The temperature-dependent tissue parameters μ_a, μ_s and g, which are defined through $z(\zeta) = z_n + (1 - \zeta)\,(z_c - z_n)$ for $z \in \{\mu_a, \mu_s, g\}$, characterize the diffusion coefficient D via $D(\zeta) = (3\,(\mu_a(\zeta) + (1 - g(\zeta))\, \mu_s(\zeta)))^{-1}$. More details about this mathematical model can be found in [6, 7].

Coarse Model The coarse model is derived from the fine model by using a combination of two different approximations chained together. The nonlinear model is first linearized, and the resulting linear model is further reduced using standard reduction techniques in order to attain a much smaller reduced order model.

Linearization In light of the nonlinear, temperature-dependent tissue parameters in (2), we simplify the model by effectively "freezing" the tissue's state by fixing a characteristic coagulation state $\bar{\zeta} \in \mathbb{R}$ and computing the associated coefficients $\bar{\mu}_a = \mu_a(\bar{\zeta})$ and $\bar{D} = D(\bar{\zeta})$. This local approximation leads to the linear parabolic-elliptic system

$$\begin{aligned} \rho\, c_p\, \partial_t T - \nabla \cdot (k\, \nabla T) &= \xi_b\, (T_b - T) + \bar{\mu}_a\, \varphi, \quad T(x,0) = T_0(x), \\ -\nabla \cdot (\bar{D}\, \nabla \varphi) &= -\bar{\mu}_a\, \varphi \end{aligned} \tag{3}$$

with appropriate boundary conditions. This simplified model behaves differently as coagulation affects the dynamics of the nonlinear model, see Fig. 1a. For example, the linear model (for $\bar{\zeta} = 1$) initially coincides with the nonlinear model for a short

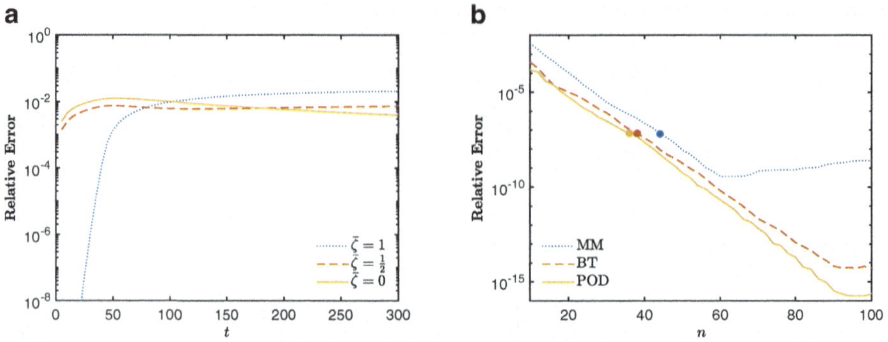

Fig. 1 (**a**) Relative error between the nonlinear and linearized models using different linearization parameters $\bar{\zeta}$. (**b**) Relative reduction error for three different model order reduction techniques. The *dots* coincide with entries in Table 1

time period before the coagulation effects cause the behavior to diverge. Different choices for $\bar{\zeta}$ result in models that are initially worse but improve slightly with time as the coagulation state in the nonlinear model nears the fixed value throughout the domain.

Model Order Reduction After introducing semi-discrete linear finite elements, (3) can be reformulated as a linear, time-invariant system (LTIS) of the form

$$\begin{aligned} \mathbf{E}\,\dot{\mathbf{x}}(t) &= \mathbf{A}\,\mathbf{x}(t) + \mathbf{B}\,\mathbf{u}(t), \quad \mathbf{x}(0) = \mathbf{0}, \\ \mathbf{y}(t) &= \mathbf{C}\,\mathbf{x}(t) + \mathbf{D}\,\mathbf{u}(t) \end{aligned} \tag{4}$$

with input $\mathbf{u} = (1, q_{\text{appl}})^{\mathsf{T}}$, state $\mathbf{x} = (\mathbf{T}, \boldsymbol{\varphi})^{\mathsf{T}} \in \mathbb{R}^N$ and output $\mathbf{y} = \mathbf{T}$. An advantage of (4) is the direct access to standard (one-sided) projection-based model reduction techniques. The projection $\mathbf{V} \in \mathbb{R}^{N \times n}$ transforms (4) into the reduced order LTIS

$$\begin{aligned} \mathbf{E}_r\,\dot{\mathbf{x}}_r(t) &= \mathbf{A}_r\,\mathbf{x}_r(t) + \mathbf{B}_r\,\mathbf{u}(t), \quad \mathbf{x}_r(0) = \mathbf{0}, \\ \mathbf{y}(t) &= \mathbf{C}_r\,\mathbf{x}_r(t) + \mathbf{D}\,\mathbf{u}(t), \end{aligned} \tag{5}$$

where $\mathbf{E}_r = \mathbf{V}^{\mathsf{T}}\mathbf{E}\,\mathbf{V}$, $\mathbf{A}_r = \mathbf{V}^{\mathsf{T}}\mathbf{A}\,\mathbf{V}$, $\mathbf{B}_r = \mathbf{V}^{\mathsf{T}}\mathbf{B}$ and $\mathbf{C}_r = \mathbf{C}\,\mathbf{V}$ with $\mathbf{x} \approx \mathbf{V}\,\mathbf{x}_r$ and $n \ll N$. The projection matrix $\mathbf{V}$ can be attained using standard methods such as moment-matching, balanced truncation and proper orthogonal decomposition. For more information on model order reduction techniques see, e.g., [1, 5, 9] and references therein.

The relative reduction error with respect to the reduced model size is shown in Fig. 1b for various reduction methods with a comparison of computation times for reduced models with similar error in Table 1. A one-sided Arnoldi algorithm is used for the moment-matching method, while the MORLAB toolbox is used for the balanced truncation method [4]. Finally, the proper orthogonal decomposition is performed using the method of snapshots.

Table 1 Comparison of the full-order model (FOM) with reduced models using moment matching (MM), balanced truncation (BT) and proper orthogonal decomposition (POD)

		Computational time[a]		
	Size	Offline [s]	Online [s]	Speedup
FOM	2596	–	7.1998e–02	–
MM	44	4.6210e–01	4.0987e–03	17.5662
BT	38	5.9567e+01	1.7777e–03	40.5014
POD	36	2.9973e–01	1.9297e–03	37.3111

For the reduced order models, the computational time is split into an offline and online phase, where the offline phase is used to precompute the projection matrix $\mathbf{V}$ and the resulting reduced order system, while the online phase only consists of the forward simulation of the reduced order model for a given input u

[a] Simulations were performed in MATLAB on an i7-6700 with 32 GB of RAM

3 Numerical Results and Discussion

The proposed space-mapping approach is used to reconstruct a given laser input. A target control q_{appl}^d is used to generate the target temperature profile T^d, which is then used in order to define the following optimal control problem

$$\min J(T, q_{\text{appl}}) = \frac{1}{2}\int_0^{t_f}\int_{\Omega}(T(x,t) - T^d(x,t))^2 \,\mathrm{d}x\,\mathrm{d}t + \frac{\lambda}{2}\int_0^{t_f} |q_{\text{appl}}(t)|^2 \,\mathrm{d}t$$

subject to (2), where the second term regularizes the optimization problem. In other words, the goal is to reconstruct the laser input q_{appl}^d from the given temperature data by solving an optimal control problem. Figure 2 shows the two-dimensional (rotationally symmetric) mesh and an exemplary temperature profile. Linear finite elements were used for the spatial discretization, while an implicit Euler method is used for the temporal discretization. Unless stated otherwise, the physical parameters from [7] are used in the simulations. The optimal controls resulting from directly optimizing the fine and coarse models, respectively, and using the space-mapping approach are shown in Fig. 3a. The coarse optimization is performed using a steepest descent method with an Armijo-type line search. The gradients are calculated with the help of adjoints. MATLAB's `fminunc` function, using finite differences to approximate the gradient, is used to solve the fine optimization problem as a reference. It can be seen that the space-mapping approach is able to account for the "lost" nonlinear effects in the coarse model. This allows for a more accurate solution of the optimal control problem. Figure 3b clearly shows the improvement achieved using the space-mapping approach in comparison to the coarse optimization, where the error in the space-mapped solution is in the same order as the fine optimal. The fine control is not expected to coincide with the targeted laser input because of the regularization term. The initial numerical results are promising. The space-mapping approach displays a great improvement

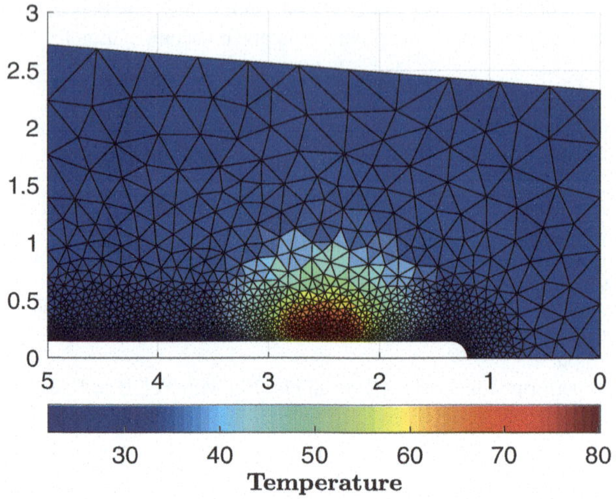

Fig. 2 Depiction of the axis-symmetrical mesh used in the computations with a temperature profile resulting from the nonlinear model. The bottom axis marks the axis of rotation, where the artificial boundary directly along the axis fulfills a symmetry boundary condition. The "pocket" along the bottom represents the cooled applicator, where the portion along which the temperature is concentrated highlights the radiating segment of the applicator

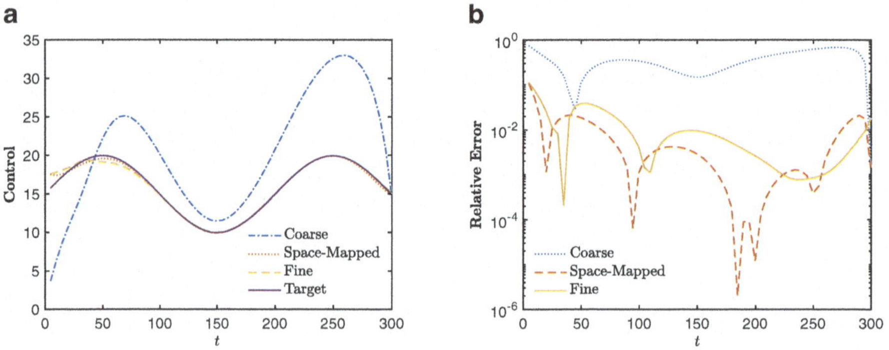

Fig. 3 (**a**) Numerical solutions of the respective optimization problems with $\lambda = 10^{-7}$. (**b**) Relative error in the optimal control with respect to the target control q^d_{appl}

in accuracy by using forward simulations of the fine model to correct the coarse optimization. After successfully demonstrating the viability of this approach for LITT problems, recent work deals with a broad performance study considering relevant parameter settings. Additionally, constraints on the control are being investigated to reflect the actual capabilities of the instruments used in the treatment.

Acknowledgements This research has been funded by the Federal Ministry of Education and Research of Germany in the framework of the project PROMT.

References

1. Antoulas, A.C.: Approximation of Large-Scale Dynamical Systems. S.I.A.M., Philadelphia (2005)
2. Bakr, M.H., Bandler, J.W., Madsen, K., Søndergaard, J.: Review of the space mapping approach to engineering optimization and modeling. Optim. Eng. **1**, 241–276 (2000)
3. Bandler, J.W., Biernacki, R.M., Chen, S.H., Grobelny, P.A., Hemmers, R.H.: Space mapping technique for electromagnetic optimization. IEEE Trans. Microwave Theory Techn. **42**, 2536–2544 (1994)
4. Benner, P., Werner, S.W.R.: MORLAB-3.0 – Model Order Reduction Laboratory. https://doi.org/10.5281/zenodo.842659 (2017)
5. Benner, P., Cohen, A., Ohlberger, M., Willcox, K. (eds.): Model Reduction and Approximation: Theory and Algorithms. S.I.A.M., Philadelphia (2017)
6. Fasano, A., Hömberg, D., Naumov, D.: On a mathematical model for laser-induced thermotherapy. Appl. Math. Model. **34**, 3831–3840 (2010)
7. Hübner, F., Leithäuser, C., Bazrafshan, B., Siedow, N., Vogl, T.J.: Validation of a mathematical model for laser-induced thermotherapy in liver tissue. Lasers Med. Sci. **32**, 1399–1409 (2017)
8. Marheineke, N., Pinnau, R.: Model hierarchies in space-mapping optimization: feasibility study for transport processes. J. Comput. Methods Sci. Eng. **12**, 63–74 (2012)
9. Schilders, W.H., van der Vorst, H.A., Rommes, J. (eds.): Model Order Reduction: Theory, Research Aspects and Applications. Springer, Heidelberg (2008)

Rational Zernike Functions Capture the Rotations of the Eye-Ball

Zoltán Fazekas, Levente Lócsi, Alexandros Soumelidis, Ferenc Schipp, and Zsolt Németh

Abstract Measurement and mathematical description of the corneal surface and of the optical properties of the human eye are actively researched topics. To enhance the mathematical tools used in the field, a novel set of orthogonal functions—called rational Zernike functions—are presented in the paper; these functions are of great promise for correcting certain types of measurement errors that adversely affect the quality of corneal maps. Such errors arise e.g., due to unintended eye-movements, or spontaneous rotations of the eye-ball. The rational Zernike functions can be derived from the well-known Zernike polynomials—the latter polynomials are used widely in eye-related measurements and ophthalmology—via an argument transformation with a Blaschke function. This transformation is a congruent transformation in the Poincaré disk model of the Bolyai-Lobachevsky hyperbolic geometry.

1 Introduction

The cornea is the primary optical structure of the human eye, contributing the greatest part to the eye's total refractive power. Since the 1980s, plenty of measurement devices for corneal topography have been developed to aid the understanding of the general and the individual optical characteristics associated with the corneal surfaces, as well as to precisely describe these surfaces: firstly only the anterior surfaces, and more recently, also the posterior surfaces. The anterior surface of the cornea is normally close to spherical, but its shape aberrations, and as a consequence its optical aberrations, may result in decrease in the visual quality. For review of devices, methods and models, see [3]. In the frame of projects, several experiments

Z. Fazekas (✉) · A. Soumelidis
Institute for Computer Science and Control (MTA SZTAKI), Budapest, Hungary
e-mail: zoltan.fazekas@sztaki.mta.hu; alexandros.soumelidis@sztaki.mta.hu

L. Lócsi · F. Schipp · Z. Németh
Department of Numerical Analysis, Faculty of Informatics, ELTE Eötvös Loránd University, Budapest, Hungary
e-mail: locsi@inf.elte.hu; schipp@inf.elte.hu; maldini@caesar.elte.hu

© Springer Nature Switzerland AG 2019
I. Faragó et al. (eds.), *Progress in Industrial Mathematics at ECMI 2018*, Mathematics in Industry 30, https://doi.org/10.1007/978-3-030-27550-1_27

and mathematical simulations were carried out by us in collaboration with other colleagues that were related to corneal measurement and shape description. These include development of an experimental multi-camera reflective cornea topographer [10], the shape description of spherical calotte-like surfaces using radial Chebyshev polynomials [9], the utilization of Zernike functions, and their discretization on the unit circle [4, 7, 8].

In Sect. 2, firstly, motivations for the presented research is given, then three shape description methods are outlined that served as precursors to the method proposed in Sect. 3. Though the new mathematical model has yet to be meticulously tested and verified, it has the potential to correct and fit the corneal images and maps that were taken in a slightly rotated position. In Sect. 4, conclusions are drawn and further work is outlined.

2 Motivations for and Precursors to the Presented Research

2.1 Motivations

In a recent study [15], the repeatability of corneal measurements was evaluated for successive topography measurements taken in follow-up of LASIK refractive surgeries. Elevation maps taken with Scheimpflug topography were included in the study. These were fitted to each other and were evaluated for repeatability within and amongst the operative stages (i.e., preoperative, 1 month and 3 months postoperative). Also, the errors due particularly to rotational and translational misalignments were calculated. The challenges posed by such longitudinal evaluations provide motivation for the development of new shape description, corneal map alignment and correction methods. The need for a simplified management and correction of misalignments is pointed out in [2]. Therein, the authors draw attention to the inconsistencies of ocular reference axes. These inconsistencies become an issue when different corneal measurement systems are used for patients, and their maps need to be compared, aligned and aggregated. A pragmatic way to unify, standardize and align these measurements would be extremely helpful. The article [1] serves as an excellent guide to ophthalmologists and biomedical engineers on the topics of elevation-based topography. It discusses and illustrates among other issues the importance of choosing the appropriate reference axis for the axial curvature calculations. The authors underline the need for corneal map corrections, if for some reason or another the choice of the reference axis were not perfect. It also discusses the role of reference surfaces, such as spherical calotte, toric ellipsoid, in locating shape aberrations, such as keratoconi, on the corneal surfaces. The latter issue is particularly relevant to the modified radial Chebyshev polynomials-based corneal shape description method outlined in next subsection.

2.2 Precursors

Zernike functions were introduced in 1934 in [14] to facilitate the mathematical description of optical systems. Zernike functions form a complete orthogonal system on the unit disk $\mathbb{D} := \{z \in \mathbb{C} : |z| < 1\}$. Since then the Zernike-based surface representation has become universally accepted in the corneal topography. One way to define the Zernike functions is via the separation of the azimuthal and radial factors as follows:

$$Z_n^\ell(\varrho e^{i\varphi}) = e^{i\ell\varphi} \cdot R_n^{|\ell|}(\varrho) \quad (\varrho \in [0,1), \varphi \in [0, 2\pi), \ell \in \mathbb{Z}, n = |\ell| + 2s, s \in \mathbb{N}), \tag{1}$$

with

$$R_{\ell+2s}^\ell = \varrho^\ell P_s^{(0,\ell)}(2\rho^2 - 1), \tag{2}$$

where $P_s^{(0,\ell)}$ denote the Jacobi polynomials. The orthogonality relation of the Zernike functions can be written as

$$\int_0^1 \int_0^{2\pi} Z_n^\ell(\rho e^{i\varphi}) \cdot \overline{Z}_m^k(\varrho e^{i\varphi}) \cdot \varrho \mathrm{d}\varphi \mathrm{d}\varrho = \frac{\pi}{n+1} \cdot \delta_{n,m}\, \delta_{\ell,k} \tag{3}$$

with the double integral serving as scalar product. In the above equation, $\delta_{i,j}$ is the Kronecker-delta symbol. From the orthogonality, it follows that a function $f \in L_2(\mathbb{D})$ can be written as the infinite sum

$$f(z) = \sum_{n=0}^{\infty} \sum_{|\ell| \le n} c_{n,\ell} \cdot Z_n^\ell(z) \quad (z \in \mathbb{D}), \tag{4}$$

where the coefficients $c_{n,\ell}$ are defined by the scalar product of f and Z_n^ℓ.

A discretization of $\mathbb{D}$ was introduced in [7, 8]. It allows the aforementioned scalar product to be computed over the proposed set of discrete points with the interpolation being guaranteed between these points. It was shown in these papers that the roots of the Legendre polynomials serve as a good discretization radially, while a uniform division—linearly dependent on n—may be used azimuthally. Several experiments and tests were carried out in conjunction with the discrete Zernike functions on corneal surface data with encouraging results [4, 11, 12].

Although the Zernike polynomials provide the most widely used and adopted way to describe functions on the unit circle, the fact that describing a simple hemisphere requires several components motivated the development of other orthogonal systems [9, 13]. It turns out that considering the even Chebyshev polynomials V_n of the second kind as radial factors allows the hemisphere to described by a single element of the approximating series (due to the weight function appearing in Eq. (5)). The orthogonality of these function can be written as

$$\int_0^1 V_n(r)V_m(r)\sqrt{1-r^2}\,\mathrm{d}r = \frac{\pi}{4}\delta_{mn} \quad . \tag{5}$$

Some useful weight function ϱ—required for a particular application—can replace the weight function in Eq. (5) via solving the following nonlinear differential equation:

$$R'(t)\sqrt{1-R^2(t)} = c \cdot \varrho(t) \ (0 \leq t \leq 1). \tag{6}$$

This way the radial Chebyshev polynomials after the argument transformation R exhibit the following orthogonality property:

$$\int_0^1 V_n(R(t))V_m(R(t))R'(t)\sqrt{1-R(t)^2}\,\mathrm{d}t = \frac{\pi}{4}\delta_{mn}. \tag{7}$$

3 The Rational Zernike Functions

An argument transformation—similar to that in Eq. (7)—was presented in [5]. Using this transformation, the authors defined a modified version of Zernike functions. In this case, however, the transformation is carried out over the entire unit disk, not just in the radial direction. To define this transform, the Blaschke functions

$$B_{a,\varepsilon}(z) := \varepsilon \cdot \frac{z-a}{1-\overline{a}z} \quad (a \in \mathbb{D}, \varepsilon \in \mathbb{T}, z \in \mathbb{D} \cup \mathbb{T}) \tag{8}$$

are used, with $\mathbb{T} := \{z \in \mathbb{C} : |z| = 1\}$. These functions are bijections both on $\mathbb{D}$ and on $\mathbb{T}$, and are analytic an $\mathbb{D} \cup \mathbb{T}$, furthermore, they form a group with respect to the composition of functions. It is the so-called Blaschke group. These functions can be considered as the analogue of the congruent transformations on the Poincaré disk model of the Bolyai–Lobachevsky hyperbolic geometry. We just note here that also the Cayley-Klein model could be used for the purpose. In practical cases, it is enough to consider a subgroup of these functions, e.g., when $z = 1$ is a fixed point.

The operators $L_{a,\varepsilon}$ $(a \in \mathbb{D}, \varepsilon \in \mathbb{T})$ define a translation of functions on $\mathbb{D}$ (and $\mathbb{T}$):

$$(L_{a,\varepsilon}f)(z) = f(B_{a,\varepsilon}^{-1}(z)) \quad (z \in \mathbb{D} \cup \mathbb{T}) \tag{9}$$

and also form a representation—actually, the regular representation—of the Blaschke group. However, this representation is not unitary. A unitary representation—i.e., when the $L_2(\mathbb{D})$ norms of the functions remain intact—can be reached as follows.

$$T_{a,\varepsilon} f = B'_{a,\varepsilon} \cdot f \circ B^{-1}_{a,\varepsilon} \tag{10}$$

With the above defined translation, the translated Zernike functions—referred to in the title as rational Zernike functions, and hereafter referred to as Zernike-Blaschke functions—$T_{a,\varepsilon} Z_n^{\ell}$, according to Eqs. (10) and (1), form a complete orthogonal system in $L_2(\mathbb{D})$ for all $a \in \mathbb{D}$ and $\varepsilon \in \mathbb{T}$. The effect of such translations on a particular Zernike polynomial is shown in Fig. 1.

Now, equipped with this mathematical tool, let us consider an image of an eye looking straight into the camera, the visual effect of a Blaschke-translation applied to it would result in an image that looks similar to an image of the eye viewing into some another direction (i.e., not straight into the camera). Now, for an image of an eye looking elsewhere, the application of the proper Blaschke-translation would result in an image with the eye looking straight into the camera. Such transformations are expected to correct corneal measurements and maps subject to the aforementioned defects.

Let us here draw attention to an important relation between Zernike and Blaschke functions based on [8]. A generating function of Zernike functions can be expressed via Blaschke functions as follows.

$$\frac{(-1)^m}{1-\overline{a}z} B^m_{a,1}(z) = \sum_{n=0}^{\infty} Z^{n-m}_{n+m}(a) \cdot z^n \quad (a = \varrho e^{i\varphi} \in \mathbb{D}) \tag{11}$$

This formula is closely related to unitary representations, and creates a bridge between the Zernike functions and the hyperbolic wavelet transformations. These relations pave the way to handle and answer questions concerning Zernike series via harmonic analysis. Furthermore, an addition formula for Zernike functions can be easily deduced.

Fig. 1 A certain Zernike function (left) and two of its translated—in hyperbolic sense—Blaschke variants over the unit disk. Shades of gray represent the magnitude of the real part of the complex function-values

4 Conclusions and Further Research

The paper considers using Zernike-Blaschke functions to correct corneal maps where the patient's eye was in a slightly rotated position. The discretization of Zernike functions—discussed briefly in Sect. 2—may be adapted to the Zernike-Blaschke functions [6]. Interpolation and approximation properties of truncated series are presently being investigated. Clearly, numerical simulation, as well as measurements on real corneal surfaces must be carried out to verify the applicability of the proposed model. It is our intention to explore the potentials of the proposed transformation, and to give more precise error bounds. Based on these expected results, we intend to formulate recommendations on its practical application.

Acknowledgements This research was supported in the frame of contract EFOP-3.6.3-VEKOP-16-2017-00001: Talent Management in Autonomous Vehicle Control Technologies. The Project is supported by the Hungarian Government and co-financed by the European Social Fund. This research was also supported by Research Funds (OTKA) in the frame of the research contract No. K115804.

References

1. Belin, M.W., Khachikian, S.S.: An introduction to understanding elevation-based topography: how elevation data are displayed – a review. Clin. Exp. Ophthalmol. **37**, 14–29 (2009)
2. Chang, D.H., Waring, G.O.: The subject-fixated coaxially sighted corneal light reflex: a clinical marker for centration of refractive treatments and devices. Am. J. Ophthalmol. **158**, 863–874 (2014)
3. Corbett, M., Rosen E.S., O'Brart, D.P.S.: Corneal Topography: Principles and Practice. BMJ, London (1999)
4. Fazekas, Z., Soumelidis, A., Schipp, F.: Utilizing the discrete orthogonality of Zernike functions in corneal measurements. In: Proceedings of the World Congress on Engineering, IAENG, Hong Kong (2009)
5. Lócsi, L., Schipp, F.: Rational Zernike functions. Annales Univ. Sci. Budapest Sec. Comp. **46**, 177–190 (2017)
6. Németh, Zs., Schipp, F.: Discrete orthogonality of Zernike–Blaschke functions. SIAM J. Numer. Anal. (to appear)
7. Pap, M., Schipp, F.: Discrete orthogonality of Zernike functions. Math. Pann. **16**, 137–144 (2005)
8. Pap, M., Schipp, F.: The voice transform on the Blaschke group II. Annales Univ. Sci. Budapest. Sec. Comp. **29**, 157–173 (2008)
9. Soumelidis, A., Fazekas, Z., Schipp, F., Csákány, B.: Description of corneal surfaces using discretised argument-transformed Chebyshev-polynomials. In: Proceedings of 18th Biennial International EURASIP Conference on Analysis of Biomedical Signals and Images, EURASIP, Darmstadt, pp. 269–274 (2006)
10. Soumelidis, A., Fazekas, Z., Bódis-Szomorú, A., Schipp, F., Németh, J.: Specular surface reconstruction method for multi-camera corneal topographer arrangements. In: Recent Advances in Biomedical Engineering, pp. 639–660. IntechOpen, London (2009)

11. Soumelidis, A., Fazekas, Z., Pap, M., Schipp, F.: Discrete orthogonality of Zernike functions and its application to corneal measurements. In: Selected Papers of the International Conference in Electronic Engineering and Computing Technology, pp. 455–469. Springer, Heidelberg (2010)
12. Soumelidis, A., Fazekas, Z., Pap, M., Schipp, F.: Generic Zernike-based surface representation of measured corneal surface data. In: Proceedings of the MeMeA, IEEE Symposium on Medical Measurements and Applications, pp. 148–153. IEEE, Piscataway (2011)
13. Soumelidis, A., Fazekas, Z., Schipp, F.: Comparison of the corneal surface representations based on Chebyshev polynomials. In: Proceedings of the MeMeA, IEEE Symposium on Medical Measurements and Applications, pp. 1–6. IEEE, Piscataway (2012)
14. Zernike, F.: Beugungstheorie des Schneidenverfahrens und seiner verbesserten Form, der Phasenkontrastmethode. Physica **7**, 689–704 (1934)
15. Zheng, X., Yang, W., Huang, L., Wang, J., Cao, S., Geraghty, B., Zhao, Y.P., Wang, Q., Bao, F., Elsheikh, A.: Evaluating the repeatability of corneal elevation through calculating the misalignment between successive topography measurements during the follow up of LASIK. Sci. Rep. **7**(1), 3122 (2017)

Multi-Obstacle Muscle Wrapping Based on a Discrete Variational Principle

Johann Penner and Sigrid Leyendecker

Abstract This work presents the integration of a discrete muscle wrapping formulation into an optimal control framework based on the direct transcription method DMOCC (discrete mechanics and optimal control for constrained systems (Leyendecker et al., Optim. Control Appl. Meth. 31(6), 505–528, 2010)). The major contribution lies in the use of discrete variational calculus to describe the entire musculoskeletal system, including the muscle path in a holistic way. The resulting coupled discrete Euler-Lagrange equations serve as equality constraints for the nonlinear programming problem, resulting from the discretisation of an optimal control problem. A key advantage of this formulation is that the structure preserving properties of the integrator enable the simulation to account for large, rapid changes in muscle paths at relativity moderate computation coasts. In particular, the derived muscle wrapping formulation does not rely on special case solutions, has no nested loops, a modular structure, and works for an arbitrary number of obstacles. A biomechanical example shows the application of the given method to an optimal control problem with smooth surfaces.

1 Introduction

One aspect of biomechanical simulations is the control of human movement, where a dynamical system must be steered from a given initial state to a predefined final state. However, there exists an infinite number of control trajectories to perform this motion. To constrain this boundary value problem, nonlinear optimal control problems are formulated, that minimise a certain objective function to find optimal control trajectories. When simulating musculoskeletal motion with multibody systems representing bones and joints and muscles acting around them, the muscle's action (three-dimensional force) is characterised by the scalar muscle force value and the

---

J. Penner (✉) · S. Leyendecker
University of Erlangen-Nuremberg, Erlangen, Germany
e-mail: Johann.Penner@fau.de

© Springer Nature Switzerland AG 2019

I. Faragó et al. (eds.), *Progress in Industrial Mathematics at ECMI 2018*,
Mathematics in Industry 30, https://doi.org/10.1007/978-3-030-27550-1_28

muscles path's (tangent) direction at the muscle origin and insertion point. Typically, most muscle paths cannot be adequately represented as straight lines because the anatomical structure of the human body forces the muscles to wrap around bones and adjacent tissue. To represent this behaviour, biomechanical simulations require methods to compute muscle paths, their lengths, and their rates of length change to determine the muscle forces. Assuming that the muscles and tendons are always under tension, they follow the path of minimum distance between origin and insertion point. In this work, we use a discrete variational principle to compute the shortest connection between two points on general smooth surfaces. The muscle path is then a G1-continuous combination of geodesics on adjacent obstacle surfaces [1, 4, 6, 7].

To simplify matters and to keep this publication short, we focus on the integration of the muscle wrapping formulation into a torque actuated multibody optimal control problem and postpone the consideration of Hill-type muscle actuation to the next (longer) publication.

2 Discrete Mechanics and Optimal with Shortest Paths

This section aims to define and solve an optimal control problem, based on the direct transcription method DMOCC [3], which consists of minimising a given discrete objective function subject to constraints that define the dynamics of the system. We therefore formulate the discrete Euler-Lagrange (DEL) equations for the multibody system and the shortest path problem.

2.1 *Discrete Euler-Lagrange Equations of the Multibody System*

In general, the DEL equations are time stepping equations derived directly from the Lagrangian. Their solution approximates the solution of the continuous Euler-Lagrange equations and inherits certain characteristic properties of there solution. Within this discrete formulation, all continuous quantities have to be approximated with discrete counterparts [2–5].

In the following, the discrete path $\boldsymbol{q}_d = \{\boldsymbol{q}_n\}_{n=0}^{N}$ is an approximation of the continuous path on a discrete time grid with constant time step $\Delta t \in \mathbb{R}$ and $N \in \mathbb{N}$ time nodes. We further choose the midpoint quadrature and finite differences to specify the discrete Lagrangian L_d. In addition, we use the discrete nullspace matrix $\boldsymbol{P}(\boldsymbol{q}_n)$ to project the DEL equations of the multibody system into the tangent space of the manifold defined by the interconnecting joint constraints in the multibody system. There only constraint fulfilling motion happens and Lagrange multipliers do not need to be determined. With these approximations, the mechanical DEL are given by

$$\boldsymbol{P}(\boldsymbol{q}_n)^T \cdot \left[D_2 L_d(\boldsymbol{q}_{n-1}, \boldsymbol{q}_n) + D_1 L_d(\boldsymbol{q}_n, \boldsymbol{F}_d(\boldsymbol{u}_{n+1}, \boldsymbol{q}_n)) + \boldsymbol{f}_{n-1}^{+} + \boldsymbol{f}_n^{-}\right] = \boldsymbol{0} \tag{1}$$

for $n = 1, \ldots, N-1$. In this equation, $D_\bullet L_d$ denotes the slot derivative with respect to the $\bullet$-th argument. Furthermore, the nodal reparametrisation $\boldsymbol{q}_{n+1} = \boldsymbol{F}_d(\boldsymbol{u}_{n+1}, \boldsymbol{q}_n)$ in term of discrete local coordinate $\boldsymbol{u}_n$ is used to reduce the system to minimum possible size. The discrete force $\boldsymbol{f}_{n-1}^{+}$ denotes the effect of the generalised joint torque $\boldsymbol{\tau}_{n-1}^{J}$ acting on $\boldsymbol{q}_n$, while $\boldsymbol{f}_n^{-}$ is coming from the effect of $\boldsymbol{\tau}_n^{J}$ acting on $\boldsymbol{q}_n$. In the discrete setting, the dimension of the joint torque vector $\boldsymbol{\tau}_d = \left\{\boldsymbol{\tau}_n^{J}\right\}_{n=0}^{N-1}$ corresponds to the degrees of freedom of the joint and the number of time steps.

2.2 Discrete Euler-Lagrange Equations of the Shortest Path Problem on Multiple Surfaces

To define the discrete shortest path problem over a given set of $I \in \mathbb{N}$ obstacles, we first assume that the muscle completely touches the surfaces, thus the solution is constrained by a scalar valued function of holonomic constraints $\phi^i(\boldsymbol{\gamma}_k^i) = 0$ ($i = 1, \ldots, I$) that define the i-th obstacle surface. Furthermore, we define the discrete geodesic curve $\boldsymbol{\gamma}_d^i = \left\{\boldsymbol{\gamma}_k^i\right\}_{k=0}^{K}$ on a discrete arc length grid with fixed arc length fraction $\Delta s \in \mathbb{R}$ and $K \in \mathbb{N}$ nodes. Depending on the start and end point on the i-th surface, a geodesic curve has to satisfy the geodesic DEL equations

$$\begin{aligned} D_2 T_d(\boldsymbol{\gamma}_{k-1}^i, \boldsymbol{\gamma}_k^i) + D_1 T_d(\boldsymbol{\gamma}_k^i, \boldsymbol{\gamma}_{k+1}^i) - \boldsymbol{\Phi}_d^i(\boldsymbol{\gamma}_k^i)^T \cdot \lambda_k^i &= \boldsymbol{0} \\ \phi^i(\boldsymbol{\gamma}_k^i) &= 0 \end{aligned} \tag{2}$$

for $k = 1, \ldots, K-1$. The term $\boldsymbol{\Phi}_d^i(\boldsymbol{\gamma}_k^i) = \Delta s\, {}^{\delta\phi^i(\gamma_k)}/_{\delta\boldsymbol{\gamma}_k^i}$ is the discrete surface Jacobian. In comparison with the mechanical DEL, the discrete Lagrangian is replaced by the discrete energy of the curve T_d.

Now, let the muscle origin point $\boldsymbol{\rho}^O$ and insertion point $\boldsymbol{\rho}^I$ be given outside the surfaces (see Fig. 1), such that the muscle path is then a G1-continuous combination of geodesics on the surface $\boldsymbol{\gamma}_d$ and adjacent straight line segments $\boldsymbol{r}^i$ and $\boldsymbol{r}^{i+1}$. G1 (geometrical) continuously joined curves share tangential direction, while the

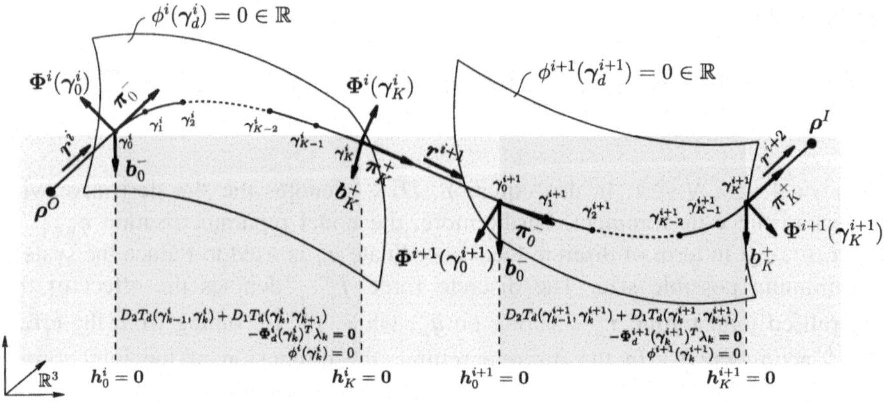

Fig. 1 Example of a discrete geodesic path (on red background) with G1-continuous transition to the straight-line segments (on green background) over $I = 2$ surfaces

length of the tangent vectors might differ. To achieve this G1-continuity, we require collinearity of the tangent and line segment by the following transition constraints

$$\boldsymbol{h}_0^i = \begin{bmatrix} \phi^i(\boldsymbol{\gamma}_0^i) \\ \boldsymbol{r}^i \cdot \boldsymbol{\Phi}^i(\boldsymbol{\gamma}_0^i) \\ \boldsymbol{r}^i \cdot \boldsymbol{b}_k^-(\boldsymbol{\gamma}_0^i, \boldsymbol{\gamma}_1^i) \end{bmatrix} = \boldsymbol{0} \quad \text{and} \quad \boldsymbol{h}_K^i = \begin{bmatrix} \phi^i(\boldsymbol{\gamma}_K^i) \\ \boldsymbol{r}^{i+1} \cdot \boldsymbol{\Phi}^i(\boldsymbol{\gamma}_K^i) \\ \boldsymbol{r}^{i+1} \cdot \boldsymbol{b}_k^+(\boldsymbol{\gamma}_{K-1}^i, \boldsymbol{\gamma}_K^i) \end{bmatrix} = \boldsymbol{0} \tag{3}$$

at the start and end point of the i-th geodesic. In the discrete setting, the tangential direction is given by the discrete momenta $\boldsymbol{\pi}_k^-(\boldsymbol{\gamma}_k^i, \boldsymbol{\gamma}_{k+1}^i) = -D_1 T_d(\boldsymbol{\gamma}_k^i, \boldsymbol{\gamma}_{k+1}^i)$ and $\boldsymbol{\pi}_k^+(\boldsymbol{\gamma}_{k-1}^i, \boldsymbol{\gamma}_k^i) = D_2 T_d(\boldsymbol{\gamma}_{k-1}^i, \boldsymbol{\gamma}_k^i)$ at the boundary points. The surface normal is given by the discrete constraint Jacobian $\boldsymbol{\Phi}^i(\boldsymbol{\gamma}_k^i)$. The binormals at the boundaries are defined as $\boldsymbol{b}_k^-(\boldsymbol{\gamma}_k^i, \boldsymbol{\gamma}_{k+1}^i) = \boldsymbol{\pi}_k^-(\boldsymbol{\gamma}_k^i, \boldsymbol{\gamma}_{k+1}^i) \times \boldsymbol{\Phi}(\boldsymbol{\gamma}_k^i)$ and $\boldsymbol{b}_k^+(\boldsymbol{\gamma}_{k-1}^i, \boldsymbol{\gamma}_k^i) = \boldsymbol{\pi}_k^+(\boldsymbol{\gamma}_{k-1}^i, \boldsymbol{\gamma}_k^i) \times \boldsymbol{\Phi}^i(\boldsymbol{\gamma}_k^i)$.

2.3 Resulting Constrained Nonlinear Optimisation Problem

Finally, the DMOCC (with shortest paths) method deals with the problem of finding the optimisation variables subject to the DEL equations given in Sect. 2, such that a certain discrete objective function J_d, or respectively the sum of a discrete cost function C_d, is minimised, i.e.

$$\min_{\boldsymbol{q}_d,\boldsymbol{\gamma}_d,\boldsymbol{\tau}_d,\Delta t} J_d(\boldsymbol{q}_d,\boldsymbol{\gamma}_d,\boldsymbol{\tau}_d,\Delta t) = \min_{\boldsymbol{q}_d,\boldsymbol{\gamma}_d,\boldsymbol{\tau}_d,\Delta t} \sum_{n=0}^{N-1} C_d(\boldsymbol{q}_n,\boldsymbol{q}_{n+1},\boldsymbol{\gamma}_n,\boldsymbol{\tau}_n^J,\Delta t) \tag{4a}$$

$$\text{subject to} \quad \begin{array}{l} \cdot \text{ mechanical DEL (1)} \\ \cdot \text{ geodesic DEL (2)} \\ \cdot \text{ transition constraints (3)} \\ \cdot \text{ boundary conditions} \end{array} \tag{4b}$$

Herein, the infinite dimensional optimal control problem is transcribed into a finite dimensional nonlinear programming problem that can be solved by any standard algorithm, e.g. Sequential Quadratic Programming (SQP) or the Interior-Point method (IP).

3 Biomechanical Example

As illustrative test scenario, the lifting of the human arm with outstretched initial configuration to a flexed elbow is examined. The simple multibody system in Fig. 2[1] consist of two rigid bodies, which represent the upper and lower arm. For simplification, a revolute joint actuated by the torque $\boldsymbol{\tau}^R \in \mathbb{R}$ is used to model the elbow and the upper arm is fixed in space (no degree of freedom). Muscle origin and insertion points according to [4] are used for the musculus triceps (TRI) and biceps (BIC). Moreover, the muscle path of the triceps is modeled around an ellipsoid which represents the elbow, and the biceps wraps over two cylinders representing the upper and lower arm. In Fig. 2, resulting muscle paths for the biceps and triceps around the elbow are shown, which are G1-continuous combinations of straight lines and geodesics on the wrapping surfaces. In this example, we use the objective function $J_d = \sum_{n=0}^{N-2}(\boldsymbol{\tau}_{n+1}^R - \boldsymbol{\tau}_n^R)^2$, which results in a smooth torque evolution. In total, the system is discretized with $I \cdot K = 3 \cdot 10$ arc length nodes and $N = 20$ time nodes, which leads to 2435 optimisation variables. For large systems such as this, it is very challenging to obtain a feasible solution at all, i.e. a sequence of configurations and joint torques satisfying the DEL equations.

[1]For the 3d bone model see https://www.thingiverse.com/thing:1543880.

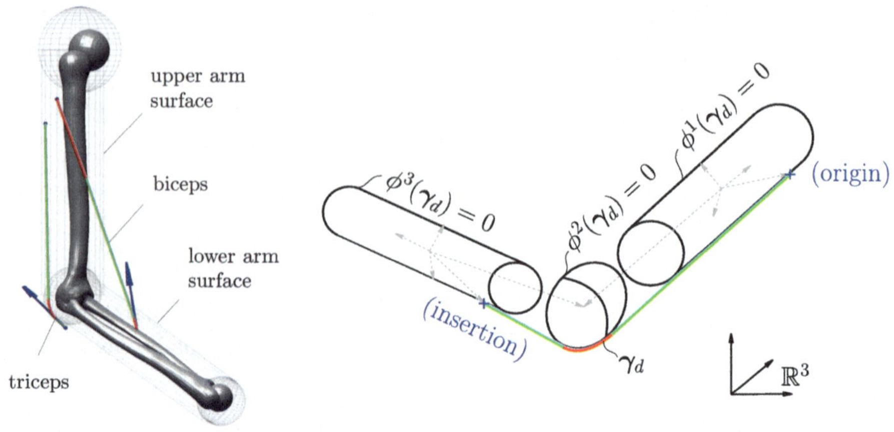

Fig. 2 Muscle paths around the elbow, with straight line segments in green and adjacent geodesics in red. The force directions acting on the lower arm are shown in blue

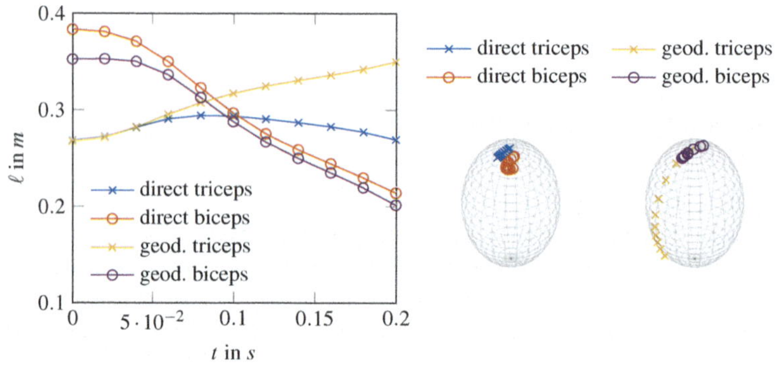

Fig. 3 Comparison of muscle length and force direction (on a unit sphere) during flexion of the elbow with a direct line connection and the geodesic muscle path formulation

3.1 Results

The simulation performs a rest-to-rest manoeuvre from an outstretched configuration to a flexed configuration. Figure 3 shows the evolution of the muscle lengths and force directions with different approaches to represent the muscle path. A direct line connection between the muscle origin and insertion points is compared to the geodesic muscle path formulation. In this comparison, the muscle path of the straight line formulation can intersect bodies, resulting in different results for the two formulations. In particular, this leads to differences in the muscle length, where the geodesic approach takes the stretching of the muscles while wrapping around obstacles into account. Consequently, when comparing the muscle length of the triceps, one sees larger values for the wrapping formulation. The reverse holds for the biceps, which is closer to the surface, resulting in smaller muscle length. Another major difference between both formulations becomes clear when investigating muscle force directions. Again, we see the sliding of the muscles around obstacles that results in a large and rapid change in force directions. While the straight line approach leads to nearly constant force direction, the force direction of the wrapping approach rotates by over 100°.

Acknowledgements The work of this paper is funded by the Federal Ministry of Education and Research (BMBF) as part of the project 05M16WEB—DYMARA.

References

1. De Sapio, V., Khatib, O., Delp, S.: Least action principles and their application to constrained and task-level problems in robotics and biomechanics. Multibody Syst. Dyn. **19**(3), 303–322 (2008)
2. Leyendecker, S., Marsden, J.E., Ortiz, M.: Variational integrators for constrained dynamical systems. ZAMM J. Appl. Math. Mech. **88**(9), 677–708 (2008)
3. Leyendecker, S., Ober-Blöbaum, S., Marsden, J.E., Ortiz, M.: Discrete mechanics and optimal control for constrained systems. Optim. Control Appl. Meth. **31**(6), 505–528 (2010)
4. Maas, R., Leyendecker, S., Biomechanical optimal control of human arm motion. Proc. Inst. Mech. Eng. Part K: J. Multi-body Dyn. **227**(4):375–389 (2013)
5. Marsden, J.E., West, M.: Discrete mechanics and variational integrators. Acta Nume. **10**, 357–514 (2001)
6. Scholz, A., Sherman, M., Stavness, I., Delp, S., Kecskeméthy, A.: A fast multi-obstacle muscle wrapping method using natural geodesic variations. Multibody Syst. Dyn. **36**(2), 195–219 (2016)
7. Thielhelm, H., Vais, A., Brandes, D., Wolte, F.E.: Connecting geodesics on smooth surfaces. Vis. Comput. **28**(6–8), 529–539 (2012)

Experimental Validation of a Mathematical Model for Laser-Induced Thermotherapy

Christian Leithäuser, Frank Hübner, Babak Bazrafshan, Norbert Siedow, and Thomas J. Vogl

Abstract Laser-induced thermotherapy (LITT) is used to treat liver cancer by inserting a laser applicator into the tumor and applying radiation to heat and destroy it. A mathematical model for the simulation of LITT is compared to experimental results with ex-vivo pig livers.

1 Introduction

Laser-induced thermotherapy (LITT) is used to treat liver cancer by inserting a laser applicator into the tumor and applying radiation to heat and destroy it. The co-talk [10] has introduced a mathematical model to simulate LITT which is based on [3]. In the following experimental results with ex-vivo pig livers are presented to validate the model. For further details on the experimental setup we refer to [6].

2 Mathematical Model

We shortly recapitulate the essential parts of the model from [10]. A water cooled laser applicator is entered into the liver. Let Ω be the computational domain (liver) without the applicator. The applicator is considered through boundary conditions. Let the boundary Γ decompose into Γ_{rad} (radiating part of applicator), Γ_{cool} (cooled part of applicator) and Γ_{amb} (ambient boundary of liver). Heat transfer is modeled

C. Leithäuser (✉) · N. Siedow
Fraunhofer Institute for Industrial Mathematics ITWM, Kaiserslautern, Germany
e-mail: christian.leithaeuser@itwm.fraunhofer.de; norbert.siedow@itwm.fraunhofer.de

F. Hübner · B. Bazrafshan · T. J. Vogl
Institut für Diagnostische und Interventionelle Radiologie, Frankfurt, Germany
e-mail: frank.huebner@kgu.de; babak.bazrafshan@kgu.de; t.vogl@em.uni-frankfurt.de

© Springer Nature Switzerland AG 2019

I. Faragó et al. (eds.), *Progress in Industrial Mathematics at ECMI 2018*,
Mathematics in Industry 30, https://doi.org/10.1007/978-3-030-27550-1_29

using the bio-heat equation

$$c_p\rho\frac{\partial T}{\partial t} = \nabla\cdot(k_h\nabla T) + \xi_b(T_b - T) + \mu_a\phi,\ T(x,0) = T_{init},$$

$$k_h\frac{\partial T}{\partial n} = \alpha_{cool}(T_{cool} - T),\ \text{on } \Gamma_{rad}\cup\Gamma_{cool},\ k_h\frac{\partial T}{\partial n} = \alpha_{amb}(T_{amb} - T),\ \text{on } \Gamma_{amb} \tag{1}$$

with temperature $T(x,t)$, specific heat c_p, thermal conductivity k_h, density ρ and blood perfusion coefficient ξ_b. The temperatures T_{init} (initial), T_{cool} (coolant), T_{amb} (ambient) and T_b (blood) are given as well as the heat exchange coefficients α_{cool} and α_{amb}.

Radiative heat transfer is modeled using the P1-approximation

$$-\nabla\cdot(D\nabla\phi(x)) + \mu_a\phi(x) = 0,\quad D = \frac{1}{3(\mu_a + (1-g)\mu_s)},$$

$$D\frac{\partial\phi}{\partial n} = \frac{q_{app}}{A_{\Gamma_{rad}}}\ \text{on } \Gamma_{rad},\ D\frac{\partial\phi}{\partial n} = 0\ \text{on } \Gamma_{cool},\ D\frac{\partial\phi}{\partial n} + \frac{1}{2}\phi = 0\ \text{on } \Gamma_{amb} \tag{2}$$

with radiative energy $\phi(x,t)$, absorption coefficient μ_a, scattering coefficient μ_s, anisotropy factor g, reduced laser power q_{app} and the area $A_{\Gamma_{rad}}$ of Γ_{rad}. The reduced laser power is derived from the actual laser power (see Table 1) by $q_{app} = (1-\beta_q)\hat{q}_{app}$ where β_q is the fraction of power which is directly absorbed by the coolant without entering the liver.

Table 1 Experimental setup for nine test cases

Case label	P22F47	P22F70	P22F92	P28F47	P28F70	P28F92	P34F47	P34F70	P34F92
Laser power [W]									
Measured $\hat{q}_{app}$	22.1	22.1	22.1	28.0	28.0	28.0	33.8	33.8	33.8
Coolant $\dot{V}$ [ml/min]	47.2	69.9	91.7	47.5	70.3	91.8	47.2	70.4	92.2
Time [s]									
Laser on t_{on}	24	30	36	18	30	60	18	24	48
Laser off t_{off}	1266	1236	684	942	1722	1098	1206	948	1182
End t_{end}	1284	1248	702	954	1734	1116	1218	972	1206
Probe position [mm]									
Radial d_r	10.1	11.4	9.2	13.5	13.7	11.1	11.2	9.9	9.6
Axis-direction d_z	12.6	25.7	20.9	21.0	7.5	10.1	23.8	26.3	35.3

Table 2 Tissue dependent parameters for pig liver (cf. [7])

Parameter	Value	Source
Optical (native)		
Absorption coefficient μ_{an} [m^{-1}]	50	[8]
Scattering coefficient μ_{sn} [m^{-1}]	8000	
Anisotropy factor g_n	0.97	
Optical (coagulated)		
Absorption coefficient μ_{ac} [m^{-1}]	60	[8]
Scattering coefficient μ_{sc} [m^{-1}]	30,000	
Anisotropy factor g_c	0.95	
Heat conductivity k_h [$W\,m^{-1}\,K^{-1}$]	0.48	[5]
Heat capacity c_p [$J\,kg^{-1}\,K^{-1}$]	3690	
Tissue density ϱ [$kg\,m^{-3}$]	1080	
Damage rate constant A [s^{-1}]	3.1×10^{98}	[9]
Damage activation energy E_a [$J\,mol^{-1}\,K^{-1}$]	6.3×10^5	
Gas constant R [$J\,mol^{-1}\,K^{-1}$]	8.31	

Tissue damage $w(x, t)$ is modeled using the Arrhenius law

$$w(x, t) = \int_0^t Ae^{-E_a/(RT(x,\tau))} d\tau, \tag{3}$$

with frequency factor A, activating energy E_a and ideal gas constant R. It is needed to model the damage dependence of the optical parameters

$$\begin{aligned} \mu_a &= \mu_{an} + (1 - e^{-w})(\mu_{ac} - \mu_{an}), \\ \mu_s &= \mu_{sn} + (1 - e^{-w})(\mu_{sc} - \mu_{sn}), \\ g &= g_n + (1 - e^{-w})(g_c - g_n) \end{aligned} \tag{4}$$

from the respective values of native and coagulated tissue (see Table 2).

2.1 Numerical Scheme

The system of partial differential equations (PDE) was solved using the finite elements method (FEM). For the heat equation (1) and the P1-approximation (2) a weak form was derived and first order Lagrangian elements were used for the discretization [2]. The Dirichlet type boundary conditions were treated as essential conditions while the Neumann and Robin boundary conditions were treated as natural conditions. The damage function (3) was also solved within the FEM scheme using a weak form of

$$\frac{\partial w}{\partial t} = Ae^{-E_a/(RT)} \tag{5}$$

and zero order elements (constant per cell). The finite elements solver GetDP [4] was used for the implementation.

3 Ex-Vivo Tests

The model was tested experimentally with ex-vivo pig livers for different laser powers and coolant flow rates (see Table 1). An applicator and a temperature probe were inserted into the liver. The relative position (d_r, d_z) of the probe with respect to the applicator tip is given in Table 1. The laser generator was switched on at time t_{on} with laser power $\hat{q}_{app}$ and it was switched off at time t_{off}. Different coolant flow rates $\dot{V}$ were used. Ambient and initial temperatures of $T_{init} = T_{amb} = 21.8\,°\mathrm{C}$ were measured as well as a coolant inflow temperature of $T_{cool} = 20\,°\mathrm{C}$. The blood perfusion rate ξ_b was set to zero (ex-vivo). The ambient heat exchange coefficient α_{amb} was also set to zero. The tissue parameters used for the model can be found in Table 2.

3.1 *Coolant Temperature*

The laser applicator is equipped with a water cooling system. The increase in coolant temperature was measured over time for all nine test cases (see Fig. 1). This data was used to derive the missing parameters β_q and α_{cool}. A coolant absorption factor of $\beta_q = 0.14$ was identified from the instant jump in coolant temperature which occurs when the laser is switched on. It was assumed that this jump originates purely from direct absorption of radiation in the coolant. A temperature exchange coefficient of $\alpha_{cool} = 250\,\mathrm{W\,K^{-1}\,m^{-2}}$ was identified such that the measured and simulated increase in coolant temperature are in good agreement. Therefore, the comparison of measured and simulated coolant temperature in Fig. 1 should be seen as a calibration of the model and not as a validation.

3.2 *Probe Temperature*

In order to validate the model a temperature probe was entered into the liver. Figure 2 shows a comparison between measured and simulated probe temperature. The curves are generally in good agreement. However, for higher temperatures (see

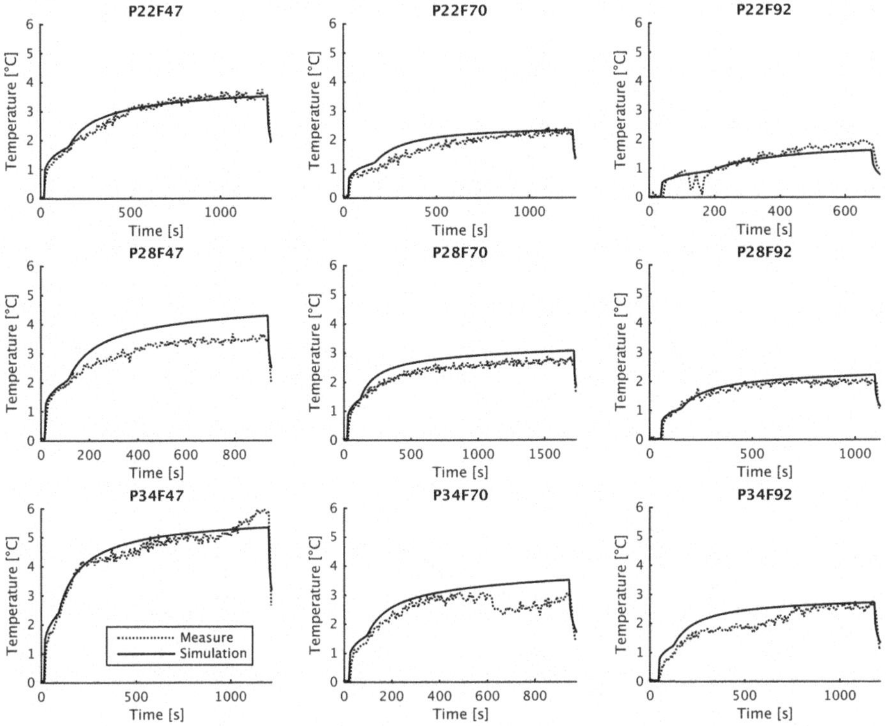

Fig. 1 Comparison of the measured and simulated increase of the coolant temperature

cases *P34F47* and *P34F70*) there is a notable deviation. Most likely this is because the model does not yet account for the energy consumed by the phase transition of water from liquid state to vapor. The model is currently being extended in this direction.

4 Outlook

The ultimate goal is to use simulations to assist surgeons during treatment (cf. [1]). First tests on analyzing patient treatment data have shown that it is important to consider the blood perfusion. However, due to the presence of larger blood vessels it is not enough to use an averaged blood perfusion rate, because the rate is highly dependent on the position of the vessels. We are currently trying to use CT- or MRI-based thermometry to identify the heat sink induced by the blood vessels at the beginning of the treatment. The simulation model can then be used to make predictions for the rest of the ongoing treatment.

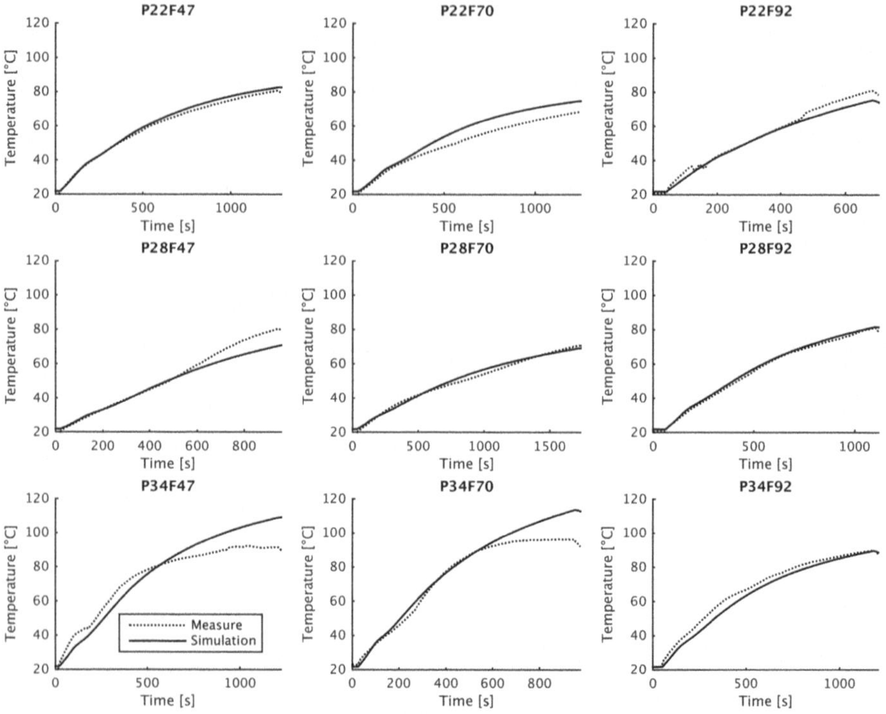

Fig. 2 Comparison of the measured and simulated probe temperature

Acknowledgements The authors acknowledge the financial support by the Deutsche Forschungsgemeinschaft (DFG) for the project *In-vitro Temperaturbestimmung und Computersimulation der Temperaturverteilung zur optimalen Planung und Steuerung der laserinduzierten Thermotherapie (LITT)* (grant no.: VO 479/10-3 and SI 1289/1-3) and by the Federal Ministry of Education and Research (BMBF) of Germany in the framework of the project *proMT: Prognostische modellbasierte online MR-Thermometrie bei minimalinvasiver Thermoablation zur Behandlung von Lebertumoren* (grant no.: 05M16AMA).

References

1. Bazrafshan, B., Koujan, A., Hübner, F., Leithäuser, C., Siedow, N., Vogl, T.J.: A thermometry software tool for monitoring laser-induced interstitial thermotherapy. Biomed. Eng. **64**(4), 449–457 (2019)
2. Ern, A., Guermond, J.L.: Theory and Practice of Finite Elements, vol. 159. Springer, Berlin (2004)
3. Fasano, A., Hömberg, D., Naumov, D.: On a mathematical model for laser-induced thermotherapy. Appl. Math. Model. **34**(12), 3831–3840 (2010)
4. Geuzaine, C.: GetDP: a general finite-element solver for the de Rham complex. In: Special Issue: Sixth International Congress on Industrial Applied Mathematics (ICIAM07) and GAMM Annual Meeting, Zürich 2007, vol. 7, pp. 1010603–1010604. Wiley, Hoboken (2008)

5. Giering, K., Minet, O., Lamprecht, I., Müller, G.: Review of thermal properties of biological tissues. In: Müller, G.J., Roggan, A. (eds.) Laser-Induced Interstitial Thermotherapy, pp. 45–65. SPIE Press, Bellingham (1995)
6. Hübner, F., Leithäuser, C., Bazrafshan, B., Siedow, N., Vogl, T.J.: Validation of a mathematical model for laser-induced thermotherapy in liver tissue. Lasers Med. Sci. **32**(6), 1399–1409 (2017)
7. Puccini, S., Bär, N.-K., Bublat, M., Kahn, T., Busse, H.: Simulations of thermal tissue coagulation and their value for the planning and monitoring of laser-induced interstitial thermotherapy (litt). Magn. Reson. Med. **49**(2), 351–362 (2003)
8. Roggan, A., Dorschel, K., Minet, O., Wolff, D., Muller, G.: The optical properties of biological tissue in the near infrared wavelength range. In: Laser-induced Interstitial Therapy, pp. 10–44. SPIE Press, Bellingham (1995)
9. Schwarzmaier, H.-J., Yaroslavsky, I.V., Yaroslavsky, A.N., Fiedler, V., Ulrich, F., Kahn, T.: Treatment planning for mri-guided laser-induced interstitial thermotherapy of brain tumors—the role of blood perfusion. J. Magn. Reson. Imaging **8**(1), 121–127 (1998)
10. Siedow, N., Leithäuser, C.: Mathematical modeling for laser-induced thermotherapy in liver tissue. In: European Consortium for Mathematics in Industry. Springer, Berlin (2018)

Adaptive Rational Transformations in Biomedical Signal Processing

Gergő Bognár, Sándor Fridli, Péter Kovács, and Ferenc Schipp

Abstract In this paper we provide a summary on our recent research activity in the field of biomedical signal processing by means of adaptive transformation methods using rational systems. We have dealt with several questions that can be efficiently treated by using such mathematical modeling techniques. In our constructions the emphasis is on the adaptivity. We have found that a transformation method that is adapted to the specific problem and the signals themselves can perform better than a transformation of general nature. This approach generates several mathematical challenges and questions. These are approximation, representation, optimization, and parameter extraction problems among others. In this paper we give an overview about how these challenges can be properly addressed. We take ECG processing problems as a model to demonstrate them.

1 Introduction

Mathematical transformation methods have a long history in signal processing, here we mention only the trigonometric Fourier-system, the wavelets, and other orthogonal systems, like the Hermite or Walsh-system. Although these methods perform generally well, their flexibility and adaptivity is usually limited. Our focus is on the adaptive transformations, where the underlying function systems have free parameters that can be adapted to the specific problem and to the signals themselves. We expect such an adapted method to provide a simpler and more concise representation for the signals that still captures the relevant behavior of them.

Our approach is to perform an adaptive transformation by means of rational functions [8]. The rational systems are especially flexible and adaptive, we have

G. Bognár (✉) · S. Fridli · P. Kovács · F. Schipp
Department of Numerical Analysis, Faculty of Informatics, ELTE Eötvös Loránd University, Budapest, Hungary
e-mail: bognargergo@caesar.elte.hu; fridli@inf.elte.hu; kovika@inf.elte.hu; schipp@numanal.inf.elte.hu

© Springer Nature Switzerland AG 2019

I. Faragó et al. (eds.), *Progress in Industrial Mathematics at ECMI 2018*, Mathematics in Industry 30, https://doi.org/10.1007/978-3-030-27550-1_30

arbitrary number of free parameters that determine their behavior. We note that they have found several areas of applications so far. Control and system theories are such important fields. In this paper we are interested in biomedical signals, in particular in ECG signals. The special motivation behind that is the observable similarity between the shapes of the basic rational functions and the natural medical segments (P, QRS, and T waves) of the ECG heartbeats. Thus, the system can be specified according to the shape of the heartbeats, and the parameters carry direct medical information about them. Easy time-domain reconstruction and time-localization of basic rational functions are just additional desirable properties. The rational transform has been successfully applied to several biomedical signal processing problems, including heartbeat modeling [4, 7], ECG compression [14, 18], heartbeat detection [9], arrhythmia classification [2, 3], geometric interpretation of heartbeats [4], EEG epileptic seizure detection [25, 26], and related parameter optimization problems [15, 17, 20]. Furthermore, a rational MATLAB toolbox have been developed [16].

2 Rational Systems

In our models we consider rational functions that are analytic on the unit disc $\mathbb{D} = \{z \in \mathbb{C} : |z| < 1\}$. In the applications in signal processing we will be interested in the real part of their restriction on the torus $\mathbb{T} = \{z \in \mathbb{C} : |z| = 1\}$. Then for any such rational function φ we obtain a real-real function by $[-\pi, \pi) \ni t \to e^{it} \to \operatorname{Re}\varphi(e^{it})$. According to the partial fraction decomposition, the building blocks of the rational functions are the basic rational functions of the form

$$r_{a,n}(z) = \frac{1}{(1 - \overline{a}z)^n} \qquad (z \in \mathbb{C},\ a \in \mathbb{D},\ n \in \mathbb{N}),$$

where the parameter a is the so-called inverse pole. Linear combinations of basic functions having the same pole will be called elementary functions or waves:

$$E_a(z) = \sum_{k=1}^{n} c_k r_{a,k}(z) = \sum_{k=1}^{n} \frac{c_k}{(1 - \overline{a}z)^k} \qquad (a \in \mathbb{D}, n \in \mathbb{N},\ c_k \in \mathbb{C},\ z \in \mathbb{C}).$$

The terminology is justified by the fact that such functions are well localized in the neighborhood of the pole.
Suppose that we have a sequence of, not necessarily distinct, inverse poles

$$a_0,\ a_1, \ldots, a_n, \ldots \qquad (a_n \in \mathbb{D})$$

with multiplicities defined as

$$m_n = \text{card}\{j \,:\, a_j = a_n,\ j \le n\}\,.$$

They generate the sequence of basic functions $\{r_{a_n,m_n} \,:\, n \in \mathbb{N}\}$. Using the Gram–Schmidt orthogonalization process for this sequence with the usual scalar product in $L_2(\mathbb{T})$, we receive an orthonormal system, called Malmquist–Takenaka (MT) system Φ_n $(n \in \mathbb{N})$. Several known classical orthogonal systems like the trigonometric or the discrete Laguerre systems can be generated this way. Generally, the orthogonalization process is computationally demanding. Fortunately, in this case there is an explicit form given by the Blaschke functions, defined as

$$B_a(z) = \frac{z-a}{1-\overline{a}z} \qquad (a \in \mathbb{D},\ z \in \overline{\mathbb{D}})\,.$$

Then

$$\Phi_n(z) = \frac{\sqrt{1-|a_n|^2}}{1-\overline{a}_n z}\prod_{j=0}^{n-1} B_{a_j}(z)\,.$$

We note that the Blaschke functions on the torus can be related to the trigonometric functions. Namely, $B_a(e^{it}) = e^{i(\alpha+\gamma_s(t-\alpha))}$, where $a = re^{i\alpha}$, $s = (1+r)/(1-r)$, and $\gamma_s(t) = 2\arctan(s\tan(t/2))$. Moreover $\gamma_s{}'(t) = \dfrac{1-r^2}{1-2r\cos t + r^2}$ is the well known Poisson kernel. This relation was utilized, for instance, in the construction of discrete rational biorthogonal systems [6].

Both the orthogonal MT and the biorthogonal systems can be expressed in an explicit form, and the projection can be performed efficiently for both representations. Moreover, the numerical approximation error is easy to calculate. In case of MT systems, the results are expressed in terms of MT functions rather than in terms of elementary waves or basic functions. The MT functions do not have the time-localization property of the elementary rational functions. The MT system derived from a given pole sequence depends on the order of the poles in this sequence while our problems are usually not sensitive to that. We can overcome these shortages by taking the biorthogonal expansions [6], which can be more useful in several applications. This is the case when the geometric properties of the ECG signal are to be extracted [4]. We note that in other cases, like features extraction for ECG arrhythmia classification [2, 3] the MT projection perform significantly better.

In practice, the proper discretization of the models is necessary, since the signals themselves are actually discrete time series, usually uniformly sampled. This leads to another interesting question, i.e. the discrete orthogonality of the system. Unlike in case of the trigonometric system, a non-uniform sampling of the torus $\mathbb{T}$ was needed, over which the signal is interpolated, and the coefficients of the projection can be effectively calculated with the fast Fourier-transform (FFT) algorithm. We note that the construction of the sample points is based on the relation between

Blaschke products and complex trigonometric functions. Namely, those products on the unit circle can be expressed by means of an argument transformation on complex trigonometric functions. Then the orthogonal and biorthogonal properties hold for a discrete scalar product including a proper weight function over the sampling points [6]. There the sampling points are defined adaptively to the poles. The drawback of this concept is that resampling of the original signal is necessary. If we have only a few poles or their multiplicities are small, then it can be more effective to use the integral approximation concept instead, like we did in case of ECG classification. For further reading about the rational systems we refer to [12].

3 Mathematical Challenges of the Application of Rational Transformation

Taking ECG processing as a model, we will show how the rational transforms can be used in biomedical applications. Here we only focus on the mathematical problems raised by them.

We emphasize that one of the greatest strength of the rational transform is the arbitrary number of free system parameters. Suppose we have a square-integrable time signal defined on $[-\pi, \pi)$. This in our case corresponds to a single heartbeat in an ECG signal. Before we apply the projection operator, a system identification step is necessary. It is a parameter optimization problem discussed later. Namely, poles and multiplicities should be specified so that the corresponding rational projection provides a good approximation of the signal. For ECG signals, good approximation means that the rational function representing the signal contains the same medical information as the original measurement. There is no purely mathematical metric that would measure the quality of approximation in medical sense. However, most of the times the classical least square approximation is used to find a representation of the signal with respect to the corresponding function system. Then the problem reduces to a well-known mathematical problem. We will however show that the problems in real applications are more complex than that described above.

3.1 Heartbeat Classification

Cardiac arrhythmia is a group of conditions where the heart shows abnormal activity or behavior. One usual way is the classification of heartbeats into predefined classes. PhysioNet [10] provides 16 classes. In this case we use rational systems for representing the ECG signal and to extract features for the classification algorithm. One possible strategy for the system optimization is when the number and multiplicities of the poles are fixed, and the optimization should be performed for the locations only. Then the problem is actually a special case of the variable

projection concept [11]. This is the case in classification problems [2, 3]. Namely, we want the coefficients of the projection and the poles themselves carry direct medical information, i.e. we want them to be related to the morphology of the P, QRS, T peaks.
The problem is not about best approximation. It is good approximation and representation at the same time.
It turned out that taking three inverse poles, corresponding to the three main waveforms of the ECG heartbeats, with fixed multiplicities 2,4,2 is reasonable. The goal is not necessarily the best approximation. Good representation of the samples that captures the relevant behavior are of equal importance. To this order, in [2] we restricted the exit conditions of the optimization algorithm, making it to rely more on its initialization. In [3] we introduced modifications on the objective function, we restricted the complex argument of the inverse poles (and thus the time locations of the rational functions) to intervals corresponding to the main ECG waveforms. Furthermore, in order to reduce intensity differences between the ECG waveforms, in [3] we performed the optimization for the three poles subsequently. We note that a good initial guess can significantly improve the efficiency of the algorithms. In ECG classification, the main peak locations are roughly estimated to this order.

The problem then reduces to constrained optimization. We not only need to keep the poles within the unit circle but within the vicinity of the corresponding peaks. Several optimization algorithms, deterministic and probabilistic are at hand. The Nelder–Mead and the particle swarm optimization (PSO) turned to be very effective. In order to satisfy the constraints in a natural way, we have developed hyperbolic versions of these algorithms [7, 17, 20]. These hyperbolic methods are the modifications of the original algorithms affecting on the unit disk following the Poincaré disk model. Based on medical properties of the ECG, we also derived constraints for the inverse poles of the P, T waves and the QRS complex in [15].

The optimization can be applied to the heartbeats individually, or to a set of heartbeats. It means a per-heartbeat, or a per-patient optimization. The per-heartbeat concept can be efficiently utilized for ECG compression, since the goal is to effectively compress each heartbeats, independently. On the other hand, the per-patient concept was proved to be more effective for ECG classification, since in this case the pole combinations are assigned to the patients instead of the individual heartbeats, and it leads to a more stable heartbeat representation. Furthermore, based on the connection between the rational systems and the shapes of the ECG signals, the poles themselves can be considered as patient descriptors [3].

We utilized the adaptive rational transform for 'class-oriented' arrhythmia classification in [2, 3]. In these studies the ECG signals were preprocessed and segmented into heartbeats, then a per-patient optimization was performed in order to find the best fitting pole combination for each patients. Morphological descriptors were extracted from the rational projections and poles, which was extended with the common RR interval features as dynamic descriptors. Finally, we used support vector machine (SVM) classifier on the feature vectors, and fusions. The evaluation on the MIT-BIH Arrhythmia Database proves that our algorithms outperform the previous ones of this kind [13, 19, 22–24, 28]. These works utilize a variety of feature

Table 1 Comparison of the proposed methods and the reference works

Method	Feature vector	Classifier	Accuracy
Work of [19]	Hermite	SOM	98.49%
Work of [23]	Wavelet + RR	ANN	96.77%
Work of [22]	Hermite + HOS	SVM	98.18%
Work of [24]	Waveform	DT	96.13%
Work of [13]	Wavelet + ICA	SVM	98.86%
Work of [28]	Wavelet + ICA + RR	SVM	99.32%
Proposed method [2]	Rational + RR	SVM	99.38%
Proposed method [3]	Rat. + Poles + RR + Fusion	SVM	**99.51%**

The best classification accuracy is shown in bold face

extraction and classification methods, like Hermite and wavelet decompositions, high-order statistics, independent and principal component analysis, combined with artificial neural network (ANN), decision tree (DT), self organizing map (SOM), and support vector machine (SVM) classifiers. The advantage of our concept is that rational variable projection is an efficient dimension reduction technique that gives good representation of ECG heartbeats using only a few coefficients. In the meantime, this representation is directly correlated to the morphology of the ECG wave components. The accuracy comparison of the reference works and the proposed method is given in Table 1.

3.2 ECG Compression

In case of ECG compression, we construct rational functions that approximate the signal in the sense described in the beginning of this section. The parameters of the rational function, i.e. the poles along with their multiplicities, and the coefficients together form the compressed data. This means that we are interested in reducing the number of parameters.

The problem is good approximation with keeping the dimension, the system complexity as low as possible.

To address this problem we developed [14, 18] an optimization method which can be considered as the generalization of the variable projection method. We note that in the variable projection method the dimension of the subspace of the projection is a priori fixed. That was the case for ECG classification above. For compressing purposes this constraint is not appropriate, therefore we added a new free parameter related to the dimension of the subspace. In the situation when increasing dimension results in nested subspaces, the optimization would terminate at the highest possible value of the dimension. On the other hand, high dimensions are not desired in real applications because it increases the complexity of the model. For controlling the dimension we introduced a penalty function $\Lambda(N)$ that is monotonically increasing with N. Then the generalized variable projection functional is of the form

$$\rho : \Gamma \times \mathbb{N}, \quad \rho(\mathbf{a}, d) = \|f - P_{\mathbf{a}}^{N} f\| + \Lambda(d), \tag{1}$$

where f is the given signal, Γ is the parameter set of the collection of systems used, d is the dimension, and $\Lambda(d)$ measures the complexity of the system with dimension d in a proper sense. We note that this setting makes sense only if all of the following conditions hold for the system: (a) it is flexible enough but easy to parametrize; (b) the complexity function Λ is properly designed; (c) an efficient optimization algorithm can be constructed. We showed that these condition hold for the rational systems. The efficient optimization algorithm that we developed for this case is the so-called multi-dimensional hyperbolic PSO (MDHPSO) algorithm [14, 18].

In a study case we demonstrated the efficiency of the generalized variable projection method for ECG compression in [18]. By performing comparison tests on 24 records of the MIT-BIH Arrhythmia Database [10] we obtained that our algorithm outperforms the previous ones [1, 21, 27]. Table 2 illustrates the results on selected records: 117 and 119. The latter contains extremely varying periods [5], which makes it ideal for compression tests. In the experiment, we segmented the first channel of the whole records into more than 3500 heartbeats. Then, we applied the proposed algorithm to solve the optimization problem in Eq. (1) for each beat. The performance is evaluated in terms of percent root-mean-square difference (PRD), compression ratio (CR), and quality score (QS):

$$\mathrm{PRD} = \frac{\|f - P_{\mathbf{a}}^{N} f\|_2}{\|f\|_2} \times 100, \quad \mathrm{CR} = \frac{\text{Size of the uncompressed data}}{\text{Size of the compressed data}}, \quad \mathrm{QS} = \frac{\mathrm{CR}}{\mathrm{PRD}},$$

where the approximation $P_{\mathbf{a}}^{N} f$ is given by the orthogonal projection of the signal f to the subspace spanned by the corresponding rational functions. Although other approaches [21, 27] also utilize rational functions, the number of different inverse poles, and their multiplicities were fixed a priori in those works. In our algorithm, these parameters are found automatically due our optimization process. Therefore, we can dynamically change the complexity of the nonlinear signal model from beat to beat.

Table 2 Experimental results of selected recordings

	Work of [21]			Work of [27]			Work of [1]			Proposed work [18]		
Rec.	PRD	CR	QS	PRD	CR	QS	PRD	CR	QS	PRD	CR	QS
117	0.62	25.64	41.36	1.57	34.78	22.15	1.60	23.00	14.38	0.39	18.15	**46.70**
119	1.17	25.64	21.91	2.37	34.78	14.68	2.20	25.00	11.36	0.54	14.14	**26.24**

The best QS of each row is shown in bold face

3.3 QRS Modeling

Another concept for system identification is the reconstruction of the poles from geometric properties of the signals. In [4] we proposed methods that provide the identification of rational models for the QRS complexes in ECG heartbeats based on medical descriptors of the signals. These can be the so-called fiducial points, or other common descriptors, like QRS duration or ventricular activation time (VAT). The idea is that if we represent the ECG wave components with rational functions, we have analytic way to find the peaks and zero crossings of the model curve, thus to derive medical descriptors. To this order, we exploited the properties of the basic rational and the Blaschke functions. The system identification is actually an inverse problem based on the analytic model. We construct synthetic model curves that complies the given descriptors, and at the same time the numerical approximation error is acceptable for the whole heartbeat. This leads to a different system identification approach compared to the previously discussed methods.
No optimization, but a reliable geometric parameter extraction method is needed. This, however, may be even more problematic than the numerical optimization, because of the ECG noises and artifacts. Namely, the peak locations may be shifted, and the zero crossing points may become uncertain, even if filtering is applied.

Acknowledgements EFOP-3.6.3-VEKOP-16-2017-00001: Talent Management in Autonomous Vehicle Control Technologies—The Project is supported by the Hungarian Government and co-financed by the European Social Fund. This research was supported by the Hungarian Scientific Research Funds (OTKA) No K115804.

References

1. Abo-Zahhad, M., Al-Ajlouni, A.F., Ahmed, S.M., Schilling, R.: A new algorithm for the compression of ECG signals based on mother wavelet parameterization and best-threshold level selection. Digit. Signal. Process **23**(3), 1002–1011 (2013)
2. Bognár, G., Fridli, S.: Heartbeat classification of ECG signals using rational function systems. In: Moreno-Díaz, R., et al. (eds.) Computer Aided Systems Theory – EUROCAST 2017. Lecture Notes in Computer Science, vol. 10672, pp. 187–195. Springer, Cham (2018)
3. Bognár, G., Fridli, S.: ECG heartbeat classification by means of variable rational projection. Biomed. Sign. Proc. Control (to appear)
4. Bognár, G., Schipp, F.: Geometric interpretation of QRS complexes in ECG signals by rational functions. Ann. Univ. Sci. Budapest. Sect. Comp. **47**, 155–166 (2018)
5. Chou, H.H., Chen, Y.J., Shiau, Y.C., Kuo, T.S.: An effective and efficient compression algorithm for ECG signals with irregular periods. IEEE Trans. Biomed. Eng. **53**(6), 1198–1205 (2006)
6. Fridli, S., Schipp, F.: Biorthogonal systems to rational functions. Ann. Univ. Sci. Budapest. Sect. Comp. **35**, 95–105 (2011)
7. Fridli, S., Kovács, P., Lócsi, L., Schipp, F.: Rational modeling of multi-lead QRS complexes in ECG signals. Ann. Univ. Sci. Budapest. Sect. Comp. **37**, 145–155 (2012)

8. Fridli, S., Lócsi, L., Schipp, F.: Rational function systems in ECG processing. In: Moreno-Díaz, R., et al. (eds.) Computer Aided Systems Theory – EUROCAST 2011. Lecture Notes in Computer Science, vol. 6927, pp. 88–95. Springer, Berlin (2012)
9. Gilián, Z., Kovács, P., Samiee, K.: Rhythm-based accuracy improvement of heart beat detection algorithms. In: Computing in Cardiology, pp. 269–272 (2014)
10. Goldberger, A.L., et al.: PhysioBank, PhysioToolkit, and PhysioNet: Components of a new research resource for complex physiologic signals. Circulation **101**(23), e215–e220 (2000)
11. Golub, G.H., Pereyra, V.: The differentiation of pseudo-inverses and nonlinear least squares problems whose variables separate. SIAM J. Numer. Anal. **10**(2), 413–432 (1973)
12. Heuberger, P.S.C., van den Hof, P.M.J., Wahlberg, B. (eds.): Modelling and identification with rational orthogonal basis functions. Springer-Verlag London Limited, London (2005)
13. Jiang, X., Zhang, L., Zhao, Q., Albayrak, S.: ECG arrhythmias recognition system based on independent component analysis feature extraction. In: TENCON 2006 – 2006 IEEE Region 10 Conference, pp. 1–4 (2006)
14. Kovács, P.: Transformation methods in signal processing. Ph.D. Thesis, ELTE Eötvös Loránd University, Budapest, Hungary (2016)
15. Kovács, P.: Rational variable projection methods in ECG signal processing. In: Moreno-Díaz, R., et al. (eds.) Computer Aided Systems Theory – EUROCAST 2017. Lecture Notes in Computer Science, vol. 10672, pp. 196–203. Springer, Cham (2018)
16. Kovács, P., Lócsi, L.: RAIT: the rational approximation and interpolation toolbox for Matlab, with experiments on ECG signals. Int. J. Adv. Telecom. Elect. Sign. Syst. **1**(2–3), 67–752 (2012)
17. Kovács, P., Kiranyaz, S., Gabbouj, M.: Hyperbolic particle swarm optimization with application in rational identification. In: Proccedings of 21st European Signal Processing Conference (EUSIPCO 2013), pp. 1–5 (2013)
18. Kovács, P., Fridli, S., Schipp, F.: Rational generalized variable projection with application in ECG processing. IEEE. Trans. Sign. Proc. (to appear)
19. Lagerholm, M., Peterson, C., Braccini, G., Edenbrandt, L., Sornmo, L.: Clustering ECG complexes using Hermite functions and self-organizing maps. IEEE Trans. Biomed. Eng. **47**(7), 838–848 (2000)
20. Lócsi, L.: A hyperbolic variant of the Nelder–Mead simplex method in low dimensions. Acta Univ. Sapientiae Mathematica **5**(2), 169–183 (2013)
21. Ma, J., Zhang, T., Dong, M.: A novel ECG data compression method using adaptive Fourier decomposition with security guarantee in e-Health applications. IEEE J. Biomed. Health Inform. **19**(3), 986–994 (2015)
22. Osowski, S., Hoai, L.T., Markiewicz, T.: Support vector machine-based expert system for reliable heartbeat recognition. IEEE Trans. Biomed. Eng. **51**(4), 582–589 (2004)
23. Prasad, G.K., Sahambi, J.S.: Classification of ECG arrhythmias using multi-resolution analysis and neural networks. In: Proccedings of Conference on Convergent Technologies for Asia-Pacific Region, vol. 1, pp. 227–231 (2003)
24. Rodriguez, J., Goni, A., Illarramendi, A.: Real-time classification of ECGs on a PDA. IEEE Trans. Info. Tech. Biomed. **9**(1), 23–34 (2005)
25. Samiee, K., Kovács, P., Gabbouj, M.: Epileptic seizure classification of EEG time-series using rational discrete short-time Fourier transform. IEEE Trans. Biomed. Eng. **62**(2), 541–552 (2015)
26. Samiee, K., Kovács, P., Gabbouj, M.: Epileptic seizure detection in long-term EEG records using sparse rational decomposition and local Gabor binary patterns feature extraction. Knowl. Based Syst. **118**, 228–240 (2017)
27. Tan, C., Zhang, L., Wu, H.: A novel Blaschke unwinding adaptive Fourier decomposition based signal compression algorithm with application on ECG signals. IEEE J. Biomed. Health Inform. **23**(2), 672–682 (2018)
28. Ye, C., Kumar, B.V.K.V., Coimbra, M.T.: Heartbeat classification using morphological and dynamic features of ECG signals. IEEE Trans. Biomed. Eng. **59**(10), 2930–2941 (2012)

Challenges in the Modelling and Control of Varicella in Hungary

Rita Csuma-Kovács, János Dudás, János Karsai, Ágnes Dánielisz, Zsuzsanna Molnár, and Gergely Röst

Abstract The introduction of varicella-zoster virus (VZV) vaccines into the routine vaccination schedule is being under consideration in Hungary. Mathematical models can be greatly useful in advising public health policy decision making by comparing predictions for different scenarios, and by quantifying the costs and benefits of immunization strategies. Here we summarize the major challenges, most of them specific to Hungary, in devising and parametrizing dynamical models of varicella transmission dynamics with vaccination policy. We gain some important insights from a simple compartmental model regarding the seasonality and intrinsic oscillation frequency of the disease dynamics, and the sensitivity to the underreporting ratio. Finally, we discuss the ideas for a more complete, realistic model.

1 Introduction

The varicella-zoster virus is a highly contagious disease that affects a huge proportion of the population, consequently the varicella incidence is of a similar magnitude to the number of births. Although most people contract the disease in their childhood, when the symptoms are generally mild, complications may occur during the infection. Furthermore, at an older age the risk of serious complications is significantly higher.

R. Csuma-Kovács · J. Dudás · J. Karsai (✉)
Bolyai Institute, University of Szeged, Szeged, Hungary
e-mail: csuma.rita@math.u-szeged.hu; dudasj@math.u-szeged.hu; karsai.janos@math.u-szeged.hu

Á. Dánielisz · Z. Molnár
National Public Health Center, Budapest, Hungary
e-mail: agnes.danielisz@nnk.gov.hu; molnar.zsuzsanna@nnk.gov.hu

G. Röst
Mathematical Institute, University of Oxford, Oxford, UK
e-mail: rost@math.u-szeged.hu

© Springer Nature Switzerland AG 2019
I. Faragó et al. (eds.), *Progress in Industrial Mathematics at ECMI 2018*, Mathematics in Industry 30, https://doi.org/10.1007/978-3-030-27550-1_31

In many developed countries, varicella vaccination programs are already implemented. Originally one-dose programs were introduced, which have been replaced by multiple-dose vaccinations in some countries by now. In Hungary, vaccination is marketed for non-routine use, and it has been made available free of charge in a few cities in recent years. There are many country-specific studies regarding the effects and cost-effectiveness of the introduction of varicella vaccination, e.g. [1, 2, 6]. However, there are hardly any studies about Hungary ([5] is a retrospective, descriptive study), where the introduction of varicella vaccination into the routine childhood vaccination program is being considered. Given the actuality and the importance of this issue, here we summarize the challenges of such a modelling work, draw some conclusions from a simple compartmental model and devise a plan for comprehensive future work.

2 Challenges in Modeling

Latency of the Virus and Reactivation as Zoster Upon recovery from the varicella infection, VZV remains in the body in latent form. In general, the individual develops lifelong immunity to VZV. This immunity usually prevents the reappearance of varicella, however the immunity can wane over time, hence the virus may reactivate causing zoster. Zoster infected people are also infectious, but at a lower rate than varicella infected persons. The length and the efficacy of VZV immunity can show a wide interindividual variability.

The Hypothesis of Exogenous Boosting The waning immunity against VZV can be boosted if the individual has a contact with a VZV infected person. Assuming exogenous boosting, it is reasonable that after introducing vaccination, the number of varicella cases decreases and consequently the zoster incidence temporarily increases [1, 6].

Age Structure Since the virus is highly contagious and appears mainly among young children, the transmission dynamics of the virus is largely age specific. Furthermore, varicella has more severe symptoms and higher risk of complications at an older age, and reactivation in the form of zoster also occurs at older age. Hence, age-structured models are necessary to capture these phenomena.

Underreporting In Hungary, monthly reporting of varicella cases to the public health authorities is obligatory. Unfortunately, the varicella incidence appears to be much higher than reported, since the annual birth number is about 2.5-times higher than the reported varicella cases; and according to most studies, these two values should be nearly equal [6]. Among others, the main reason is that not every child is taken to the pediatrician, as there is no effective medical treatment.

Seasonality Available data also reflects a seasonal behaviour in varicella incidence. It can be traced back to the high number of infected children; consequently the school term and vacation play an important role in the spread of VZV. To describe this phenomenon, time-dependent contact rates are needed in the model.

Lack of Zoster Data Contrary to varicella cases, it is not compulsory in Hungary to report the zoster cases. Therefore, there is no available data related to zoster. We need to make assumptions, based on studies from other countries.

Vaccines are Already Present Parents have the opportunity to buy the vaccine on the market in Hungary. Some cities have made the vaccine available for free for local children. Thus, a fraction of children have already been immunized.

Vaccination Efficacy and Waning Since varicella vaccination was licensed in the mid-80s in some European countries, the vaccine parameters are fairly reliable. In case of MMRV vaccine, 65% of the vaccinated population acquires full protection after one dose and 95% after the second dose. The vaccine-induced protection wanes in 15–20 years after one dose; while the two-dose vaccination provides lifelong immunity [6].

Long Term Dynamics Since we need predictions for many years ahead, an age-structured model should handle the transitions between age cohorts, which makes it more difficult than in models for single outbreaks, such as influenza with short-term behaviour [3]. Demographic changes also need to be taken into consideration.

Cost-Benefit Calculations In 2017, [5] gave a comprehensive study on the economic burden of varicella in Hungary using descriptive statistical methods. There are many uncertainties related to the introduction of VZV-vaccination, for instance, the specific program, the type of the vaccination etc. are still unknown. Hence, detailed dynamic model-based studies of the economic effects can be extremely useful.

3 Insights from a Simple Compartmental Model

Based on the known models in the literature [1, 6], we use a simple compartmental system in our studies with the compartments representing the varicella disease states: *Susceptible*, *Exposed*, *Infectious*, *Recovered*, *Susceptible to Zoster*, *Zoster*, *Zoster Immune*. Maternal immunity is not taken into account in our model. Although the real situation is different, for the sake of simplicity we assume that the birth and death rates are equal (d). Then the total population is constant and a proportional model can be used where $1 = s + e + i + r + s_z + z + r_z$. The model is as follows:

$$\begin{aligned} s' &= d - \lambda s - ds, & s_z' &= -\sigma\lambda s_z + \zeta r - \eta s_z - ds_z, \\ e' &= \lambda s - \varepsilon e - de, & i_z' &= \eta s_z - \kappa i_z - di_z, \\ i' &= \varepsilon e - \gamma i - di, & r_z' &= \kappa i_z - dr_z, \\ r' &= \gamma i + \sigma\lambda s_z - \zeta r - dr, \end{aligned} \tag{1}$$

where the force of infection is $\lambda = \beta(i + \nu i_z)$ and $(.)'$ represents time derivative. Newborns directly become susceptible, then, one can become infected by being in contact with a varicella or zoster infectious person. Having been infected, individuals go through a non-infectious latent period, and then they will be infectious. Following the recovery, individuals acquire immunity to VZV. Immunity may wane, and then individuals become susceptible to zoster. One can either be boosted through exposure to VZV and regain immunity with efficiency σ or become zoster infectious through reactivation of VZV with the rate η. Zoster recovered individuals have lifelong immunity to VZV. The average length of the exposed, infectious, temporary immunity, and zoster states are ε^{-1}, γ^{-1}, ζ^{-1} and κ^{-1}, respectively.

The basic reproduction number R_0 is a key parameter regarding the level of virulence of the disease. In [7] the basic reproduction number was determined for a slightly different model, and the usual result holds, namely that if $R_0 < 1$ then the disease-free equilibrium is asymptotically stable, but if $R_0 > 1$ then the disease will persist. With straightforward calculations, using the same method, we obtain

$$R_0 = \frac{\beta\varepsilon}{(\gamma+d)(\varepsilon+d)} + \frac{\nu\beta}{(\kappa+d)} \cdot \frac{\varepsilon\gamma\zeta\eta}{(\varepsilon+d)(\gamma+d)(\zeta+d)(\eta+d)}, \tag{2}$$

where the terms correspond to the expected number of cases generated by a typical individual during primary varicella infection or acute herpes-zoster, respectively.

3.1 Data Analysis and Model Fitting

Annual Varicella incidence data for 20 years and monthly data since 2010 in Hungary were available to us (red curves in Fig. 1 show the incidence corrected by the fitted underreporting ratio $q = 0.4$). Since zoster incidence data is not available, the related parameters were taken from the literature. Values of $(s, e, i, r, s_z, z, r_z)$ at any time are not known, hence initial values of the solutions were taken close to

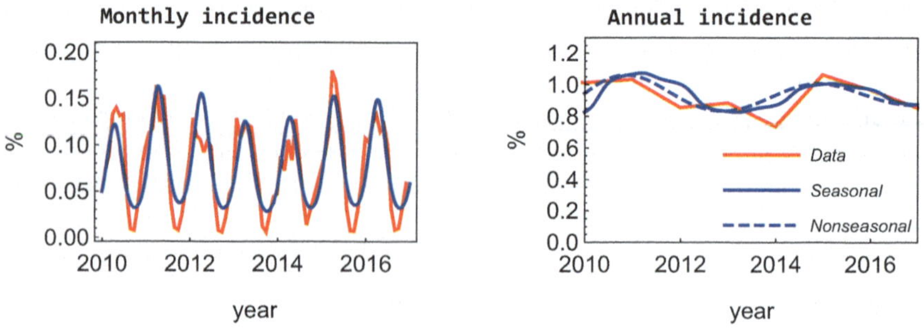

Fig. 1 Varicella incidences: data (red) and fitted model (blue)

the endemic equilibrium according to the values of parameters. Based on our former arguments, the underreporting ratio (q) is included into the fitting process.

Due to the strong seasonality of varicella, we replaced the constant β in the system by a periodic function $\hat{\beta}(t) = \beta(1 + b\cos(2\pi t - c))$ with $b = 0.25$ and $c = 0.5$ chosen by a separate fitting process. The seasonal system with parameters β (transmission rate) and q (underreporting ratio) was fitted to the monthly data. The fitting model is simple: the cumulative growth of $i(t)$ is measured by $\hat{i}(t)$ with $\hat{i}'(t) = \varepsilon e(t)$, and hence the monthly and annual incidences are modeled by $MM(t) = q(\hat{i}(t + 1/12) - \hat{i}(t))$ and $AM(t) = q(\hat{i}(t + 1) - \hat{i}(t))$, respectively.

Fitting was performed by the sophisticated and well-tested command NonlinearModelFit in Wolfram Mathematica 11.3, which can be applied to implicitly defined models such as numerical solutions of differential equations, and it can measure the goodness of the fit. Default options and $ConfidenceLevel \to 0.95$ were used.

After iteratively applied fitting and some fine-tuning, the final rounded values of fixed parameters are $d = 0.01$, $\epsilon = 26$, $\gamma = 52$, $\nu = 0.07$, $\zeta = 0.05$, $\eta = 0.003$, $\sigma = 0.7$, $\kappa = 40$. The goodness of the fit was measured by the adjusted $R^2 = 0.933$. The fitted values are $q = 0.398$ (standard error: 0.012, 95%, confidence interval: [0.374, 0.422]); $\beta = 768.94$ (standard error: 54.27, 95%, confidence interval: [660.88, 877]). The result can be seen on Fig. 1. The monthly incidence data and fitted model $MM(t)$ can be found on the left side, while the right one contains the annual data and the fitted model $AM(t)$ as well as the corresponding autonomous model with the same parameters. Finally, we emphasize that although the seasonality is very strong, both the monthly and annual incidence models show a multi-annual periodicity. The yearly peaks have maxima approximately at every 4 years. This phenomenon is known in the epidemiology of varicella and the value agrees the practice. The same period can be obtained by the autonomous model.

3.2 Sensitivity to Underreporting Ratio

According to the previous section, varicella cases are likely to be seriously underreported in Hungary ($q \approx 0.4$). The model fitting is coherent with what the serological studies suggest. In this section we investigate, how sensitive our model is to the ratio of the reported and total cases, i.e., we examine dependence of the basic reproduction number R_0 (see Eq. (2)) on this ratio q at the parameters fitted above.

Assuming that in Hungary the population is at the endemic equilibrium and using the equality $n_V/q = \gamma i^*$ (where n_V is the average annual reported varicella incidence since 2010 and i^* is the endemic equilibrium of i), we obtain the relation between q and R_0 depicted in Fig. 2. Note that in the literature a wide variety of different R_0 values can be found for the VZV. In [4], the highest value is 16.91 (Netherlands) and the lowest is 3.31 (Italy). According to these values, the

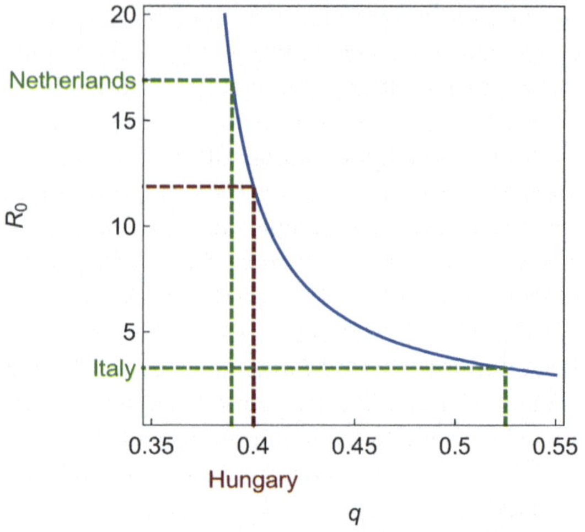

Fig. 2 Relation between the underreporting ratio q and the basic reproduction number

underreporting ratio in Hungary would change between 0.39 and 0.53. As we found above, the fitted value of q is about 0.4 and the corresponding R_0 is 11.87.

4 Conclusion

We gave an overview of the main challenges in the modelling of varicella in Hungary. We fitted a very simple model to the available data, and found that the strong seasonality of varicella infections and the underreporting are essential. The main aim of our research is to forecast the impact of vaccination in Hungary. Based on our simple model the global effects and strategic goals can be already visible. To build a realistic model which can be used to evaluate the impact of vaccination policies, the simple compartmental system should be significantly extended by vaccination, seasonal effects and age structure with age specific parameters and contact patterns.

Acknowledgements Research was supported by the projects EFOP-3.6.2-16-2017-00015, NKFI KH 125628, 20391-3/2018/FEKUSTRAT, and MSCA-IF 748193 (G. Röst).

References

1. Brisson, M., et al.: Modeling the impact of one- and two-dose varicella vaccination on the epidemiology of varicella and zoster. Vaccine **28**(19), 3385–3397 (2010)
2. Damm, O., et al.: Systematic review of models assessing the economic value of routine varicella and herpes zoster vaccination in high-income countries. BMC Public Health **15**(1), 533 (2015)

3. Knipl, D., Röst, G.: Modelling the strategies for age specific vaccination scheduling during influenza pandemic outbreaks. Math. Biosci. Eng. **8**(1), 123–139 (2011)
4. Medić, S., et al.: Varicella zoster virus transmission dynamics in Vojvodina, Serbia. PLoS One **13**(3), e0193838 (2018)
5. Meszner, Z., et al.: Economic burden of varicella in children 1–12 Years of age in Hungary, 2011–2015. BMC Infect. Dis. **17**(1), 495–595 (2017)
6. Ouwens, M.J.N.M., et al.: The impact of 2-dose routine measles, mumps, rubella, and varicella vaccination in france on the epidemiology of varicella and zoster using a dynamic model with an empirical contact matrix. Clin. Ther. **37**(4), 816–829 (2015)
7. Schuette, M.C.: A qualitative analysis of a model for the transmission of varicella-zoster virus. Math. Biosci. **182**(2), 113–126 (2003)

Modeling Neuronal Firing in Epilepsy: Fitting Hawkes Processes to Single-Unit Activity

György Perczel, Loránd Erőss, Dániel Fabó, László Gerencsér, and Zsuzsanna Vágó

Abstract Forecasting seizures based on information extracted from neuronal firing has a great potential in controlling closed-loop neurostimulators. For the description of neuronal firing patterns we use self-exiting point processes or Hawkes processes. In fitting them to simulated data, using a large variety of models, we consider both computability and reliability issues related to the maximum likelihood estimation (MLE) method. The models are classified via a single parameter related to stability regimes. The dependence of the accuracy of the individual parameter estimates on different regimes will be explored. We demonstrate the applicability of the MLE method to discriminate between different models with high confidence.

1 Introduction

1.1 A Brief Introduction to Epilepsy

With a prevalence of 0.5–1% epilepsy is one of the most common neurological disorders. Its most characteristic features are recurrent seizures. Despite that a number of causes have already been identified, including genetic or cerebrovascular disorders, brain injury and infections, 6 out of 10 cases are categorized as idiopathic, i.e. the main cause is unknown. Numerous anti-epileptic drugs are available and in

G. Perczel · L. Erőss · D. Fabó
National Institute of Clinical Neurosciences, Budapest, Hungary

Faculty of Information Technology and Bionics, Pázmány Péter Catholic University, Budapest, Hungary

L. Gerencsér (✉)
Systems and Control Lab, Institute for Computer Science and Control, Budapest, Hungary
e-mail: gerencser.laszlo@sztaki.hu

Z. Vágó
Faculty of Information Technology and Bionics, Pázmány Péter Catholic University, Budapest, Hungary

© Springer Nature Switzerland AG 2019

I. Faragó et al. (eds.), *Progress in Industrial Mathematics at ECMI 2018*, Mathematics in Industry 30, https://doi.org/10.1007/978-3-030-27550-1_32

some cases even surgical options exist, such as excision of the epileptogenic focus or the implantation of a neurostimulator. However, for approximately 30% of the patients sufficient seizure control cannot be achieved, see [14].

The unexpectedness of the seizures has such an influence on the patients' well-being that *reducing* the frequency of seizures only moderately improves their quality of life. Even in the case of rare seizures, the patient's life is determined by the fear of a forthcoming one [14]. Thus, it is generally accepted, that a system capable of forecasting seizures would ameliorate the quality of life of patients. Furthermore, it is hypothesized that combining such a system with a neurostimulator or a drug-delivery pump the development of seizures could be avoided [4].

1.2 Seizure Prediction and Closed-Loop Neurostimulators in Epilepsy

In order to forecast forthcoming seizures numerous methods have been proposed, see [4]. Most often these are based on the analysis of the electrical signals of the brain, such as the ones recorded on the scalp (electroencephalogram, EEG) or on the surface of the brain (electrocorticogram, ECoG). It is assumed that focal seizures starting from a distinct cortical area evolves as a cascade of events and thus can be theoretically predicted, though we are not aware of a specific biomarker preceding them [9]. Although some methods satisfy current statistical criteria (above chance), applicability in a clinical setting requires more rigorous standards [4].

We note that experimental and modeling studies carried out hitherto suggest that some seizures are inherently unpredictable. This is the case for primary generalized absence seizures that involve both cortical hemispheres from the very beginning of the onset of the seizure. A brain with this type of seizures is regarded as a bistable system that switches between its states stochastically [9].

To date, there is only one implantable closed-loop system available on the market, the Responsive Neurostimulation or RNS (NeuroPace, CA, USA). However, this device detects seizures, instead of predicting them, and delivers electrical stimulation to the area thought to be responsible for the seizure initiation. In contrast, the Seizure Advisory System (NeuroVista, WA, USA) aimed to achieve real seizure forecasting, but is not yet available on the market, [4].

The tools for seizure prediction mentioned above utilize the low-frequency components of brain electrical signals called local field potentials (LFP) ranging from approximately 0.1 Hz to 200–300 Hz. As an alternative the appropriate signals can be filtered using a high pass filter with a cutoff frequency around 200–500 Hz. The resulting time-series (multi-unit activity, MUA) will contain primarily the action potentials of neurons from the vicinity of the recording electrodes. With further processing, called spike-sorting, these APs can be assigned to individual units (neurons) based on their morphology. This provides us sequences of time-points indicating the APs of distinct units termed single-unit activities (SUAs) [13].

As there is evidence that epileptic activity is a result of the underlying pathological neuronal firing [10] the question arises if seizure prediction can be achieved based on the investigation of SUAs [15]. In this paper we provide a framework for the statistical analysis of SUAs using the theory of point processes, and give a summary of our extensive simulation-based investigations.

2 Modeling Single-Unit Activity via Hawkes Processes

The series of time points of the APs is modeled with a so called point process, [2]. Mathematically this is a strictly increasing sequence of random time points $0 = T_0 < T_1 < T_2 \ldots$ with no accumulation point. For the sake of mathematical convenience we will also consider two-sided point processes (T_n), $-\infty < n < \infty$, for which the range of T_n-s is $(-\infty, +\infty)$. For an excellent introduction see [1].

An alternative description of a point process is given by its counting process defined for the one-sided point process as $N_t = \#\{n : 0 < T_n \leq t\}$. More generally, we may define a counting measure for any interval $(a, b]$ by the equation $N(a, b] = \#\{n : a < T_n \leq b\}$. The internal history or the past of a point process is defined as the σ-algebra

$$\mathscr{F}_t = \sigma\{N(a, b] : a < b \leq t\}.$$

The definition of counting measure and internal history works equally for one-sided and two-sided point processes. We define the integral of a random so-called predictable function $f_t \geq 0,\ t \geq 0$ with respect to dN_t. The heuristic meaning of predictability is that f_t is the limit of left-continuous $\mathscr{F}_t$-adapted processes. Then set

$$\int_0^\infty f_t dN_t = \sum_{n \geq 0} f(T_n).$$

It can be shown that associated with dN_t is a so-called (predictable) intensity process with the property that

$$\mathrm{E}\left(\int_0^\infty f_t dN_t\right) = \mathrm{E}\left(\int_0^\infty f_t \lambda_t dt\right).$$

A prominent class of point processes in the field of neuroscience, emulating the firing pattern of a network of neurons interacting via APs in the brain, is the class of (multi-variate) mutually exiting point processes, or Hawkes processes, introduced in [8], see also [1, 2, 7] and [6].

A two-sided multivariate point process, $(T_{i,n}), i = 1, \ldots, k$, is a Hawkes process, if its counting measures $N_i(.)$ are shift invariant in time, with intensity functions

$$\lambda_{i,t} = \mu_i + \sum_{j=1}^{k} \int_{-\infty}^{t} g_{ij}(t-s) dN_{j,s}, \qquad \mu_i > 0,\ g_{ij}(u) \geq 0. \tag{1}$$

Here μ_i are the background intensities, and $g_{ij}(u)$ are non-negative impulse response functions (IRF). For the analysis of the firing pattern of a single unit we use a univariate (self-exciting) Hawkes process, implicitly defined by the feedback loop

$$\lambda_t = \mu + \int_{-\infty}^{t} g(t-s) dN_s, \qquad \mu > 0,\ g(u) \geq 0. \tag{2}$$

Taking expectation on both side, and setting $\overline{\lambda} = \mathrm{E}\lambda_t$, we get the equation $\overline{\lambda} = \mu + c\overline{\lambda}$, and the necessary (and sufficient, see [11]) condition for the existence of a Hawkes process satisfying (2):

$$c = \int_0^\infty g(t)dt < 1. \tag{3}$$

The rational behind the application of Hawkes processes in the analysis of SUA is that the burst-mode of neurons indicates a feedback-effect. The objective of the present study is to model individual neurons' firing pattern by fitting a univariate Hawkes process via the maximum-likelihood method, see [12], and to provide a summary of extensive experimental findings based on simulated data. In particular, we explore the typical configurations in the parameter space and establish confidence limits for discerning different regimes.

3 Statistical Fitting of Hawkes Processes

In order to fit Hawkes processes to real SUA data we consider a parametric class of Hawkes processes with

$$g(u) = \sigma \cdot e^{au}, u \geq 0, \qquad \text{with} \quad \sigma > 0,\ a < 0,$$

see [12]. In this case the stability criteria (3) becomes $-\sigma/a < 1$, or equivalently, $\alpha := a + \sigma < 0$. Here α is called the stability margin. Let $\eta = (\mu, a, \sigma)$, and assume that our data are in fact generated by a Hawkes process defined above with true parameter η^*. To estimate this we take an arbitrary feasible parameter η, satisfying the conditions $a + \sigma < 0 < \mu$, and define an intensity function $\lambda_t(\eta)$:

$$\lambda_t(\eta) = \mu + \int_0^t g(t-s,\eta)\, dN_s = \mu + \int_0^t \sigma \cdot e^{a(t-s)}\, dN_s. \tag{4}$$

The computation of the (conditional) log-likelihood function, under the condition that $dN_t = 0$ for $t \le 0$, is the mathematically substantiated heuristics that under minimal conditions a point process is locally a Poisson process, see [3, 12]. Thus, the negative log-likelihood function on the interval $[0, T]$ is, modulo constants,

$$L_T(\eta) = \int_0^T \lambda_t(\eta) dt - \int_0^T \log \lambda_t(\eta) dN_t. \tag{5}$$

For a rigorous foundation we refer to [12], and [3] for a more up to date reference.

The apparently cumbersome computation of the (predictable) intensity $\lambda_t(\eta)$ is actually quite simple for an exponential IRF. Namely, it follows directly from (4), by moving μ to the l.h.s. and then differentiating w.r.t. t, that on the interval $T_{n-1} < t \le T_n$, where no event occurs, we have

$$\lambda_{T_{n-1+}} = \lambda_{T_{n-1}} + \sigma, \qquad \lambda_t(\eta) - \mu = e^{a(t-T_{n-1})}(\lambda_{T_{n-1+}}(\eta) - \mu). \tag{6}$$

To bring the model closer to physiological reality we introduce an alternative parameterization using the stability margin and the average intensity as parameters, thus obtaining $\theta = (\alpha, \sigma, \bar{\lambda})$. As a measure of the precision of our estimators we use 95% confidence-ellipsoids. We note that the Fisher information matrix, for a general parameteric class of Hawkes processes, is obtained from (5) as follows:

$$I(\theta^*) = \lim_{T\to\infty} \frac{1}{T} \sum_{0<T_n\le T} \frac{\lambda_{\theta T_n}(\theta^*) \cdot \lambda_{\theta T_n}^T(\theta^*)}{\lambda_{T_n}^2(\theta^*)}, \tag{7}$$

where the subscript $_\theta$ denotes differentiation w.r.t. θ, assuming the validity of an appropriate strong law of large numbers.

4 Experimental Results

We implemented the above method in MATLAB and tested its performance. The accuracy of the MLE method was tested using simulated data generated by an improved version of the procedure presented in [12]. The length of an experiment is defined via the number of simulated events, which is in the range of 10,000 in

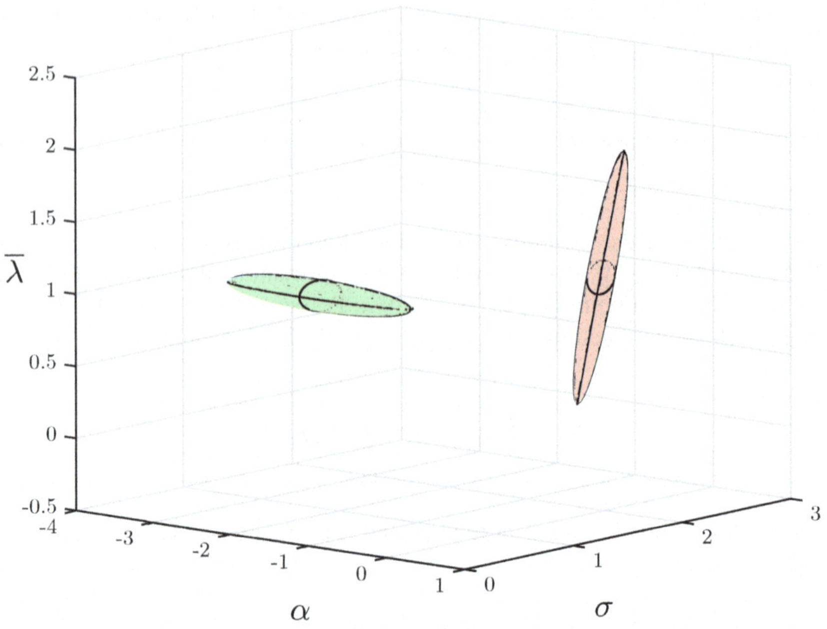

Fig. 1 95% confidence ellipsoids of the two reference examples

our case. The accuracy of the estimators are characterized by confidence ellipsoids defined for level 95%. The scope of experimental studies was focused on the sensitivity of the method w.r.t. changes in model dynamics, including changes in the orientation of the respective confidence ellipsoids.

On Fig. 1 we present the confidence ellipsoids of two simulated SUAs, taken as benchmark examples. The processes were simulated with $N = 10{,}000$ events with $\theta = (-2.1; 0.9; 1.0)$ and $\theta = (-0.6; 2.4; 1.0)$, denoted with green and red, respectively. In order to enhance our potential to discriminate between two models we can increase the number of observed events. We note that the volume of a confidence ellipsoid, denoted by V_{CE}, corresponding to a fixed model, based on N events is proportional to $N^{-3/2}$. However, in a real-life situation when estimating the dynamics during the preictal period, the number of events associated with a stationary regime is limited due to changes in the dynamics close to the onset of a seizure.

We studied the influence of different parameter setting on the accuracy of the estimation. First we note that we may chose $\overline{\lambda} = 1$ for simplicity, since the estimation of this parameter is independent from that of the others. A major characteristic of a Hawkes process is its the integral of the IRF, denoted by c, see (3), defining different regimes w.r.t. stability. A second feature that we considered is simply the attenuation determined by the parameter a.

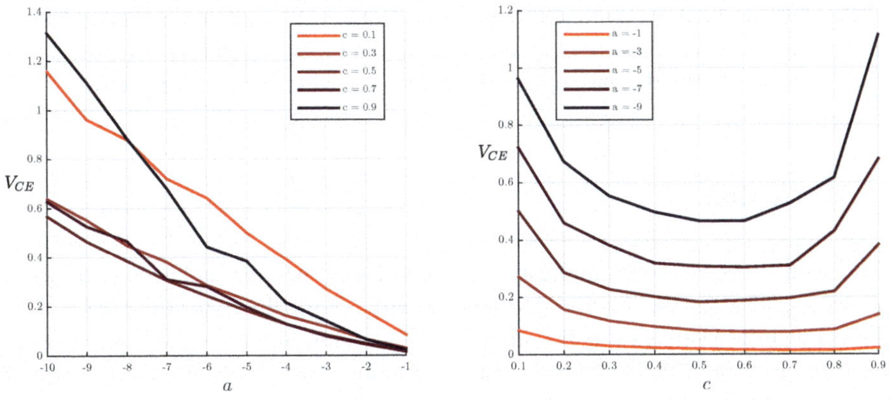

Fig. 2 Left: V_{CE} vs. a. Right: V_{CE} vs. c

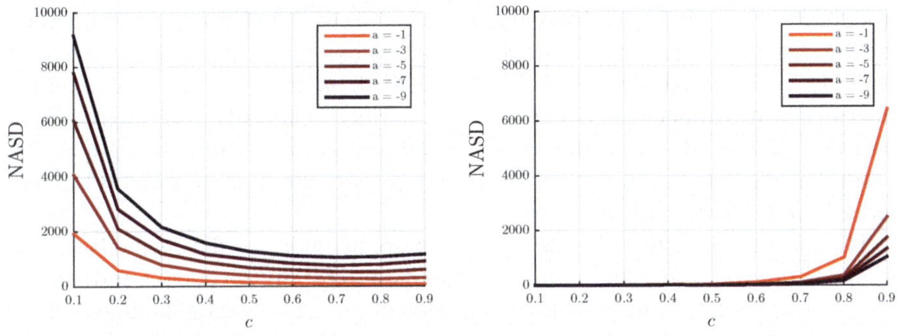

Fig. 3 Normalized asymptotic standard deviation (NASD) of α and $\overline{\lambda}$, respectively

We simulated numerous Hawkes processes (9 regimes at 10 different a-s, $N = 10{,}000$), and computed the volume of the confidence ellipsoids (V_{CE}). The dependence of V_{CE} on the particular regime c and the attenuation a is demonstrated on Fig. 2. On the left hand side it is seen that V_{CE} is a monotone decreasing function of a for each regime. On the right hand side V_{CE} is depicted as a function of c for various choices of a. It is interesting to observe that estimation problem becomes more difficult for values of c close to 1 or 0.

In order to understand the details about the increased uncertainty of the estimators when c is close to 1 or 0, we compute the asymptotic standard deviation (ASD) of individual parameters, which are simply the diagonal elements of $I^{-1}(\theta^*)$. To make different parameter-settings comparable we normalized these values by $V_{CE}^{1/3}$, see Fig. 3. These results show that the overall uncertainty, when c is close to 1 or 0, is due to the uncertainty in the estimation of α for c close 0, and that of $\overline{\lambda}$ for c close 1. The accuracy of the estimation of σ is quite satisfactory for all values of c.

The shift in the degree of uncertainty between α and $\overline{\lambda}$ indicates a change in the orientation of the confidence ellipsoid. This finding may be used to detect regime-changes more efficiently, and ultimately to detect changes in the brain-state.

5 Discussion

Forecasting seizures with application in closed-loop neurostimulators is of great need for patients with therapy-resistant epilepsy. With the expanding arsenal of clinical neurophysiology it is becoming possible to monitor patients' brain-activity at a cellular level [16]. Therefore seizure prediction based on information extracted from neuronal firing is a promising research topic.

A convenient framework to describe neuronal firing patterns in a compressed manner are self-exiting point processes or Hawkes processes. When fitting Hawkes processes to simulated or real-world data critical factors are both the computability and the statistical reliability of the MLE. In the present experimental mathematical research we explored the sensitivity of the MLE method for the class Hawkes processes with exponential IRF for a large variety of models. The models were classified via a single parameter related to stability, defining different regimes. We found that the estimation accuracy of parameters (pattern of uncertainty) highly depends on the actual regime.

The ultimate measure of accuracy is the applicability of the methods to discriminate between different brain-states based on experimental data. A further step towards real life applications is the integration of our experimental findings for the off-line MLE method into the development of a reliable on-line MLE method along the lines proposed back in [5], to be discussed in a forthcoming paper.

Acknowledgements This research has been partially supported by the European Union, co-financed by the European Social Fund (EFOP-3.6.3-VEKOP-16-2017-00002).

References

1. Brémaud, P.: Point Processes and Queues: Martingale Dynamics. Springer, New York (1981)
2. Chornoboy, E.S., Schramm, L.P., Karr, A.F.: Maximum likelihood identification of neural point process systems. Biol. Cybern. **59**, 265–75 (1988)
3. Daley, D.J., Vere-Jones, D.: An Introduction to the Theory of Point Processes. Springer Science & Business Media, New York (2013)
4. Gadhoumi, K., Lina, J.-M., Mormann, F., Gotman, J.: Seizure prediction for therapeutic devices: a review. J. Neurosci. Methods **29**, 1–13 (2015)
5. Gerencsér, L., Matias, C., Vágó, Z., Torma, B., Weiss, B.: Self-exciting point processes with applications in finance and medicine. In: Proccedings of the 18th International Symposium on Mathematical Theory of Networks and Systems (MTNS2008), Virginia Tech, Blacksburg, Virginia, USA (2008)

6. Gerhard, F., Deger, M., Truccolo, W.: On the stability and dynamics of stochastic spiking neuron models: nonlinear Hawkes process and point process GLMs. PLoS Comp. Biol. **13**, e1005390 (2017). https://doi.org/10.1371/journal.pcbi.1005390
7. Hansen, N.R., Reynaud-Bouret, P., Rivoirard, V.: Lasso and probabilistic inequalities for multivariate point processes. Bernoulli **21**, 83–143 (2015)
8. Hawkes, A.G.: Spectra of some self-exiciting and mutually exciting point processes. Biometrika **58**, 83–90 (1971)
9. Kuhlmann, L., Grayden, D.B., Wendling, F., Schiff, S.J.: Role of multiple-scale modeling of epilepsy in seizure forecasting. J. Clin. Neurophysiol. **32**, 220–226 (2015)
10. Merricks, E.M., Smith, E.H., McKhann, G.M., Goodman, R.R., Bateman, L.M., Emerson, R.G., Schevon, C.A., Trevelyan, A.J.: Single unit action potentials in humans and the effect of seizure activity. Brain **138**, 2891–2906 (2015)
11. Møller, J., Rasmussen, J.G..: Perfect simulation of Hawkes processes. Adv. Appl. Probab. **37**, 629–646 (2005)
12. Ozaki, T.: Maximum likelihood estimation of Hawkes' self-exciting point processes. Ann. Inst. Stat. Math. **31**, 145–155 (1979)
13. Quiroga, R.Q., Nadasdy, Z., Ben-Shaul, Y.: Unsupervised spike detection and sorting with wavelets and superparamagnetic clustering. Neural. Comput. **16**, 1661–1687 (2004)
14. Schulze-Bonhage, A., Kühn, A.: Unpredictability of seizures and the burden of epilepsy. In: Seizure Prediction in Epilepsy: From Basic Mechanisms to Clinical Applications, p. 1–10 (2008)
15. Truccolo, W., Donoghue, J.A., Hochberg, L.R., Eskandar, E.N., Madsen, J.R., Anderson, W.S., Brown, E.N., Halgren, E., Cash, S.S.: Single-neuron dynamics in human focal epilepsy. Nat. Neurosci. **14**, 635–641 (2011)
16. Ulbert, I., Maglóczky, Z., Erőss, L., Czirják, S., Vajda, J., Bognár, L., Tóth, S., Szabó, Z., Halász, P., Fabó, D., Halgren, E., Freund, T.F., Karmos, G.: In vivo laminar electrophysiology co-registered with histology in the hippocampus of patients with temporal lobe epilepsy. Exp. Neurol. **187**, 310–318 (2004)

Part III
Novel Numerical Methods for Industrial Mathematics

Solution of MHD Flow with BEM Using Direct Radial Basis Function Interpolation

Merve Gürbüz and Munevver Tezer-Sezgin

Abstract In this study, the two-dimensional steady MHD Stokes and MHD incompressible flows of a viscous and electrically conducting fluid are considered in a lid-driven cavity under the impact of a uniform horizontal magnetic field. The MHD flow equations are solved iteratively in terms of velocity components, stream function, vorticity and pressure by using direct interpolation boundary element method (DIBEM) in which the inhomogeneity in the domain integral is interpolated by using radial basis functions. The boundary is discretized by constant elements and the sufficient number of the interior points are taken. The interpolation points are different from the source points due to the singularities of the fundamental solution. It is found that as Hartmann number increases, the main vortex of the flow shifts through the moving top lid with a decreasing magnitude and secondary flow below it is squeezed through the main flow leaving the rest of the cavity almost stagnant. The increase in M develops side layer near the moving lid, but weakens the effect of Re in the MHD incompressible flow.

1 Introduction

The study of incompressible flow under the effect of magnetic field has many industrial and medical applications such as MHD generators, pumps, nuclear reactors and blood flow measurements. Magnetohydrodynamic (MHD) flow is governed by the hydrodynamics (Navier-Stokes equations) and electromagnetics (Maxwell's equations) through Ohm's law. Many numerical methods are developed to solve MHD equations, since the analytical solutions are restricted to simple geometry and boundary conditions due to the nonlinearities and Lorentz force

M. Gürbüz (✉)
Department of Management, Baskent University, Ankara, Turkey
e-mail: mervegurbuz@baskent.edu.tr

M. Tezer-Sezgin
Department of Mathematics, Middle East Technical University, Ankara, Turkey
e-mail: munt@metu.edu.tr

© Springer Nature Switzerland AG 2019

I. Faragó et al. (eds.), *Progress in Industrial Mathematics at ECMI 2018*,
Mathematics in Industry 30, https://doi.org/10.1007/978-3-030-27550-1_33

terms in the equations. A finite element method (FEM) is implemented to solve three-dimensional incompressible MHD flow by Salah et al. [9]. Nath et al. [7] considered the MHD flow in a lid-driven cavity using a meshless method based on fundamental and particular solution (MFS-MPS). In [3] radial basis function (RBF) approximation is employed to the two-dimensional MHD equations by using Stokes approximation ($Re << 1$).

In this study, we apply the direct interpolation boundary element method (DIBEM) to solve the MHD flow equations with or without Stokes approximation. DIBEM is a new technique to transform the domain integral due to the inhomogeneity of differential equation into the boundary integral. The only difference from DRBEM is that the complete kernel of the domain integral is interpolated directly by using the radial basis functions (RBF). The DIBEM has been implemented to solve Poisson's equations [4] and Helmholtz equation [5] using classical RBF. In [4], the analysis of better accuracy of DIBEM is carried and the numerical results are compared with the DRBEM results. In this study, the MHD flow equations are treated as Poisson's equations and solved by using DIBEM. Numerical results obtained from DIBEM are simulated in terms of streamlines, equivorticity lines and pressure contours in the lid-driven cavity for several values of Reynolds number and Hartmann number.

2 Governing Equations

The steady, two-dimensional, laminar flow of a viscous, incompressible and electrically conducting fluid is considered in the lid-driven cavity under the effect of uniform horizontal magnetic field. This problem is modelled by the dimensionless MHD equations [2, 6] in terms of stream function ψ, vorticity ω first, and then going back to the velocity components (u, v) and pressure p as

$$\nabla^2 u = -\frac{\partial \omega}{\partial y}, \qquad \nabla^2 v = \frac{\partial \omega}{\partial x}, \qquad \nabla^2 \psi = -\omega \tag{1}$$

$$\nabla^2 \omega = Re(u\frac{\partial \omega}{\partial x} + v\frac{\partial \omega}{\partial y}) + M^2\frac{\partial v}{\partial x} \tag{2}$$

$$\nabla^2 p = -2Re(\frac{\partial v}{\partial x}\frac{\partial u}{\partial y} - \frac{\partial u}{\partial x}\frac{\partial v}{\partial y}) - M^2\frac{\partial v}{\partial y} \tag{3}$$

where Reynolds number $Re = LU_0/\nu$ and Hartmann number $M = L\mu H_0\sqrt{\sigma/\rho\nu}$ are non-dimensional parameters. Here, $L, U_0, \nu, \mu, H_0, \sigma$ and ρ are characteristic length, characteristic velocity, kinematic viscosity, magnetic permeability, magnetic field intensity, electric conductivity and density of the fluid. The last terms in Eqs. (2) and (3) are due to Lorentz force. Equations (1)–(3) are supplied with the no-slip boundary condition $u = v = \psi = 0$ on $x = 0, 1$ and $y = 0$ wall, but $u = 1, v = \psi = 0$ on the moving lid $y = 1$. In case of highly viscous fluid, called MHD Stokes flow, the convective terms in Eqs. (2)–(3) are neglected due to the small values of Reynolds number ($Re << 1$). The unknown vorticity boundary conditions are obtained from the stream function equation (1) by using FD scheme. The unknown pressure boundary values are computed by a FD scheme for pressure derivatives and the coordinate matrix $\boldsymbol{F}$ for the space derivatives of the velocity in the momentum equations, [2]. $\boldsymbol{F} = (f_{ij}) = 1 + r_{ij}$, r_{ij} being the distance.

3 DIBEM Application

The direct interpolation boundary element method transforms Poisson's type equations (1)–(3) into boundary integral equations using the fundamental solution of Laplace equation, $u^* = \frac{1}{2\pi} ln(1/r)$. Weighting the equations by u^* and applying Green's identity, we get the BEM formulation [1]

$$c(\xi)K(\xi)+\int_\Gamma K(X)\frac{\partial u^*(\xi;X)}{\partial n}\,d\Gamma-\int_\Gamma \frac{\partial K(X)}{\partial n}u^*(\xi;X)\,d\Gamma = -\int_\Omega z_K(X)u^*(\xi;X)\,d\Omega \tag{4}$$

where K denotes u, v, ψ, ω or p and z_K is the right-hand side function of each corresponding Poisson equation for K. The coefficient $c(\xi)$ depends on the position of the source point ξ and given as in [1].

In the DIBEM procedure [5], the complete kernel of the domain integral is directly interpolated by using radial basis function F^i which is related to primitive interpolation function Ψ^i as $\nabla^2\Psi^i = F^i$

$$z_K(X)u^*(\xi;X) = \sum_{i=1}^{n} \alpha_i(\xi)F^i(X^i;X) = \sum_{i=1}^{n} \alpha_i(\xi)\nabla^2\Psi^i(X^i;X) \tag{5}$$

where X^i is the interpolation point and n is the number of points used in the interpolation. The undetermined coefficients $\alpha_i(\xi)$'s change for each source point $\xi, \xi = 1, \ldots, \tilde{n}$. Here, $\tilde{n}$ is the total number of N_b boundary and N_i interior nodes.

Substituting the relation (5) into the domain integral of the source term in Eq. (4) and applying Green's identity, we obtain the boundary integral equation

$$c(\xi)K(\xi)+\int_\Gamma K(X)\frac{\partial u^*(\xi;X)}{\partial n}\,d\Gamma-\int_\Gamma \frac{\partial K(X)}{\partial n}u^*(\xi;X)\,d\Gamma = -\sum_{i=1}^{n}\alpha_i(\xi)\int_\Gamma \frac{\partial\Psi^i}{\partial n}(X^i;X)\,d\Gamma. \tag{6}$$

Discretization of the boundary by using constant elements Γ_j gives matrix-vector form of Eq. (6) as

$$\boldsymbol{HK} - \boldsymbol{G}\frac{\partial \boldsymbol{K}}{\partial \boldsymbol{n}} = \bar{\boldsymbol{A}}\overline{\boldsymbol{N}} \tag{7}$$

where $\boldsymbol{H}_{ij} = c(\xi)\delta_{ij} + \int_{\Gamma_j} \frac{\partial u^*}{\partial n} d\Gamma_j$, $\boldsymbol{G}_{ij} = \int_{\Gamma_j} u^* \, d\Gamma_j$ and $\boldsymbol{G}_{ii} = \frac{1}{2\pi}(ln(2/l) + 1)$, and $\overline{\boldsymbol{N}}_i = -\sum_{j=1}^{N_b} \int_{\Gamma_j} \frac{\partial \Psi^i}{\partial n}(X^i; X_j)\, d\Gamma_j$. Here, l is the length of the element Γ_j.

Thus, the DIBEM discretized system of the equations for (1)–(3) is

$$\boldsymbol{H}\boldsymbol{u} - \boldsymbol{G}\boldsymbol{q}_u = \bar{\boldsymbol{A}}_u \overline{\boldsymbol{N}}, \qquad \boldsymbol{H}\boldsymbol{v} - \boldsymbol{G}\boldsymbol{q}_v = \bar{\boldsymbol{A}}_v \overline{\boldsymbol{N}}, \qquad \boldsymbol{H}\boldsymbol{\psi} - \boldsymbol{G}\boldsymbol{q}_\psi = \bar{\boldsymbol{A}}_\psi \overline{\boldsymbol{N}} \tag{8}$$

$$\boldsymbol{H}\boldsymbol{\omega} - \boldsymbol{G}\boldsymbol{q}_\omega = \bar{\boldsymbol{A}}_\omega \overline{\boldsymbol{N}}, \qquad \boldsymbol{H}\boldsymbol{p} - \boldsymbol{G}\boldsymbol{q}_p = \bar{\boldsymbol{A}}_p \overline{\boldsymbol{N}}. \tag{9}$$

The matrices $\bar{\boldsymbol{A}}_u$, $\bar{\boldsymbol{A}}_v$, $\bar{\boldsymbol{A}}_\psi$, $\bar{\boldsymbol{A}}_\omega$ and $\bar{\boldsymbol{A}}_p$ are constructed row-wise with the unknown vectors

$$\boldsymbol{\alpha}_u(\xi) = \boldsymbol{F}^{-1}\boldsymbol{\Lambda}(\xi)\frac{\partial \boldsymbol{F}}{\partial y}\boldsymbol{F}^{-1}(-\boldsymbol{\omega}), \quad \boldsymbol{\alpha}_v(\xi) = \boldsymbol{F}^{-1}\boldsymbol{\Lambda}(\xi)\frac{\partial \boldsymbol{F}}{\partial x}\boldsymbol{F}^{-1}\boldsymbol{\omega}, \quad \boldsymbol{\alpha}_\psi(\xi) = \boldsymbol{F}^{-1}\boldsymbol{\Lambda}(\xi)(-\boldsymbol{\omega}) \tag{10}$$

$$\boldsymbol{\alpha}_\omega(\xi) = \boldsymbol{F}^{-1}\boldsymbol{\Lambda}(\xi)(\frac{\partial \boldsymbol{F}}{\partial x}\boldsymbol{F}^{-1}(M^2\boldsymbol{v}) + Re(\boldsymbol{u}\frac{\partial \boldsymbol{F}}{\partial x}\boldsymbol{F}^{-1}\boldsymbol{\omega} + \boldsymbol{v}\frac{\partial \boldsymbol{F}}{\partial y}\boldsymbol{F}^{-1}\boldsymbol{\omega})) \tag{11}$$

$$\boldsymbol{\alpha}_p(\xi) = \boldsymbol{F}^{-1}\boldsymbol{\Lambda}(\xi)(\frac{\partial \boldsymbol{F}}{\partial y}\boldsymbol{F}^{-1}(-M^2\boldsymbol{v}) - 2Re(\frac{\partial \boldsymbol{F}}{\partial x}\boldsymbol{F}^{-1}\boldsymbol{v}\frac{\partial \boldsymbol{F}}{\partial y}\boldsymbol{F}^{-1}\boldsymbol{u} - \frac{\partial \boldsymbol{F}}{\partial x}\boldsymbol{F}^{-1}\boldsymbol{u}\frac{\partial \boldsymbol{F}}{\partial y}\boldsymbol{F}^{-1}\boldsymbol{v})) \tag{12}$$

where the diagonal matrix $\boldsymbol{\Lambda}(\xi)$ is constructed by the fundamental solution of Laplace equation u^* for each ξ.

4 Numerical Result

The DIBEM system of MHD flow equations (8)–(9) are solved iteratively with the preassigned tolerance $\epsilon = 10^{-3}$. We take $N_b = 60$ boundary nodes and $N_i = 225$ interior nodes for Reynolds number $Re = 0, 50, 100, 140$ and the Hartmann number values $M = 0, 30$. The numerical results are shown in terms of streamlines, vorticity and pressure contours in Figs. 1, 2, and 3. $Re = 0$ corresponds to MHD Stokes flow.

Figure 1 depicts the effect of the increases in Re and M on the streamlines. In the absence of the magnetic field ($M = 0$), it is observed that as Reynolds number increases, the primary vortex of the flow pushes through the center of the cavity and fluid flows in almost all parts of the cavity [8]. Secondary vortex occurs at the right lower corner of the cavity when $Re = 140$. Existence of the magnetic field ($M = 30$) destroys this behavior of the flow. Main flow is concentrated through the top lid forming boundary layer and secondary vortex appears below the main vortex with a small magnitude. The increase in Re does not change the behavior of the flow under the impact of magnetic field.

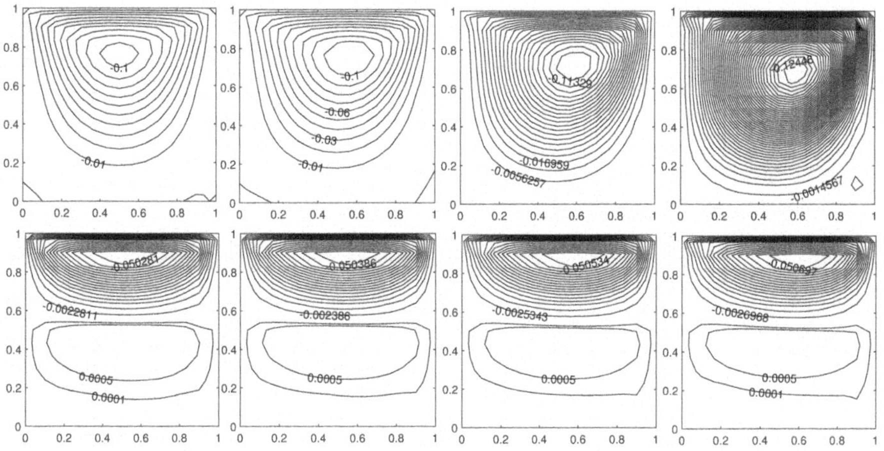

Fig. 1 Streamlines

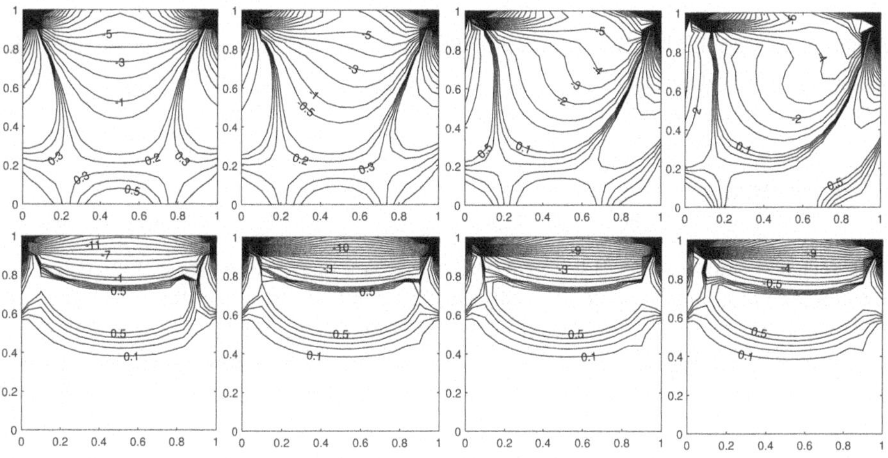

Fig. 2 Equivorticity lines

In Fig. 2, when the magnetic field is neglected, equivorticity lines are symmetric with respect to $x = 0.5$. As Re increases, they start to circulate through the right top corner. However, The increase in the intensity of the magnetic field alters the well-known behavior due to the dominance of the term containing M in the momentum equations.

When $M = 0$ and $Re = 0$, two anti-symmetric pressure profiles are observed at the upper corners in Fig. 3. As Re increases, pressure contours are enlarged through the upper right corner. A further increase in Re increases the intensity of pressure of the fluid all over the cavity reducing to zero at the right corner. Existence of the magnetic field generates secondary pressure vortices below the main vortices in

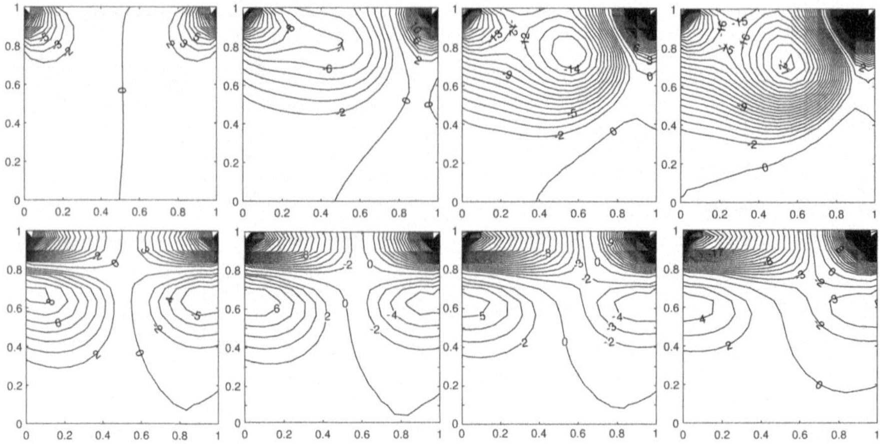

Fig. 3 Pressure contours

front of the moving lid.The increase in *Re* weakens the effect of the magnetic field on the pressure of the fluid.

5 Conclusion

The DIBEM numerical results for MHD lid-driven cavity flow are obtained for the parameter values in $0 \leq Re \leq 140$ and $0 \leq M \leq 30$ to analyze the effect of horizontal magnetic field on the flow and pressure of the fluid. It is found that the increase in Reynolds number shifts the primary vortex of the flow through the center of the cavity. However, the flow is squeezed through the moving top lid forming a boundary layer due to the direction of the magnetic field as the Hartmann number increases. The increase in Reynolds number diminishes the impact of magnetic field intensity on the pressure profiles.

References

1. Brebbia, C.A., Telles, J.C.F., Wrobel, L.C.: Boundary Element Techniques. Springer, Berlin (1984)
2. Gürbüz, M.: Radial Basis Function and Dual Reciprocity Boundary Element Solutions of Fluid Dynamics Problems. Doctoral Thesis, Middle East Technical University, Ankara, Turkey (2017)
3. Gürbüz, M., Tezer-Sezgin, M.: MHD stokes flow in lid-driven cavity and backward-facing step channel. Eur. J. Comput. Mech. **24**, 279–301 (2015)

4. Loeffler, C.F., Cruz, A.L., Bulcão, A.: Direct use of radial basis interpolation functions for modelling source terms with the boundary element method. Eng. Anal. Bound. Elem. **50**, 97–108 (2015)
5. Loeffler, C.F., Mansur, W.J., Barcelos, H., de M., Bulcão, A.: Direct radial basis function interpolation. Eng. Anal. Bound. Elem. **61**, 218–225 (2015)
6. Müller, U., Bühler, L.: Magnetofluiddynamics in Channels and Containers. Springer, New York (2001)
7. Nath, D., Kalra, M.S., Munshi, P.: Numerical simulation of time-dependent Navier-Stokes and MHD equations using a meshless method based on fundamental and particular solutions. Eng. Anal. Bound. Elem. **67**, 81–95 (2016)
8. Sahin M., Owens, R.G.: A novel fully implicit finite volume method applied to the lid-driven cavity problem – Part I: high reynolds number flow calculations. Int. J. Numer. Methods Fluids **42**, 57–77 (2003)
9. Salah, N., Soulaimani, A., Habashi, W.G.: A finite element method for magnetohydrodynamics. Comput. Methods Appl. Mech. Eng. **190**, 5867–5892 (2001)

DRBEM Solution of MHD Flow in an Array of Electromagnetically Coupled Rectangular Ducts

Munevver Tezer-Sezgin and Pelin Senel

Abstract We present the dual reciprocity boundary element method (DRBEM) solution to magnetohydrodynamic (MHD) flow in a single and two parallel ducts which are separated by conducting walls of arbitrary thickness in the direction of external magnetic field. The DRBEM discretized coupled MHD convection-diffusion equations in the ducts and the Laplace equations on the shared walls are solved as a whole by using constant boundary elements with the coupled induced current wall conditions. It is shown that, the conducting walls in the double ducts have a strong influence on the currents near the walls, and the core flow increases on the co-flow case but there is a strong reduction in the core flow in the counter-flow case. The coupling between the ducts with conducting thick walls induces reversed flow and counter current flows which may be used for the heat and mass transfer in fusion applications. The proposed numerical scheme using DRBEM captures the well-known MHD flow characteristics when Hartmann number increases.

1 Introduction

The flow of electrically conducting fluids in ducts under the influence of external magnetic field is of interest in many industrial and medical applications. Examples range from MHD generators, pumps, microfluidics to the flow of liquid metals in casting and nuclear fusion blankets. There are many numerical studies for the MHD flow in a single duct of rectangular or circular geometry if the outside medium, and the duct thin walls are insulating and/or conducting for which the analytical solutions are available [3]. Among these, solutions by using finite difference method

M. Tezer-Sezgin (✉)
Department of Mathematics, Middle East Technical University, Ankara, Turkey
e-mail: munt@metu.edu.tr

P. Senel
Department of Mathematics, Karadeniz Technical University, Trabzon, Turkey

© Springer Nature Switzerland AG 2019

I. Faragó et al. (eds.), *Progress in Industrial Mathematics at ECMI 2018*,
Mathematics in Industry 30, https://doi.org/10.1007/978-3-030-27550-1_34

(FDM) [8], by using finite element method (FEM) [4, 5], and by using boundary element method (BEM) [2, 9, 10] can be counted.

This paper considers the DRBEM solution of the MHD flow in a single and double ducts stacked in the direction of applied magnetic field and separated with arbitrarily conducting thick vertical walls. The upper and lower walls of the ducts are assumed to be thin and insulated. The flow and the induced magnetic field are greatly affected by the conductivities, magnetic permeabilities of the fluid and the walls, wall thickness and the increase in the Hartmann number. The analytical solution given by Bluck and Wolfendale [1] which is valid for thick walls and based on homogeneous solution and the FEM solution in [11] provide validation data for computational MHD flow in an array of ducts. The DRBEM has the advantage of discretizing only the boundary of the ducts, which results in small sized linear system of equations and thus, the solution is obtained at a small expense.

2 Physical Problem and Governing Equations

The non-dimensional MHD equations for an incompressible, viscous, electrically conducting fluid in an array of two square ducts Ω_i separated by conducting walls with thickness δ are given with boundary conditions shown on Fig. 1 as, [1]

$$\nabla^2 V_i + Ha\frac{\partial B_i}{\partial x} = \Delta P_i, \quad \nabla^2 B_i + Ha\frac{\partial V_i}{\partial x} = 0 \quad \text{in } \Omega_i, \ i = 1, 2 \tag{1}$$

$$\nabla^2 B_{w_i} = 0 \qquad \text{in } \Omega_{w_i}, \ i = 1, 2, 3 \tag{2}$$

where $V(x, y)$ is the velocity of the fluid and $B(x, y)$ and $B_w(x, y)$ are induced magnetic fields of the fluid and the thick walls, respectively. Subscript i denotes the corresponding ducts (Ω_1, Ω_2) and the thick walls (Ω_{w_1}, Ω_{w_2}, Ω_{w_3}). ΔP_i =

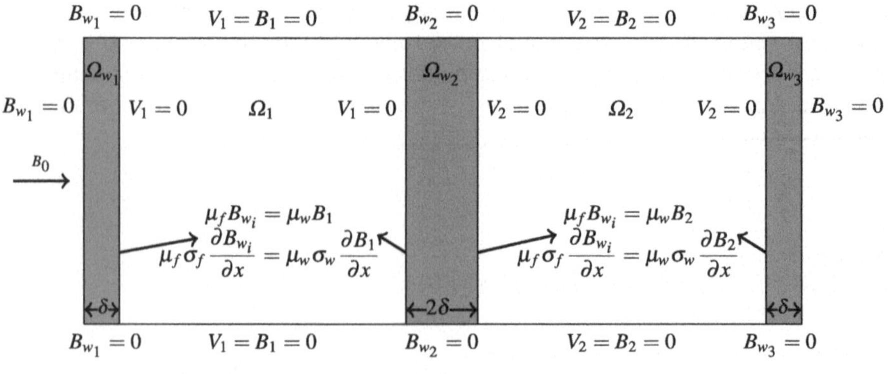

Fig. 1 Double ducts separated by conducting vertical thick walls

$\frac{\partial P_i}{\partial z} / \frac{\partial P_1}{\partial z}$ is the pressure gradient in Ω_i; μ_f, σ_f and μ_w, σ_w are the magnetic permeability and electrical conductivity of the fluid and the walls, respectively. Ha is the Hartmann number given as $Ha = B_0 L \sqrt{\sigma_f} / \sqrt{\rho_f \nu_f}$ with characteristic length L, density and kinematic viscosity ρ_f, ν_f of the fluid, and applied magnetic field intensity B_0. The same direction pressure gradients $\Delta P_1 = \Delta P_2 = 1$ is the case of co-flow in the double ducts whereas $\Delta P_1 = 1, \Delta P_2 = -1$ represents counter flow in the ducts separated by a conducting thick wall.

3 The DRBEM Formulation

The coupled MHD equations (1) and Laplace equation (2) on the vertical thick walls are converted to boundary integral equations by using the fundamental solution of Laplace equation $u^* = (1/2\pi)\ln(1/r)$. Corresponding integral equations are obtained by weighting differential equations (1), (2) with u^* and using the Green's first identity two times [7]. All the terms except the Laplacian are treated as inhomogeneity $b(x, y)$ ($b = b_{V_i}$ or $b = b_{B_i}$ in (1)) and it is approximated by radial basis functions $f_j(r) = 1 + r_j$ which are connected to particular solutions with $\nabla^2 \hat{u}_j = f_j$, r_j being the distance between the source and the field points. Then, $b(x, y) = \sum_{j=1}^{N+L} \alpha_j f_j = \sum_{j=1}^{N+L} \alpha_j \nabla^2 \hat{u}_j$ where α_j's are the undetermined coefficients.

Now, applying the Green's first identity two times also to the domain integrals due to the inhomogeneities and discretizing the boundary with constant elements we obtain (with $\partial\Omega = \Gamma$ notation)

$$c_k(V_i)_k + \sum_{m=1}^{N} \int_{\Gamma_m} V_i q^* d\Gamma - \sum_{m=1}^{N} \int_{\Gamma_m} \frac{\partial V_i}{\partial n} u^* d\Gamma = \sum_{j=1}^{N+L} \alpha_j (c_k \hat{u}_{kj} + \sum_{m=1}^{N} \int_{\Gamma_m} \hat{u}_j q^* d\Gamma - \sum_{m=1}^{N} \int_{\Gamma_m} \frac{\partial \hat{u}_j}{\partial n} u^* d\Gamma) \tag{3}$$

$$c_k(B_i)_k + \sum_{m=1}^{N} \int_{\Gamma_m} B_i q^* d\Gamma - \sum_{m=1}^{N} \int_{\Gamma_m} \frac{\partial B_i}{\partial n} u^* d\Gamma = \sum_{j=1}^{N+L} \beta_j (c_k \hat{u}_{kj} + \sum_{m=1}^{N} \int_{\Gamma_m} \hat{u}_j q^* d\Gamma - \sum_{m=1}^{N} \int_{\Gamma_m} \frac{\partial \hat{u}_j}{\partial n} u^* d\Gamma) \tag{4}$$

$$c_k(B_{w_i})_k + \sum_{m=1}^{N} \int_{\Gamma_m} B_{w_i} q^* d\Gamma - \sum_{m=1}^{N} \int_{\Gamma_m} \frac{\partial B_{w_i}}{\partial n} u^* d\Gamma = 0 \tag{5}$$

where $q^* = \partial u^* / \partial n$, V_i, B_i and B_{w_i} are nodal values of velocity, induced magnetic field of the fluid and the thick walls at the discretization points, respectively. The constant c_k is $1/2$ for boundary source point k $(k = 1, \ldots, N)$, and 1 for interior nodes $(k = N+1, \ldots, N+L)$. In the integrals on the left-hand side, V_i, B_i and B_{w_i} are the values at the boundary nodes.

Collocating b_{V_i} and b_{B_i} using the radial basis functions gives $b_{V_i} = \mathbf{F}\alpha$, $b_{B_i} = \mathbf{F}\beta$ where the matrix $\mathbf{F}$ is constructed by taking f_j's as columns and $\alpha = \mathbf{F}^{-1} b_{V_i}$, $\beta = \mathbf{F}^{-1} b_{B_i}$. In order to depict the flow and induced magnetic field behaviors inside the cavity and the thick walls, the DRBEM discretized matrix-vector equations are obtained by writing Eqs. (3)–(5) for all boundary and interior nodes

$$\mathbf{H}V_i - \mathbf{G}\frac{\partial V_i}{\partial n} = (\mathbf{H}\hat{\mathbf{U}} - \mathbf{G}\hat{\mathbf{Q}})\mathbf{F}^{-1}\{\Delta P_i - Ha\frac{\partial B_i}{\partial x}\} \tag{6}$$

$$\mathbf{H}B_i - \mathbf{G}\frac{\partial B_i}{\partial n} = (\mathbf{H}\hat{\mathbf{U}} - \mathbf{G}\hat{\mathbf{Q}})\mathbf{F}^{-1}\{-Ha\frac{\partial V_i}{\partial x}\} \tag{7}$$

$$\mathbf{H}B_{w_i} - \mathbf{G}\frac{\partial B_{w_i}}{\partial n} = 0 \tag{8}$$

where

$$\mathbf{H}_{kj} = \int_{\Gamma_j} q^* d\Gamma_j, \;\; \mathbf{H}_{kk} = c_k, \;\; \mathbf{G}_{kj} = \int_{\Gamma_j} u^* d\Gamma_j, \;\; \mathbf{G}_{kk} = \frac{l}{2\pi}(\ln(\frac{2}{l}) + 1) \tag{9}$$

and l is the length of the boundary element. The matrices $\hat{\mathbf{U}}$, $\hat{\mathbf{Q}}$ are constructed by taking each vector $\hat{u}_j$ and $\hat{q}_j$ as columns, respectively. Both $j, k = 1, 2, .., N$. The space derivatives are also approximated by using coordinate matrix $\mathbf{F}$ as $\partial V_i/\partial x = (\partial\mathbf{F}/\partial x)(\mathbf{F}^{-1}V_i)$, $\partial B_i/\partial x = (\partial\mathbf{F}/\partial x)(\mathbf{F}^{-1}B_i)$.

The DRBEM discretized matrix-vector equations (6)–(8) are solved as a whole for the velocities of the fluid and the induced magnetic fields in the ducts and the walls, and their normal derivatives after inserting the coupled boundary conditions shown on Fig. 1.

4 Numerical Results

MHD flow equations are solved numerically by using the DRBEM in a single and double (co-flow or counter-flow) ducts for several values of Ha, $c_w = \sigma_w/\sigma_f$ and δ by taking $\mu_w = \mu_f = 1$. The effects of these parameters on the flow and induced magnetic current are discussed. Figure 2 shows the effect of thickness δ of conducting walls in a single duct for $Ha = 5$ and $\sigma_f = \sigma_w = 1$.

When $\delta = 0.01$, the MHD flow represents almost the insulated thin wall case and it is in well agreement with the MHD flow in a duct with insulated walls [3, 6]. As

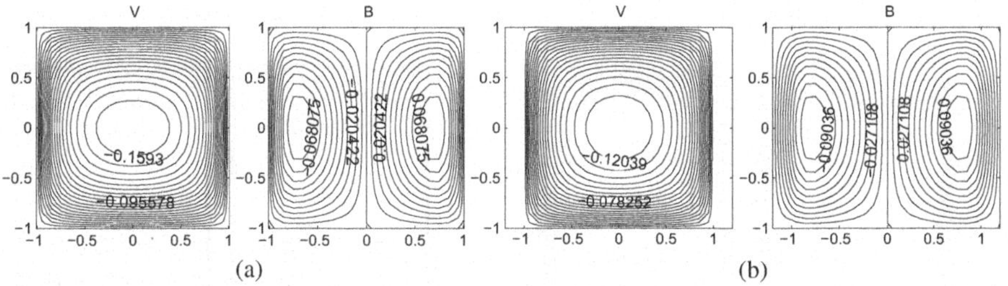

Fig. 2 Velocity and induced magnetic field profiles, $Ha = 5$, $\sigma_w = \sigma_f = 1$, $N = 140$. (**a**) $\delta = 0.01$. (**b**) $\delta = 0.2$

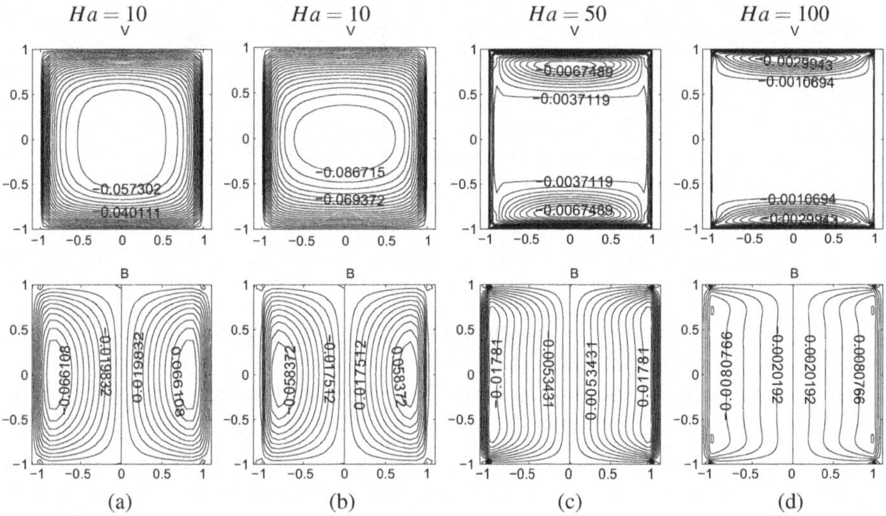

Fig. 3 Velocity and induced magnetic field, $\delta = 0.1$. (**a**) $N = 170$, $c_w = 1$. (**b**) $N = 170$, $c_w = 0.1$. (**c**) $N = 340$, $c_w = 1$. (**d**) $N = 560$, $c_w = 1$

δ increases the passage from the induced current in the fluid to the induced current of the thick walls is well observed according to the coupling of wall conditions.

Figure 3 shows the influence of an increase in Hartmann number (a), (c), (d) and the change in the wall conductance ratio c_w for $Ha = 10$ (a), (b). It is observed that as Ha increases boundary layers are developed, the velocity magnitude drops (flow is flattened). Flow is separated and concentrated in front of the side walls, the rest of the duct is stagnant. Two opposing induced current loops are formed each of which parallel to thick Hartmann walls and return through the central core. The continuation of induced currents from the duct to the thick walls is well observed. A decrease in the wall conductance ratio forces the flow and induced magnetic current to become as if the vertical walls are also insulated, since the electrical conductivity of the fluid is very large compared to the conductivity of the walls. Induced current

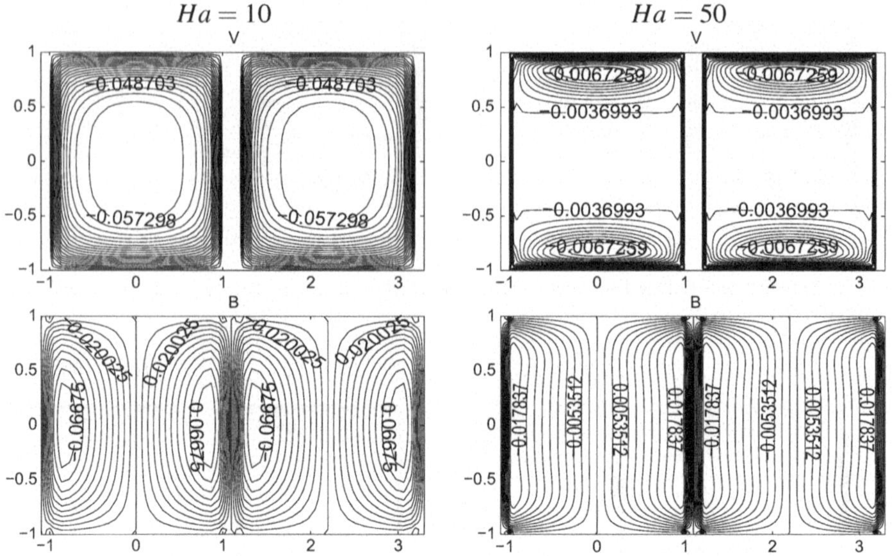

Fig. 4 Velocity and induced magnetic field in double ducts, $\delta = 0.1$, $\sigma_w = \sigma_f = 1$ co-flow

in the fluid can not connect to the currents in the walls anymore when $c_w = 1/10$. These results are in agreement with the ones given in [1].

In the double ducts, we consider MHD flow for both the co-flow and counter flow cases. In Fig. 4, Ha increase is studied for co-flow ($\Delta P_1 = \Delta P_2 = 1$) in the two ducts when $\delta = 0.1$, $\sigma_w = \sigma_f = 1$. In the co-flow case both the flow and the induced magnetic field repeat themselves in the neighboring duct with the same magnitudes. Magnetic fields of the fluid and the walls again continue on the joint walls obeying the coupling conducting wall conditions.

In the counter-flow case (Fig. 5), induced currents at the outer Hartmann layers are reduced relative to the co-flow case. At the inner conducting joint walls the current flows directly through the connecting wall into the neighboring duct. Away from the walls the flow is flattened at the cores as Ha increases.

For the counter-flow, boundary layer at the exterior conducting Hartmann walls are pronounced as Ha increases but at the center connecting walls boundary layers are diminished. Further increase in Ha pushes the flow near the side walls and results in one large current loop connected at the interface of the ducts.

5 Conclusion

The impact of electromagnetic coupling of MHD flows between the conducting walls and the fluid is demonstrated by increasing values of Hartmann number, wall thickness, conductance ratio in the co-flow and counter-flow cases. It is shown that,

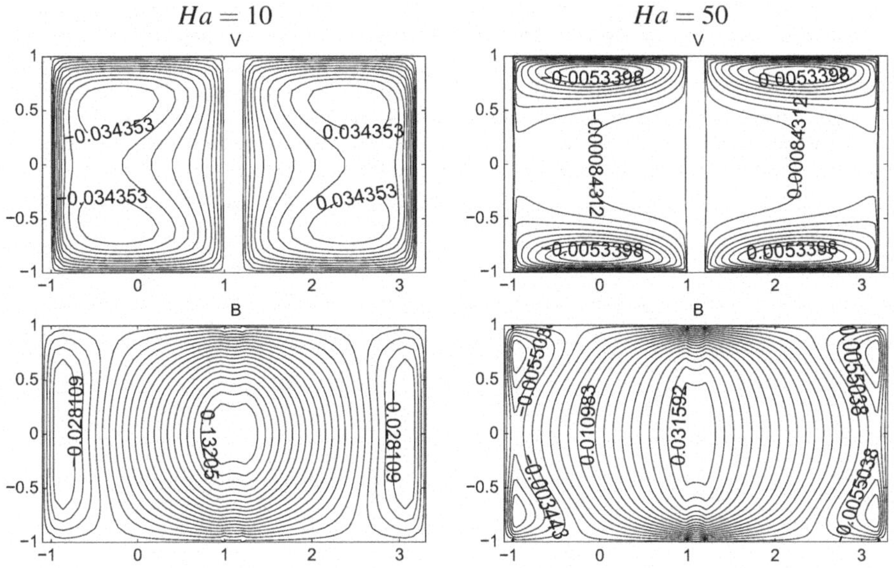

Fig. 5 Velocity and induced magnetic field in double ducts, $\delta = 0.1$, $\sigma_w = \sigma_f = 1$ counter-flow

the conducting walls in the double ducts have a strong influence on the core flow which increases in the co-flow case but there is a strong reduction in the counter-flow case. The reversal flow and counter current loops are induced which may be used for the heat and mass transfer in fusion applications.

References

1. Bluck, M.J., Wolfendale, M.J.: An analytical solution to electromagnetically coupled duct flow in MHD. J. Fluid Mech. **771**, 595–623 (2015)
2. Carabineanu, A., Dinu, A., Oprea, I.: The application of the boundary element method to the magnetohydrodynamics duct flow. Z. Angew. Math. Phys. **46**, 971–981 (1995)
3. Dragos, L.: Magnetofluid Dynamics. Abacus Press, Preston (1975)
4. Lungu, E., Pohoata, A.: Finite element-boundary element approach of MHD pipe flow. In: Proceedings of Conference in Fluid Mechanics and Technical Applications, pp. 79–88 (2005)
5. Meir, A.J.: Finite element analysis of magnetohydrodynamic pipe flow. Appl. Math. Comput. **57**, 177–196 (1993)
6. Muller, U., Buhler, L.: Magnetofluiddynamics in Channels and Containers. Springer, Berlin (2001)
7. Partridge, P.W., Brebbia, C.A., Wrobel, L.C.: The Dual Reciprocity Boundary Element Method. Computational Mechanics Publications, Sauthampton, (1992)
8. Sheu, Y.W.H., Lin, R.K.: Development of a convection-diffusion-reaction magnetohydrodynamic solver on nonstaggered grids. Int. J. Numer. Methods Fluids **45**, 1209–1233 (2004)

9. Tezer-Sezgin, M., Bozkaya, C.: Boundary element method solution of magnetohydrodynamic flow in a rectangular duct with conducting walls parallel to applied magnetic field. Comput. Mech. **41**, 769–775 (2008)
10. Tezer-Sezgin, M., Han Aydin, S.: BEM solution of MHD flow in a pipe coupled with magnetic induction of exterior region. Computing **95**, S751–S770 (2013)
11. Tezer-Sezgin, M., Han Aydin, S.: FEM solution of MHD flow in an array of electromagnetically coupled rectangular ducts. Prog. Comput. Fluid Dyn. (in press)

Application of Splitting Algorithm for Solving Advection-Diffusion Equation on a Sphere

Yuri N. Skiba and Roberto Carlos Cruz-Rodríguez

Abstract The new algorithm proposed in Skiba (Int. J. Numer. Methods Fluids (2015), https://doi.org/10.1002/fld.4016) is applied for solving linear and nonlinear advection-diffusion problems on the surface of a sphere. The discretization of advection-diffusion equation is based on the use of a spherical grid, finite volume method and the splitting of the operator in coordinate directions. The numerical algorithm is of second order approximation in space and time. It is implicit, unconditionally stable, direct (without iterations) and rapid in realization. The theoretical results obtained in Skiba (Int. J. Numer. Methods Fluids (2015), https://doi.org/10.1002/fld.4016) are confirmed numerically by simulating various linear and nonlinear advection-diffusion processes. The results show high accuracy and efficiency of the method that correctly describes the advection-diffusion processes and balance of mass of substance in the forced and dissipative discrete system, and conserves the total mass and L_2-norm of the solution in the absence of external forcing and dissipation.

1 Advection-Diffusion Problem on the Surface of a Sphere

In the present work, the implicit unconditionally stable method developed and described in detail in [5] is applied for solving linear and nonlinear advection-diffusion problems on a sphere. The numerical algorithm differs from the finite-difference method proposed in [6] and is based on the use of disjoint cells completely covering the sphere, finite volume method and the symmetrized Marchuk-Strang splitting in coordinate directions [2]. The 1D problems with periodic bound-

Y. N. Skiba (✉)
Centro de Ciencias de la Atmósfera, Universidad Nacional Autónoma de México, México City, Mexico
e-mail: skiba@unam.mx

R. C. Cruz-Rodríguez
Posgrado en Ciencias de la Tierra, Universidad Nacional Autónoma de México, México City, Mexico

© Springer Nature Switzerland AG 2019

I. Faragó et al. (eds.), *Progress in Industrial Mathematics at ECMI 2018*, Mathematics in Industry 30, https://doi.org/10.1007/978-3-030-27550-1_35

ary conditions, obtained at splitting in the longitudinal direction, are solved using the Sherman-Morrison formula [4] and the Thomas algorithm [7]. The solution of the 1D Dirichlet problems arising during the splitting in the latitudinal direction requires the application of the bordering method to the matrix of the discrete problem, and as a consequence, a preliminary determination of the values of the solution at the poles [5]. The resulting linear systems have tridiagonal matrices and are solved by the Thomas algorithm. The new numerical algorithm is implicit, unconditionally stable, of the second-order approximation in space and time and fast in implementation. One of its main advantages is that it is a direct method, the implementation of which does not require iterations. Parallel processes can be used for solving the split 1D problems in both directions. The method correctly describes the advection-diffusion processes and the balance of mass of a substance in the forced and dissipative system, and conserves the total mass and L_2-norm of the solution in the absence of external forcing and dissipation. It can be used for simulating the dispersion of a pollutant, predicting temperature and water vapor in the Earth's atmosphere, and solving various problems of linear and nonlinear diffusion, certain elliptic problems, and conjugate problems of advection-diffusion on a sphere [5].

The advection-diffusion problem is written as

$$\frac{\partial \phi}{\partial t} + \text{div}\,(\mathbf{U}\phi) - \text{div}\,(\mu \nabla \phi) + \sigma\phi = f, \quad \phi(\mathbf{x}, 0) = \phi^0(\mathbf{x}) \tag{1}$$

where $\mathbf{U} = \{u(\mathbf{x}, t), v(\mathbf{x}, t)\}$ is the known and non-divergent velocity field on a sphere:

$$\text{div}\mathbf{U} = \frac{1}{a \sin\vartheta}[u_\lambda + (v \sin\vartheta)_\vartheta] = 0 \tag{2}$$

Problem (1) describes the evolution of concentration $\phi(\mathbf{x}, t)$ of a physical substance on the sphere S of radius a, where $\mathbf{x} = (\lambda, \vartheta)$ is a point on S, λ is the longitude and ϑ is the colatitude, $\mu(\mathbf{x}, t) > 0$ is the diffusion coefficient, ∇ is the gradient in spherical coordinates, $\sigma(\mathbf{x}, t) > 0$ characterizes the rate of exponential decay of $\phi(\mathbf{x}, t)$ due to physical and chemical processes, and $f(\mathbf{x}, t)$ is a known forcing (for example, the intensity of pollution sources).

The total mass $\int_S \phi dS$ of substance satisfies the balance equation

$$\frac{\partial}{\partial t}\int_S \phi dS = \int_S f dS - \int_S \sigma\phi dS \tag{3}$$

while the evolution of L_2-norm $||\phi|| = \left(\int_S \phi^2 dS\right)^{\frac{1}{2}}$ of ϕ is governed by the integral equation

$$\frac{1}{2}\frac{\partial}{\partial t}\int_S \phi^2 dS = \int_S f\phi dS - \int_S \left(\sigma\phi^2 + \mu|\nabla\phi|^2\right) dS \tag{4}$$

where $dS = a^2 \sin\vartheta d\lambda d\vartheta$. In particular, if $f = \mu = \sigma = 0$, then the total mass and norm of solution are conserved in time.

2 Numerical Experiments

The ability of the new numerical method is tested by simulating various diffusion processes.

1. ***Linear diffusion in a spherical sector***. Let $\mathbf{U}(\mathbf{x}, t) \equiv \mathbf{0}$, $f(\mathbf{x}, t) \equiv 0$ and $\sigma(\mathbf{x}, t) \equiv 0$. The diffusion coefficient $\mu = \text{Const}$ and is nonzero only in a spherical sector, besides, $\Delta\lambda = \Delta\vartheta = 0.5°$. Figure 1 confirms that the diffusion of an initial concentration $\phi^0(\mathbf{x})$ (red spot) occurs only in the spherical sector, where μ is nonzero.

   Since the poles are singular points in spherical coordinates, the purpose of the following two experiments is to demonstrate that both diffusion and advection through the pole cells are performed correctly.
2. ***Linear diffusion of a spot from the pole***. Let $\mathbf{U}(\mathbf{x}, t) \equiv \mathbf{0}$, $\sigma(\mathbf{x}, t) \equiv 0$, $f(\mathbf{x}, t) \equiv 0$ and $\mu = \text{Const}$. Figure 2 shows that the diffusion of the initial form of the solution (the red spot centered at the pole) occurs from the pole uniformly in all directions.
3. ***Advection flux through the pole***. Let $f(\mathbf{x}, t) \equiv 0$, $\mu(\mathbf{x}, t) \equiv 0$, $\sigma(\mathbf{x}, t) \equiv 0$, and let the velocity field $\mathbf{U}(\mathbf{x}, t)$ on the sphere be directed through the pole (Fig. 3a). Figure 3b shows that the initial form of the solution (the yellow-red round spot) is not distorted after passing through the North Pole.

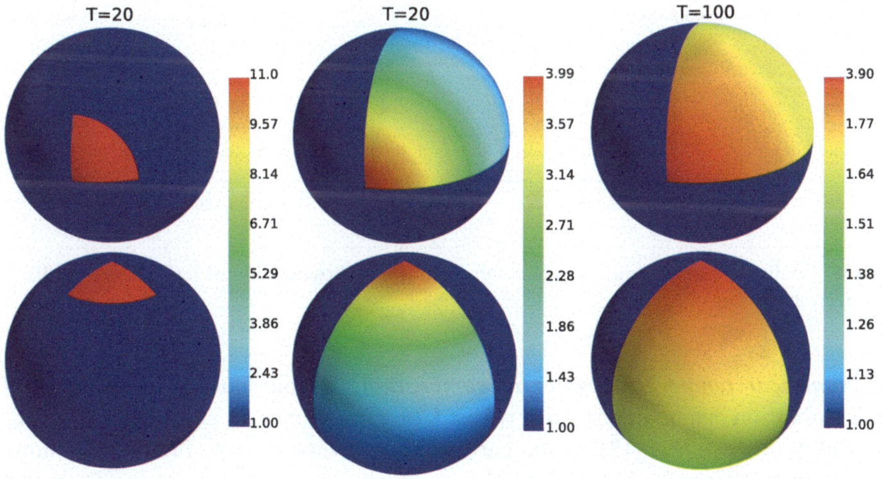

Fig. 1 Diffusion process in the spherical sector

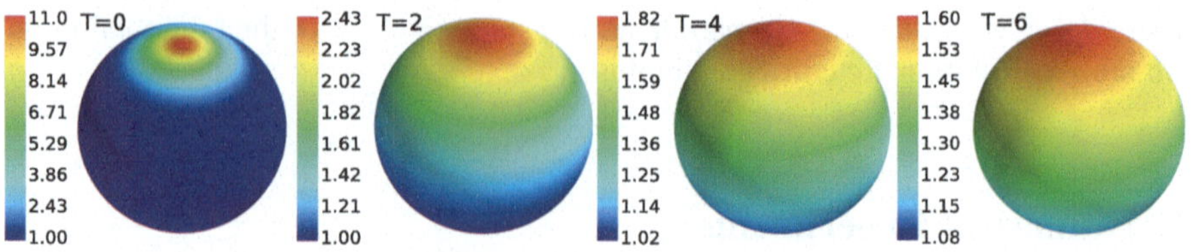

Fig. 2 Diffusion of an initial spot from the pole

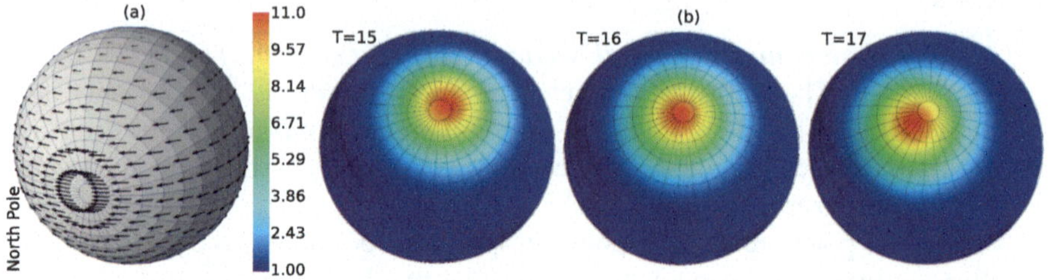

Fig. 3 Flow through the poles (**a**). Advection of initial circular spot through the North Pole (**b**)

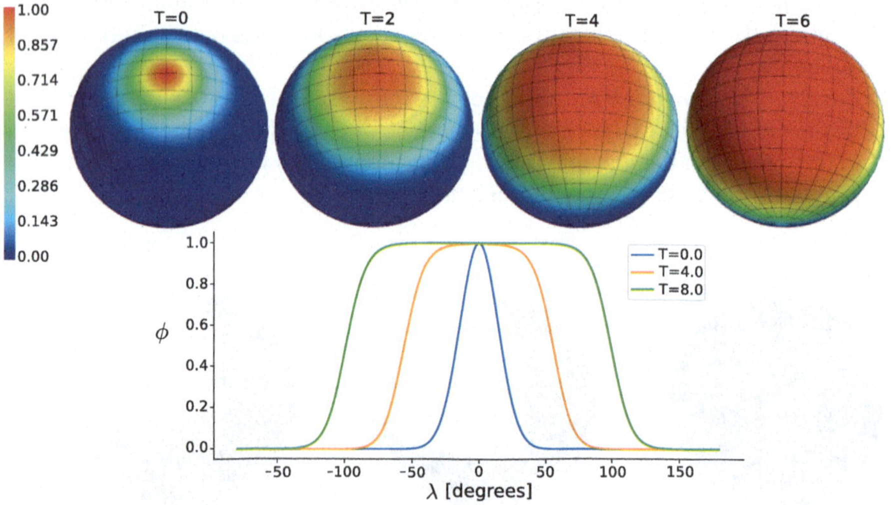

Fig. 4 Nonlinear temperature wave of combustion (top). Cross section of temperature profile in λ-direction (bottom)

4. ***Nonlinear temperature wave of combustion***. Various nonlinear processes of combustion can be described if we set $\mathbf{U}(\mathbf{x}, t) \equiv \mathbf{0}$, $\sigma(\mathbf{x}, t) \equiv 0$, and $\mu = \text{Const}$ and $f = \alpha\phi - \beta\phi^3$ [3]. In the particular case when $\alpha = \beta$, Fig. 4 (top) shows a homogeneous (in all directions) propagation of a temperature wave of constant amplitude (see Fig. 4 (bottom)), leading to an increase in the initial burning area (red spot).

5. ***Nonlinear diffusion. Blow-up combustion regimes***. Let $\mathbf{U}(\mathbf{x}, t) \equiv \mathbf{0}$, $\sigma(\mathbf{x}, t) \equiv 0$, and let both the diffusion coefficient and the forcing of the diffusion equation be nonlinear: $\mu = \mu(\phi)$ and $f = f(\phi)$. Then Eq. (1) describes a nonlinear diffusion process

$$\phi_t = \operatorname{div}(\mu(\phi)\nabla\phi) + f(\phi), \quad \phi(\mathbf{x}, 0) = \phi^0(\mathbf{x}), \tag{5}$$

and the only change that needs to be done in the new method is the linearization of discrete system in each double time interval (t_{n-1}, t_{n+1}), namely, $\mu = \mu(\phi(t_{n-1}))$ and $f = f(\phi(t_{n-1}))$.

Let $\mu = k\phi^\alpha$, $f = q\phi^\beta$ and $k, q > 0$. Then the regimes with unboundedly growing solutions in finite time (the so-called blow-up regimes) can appear due to a strong positive nonlinear feedback [1]. Such extremely growing solutions can describe rapid compression and accumulation of matter (laser fusion), as well as some important processes in chemical kinetics, magnetohydrodynamics, meteorology (tornadoes and lightning), ecology (growth of biological populations), neurophysiology, epidemiology (infectious disease outbreaks), economics (rapid economic growth), demography (world population growth), etc.

In this work, we successfully modeled three blow-up modes of combustion: the HS mode when the temperature rises rapidly in the expanding region ($\beta < \alpha + 1$, Fig. 5), the LS mode when the temperature rapidly increases in the contracting region ($\beta > \alpha + 1$, Fig. 6), and S mode when the temperature rises rapidly in a region of fixed size ($\beta = \alpha + 1$, Fig. 7).

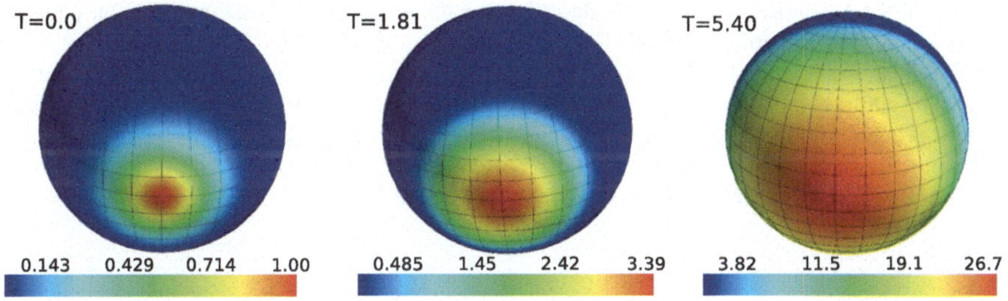

Fig. 5 The HS mode of combustion ($\alpha = 1$; $\beta = 1$)

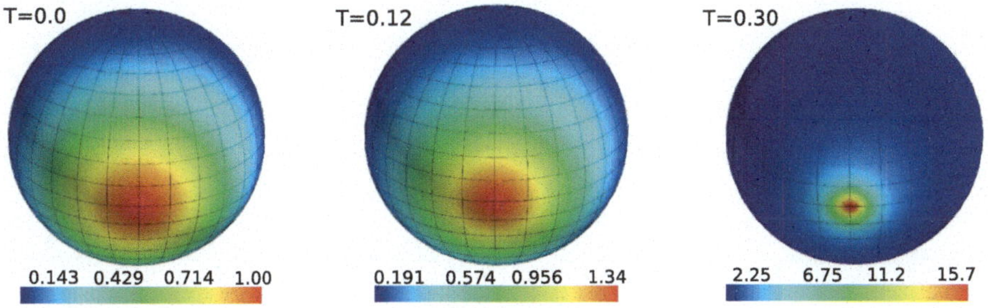

Fig. 6 The LS mode of combustion ($\alpha = 1$; $\beta = 3$)

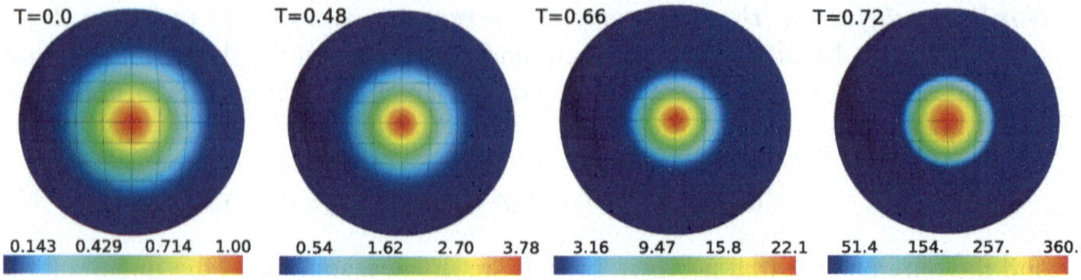

Fig. 7 The S mode of combustion ($\alpha = 1$; $\beta = 2$)

Acknowledgements The work was partially supported by the grant No. 14539 of the National System of Researchers of Mexico (SNI, CONACyT) and scholarship of CONACyT, Mexico.

References

1. Kurdyumov, S.P.: Regimes with Blow-Up. Fizmatlit, Moscow (2006) (in Russian)
2. Marchuk, G.I.: Methods of Numerical Mathematics. Springer, New York (1982)
3. Samarskii, A.A.: Nonlinear effects of blow-up and localization processes in burning problems. In: Brauner, C.M., Schmidt-Laine, C. (eds.) Mathematical Modeling in Combustion and Related Topics, pp. 217–231. Martinus Nijhoff Publishers, Leiden (1988)
4. Sherman, J., Morrison W.J.: Adjustment of an inverse matrix corresponding to changes in the elements of a given column or a given row of the original matrix. Ann. Math. Stat. **20**, 620–624 (1949)
5. Skiba, Y.N.: A non-iterative implicit algorithm for the solution of advection-diffusion equation on a sphere. Int. J. Numer. Methods Fluids (2015) https://doi.org/10.1002/fld.4016
6. Skiba, Y.N., Filatov, D.M.: Modelling of combustion and diverse blow-up regimes in a spherical shell. In: Quintela, P., et al. (eds.) Progress in Industrial Mathematics at ECMI 2016, pp. 729–735. Springer, Heidelberg (2017)
7. Thomas, L.H.: Elliptic Problems in Linear Difference Equations over a Network. Watson Sc. Comp. Lab. Rep., Columbia University, New York (1949)

Index-Preserving Model Order Reduction for Differential-Algebraic Systems Arising in Gas Transport Networks

Nicodemus Banagaaya, Peter Benner, and Sara Grundel

Abstract Gas transportation networks can be modeled by the isothermal Euler equations. Spatial discretization of these equations leads to large-scale systems of nonlinear differential-algebraic equations. Often, model order reduction is necessary for simulation of the discretized network equations under time constraints during operation. Direct reduction of such systems leads to ordinary differential equations which are very difficult to simulate especially if the index of the differential-algebraic equation is greater than one. We consider gas flow through a gas transportation network with more than one supply node which leads to differential-algebraic equations of index 2. We propose an index-aware approach which first automatically decouples the index 2 gas network into differential and algebraic parts leading to reduced-order models which are also differential-algebraic equations of the same index. This approach gives very accurate reduced order models which can be simulated using any standard ordinary differential equation numerical solver leading to accurate solutions.

1 Introduction

Modeling of gas transportation networks can be done using the isothermal Euler equations [3]. Spatial discretization of these equations leads to a nonlinear system of differential-algebraic equations (DAEs) of the form

$$\mathbf{E}\mathbf{x}' = \mathbf{H}\mathbf{x} + \mathbf{f}(\mathbf{x}) + \mathbf{B}\mathbf{u}, \quad \mathbf{E}\mathbf{x}(0) = \mathbf{E}\mathbf{x}_0, \quad \mathbf{y} = \mathbf{C}\mathbf{x} + \mathbf{h}(\mathbf{x}), \tag{1}$$

where $\mathbf{E} \in \mathbb{R}^{n\times n}$ is singular. $\mathbf{H} \in \mathbb{R}^{n\times n}$, $\mathbf{B} \in \mathbb{R}^{n\times m}$, $\mathbf{C}, \mathbf{h}(\mathbf{x}) \in \mathbb{R}^{\ell\times n}$, $\mathbf{f}(\mathbf{x}) \in \mathbb{R}^{n}$, is nonlinear and the state vector $\mathbf{x} \in \mathbb{R}^{n}$ includes the mass flux $\mathbf{x}_q \in \mathbb{R}^{n_1}$ and the gas

N. Banagaaya (✉) · P. Benner · S. Grundel
Max Planck Institute for Dynamics of Complex Technical Systems, Magdeburg, Germany
e-mail: banagaaya@mpi-magdeburg.mpg.de; benner@mpi-magdeburg.mpg.de; grundel@mpi-magdeburg.mpg.de

© Springer Nature Switzerland AG 2019
I. Faragó et al. (eds.), *Progress in Industrial Mathematics at ECMI 2018*, Mathematics in Industry 30, https://doi.org/10.1007/978-3-030-27550-1_36

density $\mathbf{x}_\rho \in \mathbb{R}^{n_2}$ within the pipe network. The input function $\mathbf{u} = \left(\mathbf{u}_p^T, \mathbf{u}_q^T\right)^T \in \mathbb{R}^m$ includes $\mathbf{u}_p \in \mathbb{R}^{m_s}$, the supply pressure, and demand mass flow $\mathbf{u}_q \in \mathbb{R}^{m_d}$, respectively. Here, we consider gas flow through a gas transportation network with more than one supply input, i.e., $m_s > 1$. The desired output vector $\mathbf{y} = \left(\mathbf{y}_p^T, \mathbf{y}_q^T\right)^T \in \mathbb{R}^\ell$ includes the supply mass flow $\mathbf{y}_p$ and the demand pressure $\mathbf{y}_q$. We are interested in a fast and stable prediction of the dynamics of natural gas transport in the pipe network. Despite the ever increasing computational power, dynamic pipeline network simulation using the system (1) is computationally expensive. Model order reduction (MOR) techniques aim to reduce the computational burden by generating reduced-order models (ROMs) that are faster and cheaper to simulate, yet accurately represent the original large-scale system behavior. MOR replaces (1) by a ROM

$$\mathbf{E}_r\mathbf{x}_r' = \mathbf{H}_r\mathbf{x}_r + \mathbf{f}_r(\mathbf{x}_r) + \mathbf{B}_r\mathbf{u}, \quad \mathbf{E}_r\mathbf{x}_r(0) = \mathbf{E}_r\mathbf{x}_{r_0}, \quad \mathbf{y}_r = \mathbf{C}_r\mathbf{x}_r + \mathbf{h}_r(\mathbf{x}_r), \tag{2}$$

where $\mathbf{E}_r, \mathbf{H}_r \in \mathbb{R}^{r\times r}$, $\mathbf{f}_r \in \mathbb{R}^r$, $\mathbf{B}_r \in \mathbb{R}^{r\times m}$ and $\mathbf{y}_r \in \mathbb{R}^{\ell\times r}$, $\mathbf{C}_r \in \mathbb{R}^{\ell\times r}$, $\mathbf{h}_r \in \mathbb{R}^{\ell\times r}$ such that the reduced order of the state vector $\mathbf{x}_r \in \mathbb{R}^r$ is $r \ll n$. A good ROM should have small approximation error $\|\mathbf{y} - \mathbf{y}_r\|$ in a suitable norm $\|.\|$ for a desired range of inputs $\mathbf{u}$. The index of the DAE is a natural number that indicates the level of difficulty of solving DAEs. There exist index concepts such as tractability index, differentiation index, etc. However, direct reduction of (1) often leads to ordinary differential equations (ODEs) which affects the choice of numerical integration schemes, especially if the index of the original DAE is greater than 1, as the reduced-order ODE is then highly stiff.

In [3], it was proved that DAEs arising from gas transportation networks are of index 1 if the gas network has only one supply node, otherwise they are of index 2. Thus, system (1) is of index 2. In [3], an index reduction approach for DAEs arising from gas transport networks was proposed leading to ODEs which can further be reduced using standard techniques such as proper orthogonal decomposition (POD). However, this approach is restricted to gas networks with special structure and can lead to very stiff ROMs. We propose an approach which first automatically decouples the index-2 gas networks into differential and algebraic parts, leading to ROMs which are also DAEs and their inherited ODE part can be simulated using any standard ODE numerical solver.

The paper is organized as follows. In Sect. 2, we present the discretized dynamic DAE model arising from gas transport networks and its linearized DAE form. In Sect. 3, we discuss the numerical difficulty of linearized DAEs arising from gas transport networks. In the final section, we discuss the proposed MOR method for gas transport networks with many supply inputs. We present also some simulations illustrating the performance of the proposed approach.

2 Discretized Model Arising from Gas Transport Networks

Here, we consider the spatial discretization approach of the isothermal Euler equations discussed in [3] which leads to a nonlinear DAE system of the form (1) with

$$\mathbf{E} = \begin{pmatrix} 0 & 0 & |\mathcal{A}_0^T| & |\mathcal{A}_S^T| \\ 0 & \mathrm{I} & 0 & 0 \\ 0 & 0 & 0 & 0 \\ 0 & 0 & 0 & 0 \end{pmatrix}, \quad \mathbf{H} = \begin{pmatrix} -\mathbf{M}_L^{-1} & 0 & 0 & 0 \\ 0 & 0 & \mathbf{M}_A\mathcal{A}_0^T & \mathbf{M}_A\mathcal{A}_S^T \\ |\mathcal{A}_0| & \mathcal{A}_0 & 0 & 0 \\ 0 & 0 & 0 & \mathrm{I} \end{pmatrix}, \quad \mathbf{f}(\mathbf{x}) = \begin{pmatrix} 0 \\ \mathbf{g}(\mathbf{x}) \\ 0 \\ 0 \end{pmatrix},$$

$$\mathbf{x} = \begin{pmatrix} \mathbf{q}_- \\ \mathbf{q}_+ \\ \boldsymbol{\rho}_d \\ \boldsymbol{\rho}_s \end{pmatrix}, \quad \mathbf{B} = -\begin{pmatrix} 0 & 0 \\ 0 & 0 \\ 0 & \mathrm{I} \\ \mathrm{I} & 0 \end{pmatrix}, \quad \mathbf{C} = \begin{pmatrix} 0 & |\mathcal{A}_S| & 0 & 0 \\ 0 & 0 & -\mathrm{I} & 0 \end{pmatrix}, \quad \mathbf{u} = \begin{pmatrix} \mathbf{s}(t) \\ \mathbf{d}(t) \end{pmatrix},$$

where $\mathbf{M}_L \in \mathbb{R}^{n_E \times n_E}$ and $\mathbf{M}_A \in \mathbb{R}^{n_E \times n_E}$ are material dependent diagonal matrices. $\boldsymbol{\rho}_s \in \mathbb{R}^{n_s}$ is the unknown density at the supply nodes and $\boldsymbol{\rho}_d \in \mathbb{R}^{n_d}$ the unknown density at the demand nodes and junctions. The other unknown vectors are $\mathbf{q}_- = \mathbf{q}_R - \mathbf{q}_L \in \mathbb{R}^{n_E}$ and $\mathbf{q}_+ = \mathbf{q}_R + \mathbf{q}_L \in \mathbb{R}^{n_E}$, where the vectors $\mathbf{q}_R$ and $\mathbf{q}_L$ are the vectors of fluxes at the end and beginning of the pipes, respectively. The dimension of the system is given by $n = 2n_E + n_d + n_s$. $\mathbf{d}(t) = (\ldots, d_i(t), \ldots)^T \in \mathbb{R}^{m_d}$ and $\mathbf{s}(t) = (\ldots, s_i(t), \ldots)^T \in \mathbb{R}^{m_s}$ are vectors for demand and supply which are considered as input functions. $\mathcal{A}_0 \in \mathbb{R}^{n_d \times n_E}$ is an incidence matrix corresponding to the demand nodes and junctions while $\mathcal{A}_S \in \mathbb{R}^{n_s \times n_E}$ is the incidence matrix corresponding to the supply nodes. The nonlinearity is defined via the state-dependent vector
$\mathbf{g}(\mathbf{x}) = \mathbf{g}(\mathbf{q}_-, \mathbf{q}_+, \boldsymbol{\rho}_d, \boldsymbol{\rho}_s) = (\ldots, g_k(\mathbf{q}_-, \mathbf{q}_+, \boldsymbol{\rho}_d, \boldsymbol{\rho}_s), \ldots)^T \in \mathbb{R}^{n_E}$, with

$$g_k(\mathbf{q}_-, \mathbf{q}_+, \boldsymbol{\rho}_d, \boldsymbol{\rho}_s) = -\frac{gA_k}{2\gamma_0}\psi_k(\boldsymbol{\rho}_d, \boldsymbol{\rho}_s)\frac{h_R^k - h_L^k}{L_k} - \frac{\lambda_k \gamma_0}{4D_k A_k}\frac{\mathbf{q}_+^k|\mathbf{q}_+^k|}{\psi_k(\boldsymbol{\rho}_d, \boldsymbol{\rho}_s)},$$

where $\psi_k(\boldsymbol{\rho}_d, \boldsymbol{\rho}_s)$ is the k-th entry of the vector-valued linear function:

$$\psi(\boldsymbol{\rho}_d, \boldsymbol{\rho}_s) = |\mathcal{A}_S^T|\boldsymbol{\rho}_s + |\mathcal{A}_0^T|\boldsymbol{\rho}_d \in \mathbb{R}^{n_E}.$$

The scalar parameters in the system are $\lambda_k, a_k, L_k, A_k, h_R^k, h_L^k$ and D_k which are known at least within some range of uncertainty. $\gamma_0 = RT_0$ is a constant, where R is the universal gas constant and T_0 is a constant temperature in time and space.

The linearization of the above nonlinear DAE model around a static point $(\mathbf{x}_s, \mathbf{u}_s)$ was discussed in [2] which leads to a linear DAE system given by

$$\mathbf{E}\bar{\mathbf{x}}' = \overline{\mathbf{A}}\bar{\mathbf{x}} + \overline{\mathbf{B}}\bar{\mathbf{u}}, \quad \mathbf{E}\bar{\mathbf{x}}(0) = \mathbf{E}\bar{\mathbf{x}}_0, \tag{3a}$$

$$\bar{\mathbf{y}} = \bar{\mathbf{C}}\bar{\mathbf{x}}, \tag{3b}$$

where $\overline{\mathbf{A}} = \mathbf{H} + \left.\frac{\partial \mathbf{f}}{\partial \mathbf{x}}\right|_{\mathbf{x}_s} \in \mathbb{R}^{n\times n}$, $\overline{\mathbf{B}} = \mathbf{B} \in \mathbb{R}^{n\times m}$, $\bar{\mathbf{C}} = \mathbf{C} \in \mathbb{R}^{\ell\times n}$, $\bar{\mathbf{x}} = \mathbf{x} - \mathbf{x}_s \in \mathbb{R}^n$ and $\bar{\mathbf{u}} = \mathbf{u} - \mathbf{u}_s \in \mathbb{R}^m$. The linearized DAE system (3a) is valid in a neighborhood of the stationary point $(\mathbf{x}_s, \mathbf{u}_s)$ for the nonlinear DAE system. In the next section, we discuss the tractability index analysis of system (3) which measures the numerical difficulty of simulating and reducing linear DAEs.

3 Index Analysis of DAEs Arising from Gas Transport Networks

In this section, we use the tractability index concept [1] to perform an index analysis of DAEs arising from gas transport networks. This can be done as follows. Assume (3) is solvable, i.e., $\det(\lambda\mathbf{E} - \bar{\mathbf{A}}) \neq 0$, then we set

$$\mathbf{E}_0 = \mathbf{E} = \begin{pmatrix} 0 & 0 & \mathbf{E}_{13} \\ 0 & \mathrm{I} & 0 \\ 0 & 0 & 0 \end{pmatrix} \quad \text{and} \quad \mathbf{A}_0 = \bar{\mathbf{A}} = \begin{pmatrix} \mathbf{A}_{11} & 0 & 0 \\ 0 & \mathbf{A}_{22} & \mathbf{A}_{23} \\ \mathbf{A}_{31} & \mathbf{A}_{32} & \mathbf{A}_{33} \end{pmatrix},$$

where $\mathbf{E}_{13} = \left(|\mathcal{A}_0^T| \; |\mathcal{A}_S^T|\right) \in \mathbb{R}^{n_E\times(n_d+n_s)}$, $\mathbf{A}_{11} = -\mathbf{M}_L^{-1} \in \mathbb{R}^{n_E\times n_E}$, $\mathbf{A}_{22} = \left.\frac{\partial \mathbf{g}}{\partial \mathbf{q}_+}\right|_{\mathbf{x}_s} \in \mathbb{R}^{n_E\times n_E}$, $\mathbf{A}_{23} = \left(\mathbf{M}_A\mathcal{A}_0^T + \left.\frac{\partial \mathbf{g}}{\partial \mathbf{p}_q}\right|_{\mathbf{x}_s} \;\; \mathbf{M}_A\mathcal{A}_S^T + \left.\frac{\partial \mathbf{g}}{\partial \mathbf{p}_s}\right|_{\mathbf{x}_s}\right) \in \mathbb{R}^{n_E\times(n_d+n_s)}$, $\mathbf{A}_{31} = \begin{pmatrix} |\mathcal{A}_0| \\ 0 \end{pmatrix} \in \mathbb{R}^{(n_d+n_s)\times n_E}$, $\mathbf{A}_{32} = \begin{pmatrix} \mathcal{A}_0 \\ 0 \end{pmatrix} \in \mathbb{R}^{(n_d+n_s)\times n_E}$, $\mathbf{A}_{33} = \begin{pmatrix} 0 & 0 \\ 0 & \mathrm{I} \end{pmatrix} \in \mathbb{R}^{n_s\times n_s}$. We choose the projectors

$$\mathbf{Q}_0 = \begin{pmatrix} \mathrm{I} & 0 & 0 \\ 0 & 0 & 0 \\ 0 & 0 & \mathbf{Q} \end{pmatrix} \in \mathbb{R}^{n\times n} \quad \text{and} \quad \mathbf{P}_0 = \mathrm{I} - \mathbf{Q}_0 = \begin{pmatrix} 0 & 0 & 0 \\ 0 & \mathrm{I} & 0 \\ 0 & 0 & \mathbf{P} \end{pmatrix} \in \mathbb{R}^{n\times n}, \tag{4}$$

such that $\operatorname{Im}\mathbf{Q}_0 = \operatorname{Ker}\mathbf{E}_0$, i.e., $\mathbf{E}_0\mathbf{Q}_0 = 0$, meaning $\mathbf{E}_{13}\mathbf{Q} = 0$, where $\mathbf{Q} \in \mathbb{R}^{(n_d+n_s)\times(n_d+n_s)}$ is the projector onto the nullspace of $\mathbf{E}_{13}$ and $\mathbf{P} \in \mathbb{R}^{(n_d+n_s)\times(n_d+n_s)}$ is its complementary projector. Then, using the definition of the tractability index, we define the matrices

$$\mathbf{E}_1 = \mathbf{E}_0 - \mathbf{A}_0\mathbf{Q}_0 = \begin{pmatrix} -\mathbf{A}_{11} & 0 & \mathbf{E}_{13} \\ 0 & \mathbf{I} & -\mathbf{A}_{23}\mathbf{Q} \\ -\mathbf{A}_{31} & 0 & -\mathbf{A}_{33}\mathbf{Q} \end{pmatrix}, \quad \mathbf{A}_1 = \mathbf{A}_0\mathbf{P}_0 = \begin{pmatrix} 0 & 0 & 0 \\ 0 & \mathbf{A}_{22} & \mathbf{A}_{23}\mathbf{P} \\ \mathbf{A}_{31} & \mathbf{A}_{32} & \mathbf{A}_{33}\mathbf{P} \end{pmatrix}.$$

According to [2], it can be proved that the matrix $\mathbf{E}_1$ is invertible if and only if

$$\mathbf{S}_0 = -\mathbf{A}_{33}\mathbf{Q} - \mathbf{A}_{31}\mathbf{A}_{11}^{-1}\mathbf{E}_{13} = \begin{pmatrix} |\mathcal{A}_0|\mathbf{M}_L|\mathcal{A}_0^T| & |\mathcal{A}_0|\mathbf{M}_L|\mathcal{A}_S^T| \\ -\mathbf{Q}_{21} & -\mathbf{Q}_{22} \end{pmatrix} \tag{5}$$

is invertible, where $\mathbf{Q}_{21} \in \mathbb{R}^{n_s \times n_d}$ and $\mathbf{Q}_{22} \in \mathbb{R}^{n_s \times n_s}$ are the entries of projector $\mathbf{Q}$. If $\mathbf{E}_1$ is nonsingular, then we say that the DAE arising from the gas transport network is of tractability index 1, otherwise it is of higher index.

Assume $\mathbf{E}_1$ is singular, then we need to construct another projector $\mathbf{Q}_1$ such that $\operatorname{Im}\mathbf{Q}_1 = \operatorname{Ker}\mathbf{E}_1$ with additional condition $\mathbf{Q}_1\mathbf{Q}_0 = 0$. Choose the projectors

$$\mathbf{Q}_1 = \begin{pmatrix} 0 & 0 & \mathbf{A}_{11}^{-1}\mathbf{E}_{13}\mathbf{Q}_S \\ 0 & 0 & \mathbf{A}_{23}\mathbf{Q}\mathbf{Q}_S \\ 0 & 0 & \mathbf{Q}_S \end{pmatrix} \quad \text{and} \quad \mathbf{P}_1 = \mathbf{I} - \mathbf{Q}_1 = \begin{pmatrix} \mathbf{I} & 0 & -\mathbf{A}_{11}^{-1}\mathbf{E}_{13}\mathbf{Q}_S \\ 0 & \mathbf{I} & -\mathbf{A}_{23}\mathbf{Q}\mathbf{Q}_S \\ 0 & 0 & \mathbf{P}_S \end{pmatrix},$$

where $\mathbf{Q}_S$ is a projector onto the nullspace of $\mathbf{S}_0$ and $\mathbf{P}_S$ is its complementary projector. Then we can compute

$$\mathbf{E}_2 = \mathbf{E}_1 - \mathbf{A}_1\mathbf{Q}_1 = \begin{pmatrix} -\mathbf{A}_{11} & 0 & \mathbf{E}_{13} \\ 0 & \mathbf{I} & \mathbf{E}_{2,23} \\ -\mathbf{A}_{31} & 0 & \mathbf{E}_{2,33} \end{pmatrix}, \quad \mathbf{A}_2 = \mathbf{A}_1\mathbf{P}_1 = \begin{pmatrix} 0 & 0 & 0 \\ 0 & \mathbf{A}_{22} & \mathbf{A}_{23} + \mathbf{E}_{2,23} \\ 0 & \mathbf{A}_{32} & \mathbf{A}_{33} + \mathbf{E}_{2,33} \end{pmatrix},$$

with $\mathbf{E}_{2,23} = -\mathbf{A}_{22}\mathbf{A}_{23}\mathbf{Q}\mathbf{Q}_S - \mathbf{A}_{23}(\mathbf{I} - \mathbf{P}\mathbf{P}_S)$, $\mathbf{E}_{2,33} = -\mathbf{A}_{32}\mathbf{A}_{23}\mathbf{Q}\mathbf{Q}_S - \mathbf{A}_{33}(\mathbf{I} - \mathbf{P}\mathbf{P}_S)$. If the matrix $\mathbf{E}_2$ is invertible, we say that the system has tractability index 2. Similarly to the index-1 case, this condition is equivalent to the invertibility of the matrix

$$\mathbf{S}_1 = -\mathbf{A}_{31}\mathbf{A}_{11}^{-1}\mathbf{E}_{13} - \mathbf{A}_{32}\mathbf{A}_{23}\mathbf{Q}\mathbf{Q}_S - \mathbf{A}_{33}(\mathbf{Q} + \mathbf{P}\mathbf{Q}_S).$$

In [3], it was shown that gas transport networks are at most of tractability index 2 and they are of tractability index 2 if and only if they have more than one supply input. According to [1], the index 2 condition implies that the solutions of (3) contain the derivatives of input data $\mathbf{u}$ which restricts the choice of the input data. Since numerical differentiations may cause considerable trouble numerically, it is very important to know the index of the DAE before applying MOR.

4 Index-Aware MOR for Gas Transport Networks

According to [1], system (3) can be re-written into an equivalent decoupled system given by

$$\mathbf{E}_p\xi_p' = \mathbf{A}_p\xi_p + \mathbf{B}_p\mathbf{u}, \quad \xi_p(0) = \xi_{p0}, \tag{6a}$$

$$-\mathcal{L}\xi_q' = \mathbf{A}_q\xi_p - \mathcal{L}_q\xi_q + \mathbf{B}_q\mathbf{u}, \tag{6b}$$

$$\mathbf{y} = \mathbf{C}_p\xi_p + \mathbf{C}_q\xi_q, \tag{6c}$$

where $\mathcal{L} \in \mathbb{R}^{n_a \times n_a}$ is of nilpotency index 2 and $\mathcal{L}_q \in \mathbb{R}^{n_a \times n_a}$, $\mathbf{E}_p \in \mathbb{R}^{n_o \times n_o}$ are non-singular matrices. The subsystems (6a) and (6b) correspond to the differential and algebraic subsystems, respectively. $\xi_p \in \mathbb{R}^{n_o}$ and $\xi_q \in \mathbb{R}^{n_a}$ are the differential and algebraic variables, and $n = n_o + n_a$. $\mathbf{A}_p \in \mathbb{R}^{n_o \times n_o}$, $\mathbf{B}_p \in \mathbb{R}^{n_o \times m}$, $\mathbf{A}_q \in \mathbb{R}^{n_o \times n_a}$, $\mathbf{B}_q \in \mathbb{R}^{n_a \times m}$, $\mathbf{C}_p \in \mathbb{R}^{\ell \times n_o}$ and $\mathbf{C}_q \in \mathbb{R}^{\ell \times n_a}$. Index-aware model order reduction (IMOR) replaces (6) by an IROM [1]

$$\mathbf{E}_{p_r}\xi_{p_r}' = \mathbf{A}_{p_r}\xi_{p_r} + \mathbf{B}_{p_r}\mathbf{u}, \quad \xi_{p_r}(0) = \xi_{p_{r_0}}, \tag{7a}$$

$$-\mathcal{L}_r\xi_{q_r}' = \mathbf{A}_{q_r}\xi_{p_r} - \mathcal{L}_{q_r}\xi_{q_r} + \mathbf{B}_{q_r}\mathbf{u}, \tag{7b}$$

$$\mathbf{y}_r = \mathbf{C}_{p_r}\xi_{p_r} + \mathbf{C}_{q_r}\xi_{q_r}, \tag{7c}$$

where $\mathbf{E}_{p_r} = \mathbf{V}_p^T\mathbf{E}_p\mathbf{V}_p, \mathbf{A}_{p_r} = \mathbf{V}_p^T\mathbf{A}_p\mathbf{V}_p \in \mathbb{R}^{r_o \times r_o}$, $\mathbf{B}_{p_r} = \mathbf{V}_p^T\mathbf{B}_p \in \mathbb{R}^{r_o \times m}$, $\xi_{p_{r_0}} = \mathbf{V}_p^T\xi_{p0} \in \mathbb{R}^{r_o \times n_o}$, $\mathcal{L}_r = -\mathbf{V}_q^T\mathcal{L}\mathbf{V}_q$, $\mathcal{L}_{q_r} = \mathbf{V}_q^T\mathcal{L}_q\mathbf{V}_q \in \mathbb{R}^{r_a \times r_a}$, $\mathbf{A}_{q_r} = \mathbf{V}_q^T\mathbf{A}_q\mathbf{V}_p \in \mathbb{R}^{r_a \times r_o}$, $\mathbf{B}_{q_r} = \mathbf{V}_q^T\mathbf{B}_q \in \mathbb{R}^{r_a \times m}$, $\mathbf{C}_{p_r} = \mathbf{C}_p\mathbf{V}_p \in \mathbb{R}^{\ell \times n_o}$, and $\mathbf{C}_{q_r} = \mathbf{C}_q\mathbf{V}_q \in \mathbb{R}^{\ell \times n_a}$. The projection matrix $\mathbf{V}_p \in \mathbb{R}^{n_o \times r_o}$ is constructed using any standard MOR method such as POD, etc applied to the ODE subsystem and $\mathbf{V}_q$ is computed using POD taking the algebraic solutions as snapshots. As a result, both the differential and algebraic subsystems are reduced. In order to illustrate the performance of this proposed approach, we used a DAE in the form (3) of dimension $n = 2023$ arising from a gas transport network with $m_s = 5$ supply inputs and $m_d = 226$ demand inputs. This system is decoupled into the form (6) with $n_o = 1341$ differential and $n_a = 682$ algebraic equations. We simulated both, the coupled and the decoupled system, using the implicit Euler scheme leading to a runtime of 29.3 s and 14.12 s, respectively. We can observe that decoupling reduces the runtime of the full order model. We obtained an IROM (7) by constructing $\mathbf{V}_p$ using POD leading to an IROM of dimension $r = r_o + r_a = 46 \ll 2023$ with $r_o = 14 \ll 1341$ and $r_a = 32 \ll 682$ at an offline cost of 10.9 s. For comparison, we applied POD directly to the DAE in the form (3) leading to an ODE ROM of dimension $r = 15$ at an offline cost 29.2 s. Hence, decoupling also reduces the offline costs of POD. Simulating the IROM leads to an output error of 3.4×10^{-5} at a speedup of 398.9, while the ODE ROM leads to an output error of 8.2×10^{-5} at a speedup of 477.6. The size of both ROMs is determined by making sure the output error is below 10^{-4}.

5 Conclusions

We have proposed an index-aware MOR method for gas transport networks with many supply inputs which leads to cheaper-to-construct ROMs than previous MOR approaches due to a lower offline cost. Moreover, numerical differentiations of the input data are no longer a problem due to the automatic decoupling which allows symbolic or explicit differentiations. This gives very good results for small to medium-size gas transport networks with many inputs.

Acknowledgements This work was supported by the German Federal Ministry for Economic Affairs and Energy, in the joint project: "**MathEnergy**—Mathematical Key Technologies for Evolving Energy Grids", sub-project: Model Order Reduction (Grant number: 0324019**B**).

References

1. Banagaaya, N.: Index-aware model order reduction methods. Ph.D. Thesis, Eindhoven University of Technology, Eindhoven, Netherlands (2014)
2. Banagaaya, N., Grundel, S., Benner, P.: Index-aware MOR for Gas Transport Networks. In: Proccedings of IUTAM Symposium on Model Order Reduction of Coupled Systems, Stuttgart, May 22–25, 2018, pp. 191–207. Springer, Cham (2020)
3. Grundel, S., Jansen, L., Hornung, N., Clees, T., Tischendorf, C., Benner, P.: Model order reduction of differential algebraic equations arising from the simulation of gas transport networks. In: Schöps, S., Bartel, A., Günther, M., ter Maten, W.J.W., Müller, P.C. (eds.) Progress in Differential-Algebraic Equations, pp. 183–205. Springer, Berlin (2014)

Double Freeform Lens Design for Laser Beam Shaping: A Least-Squares Approach

J. H. M. ten Thije Boonkkamp, Nitin Kumar Yadav, and W. L. IJzerman

Abstract The location of the surfaces of a double freeform lens, required for laser beam shaping, is governed by a Monge-Ampère type equation. We outline a least-squares solver and demonstrate the performance of the method for an example.

1 Introduction

Laser beam shaping is the technique to control the phase and irradiance of a laser beam. Typically, a laser beam is required to have a top-hat irradiance, constant over some cross section, whereas the exitance is usually Gaussian shaped. The goal of this paper is to compute a double freeform lens, i.e., a lens having two freeform surfaces, that converts the Gaussian exitance to a desired top-hat irradiance. We restrict ourselves to planar wave fronts, hence the phase is constant.

To compute the freeform surfaces, methods from illumination optics can be employed. The beam shaping problem is governed by the principles of geometrical optics and conservation of energy. These principles translate into the optical map, connecting the coordinates on source and target domain, and a relation for the location of the lens surfaces. Combining these with the energy conservation relation, we obtain a fully nonlinear elliptic PDE of Monge-Ampère type, defining the location of one lens surface. We compute the numerical solution in a two-stage algorithm. In the first stage we compute the optical map, and subsequently we compute the location of the lens surfaces. Both stages are evaluated in a least-squares sense.

J. H. M. ten Thije Boonkkamp (✉) · N. K. Yadav
Eindhoven University of Technology, Eindhoven, The Netherlands
e-mail: j.h.m.tenthijeboonkkamp@tue.nl

W. L. IJzerman
Eindhoven University of Technology, Eindhoven, The Netherlands

Signify Research, Eindhoven, The Netherlands
e-mail: wilbert.ijzerman@signify.com

© Springer Nature Switzerland AG 2019
I. Faragó et al. (eds.), *Progress in Industrial Mathematics at ECMI 2018*,
Mathematics in Industry 30, https://doi.org/10.1007/978-3-030-27550-1_37

The contents of this contribution is the following. In Sect. 2 we present the mathematical model and next, in Sect. 3, we outline the least-squares algorithm. A numerical example is presented in Sect. 4 and we end with conclusions in Sect. 5.

2 Mathematical Formulation

In this section we outline the mathematical model for a double freeform lens for laser beam shaping. A more detailed derivation, albeit in the context of illumination optics, is presented in [3]. To that purpose we first determine the optical map and subsequently apply conservation of energy.

Consider the optical system shown in Fig. 1, consisting of a source $\mathscr{S}$ (the laser) in $z = 0$, emitting a parallel beam of light, a target $\mathscr{T}$ in $z = \ell$, receiving a parallel bundle, and a lens in between. The index of refraction is n. The two freeform surfaces are defined by the relations $z = u_1(\boldsymbol{x})$ for $\boldsymbol{x} \in \mathscr{S}$ and $\ell - z = u_2(\boldsymbol{y})$ for $\boldsymbol{y} \in \mathscr{T}$. To determine the optical map $\boldsymbol{y} = \boldsymbol{m}(\boldsymbol{x})$, we combine the law of refraction (Snell's law) and the principle of equal optical path lengths [1], stating that the optical path length $L(\boldsymbol{x}, \boldsymbol{y})$ between the planar wave fronts at source and target is constant, i.e.,

$$L(\boldsymbol{x}, \boldsymbol{y}) = u_1(\boldsymbol{x}) + nd(\boldsymbol{x}) + u_2(\boldsymbol{y}) = L = \text{Const}, \tag{1}$$

where $d(\boldsymbol{x})$ is the distance between the lens surfaces, measured along the refracted ray. This way we find

$$\boldsymbol{y} = \boldsymbol{m}(\boldsymbol{x}) = \boldsymbol{x} - \frac{\beta \nabla u_1(\boldsymbol{x})}{\sqrt{n^2 + (n^2 - 1)|\nabla u_1|^2}}, \tag{2}$$

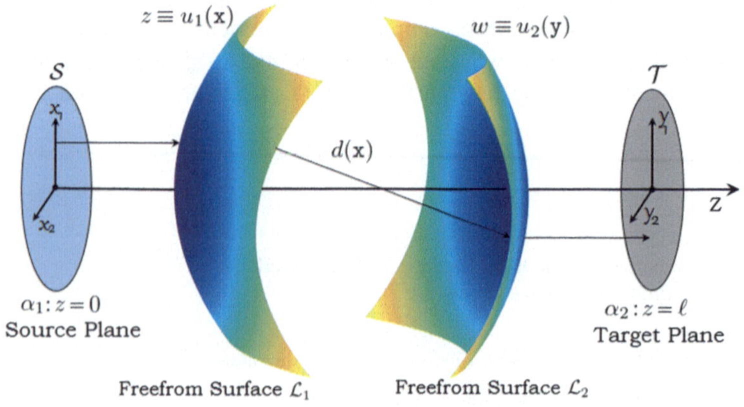

Fig. 1 Double freeform lens

where $\beta = L - \ell$ is the reduced optical path length. Moreover, eliminating d from (1), we can derive the following relation for the location of the lens surfaces

$$u_1(\boldsymbol{x}) + u_2(\boldsymbol{y}) = \ell - \frac{\beta}{n^2 - 1} - \frac{n}{n^2 - 1}\sqrt{\beta^2 - \left(n^2 - 1\right)|\boldsymbol{x} - \boldsymbol{y}|^2} = c(\boldsymbol{x}, \boldsymbol{y}), \qquad (3)$$

where $c = c(\boldsymbol{x}, \boldsymbol{y})$ is the so-called cost function in optimal transport theory [4].

Conservation of energy can be expressed as

$$\int_{\mathscr{A}} E(\boldsymbol{x})\mathrm{d}A(\boldsymbol{x}) = \int_{\boldsymbol{m}(\mathscr{A})} I(\boldsymbol{y})\mathrm{d}A(\boldsymbol{y}), \qquad (4)$$

for an arbitrary set $\mathscr{A} \subset \mathscr{S}$ and image set $\boldsymbol{m}(\mathscr{A}) \subset \mathscr{T}$, where E is the exitance of the laser and I the desired top-hat irradiance. Introducing the variable $\boldsymbol{y} = \boldsymbol{m}(\boldsymbol{x})$ in the right hand side of (4), the energy constraint becomes

$$\det\left(\mathrm{D}\boldsymbol{m}(\boldsymbol{x})\right) = \frac{E(\boldsymbol{x})}{I(\boldsymbol{m}(\boldsymbol{x}))}, \quad \boldsymbol{x} \in \mathscr{S}, \qquad (5a)$$

with $\mathrm{D}\boldsymbol{m}(\boldsymbol{x})$ the Jacobian of the optical map. The accompanying transport boundary condition reads

$$\boldsymbol{m}(\partial\mathscr{S}) = \partial\mathscr{T}, \qquad (5b)$$

implying that all light from the source arrives at the target.

The governing equations to compute u_1, u_2 and $\boldsymbol{m}$ are (2), (3) and boundary value problem (5), and allow many possible solutions. To enforce uniqueness we restrict ourselves to the c-convex solution given by

$$u_1(\boldsymbol{x}) = \max_{\boldsymbol{y}\in\mathscr{T}}\left(c(\boldsymbol{x}, \boldsymbol{y}) - u_2(\boldsymbol{y})\right), \quad u_2(\boldsymbol{y}) = \max_{\boldsymbol{x}\in\mathscr{S}}\left(c(\boldsymbol{x}, \boldsymbol{y}) - u_1(\boldsymbol{x})\right), \qquad (6)$$

which necessarily requires that $\boldsymbol{x}$ be a stationary point of $c(\boldsymbol{x}, \boldsymbol{y}) - u_1(\boldsymbol{x})$, i.e.,

$$\nabla_{\boldsymbol{x}} c(\boldsymbol{x}, \boldsymbol{y}) - \nabla u_1(\boldsymbol{x}) = \boldsymbol{0}. \qquad (7)$$

Straightforward evaluation shows that $\nabla_{\boldsymbol{x}} c(\boldsymbol{x}, \cdot)$ is injective, i.e., if $\nabla_{\boldsymbol{x}} c(\boldsymbol{x}, \boldsymbol{y}_1) = \nabla_{\boldsymbol{x}} c(\boldsymbol{x}, \boldsymbol{y}_2)$ then $\boldsymbol{y}_1 = \boldsymbol{y}_2$, implying that $\boldsymbol{y} = \left(\nabla_{\boldsymbol{x}} c(\boldsymbol{x}, \cdot)\right)^{-1} \circ \nabla u_1(\boldsymbol{x})$ is uniquely determined by (7). Indeed, solving equation (7) for $\boldsymbol{y}$, we recover $\boldsymbol{y} = \boldsymbol{m}(\boldsymbol{x})$ from (2). Conversely, for given $\boldsymbol{y}$, Eq. (2) uniquely determines ∇u_1, hence u_1 is determined up to an additive constant. Alternatively, existence of the unique solution $\boldsymbol{y} = \boldsymbol{m}(\boldsymbol{x})$ of (7) is also a consequence of the implicit function theorem, since the Jacobi matrix $\boldsymbol{C} = \mathrm{D}_{\boldsymbol{xy}} c = \left(c_{x_i y_j}\right)$ is regular for all $\boldsymbol{x}$ and $\boldsymbol{y}$. Then, assuming $\boldsymbol{y} = \boldsymbol{m}(\boldsymbol{x})$ and subsequently differentiating equation (7) with respect to $\boldsymbol{x}$, we obtain

$$\boldsymbol{C}(\boldsymbol{x})\mathrm{D}\boldsymbol{m}(\boldsymbol{x}) = \mathrm{D}^2 u_1(\boldsymbol{x}) - \mathrm{D}_{\boldsymbol{xx}} c(\boldsymbol{x}, \boldsymbol{m}(\boldsymbol{x})) = \boldsymbol{P}(\boldsymbol{x}), \qquad (8)$$

with $\mathrm{D}_{\boldsymbol{xx}}c$ and D^2u_1 the Hessian matrices of c (w.r.t. $\boldsymbol{x}$) and u_1, respectively. A sufficient condition for the c-convex solution is that the matrix $\boldsymbol{P}(\boldsymbol{x})$ defined in (8) is symmetric positive definite (SPD).

Finally, combining the energy balance (5a) with (8), we obtain the equation

$$\det(\boldsymbol{P}(\boldsymbol{x})) = \frac{E(\boldsymbol{x})}{I(\boldsymbol{m}(\boldsymbol{x}))}\det(\boldsymbol{C}(\boldsymbol{x})) = F(\boldsymbol{x}) > 0, \tag{9}$$

which is a nonlinear, second order PDE for u_1, reminiscent of the Monge-Ampère equation. Recall that the 2×2-matrix $\boldsymbol{P}$ is SPD if $\det(\boldsymbol{P}) > 0$ and $\mathrm{tr}(\boldsymbol{P}) > 0$. Only the latter condition needs to be verified.

3 Solution Strategy

The global solution strategy is as follows: first compute $\boldsymbol{m}$ from the boundary value problem (9) and (5b), next compute u_1 from (7) and finally u_2 from (3). We compute both $\boldsymbol{m}$ and u_1 in a least-squares sense. The algorithm in this section is a modification of the least-squares method detailed in [2].

First, to solve the BVP (9) and (5b) for $\boldsymbol{m}$, we subsequently compute $\boldsymbol{b}$, $\boldsymbol{P}$ and $\boldsymbol{m}$ minimizing the following functionals

$$J_{\mathrm{B}}[\boldsymbol{m}, \boldsymbol{b}] = \tfrac{1}{2}\int_{\partial\mathscr{S}} \|\boldsymbol{m} - \boldsymbol{b}\|^2\,\mathrm{d}s, \tag{10a}$$

$$J_{\mathrm{I}}[\boldsymbol{m}, \boldsymbol{P}] = \tfrac{1}{2}\int_{\mathscr{S}} \|\boldsymbol{C}\mathrm{D}\boldsymbol{m} - \boldsymbol{P}\|^2\,\mathrm{d}A, \tag{10b}$$

$$J[\boldsymbol{m}, \boldsymbol{P}, \boldsymbol{b}] = \alpha J_{\mathrm{I}}(\boldsymbol{m}, \boldsymbol{P}) + (1-\alpha)J_{\mathrm{B}}(\boldsymbol{m}, \boldsymbol{b}), \quad (0 < \alpha < 1), \tag{10c}$$

over appropriate function spaces, with $\boldsymbol{b} : \partial\mathscr{S} \to \partial\mathscr{T}$ and where $\boldsymbol{P}$ is SPD satisfying equation (9). The norms in J_{B} and J_{I} are the ℓ_2-norm and the Frobenius norm, respectively. We repeat this procedure iteratively, starting from an initial guess $\boldsymbol{m}^0$. Here, we give a brief description of the minimization of J_{I} and J; for more details see [2, 5].

Minimization Procedure for $\boldsymbol{P}$

We assume $\boldsymbol{m}$ fixed and minimize $J_{\mathrm{I}}[\boldsymbol{m}, \boldsymbol{P}]$ over all SPD matrices $\boldsymbol{P}$ that satisfy (9). Since the integrand of $J_{\mathrm{I}}[\boldsymbol{m}, \boldsymbol{P}]$ does not depend on derivatives of $\boldsymbol{P}$, the minimization procedure can de carried out point-wise. Thus, we minimize $\frac{1}{2}\|\boldsymbol{C}\boldsymbol{D} - \boldsymbol{P}\|^2$ for each grid point $\boldsymbol{x}_{i,j} \in \mathscr{S}$, where $\boldsymbol{D}$ is the central difference approximation of $\mathrm{D}\boldsymbol{m}$. This give rise to the following constrained minimization problem

$$\text{Minimize} \quad H(p_{11}, p_{22}, p_{12}) = \tfrac{1}{2}\|\boldsymbol{Q} - \boldsymbol{P}\|^2, \tag{11a}$$

$$\text{subject to} \quad \det(\boldsymbol{P}) = p_{11}p_{22} - p_{12}^2 = F(\boldsymbol{x}), \tag{11b}$$

$$\mathrm{tr}(\boldsymbol{P}) = p_{11} + p_{22} > 0, \tag{11c}$$

where $\boldsymbol{Q} = \boldsymbol{C}\boldsymbol{D}_\mathrm{S}$ with $\boldsymbol{D}_\mathrm{S}$ the symmetric part of $\boldsymbol{D}$. The solution of (11a)–(11b) is given by the stationary points of the Lagrangian function $L(p_{11}, p_{22}, p_{12}; \mu) = H(p_{11}, p_{22}, p_{12}) - \mu(\det(\boldsymbol{P}) - F)$. This way we obtain the algebraic system

$$p_{11} + \lambda p_{22} = q_{11}, \tag{12a}$$

$$\lambda p_{11} + p_{22} = q_{22}, \tag{12b}$$

$$(1 - \lambda)p_{12} = \tfrac{1}{2}(q_{12} + q_{21}), \tag{12c}$$

$$p_{11}p_{22} - p_{12}^2 = F, \tag{12d}$$

with $\lambda = \mu/\det(\boldsymbol{C})$. In [2, 5] it is shown that (12) has always a solution satisfying the inequality constraint (11c).

Minimization Procedure for *m*

We assume $\boldsymbol{b}$ and $\boldsymbol{P}$ are fixed. The minimizer for $J[\boldsymbol{m}, \boldsymbol{P}, \boldsymbol{b}]$ has to satisfy

$$\delta J[\boldsymbol{m}, \boldsymbol{P}, \boldsymbol{b}](\boldsymbol{\eta}) = 0, \tag{13}$$

where δJ represents the first variation of J with respect to $\boldsymbol{m}$ in the direction of $\boldsymbol{\eta}$. Applying the fundamental lemma of calculus of variations, we obtain the following elliptic PDE system with Robin boundary conditions

$$\nabla \cdot \big(\boldsymbol{C}^\mathrm{T}\boldsymbol{C}\mathrm{D}\boldsymbol{m}\big) = \nabla \cdot \big(\boldsymbol{C}^\mathrm{T}\boldsymbol{P}\big), \quad \boldsymbol{x} \in \mathscr{S}, \tag{14a}$$

$$(1-\alpha)\boldsymbol{m} + \alpha(\boldsymbol{C}^\mathrm{T}\boldsymbol{C}\nabla\boldsymbol{m}) \cdot \boldsymbol{\nu} = (1-\alpha)\boldsymbol{b} + \alpha\boldsymbol{C}\boldsymbol{P}\boldsymbol{\nu}, \quad \boldsymbol{x} \in \partial\mathscr{S}, \tag{14b}$$

with $\boldsymbol{\nu}$ the outward unit normal on $\partial\mathscr{S}$. For space discretization we employ the cell-centered finite volume method.

Calculation of the Freeform Surfaces

We compute the location of the first freeform lens surface $z = u_1(\boldsymbol{x})$ from Eq. (7) in the least-squares sense, i.e., we minimize the functional

$$I[u_1] = \int_{\mathscr{S}} \|\nabla u_1 - \nabla_{\boldsymbol{x}} c\|^2 \,\mathrm{d}A. \tag{15}$$

Analogous to the derivation of (14), we find the following Neumann BVP

$$\nabla^2 u_1 = \nabla \cdot \nabla_{\boldsymbol{x}} c(\cdot, \boldsymbol{m}), \quad \boldsymbol{x} \in \mathscr{S}, \tag{16a}$$

$$\nabla u_1 \cdot \boldsymbol{\nu} = \nabla_{\boldsymbol{x}} c \cdot \boldsymbol{\nu}, \quad \boldsymbol{x} \in \partial\mathscr{S}. \tag{16b}$$

We compute u_1 using standard central differences. Substituting the converged mapping $\boldsymbol{y} = \boldsymbol{m}(\boldsymbol{x})$ and $u_1(\boldsymbol{x})$ in (3), we can compute the location of the second lens surface $\ell - z = u_2(\boldsymbol{y})$.

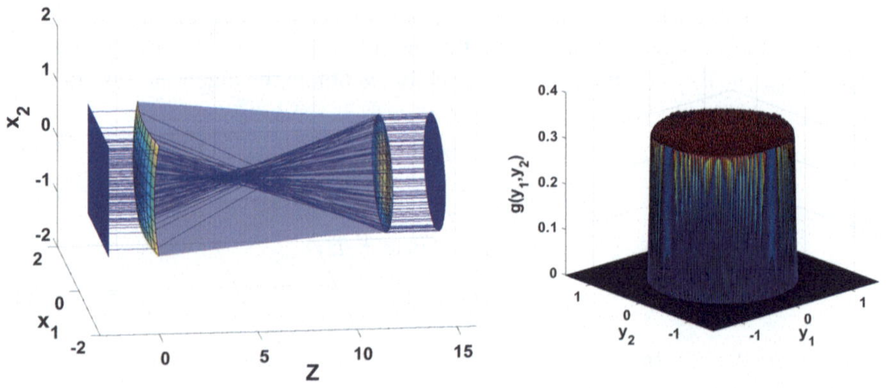

Fig. 2 Computed double freeform lens (left) and target illuminance (right). Parameter values are: $n = 1.5$, $\ell = 20$ and $\beta = 2\pi$

4 Numerical Example

As an example we compute the freeform lens that generates a circular top-hat target irradiance. The source and target domains are given by $\mathscr{S} = \mathscr{T} = [-1, 1] \times [-1, 1]$. The source has emittance $E(\boldsymbol{x}) = A\mathrm{e}^{-10|\boldsymbol{x}|^2}$, and the target plane receives the irradiance $I(\boldsymbol{y})$ given by $I(\boldsymbol{y}) = 1/\pi$ if $|\boldsymbol{y}| \leq 1$, otherwise $I(\boldsymbol{y}) = 0$. The constant A is chosen to enforce global energy conservation, i.e., relation (4) should hold for the entire source domain $\mathscr{A} = \mathscr{S}$. The numerically computed lens is shown in Fig. 2. Clearly, the lens surface $z = u_1(\boldsymbol{x})$ closest to the source is convex. To validate the result we have traced 10^7 rays through the lens to compute the target irradiance; a selected ray set is shown. The resulting irradiance is also shown in Fig. 2. We conclude that the computed target irradiance is in good approximation a circular top-hat.

5 Concluding Remarks

We have presented a nonlinear elliptic PDE of Monge-Ampère type describing a double freeform lens that converts a Gaussian exitance into a circular top-hat irradiance, relevant for laser beam shaping. Moreover, we outlined a least-squares algorithm to compute both lens surfaces. We have restricted ourselves to one single lens with c-convex surfaces, however, our least-squares algorithm can be easily adapted to compute c-concave surfaces or even freeform surfaces of a two-lens system; for more details see [5].

References

1. Born, M., Wolf, E.: Principles of Optics, 7th edn. Cambridge University Press, Cambridge (1999)
2. Prins, C.R., Beltman, R., ten Thije Boonkkamp, J.H.M., IJzerman, W.L., Tukker, T.W.: A least-squares method for optimal transport using the Monge-Ampère equation. SIAM J. Sci. Comput. **37**, B937–B961 (2015)
3. Ten Thije Boonkkamp, J.H.M., Romijn, L.B., Yadav, N.K., IJzerman, W.L.: Monge-Ampère type equations for freeform illumination optics. In: Proceedings SPIE Optical Systems Design 2018, vol. DL 10693. Illumination Optics V (2018)
4. Villani, C.: Optimal Transport, Old and New. Springer, Berlin (2008)
5. Yadav, N.K., ten Thije Boonkkamp, J.H.M., IJzerman, W.L.: Computation of double freeform optical surfaces using a Monge-Ampère solver: application to beam shaping, CASA-Report 18–03, Eindhoven University of Technology (2018)

Efficient Numerical Solution of Space-Fractional Diffusion Problems

Ferenc Izsák and Béla J. Szekeres

Abstract An efficient numerical method is introduced for the solution of space-fractional diffusion problems. We use the spectral fractional Laplacian operator with homogeneous Neumann and Dirichlet boundary conditions. The spatial discretization is based on the matrix transformation method. Using a recent algorithm for the computation of fractional matrix power-vector products and explicit time stepping, we develop a simple and efficient full discretization. The performance of our approach is demonstrated in some numerical experiments.

1 Introduction

Fractional diffusion has been detected in a wide range of real-life observations [3]. This dynamics is sometimes hard to distinguish from the conventional diffusion: a number of individuals should be tracked. At the discrete level, the fractional diffusion can be given as a Lévy process [2]. The corresponding continuous model is the space-fractional diffusion equation, which seems to be the only true model [5]. A few attempts [1] were also made to take inhomogeneous boundary data into account, but these were not related to the real-life observations. The numerical analysis of space-fractional diffusion equations was worked out in the last 15 years proposing a number of methods. Whenever a solid theoretic framework was developed for these approaches, in practice, many of them need huge computational efforts. An important observation was reported in [4], where the idea of the matrix transformation method was proposed. A fast algorithm for the approximation of

F. Izsák (✉)
Department of Applied Analysis and Computational Mathematics & ELTE-MTA Numnet Research Group, Eötvös Loránd University, Budapest, Hungary
e-mail: izsakf@cs.elte.hu

B. J. Szekeres
Eötvös Loránd University, Faculty of Informatics, Department of Numerical Analysis, Budapest, Hungary
e-mail: szpbgat@cs.elte.hu

© Springer Nature Switzerland AG 2019
I. Faragó et al. (eds.), *Progress in Industrial Mathematics at ECMI 2018*,
Mathematics in Industry 30, https://doi.org/10.1007/978-3-030-27550-1_38

the corresponding matrix powers was proposed in [9]. An alternative attempt to accelerate the computation of the conventional (large and full) finite difference matrices was given in [7].

The aim of this contribution is to propose an alternative of these approaches: we develop a simple and highly efficient algorithm for the numerical solution of space-fractional diffusion problems, which is based on existing theoretical results.

2 Mathematical Preliminaries

We investigate the efficient numerical solution of the space fractional diffusion problem

$$\begin{cases} \partial_t u(t,\mathbf{x}) = -(-\Delta_{\mathscr{D}})^\alpha u(t,\mathbf{x}) + f(t,\mathbf{x}) & t \in (0,T),\ \mathbf{x} \in \Omega \subset \mathbb{R}^d \\ u(0,\mathbf{x}) = u_0(\mathbf{x}) & \mathbf{x} \in \Omega, \end{cases} \tag{1}$$

for the unknown function $u : (0,T) \times \Omega \to \mathbb{R}$, where Ω is a Lipschitz domain, α is a positive exponent, $u_0 \in L_2(\Omega)$ is a given initial function and $f : (0,T) \times \Omega \to \mathbb{R}$ is a given function corresponding to a source term. $-\Delta_{\mathscr{D}}$ denotes the negative Laplacian operator with homogeneous Dirichlet boundary condition, which is positive and self-adjoint, so that its power makes sense. Accordingly, we will use the subscript $\mathscr{N}$ in case of homogeneous Neumann boundary conditions.

To introduce the matrix transformation method (MTM) for the numerical solution of (1), we use the notation $\mathbf{u}_h(t) \in \mathbb{R}^N$ and $\mathbf{f}_h(t) \in \mathbb{R}^N$ for the approximation of the function $u(t,\cdot) : \Omega \to \mathbb{R}$ and $f(t,\cdot) : \Omega \to \mathbb{R}$, respectively, using either a finite difference or a finite element discretization. The matrix $A_h \in \mathbb{R}^{N\times N}$ is for the corresponding discretization of $-\Delta_{\mathscr{D}}$.

According to the MTM, for the full discretization of (1), we have only to solve numerically the following system of ordinary differential equations:

$$\begin{cases} \partial_t \mathbf{u}_h(t) = -A_h^\alpha \mathbf{u}_h(t) + \mathbf{f}_h(t) & t \in (0,T) \\ \mathbf{u}_h(0) = \mathbf{u}_{0,h}, \end{cases} \tag{2}$$

where $\mathbf{u}_{0,h}$ is given. Note that A_h is positive definite and therefore, the power A_h^α makes sense. For the convergence analysis of this approach, we refer to [8].

3 Results

Whenever the semidiscretization in (2) delivers a formally simple approach, the computation of the matrix power is very expensive. To avoid this step, we note that in an explicit time stepping for solving (2), we need to compute only terms $A_h^\alpha \mathbf{w}$, where $\mathbf{w} \in \mathbb{R}^N$ is a given vector.

In the proposed procedure, we use the notation $\lambda_1 \le \lambda_2 \le \cdots \le \lambda_N$ for the eigenvalues of A_h in increasing order and $\mathbf{v}_1, \mathbf{v}_2, \ldots, \mathbf{v}_N$ for the corresponding eigenvectors.

3.1 *The Algorithm*

3.1.1 Initialization of the Algorithm

One should first compute the smallest eigenvalues $\lambda_1, \lambda_2, \ldots, \lambda_{k_1}$ and the largest ones $\lambda_N, \lambda_{N-1}, \ldots, \lambda_{N-k_2}$ along with the corresponding eigenvectors $\mathbf{v}_1, \mathbf{v}_2, \ldots, \mathbf{v}_{k_1}$ and $\mathbf{v}_N, \mathbf{v}_{N-1}, \ldots, \mathbf{v}_{N-k_2}$.

Using this, we can compute the projection matrix $P_{\min,\max} \in \mathbb{R}^{N\times N}$ to the subspace

$$\operatorname{span}\{\mathbf{v}_1, \mathbf{v}_2, \ldots, \mathbf{v}_{k_1}, \mathbf{v}_{N-k_2}, \ldots, \mathbf{v}_{N-1}, \mathbf{v}_N\},$$

so that for any $\mathbf{w} \in \mathbb{R}^N$, we have

$$P_{\min,\max}\mathbf{w} = w_1\mathbf{v}_1 + \cdots + w_{k_1}\mathbf{v}_{k_1} + w_{N-k_2}\mathbf{v}_{N-k_2} + \cdots + w_N\mathbf{v}_N.$$

3.1.2 The Subroutine for Computing $A^\alpha\mathbf{w}$

Based on [6], we approximate $A^\alpha\mathbf{w}$ as follows.

$$A^\alpha\mathbf{w} \approx \lambda_1^\alpha w_1\mathbf{v}_1 + \cdots + \lambda_{k_1}^\alpha w_{k_1}\mathbf{v}_{k_1} + \lambda_{N-k_2}^\alpha w_{N-k_2}\mathbf{v}_{N-k_2} + \cdots + \lambda_N^\alpha w_N\mathbf{v}_N$$
$$+ \left(\frac{2A}{\sigma(A)}\right)^\alpha \sum_{j=0}^{K} \binom{\alpha}{j} \left(\frac{2A}{\sigma(A)} - I\right)^j (\mathbf{w} - P_{\min,\max}\mathbf{w}), \tag{3}$$

where $\sigma(A)$ denotes the spectral radius of A, I denotes the identity matrix and the parameter K gives the length of the Taylor approximation. After the initialization, the first component in (3) can be computed quickly. Also, the sum in the second component is composed of sparse matrix-vector products.

3.1.3 Second Order Time Discretizations

Two procedures will be discussed to solve (2) numerically, where the time step is denoted with δ. In each case, we define $\mathbf{u}_h(\delta)$ with a modified Euler method as

$$\mathbf{u}_h(\delta) = \mathbf{u}_h(0) + \frac{\delta}{2}\left[f(0) - A_h^\alpha\mathbf{u}_h(0)\right.$$
$$\left. + f(\delta) - A_h^\alpha(\mathbf{u}_h(0) + \delta(-A_h^\alpha\mathbf{u}_h(0) + f(\mathbf{u}_h(0))))\right].$$

This can be continued to obtain the *modified Euler method*

$$\mathbf{u}_h((k+1)\delta) = \mathbf{u}_h(k\delta) + \frac{\delta}{2}\big[f(k\delta) - A_h^\alpha \mathbf{u}_h(k\delta) \\ + f((k+1)\delta) - A_h^\alpha(\mathbf{u}_h(k\delta) + \delta(-A_h^\alpha \mathbf{u}_h(k\delta) + f(k\delta)))\big].$$

In the second series of experiments, the two-step *Adams–Bashforth method*

$$\mathbf{u}_h((k+1)\delta) = \mathbf{u}_h(k\delta) + \frac{3\delta}{2}\left(f(k\delta) - A_h^\alpha \mathbf{u}_h(k\delta)\right) \\ - \frac{\delta}{2}\left(f((k-1)\delta) - A_h^\alpha \mathbf{u}_h((k-1)\delta)\right).$$

is applied. The corresponding results are shown in Table 1.

3.2 Implementation Issues, Numerical Results and Discussion

The performance of the algorithms in Sect. 3.1 is demonstrated in case of the model problem

$$\begin{cases} \partial_t u(t,x,y) = -(-\Delta_{\mathscr{D}})^\alpha u(t,x,y) + \sin x \sin y & (x,y) \in \Omega,\ t \in (0,T) \\ u(t,x,y) = 0 \quad (x,y) \in \partial\Omega,\ t \in (0,T) \\ u(0,x,y) = \sin x \sin 2y - \sin 2x \sin y + 2^{-\alpha} \sin x \sin y \quad (x,y) \in \Omega, \end{cases} \tag{4}$$

where $\Omega = (0,\pi) \times (0,\pi)$ is the domain, $T = 1$ and $\alpha \in \mathbb{R}^+$ is a given parameter. Note that the analytic solution of (4) is given with

$$u(t,x,y) = \exp(-5^\alpha t)(\sin x \sin 2y - \sin 2x \sin y) + 2^{-\alpha} \sin x \sin y.$$

For homogeneous Neumann boundary conditions, the model problem was given so that its solution is

$$u(t,x,y) = \exp(-5^\alpha t)(\cos x \cos 2y - \cos 2x \cos y) + 2^{-\alpha} \cos x \cos y.$$

In each case, spatially, the standard five-point second order finite difference discretization is applied on a uniform grid with $n \times n$ internal grid points.

Since the approximation $A_h^\alpha \approx (-\Delta_{\mathscr{D}})^\alpha$ is second order (see [8]), we have only tested the performance of the method using uniform bisection: taking $\delta/2$ as a time step and $(2n+1) \times (2n+1)$ internal grid points.

Table 1 Numerical results for the model problem (4) using the algorithm in Sect. 3.1

α	BC	Method	K	k_1	n	N_t	Error	Time	Order
0.8	$\mathscr{D}$	AB	200	8	40	400	4.1×10^{-4}	2.8	2.0
0.8	$\mathscr{D}$	ME	200	8	80	800	1.1×10^{-4}	22.9	2.0
0.8	$\mathscr{D}$	ME	100	8	40	400	4.2×10^{-4}	2.7	2.1
0.8	$\mathscr{D}$	ME	100	12	80	800	4.2×10^{-4}	14.2	2.1
0.8	$\mathscr{N}$	AB	200	8	40	400	4.1×10^{-4}	2.8	2.0
0.8	$\mathscr{N}$	ME	200	8	80	800	1.1×10^{-4}	20.5	2.0
0.8	$\mathscr{N}$	ME	100	8	80	800	4.1×10^{-4}	11.7	2.1
0.8	$\mathscr{N}$	ME	100	12	80	800	4.1×10^{-4}	13.1	2.0
1.2	$\mathscr{D}$	ME	200	8	40	20,000	3.6×10^{-4}	126.3	2.0
1.2	$\mathscr{D}$	ME	100	8	40	20,000	3.6×10^{-4}	66.4	2.0
1.2	$\mathscr{D}$	ME	100	8	40	20,000	3.6×10^{-4}	67.2	2.0

The main ingredient of the proposed algorithm is an efficient and accurate eigensolver. The built-in Matlab subroutine `eigs.m` is a good choice if the optional parameters are set to increase the accuracy. In case of multiple eigenvalues, one can make use of the algorithm `bchdav.m`, see [10].

For each row in Table 1, we have performed a series of numerical experiments. We give only the result with the largest grid where the stability could be maintained. The convergence order is computed using the final bisection. To summarize, the parameters in the corresponding code are the following:

- $k_1 = k_2$—number of the largest and smallest eigenvalues in the algorithm,
- K—number of the terms in the second term in (3),
- n—number of internal grid points in one direction,
- N_t—number of time steps in the final experiment,
- error—discrete L_2 norm of the error,
- time—computational time in seconds,
- $\mathscr{D}$ and $\mathscr{N}$—homogeneous Dirichlet and Neumann boundary condition,
- ME and AB—modified Euler and second order Adams–Bashforth time stepping.

Incorporating many eigenvector-eigenvalue pairs improves the efficiency of the algorithm: in this case, we have to use a moderate number of terms in the second component in (3). At the same time, small errors in the computation of eigenvectors lead to an inaccurate projection matrix and then again an overly long summation is necessary in the second component in (3). Therefore, based on our experience, as a good balance, the parameters k_1 and k_2 should be set at about 10.

In case of accurate eigenvalues, the number of terms in the summation can be set to a few hundreds, which ensures already an acceptable convergence order. By using only sparse matrix-vector products, the total computational time remains then at a moderate level.

Increasing parameters K and k_1 further does not significantly enhance accuracy.

Since the modified Euler time stepping proved to be slightly more stable, only this was applied for $\alpha = 1.2$. Even in this case, we needed a large number of time steps to ensure the stability of the method. This motivates us to develop similar algorithms for implicit time stepping as a next project.

Acknowledgements This work was completed in the ELTE Institutional Excellence Program (1783-3/2018/FEKUTSRAT) supported by the Hungarian Ministry of Human Capacities.

F. Izsák also acknowledges the financial support in the framework of the project HU-MATHS-IN Intensification of the activity of the Hungarian Industrial Innovation Mathematical Service Network (EFOP-3.6.2-16-2017-00015), which made possible to take part at the conference.

B. J. Szekeres was supported by the European Union, co-financed by the Social Fund. EFOP-3.6.1-16-2016-0023.

References

1. Abatangelo, N., Dupaigne, L.: Nonhomogeneous boundary conditions for the spectral fractional Laplacian. Ann. Inst. H. Poincaré Anal. Non Linéaire **34**(2), 439–467 (2017)
2. Baeumer, B., Meerschaert, M.M.: Tempered stable Lévy motion and transient super-diffusion. J. Comput. Appl. Math. **233**(10), 2438–2448 (2010)
3. Bucur, C., Valdinoci E.: Nonlocal diffusion and applications. In: Lecture Notes of the Unione Matematica Italiana, vol. 20. Springer, Bologna (2016)
4. Ilic, M., Liu, F., Turner, I., Anh, V.: Numerical approximation of a fractional-in-space diffusion equation I. Fractional Calculus. Appl. Anal. **8**(3), 323–341 (2005)
5. Izsák, F., Szekeres, B.J.: Models of space-fractional diffusion: a critical review. Appl. Math. Lett. **71**, 38–43 (2017)
6. Izsák, F., Szekeres, B.J.: Efficient computation of matrix power-vector products: application for space-fractional diffusion problems. Appl. Math. Lett. **86**, 70–76 (2018)
7. Jia, J., Wang, H.: Fast finite difference methods for space-fractional diffusion equations with fractional derivative boundary conditions. J. Comput. Phys. **293**, 359–369 (2015)
8. Szekeres, B.J., Izsák, F.: Finite difference approximation of space-fractional diffusion problems: the matrix transformation method. Comput. Math. Appl. **73**(2), 261–269 (2017)
9. Vabishchevich, P.N.: Numerical solution of time-dependent problems with fractional power elliptic operator. Comput. Methods Appl. Math. **18**(1), 111–128 (2018)
10. Zhou, Y.: A block Chebyshev-Davidson method with inner-outer restart for large eigenvalue problems. J. Comput. Phys. **229**(24), 9188–9200 (2010)

Black-Scholes Equation with Distributed Order in Time

Luísa Morgado and Magda Rebelo

Abstract In this work we consider a Black-Scholes model which consists of a generalization of a fractional Black-Scholes equation model proposed previously. A numerical scheme is presented to solve such type of models and some numerical results are presented for European double-knock out barrier options. In this way, we are able to conclude that this generalized model is able to describe other scenarios than the ones described with the classical (integer-order) and the fractional Black-Scholes models.

1 Introduction

In this work we investigate the pricing of double barrier options when the change in the option price with time is a fractal transmission system. In order to describe the option price we propose the modified Black-Sholes (BS) equation with distributed derivative in time which is a generalization of the modified Black-Sholes equation with a time-fractional derivative. A double barrier option is an option category with both upper (B_ℓ) and lower (B_ℓ) trigger prices, called the barriers, placed on the underlying asset. A knock-in barrier options becomes valid when the underlying exceeds either barrier. A knock-out barrier option becomes invalid, or ceases to exist, when the underlying exceeds either barrier. Barrier options can be puts or calls. Here we consider European double-knock out barrier option.

Let $C(S, t)$ be the European option fair price at the stock price S and at time t. Following a generalization of the fractional Black-Scholes equation model

L. Morgado
Center for Computational and Stochastic Mathematics (CEMAT), Lisbon and Departamento de Matemática, Universidade de Trás-os-Montes e Alto Douro, Quinta de Prados, Vila Real, Portugal
e-mail: luisam@utad.pt

M. Rebelo (✉)
Centro de Matemática e Aplicações (CMA), Departamento de Matemática, Faculdade de Ciências e Tecnologia, Universidade NOVA de Lisboa, Quinta de Prados, Caparica, Portugal
e-mail: msjr@fct.unl.pt

© Springer Nature Switzerland AG 2019
I. Faragó et al. (eds.), *Progress in Industrial Mathematics at ECMI 2018*,
Mathematics in Industry 30, https://doi.org/10.1007/978-3-030-27550-1_39

introduced in [1] and [5], we propose the following model to describe the fair price of an option:

$$\begin{cases} {}_t\mathbb{D}_T^{R,\phi} C(S,t) + \frac{1}{2}\sigma^2 S^2 \frac{\partial^2 C(S,t)}{\partial S^2} + (r-D)S\frac{\partial C(S,t)}{\partial S} - rC(S,t) = 0, \ (S,t) \in (B_\ell, B_u)\times(0,T), \\ C(B_\ell,t) = p(t), \quad C(B_u,t) = q(t), \ t \in [0,T], \\ C(S,T) = g(S), \ S \in [B_\ell, B_u] \end{cases} \tag{1}$$

where T is the expiry rate time, r is the risk-free rate, $D = \frac{1}{2}\sigma^2$ is the yield dividend, $\sigma \geq 0$ is the volatility of the returns from the stock price S and $B_u > B_\ell > 0$ are the two barriers.

${}_t\mathbb{D}_T^{R,\phi} C(S,t) = \int_0^1 \phi(\alpha) \frac{\partial_*^\alpha C(S,t)}{\partial t^\alpha} d\alpha$ is the modified Riemann-Liouville derivative with distributed-order, being:

$$\frac{\partial_*^\alpha C(S,t)}{\partial t^\alpha} = \frac{1}{\Gamma(1-\alpha)} \frac{d}{dt} \int_t^T \frac{C(S,\eta) - C(S,T)}{(\eta - t)^\alpha} d\eta, \quad 0 < \alpha < 1,$$

and ϕ is a positive function acting as a distribution of the orders of the derivative in the range [0, 1], satisfying $\int_0^1 \phi(\alpha)\, d\alpha = \mathcal{C} > 0$. Note that if ϕ is the Dirac delta function $\delta(\alpha)$, then you recover the model introduced in [5], and if you take $\alpha = 1$, then you obtain a classical (integer-order) model. The way that the payoff function g is defined depends on the cases where C is a call or a put option. Namelly,

- if C is a call option then $g(S) = C(S,T) = \max\{S - E, 0\}$,
- if C is a put option then $g(S) = C(S,T) = \max\{E - S, 0\}$,

where E is the strike price. The strike price is the price at which a derivative contract can be exercised. For call options, the strike price is where the security can be bought by the option buyer up till the expiration date. For put options, the strike price is the price at which shares can be sold by the option buyer. Proceeding as in [5], we consider the variable transformations $t = T - \tau$, $x = \ln S$ and $u(x,\tau) = C(S, T-\tau)$ and rewrite (1) as an advection-diffusion initial-boundary value problem, with distributed order in time

$${}_0\mathbb{D}_\tau^\phi u(x,\tau) = a\frac{\partial^2 u(x,\tau)}{\partial x^2} + b\frac{\partial u(x,\tau)}{\partial x} - cu(x,\tau) + f(x,\tau), \quad (x,\tau) \in (L_0, L_1)\times(0,T), \tag{2}$$

$$u(a,\tau) = p(\tau), \quad u(b,\tau) = q(\tau), \ \tau \in [0,T], \tag{3}$$

$$u(x,0) = g(x), \ x \in [L_0, L_1], \tag{4}$$

where $L_0 = \ln(B_\ell)$, $L_1 = \ln(B_u)$, $a = \frac{1}{2}\sigma^2$, $b = r - a$, $c = r$ and

$$ {}_0\mathbb{D}_\tau^\phi u(x,\tau) = \int_0^1 \phi(\alpha) \frac{\partial^\alpha u(x,\tau)}{\partial t^\alpha}\, d\alpha, \tag{5}$$

where $\frac{\partial^\alpha u(x,\tau)}{\partial \tau^\alpha}$ is the Caputo derivative of order α of the function u with respect to τ ([2]). The source function is added for the purposes of validation of the numerical method that will be presented in the next section. In the Black-Sholes model the source function $f(x,\tau) \equiv 0$.

2 Numerical Method

In this section we present an implicit numerical method for the approximation to the solution of (2)–(4). Using a quadrature rule we approximate the integral in (5) by a finite sum. Let us then consider a partition of the interval $[0, 1]$ into N subintervals, $[\beta_{j-1}, \beta_j]$, $j = 1, \ldots, N$, of equal amplitude $h = 1/N$. Using the midpoint rule, with $h = 1/N$, to approximate the integral in (2) we obtain

$$\int_0^1 \phi(\alpha) \frac{\partial^\alpha u(x,\tau)}{\partial \tau^\alpha}\, d\alpha = h \sum_{j=1}^{N} \phi(\alpha_j) \frac{\partial^{\alpha_j} u(x,\tau)}{\partial \tau^{\alpha_j}} - \frac{h^2}{24} H''(\nu), \quad \nu \in (0,1), \tag{6}$$

where $\alpha_j = \frac{\beta_{j-1} + \beta_j}{2}$, $j = 1, \ldots, N$, and H is defined by $H(\alpha) = \phi(\alpha) \frac{\partial^\alpha u(x,\tau)}{\partial \tau^\alpha}$.

In order to approximate the space derivatives, we consider a uniform space mesh, on the interval $[L_0, L_1]$, defined by the gridpoints $x_i = L_0 + i\Delta x$, $i = 0, 1, \ldots, K$, where $\Delta x = (L_1 - L_0)/K$, and we approximate the space derivatives at $x = x_i$, with the second order finite differences:

$$\left.\frac{\partial u(x,\tau)}{\partial x}\right|_{x=x_i} = \frac{u(x_{i+1},\tau) - u(x_{i-1},\tau)}{2\Delta x} - \frac{(\Delta x)^2}{6} \frac{\partial^3 u}{\partial x^3}(\eta_i, \tau), \tag{7}$$

$$\left.\frac{\partial^2 u(x,\tau)}{\partial x^2}\right|_{x=x_i} = \frac{u(x_{i+1},\tau) - 2u(x_i,\tau) + u(x_{i-1},\tau)}{(\Delta x)^2} - \frac{(\Delta x)^2}{12} \frac{\partial^4 u}{\partial x^4}(\xi_i, \tau), \tag{8}$$

with $\eta_i, \xi_i \in (x_{i-1}, x_{i+1})$. For a fixed h and Δx, denoting by $U_i(\tau)$ the approximated value for $u(x_i, \tau)$, and substituting (6), (7) and (8) (neglecting the $O\left(h^2\right)$ and

$O\left((\Delta x)^2\right)$ terms) in (2), we obtain the semi-discretised scheme:

$$h\sum_{j=1}^{N}\phi(\alpha_j)\frac{\partial^{\alpha_j}U_i(\tau)}{\partial\tau^{\alpha_j}} = b\frac{U_{i+1}(\tau)-U_{i-1}(\tau)}{2(\Delta x)} + a\frac{U_{i+1}(\tau)-2U_i(\tau)+U_{i-1}(\tau)}{(\Delta x)^2}$$
$$-cU_i(\tau)+f(x_i,\tau), \quad i=1,\ldots,K-1.$$

Note that from the boundary conditions (3):

$$U_0(\tau)=p(\tau), \quad U_K(\tau)=q(\tau), \tag{9}$$

and from the initial condition (4), we have

$$U_i(0)=g(x_i), \quad i=1,\ldots,K-1. \tag{10}$$

In order to approximate the fractional derivatives $\dfrac{\partial^{\alpha_j}u(x,\tau)}{\partial\tau^{\alpha_j}}$, we define the time gridpoints $\tau_l = l\Delta\tau, l=0,1,\ldots,n$, where $\Delta\tau = T/n$ and use the backward finite difference formula provided by Diethelm (see [3]):

$$\frac{\partial^{\alpha_j}U_i(\tau_l)}{\partial\tau^{\alpha_j}} = \frac{(\Delta\tau)^{-\alpha_j}}{\Gamma(2-\alpha_j)}\sum_{m=0}^{l}a_{m,l}^{(\alpha_j)}\left(U_i\left(\tau_{l-m}\right)-U_i(0)\right)$$
$$+c_{\alpha_j}(\Delta\tau)^{2-\alpha_j}\frac{\partial^2 u}{\partial\tau^2}(x_i,\eta_l), \quad \eta_l\in(0,\tau_l),$$

where the constants c_{α_j} do not depend on $\Delta\tau$, and the coefficients $a_{m,l}^{(\alpha_j)}$ are given by:

$$a_{m,l}^{(\alpha_j)} = \begin{cases} 1, & m=0,\\ (m+1)^{1-\alpha_j}-2m^{1-\alpha_j}+(m-1)^{1-\alpha_j}, & 0<m<l,\\ (1-\alpha_j)l^{-\alpha_j}-l^{1-\alpha_j}+(l-1)^{1-\alpha_j}, & m=l.\end{cases}$$

Substituting in (9), and denoting by $U_i^l \approx u(x_i,\tau_l)$, $f_i^l = f(x_i,\tau_l)$, we obtain the finite difference scheme:

$$h\sum_{j=1}^{N}\phi(\alpha_j)\frac{(\Delta\tau)^{-\alpha_j}}{\Gamma(2-\alpha_j)}\sum_{m=0}^{l}a_{m,l}^{(\alpha_j)}\left(U_i^{l-m}-U_i^0\right) = +a\frac{U_{i+1}^l-2U_i^l+U_{i-1}^l}{(\Delta x)^2}$$
$$+b\frac{U_{i+1}^l-U_{i-1}^l}{2(\Delta x)}-cU_i^l+f_i^l, \quad i=1,\ldots,K-1, \; l=1,\ldots,n. \tag{11}$$

Hence, in order to obtain an approximation to the solution of (2)–(4) subject to the initial condition (4) and boundary conditions (3), we need to solve the linear system of Eq. (11), and taking (9) and (10) into account:

$$U_0^l = p(\tau_l), \quad U_L^l = q(\tau_l), \quad l = 1, \dots, n, \tag{12}$$

$$U_i^0 = g(x_i), \quad i = 1, \dots, K-1. \tag{13}$$

3 Numerical Results

In this section one example with exact solution is presented to illustrate the efficiency and accuracy of the proposed method. We also use the proposed scheme to price a double barrier knock-out call European option.

In order to test the robustness of the presented numerical scheme, we first consider an example of (2)–(4) with a proper choice of the coefficients a, b, c, and functions p, q and g, so that the analytical solution is known and given by $u(x,\tau) = \tau^2 x(1-x)$.

Example 1: Advection-diffusion problem with distributed order in time

$$\begin{cases} {}_0\mathbb{D}_\tau^\phi u(x,\tau) = \dfrac{\partial^2 u(x,\tau)}{\partial x^2} + \dfrac{\partial u(x,\tau)}{\partial x} + u(x,\tau) + f(x,\tau), \quad (x,\tau) \in (0,1) \times (0,1), \\ \\ u(0,\tau) = 0, \quad u(1,\tau) = 0, \quad \tau \in [0,1], \qquad u(x,0) = 0, \quad x \in (0,1), \end{cases}$$

where $\phi(\alpha) = \dfrac{\Gamma(3-\alpha)}{2}$ and $f(x,\tau)$ is such that the exact solution is $u(x,\tau) = \tau^2(x - x^2)$.

In Table 1 we present the maximum of the absolute errors at the meshpoints $E_{K,n} = \max\limits_{i=0,1,\dots,K,\ j=0,1,\dots,n} |u(x_i,\tau_j) - U_i^j|$, where U_i^j is an approximation of $u(x_i,\tau_j)$ obtained with the stepsizes $\Delta x = h = \dfrac{1}{K}$ and $\Delta\tau = \dfrac{1}{n}$ and the experimental convergence orders, computed as it is usual (see [4]).

Table 1 Maximum of errors and experimental convergence orders

$\Delta\tau$	$h = \Delta x$	$E_{K,n}$	$p_x = p_h$	p_τ
0.25	0.5	$1.18 \cdot 10^{-3}$	–	–
0.0625	0.25	$2.19 \cdot 10^{-4}$	2.43	1.22
0.015625	0.125	$4.37 \cdot 10^{-5}$	2.33	1.16
0.00390625	0.0625	$9.04 \cdot 10^{-6}$	2.27	1.14

The results presented in Table 1 suggest that the experimental order of convergence with respect to the time variable, is approximately 1, and the space and numerical integration orders of convergence are approximately 2, as expected.

Example 2: Black-Sholes model, with distributed order in time, (BSMDT) governing a European double barrier knock-out call option:

$$\begin{cases} {}_t^{RL}\mathbb{D}_T^{\phi} C(S,t) + \frac{1}{2}\sigma^2 S^2 \frac{\partial^2 C(S,t)}{\partial S^2} + (r-D) S \frac{\partial C(S,t)}{\partial S} - rC(S,t) = 0, \ (S,t) \in (B_l, B_u) \times (0,T), \\ C(B_l,t) = C(B_u,t) = 0, \ t \in [0,T], \ C(S,T) = \max\{S-E, 0\}, \ S \in (B_l, B_u), \end{cases}$$

where $\sigma = 0.45$, $r = 0.03$, $E = 10$, $D = 0.01$, $T = 1$(year), $B_l = 3$ and $B_u = 15$.

In Figs. 1 and 2 we show the approximate solutions of Example 2, considering several choices for the weight function ϕ, obtained with the proposed method with $h = 0.1$, $K = n = 100$ and compare them with the corresponding Black Sholes solution for double barrier options under the same parameter settings.

We observe that BSMDT delivers lower prices when S is less than a critical value (close to the strike price E) and higher prices for in-the-money options ($S > E$). The distributed order model also delivers higher option prices for $S > E$, when compared with the classical Black-Sholes model. Therefore, this generalized model is able to simulate different scenarios that could not be described with previous models.

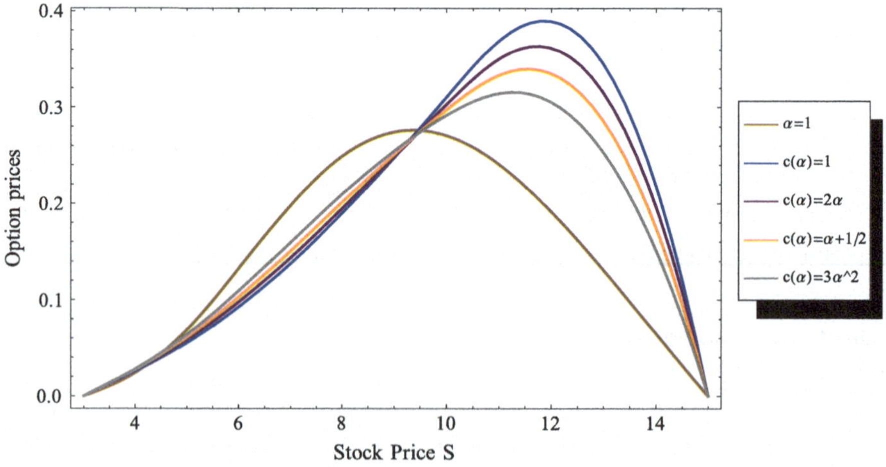

Fig. 1 Double barrier option prices obtained with several choices for the weight function $\phi(\alpha) = c(\alpha)$

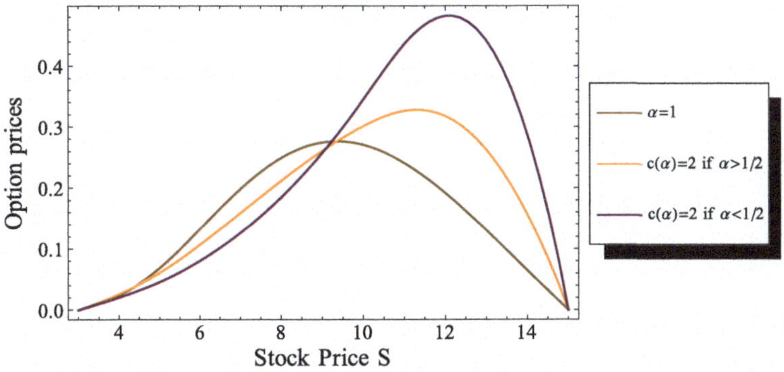

Fig. 2 Double barrier option prices obtained with several choices for the weight function $\phi(\alpha) = c(\alpha)$

From Fig. 2 we observe that the results that we obtained are in agreement with the results presented in [1]. Smaller α is, the larger the price bias becomes.

Acknowledgements The authors acknowledge the Fundação para a Ciência e a Tecnologia (Portuguese Foundation for Science and Technology) through projects UID/Multi/04621/2013 and UID/MAT/00297/2013 (Centro de Matemática e Aplicações), respectively.

References

1. Chen, W., Xu, X., Zhu, S.: Analytically pricing double barrier options based on time-fractional Black-Scholes equation. Comput. Math. Appl. **69**, 1407–1419 (2015)
2. Diethelm, K.: The analysis of fractional differential equations. In: An Application-Oriented Exposition Using Differential Operators of Caputo Type. Springer, Berlin (2004)
3. Diethelm, K., Ford, N.J., Freed, A.D., Luchko, Yu.: Algorithms for the fractional calculus: a selection of numerical methods. Comput. Methods Appl. Mech. Eng. **194**, 743–773 (2005)
4. Morgado, M.L., Rebelo, M.: Numerical approximation of distributed order reaction-diffusion equations. J. Comput. Appl. Math. **275**, 216–227 (2015)
5. Zhang, H., Liu, F., Turner, I., Yang, Q.: Numerical solution of the time fractional Black-Scholes model governing European options. Comput. Math. Appl. **71**, 1772–1783 (2016)

Periodic Homogenization of a Pseudo-Parabolic Equation via a Spatial-Temporal Decomposition

Arthur J. Vromans, Alphons Adrianus Francisca van de Ven, and Adrian Muntean

Abstract Pseudo-parabolic equations have been used to model unsaturated fluid flow in porous media. In this paper it is shown how a pseudo-parabolic equation can be upscaled when using a spatio-temporal decomposition employed in the Peszyńska-Showalter-Yi paper (Appl Anal 88(9):1265–1282, 2009). The spatial-temporal decomposition transforms the pseudo-parabolic equation into a system containing an elliptic partial differential equation and a temporal ordinary differential equation. To strengthen our argument, the pseudo-parabolic equation has been given advection/convection/drift terms. The upscaling is done with the technique of periodic homogenization via two-scale convergence. The well-posedness of the extended pseudo-parabolic equation is shown as well. Moreover, we argue that under certain conditions, a non-local-in-time term arises from the elimination of an unknown.

1 Introduction

Groundwater recharge and pollution prediction for aquifers need models for describing unsaturated fluid flow in porous media. Pseudo-parabolic equations were found to be adequate models, see eqn. 25 in [4]. In [8] a spatial-temporal decomposition

A. J. Vromans (✉)
Centre for Analysis, Scientific computing and Applications (CASA), Eindhoven University of Technology, Eindhoven, The Netherlands

Department of Mathematics and Computer Science, Karlstad University, Karlstad, Sweden
e-mail: a.j.vromans@tue.nl

A. A. F. van de Ven
Centre for Analysis, Scientific computing and Applications (CASA), Eindhoven University of Technology, Eindhoven, The Netherlands
e-mail: a.a.f.v.d.ven@tue.nl

A. Muntean
Department of Mathematics and Computer Science, Karlstad University, Karlstad, Sweden
e-mail: adrian.muntean@kau.se

© Springer Nature Switzerland AG 2019

I. Faragó et al. (eds.), *Progress in Industrial Mathematics at ECMI 2018*, Mathematics in Industry 30, https://doi.org/10.1007/978-3-030-27550-1_40

of a pseudo-parabolic system was introduced. It was shown that this decomposition made upscaling of this system rather straightforward with the use of a toy pseudo-parabolic model. In our framework, this toy model is extended with convective terms, which yield no additional problems. We want to convey the message that this decomposition can be applied not only to the physical system in [8] but also to other physical systems with pseudo-parabolic equations, such as the concrete corrosion reaction model introduced in [9]. Both these pseudo-parabolic systems are physical systems on a spatial micro scale with an intrinsic microscopic periodicity of size $\epsilon \ll 1$. Similar intrinsic microscopic periodic behaviors are found in highly active research fields using composite structures or nano-structures.

In this paper, we use this spatial-temporal decomposition to upscale our pseudo-parabolic equation by using the concept of periodic homogenization via two-scale convergence. The spatial-temporal decomposition leads to upscaled systems within the partial differential equation framework, while the upscaled pseudo-parabolic equation might not live within this framework due to the non-local term. We start in Sect. 2 with formulating our pseudo-parabolic system $(\mathbf{Q}^\epsilon)$, the decomposition system (P^ϵ) and stating our assumptions. In Sect. 3, an existence and uniqueness result for weak solutions to our problem $(\mathbf{P}^\epsilon)$ is derived. In Sect. 4, we apply the idea of two-scale convergence to a weak version of problem $(\mathbf{P}^\epsilon)$, denoted $(\mathbf{P}^\epsilon_\mathbf{w})$, that contains the microscopic information at the ϵ-level. Furthermore in this section, an upscaled system $(\mathbf{P}^\mathbf{0}_\mathbf{w})$ of the weak system $(\mathbf{P}^\epsilon_\mathbf{w})$ is derived in the limit $\epsilon \downarrow 0$, containing a non-local-in-time term.

2 Basic System and Assumptions

Our pseudo-parabolic system $(\mathbf{Q}^\epsilon)$ consists of a family of N partial differential equations for the variable vector $\mathbf{U}^\epsilon(t, \mathbf{x}, \mathbf{x}/\epsilon) = (U_1^\epsilon, \ldots, U_\alpha^\epsilon, \ldots, U_N^\epsilon)$ with $t > 0$ and $\mathbf{x} = (x_1, \ldots, x_i, \ldots, x_d) \in \Omega \subset \mathbf{R}^d$. The vectors $\mathbf{V}^\epsilon$ and $\mathbf{U}^\epsilon$ are both functions of the time coordinate t, the global or macro position coordinate $\mathbf{x}$, and also periodic functions of the micro (or nano) coordinate $\mathbf{y} \in Y$, where $\mathbf{y} = \mathbf{x}/\epsilon$, where the size of the micro domain Y is $\mathcal{O}(\epsilon)$ of the size of the macro domain Ω. For $\epsilon \in (0, \epsilon_0)$ with $\epsilon_0 > 0$, system $(\mathbf{Q}^\epsilon)$ is formulated as

$$
(\mathbf{Q}^\epsilon) \quad \begin{cases} \mathsf{M}^\epsilon \mathsf{G}^{-1}\partial_t \mathbf{U}^\epsilon - \nabla \cdot \Big((\mathsf{E}^\epsilon \cdot \nabla + \mathsf{D}^\epsilon)\mathsf{G}^{-1}(\partial_t \mathbf{U}^\epsilon + \mathsf{L}\mathbf{U}^\epsilon) \Big) & \\ \qquad = \mathbf{H}^\epsilon + (\mathsf{K}^\epsilon - \mathsf{M}^\epsilon \mathsf{G}^{-1}\mathsf{L})\mathbf{U}^\epsilon + \mathsf{J}^\epsilon \cdot \nabla \mathbf{U}^\epsilon & \text{on } \mathbf{R}_+ \times \Omega, \\ \mathbf{U}^\epsilon = \mathbf{U}_* & \text{on } \{0\} \times \Omega, \\ \partial_t \mathbf{U}^\epsilon + \mathsf{L}\mathbf{U}^\epsilon = \mathbf{0} & \text{on } \mathbf{R}_+ \times \partial\Omega. \end{cases}
$$

Our dimensionless decomposition system $(\mathbf{P}^\epsilon)$ consists of a family of N partial differential equations (PDEs) and a family of N ordinary differential equations (ODEs) for the two variable vectors $\mathbf{V}^\epsilon(t, \mathbf{x}, \mathbf{x}/\epsilon) = (V_1^\epsilon, \ldots, V_\alpha^\epsilon, \ldots, V_N^\epsilon)$ and

$\mathbf{U}^\epsilon(t, \mathbf{x}, \mathbf{x}/\epsilon)$. For $\epsilon \in (0, \epsilon_0)$ with $\epsilon_0 > 0$, it is formulated as

$$(\mathbf{P}^\epsilon) \begin{cases} \mathsf{M}^\epsilon \mathbf{V}^\epsilon - \nabla \cdot \left(\mathsf{E}^\epsilon \cdot \nabla \mathbf{V}^\epsilon + \mathsf{D}^\epsilon \mathbf{V}^\epsilon\right) = \mathbf{H}^\epsilon + \mathsf{K}^\epsilon \mathbf{U}^\epsilon + \mathsf{J}^\epsilon \cdot \nabla \mathbf{U}^\epsilon & \text{on } \mathbf{R}_+ \times \Omega, \\ \partial_t \mathbf{U}^\epsilon + \mathsf{L}\mathbf{U}^\epsilon = \mathsf{G}\mathbf{V}^\epsilon & \text{on } \mathbf{R}_+ \times \Omega, \\ \mathbf{U}^\epsilon = \mathbf{U}_* & \text{on } \{0\} \times \Omega, \\ \mathbf{V}^\epsilon = \mathbf{0} & \text{on } \mathbf{R}_+ \times \partial\Omega. \end{cases}$$

Above, the ϵ-dependent notation $c^\epsilon(t, \mathbf{x}) = c(t, \mathbf{x}, \mathbf{x}/\epsilon)$ is used for the ϵ-independent 1-, 2- and 3-tensors of assumption (A1).

(A1) For all $\alpha, \beta \in \{1, \ldots, N\}$ and for all $i, j \in \{1, \ldots, d\}$, we have

$$\begin{aligned} \mathsf{M}_{\alpha\beta},\ \mathsf{E}_{ij},\ \mathsf{D}_{i\alpha\beta},\ \mathbf{H}_\alpha,\ \mathsf{K}_{\alpha\beta},\ \mathsf{J}_{i\alpha\beta} &\in L^\infty(\mathbf{R}_+ \times \Omega; C_\#(Y)), \\ \mathsf{L}_{\alpha\beta},\ \mathsf{G}_{\alpha\beta} &\in L^\infty(\mathbf{R}_+; W^{1,\infty}(\Omega)), \\ \mathbf{U}_* &\in C^1(\Omega)^N, \end{aligned}$$

with G invertible.

(A2) Let the tensors M^ϵ and E^ϵ be in diagonal form with elements $m_\alpha^\epsilon > 0$ and $e_i^\epsilon > 0$, respectively, satisfying $1/m_\alpha^\epsilon, 1/e_i^\epsilon \in L^\infty(\mathbf{R}_+ \times \Omega; C_\#(Y))$.

(A3) The inequality

$$\|\mathsf{D}_{i\beta\alpha}^\epsilon\|^2_{L^\infty(\mathbf{R}_+\times\Omega^\epsilon; C_\#(Y))} < \frac{4}{dN^2 \left\|\frac{1}{m_\alpha^\epsilon}\right\|_{L^\infty(\mathbf{R}_+\times\Omega^\epsilon; C_\#(Y))} \left\|\frac{1}{e_i^\epsilon}\right\|_{L^\infty(\mathbf{R}_+\times\Omega^\epsilon; C_\#(Y))}}$$

holds for all $\alpha, \beta \in \{1, \ldots, N\}$, for all $i \in \{1, \ldots, n\}$, and for all $\epsilon \in (0, \epsilon_0)$.

Remark, inequality (2) implies that automatically (2) holds for the Y-averaged functions $\overline{\mathsf{D}_{i\beta\alpha}^\epsilon}$, $\overline{\mathsf{M}_{\beta\alpha}^\epsilon}$, and $\overline{\mathsf{E}_{ij}^\epsilon}$ in $L^\infty(\mathbf{R}_+ \times \Omega)$, using the notation

$$\overline{f}(t, \mathbf{x}) = \frac{1}{|Y|} \int_Y f(t, \mathbf{x}, \mathbf{y}) d\mathbf{y}.$$

3 Existence and Uniqueness of Weak Solutions to ($\mathbf{P}_\mathbf{w}^\epsilon$)

In this section, we show the existence and uniqueness of a weak solution $(\mathbf{U}, \mathbf{V})$ to ($\mathbf{P}^\epsilon$). We define a weak solution to ($\mathbf{P}^\epsilon$) for $\epsilon \in (0, \epsilon_0)$ and $T \in \mathbf{R}_+$ as a pair of sequences $(\mathbf{U}^\epsilon, \mathbf{V}^\epsilon) \in H^1((0, T) \times \Omega)^N \times L^\infty((0, T), H_0^1(\Omega))^N$ satisfying

$$(\mathbf{P}_\mathbf{w}^\varepsilon) \begin{cases} \displaystyle\int_\Omega \boldsymbol{\phi}^\top \left[\mathsf{M}^\epsilon \mathbf{V}^\epsilon - \mathbf{H}^\epsilon - \mathsf{K}^\epsilon \mathbf{U}^\epsilon - \mathsf{J}^\epsilon \cdot \nabla \mathbf{U}^\epsilon\right] + (\nabla\boldsymbol{\phi})^\top \cdot \left(\mathsf{E}^\epsilon \cdot \nabla \mathbf{V}^\epsilon + \mathsf{D}^\epsilon \mathbf{V}^\epsilon\right) d\mathbf{x} = 0, \\ \displaystyle\int_\Omega \boldsymbol{\psi}^\top \left[\partial_t \mathbf{U}^\epsilon + \mathsf{L}^\epsilon \mathbf{U}^\epsilon - \mathsf{G}^\epsilon \mathbf{V}^\epsilon\right] d\mathbf{x} = 0, \\ \mathbf{U}^\epsilon(0, \mathbf{x}) = \mathbf{U}_*(\mathbf{x}) \text{ for all } \mathbf{x} \in \Omega, \end{cases}$$

for a.e. $t \in (0, T)$, for all test-functions $\boldsymbol{\phi} \in H_0^1(\Omega)^N$ and $\boldsymbol{\psi} \in L^2(\Omega)^N$.

The existence and uniqueness can only hold when the first equation of $(\mathbf{P}_\mathbf{w}^\epsilon)$ satisfies all the conditions of Lax-Milgram. The next lemma provides the coercivity condition, while the continuity condition is trivially satisfied.

Lemma 1 *Assume assumptions (A1)–(A3) hold, then there exist positive constants* $\tilde{m}_\alpha$, $\tilde{e}_i$, $\tilde{H}$, $\tilde{K}_\alpha$, $\tilde{J}_{i\alpha}$ *for* $\alpha \in \{1, \dots, N\}$ *and* $i \in \{1, \dots, d\}$ *such that the a-priori estimate*

$$\begin{aligned} &\sum_{\alpha=1}^N \tilde{m}_\alpha \|\mathbf{V}_\alpha^\epsilon\|_{L^2(\Omega)}^2 + \sum_{i=1}^d \sum_{\alpha=1}^N \tilde{e}_i \|\partial_{x_i} \mathbf{V}_\alpha^\epsilon\|_{L^2(\Omega)}^2 \\ &\leq \tilde{H} + \sum_{\alpha=1}^N \tilde{K}_\alpha \|\mathbf{U}_\alpha^\epsilon\|_{L^2(\Omega)}^2 + \sum_{i=1}^d \sum_{\alpha=1}^N \tilde{J}_{i\alpha} \|\partial_{x_i} \mathbf{U}_\alpha^\epsilon\|_{L^2(\Omega)}^2 \end{aligned} \tag{1}$$

holds for a.e. $t \in (0, T)$.

Proof See p. 92, 93 in [9] for proof and relation with parameters of $(\mathbf{P}_\mathbf{w}^\epsilon)$. □

Theorem 1 *Assume assumptions (A1)–(A3) hold, then there exists a unique pair* $(\mathbf{U}^\epsilon, \mathbf{V}^\epsilon) \in H^1((0, T) \times \Omega)^N \times L^\infty((0, T), H_0^1(\Omega))^N$ *such that* $(\mathbf{U}^\epsilon, \mathbf{V}^\epsilon)$ *is a weak solution to* $(\mathbf{P}_\mathbf{w}^\epsilon)$.

Proof Use $\boldsymbol{\phi} = \mathbf{V}^\epsilon$ and apply Lemma 1. Then use $\boldsymbol{\psi} \in \{\mathbf{U}^\epsilon, \partial_t \mathbf{U}^\epsilon\}$. Moreover, apply a gradient to the second equation of $(\mathbf{P}^\epsilon)$ and test that equation with $\nabla \mathbf{U}^\epsilon$ and $\partial_t \nabla \mathbf{U}^\epsilon$. Application of Young's inequality, use of (1) and application of Gronwall's inequality, see [2, Thm. 1], yields the existence for $\mathbf{U}^\epsilon$. Then Lax-Milgram yields the existence for $\mathbf{V}^\epsilon$. Uniqueness follows from the bilinearity of $(\mathbf{P}_\mathbf{w}^\epsilon)$. For more details, see pages 93 and 94 in [9]. □

4 Upscaling the System $(\mathbf{P}_w^\epsilon)$ via Two-Scale Convergence

Based on two-scale convergence, see [1, 5, 7] for details, we obtain the following Lemma ensuring that the weak solution to problem $(\mathbf{P}_\mathbf{w}^\epsilon)$ has two-scale limits in the limit $\epsilon \downarrow 0$.

Lemma 2 *Assume assumptions (A0), (A1), (A2) to hold. For each* $\epsilon \in (0, \epsilon_0)$, *let the pair of sequences* $(\mathbf{U}^\epsilon, \mathbf{V}^\epsilon) \in H^1((0, T) \times \Omega) \times L^\infty((0, T); H_0^1(\Omega))$ *be the unique weak solution to* $(\mathbf{P}_\mathbf{w}^\epsilon)$. *Then this sequence of weak solutions satisfies the estimate* $\|\mathbf{U}^\epsilon\|_{H^1((0,T)\times\Omega)^N} + \|\mathbf{V}^\epsilon\|_{L^\infty((0,T),H_0^1(\Omega))^N} \leq C$, *for all* $\epsilon \in (0, \epsilon_0)$ *and there exist vector functions*

$$\begin{aligned} &\mathbf{u} \text{ in } H^1((0, T) \times \Omega)^N, && \mathscr{U} \text{ in } H^1((0, T); L^2(\Omega; H_\#^1(Y)/\mathbf{R}))^N, \\ &\mathbf{v} \text{ in } L^\infty((0, T); H_0^1(\Omega))^N, && \mathscr{V} \text{ in } L^\infty((0, T) \times \Omega; H_\#^1(Y)/\mathbf{R})^N, \end{aligned}$$

and a subsequence $\epsilon' \subset \epsilon$, for which the following two-scale convergences

$$\mathbf{U}^{\epsilon'} \xrightarrow{2} \mathbf{u}(t, \mathbf{x}), \qquad \nabla \mathbf{U}^{\epsilon'} \xrightarrow{2} \nabla \mathbf{u}(t, \mathbf{x}) + \nabla_{\mathbf{y}} \mathscr{U}(t, \mathbf{x}, \mathbf{y}),$$

$$\partial_t \mathbf{U}^{\epsilon'} \xrightarrow{2} \partial_t \mathbf{u}(t, \mathbf{x}), \qquad \partial_t \nabla \mathbf{U}^{\epsilon'} \xrightarrow{2} \partial_t \nabla \mathbf{u}(t, \mathbf{x}) + \partial_t \nabla_{\mathbf{y}} \mathscr{U}(t, \mathbf{x}, \mathbf{y}),$$

$$\mathbf{V}^{\epsilon'} \xrightarrow{2} \mathbf{v}(t, \mathbf{x}), \qquad \nabla \mathbf{V}^{\epsilon'} \xrightarrow{2} \nabla \mathbf{v}(t, \mathbf{x}) + \nabla_{\mathbf{y}} \mathscr{V}(t, \mathbf{x}, \mathbf{y})$$

hold for a.e. $t \in (0, T)$.

Proof See pages 95 and 96 of [9]. □

Using Lemma 2, we upscale $(\mathbf{P}^{\epsilon}_{\mathbf{w}})$ to $(\mathbf{P}^{\mathbf{0}}_{\mathbf{w}})$ via two-scale convergence.

Theorem 2 *Assume the conditions of Lemma 2 are met. Then the two-scale limits $\mathbf{u} \in H^1((0,T) \times \Omega)^N$, $\mathscr{U} \in H^1((0,T); L^2(\Omega; H^1_{\#}(Y)/\mathbf{R}))^N$ and $\mathbf{v} \in L^\infty((0,T); H^1_0(\Omega))^N$ introduced in Lemma 2 form the weak solution triple to*

$$(\mathbf{P}^{\mathbf{0}}_{\mathbf{w}}) \qquad \begin{cases} \displaystyle\int_\Omega \boldsymbol{\phi}^\top \left[\overline{\mathsf{M}}\mathbf{v} - \overline{\mathbf{H}} - \overline{\mathsf{K}}\mathbf{u} - \overline{\mathsf{J}} \cdot \nabla \mathbf{u} - \frac{1}{|Y|} \int_Y \mathsf{J} \cdot \nabla_{\mathbf{y}} \mathscr{U} d\mathbf{y} \right] \\ \qquad\qquad + (\nabla \boldsymbol{\phi})^\top \cdot \left(\mathsf{E}^* \cdot \nabla \mathbf{v} + \mathsf{D}^* \mathbf{v} \right) d\mathbf{x} = 0, \\ \displaystyle\int_\Omega \boldsymbol{\psi}^\top \left[\partial_t \mathbf{u} + \mathsf{L}\mathbf{u} - \mathsf{G}\mathbf{v} \right] d\mathbf{x} = 0, \\ \displaystyle\int_Y \boldsymbol{\xi}^\top \cdot \nabla_{\mathbf{y}} \left[\partial_t \mathscr{U} + \mathsf{L}\mathscr{U} - \tilde{\delta}\mathbf{v} - \tilde{\omega} \cdot \nabla \mathbf{v} \right] d\mathbf{y} = 0, \\ \mathbf{u}(0, \mathbf{x}) = \mathbf{U}_*(\mathbf{x}) \qquad \text{on } \Omega, \\ \nabla_{\mathbf{y}} \mathscr{U}(0, \mathbf{x}, \mathbf{y}) = \mathbf{0} \qquad \text{on } \Omega \times Y, \end{cases}$$

for a.e. $t \in (0, T)$, for all test-functions $\boldsymbol{\phi} \in H^1_0(\Omega)^N$, $\boldsymbol{\psi} \in L^2(\Omega)^N$, and $\boldsymbol{\xi} \in H^1_{\#}(Y)^{d \times N}$, where the effective *coefficients E^* and D^* are given by*

$$\mathsf{E}^* = \frac{1}{|Y|} \int_Y \mathsf{E} \cdot (1 + \nabla_{\mathbf{y}} \mathbf{W}) d\mathbf{y}, \qquad \mathsf{D}^* = \frac{1}{|Y|} \int_Y \mathsf{D} + \mathsf{E} \cdot \nabla_{\mathbf{y}} \delta d\mathbf{y},$$
$$\tilde{\delta} = \nabla_{\mathbf{y}}(\mathsf{G}\delta), \qquad \tilde{\omega} = \nabla_{\mathbf{y}} \mathbf{W} \otimes \mathsf{G},$$

and the tensor $\delta_{\alpha\beta} \in L^\infty((0,T) \times \Omega; H^1_{\#}(Y)/\mathbf{R})$ and vector $\mathbf{W}_i \in L^\infty((0,T) \times \Omega; H^1_{\#}(Y)/\mathbf{R})$ satisfy the cell problems

$$0 = \int_Y \Phi^\top \cdot (\nabla_{\mathbf{y}} \cdot [\mathsf{E} \cdot (1 + \nabla_{\mathbf{y}} \mathbf{W})]) d\mathbf{y}, \quad 0 = \int_Y \Psi^\top (\nabla_{\mathbf{y}} \cdot [\mathsf{D} + \mathsf{E} \cdot \nabla_{\mathbf{y}} \delta]) d\mathbf{y}$$

for all $\Phi \in C_{\#}(Y)^d$, $\Psi \in C_{\#}(Y)^{N \times N}$.

Proof In $(\mathbf{P}^{\epsilon}_{\mathbf{w}})$, we choose $\boldsymbol{\phi} = \boldsymbol{\phi}^{\epsilon} = \Phi\left(t, \mathbf{x}, \frac{\mathbf{x}}{\epsilon}\right)$ for the test-function $\Phi \in L^2((0,T); \mathscr{D}(\Omega; C^{\infty}_{\#}(Y)))^N$, and $\boldsymbol{\psi} = \boldsymbol{\psi}^{\epsilon} = \Psi(t, \mathbf{x}) + \epsilon\boldsymbol{\varphi}\left(t, \mathbf{x}, \frac{\mathbf{x}}{\epsilon}\right)$ for the test-function $\Psi \in L^2((0,T); C^{\infty}_0(\Omega))^N$ and for $\boldsymbol{\varphi} \in L^2((0,T); \mathscr{D}(\Omega; C^{\infty}_{\#}(Y)))^N$. Two-scale convergence limits in [1, 5, 7] together with standard cell-function arguments, see [6], give $(\mathbf{P}^{\mathbf{0}}_{\mathbf{w}})$. See pages 97 and 98 of [9] for details. □

We have shown that upscaling system $(\mathbf{P}^{\epsilon}_{\mathbf{w}})$ yields system $(\mathbf{P}^{\mathbf{0}}_{\mathbf{w}})$. This system contains only PDEs with respect to $(t, \mathbf{x})$. However, an extra variable $\nabla_{\mathbf{y}}\mathscr{U}$ was needed. Removing $\nabla_{\mathbf{y}}\mathscr{U}$ needs the use of continuous semi-group theory, see papers 10 and 14 of [10], for solving the third equation of system $(\mathbf{P}^{\mathbf{0}}_{\mathbf{w}})$. This leads to a non-local-in-time term as a consequence of removing $\nabla_{\mathbf{y}}\mathscr{U}$.

5 Conclusion

Our main goal of this paper is to show that the spatial-temporal decomposition, as employed in [8], allows for the straightforward upscaling of pseudo-parabolic equations, in specific for system $(\mathbf{Q}^{\epsilon})$. The upscaling procedure is here performed using the concept of two-scale convergence as reported in Sect. 4. Moreover, the decomposition is retained in the upscaled limit. A non-local-in-time term arose when an extra variable was eliminated. The spatial-temporal decoupling showed why this non-local term is non-local in time.

In future research we intend to investigate the applicability of the spatial-temporal decomposition of our pseudo-parabolic system to perforated periodic domains, corrector estimates (convergence speed estimate), high-contrast situations and the interplay between homogenization and stochastic effects.

The spatial-temporal decomposition allowed, indeed, for a straightforward homogenization process of the pseudo-parabolic structure of the system. Furthermore, the approach has the potential to be used within the framework of a multiscale discretization scheme such that homogenization limiting procedure and the convergence of the numerical scheme are done simultaneously, somewhat in a similar spirit as in [3]. An actual implementation of such numerical homogenization methodology is subject of further work.

Acknowledgements We acknowledge the Netherlands Organisation of Scientific Research (NWO) for the MPE grant 657.000.004 and we acknowledge the NWO Cluster Nonlinear Dynamics in Natural Systems (NDNS+) for funding a research stay of AJV at Karlstads Universitet.

References

1. Allaire, G.: Homogenization and two-scale convergence. SIAM J. Math. Anal. **23**(6), 1482–1518 (1992)

2. Dragomir, S.S.: Some Gronwall Type Inequalities and Applications. RGMIA Monographs. Nova Science, New York (2003)
3. Essel, E.K., Kuliev, K., Kulieva, G., Persson, L.-E.: Homogenization of quasilinear parabolic problems by the method of Rothe and two scale convergence. Appl. Math. **55**(4), 305–327 (2010)
4. Hassanizadeh S.M., Celia M.A., Dahle H.K.: Dynamical effect in the capillary pressure-saturation relationship and its impact on unsaturated flow. Vadose Zone J. **1**, 38–57 (2002)
5. Lukkassen, D., Nguetseng, G., Wall, P.: Two-scale convergence. Int. J. Pure Appl. Math. **2**(1), 35–62 (2002)
6. Muntean, A., Chalupecký, V.: Homogenization Method and Multiscale Modeling. No. 34 in COE Lecture Note. Institute of Mathematics for Industry. Kyushu University, Japan (2011)
7. Nguetseng, G.: A general convergence result for a functional related to the theory of homogenization. SIAM J. Math. Anal. **20**(3), 608–623 (1989)
8. Peszyńska, M., Showalter, R., Yi, S.Y.: Homogenization of a pseudoparabolic system. Appl. Anal. **88**(9), 1265–1282 (2009)
9. Vromans, A.J.: A Pseudoparabolic Reaction-Diffusion-Mechanics System: Modeling, Analysis and Simulation. Licentiate thesis. Karlstad University, Karlstad (2018)
10. Yosida, K.: Functional Analysis. No. 123 in Die Grundlehren der Mathematischen Wissenschaften in Einzeldarstellungen mit Besonderer Berücksichtigung der Anwendungsgebiete. Springer, Berlin (1965)

Approximating a Class of Linear Third-Order Ordinary Differential Problems

Emilio Defez, Michael M. Tung, J. Ibáñez, and Jorge Sastre

Abstract In this work, a procedure to approximate the solution of special linear third-order matrix differential problems of the type $Y^{(3)}(x) = A(x)Y(x) + B(x)$ with higher-order matrix splines is proposed. An illustrative example is given.

1 Introduction

In this paper a new spline method is developed for computing third-order linear ordinary differential equations of the form

$$Y^{(3)} = A(x)Y(x) + B(x)\ ,\ A(x) \in \mathbb{C}^{r\times r}\ ,\ B(x), Y(x) \in \mathbb{C}^{r\times q}, x \in [a, b] \tag{1}$$

with initial conditions $Y(a) = Y_a, Y'(a) = Y'_a$ and $Y''(a) = Y''_a$. This type of problem can be found in various fields of applied science and engineering, see [2, 5, 7] and references therein, specially in fluid dynamic problems [10] and also in the study of the Einstein-Weyl spaces [9].

Traditionally, the third-order ordinary differential equations can be rewritten as a first-order system of ordinary differential equations, so that standard numerical methods can be applied, but this method increases the computational cost and it can cause in numerical instability. Therefore, direct integration methods have attracted

E. Defez · M. M. Tung (✉)
Instituto de Matemática Multidisciplinar, Universitat Politècnica de València, Valencia, Spain
e-mail: edefez@imm.upv.es; mtung@mat.upv.es

J. Ibáñez
Instituto de Instrumentación para Imagen Molecular, Universitat Politècnica de València, Valencia, Spain
e-mail: jjibanez@dsic.upv.es

J. Sastre
Instituto de Telecomunicaciones y Aplicaciones Multimedia, Universitat Politècnica de València, Valencia, Spain
e-mail: jsastrem@upv.es

© Springer Nature Switzerland AG 2019
I. Faragó et al. (eds.), *Progress in Industrial Mathematics at ECMI 2018*,
Mathematics in Industry 30, https://doi.org/10.1007/978-3-030-27550-1_41

significant attention from several authors because these direct methods improve the accuracy and the speed, see [3, 8, 11]. Throughout this work, we will use the standard notation for matrix splines, see [1]. In the following section we give a description of the proposed method and conclude with a numerical example.

2 Description of the Method and Example

Let

$$Y^{(3)}(x) = A(x)Y(x) + B(x), a \leq x \leq b, \tag{2}$$

be a third-order linear matrix problem, where $Y(x) \in \mathbb{C}^{r\times q}$ and $Y(a) = Y_a, Y'(a) = Y'_a, Y''(a) = Y''_a \in \mathbb{C}^{r\times q}$ are initial conditions. The functions $A : [a, b] \to \mathbb{C}^{r\times r}$ and $B : [a, b] \to \mathbb{C}^{r\times q}$ are of differentiability class $A, B \in \mathscr{C}^s(I), s \geq 1$, with $I = (a, b)$. Let the partition of $[a, b]$ be defined by

$$\Delta_{[a,b]} = \{a = x_0 < x_1 < \ldots < x_n = b\}, \ x_k = a + kh, \ k = 0, 1, \ldots, n, \tag{3}$$

where n is a positive integer, and $h = (b - a)/n$ is the step size. In each subinterval $I_k = [a + kh, a + (k + 1)h]$ we will define a matrix spline $S(x)$ of order $m \in \mathbb{N}$ with $4 \leq m \leq s$, which will be an approximation of the solution of problem (2) so that $S(x) \in C^3(I)$. For the first interval I_0, we define the following matrix spline:

$$\begin{aligned} S\big|_{I_0}(x) &= Y(a) + Y'(a)(x-a) + \frac{1}{2!}Y''(a)(x-a)^2 + \frac{1}{3!}Y^{(3)}(a)(x-a)^3 + \\ &\cdots + \frac{1}{(m-1)!}Y^{(m-1)}(a)(x-a)^{m-1} + \frac{1}{m!}A_0(x-a)^m, \end{aligned} \tag{4}$$

where $A_0 \in \mathbb{C}^{r\times q}$ must be determined. It is easy to prove

$$S\big|_{I_0}(a) = Y_a, S'\big|_{I_0}(a) = Y'_a, S''\big|_{I_0}(a) = Y''_a, S^{(3)}\big|_{I_0}(a) = Y^{(3)}(a) = A(a)Y_a + B(a),$$

and hence the matrix spline (4) satisfies (2) at point $x = a$. We must compute the values of $Y^{(4)}(a), Y^{(5)}(a), Y^{(6)}(a), \ldots, Y^{(m-1)}(a)$ and A_0. Taking derivatives in the differential equations one gets

$$\begin{aligned} Y^{(4)}(x) &= A'(x)Y(x) + A(x)Y'(x) + B'(x) \\ &= g_1\left(x, Y(x), Y'(x)\right), \end{aligned} \tag{5}$$

where $g_1 \in \mathscr{C}^{s-1}(I)$. We are now in the position to evaluate $Y^{(4)}(a) = g_1\left(a, Y_a, Y'_a\right)$ using (5). Now, we can assume that $A, B \in \mathscr{C}^s(I)$ for $s \geq 2$. Then, the second derivatives of A and B exist and are continuous. This yields the fifth derivative $Y^{(5)}(x)$:

$$\begin{aligned} Y^{(5)}(x) &= A''(x)Y(x)+2A'(x)Y'(x)+A(x)Y''(x)+B''(x) \\ &= g_2\left(x, Y(x), Y'(x), Y''(x)\right) \in \mathcal{C}^{s-2}(I). \end{aligned} \tag{6}$$

Now we can evaluate $Y^{(5)}(a) = g_2\left(a, Y(a), Y'(a), Y''(a)\right) = g_2\left(a, Y_a, Y'_a, Y''_a\right)$ using (6). For the rest of derivatives $Y^{(6)}(x), \ldots, Y^{(m-1)}(x)$ we proceed in like manner and calculate

$$\left.\begin{aligned} Y^{(6)}(x) &= g_3\left(x, Y(x), Y'(x), Y''(x), Y'''(x)\right) \in \mathcal{C}^{s-3}(I) \\ &\vdots \\ Y^{(m-1)}(x) &= g_{m-4}\left(x, Y(x), Y'(x), \ldots, Y^{(m-4)}(x)\right) \in \mathcal{C}^{s-(m-4)}(I) \end{aligned}\right\}. \tag{7}$$

A list of all these derivatives can be easily established by employing standard computer algebra systems as *Mathematica* or *Matlab*, for example. Taking $x = a$ in (7), one gets $Y^{(6)}(a), \ldots, Y^{(m-1)}(a)$. To compute A_0, we impose that (4) is a solution of matrix differential equation (2) at $x = a + h$, i.e.

$$S^{(3)}\big|_{I_0}(a+h) = A\,(a+h)\, S\big|_{I_0}(a+h) + B(a+h). \tag{8}$$

From (8) we obtain the following implicit matrix equation:

$$\begin{aligned} A_0 = \tfrac{(m-3)!}{h^{m-3}} \Big[& A\,(a+h)\left(Y(a) + Y'(a)h + \cdots + \tfrac{h^{m-1}}{(m-1)!}Y^{(m-1)}(a) + \tfrac{h^m}{m!}A_0\right) \\ & + B(a+h) - Y^{(3)}(a) - Y^{(4)}(a)h - \cdots - \tfrac{1}{(m-4)!}Y^{(m-1)}(a)h^{m-4}\Big]. \end{aligned} \tag{9}$$

If matrix equation (9) has only one solution A_0, the matrix spline (4) is totally determined at I_0. In the following interval I_1 we take the matrices

$$\left.\begin{aligned} \overline{Y^{(3)}(a+h)} &= A\,(a+h)\, S\big|_{I_0}(a+h) + B\,(a+h) \\ \overline{Y^{(4)}(a+h)} &= g_1\left(a+h, S\big|_{I_0}(a+h), S'\big|_{I_0}(a+h)\right) \\ &\vdots \\ \overline{Y^{(m-1)}(a+h)} &= g_{m-4}\left(a+h, S\big|_{I_0}(a+h), \ldots, S^{(m-4)}\big|_{I_0}(a+h)\right) \end{aligned}\right\} \tag{10}$$

and we define the spline at I_1 as

$$\begin{aligned} S\big|_{I_1}(x) = & \sum_{i=0}^{2} \frac{S^{(i)}\big|_{I_0}(a+h)}{i!}(x-(a+h))^i + \sum_{j=3}^{m-1} \frac{\overline{Y^{(j)}(a+h)}}{j!}(x-(a+h))^j \\ & + \frac{A_1}{m!}(x-(a+h))^m, \end{aligned} \tag{11}$$

Note that matrix spline $S(x)$ defined by (4) and (11) is of differentiability class $\mathscr{C}^3(I_0 \cup I_1)$. Thus, spline (11) satisfies the differential equation (2) at point $x = a + h$, and all of its coefficients are determined except $A_1 \in \mathbb{C}^{r \times q}$. Matrix A_1 can be found imposing that the spline (11) is also the solution of (2) at $x = a + 2h$:

$$S^{(3)}\big|_{I_1}(a+2h) = A\,(a+2h)\,S\big|_{I_1}(a+2h) + B\,(a+2h)\,.$$

Developing that expression we obtain

$$A_1 = \frac{(m-3)!}{h^{m-3}}\left[A\,(a+2h)\left(\sum_{i=0}^{2}\frac{S^{(i)}\big|_{I_0}(a+h)}{i!}h^i + \sum_{j=3}^{m-1}\frac{\overline{Y^{(j)}(a+h)}}{j!}h^j + \frac{A_1 h^m}{m!}\right)\right. \tag{12}$$

$$\left. + B\,(a+2h) - \overline{Y^{(3)}(a+h)} - \overline{Y^{(4)}(a+h)}h - \cdots - \frac{h^{m-4}}{(m-4)!}\overline{Y^{(m-1)}(a+h)}\right].$$

If matrix equation (12) has only one solution, A_1, then the spline is well determined at I_1. Applying the same procedure, we can obtain the matrix spline approximation from the interval I_0 to the interval I_{k-1}. For the following subinterval I_k, the matrix spline is defined as

$$S\big|_{I_k}(x) = \sum_{i=0}^{2}\frac{S^{(i)}\big|_{I_{k-1}}(a+kh)}{i!}(x-(a+kh))^i$$
$$+ \sum_{j=3}^{m-1}\frac{\overline{Y^{(j)}(a+kh)}}{j!}(x-(a+kh))^j + \frac{A_k}{m!}(x-(a+kh))^m,$$

where

$$\left.\begin{aligned}
\overline{Y^{(3)}(a+kh)} &= A\,(a+kh)\,S\big|_{I_{k-1}}(a+kh) + B\,(a+kh)\\
\overline{Y^{(4)}(a+kh)} &= g_1\left(a+kh,\, S\big|_{I_{k-1}}(a+kh),\, S'\big|_{I_{k-1}}(a+kh)\right)\\
&\;\;\vdots\\
\overline{Y^{(m-1)}(a+kh)} &= g_{m-4}\left(a+kh,\, S\big|_{I_{k-1}}(a+kh),\ldots,\, S^{(m-4)}\big|_{I_{k-1}}(a+kh)\right)
\end{aligned}\right\}$$

Thus, the matrix spline $S(x) \in \mathscr{C}^3\left(\bigcup_{j=0}^{k} I_j\right)$ and fulfills the differential equation (2) at $x = a + kh$. As a last requirement, we impose that $S\big|_{I_k}(x)$ satisfies (2) at $x = a + (k+1)h$, and then one gets

$$S^{(3)}\big|_{I_k}(a+(k+1)h) = A\,(a+(k+1)h)\,S\big|_{I_k}(a+(k+1)h) + B\,(a+(k+1)h)\,.$$

Expanding this expression, we obtain

$$A_k = \frac{(m-3)!}{h^{m-3}} \Bigg[A\left(a+(k+1)h\right) \left(\sum_{i=0}^{2} \frac{S^{(i)}\big|_{I_{k-1}}(a+kh)}{i!} h^i + \sum_{j=3}^{m-1} \frac{\overline{Y^{(j)}(a+kh)}}{j!} h^j + \frac{A_k}{m!} h^m \right)$$

$$+B\left(a+(k+1)h\right) - \overline{Y^{(3)}(a+kh)} - \cdots - \frac{h^{m-4}}{(m-4)!} \overline{Y^{(m-1)}(a+kh)} \Bigg]. \tag{13}$$

Observe that (13) gives us the Eqs. (9) and (12), when $k = 0$ and $k = 1$. We will show that (13) has a unique solution for each $k = 0, 1, \ldots, n-1$.

Now we can write Eq. (13) as

$$\left(I_{r\times r} - \frac{h^3 A(a+(k+1)h)}{m(m-1)(m-2)} \right) A_k$$

$$= \frac{(m-3)!}{h^{m-3}} \Bigg[A\left(a+(k+1)h\right) \left(\sum_{i=0}^{2} \frac{S^{(i)}\big|_{I_{k-1}}(a+kh)}{i!} h^i + \sum_{j=3}^{m-1} \frac{\overline{Y^{(j)}(a+kh)}}{j!} h^j \right)$$

$$+B\big(a+(k+1)h\big) - \overline{Y^{(3)}(a+kh)} - \cdots - \frac{h^{m-4}}{(m-4)!} \overline{Y^{(m-1)}(a+kh)} \Bigg]. \tag{14}$$

Observe that the solvability of Eq. (14) is guaranteed if the matrix coefficients $\mathscr{C}_k = \left(I_{r\times r} - \frac{h^3 A(a+(k+1)h)}{m(m-1)(m-2)} \right)$ are invertible for $k = 0, 1, \ldots, n-1$. Let be

$$M = \max\{\|A(x)\| \ ; \ x \in [a, b]\}, \tag{15}$$

then it easy to prove that $\|I_{r\times r} - \mathscr{C}_k\| \le \frac{h^3 M}{m(m-1)(m-2)}$. Thus, if $h < \sqrt[3]{\frac{m(m-1)(m-2)}{M}}$, according to Lemma 2.3.3 in [4], it follows that $\mathscr{C}_k$ is invertible for $0 \le k \le n-1$. Hence, Eq. (13) has a unique solution A_k for each $k = 0, 1, \ldots, n-1$, and the matrix spline is determined. Summarising, the following theorem is proved:

Theorem 1 *For the third-order matrix differential equation (2), let M be the constant defined by (15). We consider the partition (3) with step size h satisfying $h < \sqrt[3]{m(m-1)(m-2)/M}$. Then, the matrix spline $S(x)$ of order m, $4 \le m \le s$ exists on each subinterval I_k, $k = 0, 1, \ldots, n-1$, as defined in the previous construction and is of differentiability class $\mathscr{C}^3(I)$.*

Table 1 Maximum error for splines of order $m = 7$ within each interval I_k, $k = 0, 1, \ldots, 9$, computed with step size $h = 0.1$ for the test problem (16)

Interval I_k	$[0, 0.1]$	$[0.1, 0.2]$	$[0.2, 0.3]$	$[0.3, 0.4]$	$[0.4, 0.5]$
max. error	7.59×10^{-11}	7.25×10^{-11}	4.36×10^{-10}	1.48×10^{-9}	3.77×10^{-9}
Interval I_k	$[0.5, 0.6]$	$[0.6, 0.7]$	$[0.7, 0.8]$	$[0.8, 0.9]$	$[0.9, 1.0]$
max. error	7.99×10^{-9}	1.50×10^{-8}	2.58×10^{-8}	4.16×10^{-8}	6.34×10^{-8}

Note that the constructed splines have a global error of $O(h^{m-1})$, see [6] for details.

As a numerical example, we consider the following system:

$$\left.\begin{array}{rl} Y^{(3)}(x) = & AY(x)\ ,\ Y(0) = \begin{pmatrix} 2 \\ -2 \\ 12 \end{pmatrix} \\ Y'(0) = & \begin{pmatrix} -12 \\ 28 \\ -33 \end{pmatrix}\ ,\ Y''(0) = \begin{pmatrix} 20 \\ -52 \\ 5 \end{pmatrix} \end{array}\right\}, A = \begin{pmatrix} \frac{817}{68} & \frac{1393}{68} & \frac{448}{68} \\ -\frac{1141}{68} & -\frac{2837}{68} & -\frac{896}{68} \\ \frac{3059}{136} & \frac{4319}{136} & \frac{1592}{136} \end{pmatrix}, 0 \le x \le 1, \tag{16}$$

whose exact solution [11, p.147] is given by $Y(x) = \begin{pmatrix} e^x - 2e^{2x} + 3e^{-3x} \\ 3e^x + 2e^{2x} - 7e^{-3x} \\ -11e^x - 5e^{2x} + 4e^{-3x} \end{pmatrix}$.

From (15) one gets $M = 90.1136$. For splines of the seventh order ($m = 7$) we have $h < \sqrt[3]{\frac{210}{90.1136}} \approx 1.32579$. We take $n = 10$ and $h = 0.1$. Table 1 displays numerical estimates for the maximum error within each subinterval I_k for $k = 0, 1, \ldots, 9$.

Acknowledgements The authors wish to thank for financial support by the Spanish *Ministerio de Economía y Competitividad* under grant TIN2017-89314-P, and by the Universitat Politècnica de València under grant PAID-06-18/SP20180016.

References

1. Defez, E., Tung, M.M., Ibáñez, J., Sastre, J.: Approximating and computing nonlinear matrix differential models. Math. Comput. Model. **55**(7), 2012–2022 (2012)
2. Duffy, B., Wilson, S.: A third-order differential equation arising in thin-film flows and relevant to Tanner's law. Appl. Math. Lett. **10**(3), 63–68 (1997)
3. Famelis, I.T., Tsitouras, C.: Symbolic derivation of Runge–Kutta–Nyström type order conditions and methods for solving $y'''(x) = f(x, y)$. Appl. Math. Comput. **297**, 50–60 (2017)
4. Golub, G.H., Loan, C.F.V.: Matrix Computations, 2nd edn. The Johns Hopkins University Press, Baltimore (1989)
5. Gregus, M.: Third Order Linear Differential Equations, vol. 22. Springer, Berlin (2012)

6. Loscalzo, F.R., Talbot, T.D.: Spline function approximations for solutions of ordinary differential equations. SIAM J. Numer. Anal. **4**(3), 433–445 (1967)
7. Momoniat, E.: Symmetries, first integrals and phase planes of a third-order ordinary differential equation from thin film flow. Math. Comput. Model. **49**(1–2), 215–225 (2009)
8. Senu, N., Mechee, M., Ismail, F., Siri, Z.: Embedded explicit Runge–Kutta type methods for directly solving special third order differential equations $y'''(x) = f(x, y)$. Appl. Math. Comput. **240**, 281–293 (2014)
9. Tod, K.: Einstein–Weyl spaces and third-order differential equations. J. Math. Phys. **41**(8), 5572–5581 (2000)
10. Tuck, E., Schwartz, L.: A numerical and asymptotic study of some third-order ordinary differential equations relevant to draining and coating flows. SIAM Rev. **32**(3), 453–469 (1990)
11. You, X., Chen, Z.: Direct integrators of Runge–Kutta type for special third-order ordinary differential equations. Appl. Numer. Math. **74**, 128–150 (2013)

An Iterative Method Based on Fractional Derivatives for Solving Nonlinear Equations

Béla J. Szekeres and Ferenc Izsák

Abstract In this work, we showed a fractional derivative based iterative method for solving nonlinear time-independent equation, where the operator is affecting on a Hilbert space. We assumed that it is equally monotone and Lipschitz-continuous. We proved that the algorithm is convergent. We also have tested our method numerically previously on a fluid dynamical problem and the results showed that the algorithm is stable.

1 Introduction

The theory of fractional order derivatives are almost as old as the integer-order [5]. There are many applications, for example in physics [1, 2, 6], finance [8, 9] or biology [3]. Our aim is to prove theoretical mathematical statements.

In this work our goal is to find a solution numerically for the equation $A(u) = f$. If we assume that u is time-dependent, then one can do this by finding a stationary solution of the equation $\partial_t u(t) = -(A(u(t)) - f)$. The numerical solution of this problem can be highly inaccurate. To avoid this we propose to replace the time derivative with a fractional one. Since the fractional order time derivative is a non-local operator, we expect that this stabilizes the time integration in the numerical solutions. Since the fractional order derivative here is defined as a limit of linear combination of past values, the time discretization will be simple. We also tested our method numerically in a fluid dynamical problem [10].

B. J. Szekeres (✉)
Department of Numerical Analysis, Eötvös Loránd University, Faculty of Informatics, Budapest, Hungary
e-mail: szekeres@inf.elte.hu

F. Izsák
Department of Applied Analysis and Computational Mathematics & ELTE-MTA Numnet Research Group, Eötvös Loránd University, Budapest, Hungary
e-mail: izsakf@cs.elte.hu

© Springer Nature Switzerland AG 2019
I. Faragó et al. (eds.), *Progress in Industrial Mathematics at ECMI 2018*, Mathematics in Industry 30, https://doi.org/10.1007/978-3-030-27550-1_42

2 Mathematical Preliminaries

The following theorem is well known, see [11].

Theorem 1 *Let H real Hilbert-space, $A : H \to H$ nonlinear operator, which satisfies the conditions below with some positive constants $M \geq m$:*

1. $\langle A(u) - A(v), u - v \rangle \geq m\|u - v\|^2$,
2. $\|A(u) - A(v)\| \leq M\|u - v\|$.

Then for any $f, u_0 \in H$ there exist a unique solution u^* of the equation $A(u) = f$. If $t \in \mathbb{R}^+$ is small enough the following iteration converges to u^*.

$$u_{n+1} = u_n - t\big[A(u_n) - f\big]. \tag{1}$$

There exist many different definitions of the fractional derivative [4, 7] we will use here the one below which is based on finite differences.

Definition 1 For the exponent $\beta \in (0, 1)$ the fractional order derivative for a given function $f : \mathbb{R}^+ \to \mathbb{R}$ is defined as

$$\frac{\partial^\beta f(t)}{\partial t^\beta} := \lim_{N\to\infty} \left\{ \sum_{k=0}^{N} \binom{\beta}{k} (-1)^k \frac{f(t - kh)}{h^\beta} \right\},$$

provided that the limit exists.

3 Results

Shortly, our objective is to find a solution for the equation $A(u) = f$ for a given nonlinear operator A, and for a given function f. The solution u is also time-dependent, our goal is to find a stationary solution for

$$-\left(A(u(t)) - f\right) = \partial_t u(t). \tag{2}$$

The method in Theorem 1 is one approach to this. Our idea was that to replace the time derivative in (2) with $\frac{\partial^\beta}{\partial t^\beta}$ for some $\beta \in (0, 1)$, according to Definition 1, and discretise the equation in time by a natural way.

We need an additional statement before we prove.

Lemma 1 (Pachpatte) *Let $(\alpha_n)_{n\in\mathbb{N}}$, $(f_n)_{n\in\mathbb{N}}$, $(g_n)_{n\in\mathbb{N}}$, $(h_n)_{n\in\mathbb{N}}$ nonnegative real sequences with the conditions below:*

$$\alpha_n \leq f_n + g_n \sum_{s=0}^{n-1} h_s \alpha_s. \tag{3}$$

Then the following inequality holds

$$\alpha_n \leq f_n + g_n \sum_{s=0}^{n-1} h_s f_s \prod_{\tau=s+1}^{n-1} \left(h_\tau g_\tau + 1\right). \tag{4}$$

The main result is a generalisation of Theorem 1. For simplicity, we will not prove the existence of the solution.

Theorem 2 *Let H be real Hilbert-space, $A : H \to H$ a nonlinear operator, which satisfies the conditions below with some positive constants $M \geq m$:*

1. $\langle A(u) - A(v), u - v\rangle \geq m\|u - v\|^2$,
2. $\|A(u) - A(v)\| \leq M\|u - v\|$.

Let u^* denote the solution of the equation $A(u) = f$. For any $f, u_0 \in H$ $\alpha \in (0, 1)$, and $t \in \mathbb{R}^+$ small enough the following iteration converges to u^*.

$$u_{n+1} = \sum_{j=1}^{n+1} \binom{\alpha}{j} (-1)^{j+1} u_{n+1-j} - t\left[A(u_{n+1}) - f\right]. \tag{5}$$

Proof We first add $t\left[A(u_{n+1}) - f\right] - u^*$ both sides of the Eq. (5) and taking their norms, we have that

$$\left\|u_{n+1} - u^* + t\left[A(u_{n+1}) - A(u^*)\right]\right\| = \left\|\sum_{j=1}^{n+1} \binom{\alpha}{j} (-1)^{j+1} u_{n+1-j} - u^*\right\|. \tag{6}$$

Using the first assumption, we get the lower estimation

$$\begin{aligned} &\|u_{n+1} - u^* + t\left[A(u_{n+1}) - A(u^*)\right]\|^2 \\ &= \|u_{n+1} - u^*\|^2 + t^2\|A(u_{n+1}) - A(u^*)\|^2 + 2t\langle A(u_{n+1}) - A(u^*), u_{n+1} - u^*\rangle \\ &\geq \|u_{n+1} - u^*\|^2 + 2tm\|u_{n+1} - u^*\|^2 \geq \|u_{n+1} - u^*\|^2. \end{aligned} \tag{7}$$

It is also known that $\sum_{j=1}^{\infty} \binom{\alpha}{j}(-1)^{j+1} = 1$ and $\binom{\alpha}{j}(-1)^{j+1} > 0$. Using this, the triangle inequality and (6) for the inequality in (7) we get

$$\begin{aligned} \|u_{n+1} - u^*\| &\leq \left\|\sum_{j=1}^{n+1} \binom{\alpha}{j} (-1)^{j+1} u_{n+1-j} - u^*\right\| \\ &= \left\|\sum_{j=1}^{n+1} \binom{\alpha}{j} (-1)^{j+1} u_{n+1-j} - \sum_{j=1}^{\infty} \binom{\alpha}{j} (-1)^{j+1} u^*\right\| \\ &\leq \sum_{j=1}^{n+1} \binom{\alpha}{j} (-1)^{j+1} \|u_{n+1-j} - u^*\| + \sum_{j=n+2}^{\infty} \binom{\alpha}{j} (-1)^{j+1} \|u^*\|. \end{aligned} \tag{8}$$

Let $\alpha_n := \|u_n - u^*\|$, $f_n := \sum_{j=n+1}^{\infty} \binom{\alpha}{j}(-1)^{j+1}\|u^*\|$ and $\beta_n = \binom{\alpha}{n}(-1)^{n+1}$. With these, we can rewrite (8) as

$$\alpha_{n+1} \le f_{n+1} + \sum_{j=1}^{n+1} \beta_j \alpha_{n+1-j}. \tag{9}$$

Also using the notation h_j instead of β_{n+1-j}, (9) can be recognised as

$$\alpha_{n+1} \le f_{n+1} + \sum_{j=0}^{n} h_j \alpha_j. \tag{10}$$

Therefore, with $g_n := 1$ we can apply Lemma 1.

$$\alpha_{n+1} \le f_{n+1} + \sum_{s=0}^{n} h_s f_s \prod_{\tau=s+1}^{n} (h_\tau + 1). \tag{11}$$

Estimate $\prod_{\tau=s+1}^{n}(h_\tau + 1)$ as

$$\prod_{\tau=s+1}^{n} (h_\tau + 1) = \prod_{\tau=s+1}^{n} (\beta_{n+1-\tau} + 1)$$

$$\le \prod_{\tau=1}^{n}(\beta_{n+1-\tau} + 1) \le \left(\frac{n + \sum_{j=1}^{n} \beta_j}{n}\right)^n \le \left(1 + \frac{1}{n}\right)^n \le e.$$

Consequently, for (11) the following holds.

$$\alpha_{n+1} \le f_{n+1} + \sum_{s=0}^{n} h_s f_s \prod_{\tau=s+1}^{n} (h_\tau + 1) \le f_{n+1} + e \sum_{s=0}^{n} h_s f_s.$$

It is clear that if $n \to \infty$ then $f_{n+1} \to 0$. We prove that $\sum_{s=0}^{n} h_s f_s \to 0$.

$$\begin{aligned}
\sum_{s=0}^{n} h_s f_s &= \|u^*\|\beta_{n+1} + \|u^*\| \sum_{s=1}^{n} \beta_{n+1-s} \sum_{j=s+1}^{\infty} \beta_j \\
&= \|u^*\|\beta_{n+1} + \|u^*\| \sum_{s=1}^{n} \beta_{n+1-s}\Big(1 - \sum_{j=1}^{s} \beta_j\Big) \qquad (12) \\
&= \|u^*\|\beta_{n+1} + \|u^*\| \sum_{s=1}^{n} \beta_{n+1-s} - \|u^*\| \sum_{s=1}^{n}\sum_{j=1}^{n} \beta_{n+1-s}\beta_j.
\end{aligned}$$

Observe first, that the last term in (12) is a Cauchy product.

$$\lim_{n\to\infty}\Big(\sum_{s=1}^{n}\sum_{j=1}^{n}\beta_{n+1-s}\beta_j\Big) = \Big(\sum_{j=1}^{\infty}\beta_j\Big)^2 = 1.$$

Therefore, the first term in (12) tends to zero, the second and the third term to $\|u^*\|$, since $\sum_{j=1}^{\infty}\beta_j = 1$. This means that $\alpha_{n+1} \to 0$ if $n \to \infty$, which has been stated.

□

4 Discussion

In this work, we solved nonlinear time-independent equations of type $A(u) = f$, where the operator A is on a Hilbert space. We assumed that it is monotone and Lipschitz-continuous and we proved that the algorithm is convergent.

Our numerical experiences show that if we replace the time-derivative operator in the equation $\partial_t u = -[A(u) - f]$ with a fractional derivative, then it stabilizes the time integration in the numerical solutions. We have tested our method numerically in a fluid dynamical problem previously [10].

Acknowledgements This work was completed in the ELTE Institutional Excellence Program (1783-3/2018/FEKUTSRAT) supported by the Hungarian Ministry of Human Capacities. The project has also been supported by the European Union, co-financed by the Social Fund. EFOP-3.6.1-16-2016-0023.

References

1. Blumen, A., Zumofen, G., Klafter, J.: Transport aspects in anomalous diffusion: Lévy walks. Phys. Rev. A **40**(7), 3964–3973 (1989)
2. Bouchaud, J., Georges, A.: Anomalous diffusion in disordered media: statistical mechanisms, models and physical applications. Phys. Rep. **195**(4), 127–293 (1990)
3. Edwards, A.M., Phillips, R.A., Watkins, N.W., Freeman, M.P., Murphy, E.J., Afanasyev, V., Buldyrev, S.V., da Luz, M.G.E., Raposo, E.P., Stanley, H.E., Viswanathan, G.M.: Revisiting Lévy flight search patterns of wandering albatrosses, bumblebees and deer. Nature **449**, 1044–1048 (2007)
4. Kwaśnicki, M.: Ten equivalent definitions of the fractional laplace operator. Fract. Calc. Appl. Anal. **20**(1), 7–51 (2017)
5. Leibniz, G.W.: Mathematische Schriften. Georg Olms Verlagsbuchhandlung, Hildesheim (1962)
6. Metzler, R., Klafter, J.: The random walk's guide to anomalous diffusion: a fractional dynamics approach. Phys. Rep. **339**(1), 1–77 (2000)
7. Podlubny, I.: Fractional Differential Equations. In: Mathematics in Science and Engineering, vol. 198. Academic, San Diego (1999)
8. Sabatelli, L., Keating, S., Dudley, J., Richmond, P.: Waiting time distributions in financial markets. Phys. Condens. Matter **27**, 273–275 (2002)

9. Scalas, E., Gorenflo, R., Mainardi, F.: Fractional calculus and continuous-time finance. Phys. A Stat. Mech. Appl. **284**(1), 376–384 (2000)
10. Szekeres, B.J., Izsák, F.: Fractional derivatives for vortex simulations. ALGORITMY 2016: 20th Conference on Scientific Computing Vysoké Tatry-Podbanské, Slovakia March 13–18, Slovak University of Technology in Bratislava, 175–182 (2016)
11. Zeidler, E.: Nonlinear Functional Analysis and Its Applications: II/B: Nonlinear Monotone Operators. Springer, New York (1990)

Homogenization of the Heat Equation with a Vanishing Volumetric Heat Capacity

T. Danielsson and P. Johnsen

Abstract This paper is a study of the homogenization of the heat conduction equation, with a homogeneous Dirichlet boundary condition, having a periodically oscillating thermal conductivity and a vanishing volumetric heat capacity. In particular, the volumetric heat capacity equals ε^q and the thermal conductivity oscillates with period ε in space and ε^r in time, where $0 < q < r$ are real numbers. By using certain evolution settings of multiscale and very weak multiscale convergence we investigate, as ε tends to zero, how the relation between the volumetric heat capacity and the microscopic structure affects the homogenized problem and its associated local problem. It turns out that this relation gives rise to certain special effects in the homogenization result.

1 Introduction

We study, by means of periodic homogenization, the heat conduction equation with a homogeneous Dirichlet boundary condition. In particular we study

$$\begin{aligned} \varepsilon^q \partial_t u_\varepsilon(x,t) - \nabla \cdot \left(a\left(\frac{x}{\varepsilon}, \frac{t}{\varepsilon^r}\right) \nabla u_\varepsilon(x,t) \right) &= f(x,t) \text{ in } \Omega \times (0,T), \\ u_\varepsilon(x,0) &= u_0(x) \text{ in } \Omega, \\ u_\varepsilon(x,t) &= 0 \text{ on } \partial\Omega \times (0,T), \end{aligned} \tag{1}$$

where $0 < q < r$, $f \in L^2(\Omega_T)$, $u_0 \in L^2(\Omega)$ and Ω is an open bounded subset of $\mathbb{R}^N$ with smooth boundary $\partial\Omega$. Here, the thermal conductivity is characterized by the function a which is continuous on $\mathbb{R}^N \times \mathbb{R}$, periodic in its arguments with respect to the unit cube $Y = (0,1)^N$ and the unit interval $S = (0,1)$ respectively,

---

T. Danielsson · P. Johnsen (✉)
Department of Mathematics and Science Education, Mid Sweden University, Östersund, Sweden
e-mail: tatiana.danielsson@miun.se; pernilla.johnsen@miun.se

© Springer Nature Switzerland AG 2019

I. Faragó et al. (eds.), *Progress in Industrial Mathematics at ECMI 2018*,
Mathematics in Industry 30, https://doi.org/10.1007/978-3-030-27550-1_43

and satisfies the coercivity condition

$$a(y,s)\xi\cdot\xi \geq C_0|\xi|^2$$

for a.e. $(y,s) \in Y \times S$, for every $\xi \in \mathbb{R}^N$ and for some $C_0 > 0$. The coefficient ε^q in front of the time derivative represents the volumetric heat capacity.

As ε tends to zero, we search for a weak limit u to the sequence of solutions $\{u_\varepsilon\}$, where u is the solution to a so-called homogenized problem, which is in turn characterized by a local problem. The matching between the scales is, up to the authors' knowledge, new and extends the study in [4]. We believe that these results are of interest both for applications regarding heat conduction in heterogeneous media and for the further development of mathematical tools in homogenization theory.

Notations Here, $\Omega_T = \Omega \times (0,T)$, $\mathcal{Y}_{n,m} = Y^n \times S^m$ with $Y^n = Y_1 \times Y_2 \times \cdots \times Y_n$ and $S^m = S_1 \times S_2 \times \cdots \times S_m$, where $Y_1 = Y_2 = \ldots = Y_n = Y = (0,1)^N$ and $S_1 = S_2 = \ldots = S_m = S = (0,1)$. We let $y^n = y_1, y_2, \ldots, y_n$, $dy^n = dy_1dy_2\cdots dy_n$, $s^m = s_1, s_2, \ldots, s_m$ and $ds^m = ds_1ds_2\cdots ds_m$. For any function space $F(\Delta)$, $\Delta \subset \mathbb{R}^M$, $F(\Delta)/\mathbb{R}$ means the subspace of functions with integral mean value zero over Δ and $F_\sharp(\Delta)$ denotes periodicity over Δ. Note that $C_\sharp(\Delta) \subset C(\mathbb{R}^M)$ and $C_\sharp^\infty(\Delta) \subset C^\infty(\mathbb{R}^M)$. We let $W^{1,2}(0,T; H_0^1(\Omega), L^2(\Omega)) = \left\{v \in L^2(0,T; H_0^1(\Omega)) : \partial_t v \in L^2(0,T; H^{-1}(\Omega)\right\}$ and $\mathcal{W} = \left\{z \in L_\sharp^2(S; H_\sharp^1(Y)/\mathbb{R}) : \partial_s z \in L_\sharp^2(S; (H_\sharp^1(Y)/\mathbb{R})')\right\}$. $\varepsilon_k(\varepsilon)$ and $\varepsilon_j'(\varepsilon)$ for $k = 1, \ldots, n$, $j = 1, \ldots, m$, are strictly positive and tend to zero as ε does and we denote lists of spatial and temporal scales by $\{\varepsilon_1, \ldots, \varepsilon_n\}$ and $\left\{\varepsilon_1', \ldots, \varepsilon_m'\right\}$, respectively.

2 Preliminaries

Our main tools in this paper are evolution multiscale and very weak evolution multiscale convergence, which are generalizations and modifications of the classical concept of two-scale convergence.

Definition 1 (Evolution Multiscale Convergence) A sequence $\{u_\varepsilon\}$ in $L^2(\Omega_T)$ is said to $(n+1, m+1)$-scale converge to $u_0 \in L^2(\Omega_T \times \mathcal{Y}_{n,m})$ if

$$\lim_{\varepsilon\to 0}\int_{\Omega_T} u_\varepsilon(x,t)\, v\left(x,t,\frac{x}{\varepsilon_1},\cdots,\frac{x}{\varepsilon_n},\frac{t}{\varepsilon_1'},\cdots,\frac{t}{\varepsilon_m'}\right) dxdt$$

$$= \int_{\Omega_T}\int_{\mathcal{Y}_{n,m}} u_0\left(x,t,y^n,s^m\right) v\left(x,t,y^n,s^m\right) dy^n ds^m dxdt$$

for all $v \in L^2(\Omega_T; C_\sharp(\mathcal{Y}_{n,m}))$. This is denoted by

$$u_\varepsilon(x,t) \overset{n+1,m+1}{\rightharpoonup} u_0\left(x,t,y^n,s^m\right).$$

A compactness result for evolution multiscale convergence is given in the theorem below. For a definition of the concept of jointly separatedness, see [5].

Theorem 1 *Let $\{u_\varepsilon\}$ be a bounded sequence in $L^2(\Omega_T)$ and suppose that the lists $\{\varepsilon_1,\ldots,\varepsilon_n\}$ and $\{\varepsilon'_1,\ldots,\varepsilon'_m\}$ are jointly separated. Then, up to a subsequence,*

$$u_\varepsilon(x,t) \overset{n+1,m+1}{\rightharpoonup} u_0\left(x,t,y^n,s^m\right)$$

where $u_0 \in L^2(\Omega_T \times \mathcal{Y}_{n,m})$.

Proof See Theorem A.1 in [2].

The idea behind the following concept originates from [3].

Definition 2 (Very Weak Evolution Multiscale Convergence) A sequence $\{w_\varepsilon\}$ in $L^1(\Omega_T)$ is said to $(n+1,m+1)$-scale converge very weakly to $w_0 \in L^1(\Omega_T \times \mathcal{Y}_{n,m})$ if

$$\lim_{\varepsilon\to 0}\int_{\Omega_T} w_\varepsilon(x,t)\, v_1\left(x,\frac{x}{\varepsilon_1},\ldots,\frac{x}{\varepsilon_{n-1}}\right) v_2\left(\frac{x}{\varepsilon_n}\right) c\left(t,\frac{t}{\varepsilon'_1},\ldots,\frac{t}{\varepsilon'_m}\right) dxdt$$
$$= \int_{\Omega_T}\int_{\mathcal{Y}_{n,m}} w_0\left(x,t,y^n,s^m\right) v_1(x,y^{n-1}) v_2(y_n)\, c(t,s^m) dy^n ds^m dxdt$$

for any $v_1 \in C_0^\infty(\Omega; C_\sharp^\infty(Y^{n-1}))$, $v_2 \in C_\sharp^\infty(Y_n)/\mathbb{R}$ and $c \in C_0^\infty(0,T; C_\sharp^\infty(S^m))$, where $\int_{Y_n} w_0(x,t,y^n,s^m)\, dy_n = 0$. We write

$$w_\varepsilon(x,t) \underset{vw}{\overset{n+1,m+1}{\rightharpoonup}} w_0\left(x,t,y^n,s^m\right).$$

Remark 1 Since the integral mean value of w_0 is zero over Y_n, the limit is unique.

We give a gradient characterization and a compactness result for very weak evolution multiscale convergence, adapted to our problem.

Theorem 2 *Assume that $\{u_\varepsilon\}$ is bounded in $L^2(0,T; H_0^1(\Omega))$ and, for any $v_1 \in C_0^\infty(\Omega)$, $c_1 \in C_0^\infty(0,T)$, $c_2 \in C_\sharp^\infty(S)$ and $r > 0$,*

$$\lim_{\varepsilon\to 0}\int_{\Omega_T} u_\varepsilon(x,t)\, v_1(x)\, \partial_t\left(\varepsilon^r c_1(t)\, c_2\left(\frac{t}{\varepsilon^r}\right)\right) dxdt = 0. \tag{2}$$

Then, for $n = m = 1$ with $\varepsilon_1 = \varepsilon$ and $\varepsilon_1' = \varepsilon^r$, up to a subsequence,

$$u_\varepsilon(x,t) \rightharpoonup u(x,t) \text{ in } L^2(0,T;H_0^1(\Omega)),$$

$$\nabla u_\varepsilon(x,t) \overset{2,2}{\rightharpoonup} \nabla u(x,t) + \nabla_y u_1(x,t,y,s)$$

and

$$\varepsilon^{-1} u_\varepsilon(x,t) \underset{vw}{\overset{2,2}{\rightharpoonup}} u_1(x,t,y,s),$$

where $u \in L^2(0,T;H_0^1(\Omega))$ and $u_1 \in L^2(\Omega_T \times S; H_\sharp^1(Y)/\mathbb{R})$.

Proof See Theorem 2.7 and Theorem 2.10 in [4].

3 Homogenization

Let us now establish a homogenization result for Eq. (1). First we state the weak form of (1): for all $v \in H_0^1(\Omega)$ and $c \in C_0^\infty(0,T)$

$$\begin{aligned}\int_{\Omega_T} -\varepsilon^q u_\varepsilon(x,t)\,v(x)\,\partial_t c(t) + a\left(\frac{x}{\varepsilon},\frac{t}{\varepsilon^r}\right)\nabla u_\varepsilon(x,t)\cdot\nabla v(x)\,c(t)\,dxdt \\ = \int_{\Omega_T} f(x,t)\,v(x)\,c(t)\,dxdt.\end{aligned} \tag{3}$$

The weak form has a unique solution for every fixed $\varepsilon > 0$, see Section 23.7 in [6].

Theorem 3 *Let $\{u_\varepsilon\}$ be a sequence of solutions to (1) in $W^{1,2}(0,T;H_0^1(\Omega), L^2(\Omega))$. Then it holds that*

$$u_\varepsilon(x,t) \rightharpoonup u(x,t) \text{ in } L^2(0,T;H_0^1(\Omega)) \tag{4}$$

and

$$\nabla u_\varepsilon(x,t) \overset{2,2}{\rightharpoonup} \nabla u(x,t) + \nabla_y u_1(x,t,y,s), \tag{5}$$

where $u \in L^2(0,T;H_0^1(\Omega))$ and $u_1 \in L^2(\Omega_T \times S; H_\sharp^1(Y)/\mathbb{R})$. Here, u is the unique solution to the homogenized problem

$$\begin{aligned}-\nabla\cdot(b\nabla u(x,t)) &= f(x,t) \text{ in } \Omega_T, \\ u(x,t) &= 0 \text{ on } \partial\Omega \times (0,T)\end{aligned} \tag{6}$$

with, for $q < r < q + 2$,

$$b\nabla u\,(x,t) = \int_{\mathcal{Y}_{1,1}} a\,(y,s)\left(\nabla u\,(x,t) + \nabla_y u_1\,(x,t,y,s)\right)dyds \tag{7}$$

where $u_1 \in L^2(\Omega_T \times S; H^1_\sharp(Y)/\mathbb{R})$ *is determined by the elliptic local problem*

$$-\nabla_y \cdot \left(a\,(y,s)\left(\nabla u\,(x,t) + \nabla_y u_1\,(x,t,y,s)\right)\right) = 0, \tag{8}$$

for $r = q + 2$, $b\nabla u\,(x,t)$ *is given by (7) where* $u_1 \in L^2(\Omega_T; \mathcal{W})$ *is determined by the parabolic local problem*

$$\partial_s u_1\,(x,t,y,s) - \nabla_y \cdot \left(a\,(y,s)\left(\nabla u\,(x,t) + \nabla_y u_1\,(x,t,y,s)\right)\right) = 0 \tag{9}$$

and, for $r > q + 2$,

$$b\nabla u\,(x,t) = \int_Y \left(\int_S a\,(y,s)\,ds\right)\left(\nabla u\,(x,t) + \nabla_y u_1\,(x,t,y)\right)dy$$

where $u_1 \in L^2(\Omega_T; H^1_\sharp(Y)/\mathbb{R})$ *is given by the elliptic local problem*

$$-\nabla_y \cdot \left(\left(\int_S a\,(y,s)\,ds\right)\left(\nabla u\,(x,t) + \nabla_y u_1\,(x,t,y)\right)\right) = 0. \tag{10}$$

Proof The sequence of solutions $\{u_\varepsilon\}$ is bounded in $L^2(0,T; H^1_0(\Omega))$ and satisfies (2), see Section 3 in [1], hence Theorem 2 gives us (4) and (5). To derive the homogenized problem we choose, in the weak form (3), the test function

$$v\,(x)\,c\,(t) = v_1\,(x)\,c_1\,(t)\,,$$

where $v_1 \in H^1_0(\Omega)$ and $c_1 \in C^\infty_0(0,T)$. Letting ε tend to zero we have

$$\int_{\Omega_T}\int_{\mathcal{Y}_{1,1}} a\,(y,s)\left(\nabla u\,(x,t) + \nabla_y u_1\,(x,t,y,s)\right)\cdot\nabla v_1\,(x)\,c_1\,(t)\,dydsdxdt$$

$$= \int_{\Omega_T} f\,(x,t)\,v_1\,(x)\,c_1\,(t)\,dxdt,$$

and by the Variational Lemma we obtain the weak form of (6).

Now we continue by finding the local problem for each of the three cases.

Case 1 $q < r < q + 2$. In (3) we choose the test function

$$v\,(x)\,c\,(t) = \varepsilon v_1\,(x)\,v_2\left(\frac{x}{\varepsilon}\right)c_1\,(t)\,c_2\left(\frac{t}{\varepsilon^r}\right),$$

where $v_1 \in C_0^\infty(\Omega)$, $v_2 \in C_\sharp^\infty(Y)/\mathbb{R}$, $c_1 \in C_0^\infty(0,T)$ and $c_2 \in C_\sharp^\infty(S)$. Carrying out the differentiations and letting $\varepsilon \to 0$, omitting terms that equal zero, we obtain

$$\lim_{\varepsilon\to 0}\left(\int_{\Omega_T} -\varepsilon^{q+1-r} u_\varepsilon(x,t)\, v_1(x)\, v_2\left(\frac{x}{\varepsilon}\right) c_1(t)\, \partial_s c_2\left(\frac{t}{\varepsilon^r}\right) dxdt\right.$$
$$\left.+\int_{\Omega_T} a\left(\frac{x}{\varepsilon},\frac{t}{\varepsilon^r}\right)\nabla u_\varepsilon(x,t)\, v_1(x)\cdot\nabla_y v_2\left(\frac{x}{\varepsilon}\right) c_1(t)\, c_2\left(\frac{t}{\varepsilon^r}\right) dxdt\right) = 0. \tag{11}$$

By Theorem 2 we have

$$\int_{\Omega_T}\int_{\mathcal{Y}_{1,1}} a(y,s)\left(\nabla u(x,t)+\nabla_y u_1(x,t,y,s)\right) v_1(x)\cdot\nabla_y v_2(y)\, c_1(t)\, c_2(s)\, dydsdxdt = 0$$

and applying the Variational Lemma we arrive at the weak form of (8).

Case 2 $r = q+2$. Taking the same test functions as in Case 1, we again arrive at (11) and, passing to the limit, Theorem 2 gives us

$$\int_{\Omega_T}\int_{\mathcal{Y}_{1,1}} -u_1(x,t,y,s) v_1(x)\, v_2(y)\, c_1(t)\partial_s c_2(s) dydsdxdt$$
$$+\int_{\Omega_T}\int_{\mathcal{Y}_{1,1}} a(y,s)\left(\nabla u(x,t)+\nabla_y u_1(x,t,y,s)\right) v_1(x)\cdot\nabla_y v_2(y) c_1(t)\, c_2(s) dyds = 0.$$

By using the Variational Lemma we get the weak form of (9).

Case 3 $r > q+2$. Before we derive the local problem for this case we establish independence of s in u_1. We choose the test function

$$v(x)\, c(t) = \varepsilon^{r-q-1} v_1(x)\, v_2\left(\frac{x}{\varepsilon}\right) c_1(t)\, c_2\left(\frac{t}{\varepsilon^r}\right),$$

where $v_1 \in C_0^\infty(\Omega)$, $v_2 \in C_\sharp^\infty(Y)/\mathbb{R}$, $c_1 \in C_0^\infty(0,T)$ and $c_2 \in C_\sharp^\infty(S)$, in the weak form (3). Carrying out the differentiations and letting $\varepsilon \to 0$, applying Theorem 2, we get

$$\int_{\Omega_T}\int_{\mathcal{Y}_{1,1}} -u_1(x,t,y,s)\, v_1(x)\, v_2(y)\, c_1(t)\, \partial_s c_2(s)\, dydsdxdt = 0.$$

From the Variational Lemma we deduce that u_1 is independent of s. Now, to find the local problem, in (3) we choose the test function

$$v(x)\, c(t) = \varepsilon v_1(x)\, v_2\left(\frac{x}{\varepsilon}\right) c_1(t),$$

where $v_1 \in C_0^\infty(\Omega)$, $v_2 \in C_\sharp^\infty(Y)/\mathbb{R}$ and $c_1 \in C_0^\infty(0,T)$. Carrying out the differentiations and letting $\varepsilon \to 0$, omitting terms that tend to zero we arrive at

$$\lim_{\varepsilon\to 0}\int_{\Omega_T} a\left(\frac{x}{\varepsilon},\frac{t}{\varepsilon^r}\right)\nabla u_\varepsilon(x,t)\, v_1(x)\cdot\nabla_y v_2\left(\frac{x}{\varepsilon}\right)c_1(t)\,dxdt = 0.$$

Theorem 2 and the fact that u_1 is independent of s give

$$\int_{\Omega_T}\int_{\mathcal{Y}_{1,1}} a(y,s)\left(\nabla u(x,t)+\nabla_y u_1(x,t,y)\right)v_1(x)\cdot\nabla_y v_2(y)\,c_1(t)\,dydsdxdt = 0$$

and by applying the Variational Lemma we arrive at the weak form of (10).

Remark 2 We refer the interested reader to an extended version of this paper, [1].

References

1. Danielsson, T., Johnsen, P.: Homogenization of the heat equation with a vanishing volumetric heat capacity (2018). arXiv: 1809.11019
2. Flodén, L., Holmbom, A., Olsson Lindberg, M., Persson, J.: Homogenization of parabolic equations with an arbitrary number of scales in both space and time. J. Appl. Math. **2014**, 16 pp. (2014)
3. Holmbom, A.: Homogenization of parabolic equations: an alternative approach and some corrector-type results. Appl. Math. **42**, 321–343 (1997)
4. Johnsen, P., Lobkova, T.: Homogenization of a linear parabolic problem with a certain type of matching between the microscopic scales. Appl. Math. **63**, 503–521 (2018)
5. Persson, J.: Homogenisation of monotone parabolic problems with several temporal scales. Appl. Math. **57**, 191–214 (2012)
6. Zeidler, E.: Nonlinear functional analysis and its applications II/A: linear monotone operators. Springer, New York (1990)

Mathematical Analysis for a Class of Partial Differential Equations with Dynamic Preisach Model

Alfredo Bermúdez, Dolores Gómez, and Pablo Venegas

Abstract This work deals with the mathematical analysis and numerical solution of a parabolic problem with dynamic hysteresis motivated by electromagnetic field equations. In this case, the values of the magnetic induction depend not only on the current values of the magnetic field, but also on the previous ones and on the velocity at which they have been attained. The hysteresis is modelled by the dynamic Preisach operator. Based upon the definition of dynamic relay, which is introduced and formalized as the solution of a multi-valued ordinary differential equation, the definition of the dynamic Preisach operator is recalled and some of their main properties established. Under suitable assumptions, the well-posedness of a weak formulation of the initial problem is shown and a numerical solution computed.

1 Introduction

The performance of electric machines is mainly influenced by energy losses that are due to the magnetic field variations in the ferromagnetic materials composing the core of the engine. These materials usually present hysteretic behaviour that is reflected in the magnetization curves describing the magnetic response of the material to an applied magnetic field. Building a mathematical model of this relation is a very important and difficult task and that is why the mathematical modelling and numerical simulation of devices involving ferromagnetic materials is still quite

A. Bermúdez · D. Gómez
Departamento de Matemática Aplicada and Instituto de Matemáticas (IMAT), Universidade de Santiago de Compostela, Santiago de Compostela, Spain

Instituto Tecnológico de Matemática Industrial (ITMATI), Edif. Instituto Investigaciones Tecnológicas, Santiago de Compostela, Spain
e-mail: alfredo.bermudez@usc.es; mdolores.gomez@usc.es

P. Venegas (✉)
GIMNAP, Department of Mathematics, Universidad del Bío-Bío, Concepción, Chile
e-mail: pvenegas@ubiobio.cl

© Springer Nature Switzerland AG 2019
I. Faragó et al. (eds.), *Progress in Industrial Mathematics at ECMI 2018*, Mathematics in Industry 30, https://doi.org/10.1007/978-3-030-27550-1_44

a challenge. From the electrical engineering point of view, having a good hysteresis model is essential to correctly estimate the losses, and mainly the so-called hysteresis and excess losses. In particular, to take into account the excess losses the classical (rate-independent) hysteresis models are not well-suited and *dynamic models* including the effect of the speed of changes of the applied field are needed, since they can reflect the dependence of the magnetic response with frequency or field waveform. In this context, this work deals with the mathematical analysis and numerical solution of a parabolic problem with dynamic hysteresis. Although the study is motivated by the electromagnetic analysis of electric machines, we will state the problem in a general abstract framework. So, first we briefly recall the definition of rate-dependent relay and that of dynamic Preisach operator and introduce some of its main properties. Next, the abstract parabolic problem with dynamic hysteresis is posed and an existence result stated. Finally, the numerical solution is computed and some results are included in order to illustrate the behaviour of the numerical solution for different configurations of the dynamic Preisach model.

2 The Dynamic Preisach Operator

The Preisach model describes the hysteresis using a superposition of elementary hysteresis operators called *relay* operators. The model assumes that the material consists of an infinite number of (magnetic) particles each one characterized by a relay, so the whole system can be modelled by a weighted parallel connections of these relays. The weight function, called Preisach density function, works as a local influence of each operator in the overall hysteresis model and it is estimated from measured data.

In the case of the *dynamic Preisach model* (see [2]) the relay is a *dynamic relay*, here denoted by η_ρ, that can be formally defined as follows: for a fixed $\rho = (\rho_1, \rho_2) \in \mathbb{R}^2$, $\rho_1 < \rho_2$, $\eta_\rho : \mathrm{L}^1(0,T) \times [-1,1] \to \mathrm{W}^{1,1}(0,T)$ such that, for any $u \in \mathrm{L}^1(0,T)$ and $\xi \in [-1,1]$, $\eta_\rho(u,\xi) : [0,T] \to [-1,1]$ is the unique function $y \in \mathrm{W}^{1,1}(0,T)$, $y(t) \in [-1,1]$ that solves the nonlinear Cauchy problem (see [1]):

$$\frac{dy}{dt}(t) = F(t, y(t)) := \begin{cases} k(u(t)-\rho_2)^+ - k(u(t)-\rho_1)^- & \text{if } -1 < y(t) < 1, \\ 0 & \text{if } y(t) = -1 \text{ and } u(t) \le \rho_2, \\ k(u(t)-\rho_2) & \text{if } y(t) = -1 \text{ and } u(t) \ge \rho_2, \\ k(u(t)-\rho_1) & \text{if } y(t) = 1 \text{ and } u(t) \le \rho_1, \\ 0 & \text{if } y(t) = 1 \text{ and } u(t) \ge \rho_1, \end{cases} \tag{1}$$

$$y(0) = \xi, \tag{2}$$

where k is a material-dependent parameter and we have used the standard notations $x^+ = \max\{x, 0\}$ and $x^- = \max\{-x, 0\}$.

Now, given $\rho_0 > 0$, we consider the *Preisach triangle* $\mathscr{T} := \{\rho = (\rho_1, \rho_2) \in \mathbb{R}^2 : -\rho_0 \le \rho_1 \le \rho_2 \le \rho_0\}$ and a Preisach density function $p \in \mathrm{L}^1(\mathscr{T})$ with $p > 0$. We introduce $Y := \left\{ v \in L^1_p(\mathscr{T}) : |v(\rho)| \le 1, \text{ a.e. } \rho \in \mathscr{T} \right\}$ where $L^1_p(\mathscr{T}) := \{\xi : \mathscr{T} \to \mathbb{R}$ Lebesgue-measurable such that $\int_{\mathscr{T}} |\xi(\rho)| p(\rho)\, d\rho < \infty\}$ endowed with the norm $\|\xi\|_{L^1_p(\mathscr{T})} := \int_{\mathscr{T}} |\xi(\rho)| p(\rho)\, d\rho$. Then, the dynamic Preisach operator $\mathscr{F}_D : \mathrm{L}^1(0, T) \times Y \to \mathrm{W}^{1,1}(0, T)$ is given by (see [2])

$$[\mathscr{F}_D(u, \xi)](t) = \int_{\mathscr{T}} [\eta_\rho(u, \xi(\rho))](t) p(\rho)\, d\rho, \tag{3}$$

where ξ contains information about the "initial state" of magnetization at each point and η_ρ is the dynamic relay.

Finally, since we are interested in the mathematical analysis and computation of distributed electromagnetic models, following [4] we introduce a space-time dependent operator $\mathscr{F} : \mathrm{L}^1(0, T; \mathrm{L}^2(\Omega)) \times \mathrm{L}^2(\Omega; Y) \to \mathrm{L}^\infty(0, T; \mathrm{L}^2(\Omega))$ as follows: given a time dependent input field $u(x, \cdot) \in \mathrm{L}^2(0, T)$ and an initial state field $\xi(x) \in Y$, we set

$$[\mathscr{F}(u, \xi)](x, t) := [\mathscr{F}_D(u(x, \cdot), \xi(x))](t), \quad \text{a.e. in } [0, T] \times \Omega, \tag{4}$$

where $\mathrm{L}^2(\Omega; Y) \subset \mathrm{L}^2(\Omega; \mathrm{L}^1_p(\mathscr{T}))$ is the space of all functions v such that $\|v\|^2_{\mathrm{L}^2(\Omega;\mathrm{L}^1_p(\mathscr{T}))} := \int_\Omega \|v\|^2_{\mathrm{L}^1_p(\mathscr{T})} < \infty$. The proof of the following lemma can be seen in [1].

Lemma 1 *The dynamic Preisach operator $\mathscr{F}$ is Lipschitz-continuous in the following sense: there exists $C > 0$ such that, for all $u, v \in \mathrm{L}^1(0, T; \mathrm{L}^2(\Omega))$ and $\xi \in \mathrm{L}^2(\Omega; Y)$, $\|\mathscr{F}(u, \xi) - \mathscr{F}(v, \xi)\|_{\mathrm{L}^\infty(0,T;\mathrm{L}^2(\Omega))} \le C \left(\|u - v\|_{\mathrm{L}^1(0,T;\mathrm{L}^2(\Omega))}\right)$.*

3 Transient Eddy Current Problem with Dynamic Hysteresis

Let $T > 0$ and $\Omega \in \mathbb{R}^d$, $d = 2, 3$ be a bounded domain with smooth boundary $\Gamma = \partial\Omega$. Let $V \subset H$ be two Hilbert spaces of scalar functions defined in Ω with continuous, dense, compact embedding. Then $V \subset H \equiv H' \subset V'$. We consider a mapping $a : (0, T) \times V \times V \to \mathbb{R}$ such that $a(t, \cdot, \cdot)$ is bilinear a.e. $t \in (0, T)$. Let $\mathscr{F}$ be the dynamic hysteresis operator defined by (4).We are interested in the mathematical analysis of the following parabolic problem:

Find $u \in \mathrm{L}^2(0, T; V) \cap \mathrm{L}^\infty(0, T; H)$ with $\partial_t u \in \mathrm{L}^2(0, T; V')$ and $w \in \mathrm{L}^2(0, T; H)$ with $\partial_t w \in \mathrm{L}^2(0, T; V')$, such that

$$\langle \partial_t u + \partial_t w, v\rangle_{V,V'} + a(t, u, v) = \langle f, v\rangle_{V,V'} \quad \forall v \in V, \quad \text{a.e. in } (0, T], \tag{5a}$$

$$w = \mathscr{F}(u, \xi) \quad \text{in } \Omega \times [0, T], \tag{5b}$$

$$(u + w)(0) = u_0 + w_0 \quad \text{in } \Omega. \tag{5c}$$

The next theorem shows the existence of solution to problem (5). The proof is carried out through three different steps: time discretization, a priori estimates and passage to the limit by using compactness (see [1] for the proof).

Theorem 1 *Let us assume that H.1, H.2 below hold true. Then, problem* (5) *has a solution.*

H.1 $a(\cdot, u, v)$ *is a continuous form in* $V \times V$ *which is Lipschitz continuous in* t *and, for some constants* $\lambda, \gamma \geq 0$, *satisfies the Gårding's inequality*

$$a(t, v, v) + \lambda \|G\|_H^2 \geq \gamma \|G\|_V^2 \quad \forall v \in V, \quad \forall t \in [0, T]. \tag{6}$$

H.2 f *belongs to* $\mathrm{H}^1(0, T; V')$, $u_0 \in V$ *and* $w_0 := \mathscr{F}(u_0) \in H$.

The previous result can be applied to the eddy current model

$$\langle \partial_t u + \partial_t w, v \rangle_{V,V'} + \left(\sigma^{-1} \nabla u, \nabla v\right) = 0 \quad \forall v \in V, \tag{7a}$$

$$w = \mathscr{F}(u, \xi) \quad \text{in } \Omega \times (0, T), \tag{7b}$$

$$u = g \quad \text{in } \partial\Omega, \tag{7c}$$

where $V = \mathrm{H}_0^1(\Omega)$ and $H = \mathrm{L}^2(\Omega)$. Here u represents the magnetic field, w is the magnetic induction, g depends on the current intensity and σ is the electrical conductivity. This problem arises in the computation of 2D electromagnetic fields in a cross-section of laminated media (see [3]) and it is important for the evaluation of the electromagnetic losses.

4 Numerical Approximation and Examples

Let $\mathscr{V}_h$ be the space of continuous piecewise linear finite elements on triangular meshes $\{\mathscr{T}_h\}_{h>0}$ of Ω and $\mathscr{V}_h^0 := \mathscr{V}_h \cap \mathrm{H}_0^1(\Omega)$, where h denotes the mesh size. We introduce a uniform partition $\{t^i := i\Delta t, i = 0, \ldots, m\}$ of $[0, T]$, with time step $t := T/m, m \in \mathbb{N}$. By using $\mathscr{V}_h$ for the spatial discretization and the backward Euler scheme for time discretization, we are led to the following approximation of problem (7): Given $u_h^0 = w_h^0 = 0$ in Ω, find $u_h^n \in \mathscr{V}_h$ and $w_h^n \in \mathscr{V}_h$, $n = 1, \ldots, m$, satisfying

$$\left(u_h^n + w_h^n, v_h\right) + \Delta t \left(\sigma^{-1} \nabla u_h^n, \nabla v_h\right) = (u_h^{n-1} + w_h^{n-1}, v_h) \quad \forall v_h \in \mathscr{V}_h^0, \tag{8a}$$

$$w_h^n = [\mathscr{F}(u_{\Delta t^n}^h, \xi)](t^n) \quad \text{in } \Omega, \tag{8b}$$

$$u_h^n = g_h^n \quad \text{on } \Gamma, \tag{8c}$$

where the piecewise linear function g_h^n is a convenient approximation of $g(t^n)$, $n = 1, \ldots, m$ and $u_{\Delta t^n}^h$ is the piecewise linear in time interpolant of $\{u_h^i\}_{i=0}^n$.

Remark 1 To solve the previous problem we need to evaluate (8b) at different quadrature nodes $P \in \Omega$. Thus, let $P := (x, y)$ and $n \in \{1, \ldots, m\}$. At time step n, the values $u_h^i(P)$, $i = 0, \ldots, n-1$ which have been previously computed, represent the history of the fully discrete problem at point P. Because of the latter, we may define the nonlinear $\mathscr{G} : \mathbb{R} \to \mathbb{R}$ function by

$$\mathscr{G}(x) := [\mathscr{F}(u^x, \xi)](t^n), \quad u^x(t^i) = u_h^i(P), i = 0, \ldots, n-1, \qquad u^x(t^n) = x,$$

with u^x a continuous piecewise linear function. As an example, we compute $\mathscr{G}(x)$ by defining the function p as the Factorized-Lorentzian distribution:

$$p(\rho_1, \rho_2) := N \left(1 + \left(\frac{\rho_2 - \omega}{\gamma\omega}\right)^2\right)^{-1} \left(1 + \left(\frac{\rho_1 + \omega}{\gamma\omega}\right)^2\right)^{-1} \tag{9}$$

with $N = 1/2000$, $\omega = 5$ and $\gamma = 4$. By taking $H(t) = 200\sin(2\pi t/t_f)$, $t \in [0, t_f]$, as history function we compute $\mathscr{G}$ for different values of t_f ($1/t_f$ being the frequency or velocity of H) and slopes values k. The corresponding $\mathscr{G}$ curves for each of these values are shown in Fig. 1. Notice that the shape of the curve highly depends on the values of k and the velocity of the history, ranging from a linear behaviour to a non-differentiable function.

To deal with the non-linear problem which must be solved at each time step of the above algorithm, a Newton-like method has been considered.

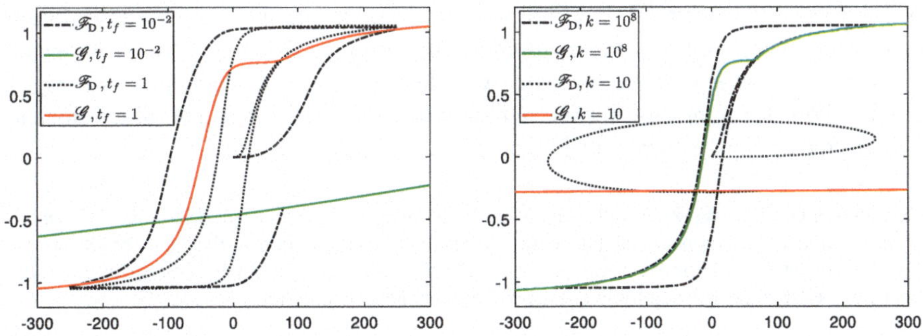

Fig. 1 $\mathscr{G}$ and $\mathscr{F}_D(H)$ for different values of $t_f \in \{10^{-2}, 1\}$ (left; $k = 50$) and slopes values $k \in \{10, 10^8\}$ (right; $t_f = 10^{-3}$). The dashed and dotted lines represent the hysteresis cycle of the history (H-$\mathscr{F}_D(H)$). The solid lines corresponds to the curve $\mathscr{G}(x)$

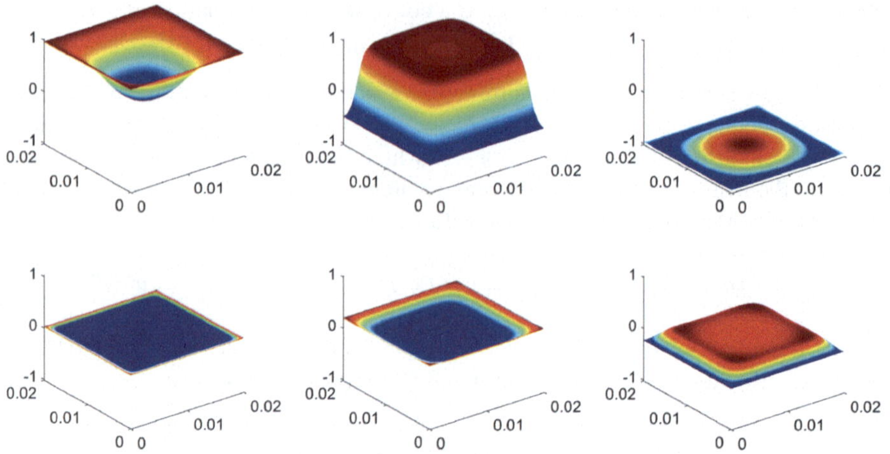

$t = 0.002$s $t = 0.0055$s $t = 0.01$s

Fig. 2 w-field solution to problem (8) for $k = 10^3$ (top) and $k = 1$ (bottom)

4.1 *Numerical Solution for Different k-Values*

In this section we illustrate the behaviour of the numerical solution to problem (8) for different configurations of the dynamic Preisach model characterized by the Factorized-Lorentzian distribution (9). The dynamic relay and, accordingly, the dynamic Preisach operator, varies with respect to the velocity of the input and the relay slope k. Problem (8) has been solved, for different values of k, in the domain $\Omega = [0, 0.02]^2$ along the time interval $[0, 0.01]$, with $\sigma = 100$ and $g(x, y, t) = 200 \sin(2\pi t/0.01)$. Field solution w is shown in Fig. 2. From this figure we deduce that changes in w are smaller when $k = 1$. This behaviour is expected as the size of the $u - w$ cycle decreases when the slope, k, decreases (see Fig. 1 (right)). This is not the case for the w field when $k = 10^3$: it reaches values close to saturation (see Figs. 2 (top) and 1 (left)).

Acknowledgements Work partially supported by Ministerio de Economía, Industria y Competitividad (MINECO), Agencia Estatal de Investigación (AEI) and Fondo Europeo de Desarrollo Regional (FEDER) under the research project MTM2017-86459-R, Xunta de Galicia (Spain) under grant 2017 GRC GI-1563 and by FONDECYT project 11160186 (Chile).

References

1. Bermúdez, A., Gómez, D., Venegas, P.: Mathematical analysis and numerical solution of models with dynamic Preisach hysteresis. J. Comput. Appl. Math. **367**, 112452 (2020). https://doi.org/10.1016/j.cam.2019.112452
2. Bertotti, G.: Dynamic generalization of the scalar Preisach model of hysteresis. IEEE Trans. Magn. **28**(5), 2599–2601 (1992)
3. Van Keer, R., Dupré, L., Melkebeek, J.A.A.: On a numerical method for 2D magnetic field computations in a lamination with enforced total flux. J. Comput. Appl. Math. **72**, 179–191 (1996)
4. Visintin, A.: Differential Models of Hysteresis. Applied Mathematical Sciences, vol. 111. Springer, Berlin (1994)

Some Embedded Pairs for Optimal Implicit Strong Stability Preserving Runge–Kutta Methods

Imre Fekete and Ákos Horváth

Abstract We construct specific embedded pairs for second and third order optimal strong stability preserving implicit Runge–Kutta methods with large absolute stability regions. These pairs offer adaptive implementation possibility for strong stability preserving (SSP) methods and maintain their inherent nonlinear stability properties, too.

1 Introduction and SSP Runge–Kutta Methods

Let us consider an initial value problem (IVP)

$$y'(t) = f(t, y(t)), \qquad y(t_0) = y_0. \tag{1}$$

The numerical solution of (1) at each time step with an implicit s-stage Runge–Kutta (RK) method RK(A, b^T) is given by

$$y_{n+1} = y_n + \Delta t \sum_{j=1}^{s} b_j f(t_n + c_j \Delta t, Y_j) \tag{2}$$

and the internal stages are computed as

$$Y_i = y_n + \Delta t \sum_{j=1}^{s} a_{ij} f(t_n + c_j \Delta t, Y_j), \qquad i = 1, \ldots, s \tag{3}$$

I. Fekete (✉)
Institute of Mathematics, Eötvös Loránd University, MTA-ELTE Numerical Analysis and Large Networks Research Group, Budapest, Hungary
e-mail: feipaat@cs.elte.hu

Á. Horváth
Institute of Mathematics, Eötvös Loránd University, Budapest, Hungary

© Springer Nature Switzerland AG 2019

I. Faragó et al. (eds.), *Progress in Industrial Mathematics at ECMI 2018*, Mathematics in Industry 30, https://doi.org/10.1007/978-3-030-27550-1_45

where y_n is an approximation to the solution of (1) at time $t_n = t_0 + n\Delta t$, $A = (a_{ij})$ and $b^T = (b_j)$ are the coefficient of the method. By using the method-of-line approach, spatial discretization of hyperbolic partial differential equations (PDEs) lead to a large system of ordinary differential equations (ODEs)

$$u_t = F(u), \tag{4}$$

where u is a vector of approximations to the exact solution of the PDE. SSP time discretization methods were designed to ensure nonlinear stability properties in (4). We assume that the semi-discretization (4) and a convex functional $||\cdot||$ (or norm, semi-norm) are given, and that there exists a Δt_{FE} such that the forward Euler condition

$$||u + \Delta t F(u)|| \leq ||u|| \text{ for } 0 \leq \Delta t \leq \Delta t_{\mathrm{FE}} \tag{5}$$

holds for all u. An implicit Runge–Kutta (IRK) method is called SSP if the estimate

$$||u_{n+1}|| \leq ||u_n||$$

holds for the numerical solution of (4), whenever (5) holds and $\Delta t \leq C\Delta t_{\mathrm{FE}}$. The constant C is called the SSP coefficient. For a complete introduction into the SSP theory we recommend monograph [2]. Below we give the main results which will be used in this paper.

Theorem 1 ([2, Theorem 3.2]) *Let us consider the matrix*

$$K = \begin{pmatrix} A & 0 \\ b^T & 0 \end{pmatrix}$$

and the SSP conditions

$$K(I + rK)^{-1} \geq 0 \tag{6a}$$

$$rK(I + rK)^{-1}e \leq e. \tag{6b}$$

Then, the SSP coefficient of the IRK method is

$$C(A, b^T) = \sup\left\{ r : (I + rK)^{-1} \textit{exists and conditions } \text{(6a)–(6b)} \textit{ hold} \right\}.$$

Theorem 2 ([2, Observation 5.2]) *Consider an IRK method. If the method has positive SSP coefficient $C(A, b^T)$, then $A \geq 0$ and $b^T \geq 0$.*

It has been showed that IRK methods with positive C cannot exist for $p > 6$ [1]. Therefore, we are interested in taking into account order conditions up to order of six.

By using embedded pairs we could allow adaptive step-size control based on local truncation error estimation [3]. The general s-stage IRK pair RK$(A, b^T, \tilde{b}^T)$ of order $p(p-1)$ has the following extended Butcher tableau.

$$\begin{array}{c|c} c & A \\ \hline & b^T \\ & \tilde{b}^T \end{array}$$

As usual, $c = (c_1, c_2, \ldots, c_s)^T$ is given by $c = A\mathbf{e}$ with $\mathbf{e} = (1, \ldots, 1)^T \in \mathbb{R}^s$. The vectors b^T, $\tilde{b}^T$ define the coefficients of the p-th and $(p-1)$-th order approximations, respectively. Motivation for providing embedded pairs for SSP methods is that several optimal implicit SSP methods have useful stability regions, small error coefficients, big absolute monotonicity radius and are frequently used even when SSP theory cannot be applied. In the next section, we give the analytical framework that enables us to construct the new family of embedded pairs and construct the embedded pairs analytically and numerically for second and third order optimal implicit SSP RK methods.

2 Embedded Pairs for Second and Third Order Implicit SSP RK Methods

We introduce the notation SSPIRK(s, p) for optimal implicit SSP RK methods, where s and p refer to the number of stages and order, respectively. We give below the desired properties for embedded pairs.

(i) The embedded method is order of $p-1$.
(ii) The embedded method is non-defective, i.e. it violates all of the p-th order conditions.
(iii) The embedded method has rational coefficients and simple structure.
(iv) The embedded method has maximum SSP coefficient $\tilde{\mathcal{C}}$, where $\tilde{\mathcal{C}}$ is the SSP coefficient of the optimal SSPIRK method; if this is not the case, then we are looking for embedded SSPIRK methods with smaller SSP coefficient or simply embedded IRK methods.

Taking into account the desired properties (i)–(iv), we seek an embedded pair $\tilde{b}^T$, with the stage coefficient A from a SSPIRK method such that these satisfy the following optimization problem

$$\text{the appropriate order conditions and property (ii) are fulfilled,} \tag{7}$$

$$\begin{pmatrix} A & 0 \\ \tilde{b}^T & 0 \end{pmatrix} \left(I + \tilde{\mathcal{C}} \begin{pmatrix} A & 0 \\ \tilde{b}^T & 0 \end{pmatrix} \right)^{-1} \geq 0, \tag{8}$$

$$\left\| \tilde{\mathcal{C}} \begin{pmatrix} A & 0 \\ \tilde{b}^T & 0 \end{pmatrix} \left(I + \tilde{\mathcal{C}} \begin{pmatrix} A & 0 \\ \tilde{b}^T & 0 \end{pmatrix} \right)^{-1} \right\|_\infty \leq 1, \tag{9}$$

where (8)–(9) are equivalent with (6a)–(6b) and $||\cdot||_\infty$ denotes the induced matrix norm. Since we fix $\tilde{C}$ therefore we have a simplified optimization problem (7)–(9). Due to Theorem 2 and the first order condition $\tilde{b}^T\mathbf{e} = 1$ we have the componentwise condition $0 \leq \tilde{b}^T \leq \mathbf{e}$. The newly constructed pairs should satisfy desired properties (i)–(iv) and should have large absolute stability regions.

2.1 Embedded Pairs for SSPIRK(s,2) Methods

The s-stage second order characterization was given by Gottlieb et al. [4]. The methods have $C = 2s$. The Butcher form of SSPIRK$(s, 2)$ methods is given in Table 1. Taking into account desired properties (i)–(iv) it turns out that for general s we cannot find embedded pairs with maximal $\tilde{C}$.

Table 1 Butcher form of SSPIRK$(s, 2)$ methods

$\frac{1}{2s}$	$\frac{1}{2s}$				
$\frac{3}{s}$	$\frac{1}{s}$	$\frac{1}{2s}$			
$\frac{5}{s}$	$\frac{1}{s}$	$\frac{1}{s}$	$\frac{1}{2s}$		
$\vdots$	$\vdots$	$\vdots$	$\ddots$	$\ddots$	
$\frac{2s-1}{2s}$	$\frac{1}{s}$	$\frac{1}{s}$	$\dots$	$\frac{1}{s}$	$\frac{1}{2s}$
	$\frac{1}{s}$	$\frac{1}{s}$	$\dots$	$\frac{1}{s}$	$\frac{1}{s}$

Theorem 3 *There is no first order embedded pair for SSPIRK*$(2, 2)$ *with properties (i)–(iv).*

Based on Theorem 3 and its generalization one can conclude that there isn't first order embedded pair with $\tilde{C} = 2s$ for SSPIRK$(s, 2)$. Therefore we are interested in giving embedd pairs with smaller $\tilde{C}$. Namely we are looking for $\tilde{C} = s$ and our numerical search suggested the following pairs satisfying the desired properties (i)–(iv).

$$\tilde{b}_1^T = \left(\frac{2}{s+1}, \dots, \frac{2}{s+1}, \frac{3}{s+1}\right)^T, \ \tilde{b}_2^T = \left(\frac{1}{s}, \dots, \frac{1}{s}, \frac{5}{4s}, \frac{3}{4s}\right)^T$$

$$\tilde{b}_3^T = \left(\frac{1}{s}, \dots, \frac{1}{s}, \frac{13}{12s}, \frac{10}{12s}, \frac{10}{12s}, \frac{15}{12s}\right)^T$$

Based on absolute stability region measurements it is obvious that embedded pair $\tilde{b}_2^T$ is recommended. Below we present a result for $s = 4$ on Fig. 1 but as we are increasing the number of stages we can see similar results

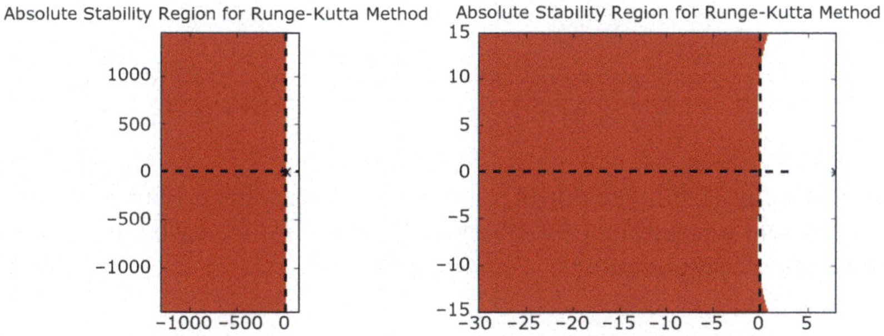

Fig. 1 The left and right plots correspond to the absolute stability region of SSPIRK(4, 2) and its $\tilde{b}_2^T$ embedded pair

2.2 *Embedded Pairs for SSPIRK(s, 3) Methods*

The s-stage third order characterization was also given by Gottlieb et al. [4]. The methods have $\mathcal{C} = s - 1 + \sqrt{s^2 - 1}$. The Butcher form of SSPIRK$(s, 3)$ methods is given in Table 2.

Table 2 Butcher form of SSPIRK$(s, 3)$ methods

β_1	β_1				
$2\beta_1 + \beta_2$	$\beta_1 + \beta_2$	β_1			
$3\beta_1 + 2\beta_2$	$\beta_1 + \beta_2$	$\beta_1 + \beta_2$	β_1		
$\vdots$	$\vdots$	$\vdots$	$\ddots$	$\ddots$	
$s\beta_1 + (s-1)\beta_2$	$\beta_1 + \beta_2$	$\beta_1 + \beta_2$	$\dots$	$\beta_1 + \beta_2$	β_1
	$\frac{1}{s}$	$\frac{1}{s}$	$\dots$	$\frac{1}{s}$	$\frac{1}{s}$

Where

$$\beta_1 = \frac{1}{2}\left(1 - \sqrt{\frac{s-1}{s+1}}\right) \text{ and } \beta_2 = \frac{1}{2}\left(\sqrt{\frac{s+1}{s-1}} - 1\right).$$

Similarly to the SSPIRK$(s, 2)$ case after tedious calculations one can see for lower stages that the desired properties (i)–(iv) cannot be satisfied with the maximal $\tilde{\mathcal{C}}$ coefficient. However, if we consider $\tilde{\mathcal{C}} = \mathcal{C}/2$ then we could give general form for SSPIRK$(s, 3)$ methods with desired properties (i)–(iv). These pairs are

$$\tilde{b}_1^T = \left(\frac{1}{\sqrt{s^2-1}}, \dots, \frac{1}{\sqrt{s^2-1}}, \frac{s - 1 - \frac{s-2}{s-1}\sqrt{s^2-1}}{2}, \frac{3 - s + \frac{s-2}{s+1}\sqrt{s^2-1}}{2}\right)$$

and

$$\tilde{b}_2^T = \left(\frac{1}{s}, \ldots, \frac{1}{s}, \frac{21s + 39 - 3\sqrt{s^2 - 1}}{16s^2 + 34s}, \frac{3s + 12 + 3\sqrt{s^2 - 1}}{8s^2 + 17s}, \frac{21s + 39 - 3\sqrt{s^2 - 1}}{16s^2 + 34s}\right).$$

Based on absolute stability region measurements we recommend embedded pair $\tilde{b}_2^T$. Here we present a result for $s = 4$ on Fig. 2. As we are increasing the number of stages we can see similar results.

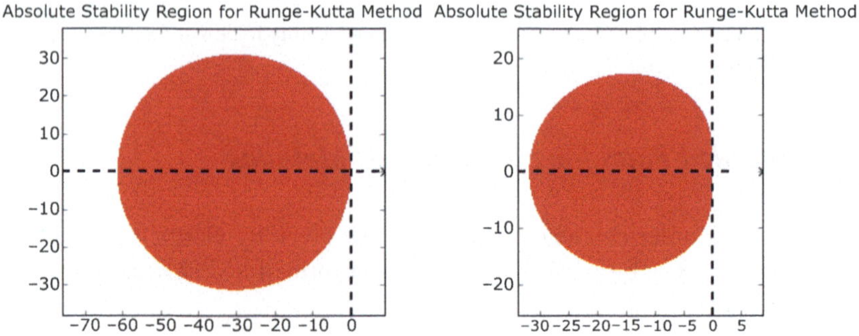

Fig. 2 The left and right plots correspond to the absolute stability region of SSPIRK(4, 3) and its $\tilde{b}_2^T$ embedded pair

Acknowledgements The project has been supported by the European Union, co-financed by the European Social Fund (EFOP-3.6.3-VEKOP-16-2017-00002).

References

1. Gottlieb, S.: Strong Stability Preserving Time Discretizations: A Review. ICOSAHOM 2014, pp. 17–30. Springer International Publishing, Cham (2015)
2. Gottlieb, S., Ketcheson, D., Shu, C.-W.: Strong Stability Preserving Runge-Kutta and Multistep Time Discretizations. World Scientific Publishing, Singapore (2011)
3. Hairer, E., Nørsett, S.P., Wanner, G.: Solving Ordinary Differential Equations I. Springer, Berlin (1993)
4. Ketcheson, D., Macdonald, C., Gottlieb, S.: Optimal Implicit Strong Stability Preserving Runge-Kutta Methods. Appl. Numer. Math. **59**(2), 373–392 (2009)

High-Order Compact Finite Difference Scheme for Option Pricing in Stochastic Volatility with Contemporaneous Jump Models

Bertram Düring and Alexander Pitkin

Abstract We extend the scheme developed in B. Düring, A. Pitkin, "High-order compact finite difference scheme for option pricing in stochastic volatility jump models", 2019, to the so-called stochastic volatility with contemporaneous jumps (SVCJ) model, derived by Duffie, Pan and Singleton. The performance of the scheme is assessed through a number of numerical experiments, using comparisons against a standard second-order central difference scheme. We observe that the new high-order compact scheme achieves fourth order convergence and discuss the effects on efficiency and computation time.

1 Introduction

The stochastic volatility with contemporaneous jump model (SVCJ) model, [3], can be seen as an extension of the Bates model [1], which combines the positive features of stochastic volatility and jump-diffusion models. In both models the option price is given as the solution of a partial integro-differential equation (PIDE), see e.g. [2]. In [5] we have presented a new high-order compact finite difference scheme for option pricing in Bates model. The implicit-explicit scheme is based on the approaches in Düring and Fournié [4] and Salmi et al. [6]. The scheme is fourth order accurate in space and second order accurate in time. In the present work we extend the scheme to the SVCJ model derived by Duffie et al. [3].

This article is organised as follows. In the next section we recall the SVCJ model for option pricing, we discuss the implementation of the implicit-explicit scheme and note the adaptations to the previously derived scheme for option pricing under the Bates model. Section 3 is devoted to the numerical experiments, where we assess the performance of the new scheme.

B. Düring · A. Pitkin (✉)
Department of Mathematics, University of Sussex, Brighton, UK
e-mail: bd80@sussex.ac.uk; a.h.pitkin@sussex.ac.uk

© Springer Nature Switzerland AG 2019
I. Faragó et al. (eds.), *Progress in Industrial Mathematics at ECMI 2018*,
Mathematics in Industry 30, https://doi.org/10.1007/978-3-030-27550-1_46

2 The SVCJ Model

The SVCJ model [3] is a stochastic volatility model which allows for jumps in both volatility and returns. Within this model the behaviour of the asset value, S, and its variance, σ, is described by the coupled stochastic differential equations,

$$dS(t) = \mu_S S(t)dt + \sqrt{\sigma(t)}S(t)dW_1(t) + S(t)dJ^S,$$
$$d\sigma(t) = \kappa(\theta - \sigma(t)) + \upsilon\sqrt{\sigma(t)}dW_2(t) + dJ^\sigma,$$

for $0 \leqslant t \leqslant T$ and with $S(0), \sigma(0) > 0$. Here, $\mu_S = r - \lambda\xi_S$ is the drift rate, where $r \geqslant 0$ is the risk-free interest rate. The two-dimensional jump process (J^S, J^σ) is a compound Poisson process with intensity $\lambda \geqslant 0$. The distribution of the jump size in variance is assumed to be exponential with mean υ. In respect to jump size z^σ in the variance process, $J + 1$ has a log-normal distribution $p(z^S, z^\sigma)$ with the mean in $\log z^s$ being $\gamma + \rho_J z^\sigma$, i.e. the probability density function is given by

$$p(z^S, z^\sigma) = \frac{1}{\sqrt{2\pi} z^S \delta \upsilon} e^{-\frac{z^\sigma}{\upsilon} - \frac{(\log z^S - \gamma - \rho_J z^\sigma)^2}{2\delta^2}}.$$

The parameter ξ_s is defined by $\xi_s = e^{\gamma + \frac{\delta^2}{2}}(1 - \upsilon\rho_J)^{-1} - 1$, where ρ_J defines the correlation between jumps in returns and variance, γ is the jump size log-mean and δ^2 is the jump size log-variance. The variance has mean level θ, κ is the rate of reversion back to mean level of σ and υ is the volatility of the variance σ. The two Wiener processes W_1 and W_2 have constant correlation ρ.

2.1 Partial Integro-Differential Equation

By standard derivative pricing arguments for the SVCJ model, obtain the PIDE

$$\frac{\partial V}{\partial t} + \frac{1}{2}S^2\sigma\frac{\partial^2 V}{\partial S^2} + \rho\upsilon\sigma S\frac{\partial^2 V}{\partial S \partial \sigma} + \frac{1}{2}\upsilon^2\sigma\frac{\partial^2 V}{\partial \sigma^2} + (r - \lambda\xi_s)S\frac{\partial V}{\partial S} + \kappa(\theta - \sigma)\frac{\partial V}{\partial \sigma}$$
$$-(r + \lambda)V + \lambda\int_0^{+\infty}\int_0^{+\infty} V(S.z^S, \sigma + z^\sigma, t)p(z^S, z^\sigma)\,\mathrm{d}z^\sigma\mathrm{d}z^S,$$

which has to be solved for $S, \sigma > 0$, $0 \le t < T$ and subject to a suitable final condition, e.g. $V(S, \sigma, T) = \max(K - S, 0)$, in the case of a European put option, with K denoting the strike price.

Through the following transformation of variables

$$x = \log S, \quad \tau = T - t, \quad y = \sigma/\upsilon \quad \text{and} \quad u = \exp(r + \lambda)V$$

we obtain

$$u_\tau = \frac{1}{2}vy\left(\frac{\partial^2 u}{\partial x^2}+\frac{\partial^2 u}{\partial y^2}\right)+\rho vy\frac{\partial^2 u}{\partial x\partial y}-\left(\frac{1}{2}vy-r+\lambda\xi_s\right)\frac{\partial u}{\partial x}$$

$$+\kappa\frac{(\theta-vy)}{v}\frac{\partial u}{\partial y}+\lambda\int_{-\infty}^{+\infty}\int_0^{+\infty}\tilde{u}(x+z^x,y+z^y,\tau)\tilde{p}(z^x,z^y)\,\mathrm{d}z^y\mathrm{d}z^x = L_D+L_I, \tag{1}$$

which is now posed on $\mathbb{R}\times\mathbb{R}^+\times(0,T)$, with
$\tilde{u}(x,y,\tau)=u(e^x,vy,\tau)$ and $\tilde{p}(z^x,z^y)=ve^{z^x}p(e^{z^x},z^y)$.

The problem is completed by suitable initial and boundary conditions. In the case of a European put option we have initial condition $u(x,y,0) = \max(1-\exp(x),0),\ x\in\mathbb{R},\ y>0$.

2.2 Implicit-Explicit High-Order Compact Scheme

For the discretisation, we replace $\mathbb{R}$ by $[-R_1,R_1]$ and $\mathbb{R}^+$ by $[L_2,R_2]$ with $R_1,R_2>L_2>0$. We consider a uniform grid $Z=\{x_i\in[-R_1,R_1]: x_i = ih_1,\ i=-N,\dots,N\}\times\{y_j\in[L_2,R_2]: y_j = L_2+jh_2,\ j=0,\dots,M\}$ consisting of $(2N+1)\times(M+1)$ grid points with $R_1=Nh_1$, $R_2=L_2+Mh_2$ and with space step $h:=h_1=h_2$ and time step k. Let $u_{i,j}^n$ denote the approximate solution of (1) in (x_i,y_j) at the time $t_n=nk$ and let $u^n=(u_{i,j}^n)$.

For the numerical solution of the PIDE we use the implicit-explicit high-order compact (HOC) scheme presented in [5]. The implicit-explicit discretisation in time is accomplished through an adaptation of the Crank-Nicholson method for which we shall define an explicit treatment for the two-dimensional integral operator, L_I.

We refer to [5] for the details of the derivation of the finite difference scheme for the differential operator L_D and the implementation of initial and boundary conditions. To form the SVCJ model the coefficients are adjusted, with constant ξ_s replacing ξ_B.

2.3 Integral Operator

After the initial transformation of variables we have the integral operator in the following form,

$$L_I=\lambda\int_{-\infty}^{+\infty}\int_0^{+\infty}\tilde{u}(x+z^x,y+z^y,\tau)\tilde{p}(z^x,z^y)\,\mathrm{d}z^y\mathrm{d}z^x,$$

We make a final change of variables $\zeta = x + z^x$ and $\eta = y + z^y$, with the intention of studying the value of the integral at the point (x_i, y_j),

$$I = \int_{-\infty}^{+\infty} \int_{0}^{+\infty} \tilde{u}(\zeta, \eta, \tau)\tilde{p}(\zeta - x_i, \eta - y_j)\, \mathrm{d}\eta \mathrm{d}\zeta \tag{2}$$

We numerically approximate the value of (2) over the rectangle $(-R_1, R_1) \times (L_2, R_2)$, with these values chosen experimentally.

$$\begin{aligned} I_{i,j} &= \int_{-\infty}^{+\infty} \int_{0}^{+\infty} \tilde{u}(\zeta, \eta, \tau)\tilde{p}(\zeta - x_i, \eta - y_j)\, \mathrm{d}\eta \mathrm{d}\zeta \\ &\approx \int_{-R_1}^{R_1} \int_{L_2}^{R_2} \tilde{u}(\zeta, \eta, \tau)\tilde{p}(\zeta - x_i, \eta - y_j)\, \mathrm{d}\eta \mathrm{d}\zeta \end{aligned} \tag{3}$$

To estimate the integral we require a numerical integration method of high order to match our finite difference scheme. We choose to use the two dimensional composite Simpson's rule. With f representing the integral in (3), we have error bounded by

$$\frac{h^4}{180}(R_2 - L_2)(2R_1) \max_{\zeta \in [-R_1, R_1], \eta \in [L_2, R_2]} |f^{(4)}(\zeta, \eta)|.$$

We evaluate the integral in (3) using the two-dimensional Simpsons rule on a equidistant grid in x, y with spacing $\Delta x = \Delta y$ and m_x grid-points in $(-R_1, R_1), (L_2, R_2)$, where each interval has length mesh-size $h/2$. We choose R_1, L_2 and R_2 such that the value of terms on the boundary can be considered negligible. Hence,

$$\begin{aligned} I_{i,j} \approx \frac{h^2}{36}\Bigg[& 16 \sum_{l=1}^{\frac{m_x}{2}} \left(\sum_{k=1}^{\frac{m_x}{2}} \tilde{u}(x_{2k-1}, y_{2l-1}, \tau)\tilde{p}(x_{2k-1} - x_i, y_{2l-1} - y_j) \right) \\ & + 4 \sum_{l=1}^{\frac{m_x-1}{2}} \left(\sum_{k=1}^{\frac{m_x-1}{2}} \tilde{u}(x_{2k}, y_{2l}, \tau)\tilde{p}(x_{2k} - x_i, y_{2l} - y_j) \right) \\ & + 8 \sum_{l=1}^{\frac{m_x-1}{2}} \left(\sum_{k=1}^{\frac{m_x}{2}} \tilde{u}(x_{2k-1}, y_{2l}, \tau)\tilde{p}(x_{2k-1} - x_i, y_{2l} - y_j) \right) \\ & + 8 \sum_{l=1}^{\frac{m_x}{2}} \left(\sum_{k=1}^{\frac{m_x-1}{2}} \tilde{u}(x_{2k}, y_{2l-1}, \tau)\tilde{p}(x_{2k} - x_i, y_{2l-1} - y_j) \right) \Bigg]. \end{aligned}$$

To avoid the construction of a dense matrix we compute this integral, as a product of the sums, at each time step.

If not mentioned otherwise, we use the following default parameters in our numerical experiments: $\kappa = 2$, $\theta = 0.01$, $v = 0.25$, $\rho = -0.5$, $\upsilon = 0.2$, $r = 0.05$, $\lambda = 0.2$, $\gamma = -0.5$, $\rho_J = -0.5$, $\delta^2 = 0.16$.

3 Numerical Experiments

We perform numerical studies to evaluate the rate of convergence and computational efficiency of the scheme. For comparison we include the results for a second-order central finite difference scheme, with the use of an appropriate two-dimensional trapezoidal rule to complete the numerical integration and the inclusion of a Rannacher-style start up to combat stability issues.

3.1 Numerical Convergence

For our convergence study we refer to both the l_2-error ϵ_2 and the l_∞-error ϵ_∞ with respect to a numerical reference solution on a fine grid with $h_{\text{ref}} = 0.025$. With the parabolic mesh ratio k/h^2 fixed to a constant value we expect these errors to converge as $\epsilon = Ch^m$ for some m and C which represent constants. From this we generate a double-logarithmic plot ϵ against h which should be asymptotic to a straight line with slope m, thereby giving a method for experimentally determining the order of the scheme.

The numerical convergence results are included in Fig. 1. We observe that the numerical convergence orders reflect the theoretical order of the schemes, with the new high-order compact scheme achieving convergence rates near fourth order.

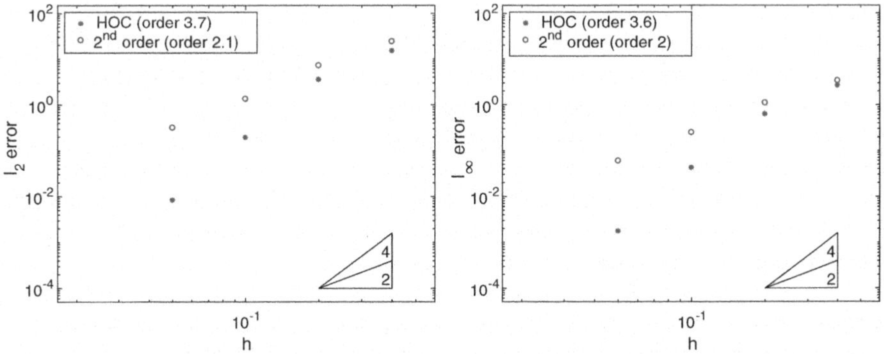

Fig. 1 l_2 and l_∞ error in option price taken at mesh-sizes $h = 0.4, 0.2, 0.1, 0.05$

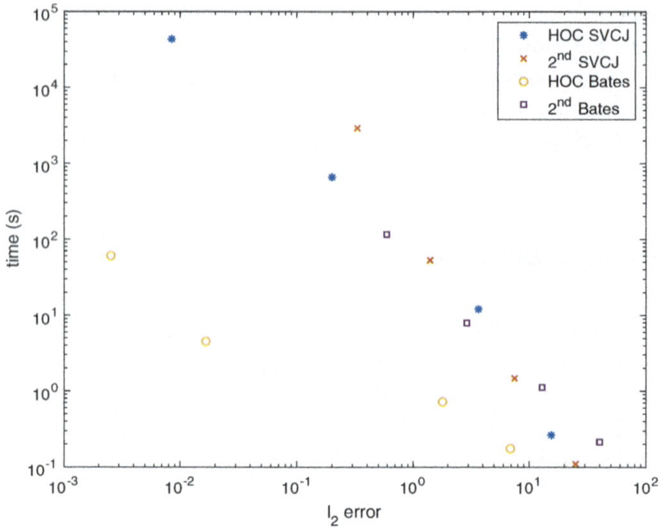

Fig. 2 Computational efficiency comparison taken at mesh-sizes $h = 0.4, 0.2, 0.1, 0.05$

3.2 Computational Efficiency Comparison

We compare the computational time of the two schemes, looking at the time to obtain a given accuracy, taking into account matrix setups, factorisation and boundary condition evaluation. The timings depend obviously on technical details of the computer as well as on specifics of the implementation, for which care was taken to avoid unnecessary bias in the results. All results were computed on the same laptop computer (2015 MacBook Air 11″).

The results are shown below in Fig. 2. The mesh-sizes used for this comparison are $h = 0.4, 0.2, 0.1$ and 0.05, with the reference mesh-size used being $h_{\text{ref}} = 0.025$.

The HOC scheme achieves higher accuracy at all mesh sizes, however, this is at the expense of computation time. We attribute this increase to the extra computational cost associated with the Simpson's rule as compared to the trapezoidal rule.

We include the results previously seen for the Bates model, [5], to indicate the increase in computation time between the two models. With access to higher memory allocation it may be possible to reduce this increase, through use of a circulant matrix and Fourier transforms to complete the numerical integration, [6]. However, it is not clear how this would be implemented with the different weightings assigned by Simpson's rule.

Acknowledgements BD acknowledges partial support by the Leverhulme Trust research project grant 'Novel discretisations for higher-order nonlinear PDE' (RPG-2015-69). AP has been supported by a studentship under the EPSRC Doctoral Training Partnership (DTP) scheme (grant number EP/M506667/1).

References

1. Bates, D.S.: Jumps and stochastic volatility: exchange rate processes implicit Deutsche mark options. Rev. Financ. Stud. **9**, 637–654 (1996)
2. Cont, R., Tankov, P.: Financial Modelling with Jump Processes. Chapman & Hall/CRC, Boca Raton (2004)
3. Duffie, D., Pan, J., Singleton, K.: Transform analysis and asset pricing for affine jump-diffusions. Econometrica **68**, 1343–1376 (2000)
4. Düring, B., Fournié, M.: High-order compact finite difference scheme for option pricing in stochastic volatility models. J. Comput. Appl. Math. **236**, 4462–4473 (2012)
5. Düring, B., Pitkin, A.: High-order compact finite difference scheme for option pricing in stochastic volatility jump models. J. Comput. Appl. Math. **355**, 201–217 (2019)
6. Salmi, S., Toivanen, J., von Sydow, L.: An IMEX-scheme for pricing options under stochastic volatility models with jumps. SIAM J. Sci. Comput. **36**(5), B817-B834 (2014)

Exploring Parallel-in-Time Approaches for Eddy Current Problems

Stephanie Friedhoff, Jens Hahne, Iryna Kulchytska-Ruchka, and Sebastian Schöps

Abstract We consider the usage of parallel-in-time algorithms of the Parareal and multigrid-reduction-in-time (MGRIT) methodologies for the parallel-in-time solution of the eddy current problem. Via application of these methods to a two-dimensional model problem for a coaxial cable model, we show that a significant speedup can be achieved in comparison to sequential time stepping.

1 Introduction

Recently, efficient and robust designs of electromechanical energy converters are gaining again in importance because of the transition towards sustainable energy in Europe ('Energiewende' in German). Electrical machinery is well understood and developed in industry close to their technical limits, but often without transient analysis or consideration of uncertainties in the design process. Such studies are only carried out late in the development process due to their high computational costs. This may lead to the fact that better or more robust designs are ruled out and not considered further on. One promising way to speed up transient analysis are parallel-in-time methods.

In contrast to classical time-integration techniques based on a time-stepping approach, i. e., solving sequentially for one time step after the other, parallel-in-time algorithms allow simultaneous solution across multiple time steps. Starting with the work of Nievergelt [1], various approaches for parallel-in-time integration have been explored; a recent review of the extensive literature in this area is [2]. The key practical aspect for choosing one of the many time-parallel methods when aiming at

S. Friedhoff (✉) · J. Hahne
Fakultät für Mathematik und Naturwissenschaften, Bergische Universität Wuppertal, Wuppertal, Germany
e-mail: friedhoff@math.uni-wuppertal.de; jens.hahne@math.uni-wuppertal.de

I. Kulchytska-Ruchka · S. Schöps
Centre for Computational Engineering, Technische Universität Darmstadt, Darmstadt, Germany
e-mail: kulchytska@temf.tu-darmstadt.de; schoeps@temf.tu-darmstadt.de

© Springer Nature Switzerland AG 2019
I. Faragó et al. (eds.), *Progress in Industrial Mathematics at ECMI 2018*, Mathematics in Industry 30, https://doi.org/10.1007/978-3-030-27550-1_47

adding parallelism to an existing application code is the level of intrusiveness, i. e., the required amount of implementation effort. There are only a few time-parallel methods that are non-intrusive. In this paper, we consider two of these approaches, the Parareal method [3] and the multigrid-reduction-in-time (MGRIT) algorithm [4] that, in a specific two-level setting, can be viewed as a Parareal-type algorithm.

2 Eddy Current Model Problem

For an open, bounded domain $\Omega \subset \mathbb{R}^3$ and $t \in \mathcal{I} = (t_0, t_{\text{end}}] \subset \mathbb{R}_{\geq 0}$, the evolution of electromagnetic fields is governed by Maxwell's equations on $\Omega \times \mathcal{I}$ [6]

$$\nabla \times \mathbf{E} = -\partial_t \mathbf{B}, \qquad \nabla \times \mathbf{H} = \partial_t \mathbf{D} + \mathbf{J}, \qquad \nabla \cdot \mathbf{B} = 0, \qquad \nabla \cdot \mathbf{D} = \rho \tag{1}$$

which are completed by the constitutive relations

$$\mathbf{D} = \varepsilon \mathbf{E}, \qquad \mathbf{J} = \sigma \mathbf{E} + \mathbf{J}_{\text{s}}, \qquad \mathbf{B} = \mu \mathbf{H}. \tag{2}$$

In these equations, $\mathbf{H}$ is the magnetic field [A/m], $\mathbf{B}$ the magnetic flux density [T], $\mathbf{E}$ the electric field [V/m], $\mathbf{D}$ the electric flux density [C/m^2], $\mathbf{J}$ and $\mathbf{J}_{\text{s}}$ are the total and source current density [A/m^2], ρ is the electric charge density [C/m^3]. All fields are functions of space $\mathbf{x} \in \Omega$ and time $t \in \mathcal{I}$. The material properties $\sigma \geq 0$, $\varepsilon > 0$ and $\mu > 0$ are the electric conductivity, the electric permittivity and the magnetic permeability, respectively. It is convenient to invert the magnetic material law, i. e., $\mathbf{H} = \nu \mathbf{B}$, using the reluctivity ν, where $\nu(B)$ can be a sufficiently smooth and bounded function of the magnitude $B = \|\mathbf{B}\|$, see [7]. In the following, we consider only devices where the displacement current can be neglected with respect to the source currents, i. e., $\|\partial_t \mathbf{D}\| \ll \|\mathbf{J}_{\text{s}}\|$. An analysis of this error can be found in [8]. Assuming $\partial_t \mathbf{D} = 0$ yields the so-called *magnetoquasistatic* approximation or *eddy current problem*. Eddy currents lead to the *skin effect*, i. e., currents through a conductor are pushed to the surface if frequency increases [6, Chapter 5.18].

One may introduce the ('modified' [9]) magnetic vector potential $\mathbf{A}$ such that $\mathbf{E} = -\partial_t \mathbf{A}$. Then, inserting the equations into each other yields

$$\sigma \partial_t \mathbf{A} + \nabla \times \big(\nu(\|\nabla \times \mathbf{A}\|)\nabla \times \mathbf{A}\big) = \mathbf{J}_{\text{s}}. \tag{3}$$

We consider the geometry shown in Fig. 1 and choose homogeneous Dirichlet conditions, i. e., $\mathbf{A} \times \mathbf{n} = 0$ with normal vector $\mathbf{n}$ on $\partial\Omega$ and the initial value $\mathbf{A}|_{t_0} \equiv 0$. The source is defined as $\mathbf{J}_{\text{s}}|_{\Omega_0}(t) = \mathbf{e}_z/(\pi r_0^2) f_n(t)$ in the inner cable Ω_0 and vanishes elsewhere, $\mathbf{e}_z$ denotes the unit vector in z-direction, and the excitation is given by

$$f_n(t) = \begin{cases} \operatorname{sign}\left[\sin\left(\frac{2\pi}{T}t\right)\right], & s_n(t) - \left|\sin\left(\frac{2\pi}{T}t\right)\right| < 0, \\ 0, & \text{otherwise}, \end{cases} \tag{4}$$

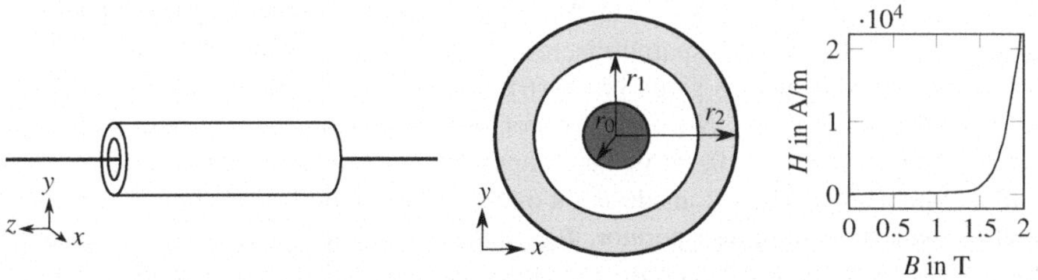

Fig. 1 Cable model, its cross section, and nonlinear material characteristic $\nu(B)$. The dark grey region Ω_0 models the copper wire, the white region the air insulation Ω_1, and the light grey annulus the conducting shield Ω_2 with nonlinear material characteristic $\nu(B)$. More details are given in [5]

where $s_n(t) = n/Tt - \lfloor n/Tt \rfloor$, $t \in [0, T]$ is the common sawtooth pattern, with $n = 1100$ teeth and period $T = 0.02$ s [14]. The reluctivity ν is modeled as vacuum $(1/\mu_0)$ in Ω_0 and Ω_1, and is given in Ω_2 by a monotone cubic spline curve, Fig. 1, the conductivity σ is only non-zero in the tube region Ω_2 (10 MS/m).

When considering a tube of very large length, it is sufficient to solve a planar 2D problem using edge shape functions that only have a z-component. They can be constructed from the nodal shape functions $N_i(\mathbf{x})$ as

$$\mathbf{A} = \sum_{i=1}^{N_{\mathrm{dof}}} \mathbf{u}_i \mathbf{w}_i(\mathbf{x}) \quad \text{with} \quad \mathbf{w}_i(\mathbf{x}) = \frac{N_i(\mathbf{x})}{l_z} \mathbf{e}_z, \tag{5}$$

where $l_z = 1$m refers to the length in z-direction and $N_{\mathrm{dof}} = 2269$ in this example. The Ritz-Galerkin approach using the first-order ansatz functions (5) yields

$$\mathbf{M}_\sigma \mathbf{u}' + \mathbf{K}_\nu(\mathbf{u})\mathbf{u} = \mathbf{j}_\mathrm{s}, \tag{6}$$

with the matrices and the right-hand side

$$M_{\sigma,i,j} = \int_\Omega \sigma \mathbf{w}_j \cdot \mathbf{w}_i \,\mathrm{d}\mathbf{x}, \;\; K_{\nu,i,j}(\cdot) = \int_\Omega \nu(\cdot) \nabla \times \mathbf{w}_j \cdot \nabla \times \mathbf{w}_i \,\mathrm{d}\mathbf{x}, \;\; j_{\mathrm{s},i} = \int_\Omega \mathbf{J}_\mathrm{s} \cdot \mathbf{w}_i \,\mathrm{d}\mathbf{x},$$

respectively. The resulting system (6) consists of differential-algebraic equations of index-1 due the vanishing entries $M_{\sigma,i,j}$ of the mass matrix $\mathbf{M}_\sigma$ in Ω_0 and Ω_1, [10].

3 Multigrid Reduction in Time

The multigrid-reduction-in-time (MGRIT) algorithm [4] is an iterative, parallel method, based on applying multigrid reduction (MGR) [11, 12] principles in time, for solving time-stepping problems of the form

$$\mathbf{u}'(t) = \mathbf{f}(t, \mathbf{u}(t)), \quad \mathbf{u}(t_0) = \mathbf{g}_0, \quad t \in (t_0, t_{\mathrm{end}}] \subset \mathbb{R}_{\geq 0}, \tag{7}$$

with initial condition, $\mathbf{g}_0$, at $t = t_0$. Note that form (7) can be a system of ODEs, arising, for example, after spatial discretization of a space-time PDE, or it can be a system of DAEs such as given in Eq. (6). Discretizing the time interval on a grid $t_i = i\Delta t$, $i = 0, 1, \ldots, N_t$, with, for notational convenience, constant time step $\Delta t = (t_{\text{end}} - t_0)/N_t > 0$, let $\mathbf{u}_i$ be an approximation to $\mathbf{u}(t_i)$ for $i = 1, \ldots, N_t$, and let $\mathbf{u}_0 = \mathbf{u}(t_0)$. Then, considering a one-step time-independent time integration method with time-stepping operator, $\Phi_{\Delta t}$, that takes a solution at time t_{i-1} to that at time t_i, the solution to (7) is defined via time-stepping, which can be represented as a (sequential) forward solve of the block-structured linear system

$$\mathbf{L}\mathbf{u} \equiv \begin{bmatrix} I & & & \\ -\Phi_{\Delta t} & I & & \\ & \ddots & \ddots & \\ & & -\Phi_{\Delta t} & I \end{bmatrix} \begin{bmatrix} \mathbf{u}_0 \\ \mathbf{u}_1 \\ \vdots \\ \mathbf{u}_{N_t} \end{bmatrix} = \begin{bmatrix} \mathbf{g}_0 \\ \mathbf{g}_1 \\ \vdots \\ \mathbf{g}_{N_t} \end{bmatrix} \equiv \mathbf{g}. \tag{8}$$

Alternatively, considering the lower block bidiagonal structure, we could apply cyclic reduction, whereby we first solve the Schur complement system,

$$\mathbf{L}_S\mathbf{u}_\Delta \equiv \begin{bmatrix} I & & & \\ -\Phi_{\Delta t}^m & I & & \\ & \ddots & \ddots & \\ & & -\Phi_{\Delta t}^m & I \end{bmatrix} \begin{bmatrix} \mathbf{u}_0 \\ \mathbf{u}_m \\ \vdots \\ \mathbf{u}_{N_t} \end{bmatrix} = \begin{bmatrix} \mathbf{g}_0 \\ \hat{\mathbf{g}}_m \\ \vdots \\ \hat{\mathbf{g}}_{N_t} \end{bmatrix} \equiv \hat{\mathbf{g}}, \tag{9}$$

for the value of the solution at every m-th temporal point, with consistently restricted forcing terms. Then define the solution at the remaining temporal points by local (and parallel) time-stepping between those points defined from the Schur complement. MGRIT is based on interpreting this cyclic reduction approach as a two-level MGR algorithm, enabling parallelism in the solution process (8). Therefore, define a coarse temporal mesh, or (using multigrid terminology) the set of C-points, to be those points included in the Schur complement system (9), with the remaining temporal points as the set of F-points. Further define "ideal" interpolation as the map which takes the solution at the C-points and yields a zero residual at the F-points, with a similar definition for "ideal" restriction. The Schur complement then arises as the standard Petrov-Galerkin coarse-grid operator with these definitions of restriction and interpolation. Cyclic reduction can be viewed as a two-level method with this Petrov-Galerkin coarse-grid operator and a block smoother (called F-relaxation) that converges in one iteration. As it is typical in the MGR setting, the MGRIT approach replaces the true Schur complement with a simpler operator (typically of the same form as the original bidiagonal system, but with a time propagator using time-step $m\Delta t$), replaces ideal restriction with simple injection, and compensates by adding relaxation. Furthermore, the two-level method can be extended to multiple levels in a simple recursive manner, and the full approximation storage (FAS) approach [13] can be applied in the same manner to accommodate nonlinear problems.

4 Numerical Results

We apply classical sequential time stepping and two MGRIT variants to the eddy current model problem (3) with the pulsed excitation (4) on the space-time domain $\Omega \times (0, 0.2]$ s, with $\Omega = \Omega_0 \cup \Omega_1 \cup \Omega_2$ depicted in Fig. 1. The spatial domain, Ω, is discretized using 2269 degrees of freedom and the backward Euler method is used on a uniform grid with 32,768 time steps for the time derivative of the space-discrete time-stepping problem (6). The time step on the finest grid, $l = 0$, is chosen to be $\Delta t = 6.1 \cdot 10^{-6}$ s, and the time step on each coarse grid, l, is given by $m^l \Delta t$, $l \geq 1$. Two MGRIT variants are considered: a two-level Parareal-type method with a coarsening factor of $m = 256$, and a five-level method that coarsens uniformly across all grids with a factor of $m = 4$. Thus, the coarsest grid consists of 128 points in time in both cases. On this coarsest temporal grid, time stepping is used. All spatial problems are solved using Newton's method with a direct LU solver.

The MGRIT algorithm was implemented in parallel using Python and Message Passing Interface (MPI). Numerical results were generated on an Intel Xeon Phi cluster consisting of 272 1.4 GHz Intel Xeon Phi processors.

Figure 2 shows convergence of the two MGRIT variants applied to the eddy current model problem. We see linear convergence for both variants. Comparing the number of spatial time-stepping solves required for the two methods to the optimal count of N_t for sequential time stepping, we note that one iteration requires about

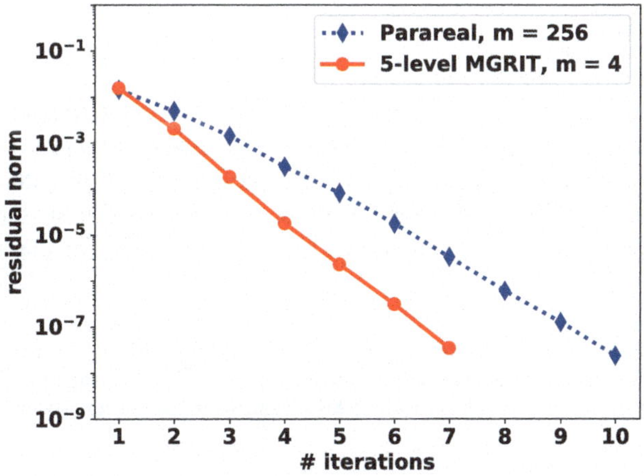

Fig. 2 Convergence of MGRIT variants applied to the eddy current model problem

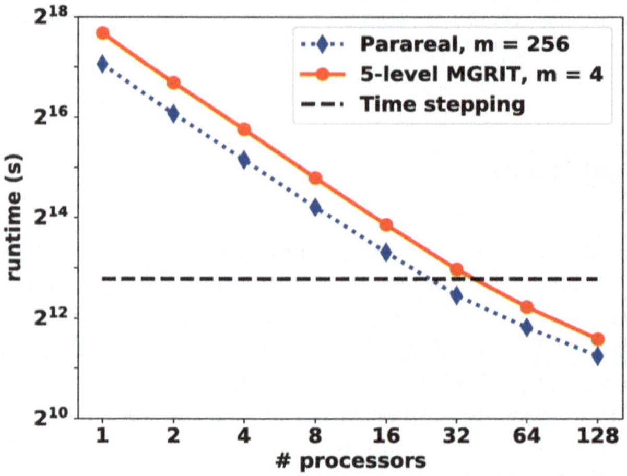

Fig. 3 Strong scaling results for MGRIT applied to the eddy current model problem

N_t or $2.5N_t$ spatial solves, respectively, when considering the two-level Parareal-type method or the five-level MGRIT scheme. This large computational overhead is demonstrated in the strong scaling results in Fig. 3. The dotted and solid lines show results for the two- and five-level methods, respectively, for increasing the number of processors in the temporal dimension only. The dashed line shows the runtime of time stepping on one processor for reference purposes. Results show that the extra work in the MGRIT variants can be effectively parallelized at high processor counts, i. e., more than 32, with good strong parallel scaling with a speedup of up to a factor of about 2.9 over sequential time stepping.

5 Conclusions

MGRIT was applied for the first time to the eddy current problem, which yields an index-1 DAE after spatial discretization. A speedup of approximately three times could be obtained. A strong scaling investigation shows that the method converges linearly with the number of processors, even for non-standard, pulsed right-hand sides, which has been shown to be problematic for classical Parareal [14].

Acknowledgements The work is supported by the Excellence Initiative of the German Federal and State Governments, the Graduate School of Computational Engineering at TU Darmstadt, and the BMBF in the framework of project PASIROM (grants 05M18RDA and 05M18PXB).

References

1. Nievergelt, J.: Parallel methods for integrating ordinary differential equations. Commun. Assoc. Comput. Mach. **7**, 731–733 (1964)
2. Gander, M.J.: 50 years of time parallel time integration. In: Carraro, T., Geiger, M., Körkel, S., Rannacher, R. (eds.) Multiple Shooting and Time Domain Decomposition, pp. 69–113. Springer, Heidelberg (2015)
3. Lions, J.-L., Maday, Y., Turinici, G.: A "parareal" in time discretization of PDEs. C. R. Acad. Sci. **332**, 661–668 (2001)
4. Falgout, R.D., Friedhoff, S., Kolev, T. V., MacLachlan, S.P., Schroder, J.B.: Parallel time integration with multigrid. SIAM J. Sci. Comput. **36**(6), C635–C661 (2014)
5. Meeker, D.C.: Finite Element Method Magnetics, Version 4.2 (28 Feb 2018 Build). http://www.femm.info
6. Jackson, J.D.: Classical Electrodynamics, 3rd edn. Wiley, New York (1998)
7. Heise, B.: Analysis of a fully discrete finite element method for a nonlinear magnetic field problem. SIAM J. Numer. Anal. **31**(3), 745–759 (1994)
8. Schmidt, K., Sterz, O., Hiptmair, R.: Estimating the eddy-current modeling error. IEEE Trans. Magn. **44**(6), 686–689 (2008)
9. Emson, C.R.I., Trowbridge, C.W.: Transient 3d eddy currents using modified magnetic vector potentials and magnetic scalar potentials. IEEE Trans. Magn. **24**(1), 86–89 (1988)
10. Nicolet, A., Delincé, F.: Implicit Runge-Kutta methods for transient magnetic field computation. IEEE Trans. Magn. **32**(3), 1405–1408 (1996)
11. Ries, M., Trottenberg, U.: MGR-Ein blitzschneller elliptischer Löser. Preprint 277 SFB 72. Universität Bonn, Bonn (1979)
12. Ries, M., Trottenberg, U., Winter, G.: A note on MGR methods. Linear Algebra Appl. **49**, 1–26 (1983)
13. Brandt, A.: Multi-level adaptive solutions to boundary-value problems. Math. Comput. **31**, 333–390 (1977)
14. Gander, M.J., Kulchytska-Ruchka, I., Niyonzima, I., Schöps, S.: A New Parareal Algorithm for Problems with Discontinuous Sources. Submitted to SISC, arXiv: 1803.05503 (2018)

Convergence of Solutions in a Mean-Field Model of Go-or-Grow Type with Reservation of Sites for Proliferation and Cell Cycle Delay

Ruth E. Baker, Péter Boldog, and Gergely Röst

Abstract We consider the mean-field approximation of an individual-based model describing cell motility and proliferation, which incorporates the volume exclusion principle, the go-or-grow hypothesis and an explicit cell cycle delay. To utilise the framework of on-lattice agent-based models, we make the assumption that cells enter mitosis only if they can secure an additional site for the daughter cell, in which case they occupy two lattice sites until the completion of mitosis. The mean-field model is expressed by a system of delay differential equations and includes variables such as the number of motile cells, proliferating cells, reserved sites and empty sites. We prove the convergence of biologically feasible solutions: eventually all available space will be filled by mobile cells, after an initial phase when the proliferating cell population is increasing then diminishing. By comparing the behaviour of the mean-field model for different parameter values and initial cell distributions, we illustrate that the total cell population may follow a logistic-type growth curve, or may grow in a step-function-like fashion.

1 Introduction

Cell proliferation and motility are key processes that govern cancer invasion or wound healing. The go-or-grow hypothesis postulates that proliferation and migration spatiotemporally exclude each other. This has been acknowledged, for example, for glioblastoma [5]. In general, two phenotypes that can be of particular importance

---

R. E. Baker
Mathematical Institute, University of Oxford, Oxford, UK

P. Boldog
Bolyai Institute, University of Szeged, Szeged, Hungary

G. Röst (✉)
Mathematical Institute, University of Oxford, Oxford, UK

Bolyai Institute, University of Szeged, Szeged, Hungary
e-mail: rost@math.u-szeged.hu

© Springer Nature Switzerland AG 2019
I. Faragó et al. (eds.), *Progress in Industrial Mathematics at ECMI 2018*,
Mathematics in Industry 30, https://doi.org/10.1007/978-3-030-27550-1_48

to progression of aggressive cancers are 'high proliferation-low migration' and 'low proliferation-high migration', and the mechanisms governing this switching are of great interest in current medical research [7]. Here we consider a strong simplification of this phenomenon by assuming that (differently from [4]) motile cells stop for a fixed period of time to complete cell division, upon which they immediately switch back into the migratory phenotype. We study the mathematical properties of a mean-field approximation of an individual based model describing this process, and this note complements our other ongoing works [1, 3] where we investigate in detail a range of biological hypotheses with the corresponding individual-based as well as mean-field, analytically tractable, models.

2 The Model

Assume that agents (representing biological cells) move and proliferate on an n-dimensional square lattice with length ℓ (in each direction), so that $K = \ell^n$ is an integer describing the number of lattice sites. We divide our agent population into two subpopulations, motile and proliferative, with the condition that a proliferative agent has to be attached to an adjacent site which is reserved until the end of proliferation. As a result, sites can either contain a motile agent, a proliferating agent, be reserved for the daughter agent of an attached proliferative agent, or be empty. At each time step, each motile agent can attempt to move into an adjacent lattice site or proliferate at its current site. However, if a motile agent attempts to move into a site that is already occupied or reserved, the movement event is aborted. Similarly, if a motile agent attempts to begin proliferation by reserving a site that is already occupied, then the proliferation event is aborted. Agents attempt to convert from being motile to proliferative at constant rate r, and the proliferative phase has length τ, upon which two motile daughter agents appear, one on the proliferating site, and one on the reserved site.

Based on the above, tracking the rate of change of the number of motile agents, $m(t)$, proliferative agents, $p(t)$, and reserved sites, $q(t)$, in time, and following the arguments and derivation in [2], we obtain the following mean-field approximation:

$$m'(t) = -rm(t)\frac{K-m(t)-p(t)-q(t)}{K} + 2rm(t-\tau)\frac{K-m(t-\tau)-p(t-\tau)-q(t-\tau)}{K},$$

$$p'(t) = rm(t)\frac{K-m(t)-p(t)-q(t)}{K} - rm(t-\tau)\frac{K-m(t-\tau)-p(t-\tau)-q(t-\tau)}{K},$$

$$q'(t) = rm(t)\frac{K-m(t)-p(t)-q(t)}{K} - rm(t-\tau)\frac{K-m(t-\tau)-p(t-\tau)-q(t-\tau)}{K}.$$

Here, the first term in the m equation expresses that m-cells attempt to proliferate with rate r, but proliferation starts only if the randomly selected target site is empty at time t, which has probability $(K - m(t) - p(t) - q(t))/K$, if we assume that there is no spatial correlation. The other terms can be interpreted in a similar way. Using the variable $u = K - m - p - q$ that accounts for empty sites, we can write

$$m'(t) = -rK^{-1}m(t)u(t) + 2rK^{-1}m(t-\tau)u(t-\tau), \tag{1}$$

$$p'(t) = rK^{-1}m(t)u(t) - rK^{-1}m(t-\tau)u(t-\tau), \tag{2}$$

$$q'(t) = rK^{-1}m(t)u(t) - rK^{-1}m(t-\tau)u(t-\tau), \tag{3}$$

$$u'(t) = -rK^{-1}m(t)u(t). \tag{4}$$

3 Long-Term Behaviour

The usual phase space for Eqs. (1)–(4) is $C = C([-\tau, 0], R^4)$, the Banach space of continuous function from the interval $[-\tau, 0]$ to R^4 equipped with the supremum norm. With the notation $x(t) = (m(t), p(t), q(t), u(t))$, our system is of the form $x'(t) = f(x_t)$ where $x_t \in C$ is defined by the relation $x_t(\theta) = x(t+\theta)$ for $\theta \in [-\tau, 0]$ and $f : C \to R^4$ is defined by the right-hand side of Eqs. (1)–(4). The standard results for delay differential equations provide existence and uniqueness of solutions from initial data $x_0 = \phi \in C$ (see, for example [6]).

Given the biological motivation, we are interested only in non-negative solutions, for which $p(t) = q(t) = rK^{-1}\int_{t-\tau}^{t} m(s)u(s)\mathrm{d}s$ holds, meaning that proliferative cells at a given time t are exactly those who started the proliferation process in the time interval $[t-\tau, t]$, and the reserved sites correspond to them. With this compatibility condition and the balance law $K = m(t) + p(t) + q(t) + u(t)$, we define the feasible phase space

$$\Omega := \left\{ \phi \in C : \phi_j(\theta) \geq 0 \text{ for all } \theta \in [-\tau, 0],\ j = 1, 2, 3, 4; \right.$$
$$\left. \sum_{j=1}^{4} \phi_j(0) = K; \quad \phi_2(0) = \phi_3(0) = rK^{-1}\int_{-\tau}^{0} \phi_1(s)\phi_4(s)\mathrm{d}s \right\}. \tag{5}$$

Lemma 1 *The set Ω is forward invariant, that is for any solution $x(t)$ with $x_0 \in \Omega$, $x_t \in \Omega$ for all $t \geq 0$.*

Proof Integrate Eq. (2) from 0 to t to obtain (similarly for $q(t)$)

$$p(t) - p(0) = rK^{-1}\int_{t-\tau}^{t} m(s)u(s)\mathrm{d}s - rK^{-1}\int_{-\tau}^{0} m(s)u(s)\mathrm{d}s.$$

From $x_0 \in \Omega$ we have $p(0) = q(0) = rK^{-1} \int_{-\tau}^{0} m(s)u(s)\mathrm{d}s$, hence

$$p(t) = q(t) = rK^{-1} \int_{t-\tau}^{t} m(s)u(s)\mathrm{d}s, \tag{6}$$

thus the third condition in the definition of Ω is preserved. The second trivially follows from summing up the equations to see $(m(t) + p(t) + q(t) + u(t))' = 0$, so $K = m(t) + p(t) + q(t) + u(t)$ is preserved. To confirm nonnegativity, note that $u(t) = u(0)\exp(-rK^{-1} \int_0^t m(s)\mathrm{d}s) \geq 0$. Assuming that $m(t) \geq 0$ for $t \leq t_0$, we have $m(t-\tau)u(t-\tau) \geq 0$ for $t \leq t_0 + \tau$, and consequently $m(t) \geq m(t_0)\exp(-rK^{-1} \int_{t_0}^t u(s)\mathrm{d}s)$ holds on $[t_0, t_0 + \tau]$. Hence, by the method of steps we obtain non-negativity of $m(t)$ for all t. Then the non-negativity of $p(t)$ and $q(t)$ follow from Eq. (6). □

Note that since solutions starting from Ω stay in this bounded set, they exist globally. Following [6], we say that a continuous functional $V : C \to R$ is a Lyapunov functional on the set Ω in C for Eqs. (1)–(4), if it is continuous on the closure of Ω, and $\dot{V} \leq 0$ on Ω. Here, $\dot{V}$ denotes the derivative of V along solutions. In our case Ω is itself closed. We also define $E := \{\phi \in \Omega : \dot{V} = 0\}$ and $M :=$ the largest set in E which is invariant with respect to Eqs. (1)–(4)).

Theorem 1 *If* $m(0) > 0$, *then* $\lim_{t\to\infty}(m(t), p(t), q(t), u(t)) = (K, 0, 0, 0)$.

Proof Consider the functional $V(\phi) = \phi_4(0)$. Then $\dot{V} = -rK^{-1}m(t)u(t) \leq 0$ for solutions in Ω, and by LaSalle's invariance principle (cf. Thm. 2.5.3 in [6]), the limit set of any solution is in M, thus on the limit set of any solution, $mu \equiv 0$ holds. Since for any solution u is always zero or always positive, we have either $m \equiv 0$ or $u \equiv 0$. In both cases, $p = q \equiv 0$ follows. Hence, the limit set can only be composed of the two equilibria $(K, 0, 0, 0)$ or $(0, 0, 0, K)$. Finally, we show that if $m(0) > 0$, then $m(t)$ can not converge to 0. Since $u(t)$ is monotone decreasing, for such a solution $m(t) + p(t) + q(t) = K - u(t) \geq K - u(0) > 0$ should hold. If $m(t) \to 0$ as $t \to \infty$, then from Eq. (6), also $p(t) = q(t) \to 0$. This contradicts $m(t) + p(t) + q(t) \geq K - u(0) > 0$ and so we can exclude $(0, 0, 0, K)$ from the limit set. Therefore $\lim_{t\to\infty}(m(t), p(t), q(t), u(t)) = (K, 0, 0, 0)$. □

Remark If $m(0) = 0$, then also $p(0) = q(0) = 0$, so $u(0) = K$ and we are on the empty lattice having the trivial solution $(0, 0, 0, K)$.

4 Simulations and Conclusion

According to the choice of the initial functions, different in vitro experiments can be modelled with Eqs. (1)–(4). One approach is to add a number of motile cells all at once at $t = 0$ to the empty cell space (e.g. a Petri dish). In this experiment the initial function ϕ_1 is given by $\phi_1(\theta) = aH(\theta)$ for $\theta \in [-\tau, 0]$, where a stands for the

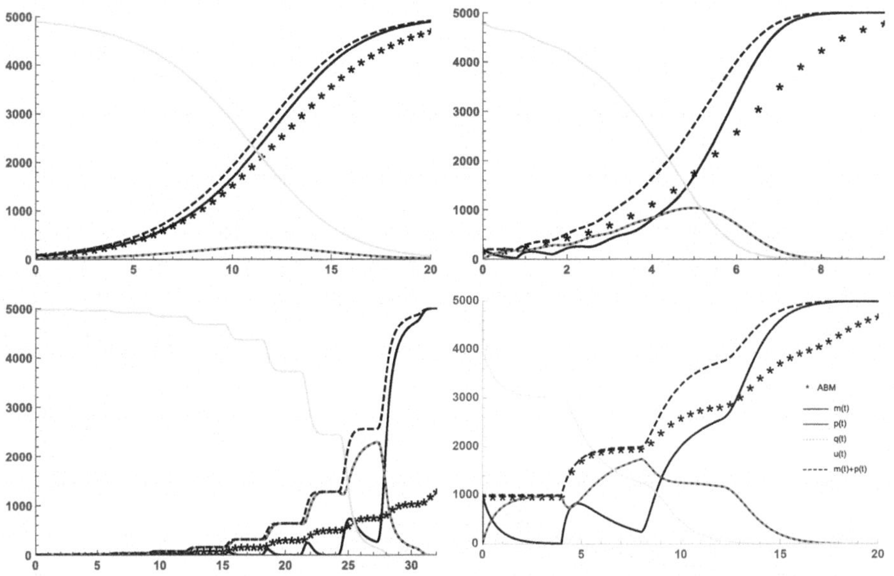

Fig. 1 Four numerical simulations, where in each realization, the initial function ϕ_1 on $[-\tau, 0]$ is given by $aH(\theta)$, where $a > 0$ is the initial number of cells, $\phi_2 = \phi_3 = 0$ and $\phi_4 = K - aH(\theta)$. This choice of the initial function models an in vitro experiment where motile cells are added to the plate at $t = 0$. The parameters are the following: *Top Left*—$r = 0.5, a = 100, \tau = 0.5$; *Top Right*—$r = 3, a = 200, \tau = 0.8$; *Bottom Left*—$r = 10.5, a = 10, \tau = 3$; *Bottom Right*—$r = 2, a = 1000, \tau = 4$; and $K = 5000$ in each case. The mean-field equations are compared to the output of averaged stochastic simulations of a corresponding agent-based model (ABM) on a 50×100 square lattice with cell motility rate 2. The legend in the bottom right figure applies to each

number of introduced cells at $t = 0$, and $H(\theta)$ is the right-continuous Heaviside-function, i.e. $H(\theta) = 0$ for $\theta < 0$ and $H(\theta) = 1$ for $\theta \geq 0$. In this setting, we take $\phi_2(\theta) = \phi_3(\theta) = 0$ and $\phi_4(\theta) = K - aH(\theta)$. While such initial data is not from C, they satisfy Eq. (5) and generate a continuous solution for $t > 0$. Some of such simulations are shown in Fig. 1.

A more elaborate in vitro experiment is the following. Instead of adding motile cells all at once, we add them in to the assay with a constant rate a for a time interval of length τ. After this, we leave the cell population intact. The initial data corresponding to this experiment can be obtained by solving a modification of Eqs. (1)–(4) with an additive forcing term $+a$ to the m-equation (and $-a$ to the u-equation), representing the gradual addition of m-cells, on an interval of length τ, starting from the state $(0, 0, 0, K)$. Then we start solutions of Eqs. (1)–(4) with such initial functions, which satisfy Eq. (5). Four realizations of this experimental setting are shown in Fig. 2.

The point of considering these two setups is that in the first we have only motile cells at $t = 0$, while in the second at $t = 0$ we have a distribution of cells in different phases of the cell cycle. This has a profound impact on the behaviour of solutions.

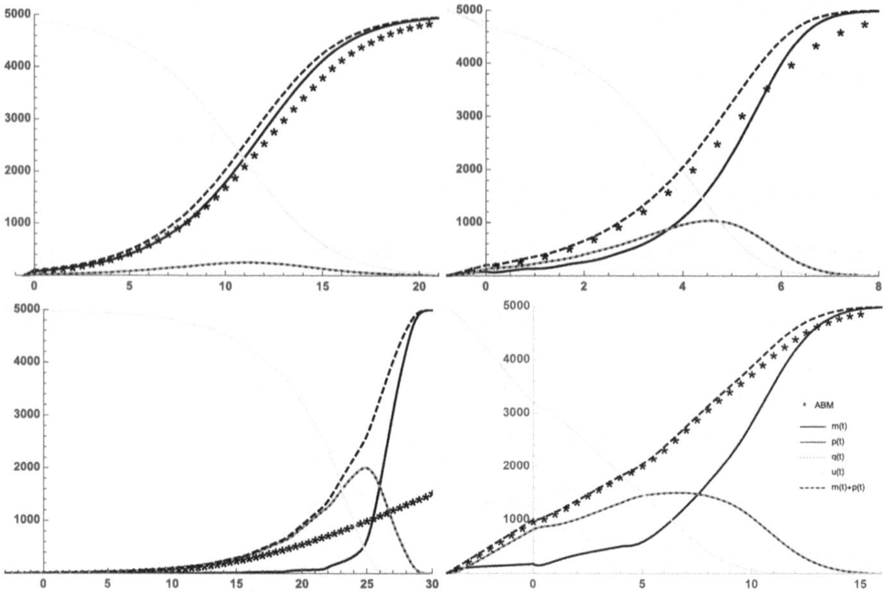

Fig. 2 Four simulations, where in each realization motile cells are added with rate a in the initial interval. The parameters are the following: *Top Left*—$r = 0.5, a = 200, \tau = 0.5$; *Top Right*—$r = 3, a = 250, \tau = 0.8$; *Bottom Left*—$r = 10.5, a = 3.3, \tau = 3$; *Bottom Right*—$r = 2, a = 250, \tau = 4$ and $K = 5000$ in each case. The ABM is the same as in Fig. 1, but the initial cell distribution has also been simulated here. The legend in the bottom right figure applies to each

For the sake of easier comparison, in both experimental settings the number of cells at $t = 0$ is exactly the same in the corresponding simulations whenever the parameter r and τ are the same. For both experimental settings, we can observe that the smaller the proliferation rate, the better the agreement between the mean-field model and the output of the agent based model (ABM). This is intuitively clear, as for smaller proliferation rate the cells in the ABM has more time to move around between proliferation events, hence the cell population becomes more well-mixed. While in Sect. 3 we proved that all solutions settle eventually at the state $(K, 0, 0, 0)$, there are distinctive features of solutions in different scenarios. Figure 1 shows that when the cell cycle delay is small, the solutions resemble logistic growth. In contrast, when the delay is large relative to the average time between individual cells attempting enter the proliferative state, the initially motile cells enter the proliferative state more or less together, and hence complete cell division more or less together too, resulting in a step-function-style growth curve in the total cell count. The sudden switching between phenotypes causes non-monotonic behaviours in $m(t)$ and $p(t)$ also. When we add motile cells continuously rather than adding them all at once, the solutions are much more similar to the expected logistic growth curve, and a different characteristic can be observed only for high proliferation rates or large numbers of initially added cells. In conclusion, an intermittent growth of a cell population can be an indication that the cell cycle length is relatively large (relative to inter-proliferation times), while its variance is small.

Acknowledgements REB is a Royal Society Wolfson Research Merit Award holder and would like to thank the Leverhulme Trust for a Research Fellowship. GR was supported by Marie Skłodowska-Curie Grant No. 748193. PB was supported by NKFI FK 124016 and EFOP-3.6.1-16-2016-00008.

References

1. Baker, R.E., Röst, G.: Global dynamics of a novel delayed logistic equation arising from cell biology. J. Nonlinear Sci. (2019). https://doi.org/10.1007/s00332-019-09577-w
2. Baker, R.E., Simpson, M.J.: Correcting mean-field approximations for birth-death-movement processes. Phys. Rev. E **82**(4), e041905 (2010)
3. Boldog P., Baker R.E., Röst, G.: Go-or-grow type models with explicit cell cycle length. In preparation
4. Gerlee, P., Nelander, S.: The impact of phenotypic switching on glioblastoma growth and invasion. PLoS Comput. Biol. **8**(6), e1002556 (2012)
5. Giese, A., Bjerkvig, R., Berens, M.E., Westphal, M.: Cost of migration: invasion of malignant gliomas and implications for treatment. J. Clin. Oncol. **21**, 1624–1636 (2003)
6. Kuang, Y.: Delay differential equations: with applications in population dynamics, vol. 191. Academic, Cambridge (1993)
7. Noren, D.P., Chou, W.H., Lee, S.H., Qutub, A.A., Warmflash, A., Wagner, D.S., Popel, S.P., Levchenko, A.: Endothelial cells decode VEGF-mediated Ca2+ signaling patterns to produce distinct functional responses. Sci. Signal. **9**(416), ra20-ra20 (2016)

Poroelasticity with Deformation Dependent Permeability

Sílvia Barbeiro

Abstract The poroelasticity theory that was originally developed in the context of geophysical applications has been successfully used to model the mechanical behavior of fluid-saturated living bone tissue. In this paper we focus on the numerical solution of the coupled fluid flow and mechanics in Biot's consolidation model of poroelasticity. The method combines mixed finite elements for Darcy flow and Galerkin finite elements for elasticity. The permeability tensor in the model is allowed to be a nonlinear function on the deformation, since this influence has relevance in the case of biological tissues like bone. We deal with the nonlinear term by considering a semi-implicit in time scheme. We provide the a priori error estimates for the numerical solution of the fully discretized model. For efficiency, we also explore an operator splitting strategy where the flow problem is solved before the mechanical problem, in an iterative process.

1 Introduction

The concept of mechanically stimulated bone adaptation has been discussed extensively in the literature. In the existing models, different approaches are considered. While some authors describe this process at a cellular (microscopic) level (e.g. [17]), others investigate the bone's (macroscopic) poroelastic structure (e.g. the survey article [6]), where poroelasticity is established as an effective model for deformation-driven bone fluid movement in bone tissue.

Poroelasticity refers to fluid flow within a deformable porous medium under the assumption of relatively small deformations, and models the influence of solid deformation to fluid flow and vice versa. The general mathematical description of this interaction is know as Biot theory [3].

S. Barbeiro (✉)
CMUC, Department of Mathematics, University of Coimbra, Coimbra, Portugal
e-mail: silvia@mat.uc.pt

© Springer Nature Switzerland AG 2019
I. Faragó et al. (eds.), *Progress in Industrial Mathematics at ECMI 2018*,
Mathematics in Industry 30, https://doi.org/10.1007/978-3-030-27550-1_49

Recent mathematical models account complex fluid/medium interactions which often lead to coupled systems of time dependent nonlinear partial differential equations. In our coupled model, the permeability tensor is a nonlinear function dependent on the deformation (see e.g. [1, 2]). The role of this dependence has relevance for modeling coupled mechanics and flow in porous media in different areas. For example, in petroleum industry, the reduction in permeability caused by the change in the stress state may significantly reduce the expected reservoir productivity [15]. In the case of the bone, the relation between porosity and permeability is discussed for instance in [5]. Other examples of deformation dependent permeability tensors for biological tissues can be found in [11] and [12].

The fully coupled approach consists in solving flow and elasticity equations simultaneously [1, 14]. Alternatively, operator splitting techniques which decouple the pressure equation from the equation for the displacement [8, 9, 18] can be used to solve the discrete system.

2 The Semi-Implicit Fully Discrete Formulation

The domain of interest is a polygonal or polyhedral domain $\Omega \subset \mathbb{R}^d$, when $d = 2$ or $d = 3$, respectively, with boundary $\partial\Omega$.

The physical parameters of the model are: λ, μ, the Lamé constants, c_o, the constrained specific storage coefficient, α, the Biot-Willis constant, μ_f, the fluid viscosity, ρ_f, the fluid mass density and $\mathbf{g}$, the body force per unit of mass. The values of the poroelasticity coefficients for different kinds of bone can be found in the literature (e.g. [7] for human femoral cortical bone data).

The primary variables are the pressure p and the displacement $\mathbf{u}$. In the context of linearized strains, the effective stress is given by $\sigma(\mathbf{u}) = 2\mu\epsilon(\mathbf{u}) + \lambda \mathrm{tr}(\epsilon(\mathbf{u}))I$, where $\epsilon(\mathbf{u}) = \frac{1}{2}\left(\mathrm{grad}\,\mathbf{u} + (\mathrm{grad}\,\mathbf{u})^t\right)$, and I is the identity matrix. By σ_m we denote the effective mean stress, $\sigma_m = \frac{1}{d}\mathrm{tr}(\sigma(\mathbf{u}))$. The total stress, $\tilde{\sigma}$, that must account for the usual material stress and for the fluid pressure, is given by $\tilde{\sigma}(\mathbf{u}, p) = \sigma(\mathbf{u}) - \alpha p I$. By K we denote the symmetric permeability tensor which is stress dependent. To introduce the mixed formulation for the Darcy flow, we consider the variable for the flux $\mathbf{z} = -\frac{1}{\mu_f} K(\sigma_m)(\nabla p - \rho_f \mathbf{g})$.

In order to define two sets of boundary conditions, one corresponding to the pressure and flux and another corresponding to the deformation, the boundary is decomposed in two unrelated ways, $\partial\Omega = \Gamma_p \cup \Gamma_f$ and $\partial\Omega = \Gamma_0 \cup \Gamma_N$, with $\mathrm{meas}(\Gamma_0) > 0$.

We summarize below the governing equations, together with the boundary and initial conditions:

$$-(\lambda + \mu)\nabla(\nabla \cdot \mathbf{u}) - \mu \nabla^2 \mathbf{u} + \alpha \nabla p = \mathbf{f} \text{ in } \Omega \times (0, T],$$
$$\frac{\partial}{\partial t}(c_o p + \alpha \nabla \cdot \mathbf{u}) - \frac{1}{\mu_f} \nabla \cdot K(\sigma_m)(\nabla p - \rho_f \mathbf{g}) = s_f \text{ in } \Omega \times (0, T],$$

$$
\begin{aligned}
p &= p_D \text{ on } \Gamma_p \times [0, T], \\
-\frac{1}{\mu_f} K(\sigma_m)(\nabla p - \rho_f \mathbf{g}) \cdot \boldsymbol{\eta} &= q \text{ on } \Gamma_f \times [0, T], \\
\mathbf{u} &= \mathbf{u}_D \text{ on } \Gamma_0 \times [0, T], \\
\tilde{\boldsymbol{\sigma}} \boldsymbol{\eta} &= \mathbf{r}_N \text{ on } \Gamma_N \times (0, T], \\
p(0) &= p_0 \text{ in } \Omega, \\
\mathbf{u}(0) &= \mathbf{u}_0 \text{ in } \Omega.
\end{aligned}
$$

The symbol η represents the outward normal vector on $\partial\Omega$.

Let us consider the spaces $\mathbf{H}(\text{div}) = \{\mathbf{s} \in (L^2(\Omega))^d : \nabla \cdot \mathbf{s} \in L^2(\Omega)\}$, $\mathbf{S}_0 = \{\mathbf{s} \in \mathbf{H}(\text{div}) : \mathbf{s} \cdot \eta|_{\Gamma_f} = 0\}$ and $\mathbf{V}_0 = \{\mathbf{v} \in H^1((\Omega))^d : \mathbf{v}|_{\Gamma_0} = 0\}$.

Since the boundary conditions are allowed to be inhomogeneous, we need to select, for each $t \in [0, T]$, a function $\mathbf{u}_d(., t) \in (H^1(\Omega))^d$ such that $\mathbf{u}_d(., t)|_{\Gamma_0} = \mathbf{u}_D(., t)$ and a function $\mathbf{z}_d(., t) \in \mathbf{H}(\text{div})$ such that $(\mathbf{z}_d(., t) \cdot \eta)|_{\Gamma_f} = q(., t)$.

The variational problem becomes: find $\mathbf{u} \in \mathbf{u}_d + H^1([0, T]; \mathbf{V}_0)$, $p \in H^1([0, T]; L^2(\Omega))$ and $\mathbf{z} \in \mathbf{z}_d + L^2([0, T]; \mathbf{S}_0)$ such that

$$a_{\mathbf{u}}(\mathbf{u}, \mathbf{v}) - \alpha(\nabla \cdot \mathbf{v}, p) = \ell_1(\mathbf{v}), \tag{1}$$

$$\left(c_0 \frac{\partial p}{\partial t}, w\right) + \alpha\left(\frac{\partial}{\partial t} \nabla \cdot \mathbf{u}, w\right) + (\nabla \cdot \mathbf{z}, w) = \ell_2(w), \tag{2}$$

$$\mu_f (K^{-1}(\sigma_m)\mathbf{z}, \mathbf{s}) - (p, \nabla \cdot \mathbf{s}) = \ell_3(\mathbf{s}) \tag{3}$$

holds for all $(\mathbf{v}, w, \mathbf{s}) \in (\mathbf{V}_0, L^2(\Omega), \mathbf{S}_0)$, $t \in (0, T]$ and $p(0) = p_0$, $\mathbf{u}(0) = \mathbf{u}_0$ in Ω, where

$$a_{\mathbf{u}}(\mathbf{u}, \mathbf{v}) = \int_\Omega \boldsymbol{\sigma}(\mathbf{u}) : \boldsymbol{\epsilon}(\mathbf{v})\, d\mathbf{x}.$$

The right hand side of the Eqs. (1), (2) and (3) is defined by the functionals

$$
\begin{aligned}
\ell_1(\mathbf{v}) &= \int_\Omega \mathbf{f} \cdot \mathbf{v} + \int_{\Gamma_N} \mathbf{r}_N \cdot \mathbf{v}, \quad \mathbf{v} \in \mathbf{V}_0, \\
\ell_2(w) &= \int_\Omega s_f w, \quad w \in L^2(\Omega), \\
\ell_3(\mathbf{s}) &= -\int_{\Gamma_p} p_D \mathbf{s} \cdot \boldsymbol{\eta} + \int_\Omega \rho_f \mathbf{g} \cdot \mathbf{s}, \quad \mathbf{s} \in \mathbf{S}_0.
\end{aligned}
$$

Let $\mathcal{E}_h$ be a nondegenerate partition of the domain Ω into non-overlapping triangles or tetrahedra with maximal element diameter h. We denote by $\mathbf{V}_h$ the space of continuous piecewise polynomials of degree $r \geq 1$ defined on $\mathcal{E}_h$ and

$$\mathbf{V}_{h,0} = \{\mathbf{v} \in \mathbf{V}_h : \mathbf{v}|_{\Gamma_0} = 0\}.$$

Let $(W_h, \mathbf{S}_h) \subset (L^2(\Omega) \times \mathbf{H}(\mathrm{div}))$ represent the Raviart-Thomas (RT) or Raviart-Thomas-Nedelec (RTN) spaces ([13, 16]) on $\mathcal{E}_h$, of order κ, which are standard mixed finite element spaces (see e.g. [4]), and let

$$\mathbf{S}_{h,0} = \{\mathbf{s} \in \mathbf{S}_h : \mathbf{s} \cdot \eta_{|\Gamma_f} = 0\}.$$

Let $\Delta t = T/N$, where N denotes the number of time steps and $t^n = n\Delta t$, $n = 0, 1, \dots, N$. We use the following notation $g^n = g(., t^n)$.

We define $\mathbf{u}^n_{h,d} \in \mathbf{V}_h$ such that $a_{\mathbf{u}}(\mathbf{u}^n_{h,d} - \mathbf{u}_d(., t^n), \mathbf{v}) = 0,\ \forall \mathbf{v} \in \mathbf{V}_h$, and $\mathbf{z}^n_{h,d} \in \mathbf{S}_h$ such that $(\nabla \cdot (\mathbf{z}^n_{h,d} - \mathbf{z}_d(., t^n)), w) = 0,\ \forall w \in W_h$.

The semi-implicit fully discrete formulation becomes: find $\mathbf{u}^n_h \in \mathbf{u}^n_{h,d} + \mathbf{V}_{h,0}$, $p^n_h \in W_h$, $\mathbf{z}^n_h \in \mathbf{z}^n_{h,d} + \mathbf{S}_{h,0}$ such that

$$a_{\mathbf{u}}(\mathbf{u}^n_h, \mathbf{v}) - \alpha(p^n_h, \nabla \cdot \mathbf{v}) = \ell^n_1(\mathbf{v}), \tag{4}$$

$$\left(c_0 \frac{p^n_h - p^{n-1}_h}{\Delta t}, w\right) + \alpha\left(\nabla \cdot \frac{\mathbf{u}^n_h - \mathbf{u}^{n-1}_h}{\Delta t}, w\right) + (\nabla \cdot \mathbf{z}^n_h, w) = \ell^n_2(w), \tag{5}$$

$$\mu_f((K(\sigma^{n-1}_{m,h}))^{-1}\mathbf{z}^n_h, \mathbf{s}) - (p^n_h, \nabla \cdot \mathbf{s}) = \ell^n_3(\mathbf{s}), \tag{6}$$

for all $(\mathbf{v}, w, \mathbf{s}) \in (\mathbf{V}_{h,0}, W_h, \mathbf{S}_{h,0})$, $n = 1, \dots, N$. Here

$$(\sigma_{m,h})^{n-1} = \frac{1}{d}\mathrm{tr}(\sigma(\mathbf{u}^{n-1}_h)).$$

Additionally, we consider the initial conditions $\mathbf{u}^0_h \in \mathbf{V}_h$, $p^0_h \in W_h$, such that $a_{\mathbf{u}}(\mathbf{u}^0_h - \mathbf{u}_0, \mathbf{v}) = 0,\ \forall \mathbf{v} \in \mathbf{V}_h$, and $(p^0_h - p_0, w) = 0,\ \forall w \in W_h$. The fully coupled scheme involves calculating $\mathbf{u}^n_h$, p^n_h and $\mathbf{z}^n_h$ simultaneously.

The prove of the next convergence result for the semi-implicit in time scheme (4)–(6) can be derived in a similar fashion of the corresponding result for the implicit in time method stated in Theorem 4 of [1].

Theorem 1 *Let $(\mathbf{u}, p, \mathbf{z})$ be the solution of (1)–(3) and $(\mathbf{u}_h, p_h, \mathbf{z}_h)$ be the solution of (4)–(6). Let us consider the same smoothness assumptions for the exact solution and for the permeability tensor stated in [1]. Then, if Δt small enough, there exists $C > 0$ such that*

$$\|\mathbf{u} - \mathbf{u}_h\|^2_{L^\infty(H^1)} + \|p - p_h\|^2_{L^\infty(L^2)} + \|\mathbf{z} - \mathbf{z}_h\|^2_{L^2(L^2)} \le C(h^{2r} + h^{2\kappa+2}) + \mathcal{O}(\Delta t^2), \tag{7}$$

where C depends on the model parameters, but is not dependent on h and Δt.

3 Iteratively Coupled Scheme

The choice of the coupling scheme affects the stability and accuracy of the numerical solutions as well as the computational efficiency. In [10], four different operator-split strategies are analyzed. Following those conclusions therein discussed, we decouple our system using the fixed-stress split method ([10, 18]).

We write the volumetric strain $\nabla \cdot \mathbf{u}$ in terms of fluid pressure p and the total mean stress $\tilde{\sigma}_m$,

$$\alpha \nabla \cdot \mathbf{u} = c_r p + \frac{c_r}{\alpha} \tilde{\sigma}_m, \tag{8}$$

where $c_r = \frac{d\alpha^2}{d\lambda + 2\mu}$.

Let k be inner loop iteration number and $p_h^{n,k}$, $\mathbf{z}_h^{n,k}$ and $\mathbf{u}_h^{n,k}$ denote the solutions of each inner loop iteration. We use equality (8) to obtain the following decoupled problems:

$$\Big((c_o + c_r)(p_h^{n,k} - p_h^{*n-1}), w\Big) + \Delta t(\nabla \cdot \mathbf{z}_h^{n,k}, w) = \Delta t \ell_2^n(w)$$
$$- \Big(\frac{c_r}{\alpha}(\tilde{\sigma}_{m,h}^{n,k-1} - \tilde{\sigma}_{m,h}^{*n-1}), w\Big), \tag{9}$$

$$\mu_f((K(\sigma_{m,h}^{*n-1}))^{-1} \mathbf{z}_h^{n,k}, \mathbf{s}) - (p_h^{n,k}, \nabla \cdot \mathbf{s}) = \ell_3^n(\mathbf{s}), \tag{10}$$

$$a_{\mathbf{u}}(\mathbf{u}_h^{n,k}, \mathbf{v}) - \alpha(p_h^{n,k}, \nabla \cdot \mathbf{v}) = \ell_1^n(\mathbf{v}), \tag{11}$$

for all $(\mathbf{v}, w, \mathbf{s}) \in (\mathbf{V}_{h,0}, W_h, \mathbf{S}_{h,0})$, where we solve first Eqs. (9) and (10) for flow, and then Eq. (11) for mechanics. Inside the outer loop for time steps, the two subproblems are solved in a staggered way until the convergence criterion

$$\|\tilde{\sigma}_{m,h}^{n,k} - \tilde{\sigma}_{m,h}^{n,k-1}\|_{L^\infty(\Omega)} < \text{Tol}$$

is satisfied. We write p_h^{*n}, $\mathbf{z}_h^{*n}$ and $\mathbf{u}_h^{*n}$ to denote, respectively, the solutions of pressure, velocity and displacement, at time n, resulting from this iterative coupling process.

The following lemma gives upper bounds for the difference of two consecutive solutions of the iterative process.

Lemma 1 *Let* $\delta_k p_h^{n,k} = p_h^{n,k} - p_h^{n,k-1}$, $\delta_k \mathbf{z}_h^{n,k} = \mathbf{z}_h^{n,k} - \mathbf{z}_h^{n,k-1}$ *and* $\delta_k \mathbf{u}_h^{n,k} = \mathbf{u}_h^{n,k} - \mathbf{u}_h^{n,k-1}$. *Then*

$$\|(\frac{c_r}{2} - \frac{\alpha^2}{2\lambda} + c_o)^{1/2} \delta_k p_h^{n,k}\|_{L^2(\Omega)}^2 + \Delta t \|K(\sigma_{m,h}^{*n-1}))^{-1/2} \delta_k \mathbf{z}_h^{n,k}\|_{L^2(\Omega)}^2$$
$$\leq \|(\frac{c_r}{2} + \frac{\alpha^2}{2\lambda})^{1/2} \delta_k p_h^{n,k-1}\|_{L^2(\Omega)}^2 \tag{12}$$

and

$$\|\nabla \cdot \delta_k \mathbf{u}_h^{n,k}\|_{L^2(\Omega)} \leq \frac{\alpha}{\lambda}\|\delta_k p_h^{n,k}\|_{L^2(\Omega)}. \tag{13}$$

The inequalities (12) and (13) can be obtained from the system (9)–(11), by setting $\mathbf{v} = \delta_k \mathbf{u}_h^{n,k}$, $w = \delta_k p_h^{n,k}$ and $\mathbf{s} = \delta_k \mathbf{z}_h^{n,k}$. Taking into account the estimates of Lemma 1, the next result follows straightforward.

Theorem 2 *If* $c_o(\mathbf{x}) > \frac{\alpha^2}{\lambda}$ $\forall \mathbf{x} \in \bar{\Omega}$ *holds then the iterative coupling scheme converges.*

Theorem 1 establishes the order of convergence of the semi-implicit in time fully coupled method. Together with the result of Theorem 2, we may conclude that the same type of convergence occurs for the numerical solution of the fixed-stress split iteratively coupled scheme.

Acknowledgements This work was partially supported by the Centre for Mathematics of the University of Coimbra—UID/MAT/00324/2019, funded by the Portuguese Government through FCT/MEC and co-funded by the European Regional Development Fund through the Partnership Agreement PT2020.

References

1. Barbeiro, S., Wheeler, M.F.: A priori error estimates for the numerical solution of a coupled geomechanics and reservoir flow model with stress-dependent permeability. Comput. Geosci. **14**, 755–768 (2010)
2. Berger, L., Bordas, R., Kay, D., Tavener, S.: A stabilized finite element method for finite-strain three-field poroelasticity. Comput. Mech. **60**(1), 51–68 (2017)
3. Biot, M.A.: General theory of three-dimensional consolidation. J. Appl. Phys. **12**, 155–164 (1941)
4. Brezzi, F., Fortin, M.: Mixed and Hybrid Finite Element Methods. Springer Series in Computational Mathematics, vol. 15, p. 350. Springer, New York (1991)
5. Cardoso, l., Fritton, S.P., Gailani, G., Benalla, M., Cowin, S.C.: A review of recent advances in the assessment of bone porosity, permeability, and interstitial fluid flow. J. Biomech. **46**(2), 253–265 (2013)
6. Cowin, S.C.: Bone poroelasticity. J. Biomech. **32**, 217–238 (1999)
7. Cowin, S.C., Sadegh, A.M.: Non-interacting modes for stress, strain and energy in anisotropic hard tissue. J. Biomech. **24**(9), 859–67 (1991)
8. Dana, S., Wheeler, M.F.: Convergence analysis of two-grid fixed stress split iterative scheme for coupled flow and deformation in heterogeneous poroelastic media. Comput. Methods Appl. Mech. Eng. **341**, 788–806 (2018)
9. Goulet, G.C., Cooper, D.M.L., Coombe, D., Zernicke, R.F.: Validation and application of iterative coupling to poroelastic problems in bone fluid flow. Bull. App. Mech. **5**(17), 6–17 (2009)
10. Kim, J., Tchelepi, H., Juanes, R.: Stability, Accuracy and efficiency of sequential methods for coupled flow and geomechanics. SPE J. **16**(02), (2009)
11. Kowalczyk, P., Kleiber, M.: Modelling and numerical analysis of stresses and strains in the human lung including tissue-gas interaction. Eur. J. Mech. A. Solids **13**(3), 367–393 (1994)

12. Lai, W., Mow, V.: Drag-induced compression of articular cartilage during a permeation experiment. Biorheology **17**(1–2), 111 (1980)
13. Nedelec, J.C.: Mixed finite elements in $\mathbb{R}^3$. Numer. Math. **35**(3), 315–341 (1980)
14. Phillips, P.J., Wheeler, M.F.: A coupling of mixed and continuous Galerkin finite element methods for poroelasticity II: The discrete-in-time case. Comput. Geosci. **11**(4), 145–158 (2007)
15. Raghavan, R., Chin, L.Y.: Productivity changes in reservoirs with stress-dependent permeability. SPE Reserv. Eval. Eng. **7**(4), 308–315 (2004)
16. Raviart, P.A., Thomas, J.M.: A mixed finite element method for second order elliptic problems. Lect. Notes Math. Springer **606**, 292–315 (1977)
17. Ryser, M.D., Komarova, S.V., Nigam, N.: The cellular dynamics of bone remodeling: a mathematical model. SIAM J. Appl. Math. **70**(6), 1899–1921 (2010)
18. Wheeler, M.F., Gai, X.: Iteratively coupled mixed and Galerkin finite element methods for poro-elasticity. Numer. Methods Partial Differ. Eq. **23**, 785–797 (2007)

Local Time Stepping Method for District Heating Networks

Matthias Eimer, Raul Borsche, and Norbert Siedow

Abstract In this article, we present a numerical solver for simulating district heating networks. The method applies a local time stepping to networks of linear advection equations. Numerical diffusion as well as the computational effort on each edge is reduced significantly. The combination with high order coupling and reconstruction techniques leads to a very efficient scheme.

1 Introduction

District heating is an efficient alternative to conventional heating systems, especially in urban regions. The transport medium water is heated in a central plant and distributed to the consumers through a network of pipes. There are systems for any common energy source e.g. fossil fuel, biomass and solar energy. In combination with a power generator in so called CHPs (combined heat and power) the systems have much higher energy efficiency and less pollution than local boilers. In order to find an optimal control for such systems, fast and accurate simulations are needed. In the following we present a local time stepping scheme for district heating networks. In Sect. 2 the full model for the system is presented. After restricting ourselves to the computation of the energy transport, we present the new scheme in Sect. 3. Additional insight is given to some special cases such as high order extension and the incorporation of source terms. In Sect. 4 we discuss the results of the new scheme compared to a high order ADER scheme [1, 7].

M. Eimer (✉) · R. Borsche
TU Kaiserslautern, Kaiserslautern, Germany
e-mail: matthias.eimer@itwm.fraunhofer.de

N. Siedow
Fraunhofer Institute for Industrial Mathematics ITWM, Kaiserslautern, Germany

© Springer Nature Switzerland AG 2019
I. Faragó et al. (eds.), *Progress in Industrial Mathematics at ECMI 2018*,
Mathematics in Industry 30, https://doi.org/10.1007/978-3-030-27550-1_50

2 Model

The behavior of density ρ, velocity v and energy density e of the transport medium water inside a pipe is described by the Euler equations. Since the pressure level always keeps the water in its liquid phase, we can assume incompressibility. The remaining system then reads

$$\begin{aligned} \partial_x v &= 0 \\ \partial_t v + \frac{1}{\rho}\partial_x p &= -\frac{\lambda}{2d} v|v| - g\partial_x h \\ \partial_t e + v\partial_x e &= -\frac{4k}{d}\left(T(e) - T_\infty\right) . \end{aligned} \tag{1}$$

Here, p is the pressure inside the pipe, d its diameter and λ the friction factor according to the Darcy-Weisbach friction law. The term $g(\partial_x h)$ takes the vertical elevation h into account with the gravitational acceleration g. The right hand side of the third equation models the energy loss to the environment, where k is the heat transmission coefficient of the pipe, $T(e)$ is the fluid temperature depending on its energy density e and T_∞ is the external temperature. In a district heating network, the water is distributed through a system of pipes. This is modeled by connecting above equations via suitable coupling conditions. They state the conservation of mass and energy in every node of the network. Furthermore, we assume a perfect mixture of incoming flows in a node resulting in all outflows to have the same temperature. Finally, the pressure level in a node has to coincide for all adjacent pipes. In order to close the model, as boundary conditions we set inflow temperature $T_0(t)$ and pressure level $p_0(t)$ at the CHP. The full system then has the following form

$$\text{HYD}\begin{cases} \text{ODE}\begin{cases} \partial_x v^i = 0 \\ \partial_t v^i + \frac{1}{\rho}\partial_x p^i = -\frac{\lambda}{2d^i} v^i|v^i| - g(\Delta b^i) \end{cases} \\ \text{CC}\begin{cases} \sum_{j\in J} A^j v^j \rho^j = 0 \\ p^i = p^j, \quad \text{for } i, j \text{ adjacent} \end{cases} \\ \text{BC}\{ \quad p(0) = p_0(t) \end{cases} \tag{2}$$

$$\text{CONSUMER}\left\{ A^k v^k (e(T^k) - e(T_{out}) = Q_k(t) \right. \tag{3}$$

$$\text{ENERGY}\begin{cases}\text{PDE}\left\{\ \partial_t e^i + v^i \partial_x e^i = -\dfrac{4k}{d^i}\left(T(e^i) - T_\infty\right)\right.\\[2ex] \text{CC}\left\{\begin{array}{l}\displaystyle\sum_{j\in J} A^j v^j e^j = 0\\[1ex] \qquad\qquad T(e^i) = T(e^j), \quad \text{for } i, j \text{ outgoing flows}\end{array}\right.\\[2ex] \text{BC}\left\{T(e^1(x=0,t)) = T_0(t)\right.\end{cases} \tag{4}$$

Equation (2) describes the hydraulics of the system. The first equation states the incompressibility, the second one is the balance of momentum. The coupling conditions (CC) state conservation of mass in every node as well as equality of pressure for all connected edges to a node. The superscript $i \in I$ indicates the specific edge, the quantities belong to, where I is the set of all edges. As boundary condition (BC) the pressure at the CHP is prescribed. Equation (3) describes the coupling between the hydraulics and the energy transport at consumer site, where A is the pipe's cross section, T_{out} is a fixed temperature level to which the water is cooled down and $Q_k(t)$ is the power demand of the consumer k. Equation (4) formulates the energy transport in the network with the advection PDE. The coupling conditions state conservation of energy inside the nodes and perfect mixture of energies therein. As boundary condition, the temperature at the inflow is prescribed.

3 Numerical Method

The full model for district heating networks (2)–(4) is a complex system of algebraic and partial differential equations. For its numerical solution, we use a splitting algorithm, i.e. for a given time t we first compute the flow with (2) and (3) for fixed temperature in time. Afterwards we update the energy using the new velocities. Such splitting reduces the accuracy of the full system to first order. By exploiting the special structure of the network, the flow can be solved efficiently. In the following, we focus on solving the energy Eq. (4). When a flow solver is needed, the method of [4] is used. The time step of the splitting is chosen according to the fastest waves in the energy model.

The evolution of energy density in the network is described by a network of linear scalar balance laws. For solving this kind of problems the Godunov scheme is commonly used, which in the linear case coincides with the Upwind scheme. Furthermore, there are some recent extensions to higher order methods for network of hyperbolic conservation laws [1]. All these classical schemes have in common, that they use a global time step for the whole system. This time step is determined by the minimal CFL bound on all edges

$$\Delta t_{net} = \min_i \Delta t^i \tag{5}$$

where Δt^i is the maximal time step of edge i. However, since the error of the schemes scales with the local CFL number, and the relative flow velocities between different pipes can strongly vary this may lead to large numerical diffusion.

3.1 Local Time Stepping

Motivated by above consideration we construct a Upwind-like scheme that decouples the time steps of every edge, such that the locally optimal time steps can be chosen [2, 6]. Note that an optimal CFL number can also be achieved with adaptive spacial discretization, however the remeshing and interpolation would be very costly, especially in the context of high order methods. Furthermore, we restrict ourselves to the homogeneous case, the extension to source terms is treated in Sect. 3.3.

The time step of an edge is chose according to

$$\int_t^{t+\Delta t} v(\tau)\mathrm{d}\tau = \Delta x \ . \tag{6}$$

Therefore the CFL number is equal to 1 for every local time step. Note that the velocity is piecewise constant in the considered time interval due to the different time step of the splitting. As with this definition the solution travels exactly one cell each time step, no computation for inner cells is needed. However, the update can only be performed, if the fluxes over the edge's boundaries are known for this time interval. In other words, the future time level of the current pipe can not exceed the future time levels of all adjacent edges, or to be more precise, current edge e_i has to fulfill

$$t^i + \Delta t^i \leq t^k + \Delta t^k \tag{7}$$

for all neighboring edges e_k. Whenever a pipe fulfills this condition, the numerical flux over its boundaries have to be computed. Note that the adjacent edges have already computed parts of the flux for some subinterval $[t^i, t^k] \subseteq [t^i, t^i + \Delta t^i]$. In this case we store the flux (or its polynomial coefficients in the high order case, see Sect. 3.2) in memory variables. The current pipe then just computes the remaining flux for the interval $[t^k, t^i + \Delta t^i]$ and the cell values are shifted in the direction of the flow, while the memory variable is emptied into the first cell. This procedure is continued, until a final time is reached. Since in general, the time steps do not exactly add up to the final time a classical upwind is performed in the last step. The procedure is schematically shown in Fig. 1 for a node connecting 3 edges.

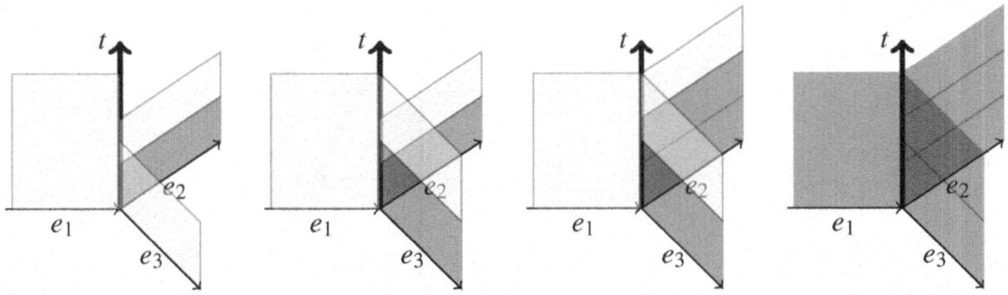

Fig. 1 Illustration of the local time steps at a single node

3.2 High Order Coupling

We can further increase the accuracy by incorporating a high order coupling at the nodes of the network. Therefore, we use a high order WENO reconstruction [5] and instead of cell means, we store the polynomial coefficients. The coupling conditions are formulated such that not only the energy density, but also its moments are conserved as well as the equality of all moments of the outgoing temperature up to a given order. In an update step, we then shift the polynomial coefficients instead of means. When the memory variables are cleared into the first cell of a pipe, we need to get a single polynomial representation out of several piecewise polynomials. This is done by solving a least squares problem for the new coefficients under the condition that the total mass must be contained [3]. In most cases during the computation only few (1–3) piecewise polynomials have to be combined, such that the numerical effort is lower than for storing means and using WENO reconstructions in each step.

3.3 Source Term

When incorporating the source term, we can exploit the fact that its coefficients are constant in space. When tracing a characteristic of the energy evolution, the change of the energy follows an ODE of the form $e_\tau = S(e)$. The evaluation of the source term does not necessarily be performed in every local update, but only when a given cell leaves the edge. We therefore keep track of the timespan each cell spends inside the given pipe (by integrating v) and solve the ODE for this whole interval just once.

4 Results and Conclusion

In several simulations, the local time stepping scheme has been compared to a high order ADER scheme with global time steps (see Fig. 2). When the velocity is set constant, the expected convergence rates of the schemes are reached. The overall

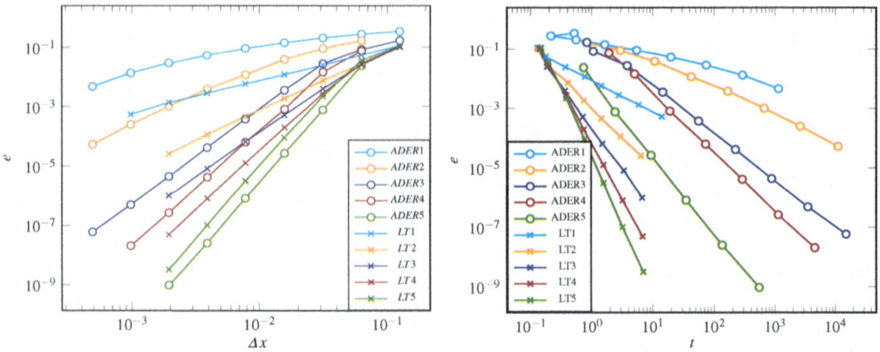

Fig. 2 Convergence plot for different schemes

error of the local time stepping scheme is lower because there is no numerical diffusion inside a pipe. The error only consists of the part arising from the coupling of different pipes and the part that depends on the resolution of the input signal. When comparing computation times, we notice a large difference between the two schemes. The advantage of the local time stepping is that in one pipe update, only $O(1)$ operations are needed, while for classical finite volume schemes you need $O(n)$, for the number of cells n. The overall computation time of the local time stepping therefore only scales linearly with the number of cells vs. quadratically for the ADER scheme. For more realistic settings, where a flow solver is involved only first order convergence can be expected. The errors of the local time stepping scheme are below those of the ADER scheme, but the gain in terms of computation time is not as big since the flow solver, which is identical for both schemes takes significant amount of time. To conclude, we constructed a numerical scheme that applies local time steps on the edges of the network. The advection inside the pipes is solved exactly which results in increasing accuracy and computational efficiency.

Acknowledgements This research was supported by Verbundprojekt 05M2018-EiFer: Energieeffizienz durch intelligente Fernwärmenetze. 05M18AMB-810303892568

References

1. Borsche, R., Kall, J.: ADER schemes and high order coupling on networks of hyperbolic conservation laws. J. Comput. Phys. **273**, 658–670 (2014)
2. Dumbser, M., Käser, M., Toro, E.F.: An arbitrary high-order Discontinuous Galerkin method for elastic waves on unstructured meshes-V. Local time stepping and p-adaptivity. Geophys. J. Int. **171** 695–717 (2007)
3. Dumbser, M., Zanotti, O., Loubère, R., Diot, S.: A posteriori subcell limiting of the discontinuous Galerkin finite element method for hyperbolic conservation laws. J. Comput. Phys. **278** 47–75 (2014)

4. Jansen, L., Pade, J.: Global unique solvability for a quasi-stationary water network model. In: Preprint series: Institut für Mathematik. Humboldt-Universität zu, Berlin (2013) https://www.mathematik.hu-berlin.de/de/forschung/pub/P-13-11
5. Jiang, G., Shu, C.: Efficient implementation of weighted ENO schemes. J. Comput. Phys. **126**, 202–228 (1996)
6. Müller, L.O., Blanco, P.J., Watanabe, S.M, Feijóo, R.A.: A high-order local time stepping finite volume solver for one-dimensional blood flow simulations: application to the ADAN model. Int. J. Numer. Methods Biomed. Eng. **32**, e02761, 36 (2016)
7. Toro, E.F., Millington, R.C., Nejad, L.A.M.: Towards very high order Godunov schemes. In: Godunov methods (Oxford, 1999), pp. 907–940. Kluwer/Plenum, New York (2001)

Stability Preserving Model Order Reduction for District Heating Networks

Markus Rein, Jan Mohring, Tobias Damm, and Axel Klar

Abstract Stability is one of the key properties when modeling a physical system on all model hierarchies. We focus on the case of hyperbolic differential algebraic equations dominated by advection at the example of district heating networks. For the transport dynamics, a solution of the corresponding Lyapunov inequality is presented ensuring stability. At the example of an existing network, we numerically demonstrate that stability also translates to the reduced order model (ROM).

1 Introduction

District heating denotes the transport of thermal energy from a centralized power plant to consumers using a network of transport pipelines. For each of the connected houses, a heat exchanger covers the time dependent power demand of customers by regulating the volume flow based on the currently available thermal energy. Due to its high flexibility towards the injection of different forms of energy, district heating has gained increasing importance for the supply with renewable energies [7]. Towards an efficient operation, finding an optimal control of such networks is a demanding mathematical and computational task. The high number of nodes and edges lead to state space dimensions in the order 10^6 making them large scale dynamical systems. Using model predictive control requires the simulation of the transport dynamics many times explaining the need for an efficient surrogate model. To this end, the reduced model should maintain desired properties such as stability from the full order model. Here port-Hamiltonian systems proved to be useful being formulated close to the underlying physical conservation laws [8]. Such systems automatically incorporate stability and passivity. Moreover it can be shown that

M. Rein (✉) · T. Damm · A. Klar
TU Kaiserslautern, Kaiserslautern, Germany
e-mail: markus.rein@itwm.fraunhofer.de

J. Mohring
Fraunhofer Institute for Industrial Mathematics ITWM, Kaiserslautern, Germany

© Springer Nature Switzerland AG 2019

I. Faragó et al. (eds.), *Progress in Industrial Mathematics at ECMI 2018*,
Mathematics in Industry 30, https://doi.org/10.1007/978-3-030-27550-1_51

these properties are passed to a reduced model obtained by Galerkin projections if the Hamiltonian energy matrix is included in the reduction process [3].

After presenting the mathematical model in Sect. 1, we sketch the derivation of a global Lyapunov matrix Q in Sect. 3. Subsequently, a reduced model is derived and its effectiveness is demonstrated for an existing heating network.

2 Model for District Heating

The transport of the energy density φ within a pipeline is modeled by one dimensional Euler equations. Since water in the liquid phase is the transport medium, the incompressible limit is assumed, simplifying the conservation of mass to $v_x = 0$. The remaining Euler equations for conservation of momentum and internal energy density read

$$0 = p_x + \frac{\lambda\rho}{2d}|v|v + \rho g h_x \tag{1}$$

$$\dot{\varphi} = -v\varphi_x - \frac{4k}{d}(T(\varphi) - T_e). \tag{2}$$

The change of pressure p_x over a pipeline is modeled by frictional forces according to the Darcy-Weisbach equation, where λ is a dimensionless friction factor, d is the pipeline diameter, v the advection velocity, and ρ the density. The quantities ρ and λ are assumed to be constants within this contribution. Gravitational forces are captured by the height difference h_x, and the gravitational constant g. For the typical dynamics of heating networks, acceleration is small compared to friction and gravitation, which is why it is neglected here. This makes (1) an algebraic equation after integration over the pipeline length. The advection of the energy density φ in (2) incorporates an additional sink term due to conduction of heat with transfer coefficient k to the environment with temperature T_e.

To allow for a numerical treatment of the partial differential equation (PDE), we perform a spatial discretization of (2) employing the upwind scheme yielding a total number of n finite volume cells. For the description of the network, we introduce the set $\mathscr{E}$ containing E edges which represent all pipelines. More specifically, pipeline $i \in \mathscr{E}$ contains the local set of cells $\mathscr{N}_i = [1, .., n_i]$ with cardinal number n_i. In the following the resulting system of ordinary differential equations is considered. To connect incoming and outgoing pipelines within the network, additional algebraic constraints at the junctions have to be posed. A prominent choice is the conservation of energy over node N yielding

$$\sum_{i \in N^-} q_i \varphi_{i,1} = \sum_{j \in N^+} q_j \varphi_{j,n_j}, \tag{3}$$

where $q_i = \Phi_i v_i$ is the volume flow on pipeline i formed by its cross section Φ_i and velocity v_i. N^- and N^+ denote edges exiting and entering node N. By instantaneous mixing of energy flows within N, the energy density is identical for all outgoing pipes, $\varphi_{i,1} = \varphi^N, \forall i \in N_-$. Here $\varphi_{e,c}$ is the finite volume cell c of edge e in flow direction. Hence, the energy density φ^N adjacent to pipe i is given by

$$\varphi^N = \frac{\sum_{j\in N_+} \Phi_j v_j \varphi_{j,n_j}}{\sum_{i\in N_-} \Phi_i v_i}. \tag{4}$$

Similarly, volume conservation over node N is assumed, yielding

$$\sum_{j\in N_+} \Phi_j v_j = \sum_{k\in N_-} \Phi_k v_k. \tag{5}$$

This allows to write the network dynamics as

$$\dot{\varphi} = A(v)\varphi + B_T(v)u_T(t), \tag{6}$$

$$y = C\varphi, \tag{7}$$

$$0 = v - Kq, \tag{8}$$

$$0 = G[v_i \cdot |v_i|]_{i\in\mathscr{E}}, \tag{9}$$

$$0 = [q_i \cdot y_i]_{i\in\mathscr{H}} - u_H(t), \tag{10}$$

where (6) mirrors the advection of the energy density subject to the injected energy $u_T(t)$, with $B_T(v) \in \mathbb{R}^{n\times 1}$. Both the upwind scheme and the conservation of energy are encoded in the velocity dependent matrix $A(v) \in \mathbb{R}^{n\times n}$. Equation (8) uses the solution $K \in \mathbb{R}^{E\times L}$ of (5), to describe the pipeline velocities by L independent volume flows q. Kirchhoff's circuit law presented in (9) claims that the sum of pressure differences over a loop within the network equals 0, where $G \in \mathbb{R}^{(L-H)\times E^2}$. Finally in (10), the demanded power consumption u_H is provided by energy density and volume flow at houses $\mathscr{H}$ with cardinal number H. Consequently the velocity changes dynamically with the energy density at the houses and their time dependent consumption $u_H(t)$.

3 Stability of the Discretized Model

In the following, we solely focus on the adjective transport on the network (6) and consider the remaining algebraic equations as generators for the velocity field v acting as a parameter to the transport system. As a consequence, the structure of the corresponding Jacobian bases on the upwind discretization mentioned in the previous chapter combined with the energy conservation (3). This results in two possible types of cell coupling on a network of pipelines and junctions. Among these are coupling with neighboring cells in the pipeline or with border cells coupling to incoming pipelines at junctions,

$$\dot{\varphi}_{i,j} = -\frac{v_i}{h_i}(\varphi_{i,j} - \varphi_{i,j-1}), \quad j \in \mathcal{N}_i \tag{11}$$

$$\dot{\varphi}_{i,1} = -\frac{v_i}{h_i}(\varphi_{i,1} - \varphi^N(\{\varphi_{j,n_j} | j \in N_+\})). \tag{12}$$

Here $\varphi_{i,j}$ is the finite volume cell j in flow direction of edge i. For a fixed velocity field $v = \bar{v}$, A is considered as the Jacobian of the ODE system (6),

$$A_{f(i,j),f(k,l)}(\bar{v}) \equiv \frac{\partial \dot{\varphi}_{i,j}}{\partial \varphi_{k,l}}(\bar{v}). \tag{13}$$

Rows and columns of the matrix A are mapped to the edge- and cell indices i, j by the ordering function

$$f(e, c) \equiv c + \sum_{k=1}^{e-1} n_k, \quad e \in \mathcal{E}, \quad c \in \mathcal{N}_e. \tag{14}$$

This allows to formulate the following theorem.

Theorem 1 *For every fixed velocity field $\bar{v}$ satisfying volume conservation (5), there exists a global, diagonal, positive definite energy matrix $Q \in \mathbb{R}^{n \times n}$, such that*

$$M = (QA)^T + (QA) \leq 0. \tag{15}$$

Remark: Q can be constructed with positive diagonal elements $Q_i \equiv diag(Q) = Q_{f(i,j),f(i,j)} = \Phi_i h_i$, $i \in \mathcal{E}$, $j \in \mathcal{N}_i$. The latter carry the volume $\Phi_i h_i$ of each of the discretization cells on edge i.

For a detailed proof we refer to [6], Theorem 1. A sketch of the proof starts by noting that the diagonal elements of M are negative,

$$M_{f(i,j),f(i,j)} = (QA)_{f(i,j),f(i,j)} + (A^T Q)_{f(i,j),f(i,j)} = -2Q_i \frac{v_i}{h_i}, \tag{16}$$

for $i \in \mathcal{E}$, $j \in \mathcal{N}_i$. Subsequently weak diagonal dominance is shown for each row of a representative edge $i \in \mathcal{E}$ of the symmetric matrix M. This is done by decomposing the cells on i into the three classes of incoming, inner, and outgoing ones. For the row describing the first cell $M_{f(i,1),k}$, $k \in \mathcal{N}_i$ volume conservation is sufficient for proving diagonal dominance. For the last cell on each pipeline, the choice $Q_{f(i,j),f(i,j)} = \Phi_i h_i$, $\forall i \in \mathcal{E}$, $j \in \mathcal{N}_i$, is sufficient for diagonal dominance. In case of inner cells no further assumption is necessary. Finally Sylvester's criterion concludes that M is a negative semi-definite matrix.

4 Derivation of the Reduced Model and Numerical Results

To numerically demonstrate that stability also translates to the ROM, we simulate a part of an existing heating network shown in Fig. 1a. To obtain the ROM, we transform the system to coordinates in which Q equals identity [5]. To account for the influence of the transport velocity changing in time, the reduction is performed at representative realizations of v. To this end, a moment-matching technique in frequency space is used. Realizations of v are picked exploiting a greedy strategy [2, 4]. Afterwards, the local projection matrices are combined using a singular value decomposition [1] forming the global Galerkin projection $V \in \mathbb{R}^{n \times r}$ inducing the reduced dimension r. After application to the full order model, the ROM reads

$$\dot{\varphi}^r = V^T A(v) V \varphi^r + V^T B_T(v) u_T(t) \tag{17}$$

$$y^r = C V \varphi^r \tag{18}$$

$$0 = g(u_H, v, y^r), \tag{19}$$

where (19) abbreviates the constraints (8–10). To ensure stability, the algebraic equations remain unreduced, and act as generators for the volume flows steering the advection on the network. Retaining the algebraic equations plays a key role, since it ensures that every currently active reduced model is the reduction of a stable, full order system. Initializing a constant energy on the network, we simulate the input signal $u_T(t) = 0.4 - 0.2\cos(\omega t)$ for constant consumption at the houses. The

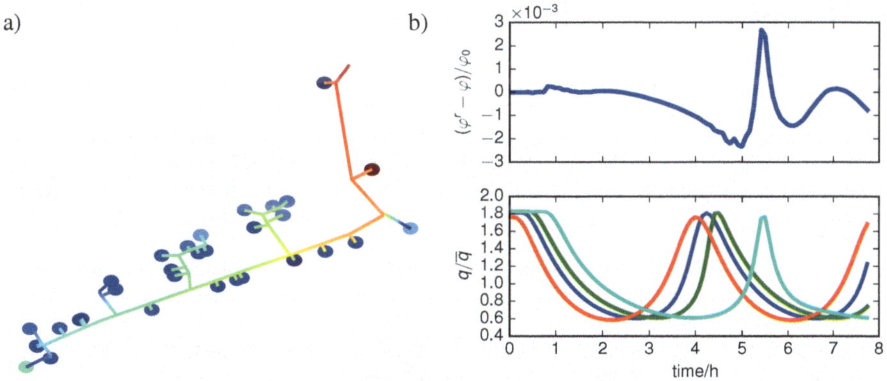

Fig. 1 (**a**) Visualization of the considered network including pipelines (colored lines), and 32 consumers (circles). Colors on pipelines show the energy density, and colored circles visualize the current volume flow. The top part in (**b**) shows the in-sample error of the reduced model φ^r compared to the PDE solution φ at the observable with the largest approximation error. φ_0 denotes the energy density of the cooled fluid returning to the power plant. The lower part shows the normalized volume flow at representative consumers emphasizing the nonlinear character of the transport

frequency $\omega = 4.49\times10^{-4}$ s is an upper bound for the typical operation frequencies. To define the fidelity of the ROM, the PDE solution is estimated by extrapolating the results of full order models with decreasing h. Hereafter an error bound δ is set, which is defined as the largest relative l_2 error of all 32 observables located at the consumers. An upwind discretization including 2269 cells allows for an acceptable error $\delta = 2.5\times10^{-3}$ and is the basis for the ROM. The given problem is challenging in terms of reduction due to the nonlinearity introduced by changing volume flows and the high number of observables. Despite this fact, the ROM suffices to use $r = 53$ states to reproduce the full dynamics to $\delta = 2.5 \times 10^{-3}$, cf. Fig. 1b.

5 Summary

We presented a global, diagonal solution of the Lyapunov inequality for the advection problem on a network discretized in space. Local stability is ensured for the dynamically changing velocity field by claiming volume conservation. Preserving this algebraic constraint in the reduction translates stability to the reduced model.

Acknowledgements We acknowledge the financial support by the Federal Ministry for Economic Affairs and Energy of Germany in the framework of the project *"Verbundvorhaben:EnEff:Wärme - DYNEEF: Dynamische Netzsimulation zur Effizienzsteigerung und Emissionsreduzierung in der Fernwärmeerzeugung - Schwerpunkt: Optimierung" (Förderkennzeichen: 03ET1346B).*

References

1. Benner, P., Gugercin, S., Willcox, K.: A survey of projection-based model reduction methods for parametric dynamical systems. SIAM Rev. **57**(4), 483–531 (2015)
2. Bui-Thanh, T., Willcox, K., Ghattas, O.: Model reduction for large-scale systems with high-dimensional parametric input space. SIAM J. Sci. Comput. **30**(6), 3270–3288 (2008)
3. Gugercin, S., Polyuga, R.V., Beattie, C.A., van der Schaft, A.: Structure-preserving tangential interpolation for model reduction of port-Hamiltonian systems. Automatica **48**(9), 1963–1974 (2012)
4. Haasdonk, B., Ohlberger, M.: Reduced basis method for finite volume approximations of parametrized linear evolution equations. ESAIM Math. Model. Numer. Anal. **42**, 277–302 (2008)
5. Polyuga, R.V., van der Schaft, A.: Structure preserving model reduction of port-Hamiltonian systems by moment matching at infinity. Automatica **46**(4), 665–672 (2010)
6. Rein, M., Mohring, J., Damm, T., Klar, A.: Model order reduction of hyperbolic systems at the example of district heating networks (2019). arXiv:1903.03342v1
7. Rezaie, B., Rosen, M.A.: District heating and cooling: review of technology and potential enhancements. Appl. Energy **93**, 2–10 (2012)
8. van der Schaft, A.J., Maschke, B.M.: Port-Hamiltonian systems on graphs. SICON **51**(2), 906–937 (2013)

Numerical Simulation of Heterogeneous Steady States for a Reaction-Diffusion Degenerate Keller-Segel Model

Georges Chamoun, Moustafa Ibrahim, Mazen Saad, and Raafat Talhouk

Abstract In this paper, a linear instability criterion is carried out to show the existence of heterogenous spatial patterns for a degenerate Keller-Segel model. We show that the nonlinear system behaves asymptotically as a linear combination of eigenvectors associated to highest eigenvalues. Finite volume method is implemented to investigate numerically the appearance of heterogeneous spatial patterns in a two-dimensional space for the given model. The nonlinear solution is compared to the predicted nonhomogeneous steady solution obtained by the analysis of the linear instability.

1 The Degenerate Keller-Segel Model

Let Ω be an open bounded polygonal and connected subset of $\mathbb{R}^n$ defined by $\Omega = \mathbb{T}^n = \prod_{i=1}^{n}]0, \pi[$ for $n = 1, 2$ or 3, and let $t_f > 0$ be a fixed finite time. We are interested in the degenerate Keller-Segel system [10] modeling the chemotaxis process, in the context of volume-filling phenomenon [13] given by the

G. Chamoun
Lebanese University, Faculty of Science IV. Laboratory of Mathematics-EDST, Hadath, Lebanon
e-mail: georgeschamoun@usek.edu.lb

M. Ibrahim (✉)
Department of Mathematics, College of Engineering and Technology, American University of the Middle East, Eqaila, Kuwait
e-mail: moustafa.ibrahim@aum.edu.kw

M. Saad
Ecole Centrale de Nantes, Laboratoire de Mathématiques Jean Leray, Nantes, France
e-mail: mazen.saad@ec-nantes.fr

R. Talhouk
Lebanese University, Faculty of Science I. Laboratory of Mathematics-EDST, Hadath, Lebanon
e-mail: rtalhouk@ul.edu.lb

© Springer Nature Switzerland AG 2019
I. Faragó et al. (eds.), *Progress in Industrial Mathematics at ECMI 2018*, Mathematics in Industry 30, https://doi.org/10.1007/978-3-030-27550-1_52

set of parabolic equations:

$$\begin{cases} \partial_t U - \nabla \cdot (a(U) \nabla U) + \nabla \cdot (\chi(U) \nabla V) = 0, & \text{in } Q_T = \Omega \times (0, T), \\ \partial_t V - d\, \Delta V = \alpha U - \beta V, & \text{in } Q_T = \Omega \times (0, T), \end{cases} \tag{1}$$

This system of equations is supplemented with the following homogeneous Neumann boundary conditions on $\Sigma_T = \partial\Omega \times (0, T)$,

$$a(U) \nabla U \cdot \eta = 0, \ \nabla V \cdot \eta = 0, \tag{2}$$

where η is the exterior unit normal vector to $\partial\Omega$ outward to Ω. The initial conditions on Ω are given by,

$$U(\mathbf{x}, 0) = U_0(\mathbf{x}), \qquad V(\mathbf{x}, 0) = V_0(\mathbf{x}). \tag{3}$$

In the above model, the density of the cell-population and the chemoattractant concentration are represented by $U = U(\mathbf{x}, t)$ and $V = V(\mathbf{x}, t)$ respectively. Next, $a(U)$ is a density-dependent diffusion coefficient. Furthermore, the function $\chi(U)$ is the chemoattractant sensitivity. Finally, $d > 0$ represents the diffusion coefficient of V while the positive constants α and β describe, respectively, the rate of production and degradation of the chemoattractant.

We give the main assumptions made about the system:

(A1) The cell-density diffusion $a : [0, 1] \longrightarrow \mathbb{R}^+$ is a continuous function satisfying: $a(0) = a(1) = 0$, and $a(U) > 0$ for $0 < U < 1$.
(A2) The chemosensitivity $\chi : [0, 1] \longrightarrow \mathbb{R}^+$ is a continuous function satisfying: $\chi(0) = \chi(1) = 0$, and $\chi(U) > 0$ for $0 < U < 1$.
(A3) The initial function U_0 and V_0 are two functions in $L^2(\Omega)$ such that, $0 \leq U_0 \leq 1$ and $V_0 \geq 0$.

The degenerate Keller-Segel system has been studied numerically in the past years to show the isotropic and anisotropic chemotaxis process (see e.g. [1–3, 9]. In this paper, we show an alternate avenue potentially leading to pattern formation via chemotaxis, inspired by the methods used by the authors of [7, 8, 12].

2 Growing Modes in the Keller-Segel Model

In this section, we summarize the classical linear instability criterion in order to show the existence of heterogenous patterns for the degenerate Keller-Segel model (1)–(3). When the diffusion terms are ignored, a uniform constant solution

$$U(\mathbf{x}, t) \equiv \overline{U}, \qquad V(\mathbf{x}, t) \equiv \overline{V}$$

forms a nontrivial homogeneous steady state provided

$$\alpha \overline{U} = \beta \overline{V}; \qquad \overline{U} \in]0, 1[. \tag{4}$$

In this paper, we study the nonlinear evolution of a perturbation

$$u(\mathbf{x}, t) = U(\mathbf{x}, t) - \overline{U}, \qquad v(\mathbf{x}, t) = V(\mathbf{x}, t) - \overline{V}$$

around $[\overline{U}, \overline{V}]$, which satisfies the equivalent system:

$$\begin{cases} \partial_t u - \nabla \cdot \left(a\left(u + \overline{U}\right) \nabla u\right) + \nabla \cdot \left(\chi\left(u + \overline{U}\right) \nabla v\right) = 0 \\ \partial_t v - d\,\Delta v = \alpha u - \beta v \end{cases} \tag{5}$$

Then, the system can be written into a matrix form:

$$\partial_t W = \underbrace{\mathcal{L}(W)}_{\text{linear term}} + \underbrace{\mathcal{N}(W)}_{\text{nonlinear term}} \tag{6}$$

where the corresponding **linearized** Keller-Segel system then takes the form:

$$\begin{cases} \partial_t u = a\left(\overline{U}\right) \Delta u - \chi\left(\overline{U}\right) \Delta v, \\ \partial_t v = d \Delta v + \alpha u - \beta v. \end{cases} \tag{7}$$

We know that the unique solution of system (7) (see e.g. [12]) is given by

$$\begin{aligned} W(\mathbf{x}, t) = [u(\mathbf{x}, t), v(\mathbf{x}, t)] &= \sum_{q \in \mathbb{N}^n} \left\{ w_q^- r_q^- \exp\left(\lambda_q^- t\right) + w_q^+ r_q^+ \exp\left(\lambda_q^+ t\right) \right\} e_q(\mathbf{x}) \\ &\equiv e^{\mathcal{L}t} W(\mathbf{x}, 0) \end{aligned} \tag{8}$$

where $q = (q_1, \ldots, q_n) \in \mathbb{N}^n$, $e_q(\mathbf{x}) = \Pi_{i=1}^n \cos(q_i x_i)$, and r_q^+ (resp. r_q^-) are the positive (resp. negative) eigenvectors corresponding to the positive eigenvalues λ_q^+ (resp. negative eigenvalues λ_q^-) of the stability matrix A (i.e. the Jacobian matrix of system (7) computed at the steady state $\left(\overline{U}, \overline{V}\right)$.

2.1 Main Result

The main result in this paper is given by the upcoming theorem. It interprets the behavior of the nonlinear solution compared to a heterogeneous stationary solution, which gives finally a mathematical description of the pattern formation for the Keller-Segel model.

Theorem 1 *Consider the instability criterion* $q^2\{a\left(\overline{U}\right)\left(dq^2+\beta\right)-\chi\left(\overline{U}\right)\alpha\}<0$, *and let:*

$$W_0(\mathbf{x}) = \sum_{q\in\mathbb{N}^n}\left\{w_q^- r_q^- + w_q^+ r_q^+\right\} e_q(\mathbf{x}) \in L^2(\Omega). \tag{9}$$

If the initial perturbation of the steady state $[\overline{U}, \overline{V}]$ *is* $W(\mathbf{x}, 0) = W_0$, *then its nonlinear evolution* $W(t, \mathbf{x})$ *satisfies*

$$\left\| W(\mathbf{x}, t) - e^{\lambda_{max} t} \sum_{q\in Q_{\max}} w_q^+ r_q^+ e_q(\mathbf{x}) \right\|_{L^2(\Omega)} \leq C \left\|W_0\right\|_{L^2(\Omega)} \left\{e^{-\nu t} + e^{K_1 t}\right\} e^{\lambda_{\max} t}, \tag{10}$$

where C, K_1, ν *are positive constants, and* $Q_{max} = \{q \in \mathbb{N}^n$ *such that* $\lambda_q^+ = \lambda_{\max}\}$.

Proof The proof of Theorem 1 is inspired from [7, 8] and using similar estimates developed in [4].

3 Pattern Formation for Keller-Segel Model in Two Dimensions

In this section, we will perform a numerical simulation in order to investigate the pattern formation of the Keller-Segel model (1)–(3). Our aim is to show that a solution of the nonlinear degenerate Keller-Segel model behaves asymptotically as $W_\infty(\mathbf{x}, t) = e^{\lambda_{max} t} \sum_{q\in Q_{\max}} w_q^+ r_q^+ e_q(\mathbf{x})$. The computation of spatial distribution of the solution W_∞ is given explicitly since the eigenvalues of the matrix A are known and given by $r_q^\pm = \left[\frac{\lambda_q^\pm + dq^2 + \beta}{\alpha}, 1\right]$ and the coefficients w_q^+ are also known and given by the initial condition. However, the numerical computation of the solution of the nonlinear Keller-Segel model, namely $W(\mathbf{x}, t)$, needs a sophisticate numerical scheme to handle with the asymptotic behavior and to reach the spatially nonhomogeneous steady solution. To approximate the nonlinear solution $W(\mathbf{x}, t)$, the numerical scheme adopts a method of Finite Volume (see [5]) for the diffusion terms, and the subsequent system of ODEs is then discretized in time using forward Euler method [6]. This choice of finite volume scheme is essential to ensure the discrete maximum principle on the solutions of system (1)–(3) (e.g. see [1]). On the other hand, it is well known that upwind technique for the chemoattractant term preserves the local conservativity of the numerical fluxes [1, 11], i.e. the numerical flux is conserved from one discretization cell to its neighbor. Finally, for the upwind technique, we use Engquist–Osher's scheme where the numerical flux

$\mathcal{F}$ is defined by

$$\mathcal{F}(a, b; c) = c^{+}\left(\chi_{\uparrow}(a) + \chi_{\downarrow}(b)\right) - c^{-}\left(\chi_{\uparrow}(b) + \chi_{\downarrow}(a)\right),$$

where $s^{+} = \max(s, 0)$, $s^{-} = \max(-s, 0)$, $\chi_{\uparrow}(z) := \int_{0}^{z} \left(\chi'(s)\right)^{+} ds$, and $\chi_{\downarrow}(z) := -\int_{0}^{z} \left(\chi'(s)\right)^{-} ds$.

In what follows, we consider a numerical test to investigate the pattern formation for system (1)–(3). We focus on the pattern formation for the first component $u(\mathbf{x}, t)$ of the function $W(\mathbf{x}, t)$. We perform our test on an unstructured triangular mesh of the space domain $\Omega = (0, 1) \times (0, 1)$ (a normalization of the space by considering the change of variable $\mathbf{x} := \mathbf{x}/\pi$ in system (1)–(3)). We verify that the space domain satisfies the orthogonality condition and finally we assume zero-flux boundary conditions.

In this numerical test, we fix $\alpha = 5$, $\beta = 11$, $d = 0.01$, $\overline{U} = 0.5$, $\Delta t = 0.01$ (time step) $a(U) = d_U(U(1-U))^2$, $d_U = 0.01$, $\chi(U) = \zeta(U(1-U))$, and $\zeta = 0.01$, and for the numerical flux $\mathcal{F}$ we set: $\chi_{\uparrow}(z) = \chi\left(\min\left\{z, \frac{1}{2}\right\}\right)$ and $\chi_{\downarrow}(z) = \chi\left(\max\left\{z, \frac{1}{2}\right\}\right) - \chi\left(\frac{1}{2}\right)$.

Figure 1 shows the plot of the initial condition $U_0(\mathbf{x})$ with a small perturbation with an order of magnitude equal to 3×10^{-3} around the steady state $\overline{U}$ for the function of $U(\mathbf{x}, t)$ solution to system (1)–(3).

Figure 2 shows the spatial nonlinear evolution with respect to the time, we remark at the beginning that the initial condition leads to some diffusions in the space which gives rise to some aggregations of densities that start the merging process

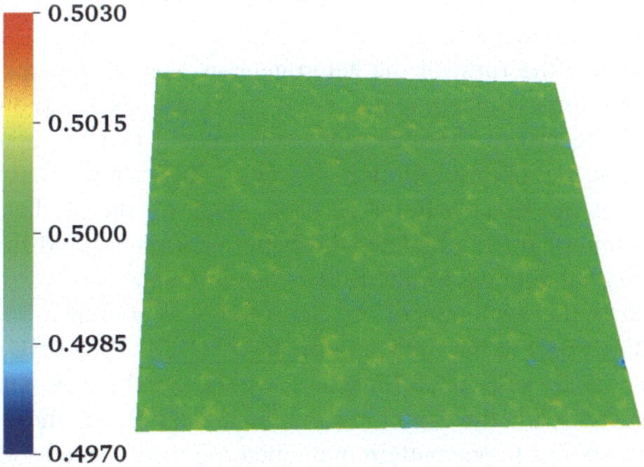

Fig. 1 Initial condition $U_0(\mathbf{x})$ with a small perturbation around $\overline{U}$, $U_0(\mathbf{x}) = \overline{U} \pm 10^{-3}$

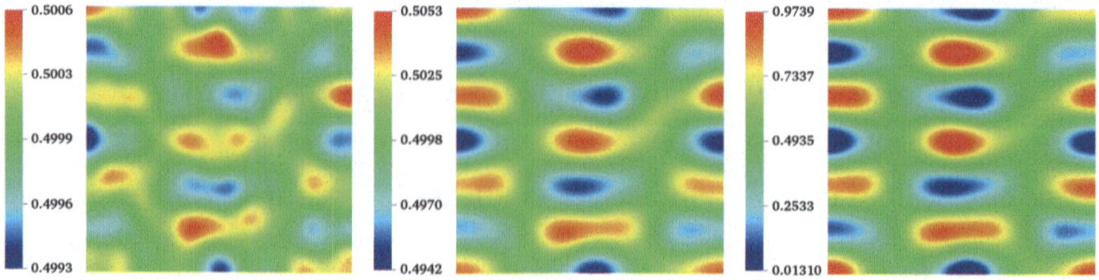

Fig. 2 *From left to right:* Nonlinear evolution of the nonlinear evolution $U(\mathbf{x}, t)$ at $t = 2.5$, $t = 325$, and $t = 997.5$

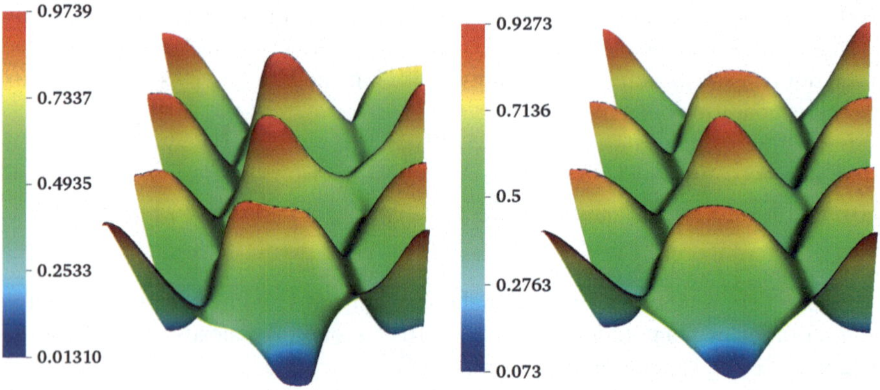

Fig. 3 Spatial patterns for $t = 997.5$ in 3D for the nonlinear approximated evolution $U(\mathbf{x}, t)$ (to the left) and for the asymptotic solution $W_\infty(\mathbf{x}, t)$ (to the right)

in such a way to generate spatial patterns which are nothing than a description of the heterogeneous stationary state.

In Fig. 3, we give the form of the heterogeneous spatial patterns associated to approximated solution at the moment $t = 997.5$. As well as, we show for the same moment of time the form of the heterogeneous spatial patterns associated to the computation of the asymptotic solution $W_\infty(\mathbf{x}, t)$. It is shown that the numerical solutions produces the same patterns as these given by the predicted asymptotic solution. This numerical test validates the pattern formation and the existence of heterogeneous spatial patterns for the Keller-Segel model.

Figure 4 shows the evolution of the approximated nonlinear solution of system (1)–(3) with respect to the time at point $P(0.5, 0.5)$. The curve shows that we have reached the steady solution; for instance, we see that the approximated solution at the point P increased at the beginning and then it stabilized after $t = 700$ so that the point P corresponds to one pattern in the heterogeneous steady state.

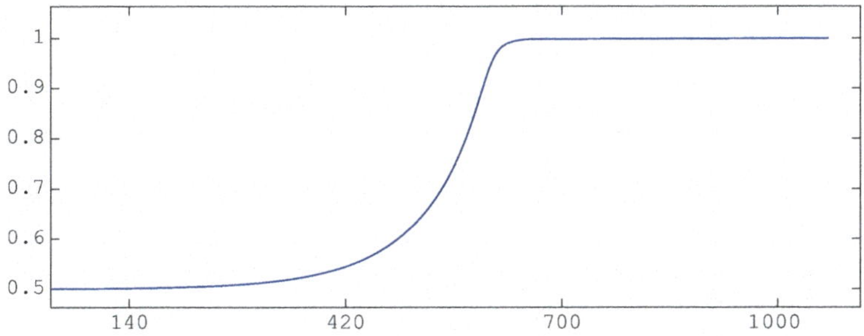

Fig. 4 Time evolution of the approximated nonlinear solution $U(\mathbf{x}, t)$ at point $P(0.5, 0.5)$

References

1. Andreianov, B., Bendahmane, M., Saad, M.: Finite volume methods for degenerate chemotaxis model. J. Comput. Appl. Math. **235**(14), 4015–4031 (2011)
2. Cancès, C., Ibrahim, M., Saad, M.: Positive nonlinear CVFE scheme for degenerate anisotropic Keller-Segel system. SMAI J. Comput. Math. **3**, 1–28 (2017)
3. Chamoun, G., Saad, M., Talhouk, R.: Monotone combined edge finite volume–finite element scheme for anisotropic Keller–Segel model. Numer. Methods Partial Differ. Equ. **30**(3), 1030–1065 (2014)
4. Chamoun, G., Saad, M., Talhouk, R.: A coupled anisotropic chemotaxis-fluid model: the case of two-sidedly degenerate diffusion. Comput. Math. Appl. **68**(9), 1052–1070 (2014)
5. Eymard, R., Gallouët, T., Herbin, R.: Finite volume methods. Handb. Numer. Anal. **7**, 713–1018 (2000)
6. Griffiths, D.F., Higham, D.J.: Numerical Methods for Ordinary Differential Equations: Initial Value Problems (Springer, Berlin, 2010)
7. Guo, Y., Hwang, H.J.: Pattern formation (II): the Turing instability. Proc. Am. Math. Soc. **135**(9), 2855–2866 (2007)
8. Guo, Y., Hwang, H.J.: Pattern formation (I): the Keller–Segel model. J. Differ. Equ. **249**(7), 1519–1530 (2010)
9. Ibrahim, M., Saad, M.: On the efficacy of a control volume finite element method for the capture of patterns for a volume-filling chemotaxis model. Comput. Math. Appl. **68**(9), 1032–1051 (2014)
10. Keller, E.F., Segel, L.A.: Model for chemotaxis. J. Theor. Biol. **30**(2), 225–234 (1971)
11. LeVeque, R.J.: Conservative methods for nonlinear problems. In: Numerical Methods for Conservation Laws, pp. 122–135 (Birkhäuser, Basel, 1990)
12. Murray, J.D.: Mathematical biology II: spatial models and biomedical applications. In: Interdisciplinary Applied Mathematics (Springer, New York, 2001)
13. Painter, K.J., Hillen, T.: Volume-filling and quorum-sensing in models for chemosensitive movement. Can. Appl. Math. Q. **10**(4), 501–543 (2002)

Part IV
Optimization and Control of Industrial Problems

Simulation and Multi-Objective Optimization of Thermal Distortions for Milling Processes

Alfred Schmidt, Carsten Niebuhr, Jonathan Montalvo-Urquizo, and Maria G. Villarreal-Marroquin

Abstract During a machining process, the produced heat results in thermomechanical deformation of the workpiece and thus an incorrect material removal by the cutting tool, which may exceed given tolerances.

We present a numerical model based on an adaptive finite element simulation for thermomechanics, which takes into account both the approximation of the temperature field as well as the approximation of the time dependent domain.

Control of the milling parameters and tool path can be used to minimize the final shape deviation. A multi-objective approach can try to additionally reduce the tool wear. We present results from a simulation-based optimization approach for a simplified workpiece.

1 Introduction

During a milling process, heat introduced by the cutting into the workpiece leads to thermoelastic deformation of the workpiece. As a consequence, the milling tool removes not the desired amount of material, but more or less. This can lead to a substantial shape error.

Mathematical modelling, simulation, and optimization can be used in order to predict and reduce this shape error. The time dependent shape of the workpiece adds another challenge to models and numerical methods. We present an approach based on a hybrid dexel/adaptive finite element model (Sect. 2) and a simulation-based multi-objective optimization method (Sect. 3). Applied to a model problem, we are able to reduce the shape error while additionally paying attention to the milling tool wear.

A. Schmidt (✉) · C. Niebuhr
University of Bremen, Center for Industrial Mathematics, Bremen, Germany
e-mail: schmidt@math.uni-bremen.de

J. Montalvo-Urquizo · M. G. Villarreal-Marroquin
Modeling Optimization and Computing Technology SAS de CV, Monterrey, Mexico
e-mail: jmontalvo@moctech.com.mx; mvillarreal@moctech.com.mx

© Springer Nature Switzerland AG 2019
I. Faragó et al. (eds.), *Progress in Industrial Mathematics at ECMI 2018*,
Mathematics in Industry 30, https://doi.org/10.1007/978-3-030-27550-1_53

2 Model and Numerical Method

With stress tensor σ depending on displacements u and temperature θ, $\sigma(u,\theta) = 2\mu(\theta)(\epsilon(u) + (\lambda(\theta)tr(\epsilon(u)) - 3\alpha(\theta-\theta_0))\,I$ with strain tensor $\epsilon(u) = \frac{1}{2}(\nabla u + \nabla u^T)$ and thermal expansion determined by α, the thermomechanical problem with quasi-stationary mechanics is given in strong form as: Find temperature θ and deformation u such that

$$\dot{\theta} - \operatorname{div}(\kappa\nabla\theta) = 0 \ \text{ in } \Omega(t), \tag{1}$$

$$-\operatorname{div}\sigma(u,\theta) = 0 \ \text{ in } \Omega(t) \tag{2}$$

for $t \in (0, t_{end})$ with initial condition $\theta(0) = \theta_0$ and boundary conditions

$$\kappa\nabla\theta\cdot\nu = g_N \text{ on } \partial\Omega(t), \quad \sigma(u,\theta)\cdot\nu = f_N \text{ on } \Gamma_N(t), \quad u = 0 \text{ on } \Gamma_D. \tag{3}$$

The heat flux g_N over the boundary is given by a cooling condition or by the flux produced during the milling process, as are the forces f_N. The workpiece is clamped at Γ_D and $\partial\Omega(t) = \Gamma_D \cup \Gamma_N(t)$.

The model is based on small deformations on a reference configuration with $\Omega(t) \subset \Omega(0)$ and uses the linearized elasticity tensor. The time dependent domain $\Omega(t)$ and its moving boundary $\Gamma_N(t)$ are given via the engagement of the milling tool. A macroscopic model is used here, where a boolean operation cuts in every time step the rotated tool sweep volume from the current, deformed workpiece geometry. For an efficient implementation, a dexel model is used here [2]. This results in a description of the domain $\Omega(t)$ which is independent of the numerical mesh of the finite element method. As microscopic effects like the creation and removal of single chips are not included in the macroscopic model, a suitable process model has to be used to compute heat fluxes and forces acting on the workpiece, resulting in the Neumann data g_N and f_N in (3).

Approximation of the Time-Dependent Workpiece Geometry A finite element discretization of the equations is used, based on a tetrahedral mesh and piecewise linear finite element spaces. Time discretization of the problem is done using an implicit Euler discretization, use of higher order methods would need further investigation due to the time dependent domain. As the model equations are given in the reference configuration, the domain is only getting smaller, $\Omega(t_2) \subseteq \Omega(t_1)$ for $t_2 \geq t_1$. Thus, data from the last time step t_{n-1} are always available in $\Omega(t_n)$. We approximate the time dependent domain $\Omega(t_n)$ by a discrete domain $\Omega_h(t_n)$ which is given by the union of all elements of a given triangulation $S_h(t_n)$ of $\Omega(0)$ which have a nonzero intersection with $\Omega(t_n)$,

$$\bar{\Omega}_h(t_n) := \bigcup\{S \in S_h(t_n) : S \cap \Omega(t_n) \neq \emptyset\}.$$

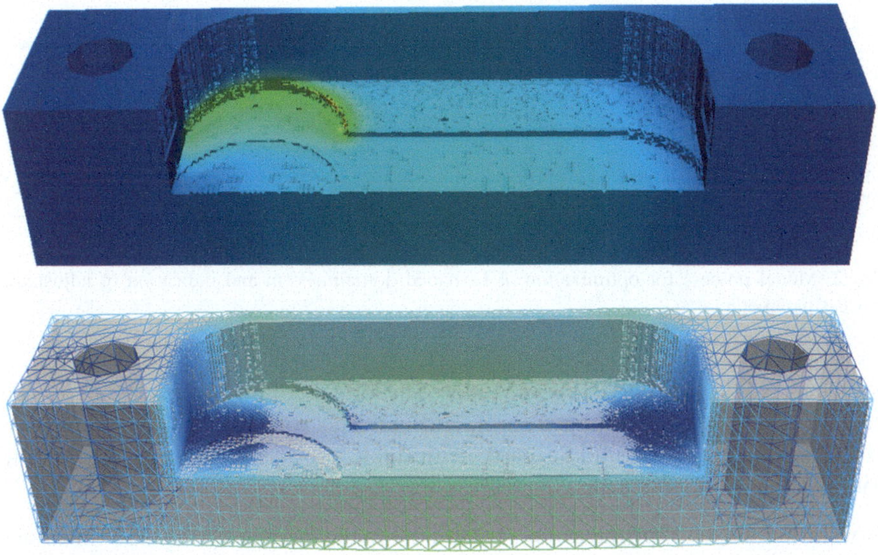

Fig. 1 Simulation of milling of a thin-walled reference workpiece. Temperature (top) and deformation (bottom, amplified by factor 100)

In order to get a convergent approximation of the solution, boundary conditions for flux and forces (3) need to be transferred suitably from $\partial\Omega(t_n)$ to the possibly rough discrete boundary $\partial\Omega_h(t_n)$. An adaptive mesh refinement for $S_h(t_n)$, based on error indicators for the solution of (1)–(3), combined with local refinement near $\partial\Omega_h(t)$, results in a good approximation of the domain as well as temperature and deformation [4, 5]. The numerical method for thermomechanics on the timedependent domain was implemented in the finite element toolbox ALBERTA [6].

Simulation of Milling Processes The numerical method was applied successfully to various milling and drilling scenarios [2, 4]. Figure 1 shows temperature, deformation and mesh at one timestep during the milling of a thin-walled workpiece from a rectangular block with holes for fixation. For this workpiece, comparisons with experiments were conducted by engineers from IFW Hannover, which show a good agreement of simulated temperatures, deformations and shape errors with experimental data [1].

3 Optimization

Based on the process simulation for given process parameters, we want to optimize the share error and tool wear with respect to some of the process' input parameters. As the simulation of the whole workpiece is rather computationally expensive, we restrict our model process further to a small part of the workpiece.

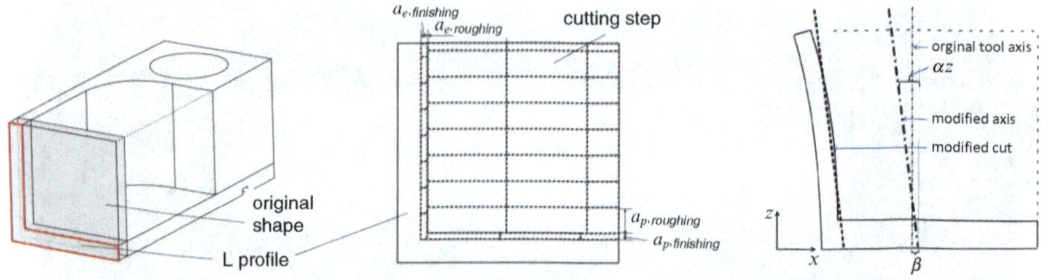

Fig. 2 Model process for optimization of L-shaped domain. Form and indication of adjustable process parameters

Reduced Workpiece and Process Model The left part of Fig. 2 depicts the reduced geometry, with a final L-shaped form, and indicates roughing and finishing steps of the milling process (middle part). Adjustable process parameters are milling parameters and the tool path.

Simulation-Based Optimization Traditional mathematical optimization methods try to use gradients of the cost functionals with respect to the adjustable parameters in order to find a descent direction. Computation of gradients can be done numerically, which results in correspondingly many evaluations of the cost functional, or by an analytic procedure which typically involves the additional solution of adjoint problems. Both approaches are very time consuming, when the control-to-state operator involves time dependent, nonlinear PDEs. Thus, a cheap approximation of the control-to-cost operator can save a lot of computing time. The "simulation-based optimization" method is able to derive approximations of the operator with only very few evaluations of the cost function [7]. It is used here in the following context for multi-objective optimization of the milling process.

Minimization of Shape Error and Tool Wear We choose the axial cutting depth $a_{p,roughing}$ as adjustable parameter, together with cutting velocity v_c, and an additional inclination α and displacement β of the tool axis for the finishing cuts, see Fig. 2 (middle and right). For all four adjustable parameters, suitable admissible ranges were selected.

We want to reduce the final shape error while having the tool wear under control. Thus, our objective functions for multi-objective optimization are given by the shape error

$$\delta_x := \max_{i,j} |L(d_{ij}) - L_d(i, j)|,$$

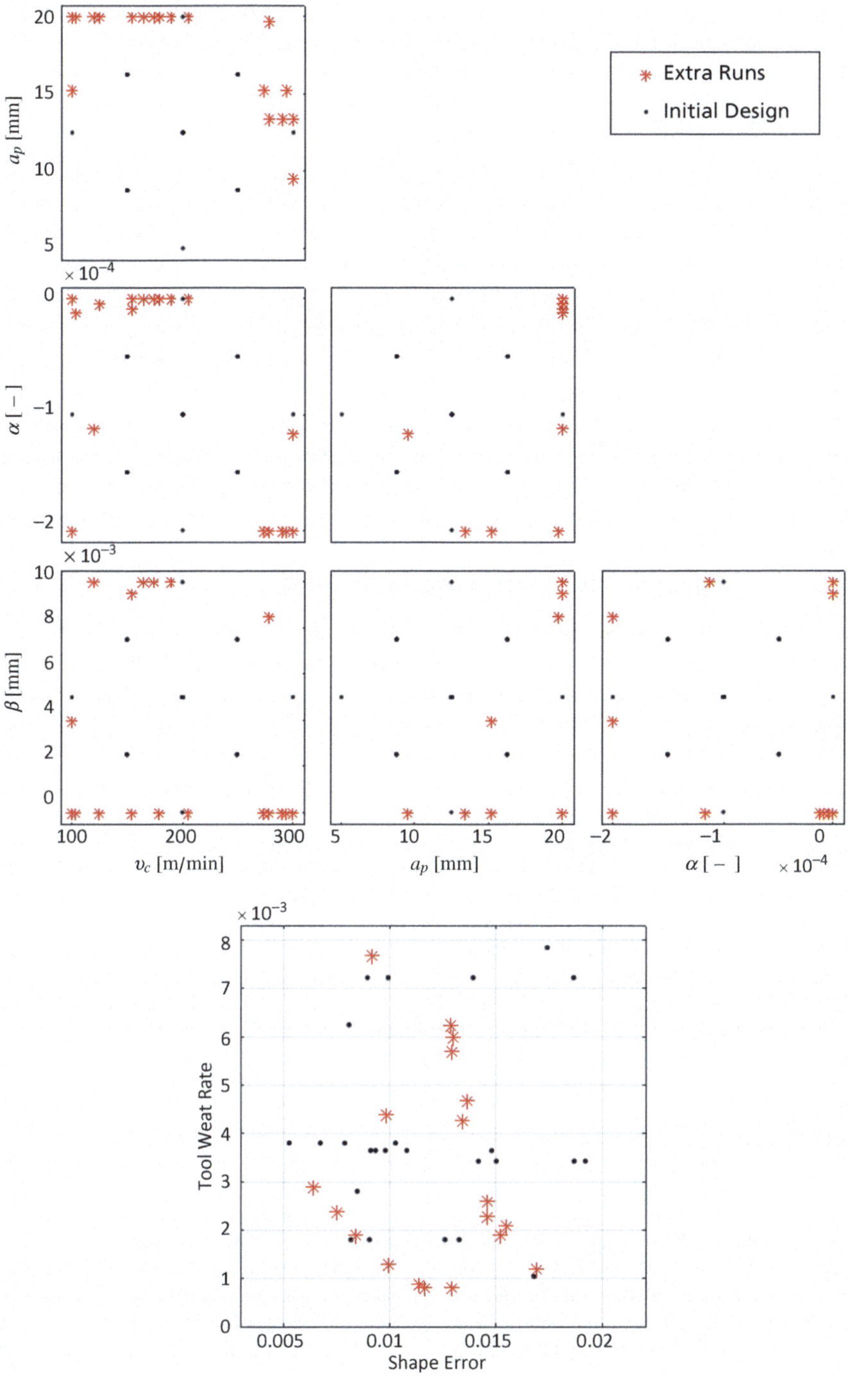

Fig. 3 2D projections of evaluated controllable process variables (top) and performance measure values of evaluated runs (bottom). Initial design (black dots) and extra runs (red stars)

computed by comparing the length of dexel d_{ij} and the corresponding desired length L_d for a 2-dimensional dexel field, and the tool wear rate is modeled by

$$\mathrm{TWR} = \frac{n_{cuts} L_{cut}}{L_f} = \frac{v_c - B}{A}\left(\left\lceil \frac{H}{a_p} \right\rceil \cdot \left(\left\lceil \frac{W}{a_e} \right\rceil + 1\right) + \left\lceil \frac{W}{a_e} \right\rceil - 1\right).$$

which is the inverse of the number of producible workpieces during tool life L_f, where H and W are the height and width of the removed pocket, from which the number of roughing and finishing steps are computed, and A, B are parameters.

Figure 3 shows control variables and results from the application of the simulation-based multi-objective optimization procedure. Black dots indicate initial parameter combinations which are selected in order to explore the set of admissible control parameters. Red stars indicate additional parameter combinations which were selected by the method in order to identify the Pareto set of parameters and Pareto front of objectives with values in the lower left corner of performance measure values in Fig. 3. These can be used to choose parameter sets for small shape error with acceptable tool wear, or small tool wear with acceptable shape error.

Details of the optimization method and the results are given in [3].

Conclusion The results presented above show, that a combination of modern and efficient approaches to simulation and optimization is able to improve rather complex production processes and thus is an important aspect of a digital factory environment.

Acknowledgements We thank Dietmar Hömberg for organizing MS26 "Maths for the Digital Factory" where we could report about the subject.

The presented results have been obtained within the research project "Thermomechanical Deformation of Complex Workpieces in Drilling and Milling Processes" (DE 447/90-3, MA 1657/21-3) within the DFG Priority Program 1480 "Modeling, Simulation and Compensation of Thermal Effects for Complex Machining Processes" as well as in the DAAD-CONACYT PROALMEX project "Numerical Simulation and Optimization of Time Dependent Processes in Engineering and Materials Science". The authors would like to thank DFG, DAAD, and CONACYT for the financial and organizational support of the projects. Furthermore, we thank our project partners from ZeTeM Bremen and IFW Hannover for cooperation.

References

1. Denkena, B., Schmidt, A., Maaß, P., Niederwestberg, D., Niebuhr, C., Vehmeyer, J.: Prediction of temperature induced shape deviations in dry milling. Procedia CIRP **31**, 340–345 (2015)
2. Denkena, B., Maaß, P., Schmidt, A., Niederwestberg, D., Vehmeyer, J., Niebuhr, C., Gralla, P.: Thermomechanical deformation of complex workpieces in milling and drilling processes. In: Biermann, D., Hollmann, F. (eds.) Thermal Effects in Complex Machining Processes - Final Report of the DFG Priority Program 1480, pp. 219–250. Springer LNPE Series (Springer, Cham, 2017)

3. Montalvo-Urquizo, J., Niebuhr, C., Schmidt, A., Villarreal-Marroquin, M.G.: Reducing deformation, stress, and tool wear during milling processes using simulation-based multiobjective optimization. Int. J. Adv. Manuf. Technol. **96**, 1859–1873 (2018)
4. Niebuhr, C.: FE-CutS – Finite Elemente Modell für makroskopische Zerspanprozesse: Modellierung, Analyse und Simulation. PhD thesis, University of Bremen (2017)
5. Niebuhr, C., Schmidt, A.: Finite element methods for parabolic problems with time-dependent domains – application to a milling simulation, 9 p. In: Radu, F.A., Kumar, K., Berre, I., Nordbotten, J.M., Pop, I.S. (eds.) Numerical Mathematics and Advanced Applications - ENUMATH 2017. Lecture Notes in Computational Science and Engineering, vol. 126 (Springer, Berlin, 2018)
6. Schmidt, A., Siebert, K.G.: Design of Adaptive Finite Element Software – The Finite Element Toolbox ALBERTA. Lecture Notes in Computational Science and Engineering, vol. 42 (Springer, Berlin, 2005)
7. Villarreal-Marroquin, M.G., Cabrera-Rios, M., Castro, J.M.: A multicriteria simulation optimization method for injection molding. J. Polym. Eng. **31**(5), 397–407 (2011)

Shape Optimization of Liquid Polymer Distributors

Raphael Hohmann and Christian Leithäuser

Abstract We consider the optimal shape design of a distributor geometry in the context of industrial fiber spinning. In this process a molten polymer is routed from a pipe to a spinneret plate with a larger cross section, where thin fibers, which are then further processed, are spun from the fluid. The residence time or material age of the polymer in the distributor, which is modeled through an additional advection-diffusion-reaction equation, has to be controlled such that fluid stagnation is prevented, since this would cause material degradation and a decrease in the quality of the fibers. In order to optimize the geometry, we formally derive the adjoint equations and the volume formulation of the shape derivative and apply them within a gradient descent method.

1 Introduction

Spin packs as depicted in Fig. 1 are used to distribute liquid polymer from a tube onto a spinneret plate, where capillary nozzles spin the material to fibers. Since long residence times and fluid stagnation within the distributor negatively affect the fiber quality, we improve the initial design of the distributing first cavity through shape optimization. The previous works [5, 6] optimized the shape of a spin pack with respect to the wall shear stress, which indirectly controls the time needed until the fluid reaches the nozzles. By solving an additional advection-diffusion-reaction equation however, the wall shear stress as an objective can be omitted and the geometry can be controlled directly with respect to the material age at the outflow.

After giving the mathematical model in Sect. 2 we state the distributed shape gradient of our problem, which are derived using a formal shape Lagrangian approach [2], in Sect. 3 and present numerical results in Sect. 4.

R. Hohmann (✉) · C. Leithäuser
Fraunhofer Institute for Industrial Mathematics, Kaiserslautern, Germany
e-mail: raphael.hohmann@itwm.fraunhofer.de; christian.leithaeuser@itwm.fraunhofer.de

© Springer Nature Switzerland AG 2019

I. Faragó et al. (eds.), *Progress in Industrial Mathematics at ECMI 2018*,
Mathematics in Industry 30, https://doi.org/10.1007/978-3-030-27550-1_54

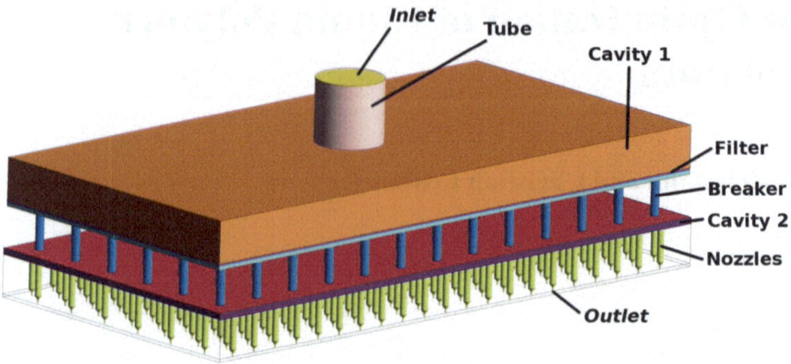

Fig. 1 Schematic diagram of a spin pack

2 Problem Formulation

2.1 *Fluid Equations and Material Age*

The fluid we are dealing with can be modeled with the stationary incompressible Stokes equations due to its high viscosity, whose non-dimensionalized form reads

$$-Re^{-1}\Delta \mathbf{u} + \nabla p = 0 \text{ in } \Omega, \qquad \nabla \cdot \mathbf{u} = 0 \text{ in } \Omega. \tag{1}$$

Here $\mathbf{u} \in \mathbb{R}^n$ denotes the fluid velocity, $p > 0$ the pressure and $Re \ll 1$ the Reynolds number. The boundary $\Gamma := \partial\Omega$ is subdivided into inflow boundary Γ^{in}, wall Γ^{w} and outflow boundary Γ^{out}. On Γ^{in} a velocity profile and on Γ^{w} no-slip conditions are prescribed. To simplify the computations we summarize the effect of the spin pack consisting of filter, breaker, second cavity and nozzles in Fig. 1 using the Darcy boundary condition of a porous medium at Γ^{out}, the bottom of the first cavity. The boundary conditions read

$$\mathbf{u} \cdot \mathbf{n} = u_{in} \text{ on } \Gamma^{in}, \ \mathbf{u} \cdot \mathbf{n} = 0 \text{ on } \Gamma^{w}, \ \eta(\mathbf{u} \cdot \mathbf{n}) - p = 0 \text{ on } \Gamma^{out}, \ \mathbf{u} \times \mathbf{n} = 0 \text{ on } \Gamma \tag{2}$$

with $\eta > 0$ being the spin pack's non-dimensionalized porosity. The inflow profile u_{in} is that of a laminar pipe flow with a given mass flow rate.

The *residence time* or *material age* τ of the fluid can be modeled using a stationary advection-diffusion-reaction equation together with a homogeneous Dirichlet condition at the inflow [4], which reads

$$\mathbf{u} \cdot \nabla \tau - Pe^{-1}\Delta \tau = f(\mathbf{u}) \text{ in } \Omega, \quad \tau = 0 \text{ on } \Gamma^{in}. \tag{3}$$

Here $f(\mathbf{u})$ is a differentiable approximation of $\chi_{\{\mathbf{x}\in\Omega \mid ||\mathbf{u}(\mathbf{x})||_2 > \epsilon\}}$ with χ being the indicator function. The influence of the regularizing second order term and the indicator function constant is kept small by choosing a Péclet number $Pe = \mathcal{O}(10^5)$ and $\epsilon > 0$ to be two orders smaller than $\max_{\mathbf{x}\in\Gamma^{in}}\{u_{in}(\mathbf{x})\}$. With this definitions it can on the one hand be shown, that (3) possesses a unique solution and the material age does not tend to infinity at stagnation zones, and on the other hand, that relatively high local variations of τ indicate slow material transport.

2.2 Optimization Problem

Given the deformable boundary $\Gamma^d \subset \Gamma^w$ we denote by $\mathscr{A} \subset \mathscr{P}(\mathbb{R}^3)$ the set of admissible shapes, which is defined as the set of images of sufficiently smooth transformations of the initial distributor Ω_0 keeping $\Gamma \setminus \Gamma^d$ fixed. Given $\Omega \in \mathscr{A}$ we define the spaces

$$
\begin{aligned}
&X(\Omega) := \{\mathbf{v} \in (H^1(\Omega))^3 \mid \mathbf{v}|_{\Gamma^w} = 0,\ \mathbf{v} \times \mathbf{n}|_{\Gamma^{in}\cup\Gamma^w} = 0\},\\
&X_0(\Omega) := \{\mathbf{v} \in X \mid \mathbf{v}\cdot\mathbf{n}|_{\Gamma^{in}} = 0\}, \quad M(\Omega) := L^2(\Omega), \quad R(\Omega) := H^1(\Omega),\\
&R_0(\Omega) := \{\sigma \in R \mid \sigma|_{\Gamma^{in}} = 0\}, \quad B(\Omega) := H^{1/2}(\Gamma^{in}),
\end{aligned}
$$

equipped with the usual L^2-norms as well as

$$
\begin{aligned}
a(\Omega, \mathbf{u}, \mathbf{v}) &:= \int_\Omega (\nabla \times \mathbf{u}) \cdot (\nabla \times \mathbf{v}) d\mathbf{x} + \int_{\Gamma^{in}} \eta(\mathbf{u}\cdot\mathbf{n})(\mathbf{v}\cdot\mathbf{n}) ds,\\
b(\Omega, \mathbf{v}, q) &:= \int_\Omega q\nabla\cdot\mathbf{v} d\mathbf{x}, \quad c(\Omega, \mathbf{u}, \tau, \sigma) := \int_\Omega (\mathbf{u}\cdot\nabla\tau)\sigma + Pe^{-1}\nabla\tau\cdot\nabla\sigma d\mathbf{x}\\
h(\Omega, \mathbf{v}, \sigma) &:= \int_\Omega f(\mathbf{v})\sigma ds.
\end{aligned}
$$

Using these definitions the weak formulation associated with the strong problem formulation from Sect. 2.1 reads

$$
\begin{aligned}
&\text{Given } \Omega \in \mathscr{A} \text{ seek } (\mathbf{q_u}, q_p, q_\tau) \in X(\Omega) \times M(\Omega) \times R_0(\Omega) \text{ s.t.}\\
&a(\Omega, \mathbf{q_u}, \boldsymbol{\psi}_\mathbf{u}) - b(\Omega, \boldsymbol{\psi}_\mathbf{u}, q_p) = 0 \qquad \forall \boldsymbol{\psi}_\mathbf{u} \in X_0(\Omega),\\
&\qquad b(\Omega, \mathbf{q_u}, \psi_p) = 0 \qquad \forall \boldsymbol{\psi}_\mathbf{p} \in M(\Omega),\\
&\qquad \mathbf{q_u}\cdot\mathbf{n} = u_{in} \qquad \text{on } \Gamma^{in},\\
&c(\Omega, \mathbf{q_u}, q_\tau, \psi_\tau) = h(\Omega, \mathbf{q_u}, \psi_\tau) \quad \forall \psi_\tau \in R(\Omega).
\end{aligned}
\tag{4}
$$

Problem (4) possesses a unique solution due to the ellipticity of the operators $a(\Omega, \cdot, \cdot)$, see [3], and $c(\Omega, \mathbf{q_u}, \cdot, \cdot)$ between the given spaces. Therefore we can define the reduced cost functional

$$J(\Omega) := \tilde{J}(\Omega, q_\tau(\Omega), \mathbf{q_u}(\Omega)) := \int_{\Gamma^{out}} g(q_\tau)\, ds + c_1 ||\nabla \mathbf{q_u}||^2_{(L^2(\Omega))^3} + c_2 \int_{\Gamma^d} 1 ds,$$

where $c_1, c_2 > 0$ are regularization parameters. We choose $g(\tau) :=^\gamma (\max\{\tau, \tau^*\})$ with $\tau^* > 0$ and $^\gamma(\cdot)$ being the Moreau envelope, see e.g. [1], and consider the problem

$$\begin{aligned} &\text{Minimize } J(\Omega) \text{ for } \Omega \in \mathscr{A} \\ &\qquad\qquad \text{subject to } (4). \end{aligned} \tag{5}$$

We note that if the solution of ((1),(2)) fulfills $(\mathbf{u}, p) \in H^2(\Omega) \times H^1(\Omega)$, we have

$$\Delta E_p := -\int_{\Gamma^{in} \cup \Gamma^{out}} p(\mathbf{u} \cdot \mathbf{n}) d\mathbf{x} = Re^{-1} ||\nabla \mathbf{u}||^2_{(L^2(\Omega))^3},$$

which is the fluid power loss from inflow to outflow. The first regularization term therefore penalizes the transformation of pressure energy into kinetic energy, which avoids that the material age at the outflow is minimized by a geometry with a small volume and thereby causing an excessive acceleration of the fluid.

3 Adjoint Equations and Shape Derivative

In order to obtain the distributed shape derivative of J we use the perturbation of the identity approach with sufficiently smooth deformation velocities $\mathbf{V} : \mathbb{R}^3 \to \mathbb{R}^3$ and apply a shape Lagrangian approach [2]. Given the unique solution of (4) as well as

$$d(\Omega, \mathbf{q_u}, \lambda_\tau, \phi_\tau) := c(\Omega, -\mathbf{q_u}, \lambda_\tau, \phi_\tau) + \int_{\Gamma^{out}} (\mathbf{q_u} \cdot \mathbf{n}) \lambda_\tau \phi_\tau \mathbf{ds},$$

$$k(\Omega, q_\tau, \phi_\tau) := \int_{\Gamma^{out}} -\partial_{\phi_\tau} g(q_\tau) \phi_\tau \mathbf{ds}, \quad z(\Omega, \boldsymbol{\phi}_\mathbf{u}, \lambda_{in}) := \int_{\Gamma^{in}} (\boldsymbol{\phi}_\mathbf{u} \cdot \mathbf{n}) \lambda_{in} \mathbf{ds},$$

$$l(\Omega, \mathbf{q_u}, q_\tau, \lambda_\tau, \boldsymbol{\phi}_\mathbf{u}) := \int_\Omega \left(-\nabla q_\tau \lambda_\tau + \partial_{\boldsymbol{\phi}_\mathbf{u}} f(\mathbf{q_u}) \lambda_\tau \right) \cdot \boldsymbol{\phi}_\mathbf{u} d\mathbf{x} - 2c_1 \int_\Omega \nabla \mathbf{q_u} : \nabla \boldsymbol{\phi}_\mathbf{u} d\mathbf{x},$$

the adjoint state is the solution of the problem

$$
\begin{aligned}
&\text{Given } \Omega \in \mathscr{A} and (\mathbf{q_u}, q_p, q_\tau) \in X(\Omega) \times M(\Omega) \times R_0(\Omega) \text{ seek} \\
&\qquad (\boldsymbol{\lambda}_\mathbf{u}, \lambda_p, \lambda_\tau, \lambda_{in}) \in X_0(\Omega) \times M(\Omega) \times R(\Omega) \times B(\Omega) \text{ s.t.} \\
&\qquad d(\Omega, \mathbf{q_u}, \lambda_\tau, \phi_\tau) = k(\Omega, q_\tau, \phi_\tau) \qquad \forall \phi_\tau \in R_0(\Omega), \\
&a(\Omega, \boldsymbol{\lambda}_\mathbf{u}, \boldsymbol{\phi}_\mathbf{u}) + b(\Omega, \boldsymbol{\phi}_\mathbf{u}, \lambda_p) = l(\Omega, \mathbf{q_u}, q_\tau, \lambda_\tau, \boldsymbol{\phi}_\mathbf{u}) \qquad \forall \boldsymbol{\phi}_\mathbf{u} \in X(\Omega), \\
&\qquad b(\Omega, \boldsymbol{\lambda}_\mathbf{u}, \phi_p) = 0 \qquad \forall \phi_p \in M(\Omega), \\
&\qquad z(\Omega, \boldsymbol{\phi}_\mathbf{u}, \lambda_{in}) = 0 \qquad \forall \boldsymbol{\phi}_\mathbf{u} \in X(\Omega).
\end{aligned}
\tag{6}
$$

From this weak formulation we see that the adjoint residence time λ_τ is transported from the outflow to the inflow boundary along the negative fluid streamlines and contributes to the adjoint fluid equations through the source term l.

With the solutions of (4) and (6) and the summation convention the shape derivative of J in direction $\mathbf{V}$ is given by

$$
\begin{aligned}
dJ(\Omega)[\mathbf{V}] =& c_1 \int_\Omega (\mathbf{Q}(\mathbf{V})\, \nabla \mathbf{q_u}) : \nabla \mathbf{q_u} \mathrm{d}\mathbf{x} + c_2 \int_{\Gamma^d} \nabla_\Gamma \cdot \mathbf{V} \mathrm{d}\mathbf{s} \\
& - Re^{-1} \int_\Omega (\epsilon_{ijk} (\mathbf{Dq_u}\, \mathbf{DV})_{jk} \mathbf{e_i}) \cdot (\nabla \times \boldsymbol{\lambda}_\mathbf{u}) \\
& \qquad + (\nabla \times \mathbf{q_u}) \cdot (\epsilon_{ijk} (\mathbf{D}\boldsymbol{\lambda}_\mathbf{u}\, \mathbf{DV})_{jk} \mathbf{e_i}) \mathrm{d}\mathbf{x} \\
& - \int_\Omega \lambda_p \mathrm{tr}(\mathbf{Dq_u}\, \mathbf{DV}) - q_p \mathrm{tr}(\mathbf{D}\boldsymbol{\lambda}_\mathbf{u}\, \mathbf{DV}) \mathrm{d}\mathbf{x} \\
& - \int_\Omega (\mathbf{q_u} \cdot (\mathbf{DV}^T\, \nabla q_\tau)) \lambda_\tau \mathrm{d}\mathbf{x} \\
& - \int_\Omega Pe^{-1} \Big((\mathbf{DV}^T\, \nabla q_\tau) \cdot \nabla \lambda_\tau + (\mathbf{DV}^T\, \nabla \lambda_\tau) \cdot \nabla q_\tau \Big) \mathrm{d}\mathbf{x} \\
& + \int_\Omega \Big(Re^{-1} (\nabla \times \mathbf{u}) \cdot (\nabla \times \boldsymbol{\lambda}_\mathbf{u}) - q_p (\nabla \cdot \boldsymbol{\lambda}_\mathbf{u}) + \lambda_p (\nabla \cdot \mathbf{q_u}) \\
& \qquad + \mathbf{q_u} \cdot \nabla q_\tau\, \lambda_\tau + Pe^{-1} \nabla q_\tau \cdot \nabla \boldsymbol{\lambda}_\tau - f(\mathbf{q_u}) \lambda_\tau \Big) (\nabla \cdot \mathbf{V}) \mathrm{d}\mathbf{x}
\end{aligned}
\tag{7}
$$

using $\mathbf{Q}(\mathbf{V}) := (\nabla \cdot \mathbf{V})\mathbb{I} - \mathbf{DV}^T - \mathbf{DV}$, where $\mathbb{I} \in \mathbb{R}^{3\times3}$ is the identity and $\mathbf{DV}$ the Jacobian. Here ϵ_{ijk} is the Levi-Civita symbol, $\mathbf{e_i} \in \mathbb{R}^3$ are the standard unit vectors and $\nabla_\Gamma \cdot \mathbf{V}$ denotes the tangential divergence [7].

4 Numerical Results

For the numerical investigation of our approach we consider a three dimensional test case and apply a gradient descent method based on the projection

$$\langle \mathbf{G}, \mathbf{V} \rangle_{(H^1(\Omega))^3} = dJ(\Omega)[\mathbf{V}] \quad \forall \mathbf{V} \in (H^1(\Omega))^3, \qquad \mathbf{G} = 0 \text{ on } \Gamma \setminus \Gamma^d,$$

of (7) together with the Armijo step size rule. Here we choose the model parameters $Re = 7.1 \times 10^{-2}$, $\eta = 3.5 \times 10^7$, $\bar{\tau} = 15$, $\gamma = 2$, $c_1 = 1 \times 10^{-3}$ and $c_2 = 2 \times 10^{-2}$ and use a mesh consisting of 9163 tetrahedral elements.

Figure 2 shows the decrease of the cost function and Fig. 3 the relative gradient norms during the iterations, where we used the norm

$$||\mathbf{G}||_{\Gamma^d} := \left(\int_{\Gamma^d} (\mathbf{G} \cdot \mathbf{n})^2 \mathbf{dx} \right)^{\frac{1}{2}},$$

since only the normal component of the gradient leads to changes in the domain. No further decrease of J could be obtained after 15 iterations, where the relative

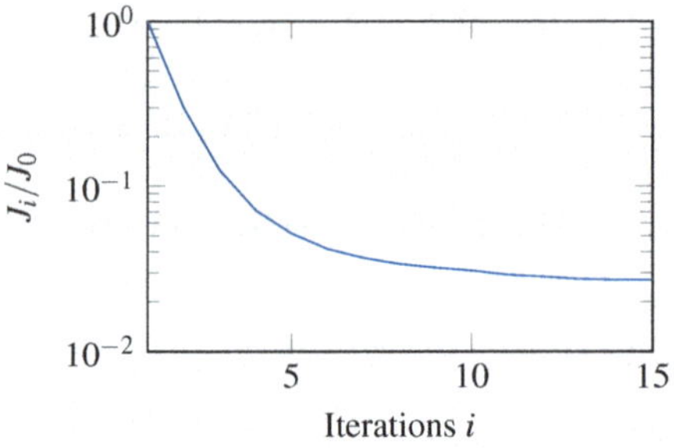

Fig. 2 Relative objective function values

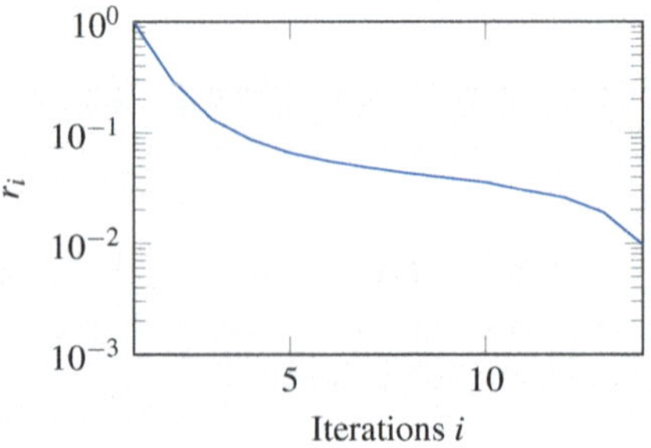

Fig. 3 Gradient norms $r_i := ||\mathbf{G_i}||_{\Gamma^d} / ||\mathbf{G_0}||_{\Gamma^d}$

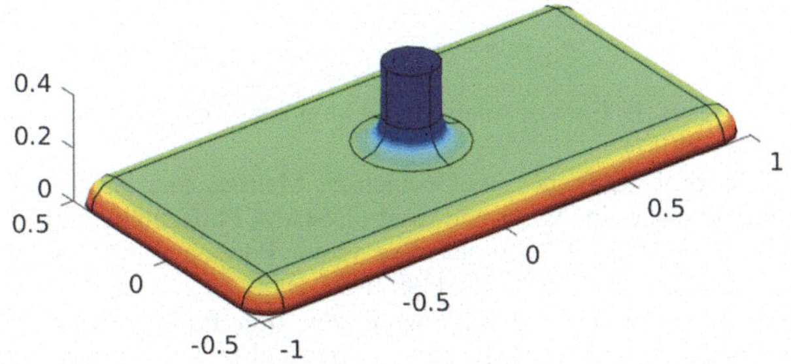

Fig. 4 Height profile of the initial geometry

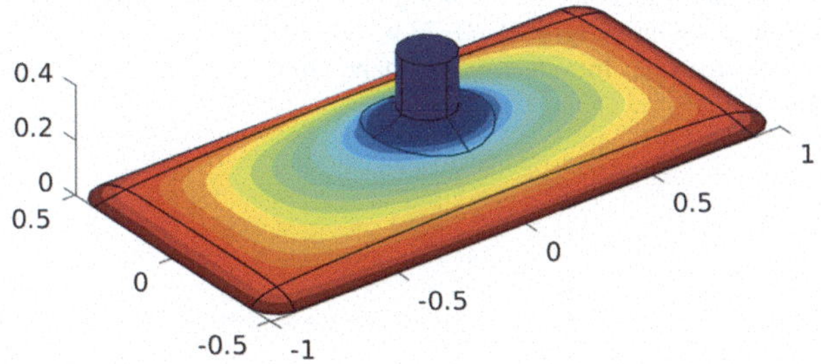

Fig. 5 Height profile of the final geometry

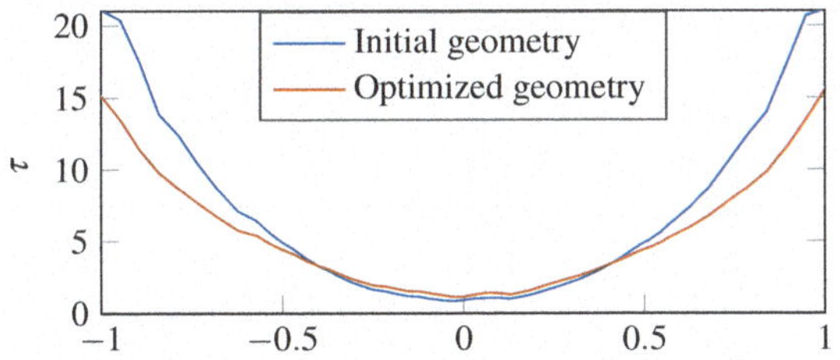

Fig. 6 Material ages τ of the initial and optimized geometry along the longitudinal axis of the outflow Γ^{out}

gradient norm falls below 1%. During the optimization Ω_0 is flattened around the corners and slightly lifted close to the non-deformable inflow tube, which can be seen from the height profiles in Figs. 4 and 5. Figure 6 shows, that the main objective of reaching material ages below $\bar{\tau} = 15$ can be obtained at most of Γ^{out}. The loss of pressure energy ΔE_p for the optimized cavity is 1.6% higher than for the initial design, which is acceptable for this application.

References

1. Bauschke, H., Combettes, P.: Convex Analysis and Monotone Operator Theory in Hilbert Spaces (Springer, New York, 2017)
2. Delfour, M.C., Zolésio, J.-P.: Shapes and Geometries: Metrics, Analysis, Differential Calculus, and Optimization. Advances in Design and Control. SIAM, Philadelphia (2011)
3. Guermond, J.-L., Quartapelle, L.: On sensitive vector Poisson and Stokes problems. Math. Models Methods Appl. Sci. **7**(5), 681–698 (1997)
4. Józsa, J., Krámer, T: Modelling residence time as advection-diffusion with zero-order reaction kinetics. In: Proceedings of the Hydrodynamics 2000 Conference, International Association of Hydraulic Engineering and Research, pp. 23–27 (2000)
5. Leithäuser, C., Pinnau, R., Feßler, R.: Approximate controllability of linearized shape-dependent operators for flow problems. ESAIM Control Optim. Calc. Var. **23**(3), 751–771 (2017)
6. Leithäuser, C., Pinnau, R., Feßler, R.: Designing polymer spin packs by tailored shape optimization techniques. Optim. Eng. **19**(3), 733–764 (2018)
7. Sokolowski, J., Zolesio, J.-P.: Introduction to Shape Optimization: Shape Sensitivity Analysis (Springer, Berlin, 1992)

Modeling and Simulation of Macroscopic Pedestrian Flow Models

Naveen Kumar Mahato, Axel Klar, and Sudarshan Tiwari

Abstract Mathematical modeling and numerical simulation of human crowd motion have become a major subject of research with a wide field of applications. A variety of models for pedestrian behavior have been proposed on different levels of description in recent years. Macroscopic pedestrian flow model involving equations for density and mean velocity of the flow is derived in Bellomo and Dogbe (Math. Models Methods Appl. Sci. 18:1317-1345, 2008), Burger et al. (Discrete Contin. Dynam. Systems Ser. B: J. Bridging Math. Sci. 19:1311-1333, 2014), Hughes (Transp. Res. B Methodol. 36:507-535, 2000) and Mahato et al. (Appl. Math. Model. 53:447-461, 2018; Int. J. Adv. Eng. Sci. Appl. Math. 10:41-53, 2018).

1 Introduction

Mathematical modeling and numerical simulation of human crowd motion have become a major subject of research with a wide field of applications. A variety of models for pedestrian behavior have been proposed on different levels of description in recent years. Macroscopic pedestrian flow model involving equations for density and mean velocity of the flow is derived in Refs. [1, 4, 5, 7, 8].

In this work, we analyze numerically some macroscopic models of pedestrian motion such as the classical Hughes model [5] and a mean field game with nonlinear mobilities [4], modeling fast exit scenarios in pedestrian crowds. A model introduced by Hughes consists of a non-linear conservation law for the density of pedestrians coupled with an Eikonal equation for a potential modeling the common sense of the task. Mean field game with nonlinear mobilities is obtained by an optimal control approach, where the motion of every pedestrian is determined by minimizing a cost functional, which depends on the position, velocity, exit time and the overall density of people. We consider a parabolic optimal control problem

N. K. Mahato (✉) · A. Klar · S. Tiwari
Technische Universität Kaiserslautern, Department of Mathematics, Kaiserslautern, Germany
e-mail: mahato@mathematik.uni-kl.de; klar@mathematik.uni-kl.de;
tiwari@mathematik.uni-kl.de

© Springer Nature Switzerland AG 2019
I. Faragó et al. (eds.), *Progress in Industrial Mathematics at ECMI 2018*,
Mathematics in Industry 30, https://doi.org/10.1007/978-3-030-27550-1_55

of nonlinear mobility in pedestrian dynamics, which leads to a mean field game structure. We show how optimal control problem related to the Hughes model for pedestrian motion. Furthermore, we provide several numerical results which relate both models in one and two dimensions.

2 Optimal Control Problem of Pedestrian Flow from [4]

For completeness of the presentation up to higher dimensions, we review the macroscopic optimal control problem for pedestrian flow, see Refs. [2–4]. There, denoting the (normalized) density function of the pedestrians by $\rho(t,x)$ and the momentum (or the flux density) by $m = F(\rho)v$ at position $x \in \Omega$, velocity $v \in \Omega$ and time t, where the function $F(\rho(t,x))$ describing the non-linear mobility of the pedestrians (or the costs created by large densities) and $\Omega \in \mathbb{R}^d$, $d = 1, 2$ is a bounded domain representing the pedestrian area. We assume the boundary $\partial\Omega$ is split into a Neumann part $\Gamma_N \subseteq \partial\Omega$ modeling walls or obstacles, $\Gamma_E \subseteq \partial\Omega$ modeling the exits such that $\partial\Omega = \Gamma_N \bigcup \Gamma_E$ and $\Gamma_N \bigcap \Gamma_E = \phi$. If we denote the rate of passing the exit by β, then we have an outflow proportional to $\beta\rho$. Hence, for a stochastic particle and a final time T sufficiently large, the minimization functional is given by the following parabolic optimal control problem:

$$\min_{(\rho,m)} I_T(\rho,m) = \min_{(\rho,m)} \frac{1}{2}\int_0^T \int_\Omega \frac{|m(t,x)|^2}{F(\rho(t,x))} dxdt + \frac{\alpha}{2}\int_0^T \int_\Omega \rho(t,x)dxdt, \tag{1a}$$

subject to

$$\partial_t \rho + \nabla \cdot m = \frac{\sigma^2}{2}\Delta\rho, \qquad \text{in } \Omega \times (0,T), \tag{1b}$$

$$\left(m - \frac{\sigma^2}{2}\nabla\rho\right) \cdot n = 0, \qquad \text{on } \Gamma_N \times (0,T), \tag{1c}$$

$$\left(m - \frac{\sigma^2}{2}\nabla\rho\right) \cdot n = \beta\rho, \qquad \text{on } \Gamma_E \times (0,T), \tag{1d}$$

$$\rho(0,x) = \rho_0(x), \qquad \text{in } \Omega. \tag{1e}$$

This optimality system can be seen as the mean field game structure, see Ref. [6]. We start by defining the Lagrangian with dual variable $\Phi = \Phi(t,x)$ as

$$L_T(\rho, m, \Phi) = I_T(\rho, m) + \int_0^T \int_\Omega (\partial_t \rho + \nabla \cdot m - \frac{\sigma^2}{2}\Delta\rho)\Phi dxdt$$
$$= I_T(\rho, m) + \left[\int_\Omega \rho\Phi dx\right]_0^T + \int_0^T \int_\Omega \left[\rho\left(-\partial_t\Phi - \frac{\sigma^2}{2}\Delta\Phi\right) - m\cdot\nabla\Phi\right]dxdt$$
$$+ \int_0^T \int_{\partial\Gamma_E} \left(\underbrace{-\frac{\sigma^2}{2}\nabla\rho\cdot n\Phi + m\cdot n\Phi}_{\beta\rho\Phi} + \frac{\sigma^2}{2}\rho\nabla\Phi\cdot n\right)dsdt + \int_0^T \int_{\partial\Gamma_N} \frac{\sigma^2}{2}\rho\nabla\Phi\cdot ndsdt.$$

The optimality condition with respect to m and ρ, yields the following equations

$$0 = \partial_m L_T(\rho, m, \Phi) = \frac{m(t,x)}{F(\rho(t,x))} - \nabla\Phi,$$
$$0 = \partial_\rho L_T(\rho, m, \Phi) = -\frac{1}{2}\frac{|m(t,x)|^2 F'(\rho)}{F^2(\rho)} + \frac{\alpha}{2} - \partial_t\Phi - \frac{\sigma^2}{2}\Delta\Phi.$$

Inserting $m = F(\rho(t,x))\nabla\Phi$ we obtain the following system of equations

$$\partial_t\rho + \nabla\cdot(F(\rho)\nabla\Phi) - \frac{\sigma^2}{2}\Delta\rho = 0, \qquad \text{in } \Omega\times(0,T), \tag{2a}$$

$$\partial_t\Phi + \frac{F'(\rho)}{2}|\nabla\Phi|^2 + \frac{\sigma^2}{2}\Delta\Phi = \frac{\alpha}{2}, \qquad \text{in } \Omega\times(0,T), \tag{2b}$$

$$\left(F(\rho)\nabla\Phi - \frac{\sigma^2}{2}\nabla\rho\right)\cdot n = 0, \quad \frac{\sigma^2}{2}\nabla\Phi\cdot n = 0, \qquad \text{on } \Gamma_N\times(0,T), \tag{2c}$$

$$\left(F(\rho)\nabla\Phi - \frac{\sigma^2}{2}\nabla\rho\right)\cdot n = \beta\rho, \quad \frac{\sigma^2}{2}\nabla\Phi\cdot n + \beta\rho = 0, \qquad \text{on } \Gamma_E\times(0,T), \tag{2d}$$

$$\rho(0,x) = \rho_0(x), \quad \Phi(T,x) = 0, \qquad \text{in } \Omega. \tag{2e}$$

System (2) has the structure of a mean field game for pedestrian dynamics, which contains the Fokker-Planck equation (2a) has to be solved forward in time and the Hamilton-Jacobi equation (2b) that has to be solved backward in time.

3 Relation to the Classical Hughes Model [5]

In this section we discuss the relation which shows that for vanishing viscosity $\sigma = 0$ of the optimality system (1) has a similar structure as the classical Hughes model for pedestrian flow. Hughes proposed that pedestrians seek the fastest path to the exit, but at the same time try to avoid congested areas, for details see Ref. [5]. Let us consider the governing equations of Hughes model for pedestrian flow,

$$\partial_t \rho - \nabla \cdot (\rho f^2(\rho) \nabla \Phi) - \frac{\sigma^2}{2} \Delta \rho = 0, \qquad \text{in } \Omega \times (0, T), \tag{3a}$$

$$|\nabla \Phi| = \frac{1}{f(\rho)}, \qquad \text{in } \Omega \times (0, T), \tag{3b}$$

$$\left(\rho f^2(\rho) \nabla \Phi - \frac{\sigma^2}{2} \nabla \rho\right) \cdot n = 0, \quad \Phi = \infty, \qquad \text{on } \Gamma_N \times (0, T), \tag{3c}$$

$$\left(\rho f^2(\rho) \nabla \Phi - \frac{\sigma^2}{2} \nabla \rho\right) \cdot n = \beta \rho, \quad \Phi = 0, \qquad \text{on } \Gamma_E \times (0, T), \tag{3d}$$

$$\rho(0, x) = \rho_0(x), \qquad \text{in } \Omega, \tag{3e}$$

where the function $f(\rho) = \rho_{max} - \rho$ with ρ_{max} denote the maximum density and models how pedestrians change their direction and velocity due to the surrounding density, i.e. provides a weighting or cost with respect to high densities. Saturation effects are included via the function $f(\rho)$ for $\rho \longrightarrow \rho_{\max}$.

On the other hand, if we choose the mobility/penalization function for high densities such as $F(\rho) = \rho f(\rho)^2$, then the optimality system (2) for vanishing viscosity can be written as

$$\partial_t \rho + \nabla \cdot (\rho f(\rho)^2 \nabla \Phi) = 0, \qquad \text{in } \Omega \times (0, T), \tag{4a}$$

$$\partial_t \Phi + \frac{f(\rho)}{2} (f(\rho) + 2\rho f'(\rho)) |\nabla \Phi|^2 = \frac{\alpha}{2}, \qquad \text{in } \Omega \times (0, T), \tag{4b}$$

where the initial, terminal and boundary conditions are same as in system (2). Furthermore, one can expect the equilibration of Φ backward in time for large T. Then for time t of order one the limiting model becomes

$$\partial_t \rho + \nabla \cdot (\rho f(\rho)^2 \nabla \Phi) = 0, \qquad \text{in } \Omega \times (0, T), \tag{5a}$$

$$(f(\rho) + 2\rho f'(\rho)) |\nabla \Phi|^2 = \frac{\alpha}{f(\rho)}, \qquad \text{in } \Omega \times (0, T). \tag{5b}$$

Hence, if we set $\alpha = 1$, the system (5) is almost equivalent to the Hughes model (3) for vanishing viscosity. Note that the sign difference in Eqs. (3a) and (5a) is not an actual, since due to the signs in the backward equation we shall obtain Φ as the negative of the distance function used in the Hughes model.

4 Numerical Results

In this section we present a series of numerical experiments for the equations from both proposed models. We compare the relation between the models for different parameters in one and two dimensions. We use finite difference scheme for solving the classical Hughes model, where central difference in space and the forward difference in time, i.e. forward time centered space (FTCS) scheme for the nonlinear conservation law and an upwind Godunov scheme for the Eikonal equation. We follow the steepest descent algorithm from Ref. [4], to solve the mean field game structure, in which we use FTCS finite difference scheme to solve both forward and backward equations.

We consider a one-dimensional domain $\Omega = [-1, 1]$ with exits located at $x = \pm 1$ for the numerical simulation as a configuration defined in Ref. [4]. We choose the maximum density ρ_{max}, the weighting parameter α and the flow rate parameter β as 1. Furthermore, we consider the time interval as $t \in [0, 3]$. The time step is set to $\Delta t = 10^{-4}$ for Hughes and $\Delta t = 10^{-3}$ for MFG. We use the spatial discretization $h = 10^{-2}$, the diffusion coefficient $\frac{\sigma^2}{2} = h$ and the initial density $\rho_0 = \frac{1}{3}$ in both models.

Figure 1 shows the evolution of solutions at different times for both mean field type structures. One observes that the non-stationary Eikonal solution of the MFG

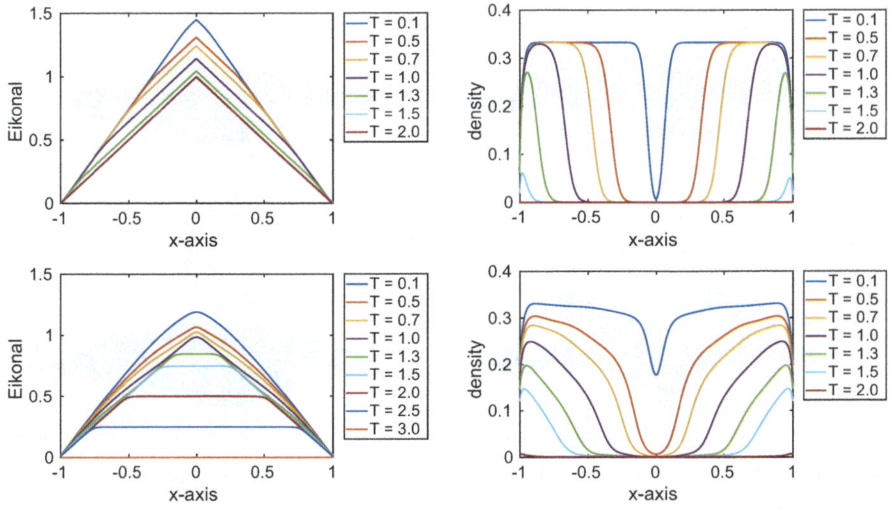

Fig. 1 Evolution of solutions at different times for the Hughes model (top) and for the MFG structure (bottom)

structure has a similar behavior as the stationary Eikonal solution of the classical Hughes model until the density is not zero, as we expected equilibration of Φ in the Eq. (4b). One also observes from the density solution that both models have similar behavior as pedestrians start in immediate vacuum formation at the center $x = 0$. Although the models have a very similar structure, pedestrians wait for a little while at the center and then start to move at a higher speed in the case of the mean field game compare to the Hughes model.

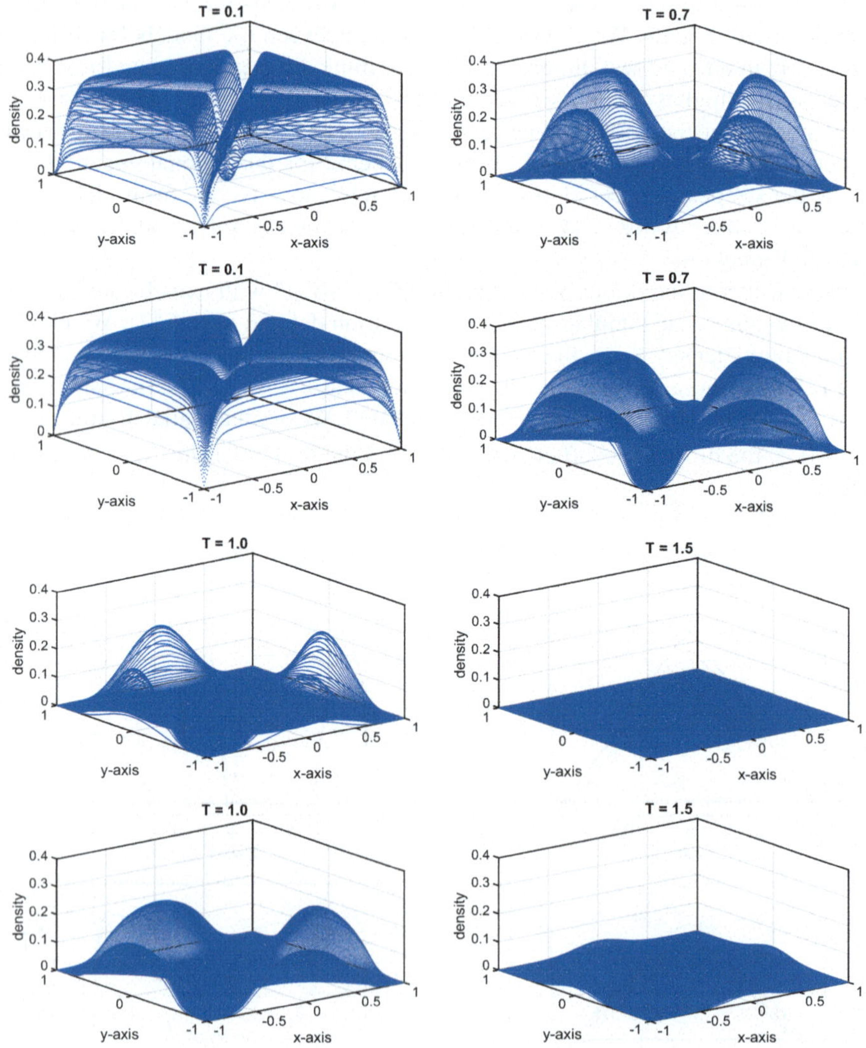

Fig. 2 Evolution of the density solution at different times for the Hughes model (top) and the MFG structure (bottom)

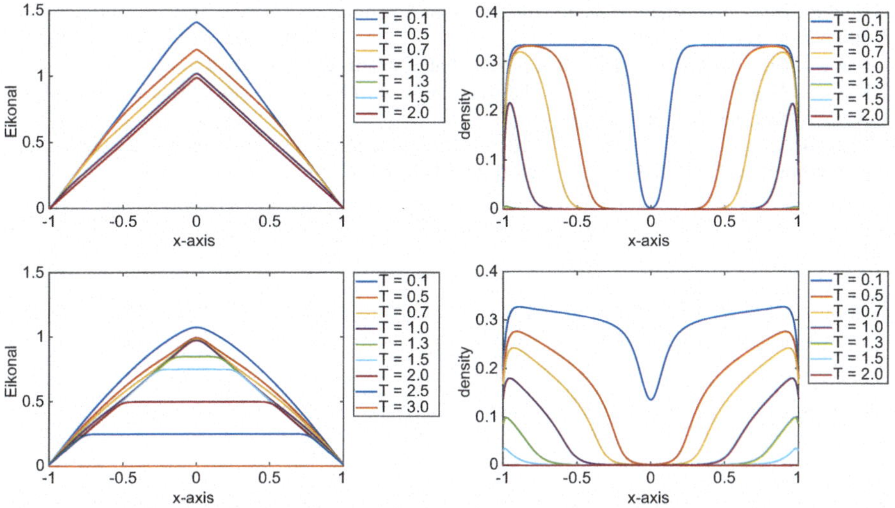

Fig. 3 Evolution of solutions through the center along the x-axis at different times for the Hughes model (top) and for the MFG structure (bottom)

The extension of the above method into higher dimensions is straight forward. Here, we restrict ourselves to two-dimensional problem. Suppose the geometry for numerical experiment is taken as $\Omega = [-1, 1] \times [-1, 1]$ with exits located at $(\pm 1, \pm 1)$. Furthermore, we choose all parameters as for one dimension. Figure 2 shows the evolution of density for both the classical Hughes model and the MFG structure at different times. Figure 3 shows the evolution of solutions through the center along the x-axis for both models at different times. One observes that the solutions in two dimensional cases have a similar behavior as the solutions in one dimensional case, see Fig. 1.

Acknowledgements This work is supported by the German research foundation, DFG grant KL 1105/27-1, by RTG GrK 1932 "Stochastic Models for Innovations in the Engineering Sciences", project area P1 and by the DAAD PhD program MIC "Mathematics in Industry and Commerce".

References

1. Bellomo, N., Dogbe, C.: On the modeling crowd dynamics from scaling to hyperbolic macroscopic models. Math. Models Methods Appl. Sci. **18**, 1317–1345 (2008)
2. Burger, M., Markowich, P.A., Pietschmann, J.-F.: Continuous limit of a crowd motion and herding model: analysis and numerical simulations. Kinet. Relat. Model. **4**, 1025–1047 (2011)
3. Burger, M., Schlake, B., Wolfram, M.-T.: Nonlinear Poisson-Nernst-Planck equations for ion flux through confined geometries. Nonlinearity, **25**, 961–990 (2012)
4. Burger, M., Di Francesco, M., Markowich, P.A., Wolfram, M.-T.: Mean field games with nonlinear mobilities in pedestrian dynamics. Discrete Contin. Dynam. Systems Ser B: J Bridging Math. Sci. **19**, 1311–1333 (2014)

5. Hughes, R.L.: A continuum theory for the flow of pedestrians. Transp. Res. B Methodol. **36**, 507–535 (2000)
6. Lasry, J.-M., Lions, P.-L.: Mean field games. Jpn. J. Math. **2**, 229–260 (2007)
7. Mahato, N.K., Klar, A., Tiwari, S.: Particle methods for multi-group pedestrian flow. Appl. Math. Model. **53**, 447–461 (2018)
8. Mahato, N.K., Klar, A., Tiwari, S.: A meshfree particle method for a vision-based macroscopic pedestrian model. Int. J. Adv. Eng. Sci. Appl. Math. **10**, 41–53 (2018)

Non-renewable Fishery Resource Management Under Incomplete Information

Hidekazu Yoshioka, Yuta Yaegashi, Yumi Yoshioka, and Kentaro Tsugihashi

Abstract In this brief paper, stochastic control theory under incomplete information is applied to mathematical modeling of inland fishery management. The inland fishery resource to be managed is non-renewable in the sense that its reproduction is unsuccessful. The incomplete information comes from the uncertain body growth rate of the individuals due to temporal regime-switching of their foods. We show that finding the most cost-effective harvesting policy of the non-renewable fishery resource reduces to solving a Hamilton-Jacobi-Bellman equation. The equation is numerically solved via a simple finite difference scheme focusing on the major inland fishery resource *Plecoglossus altivelis* (*P. altivelis*: Ayu) in Japan.

1 Introduction

Resource management in natural environment is always subject to uncertainties, due mainly to inherent stochasticity of the resource dynamics and our lack of knowledge on the dynamics. Inland fishery resource management is such an example where the body growth rate of the individuals is one of the most important biological parameters, which highly depends on their living environmental conditions that are usually only partially-observable. A mathematical framework for describing and

H. Yoshioka (✉)
Graduate School of Natural Science and Technology, Shimane University, Matsue, Japan
e-mail: yoshih@life.shimane-u.ac.jp

Y. Yaegashi
Graduate School of Agriculture, Kyoto University, Sakyo-ku, Kyoto, Japan
e-mail: yaegashi.yuta.54s@st.kyoto-u.ac.jp

Y. Yoshioka
Faculty of Agriculture, Tottori University, Tottori, Japan
e-mail: yoshioka@tottori-u.ac.jp

K. Tsugihashi
Graduate School of Life and Environmental Science, Shimane University, Matsue, Japan
e-mail: a179806@matsu.shimane-u.ac.jp

© Springer Nature Switzerland AG 2019
I. Faragó et al. (eds.), *Progress in Industrial Mathematics at ECMI 2018*,
Mathematics in Industry 30, https://doi.org/10.1007/978-3-030-27550-1_56

controlling such dynamics should be established for attaining better inland fisheries; however, such approaches are still rare to our knowledge. The stochastic control theory under incomplete information [2, 4] can be an effective candidate for this purpose.

The objective of this paper is to propose a prototype mathematical model for cost-effective management of a non-renewable inland fishery resource under uncertainty, focusing on its application to *Plecoglossus altivelis* (*P. altivelis*: Ayu) living in Japanese rivers. The fish has an annual life cycle, but its reproduction has not been always successful in the country [9]. Their population dynamics are therefore reasonably considered to be non-renewable. The main source of the partial observability is the regime-switching phenomenon of their body growth rate caused by temporal dynamics of the benthic algae, the main food source of the fish. The model here employs the formalism of stochastic control under incomplete information where the population dynamics are described as a system of stochastic differential equations (SDEs) controlled so that a performance index as a total profit to be maximized. Finding the optimal management policy of the fish reduces to solving a Hamilton-Jacobi-Bellman (HJB) equation: a degenerate nonlinear parabolic equation. The non-renewability assumption combined with the functional form of the performance index leads to an equivalent reduced HJB equation that is easier to handle. The reduced HJB equation is discretized with a simple finite difference scheme based on the central differencing as much as possible strategy [7]. A demonstrative computational example of the model to management of the fish in a Japanese river is finally presented.

2 Mathematical Model

The mathematical model for the fishery resource management, namely the system of SDEs to be controlled, the performance index to be maximized, and the HJB equation to be solved, are presented. Detailed mathematical setting follows that of standard stochastic control approach [4] and is therefore not presented here.

2.1 System of SDEs

The system of SDEs describing the population dynamics of the fishery resource is similar to that of Yoshioka and Yaegashi [8] for aquaculture management since both handle non-renewable population dynamics. A main difference between their and present formulations is that the former considers that the body growth rate of the fish is observable, while it is not in the latter. Another difference is the drift and diffusion coefficients. This paper employs the coefficients [3] so that the HJB equation presented later is defined in a compact domain.

The time is denoted as t and the standard Brownian motion defined through the nonlinear filtering formalism under incomplete information (Chapter 8 of [2]) is represented as B_t. The representative body weight of the individuals is denoted as W_t. The growth rate is assumed to follow a two-state Markov chain, which is not directly observable. The population of the individuals in a habitat, which is assumed to be closed, is denoted as N_t. The two-state Markov chain has the states High and Low, with the higher μ_{H} and lower body growth rates μ_{L} such that $\mu_{\mathrm{H}} > \mu_{\mathrm{L}} > 0$. The transition rate from the state High (Low) to the state Low (High) is denoted as $\lambda_{\mathrm{HL}} > 0$ ($\lambda_{\mathrm{LH}} > 0$). The conditional probability based on the filtration G_t generated by the processes X_t and Y_t is denoted as $Y_t = \Pr(\mu_t = \mu_{\mathrm{H}} | G_t)$ where μ_t is the optimal estimation [2] of the body growth rate. The measurable process c_t adapted to B_t represents the harvesting rate, which is the control variable to be optimized in the present model. This c_t is valued in the compact set $[0, c_{\max}]$ with a constant $c_{\max} > 0$.

The harvesting is assumed to be allowed after a fixed time $\tau > 0$. The indicator function for the set $t \geq \tau$ is denoted as z_t. Namely, $z_t = 1$ if $t \geq \tau$ and $z_t = 0$ otherwise.

The system of SDEs for the population dynamics is then formulated through observable quantities as follows:

$$\mathrm{d}N_t = -(z_t c_t + R) N_t \mathrm{d}t, \quad t > 0, \quad 0 \leq N_0, \tag{1}$$

$$\mathrm{d}X_t = g(X_t)\left(\hat{\mu}(Y_t)\mathrm{d}t + \sigma \mathrm{d}B_t\right), \quad X_t = K^{-1} W_t, \quad t > 0, \quad 0 \leq X_0 \leq 1, \tag{2}$$

$$\mathrm{d}Y_t = (\lambda_{\mathrm{LH}} - (\lambda_{\mathrm{LH}} + \lambda_{\mathrm{HL}}) Y_t) + \sigma^{-1} \Delta\mu g(Y_t) B_t, \quad t > 0, \quad 0 \leq Y_0 \leq 1. \tag{3}$$

Here, the function g is defined as $g(x) = x(1-x)$ for $0 \leq x \leq 1$ and the notation $\hat{\mu}(Y_t) = Y_t \mu_{\mathrm{H}} + (1 - Y_t)\mu_{\mathrm{L}}$ is employed. The constant $R > 0$ represents the mortality rate per unit time of the individuals, $K > 0$ represents the maximum body weight, $\sigma > 0$ represents the magnitude of inherent stochasticity involved in the dynamics of X_t, and $\Delta\mu = \mu_{\mathrm{H}} - \mu_{\mathrm{L}} > 0$. Without significant loss of generality, the process c_t is assumed to be a Markov control.

2.2 *Performance Index and Value Function*

The performance index J to be maximized with respect to the control variable c_t ($0 \leq t \leq T$) with a fixed terminal time $T > 0$ subject to the above-presented population dynamics is formulated as

$$J(t, n, x, y; c) = \mathrm{E}\left[\int_t^T \alpha z_s c_s N_s X_s \mathrm{d}s - \int_t^T \beta p N_s X_s \mathrm{d}s\right]. \tag{4}$$

Here, E represents the expectation conditioned on G_t and the notations $N_t = n$, $X_t = x$, and $Y_t = y$ have been used. $\alpha > 0$ and $\beta > 0$ are weighting constants, and $p > 0$ is the management cost per unit time. The performance index J simply measures the expectation of the profit by harvesting (the first term) minus the cost of management (the second term), the latter comes from then maintenance cost of the environment where the fishery resource lives [9].

The value function $u = u(t, n, x, y)$ is the maximum of J:

$$u(t, n, x, y) = \sup_c J(t, n, x, y; c). \tag{5}$$

The optimal c_t that achieves the maximum in (4) is denoted as c_t^*. Finding this c_t^* is the goal of the present stochastic control problem.

2.3 HJB Equation

The formal linearity of the SDE (1) and the performance index J with respect to the variable N_t allows us to decompose the value function u as $u = n\Psi$ with some function $\Psi = \Psi(t, x, y)$. The conventional dynamic programming principle then leads to the (reduced) HJB equation that governs Ψ as

$$\begin{aligned}
&\frac{\partial \Psi}{\partial t} - R\Psi + \hat{\mu}(y) g(x) \frac{\partial \Psi}{\partial x} + \frac{\sigma^2}{2} g^2(x) \frac{\partial^2 \Psi}{\partial x^2} + \Delta\mu g(x) g(y) \frac{\partial^2 \Psi}{\partial x \partial y} \\
&+ (\lambda_{\mathrm{LH}} - (\lambda_{\mathrm{LH}} + \lambda_{\mathrm{HL}}) y) \frac{\partial \Psi}{\partial y} + \frac{(\Delta\mu)^2 g^2(y)}{2\sigma^2} \frac{\partial^2 \Psi}{\partial y^2} \\
&- \beta p x - \min_{0 \le c \le c_{\max}} (z_t c (\Psi - \alpha x)) = 0
\end{aligned} \tag{6}$$

for $0 \le t < T$, $0 \le x \le 1$, and $0 \le y \le 1$ subject to the terminal condition $\Psi = 0$ at $t = T$. The boundary conditions along $x = 0, 1$ and $y = 0, 1$ are specified by directly considering (6) on these boundaries. The above-presented approach for the boundaries is justified by the characteristics argument for second-order partial differential equations degenerated on boundaries [5].

Hereafter, the condition $\alpha c_{\max} - \beta p > 0$ is assumed, meaning that there is a possibility to make the profit be larger than the cost by choosing some c_t. This is a quite natural assumption from a fisheries viewpoint. Through the HJB equation (6), the optimal c_t^* is found as the maximizer of its last term, which is expressed with an abuse of notation as $c^*(t, X_t, Y_t)$. This c^* is expressed through Ψ as

$$c^*(t, X_t, Y_t) = z_t \operatorname*{argmin}_{0 \le c \le c_{\max}} (c(\Psi(t, X_t, Y_t) - \alpha X_t)) \tag{7}$$

Since this c^* does not explicitly depend on the total population N_t, the optimal harvesting strategy is based on the growth and its information uncertainty.

In the above-mentioned sense, finding the optimal management policy of the fishery resource reduces to solving the HJB equation (6), which is carried out numerically in this paper. This is because we have no hope to solve the equation analytically due to its multi-dimensional and highly nonlinear nature. Nevertheless, we have several qualitative results on the solution Ψ. although the proofs are not presented here, we can show that Ψ is non-negative and bounded from the above, and asymptotically behaves polynomially near the boundary $x = 0$. In addition, Ψ is increasing and decreasing with respect to x and t, respectively, agreeing well with our intuition that the total profit would increase when the fishery resource is larger and when the harvesting is carried out in a longer period.

3 Application

The HJB equation (5) is discretized with a finite difference scheme based on the central differencing as much as possible strategy [7] with a fully-implicit time discretization. The mixed derivative term is discretized with a conventional central difference discretization, which is not convergent in the viscosity sense and may violate maximum principles numerically. Nevertheless, it has been found that the employed numerical discretization strategy can handle HJB equations without numerical instability. We conjecture that this may be due to that solutions to HJB equations are actually classical $C^{1,2}$-class solutions almost everywhere in the domain. The spatio-temporal domain is uniformly discretized into 256 cells in each direction.

The parameter values are specified as follows: $T = 180$ (day), $\tau = 45$(day), $R = 0.010$ (1/day), $c_{\max} = 0.030$ (1/day), $\mu_{\mathrm{L}} = 0.016$ (1/day), $\sigma^2 = 0.004$ (1/day), $\mu_{\mathrm{H}} = 0.040$ (1/day), $\lambda_{\mathrm{HL}} = \lambda_{\mathrm{LH}} = 0.1$ (1/day), $p = 0.015$ (1/day), $\alpha = 1.0$, $\beta = 0.2$. These parameter values are based on our field survey results in Hii River, Japan, where $P.\ altivelis$ is the most important inland fishery resource. Reproduction of the fish in the river is considered to be unsuccessful, and its population is maintained annual release of farmed juveniles in spring.

Computationally, for $t \geq \tau$, the optimal harvesting policy is as follows: there is a continuous and non-negative bi-variate function b, namely a free boundary such that $c^*(t, X_t, Y_t) = c_{\max}$ if $X_t \geq b(t, Y_t)$ and $c^*(t, X_t, Y_t) = 0$ if $X_t < b(t, Y_t)$. Figure 1 shows the computed b. In this sense, the free boundary completely determines the optimal harvesting policy. Qualitatively the same results have been obtained for different parameter values, supporting this conjecture. The results suggest that harvesting the fish is not always optimal just after $t = \tau$. This is a theoretical, but possible policy.

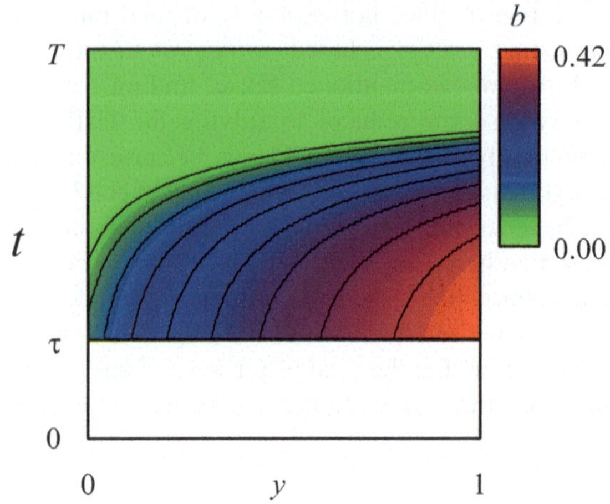

Fig. 1 Computed free boundary $b = b(t, y)$ for $t \geq \tau$

4 Concluding Remark

In this paper, we focused on formulating the problem. Mathematical analysis of the model, especially the HJB equation, was not discussed. A major issue is identifying the function space where the value function Ψ belongs to. Several literature (For example, see [6]) implied that solutions to degenerate parabolic equations, even when they are linear, do not belong to standard Sobolev spaces like $W^{1,2}$, but to some weighted Sobolev spaces: the latter are able to handle the degenerate coefficients in domains and on boundaries. Approaches from the viewpoint of viscosity solutions [1] would effectively work as well. Further research is necessary from both mathematical and practical viewpoints to achieve more cost-effective inland fisheries.

Acknowledgements This work was supported by The River Foundation under grant The River Fund No. 285311020, The Japan Society for the Promotion Science under grant KAKENHI No. 17K15345 and No. 17J09125, and Water Resources Environment Center under grant The WEC Applied Ecology Research Grant No. 2016-02. The authors thank the officers of Hii River Fisheries Cooperatives for providing valuable data and comments for this research.

References

1. Fleming, H.W., Soner, H.M.: Controlled Markov Processes and Viscosity Solutions. Springer, Berlin (2006)
2. Liptser, R.S., Shiryaev, A.N.: Statistics of Random Processes: I. General Theory. Springer, Berlin (2013)
3. Lungu, E.M., Øksendal, B.: Optimal harvesting from a population in a stochastic crowded environment. Math. Biosci. **145**, 47–75 (1997)
4. Øksendal, B.: Stochastic Differential Equations. Springer, Berlin (2003)

5. Oleinik, O.A., Radkevic, E.V.: Second-Order Equations with Nonnegative Characteristic Form. Springer, Berlin (2006)
6. Valkov, R.: Convergence of a finite volume element method for a generalized Black-Scholes equation transformed on finite interval. Numer. Algorithms **68**, 61–80 (2015)
7. Wang, J., Forsyth, P.A.: Maximal use of central differencing for Hamilton-Jacobi-Bellman PDEs in finance. SIAM J. Numer. Anal. **46**, 1580–1501 (2008)
8. Yoshioka, H., Yaegashi, Y.: Optimization model to start harvesting in stochastic aquaculture system. Appl. Stoch. Model. Bus. Ind. **33**, 476–493 (2017)
9. Yoshioka, H., Yaegashi, Y.: Mathematical analysis for management of released fish. Optimal Control Appl. Methods **39**, 1141–1146 (2018)

Environmentally Optimized Management of Urban Road Networks

Lino J. Álvarez-Vázquez, Néstor García-Chan, Aurea Martínez, and Miguel Ernesto Vázquez-Méndez

Abstract In this paper we address the optimal management of an urban road network by combining optimal control of partial differential equations, numerical simulation and optimization techniques. Specifically, we are interested in analyzing the optimal management of the intersections of an urban road network, in order to reduce both atmospheric pollution and traffic congestion. To optimize the network management, we consider a multi-objective optimal control problem, balancing—within a cooperative Pareto framework—a traffic cost function involving travel times and outflows, and a pollution cost function related to contaminant concentrations. In the second part of this work we present numerical tests for a real-world example of ecological interest, posed in the Guadalajara Metropolitan Area (Mexico), where the possibilities of our approach are shown.

1 Introduction

Two of the main urban problems in modern cities are atmospheric pollution and traffic congestion. In order to deal with these problems (that are closely related to each other), a careful urban planning has shown to be of crucial importance. Urban planners are involved in design and management of the urban road network, taking into account several issues: network topology (number and location of roads, intersections...), roads characteristics (capacity, length, number of lanes, construction costs...), intersections characteristics (existence of level crossing,

L. J. Álvarez-Vázquez (✉) · A. Martínez
Universidade de Vigo, Vigo, Spain
e-mail: lino@dma.uvigo.es; aurea@dma.uvigo.es

N. García-Chan
Universidad de Guadalajara, Guadalajara, Mexico
e-mail: nestor.gchan@academicos.udg.mx

M. E. Vázquez-Méndez
Universidade de Santiago de Compostela, Lugo, Spain
e-mail: miguelernesto.vazquez@usc.es

© Springer Nature Switzerland AG 2019

I. Faragó et al. (eds.), *Progress in Industrial Mathematics at ECMI 2018*,
Mathematics in Industry 30, https://doi.org/10.1007/978-3-030-27550-1_57

presence of traffic lights...) and so on. Nevertheless, the behaviour of traffic network users (whose main interests usually tend to optimize travel times and costs) commonly makes difficult the effectiveness of adopted measures (mainly directed to a better functioning of the whole road network and to a reduction in pollution).

The analysis of traffic flow in urban networks coupled with the study of air pollution has been dealt with the use of partial differential equations models [4, 7, 9], but the optimal control of such coupled problem has been much less addressed. In previous works [1, 2] the authors proposed a novel methodology that brings together a 1D model for vehicular flow with a 2D model for pollutant dispersion, in order to apply optimal control techniques to the interactions between road network flow and air pollution (obtaining an explicit relation between the design variables of the problem and the corresponding states: traffic flow and air pollution).

The present work is devoted to set, analyze and solve a multi-objective cooperative (Pareto) optimal control problem, related to the optimal management of the network intersections, looking for environmental and operational interests (that is, minimizing air pollution and optimizing traffic flow). We also present and discuss some numerical experiences obtained when applying our methodology to a simple, real-world case posed in the Guadalajara Metropolitan Area (Mexico), although it can be applied to any other, more complex scenario.

2 The Optimal Control Problem

Within an urban domain $\Omega \subset \mathbb{R}^2$ we consider a road network composed of N_R unidirectional avenues (segments) meeting at a number N_J of junctions (intersections). Each segment $A_i \subset \Omega$, $i = 1, \ldots, N_R$, is modelled by an interval $[a_i, b_i]$ and a parametrization $\sigma_i : s \in [a_i, b_i] \rightarrow \sigma_i(s) = (x_i(s), y_i(s)) \in A_i$ preserving the sense of motion on the avenue. We denote by $\mathcal{J}^{in}$, $\mathcal{J}^{out} \subset \{1 \ldots, N_R\}$ the sets of indices corresponding to incoming and outgoing roads in the network, respectively. Finally, for each junction $j = 1 \ldots, N_J$, we denote $\mathcal{J}_j^{in}$, $\mathcal{J}_j^{out} \subset \{1 \ldots, N_R\}$ the sets of indices corresponding to avenues incoming and outgoing in that junction, respectively.

To model traffic flow in the road network we consider the classical LRW model coupled with a queue model. We denote by $\rho_i(s,t) \in [0, \rho_i^{max}]$ the density of cars in the avenue A_i, (ρ_i^{max} standing for the maximum allowed value). We suppose known the static relations $f_i : [0, \rho_i^{max}] \rightarrow \mathbb{R}$ giving the flow rate on the avenue A_i in terms of the density ($f_i(\rho_i) = \rho_i v_i$, with $v_i(s,t)$ the velocity on the avenue A_i). Finally, for each $y \in \mathcal{J}^{in}$, we denote by $q_y(t) \geq 0$ the queue length downstream the avenue A_y. Then, the traffic flow in the road network can be modelled by the following system [2]: for $i = 1, \ldots, N_R$, $y \in \mathcal{J}^{in}$, $z \in \mathcal{J}^{out}$, $j = 1, \ldots, N_J$,

$k \in \mathcal{I}_j^{in}$, and $l \in \mathcal{I}_j^{out}$:

$$\frac{\partial \rho_i}{\partial t} + \frac{\partial f_i(\rho_i)}{\partial s} = 0 \quad \text{in } (a_i, b_i) \times (0, T), \tag{1}$$

$$f_k(\rho_k(b_k, .)) = \sum_{l \in \mathcal{I}_j^{out}} \min\left\{\alpha_{lk}^j D_k(\rho_k(b_k, .)),\ \beta_{kl}^j S_l(\rho_l(a_l, .))\right\} \quad \text{in } (0, T), \tag{2}$$

$$f_l(\rho_l(a_l, .)) = \sum_{k \in \mathcal{I}_j^{in}} \min\left\{\alpha_{lk}^j D_k(\rho_k(b_k, .)),\ \beta_{kl}^j S_l(\rho_l(a_l, .))\right\} \quad \text{in } (0, T), \tag{3}$$

$$f_z(\rho_z(b_z, .)) = \min\{f_z^{out}, D_z(\rho_z(b_z, .))\} \quad \text{in } (0, T), \tag{4}$$

$$f_y(\rho_y(a_y, .)) = \min\{D_y^{in}(q_y, .), S_y(\rho_y(a_y, .))\} \quad \text{in } (0, T), \tag{5}$$

$$\frac{dq_y}{dt} = f_y^{in} - f_y(\rho_y(a_y, .)) \quad \text{in } (0, T), \tag{6}$$

with corresponding initial conditions. In above system, $D_i,\ S_i : [0, \rho_i^{max}] \to \mathbb{R}$ denote, respectively, the demand and supply functions given by:

$$D_i(\rho) = \begin{cases} f_i(\rho) & \text{if } 0 \le \rho \le \rho_{C_i}, \\ C_i & \text{if } \rho_{C_i} \le \rho \le \rho_i^{max}, \end{cases} \qquad S_i(\rho) = \begin{cases} C_i & \text{if } 0 \le \rho_i \le \rho_{C_i}, \\ f_i(\rho) & \text{if } \rho_{C_i} \le \rho \le \rho_i^{max}, \end{cases} \tag{7}$$

for C_i the road capacity for avenue A_i, and ρ_{C_i} its critical density.

Parameters α_{lk}^j represent the preferences of drivers arriving to a junction, that is, α_{lk}^j gives the percentage of drivers that, arriving to junction j from the incoming avenue A_k, try to take the outgoing avenue A_l. Consequently, they should verify:

$$0 \le \alpha_{lk}^j \le 1, \qquad \sum_{l \in \mathcal{I}_j^{out}} \alpha_{lk}^j = 1. \tag{8}$$

On the other hand, parameters β_{kl}^j represent the ingoing capacities in outgoing avenues, that is, β_{kl}^j gives the percentage of vehicles that, at a junction j and coming from the avenue A_k, can enter the outgoing avenue A_l. In a similar way to previous case, these parameters should verify:

$$0 \le \beta_{kl}^j \le 1, \qquad \sum_{k \in \mathcal{I}_j^{in}} \beta_{kl}^j = 1. \tag{9}$$

Finally, term $D_y^{in}(q_y, t)$ represents the demand of queue q_y at time t, given by:

$$D_y^{in}(q_y, t) = \begin{cases} \min\{f_y^{in}(t), C_y^{in}\} & \text{if } q_y = 0, \\ C_y^{in} & \text{if } q_y > 0. \end{cases} \tag{10}$$

In order to simulate air pollution due to vehicular traffic, we consider a mathematical model similar to that proposed in [1]. We denote by $\phi(x, t)$ the CO concentration at point $x \in \Omega$ and at time $t \in [0, T]$, obtained as the solution of the following initial/boundary value problem:

$$\frac{\partial \phi}{\partial t} + \mathbf{v} \cdot \nabla \phi - \nabla \cdot (\mu \nabla \phi) + \kappa \phi = \sum_{i=1}^{N_R} \xi_{A_i} \quad \text{in } \Omega \times (0, T), \tag{11}$$

$$\mu \frac{\partial \phi}{\partial n} - \phi \mathbf{v} \cdot \mathbf{n} = 0 \quad \text{on } S^-, \tag{12}$$

$$\mu \frac{\partial \phi}{\partial n} = 0 \quad \text{on } S^+, \tag{13}$$

with corresponding initial condition, and where $\mathbf{v}(x, t)$ represents the wind velocity field, $\mu(x, t)$ is the CO molecular diffusion coefficient, $\kappa(x, t)$ is the CO extinction rate, and $\mathbf{n}$ denotes the unit outward normal vector to the boundary $\partial\Omega = S^- \cup S^+$, with $S^- = \{(x, t) \in \partial\Omega \times (0, T) \ : \ \mathbf{v} \cdot \mathbf{n} < 0\}$ the inflow boundary, and $S^+ = \{(x, t) \in \partial\Omega \times (0, T) \ : \ \mathbf{v} \cdot \mathbf{n} \geq 0\}$ the outflow boundary. Finally, each second member term ξ_{A_i} stands for pollution sources due to vehicular traffic on road A_i, given by the following Radon measure, with $\gamma_i, \ \eta_i$ weight parameters for contamination rates:

$$\langle \xi_{A_i}(t), v \rangle = \int_{a_i}^{b_i} (\gamma_i f_i(\rho_i(s, t)) + \eta_i \rho_i(s, t)) \ v(\sigma_i(s)) \, \|\sigma_i'(s)\| \, ds, \quad \forall v \in \mathscr{C}(\overline{\Omega}),$$

Two different objectives will be simultaneously considered here: the former involving traffic flow, and the latter related to air pollution. So, regarding the optimization of traffic flow, we try to minimize the length of queues and the total travel time, and to maximize the outflow of the system. Thus, the following functional will be minimized [8]:

$$J_T = \int_0^T \left(\sum_{y \in \mathscr{J}^{in}} \epsilon_y^q q_y(t) + \sum_{i=1}^{N_R} \epsilon_i \int_{a_i}^{b_i} \rho_i(x, t) dx - \sum_{z \in \mathscr{J}^{out}} \epsilon_z^{out} f_z(\rho_z(b_z, t)) \right) dt, \tag{14}$$

with $\epsilon_y^q, \ \epsilon_i, \ \epsilon_z^{out} \geq 0$ weight parameters (alternative formulations -related to strategic decisions from the viewpoint of game theory- can be seen, for instance, in [5] and the references therein). With respect to contamination, we try to keep the mean CO concentration is as low as possible. So, we will minimize the functional:

$$J_P = \frac{1}{T\,|\Omega|}\int_0^T \int_\Omega \phi(x,t)\,dx\,dt, \tag{15}$$

where $|\Omega|$ denotes the Euclidean measure of set Ω.

Finally, with respect to the design variables (controls) that can be managed within the network, we will focus on the optimal management of the intersections, trying to look for the values of the parameters that are the most suited to our objectives.

So, we can consider the following multi-objective cooperative problem: Let us suppose that the preferences of the drivers α_{lk}^j are known. In this case, the control of the problem will be the vector $\boldsymbol{\beta} = (\beta_{kl}^j)$, and, assuming a unique organization managing the entire network, we will try to solve the multi-objective problem:

$$\begin{aligned} &\min \mathbf{J}(\boldsymbol{\beta}) = (J_T(\boldsymbol{\beta}), J_P(\boldsymbol{\beta})) \\ &\text{subject to (9)} \end{aligned} \tag{16}$$

within a cooperative framework, that is, looking for Pareto-optimal solutions.

From the computational viewpoint, in order to find the Pareto-optimal frontier of problem (16), the use of an elitist genetic algorithm is proposed (in particular, a well-known variant of the multi-objective non-dominated sorting-based evolutionary algorithm NSGA-II [6], that is included in the Global Optimization Toolbox of MATLAB R2017a).

3 Numerical Results

We discuss here some numerical results obtained when applying our methodology to a real-world case posed in Guadalajara Metropolitan Area (Mexico) with almost 5 million inhabitants and more than 2 million vehicles. So, domain Ω shown in Fig. 1 was taken and, for the sake of brevity, a main road network of only $N_R = 15$ avenues and $N_J = 9$ junctions and a time interval of $T = 24$ h was chosen.

In relation to the traffic model, all the avenues of the network were assumed to present the same theoretical characteristics. Moreover, we took the same downstream road capacity for the three incoming roads, and also equal maximum outflow rates for the three outgoing roads. Finally, we considered weight parameters with values $\epsilon_i = 0.5$, $\epsilon_y^q = 0.1$ and $\epsilon_z^{out} = 0.5$. With respect to the pollution model, a characteristic wind field was considered (see Fig. 1), and typical values for CO ($\mu = 3.5 \times 10^{-8}$, $\kappa = 0.6 \times 10^{-2}$, $\gamma_i = 10^{-6}$, $\eta_i = 3.16 \times 10^{-5}$) were taken.

The Pareto-optimal frontier obtained can be seen in Fig. 2, where three distinguished points were chosen for its analysis: β_T, the best solution from the viewpoint of traffic flows and travel times; β_P, the best solution for pollution minimization; and β_C, an intermediate compromise solution.

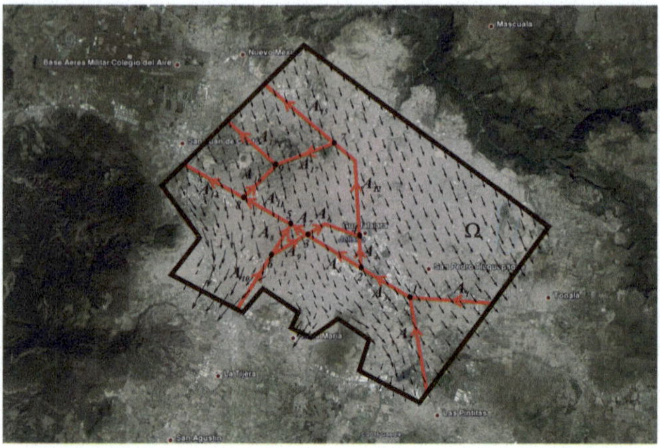

Fig. 1 Satellite photo of Guadalajara metropolitan area (Mexico). The domain Ω considered for pollution simulation is drawn in black, the road network is sketched in red, and vectors show the wind velocity field corresponding to the numerical experiment

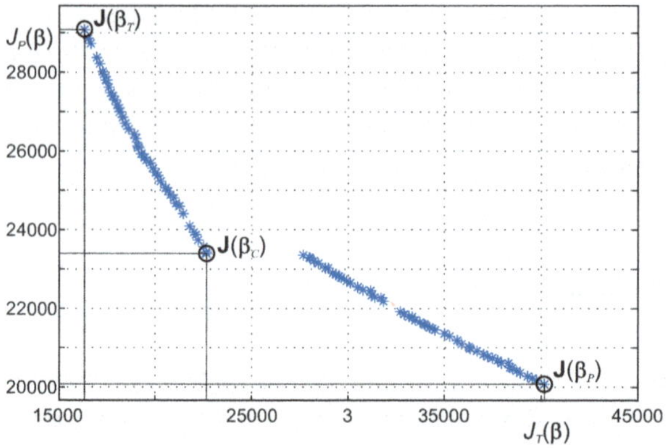

Fig. 2 Pareto-optimal frontier obtained for the multi-objective optimal control problem, where three particular solutions are emphasized: the optimal solution for the optimization of traffic flow (β_T), the optimal solution for the minimization of pollution (β_P), and a compromise solution corresponding to an intermediate balance choice (β_C)

Finally, Fig. 3 compares, for these three Pareto-optimal solutions, the values (along the whole simulation time interval) of two variables for the traffic model. In this case, for the sake of conciseness, only queue lengths at entry road A_1 and outflow rates at exit avenue A_{15} are shown. The full numerical algorithm and more detailed computational results can be found in the recent paper of the authors [3]. Results show that, as was expected, solution β_T results in a greater outflow rate at outgoing avenues, and smaller queues for incoming roads. On the contrary, solution β_P causes lower CO concentration (since car density and flow rate decrease), but

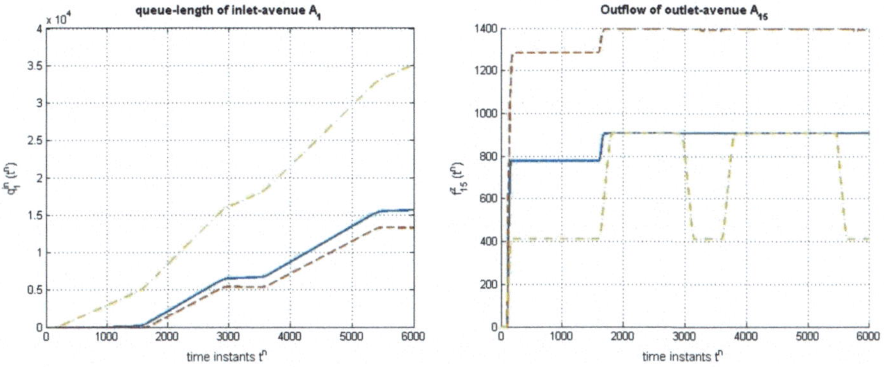

Fig. 3 Values for the three Pareto-optimal solutions β_T (dashed lines), β_P (dash-dot lines), and β_C (solid lines) of queue length for entry road A_1 (left), and outflow rate for exit road A_{15} (right)

queue lengths increase. Finally, solution β_C reaches a balancing situation that, under satisfactory traffic conditions, reduces pollution to acceptable levels.

Acknowledgements The authors thank the support from the project MTM2015-65570-P of MINECO/FEDER (Spain), and also from SNI-52768 and PRODEP/103.5/16/8066 (Mexico).

References

1. Alvarez-Vázquez, L.J., García-Chan, N., Martínez, A., Vázquez-Méndez, M.E.: Numerical simulation of air pollution due to traffic flow in urban networks. J. Comput. Appl. Math. **326**, 44–61 (2017)
2. Alvarez-Vázquez, L.J., García-Chan, N., Martínez, A., Vázquez-Méndez, M.E.: Optimal control of urban air pollution related to traffic flow in road networks. Math. Control Rel. Fields **8**, 177–193 (2018)
3. Alvarez-Vázquez, L.J., García-Chan, N., Martínez, A., Vázquez-Méndez, M.E.: Optimal management of an urban road network with an environmental perspective. Comput. Math. Appl. **77**, 1786–1797 (2019)
4. Berrone, S., de Santi, F., Pieraccini, S., Marro, M.: Coupling traffic models on networks and urban dispersion models for simulating sustainable mobility strategies. Comput. Math. Appl. **64**, 1975–1991 (2012)
5. Csercsik, D., Sziklai, B.: Traffic routing oligopoly. Cent. Eur. J. Oper. Res. **23**, 743–762 (2015)
6. Deb, K.: Multiobjective Optimization Using Evolutionary Algorithms. Wiley, New York (2001)
7. García-Chan, N., Alvarez-Vázquez, L.J., Martínez, A., Vázquez-Méndez, M.E.: Numerical simulation for evaluating the effect of traffic restrictions on urban air pollution. In: Quintela, P. et al. (eds.) Progress in Industrial Mathematics at ECMI 2016, pp. 367–373. Springer, Berlin (2017)
8. Goatin, P., Goettlich, S., Kolb, O.: Speed limit and ramp meter control for traffic flow networks. Eng. Optim. **48**, 1121–1144 (2016)
9. Skiba, Y.N., Parra-Guevara, D.: Control of emission rates. Atmosfera **26**, 379–400 (2013)

Optimal Control of Heavy Metals Phytoremediation

Aurea Martínez, Lino J. Alvarez-Vázquez, Carmen Rodríguez, Miguel E. Vázquez-Méndez, and Miguel A. Vilar

Abstract Heavy metals enter aquatic systems as a result of very different human activities involving the mining, processing and use of substances containing metal pollutants. Phytoremediation is a cost-effective plant-based approach of remediation for heavy metal-contaminated bodies of water, that takes advantage of the ability of algae to concentrate elements from the environment in their tissues. This paper deals with the optimization of phytoremediation methods, by combining mathematical modelling, optimal control and numerical optimization. In particular, we propose a 2D mathematical model coupling partial differential equations modelling the concentrations of heavy metals, algae and nutrients in large waterbodies. Questions related to determining the minimal quantity of algae to be used, and also to locating the optimal place for such algal mass, are formulated as an optimal control problem for this scenario, and several numerical results for a realistic case are presented.

1 Introduction

Heavy metals pollution is nowadays one of the major environmental engineering problems related to the quality of coastal water in highly industrialized areas, since its threats to human health are well known from many decades ago [6]. The main difficulty in the remediation of heavy metals (cooper, lead, cadmium, mercury, chromium and so on) arises from the fact that they cannot be biodegraded and, consequently, they persist indefinitely in water.

A. Martínez (✉) · L. J. Alvarez-Vázquez
Universidade de Vigo, Vigo, Spain
e-mail: aurea@dma.uvigo.es; lino@dma.uvigo.es

C. Rodríguez
Universidade de Santiago de Compostela, Santiago, Spain
e-mail: carmen.rodriguez@usc.es

M. E. Vázquez-Méndez · M. A. Vilar
Universidade de Santiago de Compostela, Lugo, Spain
e-mail: miguelernesto.vazquez@usc.es; miguel.vilar@usc.es

© Springer Nature Switzerland AG 2019

I. Faragó et al. (eds.), *Progress in Industrial Mathematics at ECMI 2018*,
Mathematics in Industry 30, https://doi.org/10.1007/978-3-030-27550-1_58

Among the several techniques for treating heavy metals pollution (ranging from physical removal or detoxification to bioleaching and bioremediation) the most commonly used in present times is phytoremediation (which refers to the use of the adsorption capacity of plants—in this case, algae—in order to reduce the concentration of heavy metals in water), due to its effectiveness and low cost [10].

Usual phytoremediation techniques are based on the placement of a quantity $\tilde{a}$ of algal mass in a subdomain K of the water region under study, in such a way that its bioadsorbent capacity may reduce the concentration levels of heavy metals inside its influence zone. Within this framework, the simplest example of optimal control problem would be the determination of the minimal quantity of algae to be placed and of the optimal location of this algal mass. So, following techniques similar to those previously used by the authors in different environmental control problems [2–4], in below sections we set a rigorous mathematical formulation of the environmental problem, deal with its numerical resolution and, finally, present several computational examples showing the validity of our method.

2 Mathematical Formulation of the Environmental Problem

To fix ideas, we consider a bounded domain $\Omega \subset \mathbb{R}^2$ (for instance, an estuary) corresponding to a shallow water region, where a heavy metals pollution problem is detected, due to industrial wastewater discharges causing a contamination of the body of water by surpassing allowed thresholds.

As above commented, our main aim is related to the determination of the minimal quantity of algae $\tilde{a}$ to be placed inside Ω and of its optimal location $K \subset \Omega$. A first necessary step in this direction is the setting of a suitable mathematical model in order to simulate the evolution of the interactions between the main species involved in the process (water, heavy metals, algae and, eventually, nutrients) along a time interval $[0, T]$. So, following the notations previously introduced by the authors in [7] (where full details and definitions can be found), we consider the following concentrations (height averaged in the water column): $c(x, t)$ representing the concentration of heavy metal in water; $q(x, t)$, the concentration of heavy metal deposited in algae; $a(x, t)$, the concentration of algae in water; and $p(x, t)$, the concentration of nutrients (mainly nitrogen and phosphorus) in water.

To model the bioadsorption capacity of algae for c and q we use the classical Lagergren's kinetic model combined with the well-known Langmuir equilibrium model for characterizing the adsorption isotherm [5]. To model the interactions of algae and nutrients for a and p we use a convection-reaction-diffusion equation with nonlinear Michaelis-Menten kinetics [3]. Thus, the proposed state system simulating the full process reads, for $x \in \Omega \subset \mathbb{R}^2$ and for $t \in (0, T)$, as the following well-posed, coupled system of partial differential equations [7]:

$$\begin{aligned}
&\frac{\partial c}{\partial t} + v \cdot \nabla c - \mu_c \, \Delta c + \kappa_c \, a \, \frac{\partial q}{\partial t} = F \quad \text{in } \Omega \times (0, T),\\
&\frac{\partial q}{\partial t} = \kappa_q \left(\frac{Q_{max} \, b \, c}{1 + b \, c} - q \right) \quad \text{in } \Omega \times (0, T),\\
&\frac{\partial a}{\partial t} + w \cdot \nabla a - \mu_a \, \Delta a - \lambda \, \frac{p}{\kappa_p + p} \, a + \gamma \, a = 0 \quad \text{in } \Omega \times (0, T),\\
&\frac{\partial p}{\partial t} + v \cdot \nabla p - \mu_p \, \Delta p + \beta \, \lambda \, \frac{p}{\kappa_p + p} \, a = G \quad \text{in } \Omega \times (0, T),
\end{aligned} \tag{1}$$

with initial conditions:

$$\begin{aligned}
&c(0, x) = c^0(x), \quad q(0, x) = q^0(x) \quad \text{in } \Omega,\\
&a(0, x) = a^0(x) + \tilde{a} \, 1_K(x), \quad p(0, x) = p^0(x) \quad \text{in } \Omega,
\end{aligned} \tag{2}$$

and boundary conditions:

$$\frac{\partial c}{\partial n} = \frac{\partial a}{\partial n} = \frac{\partial p}{\partial n} = 0 \quad \text{on } \partial\Omega \times (0, T), \tag{3}$$

where vector field $v(x, t)$ represents the velocity of water, averaged in height $h(x, t)$, being both solutions of the classical shallow water equations; μ_c, μ_a, μ_p are the diffusion coefficients of metal, algae, and nutrient, respectively; κ_c is the mass transfer coefficient; $F(x, t)$ represents the source term of heavy metal; κ_q is the Lagergren constant; Q_{max} represents the maximum adsorption capacity; b is the Langmuir constant; $w(x, t)$ represents the velocity of algae; λ represents the luminosity coefficient; κ_p is the nutrient semi-saturation constant; γ is the mortality rate of algae; β is the nutrient-carbon stoichiometric coefficient; $G(x, t)$ represents the source term of nutrients; and $1_K(x)$ represents the indicator function of the region $K \subset \Omega$ where algae are initially added with a mean concentration $\tilde{a} \geq 0$.

As already commented, we try to find the optimal location K where the algal mass can be placed, and also its minimal quantity $\tilde{a}$, in such a way that water quality, given by heavy metals level, is maximized. Control variables K and $\tilde{a}$ enter the state system *via* the initial condition (2) for state variable a. Regarding the first control variable K, and based on geopolitical reasons, we impose some constraints related to the optimal location of the mass of algae. In particular, algal area K is defined from a central point p that we demand to be located inside an admissible region $\mathcal{K}_{ad} \subset \Omega$ (assumed convex for simplicity). So, area K is formed by the triangle in which the point p lies and the three adjacent triangles (that is, region K is composed by four contiguous elements). On the other hand, due to economic causes, for the second control variable $\tilde{a}$, which is nonnegative by definition, we require it not to exceed maximal and minimal thresholds $A_{max} > A_{min} \geq 0$. That is, we are actually imposing the following control constraints:

$$p \in \mathcal{K}_{ad}, \qquad \tilde{a} \in [A_{min}, A_{max}]. \tag{4}$$

Finally, as cost function J to be minimized we take the mean concentration of heavy metals in water inside a sensitive region to be protected $S \subset \Omega$ (which can be even the whole domain Ω). For instance, the objective function could be:

$$J(p, \tilde{a}) = \frac{1}{T\,|S|} \int_0^T \int_S c(x, t)\, dx\, dt \tag{5}$$

where $|S|$ denotes the measure of S, and c is the solution of the state system (1)–(3) with initial conditions (2) posed for K and $\tilde{a}$, derived from design variables $(p, \tilde{a})$.

Then, the optimal control problem consists of finding the optimal location K and the optimal amount $\tilde{a}$ of algae such that verify state system (1)–(3), minimize cost function (5) and satisfy constraints (4). Thus, the problem can be written as:

$$\min_{\substack{p \in \mathcal{K}_{ad} \\ A_{min} \leq \tilde{a} \leq A_{max}}} \quad J(p, \tilde{a}) \tag{6}$$

To solve this optimization problem (6), and given its essentially geometric nature, we propose a direct search algorithm: the Nelder-Mead simplex method [9]. This classical algorithm is a derivative-free method, based on the mere comparison of values of the objective function, that constructs a sequence of simplices (sets of sample points) as an approximation to the optimal point, and that has been successfully used by the authors in other related environmental problems (see, for instance, previous work [1], where a full description of the algorithm can be also found).

Moreover, although Nelder-Mead algorithm is not guaranteed to converge in the general case, it presents good convergence properties for low dimensions (and in our problem design variable $(p, \tilde{a}) \in \mathbb{R}^3$). However, since Nelder-Mead algorithm has been designed for unconstrained minimization problems, in order to apply it to the control-constrained optimization problem (6) we need first to modify our cost function (5) by adding a penalty term related to the fulfilling of the control constraints (4), which can be made here in a very simple and straightforward way.

3 Computational Results

We present here some numerical tests for a realistic case posed in the estuary *Ría de Vigo* (Galicia, NW Spain), in the area surrounding the harbour in Cangas, corresponding to a lead discharge from several shipyards located in the region. Numerical results have been obtained by using the module *Heavy metals* of the 2D hydrodynamic model MIKE 21 [8], developed by DHI (Danish Institute of Technology) and widely employed in the study of environmental issues.

From the large collection of computational experiences performed to analyze the effects of the placement of a mass of algae (in this case, green alga *Ulva*) in the vicinity of the discharge zone, only two figures are presented here.

For the optimization algorithm, the admissible region $\mathcal{K}_{ad}$ is given by a rectangle surrounding the wastewater discharge point, and the lower and upper bounds are taken as $A_{min} = 0$ and $A_{max} = 60$. Then, we start from a set of four initial simplices: $(p_1, \tilde{a}_1) = ((518376.99, 4677839.17), 20.00)$, $(p_2, \tilde{a}_2) = ((518587.76, 4677845.97), 25.00)$, $(p_3, \tilde{a}_3) = ((518465.38, 4678070.33), 15.00)$, and $(p_4, \tilde{a}_4) = ((518492.57, 4677920.76), 30.00)$, whose cost values range from $J = 0.0146$ to $J = 0.0151$. After 173 cost function evaluations, we arrive to optimal control $(p, \tilde{a}) = ((518864.62, 4677937.99), 59.99)$, where the objective function takes the value $J = 0.0106$ (representing a reduction in lead concentration of a 30% with respect to initial situation).

The two figures correspond to zooms on the area of interest in the *Ría de Vigo* (Cangas harbour region). So, Fig. 1 shows lead concentration c at low tide (after two tidal cycles: about $T = 24.8$ h) for the initial (uncontrolled) configuration, obtained by the numerical resolution of the problem in a spatial triangular mesh of 1941 elements and 1129 vertices, with a time step of 30 s. We can appreciate there the protected area (formed by two regions delimited by a white thick line), and the original location K for algae (region composed of four triangles depicted by a white thin line). Figure 2 shows c at same time for the optimized (controlled) configuration, with optimal location K depicted by a coloured thick line. It can be easily noted how lead concentration is significantly reduced by placing the optimal amount of algae in the optimal area near to the source of pollution (point close to the most highly contaminated vertex in Fig. 1). It can be also observed how the reduction of lead appears not only in the area where algae are placed, but also affects to the surrounding zones of the estuary, due to natural tide effects. Finally, from a quantitative viewpoint, we can see how the optimized quantity of added algae $\tilde{a}$

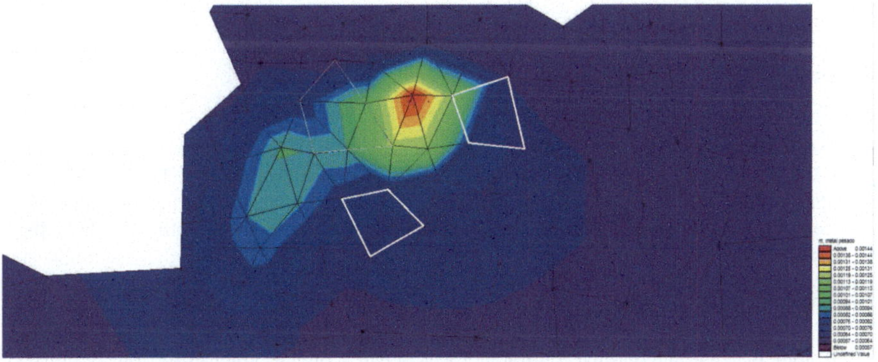

Fig. 1 Lead concentration levels for the uncontrolled situation posed in a region of the estuary *Ría de Vigo*. Discharge zone corresponds to black point close to concentration peak, protected areas S are depicted by white thick lines, and initial location K is delimited by a white thin line

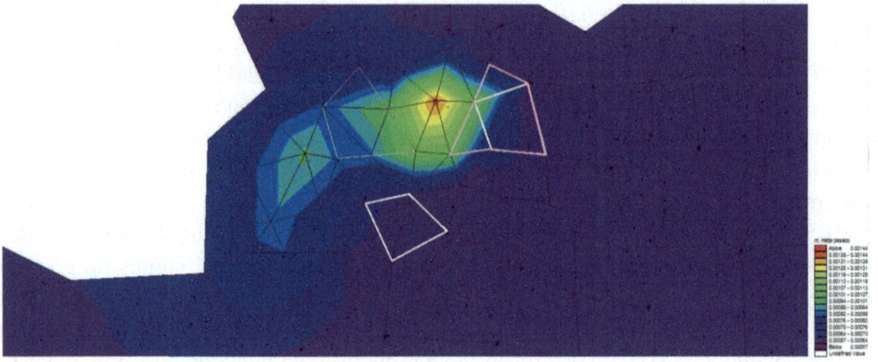

Fig. 2 Lead concentration levels, corresponding to the same scenario, after the controlled placement of a mass of algae $\tilde{a} = 59.99$ g/m^3 in the optimal area K depicted by a coloured thick line

approaches the maximum allowed quantity A_{max}, and how, also as expected, the optimal region K moves closer to the protected region S and to the wastewater discharge point, in order to intensify phytoremediation benefits.

Alternative numerical results have been presented by the authors in a recent paper [5], by using the finite element method for the discretization of the state system (1)–(3) and an interior-point algorithm for the resolution of the nonlinear, constrained optimization problem (6). These alternative results are very similar to those given here, both qualitatively and quantitatively, which shows the robustness of our approach: a novel and reliable combination of optimal control theory of partial differential equations, mathematical modelling and numerical optimization.

Acknowledgements The authors thank funding from project MTM2015-65570-P of MINECO/FEDER (Spain), and from project ED431C 2018/50 of Xunta de Galicia.

References

1. Alvarez-Vázquez, L.J., Martínez, A., Rodríguez, C., Vázquez-Méndez, M.E.: Numerical optimization for the location of wastewater outfalls. Comput. Optim. Appl. **22**, 399–417 (2002)
2. Alvarez-Vázquez, L.J., Martínez, A., Muñoz-Sola, R., Rodríguez, C., Vázquez-Méndez, M.E.: The water conveyance problem: optimal purification of polluted waters. Math. Models Methods Appl. Sci. **15**, 1393–1416 (2005)
3. Alvarez-Vázquez, L.J., Fernández, F.J., Martínez, A.: Optimal management of a bioreactor for eutrophicated water treatment: A numerical approach. J. Sci. Comput. **43**, 67–91 (2010)
4. Alvarez-Vázquez, L.J., Martínez, A., Rodríguez, C., Vázquez-Méndez, M.E.: Sediment minimization in canals: an optimal control approach. Math. Comput. Simul. **149**, 109–122 (2018)
5. Alvarez-Vázquez, L.J., Martínez, A., Rodríguez, C., Vázquez-Méndez, M.E., Vilar, M.A.: Optimal control of phytoremediation techniques for heavy metals removal in shallow water. In: Rodrigues, H.C., et al. (eds.) EngOpt 2018 Proceedings, pp. 352–360. Springer, Berlin (2019)

6. Mani, D., Kumar, C.: Biotechnological advances in bioremediation of heavy metals contaminated ecosystems: an overview with special reference to phytoremediation. Int. J. Environ. Sci. Technol. **11**, 843–872 (2014)
7. Martínez, A., Alvarez-Vázquez, L.J., Rodríguez, C., Vázquez-Méndez, M.E., Vilar, M.A.: Heavy metals phytoremediation: first mathematical modelling results. In: Radu, F.A. et al. (eds.) Numerical Mathematics and Advanced Applications - ENUMATH 2017, pp. 819–827. Springer, Berlin (2019)
8. MIKE 21: User Guide and Reference Manual. Danish Hydraulic Institute, Horsholm (2001)
9. Nelder, J.A., Mead, R.: A simplex method for function minimization. Comput. J. **7**, 308–313 (1965)
10. Perales-Vela, H., Peña-Castro, J., Cañizares-Villanueva, R.: Heavy metal detoxification in eukaryotic microalgae. Chemosphere **64**, 1–10 (2006)

Intraday Renewable Electricity Trading: Advanced Modeling and Optimal Control

Silke Glas, Rüdiger Kiesel, Sven Kolkmann, Marcel Kremer, Nikolaus Graf von Luckner, Lars Ostmeier, Karsten Urban, and Christoph Weber

Abstract This paper is concerned with a new mathematical model for intraday electricity trading involving both renewable and conventional generation. The model allows us to incorporate market data e.g. for half-spread and immediate price impact. The optimal trading and generation strategy of an agent is derived as the viscosity solution of a second-order Hamilton-Jacobi-Bellman (HJB) equation for which no closed-form solution can be given. We thus construct a numerical approximation allowing us to use continuous input data. Numerical results for a portfolio consisting of three conventional units and wind power are provided.

1 Introduction

Due to the extensive rise of renewable power supply as a response to the global climate change, electricity short-term markets like EPEX SPOT, in particular continuous intraday trading, gained more importance. This, in turn, motivates the interest in mathematical modeling of such trading as a basis for deeper understanding and optimization. Early work in that direction can be found in [3]. In [1], the authors derive a Hamilton-Jacobi-Bellman (HJB) equation for determining an optimal trading strategy by modeling the dynamics of the electricity market by *stochastic differential equations* (SDEs) and formulating a corresponding *value function* to be optimized. The specific market model in [1] allows to solve the arising HJB analytically, i.e., the authors derive a solution formula. The starting point of this paper is a statistical analysis of EPEX SPOT data, which shows that some of the model assumptions in [1] are not satisfied under real market conditions. Thus, we

S. Glas · K. Urban (✉)
Ulm University, Institute for Numerical Mathematics, Ulm, Germany
e-mail: silke.glas@uni-ulm.de; karsten.urban@uni-ulm.de

R. Kiesel · M. Kremer · N. G. von Luckner · S. Kolkmann · L. Ostmeier · C. Weber
University of Duisburg-Essen, Essen, Germany
e-mail: ruediger.kiesel@uni-due.de; sven.kolkmann@uni-due.de; marcel.kremer@uni-due.de; nikolaus.graf-von-luckner@uni-due.de; lars.ostmeier@uni-due.de; christoph.weber@uni-due.de

© Springer Nature Switzerland AG 2019

I. Faragó et al. (eds.), *Progress in Industrial Mathematics at ECMI 2018*, Mathematics in Industry 30, https://doi.org/10.1007/978-3-030-27550-1_59

introduce a more sophisticated model. The arising HJB equation can no longer be solved analytically; the value function is shown to be the unique *viscosity solution* of this HJB equation. Thus, we need an appropriate numerical scheme.

From an economical point of view, the main new ingredients of our model are: (1) Portfolio of renewable and conventional energy represented by a cost function that reflects the stepwise merit order of a portfolio rather than a systemwide quadratic function; (2) Pricing model using time-varying half-spread and being capable of representing time-varying liquidity; (3) Approximation of market data for half-spread and instantaneous price impact; (4) Variable penalty depending on the state of the market at final time. The main focus of this paper is a novel application-related modeling of the intraday trading and the determination of a numerical approximation for this problem. We show an example of a real-world problem and compute the optimal trading strategy. The remainder of this paper is organized as follows: In Sect. 2, we introduce the new model and the arising HJB equation, Sect. 3 is devoted to the presentation of numerical experiments and we finish by an outlook.

2 A New Mathematical Model

In order to take both renewable and conventional generation into account, our model is based upon the consideration of an agent owning both kinds of power plants and aiming at selling a combination of renewable[1] and additionally conventionally produced electricity. In detail, depending on the weather forecast and the expected price at the final time, a combination of conventional and renewable electricity is sold at the day-ahead market. With this sold amount, the agent starts the continuous intraday trading aiming at maximizing her profit by determining an optimal trading strategy as well as an optimal production of conventional power.[2] We are now going to describe both involved frameworks, namely the trading model including day-ahead as well as intraday trading and the stochastic model of the dynamics.

Day-Ahead and Intraday Electricity Trading Consider a delivery hour h on day d. The day before, the *day-ahead auction* takes place with gate closure at 12 pm. In this auction, each participant can offer (ask) or request (bid) a certain demand of electricity at a specific price. Then, a clearing price is set and power is exchanged accordingly. Next, the *continuous intraday trading* starts at 3 pm on day $d - 1$ and closes half an hour before the actual delivery hour h, see Fig. 1.

Dynamics of the Electricity Market The dynamics of the market includes the forecasted renewable power production and the price process. The latter one is

[1] In our numerical experiments, we consider wind energy.

[2] That means that we do not optimize day-ahead and intraday trading at the same time.

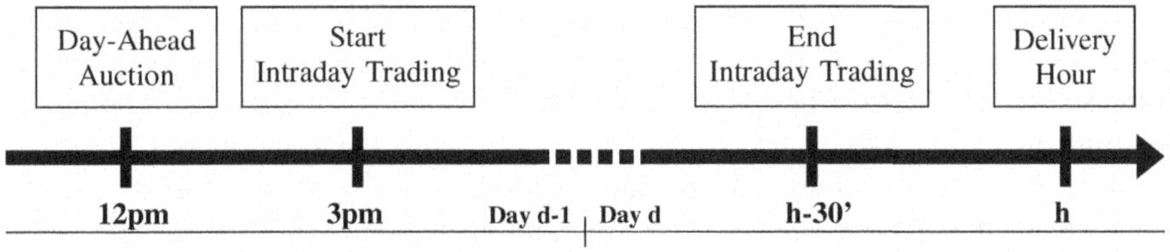

Fig. 1 Scheme of continuous intraday trading

influenced by the current trading activity of the agent. We use stochastic processes and derive stochastic differential equations (SDEs).

Forecast Model for the Renewable Power By $D = (D_t)_{0 \le t \le T}$ we denote the *forecasted production* of renewable electricity during the trading session. The uncertainty is modeled by means of the dynamics $dD_t = \mu_D\, dt + \sigma_D\, dW_{t,D}$, where μ_D is the drift, σ_D is the volatility and $(W_{t,D})_{0 \le t \le T}$ is a standard Brownian motion. For the sake of simplicity, this variable is unbounded, whereas in the real world, there are restrictions by zero (no wind) and the maximum capacity of the wind farm.

Agent's Position The position resulting from the agent's trading activity is denoted by $X = (X_t)_{0 \le t \le T}$, which resembles the agents current amount of electricity. The agent participates in the intraday market with continuous trading at rate $q_t \in \mathcal{Q} \subset \mathbb{R}$ ($q_t > 0$ means buying, $q_t < 0$ selling), i.e., $dX_t = q_t\, dt$ and we denote by $X^{q,t,x}$ the solution of the SDE starting from x at t (for $t = 0$, x_0 is the amount of electricity sold on the day-ahead market).

Price Model The *execution price* is the price the agent pays (receives) when actually buying (selling). We require a more advanced approach of the pricing model as in [1], where the half-spread and its time variability as well as the time variability of the immediate price impact are ignored. Incorporating these effects, the execution price depends on a number of quantities to be introduced now. First, we denote by $Y = (Y_t)_{0 \le t \le T}$ the sum of the *mid price* of energy and the *permanent impact* of the agent's trading modeled by some function $\psi : \mathbb{R} \to \mathbb{R}$. Its dynamics is modeled by the SDE $dY_t = (\mu_Y + \psi(q_t))dt + \sigma_Y\, dW_{t,Y}$, where μ_Y is the drift, σ_Y is the volatility and $(W_{t,Y})_{0 \le t \le T}$ is a standard Brownian motion. We denote by $Y^{t,y}$ the solution of the SDE starting from y at t. The next ingredient is the *half-spread* $h : [0, T] \to \mathbb{R}$, i.e., the half of the distance between the best ask and the best bid price. This is data which can be retrieved from the market. With all these quantities at hand, the execution price $P^{q,t,y} = (P_s^{q,t,y})_{0 \le s \le T}$ is modeled as

$$P_s^{q,t,y} := Y_s^{t,y} + \frac{|q_s|}{q_s} h(s) + \varphi(t, q_s), \qquad (1)$$

i.e., the permanently impacted mid price plus (minus) the half-spread and the instantaneous price impact $\varphi : [0, T] \times \mathcal{Q} \to \mathbb{R}$.

Conventional Production/Payoff At the end of the trading session T, the agent chooses how much electricity $\xi \in \mathbb{R}_0^+$ she will produce during the delivery period. In doing so, she also has the option to place a final buy or sell market order, potentially resulting in $\xi \neq -Z_T$, with $Z_t := X_t + D_t$ being the sum of the forecasted production from renewables and what has been sold by the agent so far. For example, she could further increase her sell position and production. The final market order goes along with costs due crossing the half-spread $h(T)$ and potentially executing limit orders whose prices are worse than the best bid/ask price due to a penalty $\alpha : \mathbb{R} \to \mathbb{R}_0^+$. The arising cost per unit depends on the state of the market at T. The terminal payoff is

$$g(\xi, Y_T, Z_T) := -c(\xi) + (\xi + Z_T)\left(Y_T - (h(T) + \alpha)\frac{|\xi + Z_T|}{\xi + Z_T}\right), \tag{2}$$

where $c : \mathbb{R}_0^+ \to \mathbb{R}_0^+$ models the cost of the conventional generation.

Value Function The value function corresponds to the agent's cash, so that an optimal strategy yields maximal cash. The total running profit from the continuous trading in the intraday market is given by $f^q(s; t, y) := -q_s P_s^{q,t,y}$. Denoting by $Z^{q,t,z}$ the solution of the SDE $dZ_t = dD_t + dX_t = (q_t + \mu_D)\,dt + \sigma_D\,dW_{t,D}$ starting from z at t, the resulting value function $V : [0, T] \times \mathcal{U} \to \mathbb{R}$ reads

$$V(t, y, z) := \sup_{(q,\xi)\in\mathcal{Q}\times\mathbb{R}} \mathbb{E}\left[\int_t^T f^q(s; t, y)\,ds + g(\xi, Y_T^{t,y}, Z_T^{q,t,z})\right], \tag{3}$$

where $\mathcal{U} := \mathcal{Y} \times \mathcal{Z} \subset \mathbb{R}^2$ is a rectangle (in order to ensure well-posedness of the optimization in (3), [2]). We prescribe Dirichlet conditions on the boundary $\partial\mathcal{U}$.

Hamilton-Jacobi-Bellman (HJB) Equation Following the well-known *dynamic programming principle* (e.g. [5, Ch. 4]), we derive the HJB equation: Find $W : [0, T] \times \mathcal{U} \to \mathbb{R}$, $W = W(t, y, z)$, such that

$$\begin{aligned} &\partial_t W + \mu_Y \partial_y W + \mu_D \partial_z W + \frac{1}{2}\sigma_Y^2\,\partial_{yy} W + \frac{1}{2}\sigma_D^2\,\partial_{zz} W \\ &+ \sup_{q(t)\in\mathcal{Q}} \left\{ -\left(y + h(t)\frac{|q(t)|}{q(t)} + \varphi(t, q(t))\right) q(t) + q(t)\,\partial_z W + \psi(q(t))\,\partial_y W \right\} = 0, \end{aligned} \tag{4}$$

for $(t, y, z) \in [0, T) \times \mathcal{U}$ with terminal condition $W(T, y, z) = g(T, y, z)$, $(y, z) \in \mathcal{U}$. One can show that this problem is well-posed and that the unique *viscosity solution* W is the value function V in (3). Due to the form of (4), we cannot expect a first-order condition for the control $q(t)$ and we have to resort to numerical solvers.

3 Numerical Experiment

Finally, we report on results of a numerical experiment concerning (4) using the following data: $\mathcal{U} := [-50, 250] \times [-1645, 145] \subset \mathbb{R}^2$ and $T = 17.5\,\text{h}$. We use a finite difference discretization from [4] with 56×301 points in space and 100 points in time. In particular, central differences are used for the approximation of the first-order terms with additional artificial diffusion, which results in a stable, consistent and monotone scheme converging to the viscosity solution, [4]. We use the well-known *policy iteration* in every time-step and the control is maximized over a discrete set (as no first-order conditions are available). Finally, the optimal conventional generation is computed as the maximum value of (2) w.r.t. ξ using Matlab's intlinprog with the interior point method.

Boundary Conditions Similar to option pricing, the choice of appropriate boundary conditions (here for y and z) is delicate. Here, we use a similar but easier HJB allowing for a closed-form solution on some $\mathcal{U} \subset \mathbb{R}^2$. Then, we prescribe the boundary values of this function as Dirichlet conditions on $\partial\mathcal{U}$.

Data We use the data $\mu_D := \mu_Y := 0.0$, $\sigma_D := \sigma_Y := 0.1$. The functions $\varphi(\cdot, \cdot)$ and $h(\cdot)$ are least-squares fifth order polynomial approximations of market data from Q2/2015 ($\psi(t) = 0$). The penalty is given by market data as $\alpha(x) := 0.5 \cdot (|x| - 20)\chi_{20<|x|\leq 45} + ((|x| - 45) + 12.5)\chi_{45<|x|\leq 145}$. We consider three conventional units, namely a hard coal plant with 25 €/MWh variable cost and min-max capacity of 250–500 MW, one combined cycle gas turbine (CCGT) unit (35 €/MWh, 100–400 MW) and open cycle gas turbine (OCGT) unit (60 €/MWh, 60–600 MW).

Our results for the **optimal conventional generation** ξ are displayed in Fig. 2. Let us comment on the case where $Z_T = -500$ MWh. As long as the final mid price is below 25 €/MWh, the agents buys the maximal amount of 145 MWh (recall, that $y \in [-1645, 145]$) and uses the power plant with the lowest marginal costs (hard

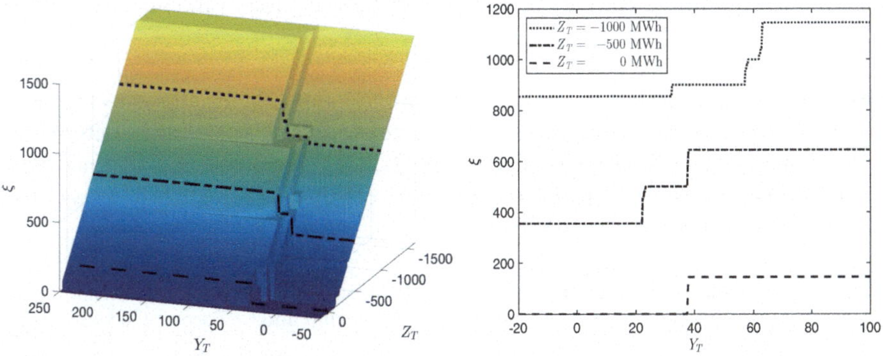

Fig. 2 Optimal conventional generation ξ as a function of Y_T and Z_T (left) as well as for some values of Z_T (right; the lines correspond to those on the left graph)

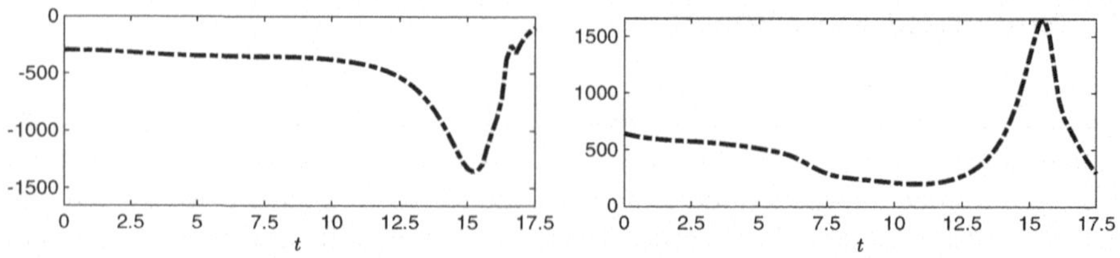

Fig. 3 Optimal trading rate over the trading window $t \in [0, 17.5]$ for $Z_t \equiv -499.4$ MWh and $Y_t \equiv 59.25$ €/MWh (left) as well as $Y_t \equiv 13.98$ €/MWh (right)

coal) accordingly, i.e. the remaining 355 MWh. Once the final mid price is 25–35 €/MWh (i.e., above the marginal cost of hard coal, but below the marginal cost of CCGT) it is optimal to produce at maximum capacity with the cheapest conventional power plant (i.e. 500 MWh by hard coal) and no final market order is required. If the final mid price exceeds 35 €/MWh, the agent sells as much electricity as possible (145 MWh) and produces exactly that amount with the CCGT plant at 35 €/MWh, which is possible because its capacity is 100–400 MW. Finally, no matter how high the final mid price is, the OCGT unit with the highest marginal cost is not used, since there is not enough sell volume on the market. These results are clearly reasonable.

Trading Rate Figure 3 shows the optimal trading rate over the trading window $t \in [0\,\text{h}, 17.5\,\text{h}]$. In both cases, we fix $Z_t \equiv -499.4$ MWh (the non-integer numbers arise from the discretization w.r.t. y and z). For the mid price, we choose $Y_t \equiv 59.25$ €/MWh (left) and $Y_t \equiv 13.98$ €/MWh (right). In the left plot, the trading rate is negative (selling), which is reasonable since $Z_t \equiv -499.4$ MWh means that the agent has only marketed the cheapest power plant and $Y_t \equiv 59.25$ €/MWh means that the execution price is above the marginal costs of the second cheapest power plant. Note, that the absolute value of the trading rate substantially increases around 15 h, since half-spread and immediate price impact are minimal there. In the right plot, the execution price is below the marginal costs of the cheapest power plant, the agent buys electricity and reduces the production of the marketed power plant.

Outlook The availability of a numerical approximation scheme allows us to extend our model to all market participants, so that regulatory constraints can be determined e.g. for reaching desired environmental goals. Ongoing work is concerned with model order reduction to make the scheme real-time efficient 24 h a day with continuous incoming data (market and forecast).

Acknowledgements This work was funded by the German *Federal Ministry for Economic Affairs and Energy* within the project *AEIT*. We are grateful to C. Greif (Ulm) for cooperation within *AEIT*.

References

1. Aïd, R., Gruet, P., Pham, H.: An optimal trading problem in intraday electricity markets. Math. Financ. Econ. **10**(1), 49–85 (2016)
2. Fleming, W.H., Soner, M.: Controlled Markov Processes and Viscosity Solutions. Springer, New York (2006)
3. Garnier, E., Madlener, R.: Balancing forecast errors in continuous-trade intraday markets. Energy Syst. **6**(3), 361–388 (2015)
4. Steck, S., Urban, K.: A Reduced Basis Method for the Hamilton-Jacobi-Bellmann Equation with Application to the European Union Emission Trading Scheme. In: Dante, K., Kunisch, K., Zhiping, R. (eds.) Hamilton-Jacobi-Bellmann Equation, pp. 175–196. de Gruyter, Berlin (2018)
5. Yong, J., Zhou, X.Y.: Stochastic Controls, Hamiltonian Systems and HJB Equations. Springer, New York (1999)

Surrogate Models for Coupled Microgrids

Sara Grundel, Philipp Sauerteig, and Karl Worthmann

Abstract We consider the operation of coupled microgrids. Each microgrid consists of a number of residential energy systems, each including an energy storage device. The goal is to determine an optimal energy exchange between the microgrids, which results in a two-level optimization problem. On the lower level, within each microgrid, a grid operator sets up an optimization scheme to coordinate the individual subsystems. We propose a surrogate model based on radial basis functions to approximate this optimization based process and investigate its applicability in the higher level by conducting a case study based on an Australian data set.

1 Introduction

The successful integration of renewable energy sources into the electricity grid is the key factor to master the energy transition. Herein, the control of locally distributed energy storage devices, e.g. batteries, plays a major role, see, e.g. [4]. While there are approaches for the control of a single microgrid, see, e.g. [6, 7], coupled microgrids incur additional challenges, see [2] and the references therein. Among them is the need to rapidly solve optimization problems at the microgrid layer such that a negotiation process w.r.t. energy exchange between microgrids can be set up, see, e.g. [8]. To this end, we propose a technique based on Radial Basis Functions (RBFs) in order to derive a surrogate model replacing the outcome of an optimization scheme.

The paper is structured as follows: In Sect. 2, we briefly recall operation of a single and of coupled microgrids. Then, we introduce the technique based on RBFs to construct a surrogate model in Sect. 3 before its suitability is investigated

S. Grundel
Max Planck Institute for Dynamics of Complex Technical Systems, Magdeburg, Germany
e-mail: grundel@mpi-magdeburg.mpg.de

P. Sauerteig (✉) · K. Worthmann
Technische Universität Ilmenau, Institut für Mathematik, Ilmenau, Germany
e-mail: Philipp.Sauerteig@tu-ilmenau.de; Karl.Worthmann@tu-ilmenau.de

© Springer Nature Switzerland AG 2019
I. Faragó et al. (eds.), *Progress in Industrial Mathematics at ECMI 2018*,
Mathematics in Industry 30, https://doi.org/10.1007/978-3-030-27550-1_60

in a numerical case study based on data provided by an Australian distribution network. Here, we consider an optimization loop based on the Alternating Direction Method of Multipliers (ADMM), see [1] for details. Finally, coupled microgrids are investigated in Sect. 4 before conclusions are drawn in Sect. 5.

Notation $\mathbb{N}$ denotes the natural numbers while $\mathbb{N}_0$ stands for $\mathbb{N} \cup \{0\}$. For $k, \ell \in \mathbb{N}_0$ with $k \leq \ell$, the set $\{k, k+1, \ldots, \ell\}$ is denoted by $[k : \ell]$.

2 Coupled Microgrids

In the following we revive the idea of coupled microgrids presented in [2]. We consider Ξ microgrids (MGs), $\Xi \in \mathbb{N}$, that are coupled through a network of transmission lines. In our numerical case study, we consider four microgrids coupled as depicted in Fig. 1.

Each MG $\kappa \in [1 : \Xi]$ consists of a number of $\mathscr{I}_\kappa \in \mathbb{N}$ residential energy systems, residential energy systems, which can be defined as follows.

Optimal Control of a Single Microgrid We consider $\mathscr{I}_\kappa$ residential energy systems; each equipped with a battery, which could be replaced by another energy storage device. The battery dynamics of subsystem i and its power demand are given by

$$x_i(n+1) = \alpha_i x_i(n) + T(\beta_i u_i^+(n) + u_i^-(n)), \tag{1a}$$

$$z_i(n) = w_i(n) + u_i^+(n) + \gamma_i u_i^-(n). \tag{1b}$$

Here, $x_i(n)$ represents the State of Charge (SoC), $u_i^+(n)$ and $u_i^-(n)$ the charging and discharging rate, $z_i(n)$ the power drawn from/supplied to the grid, and $w_i(n)$ the net energy consumption (load minus generation) at time n, $n \in \mathbb{N}_0$. The parameters α_i, β_i, and γ_i are losses due to self-discharge of the battery and energy conversion while

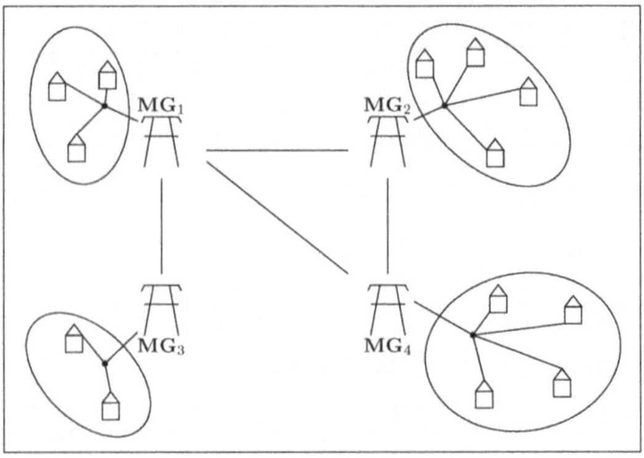

Fig. 1 Schematic representation of four coupled MGs

T denotes the length of the sampling interval in hours. In addition, the following constraints are imposed to model battery capacity limits (2a), bounds on the (dis-) charging rate (2b), (2c), and to avoid simultaneous charging and discharging (2d):

$$0 \leq \quad x_i(n) \quad \leq C_i \tag{2a}$$

$$\underline{u}_i \leq \quad u_i^-(n) \quad \leq 0 \tag{2b}$$

$$0 \leq \quad u_i^+(n) \quad \leq \overline{u}_i \tag{2c}$$

$$0 \leq \frac{u_i^-(n)}{\underline{u}_i} + \frac{u_i^+(n)}{\overline{u}_i} \leq 1. \tag{2d}$$

All systems are coupled via a common point of coupling, at which a grid operator (Central Entity; CE) is located. The CE has to provide the desired power demand—independently of its sign, i.e. whether there is an excess or a need of energy. Therefore, the CE has to retain balancing energy. Hence, an objective is peak shaving, i.e. minimizing the deviation from the average net consumption $\bar{\zeta}_\kappa(k) = (\mathcal{I}_\kappa \cdot \min\{N, k+1\})^{-1} \sum_{n=k-\min\{k,N-1\}}^{k} \sum_{i=1}^{\mathcal{I}_\kappa} w_i(n)$. In summary, the optimization problem reads as

$$\begin{aligned} &\underset{\mathbf{u}=(\mathbf{u}^+,\mathbf{u}^-)}{\text{Minimize}} \quad \frac{1}{N} \sum_{n=k}^{k+N-1} \left(\frac{1}{\mathcal{I}_\kappa} \sum_{i=1}^{\mathcal{I}_\kappa} \left[w_i(n) + u_i^+(n) + \gamma_i u_i^-(n) \right] - \bar{\zeta}_\kappa(n) \right)^2 \\ &\text{s.t.} \quad \text{battery dynamics (1a) and constraints (2),} \end{aligned} \tag{3}$$

where $\mathbf{u}^+ = \left(\mathbf{u}^+(n)\right)_{n=k}^{k+N-1}$ with $\mathbf{u}^+(n) = (u_1^+(n), u_2^+(n), \ldots, u_{\mathcal{I}_\kappa}^+(n))^\top$ and $\mathbf{u}^-$ is analogously defined. The optimization problem (3) can be solved via the Alternating Direction Method of Multipliers (ADMM), see, e.g. [1], which may be interpreted as follows: Compute the power demand $\bar{z}_\kappa(n) := \frac{1}{\mathcal{I}_\kappa} \sum_{i=1}^{\mathcal{I}_\kappa} z_i(n)$ for $n \in [k : k+N-1]$ based on the average net consumption, i.e.

$$(\bar{\zeta}_\kappa(k), \ldots, \bar{\zeta}_\kappa(k+N-1))^\top \quad \overset{\text{ADMM}}{\rightsquigarrow} \quad (\bar{z}_\kappa(k), \ldots, \bar{z}_\kappa(k+N-1))^\top. \tag{4}$$

Negotiation Between Microgrids Microgrids may benefit from an energy exchange even if this exchange involves losses as numerically shown in [2]. To this end, we solve the minimization problem

$$\underset{\delta(k),\ldots,\delta(k+N-1)}{\text{Minimize}} \sum_{n=k}^{k+N-1} \left(\sum_{\kappa=1}^{\Xi} \left(\mathcal{I}_\kappa \bar{\zeta}_\kappa(n) - \sum_{\nu=1}^{\Xi} (\delta_{\nu,\kappa}(n)\eta_{\nu,\kappa}) \mathcal{I}_\nu \bar{\mathbf{z}}_\nu(n) \right)^2 \right) \tag{5}$$

$$\text{s.t.} \sum_{\nu=1}^{\Xi} \delta_{\kappa,\nu}(n) = 1 \text{ and } \delta_{\kappa,\nu}(n) \cdot \delta_{\nu,\kappa}(n) \leq 0,\ \kappa \in [1:\Xi],\ n \in [k : k+N-1]$$

for given $\bar{\mathbf{z}}_\nu(n)$, $\nu \in [1 : \Xi]$ and $n \in [k : k + N - 1]$. Here, the matrices $\delta(n) \in [0, 1]^{\Xi \times \Xi}$ and the symmetric matrix $\eta \in [0, 1]^{\Xi \times \Xi}$ denote the rate of exchange between the microgrids at time instant n and the corresponding losses resp. If there is no transmission line between MG_ν and MG_κ, we set $\eta_{\kappa,\nu} = \eta_{\nu,\kappa} = 0$, $\kappa, \nu \in [1 : \Xi]$. Furthermore, we assume no losses without exchange, i.e. $\eta_{\kappa,\kappa} = 1$, $\kappa \in [1 : \Xi]$. To ensure that exactly the available amount of energy of each MG is distributed the linear constraints $\sum_{\nu=1}^{\Xi} \delta_{\kappa,\nu}(n) = 1$ are introduced. The nonlinear constraints $\delta_{\kappa,\nu}(n) \cdot \delta_{\nu,\kappa}(n) \leq 0$, $\kappa, \nu \in [1 : \Xi]$ with $\kappa \neq \nu$, model that power cannot be exchanged in both directions of a transmission line at one time instant $n \in [k : k + N - 1]$.

Numerical Case Study We consider four MGs with $\mathscr{I}_1 = 50$ and $\mathscr{I}_2 = \mathscr{I}_3 = \mathscr{I}_4 = 10$ and the corresponding net topology and losses given by $\eta_{12} = \eta_{21} = 0.7$, $\eta_{13} = \eta_{31} = 0.8$, $\eta_{14} = \eta_{41} = 0.9$, $\eta_{24} = \eta_{42} = 0.8$, i.e., there is a large MG connected to three smaller MGs, see also Fig. 1. Then, we consider 25 randomly selected time instants and solve (the large scale optimization) Problem (5) using ADMM [2] in a distributed manner, see Fig. 3 (left) for the results.

3 Surrogate Model: Radial Basis Function Approximations

For the negotiation process mentioned in the previous section, each MG needs to solve its optimization problem, which may be very time-consuming. Here, we propose a technique to replace the *input-output-map* (4) by a surrogate model in form of a radial basis function approximation in order to speed up this computation and, thus, to facilitate the optimization of coupled MGs. To be more precise, we replace the map (4) by a function $f : \mathbb{R}^N \to \mathbb{R}^N$, which is given as a Radial Basis Function interpolation of a certain training set. Assume the training set is given by M samples of $\bar{\zeta}(n)$, $n \in [k : k + N - 1]$, to $\bar{z}(n)$, $n \in [k : k + N - 1]$. We stack the N values of the input $\bar{\boldsymbol{\zeta}}_1, \ldots, \bar{\boldsymbol{\zeta}}_M \in \mathbb{R}^N$ and the output $\bar{z}_1, \ldots, \bar{z}_M \in \mathbb{R}^N$. Then, we set up an interpolation function f as a sum of basis functions $\phi_i : \mathbb{R}^N \to \mathbb{R}$ and a regularization term $q : \mathbb{R}^N \to \mathbb{R}^N$, i.e.

$$f(\bar{\zeta}) = \sum_{i=1}^{M} \phi_i(\bar{\zeta})\alpha_i + q(\bar{\zeta})$$

for some $\alpha_1, \ldots, \alpha_M \in \mathbb{R}^N$. This approximation is based on a particular function $\phi : \mathbb{R} \to \mathbb{R}$, which is used to define ϕ_i by $\phi_i(\bar{\zeta}) = \phi(\|\bar{\zeta} - \bar{\zeta}_i\|)$. The function q is typically a short sum of basis functions that are some polynomials. We use a linear regularization term $q(\bar{\zeta}) = \beta_0 + B\bar{\boldsymbol{\zeta}}$, with $\beta_0 \in \mathbb{R}^N$, $B \in \mathbb{R}^{N \times N}$. It is obviously important and possibly difficult to choose the right radial basis function ϕ. Classically this is done by dividing your data into a training and a validation set.

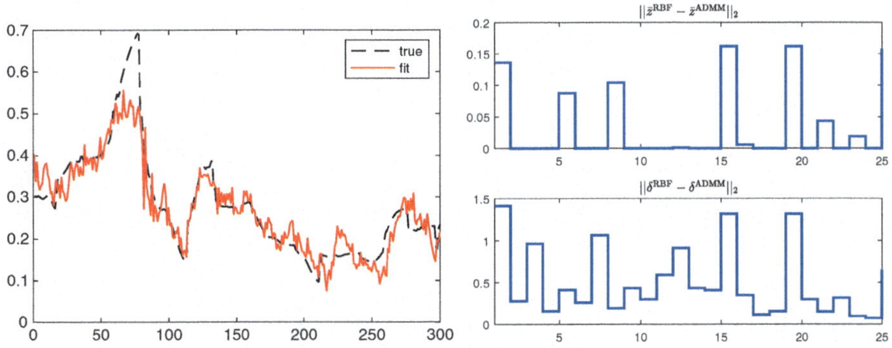

Fig. 2 Comparison of the output of the first component of $\bar{z}$ for a random selection of 300 points which are not used to create the surrogate model (left) and for the 25 randomly selected samples both for the output and w.r.t. its impact on the energy exchange (right)

In order to determine the particular approximation function we need to specify the $MN+N+N^2$ unknowns α_i, $i \in [1 : M]$, β_0, and B. This is done by imposing the NM interpolation conditions $f(\bar{\zeta}_i) = \bar{z}_i$, $i \in [1 : M]$ as well as the following $N + N^2$ additional constraints: $\sum_{i=1}^{M} \alpha_i = 0$ and $\bar{Z}A^T = 0$ where $\bar{Z} = [\bar{\zeta}_1, \ldots, \bar{\zeta}_M]$ and $A = [\alpha_1, \ldots, \alpha_M]$, see [3] for further details.

Numerical Case Study We use 400 samples for input and output, a linear regression term $q(\bar{\zeta}) = \beta_0 + B\bar{\zeta}$, and a spherical basis function $\phi : \mathbb{R}_{\geq 0} \to [0, 1]$, $x \mapsto 1 - 1.5 \min\{1, \theta x\} + 0.5 \min\{1, \theta x\}^3$ with $\theta = 0.01$. With this we get an average error of 0.03, a median of 0.02, and a standard deviation of 0.03 overall the tested 16,800 samples where we used the MatLab toolbox DACE [5] for calculations, see also Fig. 2 (left). The type of function ϕ used and the value of θ are not optimized yet.

4 Case Study

The large MG is once optimized using ADMM, see (4), and once with the help of its approximation based on RBFs while the other MGs are computed only via ADMM. Then, the exchange matrices δ^{RBF} and δ^{ADMM} based on Problem (5) are computed. The deviations are depicted in Fig. 2 on the right. Here, the three main peaks in both plots occur at the same sampling time. However, the error in $\bar{z}$ is no proper indicator for the error in δ. The mean value and the standard deviation are given by 0.0352 and 0.0598 for the approximation error w.r.t. the input-output-map (4) and 0.5084 and 0.4135 w.r.t. the deviations of the coefficients δ for the energy exchange between the MGs. In Fig. 3 the impact of the energy exchange is shown for each of the four MGs.

In the ADMM case, it can be observed that not every MG benefits at every sampling time. Since MG_1 consists of 50 residential energy systems, its performance is

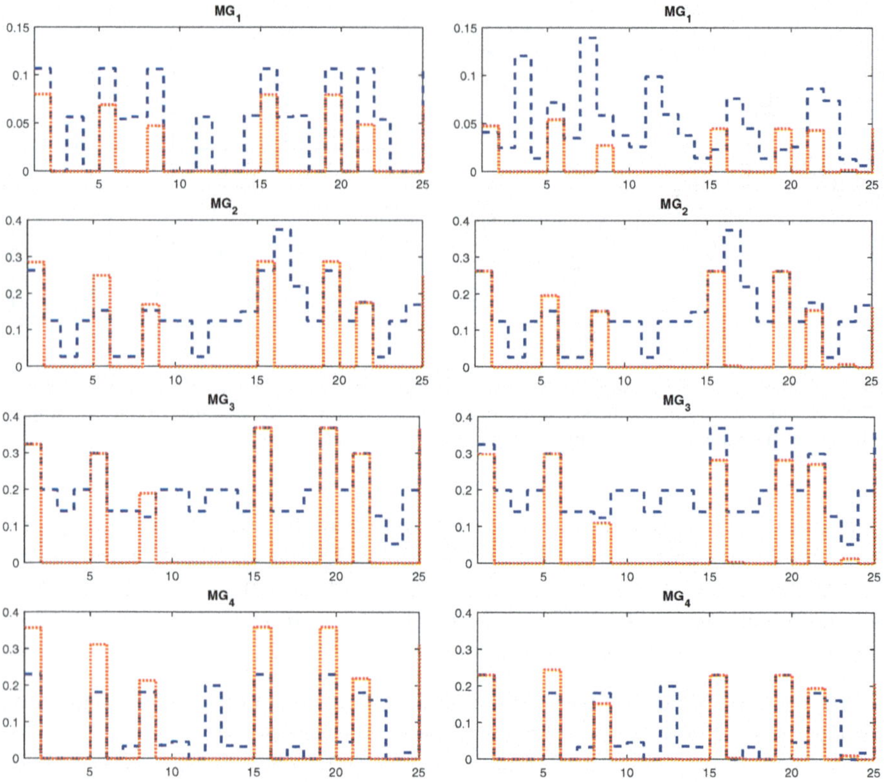

Fig. 3 Deviation from the reference value $\bar{\zeta}_\kappa$, $\kappa \in [1 : 4]$, using ADMM (left) and RBF (right) before (dashed blue line) and after the energy exchange (dotted red line)

weighted five times the performance of the other MGs, compare with (5). Hence, in our example it always benefits from the exchange. The results are similar using the approximation based on RBFs. Here, however, one can observe that MG_1 does not improve its performance for every sample.

5 Conclusions and Outlook

In this paper, we introduced surrogate models for the optimization of microgrids in order to save computation time. To this end, we briefly recapped the idea of coupled microgrids and calculated radial basis function approximations based on samplings. Our numerical case study showed promising results that the exchange coefficients are well approximated—even if individual elements may differ. Moreover, there seems to be a great potential for the use in Model Predictive Control (MPC) as numerically shown in Fig. 2 (left), in which only the first element of the computed vectors is applied, see, e.g. [7] for the use of MPC in smart grids.

Acknowledgements The authors are indebted to the German Research Foundation (DFG; grants WO 2056/4-1 and WO 2056/6-1) and the Federal Ministry for Education and Research (BMBF; grants 05M18EVA and 05M18SIA).

References

1. Braun, P., Faulwasser, T., Grüne, L., Kellett, C.M., Weller, S.R., Worthmann, K.: Hierarchical distributed ADMM for predictive control with applications in power networks. IFAC J. Syst. Control **3**, 10–22 (2018)
2. Braun, P., Sauerteig, P., Worthmann, K.: Distributed optimization based on the example of microgrids. In: Blondin, M.J., Pardalos, P.M., Saéz, J. (eds.) Computational Intelligence and Optimization Methods for Control Engineering. Springer Optimization and Its Applications, vol. 150, pp. 173–200. Springer, Cham (2019)
3. Buhmann, M.D.: Radial Basis Functions: Theory and Implementations, vol. 12. Cambrigde University Press, Cambrigde (2003)
4. Hashemi, S, Østergaard, J.: Efficient control of energy storage for increasing the PV hosting capacity of LV grids. IEEE Trans. Smart Grid **9**(3), 2295–2303 (2018)
5. Lophaven, S.N., Nielsen, H.B., Søndergaard, J.: DACE-A Matlab Kriging toolbox, version 2.0, Technical Report (2002)
6. Morstyn, T., Hredzak, B., Agelidis, V.G.: Control strategies for microgrids with distributed energy storage systems: an overview. IEEE Trans. Smart Grid **9**(4), 3652–3666 (2018)
7. Worthmann, K., Kellett, C.M., Braun, P., Grüne, L., Weller, S.R.: Distributed and decentralized control of residential energy systems incorporating battery storage. IEEE Trans. Smart Grid **6**(4), 1914–1923 (2015)
8. Xiao, L., Mandayam, N.B., Poor, H.V.: Prospect theoretic analysis of energy exchange among microgrids. IEEE Trans. Smart Grid **6**(1), 63–72 (2015)

Optimal Inflow Control Penalizing Undersupply in Transport Systems with Uncertain Demands

Simone Göttlich, Ralf Korn, and Kerstin Lux

Abstract We are concerned with optimal control strategies subject to uncertain demands. An Ornstein-Uhlenbeck process describes the uncertain demand. The transport within the supply system is modeled by the linear advection equation. We consider different approaches to control the produced amount at a given time to meet the stochastic demand in an optimal way. In particular, we introduce an undersupply penalty and analyze its effect on the optimal output in a numerical simulation study.

1 Introduction

In many real-world situations, taking uncertainty into account becomes more and more important. In the context of supply chain management, a need for appropriate control strategies under uncertainty naturally arises when it comes to production planning. The size and timing of product orders is often not known in advance. However, for a delivery on time, the production process needs to be started in advance. In this work, we tackle the challenging question of when to feed how many goods into a supply system to meet the stochastic demand. We use the framework of [3], where a corresponding stochastic optimal control problem is set up in the context of electricity injection and extend it by introducing a penalty term into the cost function. This term penalizes a production not leading to demand satisfaction, i.e. an undersupply.

The main contribution of this work is to provide insight into the effect of an undersupply penalty on the optimal production plan. In a numerical simulation study, we highlight the effect for different penalty parameters.

S. Göttlich · K. Lux (✉)
University of Mannheim, Department of Mathematics, Mannheim, Germany
e-mail: goettlich@uni-mannheim.de; klux@mail.uni-mannheim.de

R. Korn
TU Kaiserslautern, Department of Mathematics, Kaiserslautern, Germany
e-mail: korn@mathematik.uni-kl.de

© Springer Nature Switzerland AG 2019
I. Faragó et al. (eds.), *Progress in Industrial Mathematics at ECMI 2018*, Mathematics in Industry 30, https://doi.org/10.1007/978-3-030-27550-1_61

2 Stochastic Optimal Control Model for Transport Systems

An analysis of optimal control strategies for a supply problem in a deterministic demand setting can be found in [1, 2]. Here, we focus on the stochastic nature of the demand and start from the stochastic optimal control framework originally set up in [3]. We consider a supply system consisting of only one production line. Goods are fed into the system at $x = 0$, and leave the system at $x = 1$. Within a finite time interval $[0, T]$, the aim is to optimally match the externally given customers' demand Y_t located at $x = 1$ by determining the inflow control $u(t) \in L^2$ of goods at $x = 0$. Thereby, the transport of goods $z = z(x, t)$ along the production line is governed by the linear advection equation with constant transport velocity $\lambda > 0$ and the following initial and inflow conditions:

$$\begin{aligned} z_t + \lambda z_x &= 0, \quad x \in (0, 1),\ t \in [0, T] \\ z(x, 0) &= 0, \quad z(0, t) = u(t). \end{aligned} \tag{1}$$

The explicit solution of (1) given by $z(x, t) = u(x - \frac{1}{\lambda}t)$ is well-known (note the boundary control instead of controlling the initial state). As in real-life often more complex dynamics than a pure advection occur (see e.g. [1, 2]), our goal is to set up a more general stochastic optimal control framework. For simplicity, we start our investigations assuming linear transport only. We denote by $y(t) = z(1, t)$ the output of the system. It is intended to match the externally given demand Y_t.

The uncertainty about the height and timing of the orders entails the stochasticity of Y_t. As in [3], we assume that the demand process fluctuates around a given time-dependent mean demand level $\mu(t)$. The latter can be seen as a forecast that is based on historical demand data. In this demand setting, one possible model choice is the Ornstein-Uhlenbeck process (OUP). Let W_t be a one-dimensional Brownian motion, $\sigma > 0$, $\kappa > 0$ be constant parameters, and denote the initial demand by y_0. Then, the OUP is the unique strong solution of the stochastic differential equation (SDE)

$$dY_t = \kappa\,(\mu(t) - Y_t)\,dt + \sigma dW_t, \qquad Y_0 = y_0. \tag{2}$$

The OUP possesses a mean-reverting property, i.e., whenever the process is away from its mean demand level, it is attracted back to it. The parameter κ describes how strong this attraction is, and σ determines how large the fluctuations are.

In this work, we make use of the known distribution of Y_t, which is given by the following normal distribution:

$$Y_t \sim N\left(y_0 e^{-\kappa t} + \kappa \int_0^t e^{-\kappa(t-s)} \mu(s)\, ds,\ \sigma^2 \int_0^t e^{-2\kappa(t-s)} ds \right). \tag{3}$$

We refer the reader to [3] for more details on the demand process and the possibility to include jumps in the demand.

The problem of interest is the arising constrained stochastic optimal control (SOC) problem

$$\min_{u\in L^2([0,T-1/\lambda])} \int_{1/\lambda}^{T} \mathrm{OF}(Y_s, t_0, y_{t_0}, y(s))ds \text{ subject to (1) and (2),} \tag{4}$$

Thereby, $1/\lambda$ is the time that one good needs to pass the production line, and $\mathrm{OF}(Y_s, t_0, y_{t_0}, y(s))$ denotes the loss function.

In [3], a possible choice of an objective function as a tracking-type function $\mathrm{OF}_{\mathrm{track}}(Y_s, t_0, y_{t_0}, y(s)) = \mathbb{E}\left[(Y_s - y(s))^2 | Y_{t_0} = y_{t_0}\right]$ has been introduced. The loss is measured in terms of the quadratic deviation between the output at the end of the line and the actual demand. In this work, we focus on an extended loss quantification including an undersupply penalty. This is of interest for companies where a supply guarantee is of crucial importance and short-term external purchase is very costly. For them, it might be more harmful to generate an output that does not lead to demand satisfaction compared to an overproduction. Therefore, we introduce a new term into the objective function that penalizes undersupply. Thereby, α regulates the intensity of penalization.

$$\begin{aligned}\mathrm{OF}_{\mathrm{pen}}(Y_s, t_0, y_{t_0}, y(s)) =& \mathbb{E}\left[(Y_s - y(s))^2 | Y_{t_0} = y_{t_0}\right] \\ &+ \alpha \mathbb{E}\left[(Y_s - y(s))^2 | Y_s > y(s) \wedge Y_{t_0} = y_{t_0}\right].\end{aligned} \tag{5}$$

According to [4, Def. 8.9], the second conditional expectation in (5) reads as

$$\begin{aligned}&\mathbb{E}\left[(Y_s - y(s))^2 | Y_s > y(s) \wedge Y_{t_0} = y_{t_0}\right] \\ &= \begin{cases} \frac{\mathbb{E}\left[(Y_s - y(s))^2 \mathbb{1}_{\{Y_s > y(s)\}} | Y_{t_0} = y_{t_0}\right]}{P(Y_s > y(s))} & \text{if } P(Y_s > y(s)) > 0 \\ 0 & \text{else.} \end{cases}\end{aligned}$$

Thus, both conditional expectations in (5) can be expressed in terms of the known demand density $\rho_{Y_t | Y_{t_0} = y_{t_0}}$ at time t given by (3). Hence, for the evaluation of the objective functions $\mathrm{OF}_{\mathrm{track}}$ and $\mathrm{OF}_{\mathrm{pen}}$, this information on the demand density is sufficient. As the objective function is the only part of the SOC problem where the stochastic demand dynamics (2) come into play, we can replace the SDE constraint (2) in (4) by the condition that Y_t has demand density (3), which is used to calculate the expectations in the objective function (5). We are left with

$$\min_{u\in L^2([0,T-1/\lambda])} \int_{1/\lambda}^{T} \mathrm{OF}_{\mathrm{pen}}(Y_s, t_0, y_{t_0}, y(s))ds \text{ subject to (1) and (3).} \tag{6}$$

We are now able to apply deterministic optimization algorithms to the SOC problem (6).

However, we still need to make assumptions on the demand information that is used to determine the optimal inflow $u(t)$. Those assumptions result in different control methods due to the measurability assumptions on the inflow control $u(t)$. We focus on two of the three presented control methods (CM) in [3] corresponding to two information scenarios that are shortly summarized here for the sake of completeness:

- **CM1**: The only available demand information is the initial demand y_0 and the demand dynamics (2). No updates on the actual evolution of the demand can be used to determine the inflow control over the optimization horizon $[0, T]$. Thus, we assume that $u(t)$ is $\mathscr{F}_t$-measurable, where $\mathscr{F}_t = \sigma(Y_s; 0 \le s \le t)$.
- **CM2**: We prespecify update times $0 = \hat{t}_0 < \hat{t}_1 < \cdots < \hat{t}_n \le T - 1/\lambda$, where $\hat{t}_i = i \cdot \Delta t_{\text{up}}$, $i \in \{0, 1, \cdots, T-1/\lambda/\Delta t_{\text{up}}\}$, and update frequency $\Delta t_{\text{up}} \in [0, T - 1/\lambda]$. At those points in time, the initial demand and the demand dynamics (2) are supplemented by the actually realized demand. The forecast is updated accordingly and the optimal inflow control is calculated based on the updated demand forecast. Hence, we assume $u(t)$ is $\mathscr{F}_{\hat{t}_i}$-measurable for $t \in [\hat{t}_i, \hat{t}_{i+1}]$.

CM1 is directly applicable to (6). For CM2, we divide the optimization period $[0, T]$ into smaller subperiods $[\hat{t}_i, \hat{t}_{i+1}]$ according to the prespecified update times $\hat{t}_i$ and solve our SOC problem thereon.

$$\min_{u \in L^2([\hat{t}_i, \hat{t}_{i+1}])} \int_{\hat{t}_i + 1/\lambda}^{\min\{\hat{t}_{i+1} + 1/\lambda, T\}} \text{OF}_{\text{pen}}(Y_s, \hat{t}_i, y_{\hat{t}_i}, y(s))ds$$

$$\text{subject to (3) and } z_t + \lambda z_x = 0, \quad z(0, t) = u(t), \quad z(x, \hat{t}_i) = z_{\text{old}}(x, \hat{t}_i),$$
$$x \in (0, 1), \; t \in [\hat{t}_i, \min\{\hat{t}_{i+1} + 1/\lambda, T\}], \tag{7}$$

where $z_{\text{old}}(x, \hat{t}_i)$ denotes the state of the production line at update time $\hat{t}_i$ ensuring that the SOC problems on the subintervals are correctly linked to each other.

Note that the usage of the demand density (3) enables us to tackle both the SOC problem (6) and the subproblems (7) with methods from deterministic optimization, which will be done in the next section.

3 A Case Study: The Effect of an Undersupply Penalty

In this section, we numerically analyze the effect of an undersupply penalty for different intensities α for control methods CM1 and CM2. Using the reformulations (6) and (7) of the original SOC problem (4), the nonlinear optimization solver *fmincon*

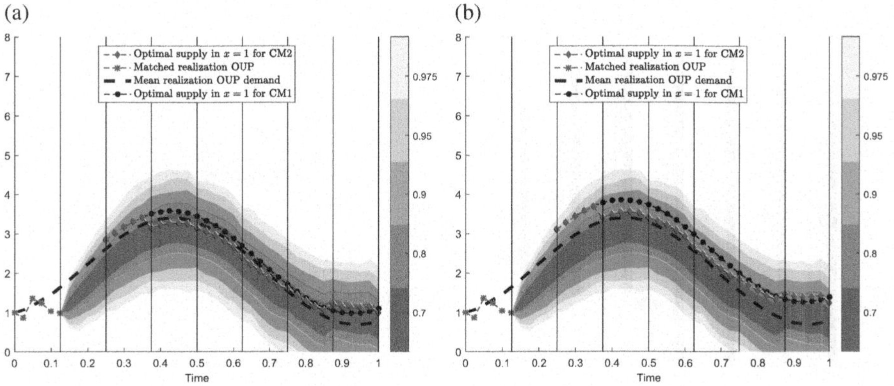

Fig. 1 Influence of penalty parameter α on optimal output. **(a)** $\alpha = 1$. **(b)** $\alpha = 3$

from MATLAB R2015b[1] is applicable (chosen initialization for CM1 and CM2: constant inflow of $u(t) \equiv \mu(0)$).

The numerical implementation to solve the SOC problem can be extended to the case of transport dynamics with non-constant, but positive transport velocity. Thus, a left-sided upwind scheme [5], i.e. $\frac{z(x_j,\tau_{i+1})-z(x_j,\tau_i)}{\Delta\tau} + \lambda\frac{z(x_j,\tau_i)-z(x_{j-1},\tau_i)}{\Delta x} = 0$, is chosen to discretize the linear advection equation (1). The applied step sizes $\Delta x = 0.1$, and $\Delta\tau = \Delta x/\lambda$ fulfill the CFL-condition. For our numerical simulations, we use 10^3 Monte Carlo repetitions with the following parameter setting for the demand process: $T = 1$, $\lambda = 4$, $\mu(t) = 2 + 3 \cdot \sin(2\pi t)$, $\kappa = 3$, $\sigma = 2$, $y_0 = 1$.

In Fig. 1, we are concerned with the influence of the penalty parameter α on the optimal output $y(t)$ for control methods CM1 and CM2. Thereby, we depict the updated confidence levels of the demand process in grey scale, the original mean realization of the demand (dashed line), the optimal CM1-output (dotted line), the optimal CM2-output (line marked by diamonds), and the tracked demand path until the first update time (line with asterisks). The vertical lines indicate the update times. For both control methods, the penalty leads to an output above the (updated) mean demand. However, the CM1-output follows well the course of the original mean demand, and the CM2-output lies well within the upper part of the updated confidence intervals. Consistent with our intuition, a higher penalty parameter α leads to an output higher above the (updated) mean demand.

In a next step, we want to quantify the number of undersupply cases, i.e., for each point in time, we count how many of the 10^3 simulated paths lie above the output (see Fig. 2a). By increasing the penalty parameter from $\alpha = 1$ to $\alpha = 3$, we are able to drastically reduce the number of undersupply cases. Based on this information, it is not clear whether CM2 is preferable over CM1 or not. Note that deciding on an undersupply is a binary decision. However, in the objective function,

[1] https://de.mathworks.com/help/optim/ug/fmincon.html, last checked: Sept 21, 2018.

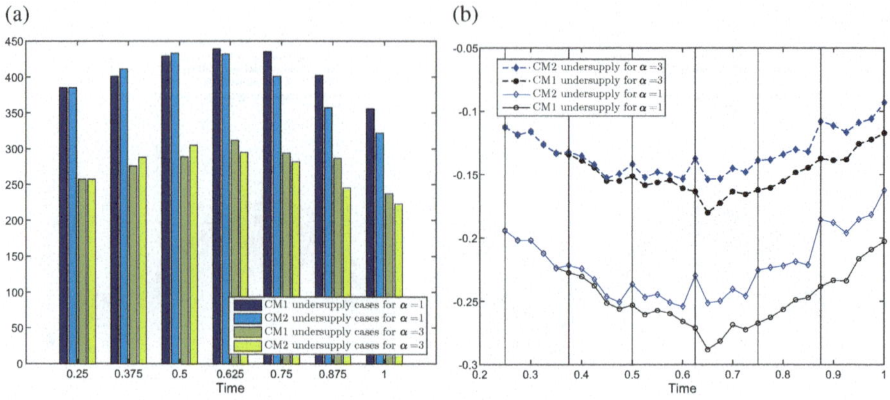

Fig. 2 (**a**) Number of undersupply cases and (**b**) average undersupply

the height of the deviation plays an important role. As there is a tradeoff between not realizing an undersupply but at the same time providing an adequate tracking of the demand, it might pay off to accept a small undersupply. However, with respect to the average undersupply, Fig. 2b shows that updates help to enhance the performance. To see this, at each point in time, we consider only those realizations where an undersupply occurs and plot the average height of the realized undersupply. The average undersupply for CM2 (lines marked by diamonds) is less or equal to the average undersupply for CM1 (dotted lines). Furthermore, there is less average undersupply for a higher penalty parameter. Finally, we can conclude that the introduction of a penalty parameter in the cost function leads to a reduction of both the undersupply cases as well as the average height of the undersupply.

Acknowledgements The authors are grateful for the support of the German Research Foundation (DFG) within the project "*Novel models and control for networked problems: from discrete event to continuous dynamics*" (GO1920/4-1) and the BMBF within the project ENets.

References

1. Göttlich, S., Teuber, C.: Space mapping techniques for the optimal inflow control of transmission lines. Optim. Methods Softw. **33**, 120–139 (2018)
2. Göttlich, S., Herty, M., Schillen, P.: Electric transmission lines: control and numerical discretization. Optim. Control Appl. Methods **37**, 980–995 (2016)
3. Göttlich, S., Korn, R., Lux, K.: Optimal control of electricity input given an uncertain demand. Math. Methods Oper. Res. (2019). https://doi.org/10.1007/s00186-019-00678-6
4. Klenke, A.: Probability Theory: A Comprehensive Course. Springer, London (2008)
5. LeVeque, R.J.: Numerical methods for conservation laws. Lectures in Mathematics ETH Zürich. Birkhäuser, Basel (1990)

A Production Model with History Based Random Machine Failures

Stephan Knapp and Simone Göttlich

Abstract In this paper, we introduce a time-continuous production model that enables random machine failures, where the failure probability depends historically on the production itself. This bidirectional relationship between historical failure probabilities and production is mathematically modeled by the theory of piecewise deterministic Markov processes (PDMPs). On this way, the system is rewritten into a Markovian system such that classical results can be applied. In addition, we present a suitable solution, taken from machine reliability theory, to connect past production and the failure rate. Finally, we investigate the behavior of the presented model numerically in examples by considering sample means of relevant quantities and relative frequencies of number of repairs.

1 Modeling Equations

We briefly recall the production network model from [1, 6] first, and according to [4], we present the stochastic extension to a load-dependent production model with machine failures. To keep the notation well-arranged, we consider a production network consisting of a single queue processor unit. We assume a processor, which is represented by an interval $(a, b) \subset \mathbb{R}$, i.e., with length $L = b - a$, where $\rho(x, t)$ describes the density of production goods at $x \in (a, b)$ and time $t \geq 0$. The dynamics of the density, and consequently of the production, is given by the following nonlinear hyperbolic partial differential equation

$$\partial_t \rho(x, t) + \partial_x \min\{v\rho(x, t), c\} = 0, \tag{1}$$

where $c \geq 0$ is the production capacity and $v > 0$ the constant production velocity. In front of the processor a storage, also called queue, is assumed and for an

S. Knapp (✉) · S. Göttlich
University of Mannheim, Mannheim, Germany
e-mail: stknapp@mail.uni-mannheim.de; goettlich@uni-mannheim.de

© Springer Nature Switzerland AG 2019
I. Faragó et al. (eds.), *Progress in Industrial Mathematics at ECMI 2018*,
Mathematics in Industry 30, https://doi.org/10.1007/978-3-030-27550-1_62

externally given time-dependent inflow $G_{\text{in}}(t)$ into the production, the queue length q follows the ordinary differential equation

$$\partial_t q(t) = G_{\text{in}}(t) - g_{\text{out}}(t), \tag{2}$$

with

$$g_{\text{out}}(t) = \begin{cases} \min\{G_{\text{in}}(t), c\}, & \text{if } q(t) = 0, \\ c, & \text{if } q(t) > 0. \end{cases}$$

The processor is coupled to the queue by a boundary condition in the form of $\rho(a,t) = \frac{g_{\text{out}}(t)}{v}$ and initial conditions $\rho(x,0) = \rho_0(x) \in L^1((a,b))$, $q(0) = q_0 \in \mathbb{R}_{\geq 0}$ are prescribed. This deterministic model is well-defined, see, e.g. [1]. The theory of piecewise deterministic Markov processes; see, e.g. [2, 7], has been used to define an appropriate production model with stochastic machine failures in [4], where the probabilities of machine failures depend on the actual workload of the processor. Since this construction only allows for a dependence on the current workload, we can not use the amount of goods produced since the last machine failure as a measure for the next failure. Our new idea lies in adding a variable w governing the workload since the last repair. To do so, we use the time-dependent variable $r(t) \in \{0, 1\}$, and set the capacity as $\mu(t) = r(t)c$ for a maximal capacity $c > 0$. This means that $r(t) = 0 \Rightarrow \mu(t) = 0$ is a down and $r(t) = 1 \Rightarrow \mu(t) = c$ a working processor at time t and we define

$$\text{WIP}(t_0, t_1) = \int_{t_0}^{t_1} \int_a^b \rho(x,t)dxdt$$

as the cumulative work-in-progress of the processor between time t_0 and t_1. The variable w should therefore satisfy

$$\partial_t w(t) = r(t) \int_a^b \rho(x,t)dx, \quad w(t_0) = w_0 = \int_a^b \rho(x,t_0)dx. \tag{3}$$

Altogether, we define the state space

$$E = \mathbb{R}_{\geq 0} \times \{0, 1\} \times \mathbb{R}_{\geq 0} \times L^1((a,b)),$$

which is a measurable space together with the σ-algebra $\mathscr{E}$ generated by the open sets induced by the metric

$$d((w,r,q,\rho), (\tilde{w}, \tilde{r}, \tilde{q}, \tilde{\rho})) = |w - \tilde{w}| + |r - \tilde{r}| + |q - \tilde{q}| + \|\rho - \tilde{\rho}\|_{L^1((a,b))}.$$

Since we construct a piecewise deterministic Markov process, we define the deterministic dynamics between jump times as

$$\Phi_{st}\colon E \to E, \quad (w_0, r_0, q_0, \rho_0) \mapsto (w(t), r(t), q(t), \rho(t)),$$

i.e., Φ_{st} is the solution to Eqs. (1), (2), (3), and $r(t) = r_0$ with initial conditions $(w_0, r_0, q_0, \rho_0) \in E$. In between the jump-times, where the capacity changes, we have a capacity, which is given by cr_0 and independent of time. This allows us to apply the theory of the deterministic model (1)–(2) to obtain continuity properties of Φ. To characterize the stochastic part, we introduce

$$\psi(t, y) = \lambda_{r,r}(t, w), \quad \eta(t, y, B) = \frac{\lambda_{r,(1-r)}(t, w)}{\psi(t, y)} \epsilon_{(rw,(1-r),q,\rho)}(B)$$

for every $y = (w, r, q, \rho) \in E$ and $B \in \mathscr{E}$, where $\lambda_{i,j}(t, w)$ describes the transition rate from capacity i to j at time t and actual workload w, $i, j \in \{0, 1\}$ and ϵ_x is the Dirac measure with unit mass in x. The function ψ is the total intensity determining whether a jump occurs, or not, and the function η describes the probability distribution of the systems jump given the system changes at time t. For example, given the state $y = (w, 1, q, \rho)$ at the time of a jump, the system jumps to $(w, 0, q, \rho)$ and, vice versa, given the state $y = (w, 0, q, \rho)$ the system jumps to $(0, 1, q, \rho)$, i.e., the workload has been "reset". The open question is whether this model can be represented by a piecewise deterministic Markov process. Following [4], it is straightforward to show

Theorem 1 *Let* $\lambda_{i,j}\colon [0, T] \times \mathbb{R}_{\geq 0} \to \mathbb{R}_{\geq 0}$ *be uniformly bounded, continuous and satisfy* $\lambda_{i,i} = \lambda_{i,i-1}$ *for* $i \in \{0, 1\}$. *Then for all initial data* $x_0 \in E$ *there exists a Markov process*

$$X = ((w(t), r(t), q(t), \rho(r)), t \in [0, T]) \subset E$$

on some probability space $(\Omega, \mathscr{A}, P)$, *satisfying*

1. $X(0) = x_0$ *P-almost surely,*
2. *for every* $t \in (0, T)$, $(w, r, q, \rho) \in E$ *and* $j \in \{0, 1\}$, *it holds that*

$$P(r(t + \Delta t) = j | X(t) = (w, r, q, \rho)) = \big(1 - \Delta t \lambda_{r,r}(t, w)\big)\mathbb{1}_r(j) + \Delta t \lambda_{r,(1-r)}(t, w)\mathbb{1}_{1-r}(j) + o(\Delta t),$$

3. *there exists a P-null set* $\mathscr{N} \in \mathscr{A}$ *such that for every* $\omega \in \Omega \setminus \mathscr{N}$, *there exist times* $T_0 = 0 \leq T_1 \leq \cdots \leq T_M = T$ *such that for every* $k = 0, \ldots, M - 1$, $X(t) = \Phi_{T_k,t}(X(T_k))$ *for* $t \in [T_k, T_{k+1})$ *with capacity* $\mu(r(T_k, \omega))$, *i.e.,* X *behaves deterministic between jump times.*

The main and new ingredient is the mapping $t \mapsto w(t)$, which is a continuous mapping since $t \mapsto \rho(t)$ is continuous.

2 Computational Results

Due to the fact that solutions to (1) move with non-negative velocities only, we can use the first order left-sided upwind scheme for a numerical approximation of the density ρ. Furthermore, we use the explicit Euler scheme to approximate the queue length q given by (2) and w given by (3), where we use a rectangular rule for the integration. This yields an approximation of the deterministic dynamics between the jump times. The simulation of the jump times is done with the thinning algorithm presented in [4]. Its basic idea is to use the uniform bound on the rate functions and generate exponentially distributed times with high intensity, representing the times between jumps, and thin these times during the numerical simulation of the whole system with an appropriate acceptance rejection procedure.

The choice of the rate functions $\lambda_{i,j}(t, w)$ is a crucial point in numerical examples. Here, we make use of the choice in [9] and set for $\theta_1, \theta_2 > 0$ the rate function as

$$\lambda_{1,0}(t, w) = \lambda_{1,0}^{\min} + (\lambda_{1,0}^{\max} - \lambda_{1,0}^{\min})(1 - e^{-(\theta_1 w)^{\theta_2}}),$$

which is a scaled version of the cumulative distribution function of a Weibull distribution, i.e., $F(t) = 1 - e^{-(\theta_1 t)^{\theta_2}}$. The classical interpretation of t in the latter expression is the lifetime of a machine and $F(t)$ is the probability that a failure happens after time t, see, e.g. [8]. In our case we use the variable w, which measures the amount of goods produced since the last repair happened. Therefore, if $w = 0$, then $\lambda_{1,0}(t, 0) = \lambda_{1,0}^{\min}$, which corresponds to the minimal failure rate and $\lim_{w\to\infty} \lambda_{1,0}(t, w) = \lambda_{1,0}^{\max}$. The function $\lambda_{1,0}(t, w)$ is monotonically increasing in w and incorporates the idea of an increasing failure rate depending on past workloads. On the other hand, we assume $\lambda_{0,1}(t, w) = \lambda_{0,1}$ because repair times do not dependent on the amount of goods produced.

In the following, we examine the presented model using numerical examples. Here, we assume a production velocity of $v = 1$, the interval $a = 0$, $b = 1$, and the capacity is given as $\mu(t) = 2r(t)$. We use a spatial discretization with step-size $\Delta x = 10^{-1}$ and a temporal step-size that satisfies the Courant-Friedrichs-Lewy condition, which reads as $\Delta t \leq \Delta x$ for the chosen parameters. The simulation results are based on samples of the stochastic process X and we use the classical Monte-Carlo estimator to evaluate moments or probabilities of the samples. We used a sample size of 10^5 for all following results.

We analyze the expected queue length, capacity and the distribution of the number of repairs within a time horizon $[0, 50]$ for two different constant inflow profiles. We denote by $G_{\text{in}}^1(t) \equiv 0.5$ and by $G_{\text{in}}^2(t) \equiv 1.5$ as inflow profiles and use the parameters

$$\lambda_{0,1}(t, w) = \frac{1}{0.5}, \quad \lambda_{1,0}^{\min} = \frac{1}{10}, \quad \lambda_{1,0}^{\max} = \frac{1}{0.5}, \quad \theta_1 = \frac{1}{10}, \quad \theta_2 = 5.$$

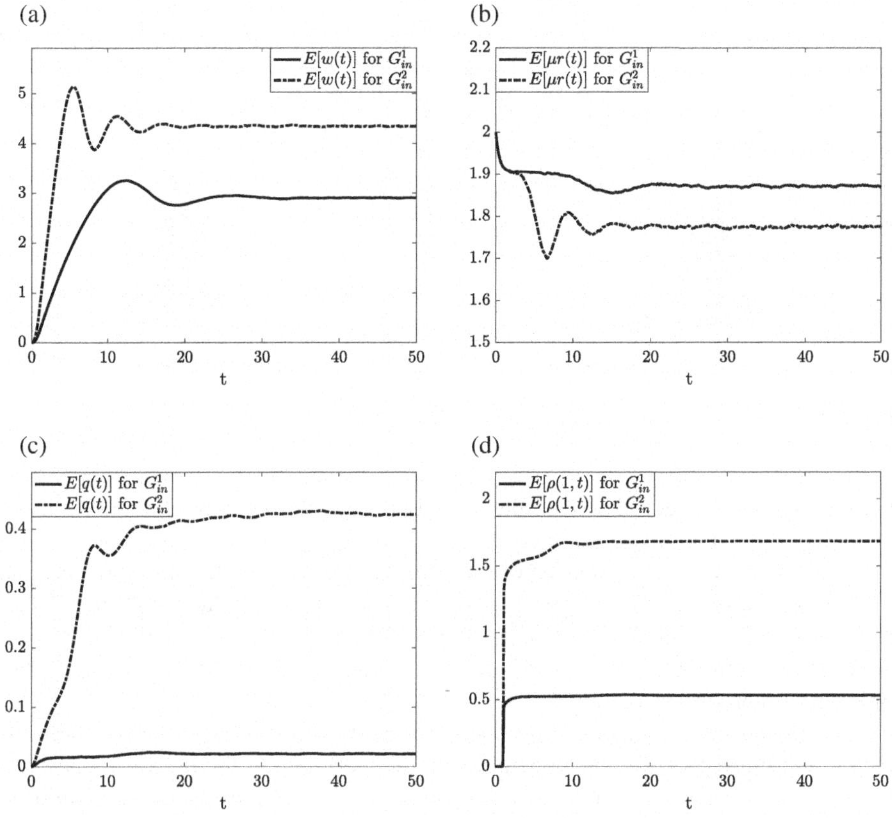

Fig. 1 First order moments of w, the capacity, queue-length and density. (**a**) Expected w. (**b**) Expected capacity $\mu(t)$. (**c**) Expected queue-length $q(t)$. (**d**) Expected density at $x = 1$

In Fig. 1, first order moment estimations are shown. In detail, Fig. 1a shows the expected value of the variable w, Fig. 1b the expected capacity, Fig. 1c the expected queue length and Fig. 1d the expected density at the end of the processor. The dynamics is quite interesting: the expected capacity decreases approximately until time $t = 6$ for the second inflow, then increases and decreases again. Indeed, the mean time to failure is given by $\Gamma(1 + \frac{1}{\theta_2})\theta_1^{-1}$, see e.g. [8]. If w corresponds to the lifetime in our model, we see that an intact system with constant inflow G_{in} is more likely to fail around time $\Gamma(1 + \frac{1}{\theta_2})(\theta_1 G_{\text{in}})^{-1}$. In our case, this leads to time 18.4 for the first and time 6.1 for the second inflow profile, which is close to the times at which the shape of the expected capacity changes. We observe these characteristic times also in the other graphs in Fig. 1. In contrast to the models presented in [3–5], where quantities monotonically converge, we obtain an oscillatory behavior of the quantities for constant inputs. The oscillatory effects are natural and caused by the history we incorporate in w. This means, the first machine failures are likely around time 18.4(6.1), the second around 36.8(12.2) and so on. At the same time

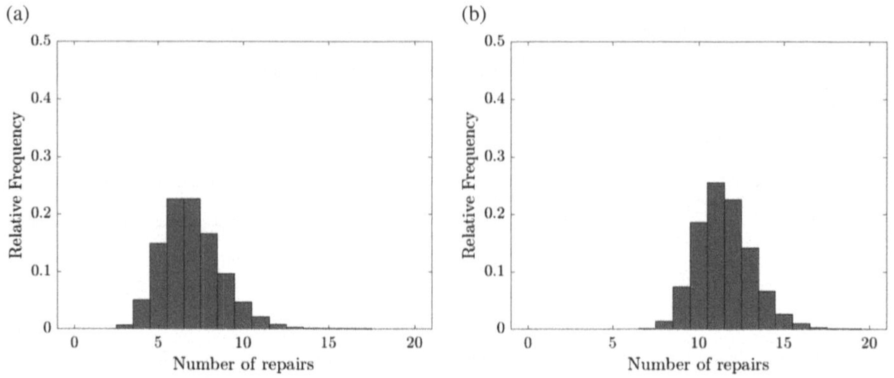

Fig. 2 Distribution of the number of repairs within [0, 50]. (**a**) Inflow G^1_{in}. (**b**) Inflow G^2_{in}

the failures, which occur between these likely times, smooth this effect out as time evolves and the quantities converge.

Figure 2 shows the distribution of the number of repairs within the time horizon [0, 50] and emphasizes the impact of the chosen inflow on the reliability of the processor. In Fig. 2a the case of G^1_{in} is shown, where mostly 5–9 repairs have been done. The situation for inflow profile G^2_{in} is different, where 9–14 repairs during the time horizon are more likely.

To conclude, we deduced a production model with random machine failures including failure probabilities depending on the workload of the machine since the last repair occurred. The extension of the model to complex production networks is straightforward, see, e.g. [4]. Simulation results showed a big impact of the history on expected workload, capacity, queue length and density. These effects are not negligible for production planning and control and must be taken into account.

Acknowledgements This work has been financially supported by the BMBF project ENets (05M18VMA).

References

1. D'Apice, C., Göttlich, S., Herty, M., Piccoli, B.: Modeling, Simulation, and Optimization of Supply Chains. SIAM, Philadelphia (2010)
2. Davis, M.H.A.: Piecewise-deterministic Markov processes: a general class of nondiffusion stochastic models. J. Roy. Stat. Soc. Ser. B **46**, 353–388 (1984)
3. Göttlich, S., Knapp, S.: Semi-Markovian capacities in production network models. Discrete Contin. Dyn. Syst. Ser. B **22**, 3235–3258 (2017)
4. Göttlich, S., Knapp, S.: Load-dependent machine failures in production network models (2018). arXiv:1806.03091v1
5. Göttlich, S., Martin, S., Sickenberger, T.: Time-continuous production networks with random breakdowns. Netw. Heterog. Media **6**, 695–714 (2011)

6. Göttlich, S., Herty, M., Klar, A.: Network models for supply chains. Commun. Math. Sci. **3**, 545–559 (2005)
7. Jacobsen, M.: Point Process Theory and Applications. Birkhäuser, Boston (2006)
8. Jiang, R., Murthy, D.N.P.: A study of Weibull shape parameter: properties and significance. Reliab. Eng. Syst. Saf. **96**, 1619–1626 (2011)
9. Rivera-Gómez, H., Montaño-Arango, O., Corona-Armenta, J.R., Garnica-González, J., Hernández-Gress, E.S., Barragán-Vite, I.: Production and maintenance planning for a deteriorating system with operation-dependent defectives. Appl. Sci. **8**, 165 (2018)

Optimization of Buckling for Textiles

Stephan Wackerle, René Pinnau, and Julia Orlik

Abstract Textiles are present in many applications and are an interesting yet complicated subject. For the industry mostly the macroscopic behavior of textiles is important.

In the following article we deal with the buckling behavior of a textile shell under uniaxial tension in the nonlinear regime. The nonlinearity redirects the applied tensional force into bending of the plate in the third direction. To model this behavior, we assume a homogenized shell of von-Kármán type, achieved via homogenization of the textile with given micro-structure. Furthermore, a careful reduction to 1D for the case of a belt-like geometry, i.e., narrow in the second direction, gives a buckling model, which can be optimized with respect to both its shape and retardation. The resulting macroscopic optimization problem with PDE-constraints yields a Pareto-optimization with local minima.

1 Introduction

Textiles as versatile materials are widely used for very different purposes. During the production processes in industry the textiles are often subjected to tensional forces. The resulting deformations may be critical for the further treatment or even the final product. While the in-plane deformations are usually no threat, the bending-type deformations induce visible folds and wrinkles and produce faults or even failures. Thus it is necessary to counteract and reduce or even inhibit such behaviors. Instead of adapting the whole infrastructure the following article presents a way for an optimization of such behavior of the textile itself under some constraints preserving certain crucial properties.

S. Wackerle (✉) · J. Orlik
Fraunhofer ITWM, Kaiserslautern, Germany
e-mail: wackerle@itwm.fraunhofer.de; orlik@itwm.fraunhofer.de

R. Pinnau
TU Kaiserslautern, Kaiserslautern, Germany
e-mail: pinnau@mathematik.uni-kl.de

© Springer Nature Switzerland AG 2019

I. Faragó et al. (eds.), *Progress in Industrial Mathematics at ECMI 2018*,
Mathematics in Industry 30, https://doi.org/10.1007/978-3-030-27550-1_63

2 Mathematical Approach

In the following we present an optimization problem with PDE-constraints for the buckling of a belt-shaped textile. The goal is to achieve a design for the textile, where the buckling is delayed as well as the buckling-shape is adapted to a given desired profile. While the desired profile depends heavily on the application and the product design, the delay of the buckling is equivalent to the maximization of the necessary critical force to reach the buckling regime. Note, buckling is nothing but a stability loss of a structure and can be characterized by the loss of coercivity in the problem. Furthermore it is a bifurcation problem. The loss of stability in plates is investigated, e.g., in [1, 3, 6] with the help of the von-Kármán model for a plate. The used von-Kármán plate is a non-linear model which is controversially discussed [4, 5], but widely used for buckling phenomena [1–3, 6]. Henceforth, we assume a one dimensional setting, which arises by a dimension reduction of the two dimensional von-Kármán plate for the given geometry, where $h \ll \ell_1 \ll \ell_2$ with the height $2h$ the width $2\ell_1$ and the length $2\ell_2$. This reduction eliminates the nonlinearities coming from a coupling between the bending and tension. Consequently, the problem on the interval $\Omega = (-\ell, \ell)$ with $\ell = \ell_1$ fulfills the reduced von-Kármán equation

$$\Delta \left(B(x)\Delta u_B - F u_B\right) = 0,$$

where Δ denotes the Laplace operator, $B = B(x)$ the bending stiffness, $F > 0$ is a compressive force and u_B is the outer-plane deflection of the middle line.

Note, that the buckling problem of a plate is a kind of an eigenvalue problem. Its shape is an eigenmode and its occurrence is characterized via a related eigenvalue. Indeed, the loss of coercivity leading to the buckling is observable by the vanishing spectral gap, i.e., the difference between the first eigenvalue and zero.

Thus introduce the generalized Rayleigh-quotient

$$R(u_B) = \frac{\langle u_B, \Delta \left(B(x)\Delta u_B - F u_B\right)\rangle_{L^2(\Omega)}}{\langle u_B, \Delta u_B\rangle_{L^2(\Omega)}},$$

to characterize the buckling problem. In fact, this generalized eigenvalue yields the spectral gap of the plate. Consequently, observe as long as $R(u_B) > 0$ the problem is coercive and the only shape the plate can assume is $u_B \equiv 0$. Hence, the buckling of the plate appears as soon as $R(u_B) \leq 0$.

For further explanation of $R(u_B)$ assume a constant bending stiffness B. Then a calculation involving the Poincaré-Wirtinger inequality (cf. [1, Sect. 14.3]) yields the analytic expression

$$R(u_B) = \frac{4\pi^2 B}{\ell^2} - F.$$

Thus, the necessary condition $R(u_B) = 0$ yields the critical force $F_{crit} = \frac{4\pi^2 B}{\ell^2}$. Observe, that the maximization of $R(u_B)$ for a constant F increases the spectral gap and thereby a larger force is necessary to induce the buckling.

The second part of the objective functional concerns the eigenmode shape. This is directly modeled by the typical tracking term

$$\|u_B - u_g\|_{L^2(\Omega)},$$

with a predefined goal function u_g. Setting both objectives together in one functional gives

$$J(u_B) = \gamma \|u_B - u_g\|_{L^2(\Omega)} - (1 - \gamma)\lambda_B,$$

where $\gamma \in [0, 1]$ is a weighting factor for both objectives: the shape-optimization and the appearance-delay of the first buckling. Also γ can be seen as the parametrization of the Pareto front. The pareto front naturally arises in multi-criteria optimization and describes the ambiguity of optima.

Furthermore, there are two more sensible constraints to the problem. First, the bending stiffness as design parameter is only allowed to take values in an interval $[a, b]$ with $0 < a \leq b$. Secondly, the mean bending stiffness of the whole problem, i.e., the mean $\frac{1}{2\ell} \int_\Omega B(x)\, dx$ can only differ up to 10% from its initial value $M = \frac{1}{2\ell} \int_\Omega B_{init}(x)\, dx$. The last constraint takes into account that a given product can be optimized without loosing to much from its original properties.

Combining the above considerations and constraints the final optimal control problem reads as follows:

$$\begin{aligned} \min_B \quad & J(u_B) \\ s.t. \quad & \lambda_B = R(u_B), \\ & a \leq B(x) \leq b, \\ & \frac{1}{2\ell M} \int_\Omega B(x) dx \in [0.9, 1.1]. \end{aligned}$$

Finally, we restrict the bending stiffness B to be piecewise constant function, which is symmetric wrt. the midpoint. The piecewise constant character corresponds to sections with the same bending stiffness, which originates from the production process.

3 Numerics

In this section we show the numerical results for the optimal control problem. The problem is implemented in Matlab. Due to the low dimensionality of the problem the optimization is done via the Matlab-function `fminsearch`. To point out the

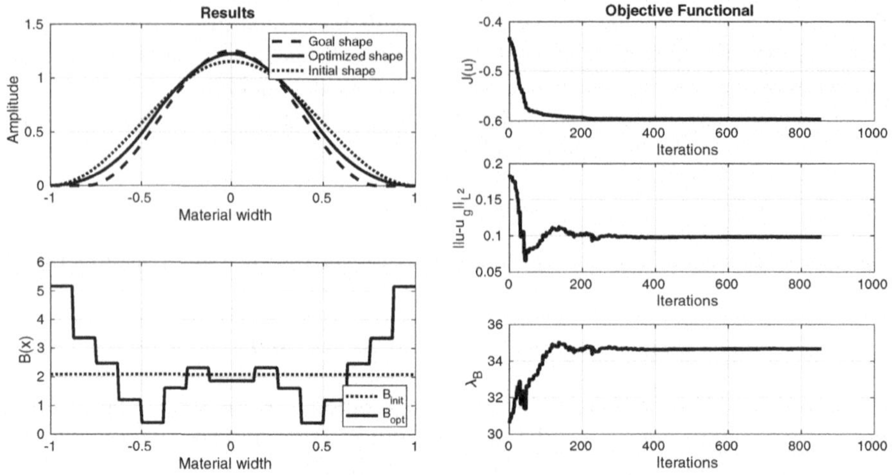

Fig. 1 The optimization problem for $\gamma = 0.98$. The left figure depicts in the upper picture mode-shapes and in the lower distributions of the piecewise constant bending stiffness in different states. On the right the evolution of the objective functional and its two components are represented

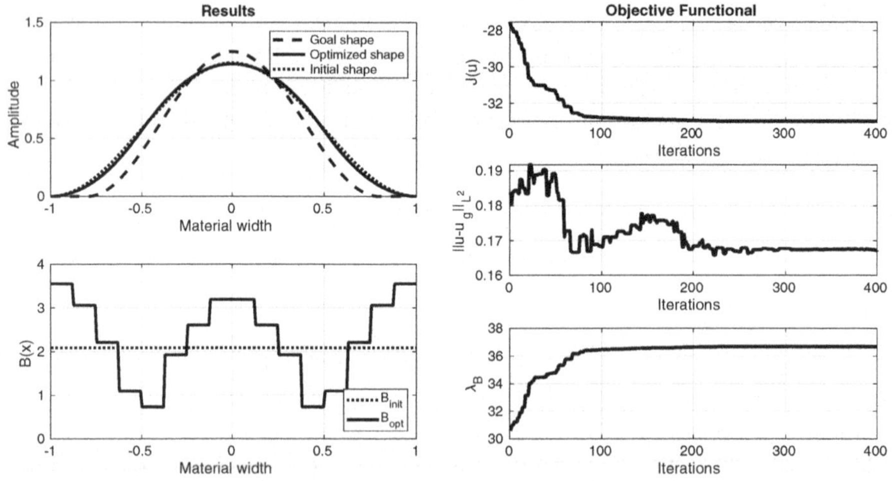

Fig. 2 The optimization problem for $\gamma = 0.1$. The left figure depicts in the upper picture mode-shapes and in the lower distributions of piecewise constant the bending stiffness in different states. On the right the evolution of the objective functional and its two components are represented

problematic of the Pareto front multiple choices of γ are depicted. Here, $\gamma = 0.98$ represents the factor for which both objectives are almost in balance (see Fig. 1). The extreme cases $\gamma = 0.1$ and $\gamma = 0.995$ correspond to a strong weighting for the delay of the buckling or the shape-optimization of the mode, respectively (Figs. 2 and 3).

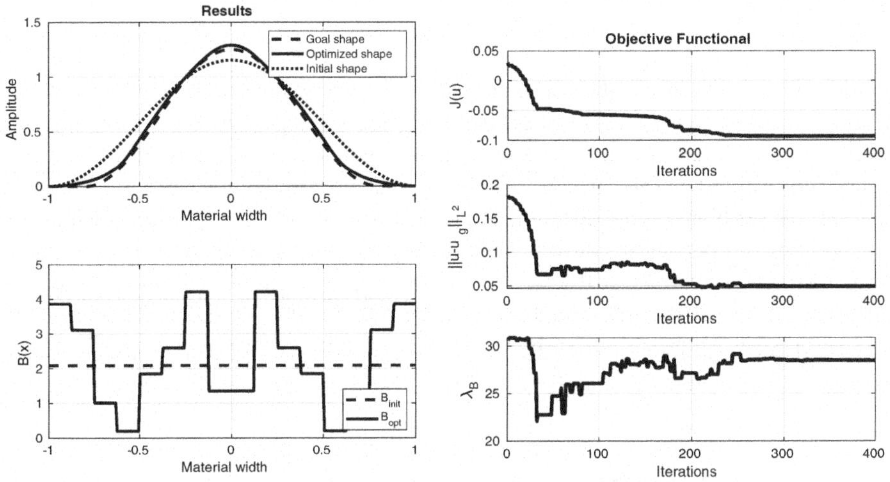

Fig. 3 The optimization problem for $\gamma = 0.995$. The left figure depicts in the upper picture mode-shapes and in the lower distributions of the piecewise constant bending stiffness in different states. On the right the evolution of the objective functional and its two components are represented

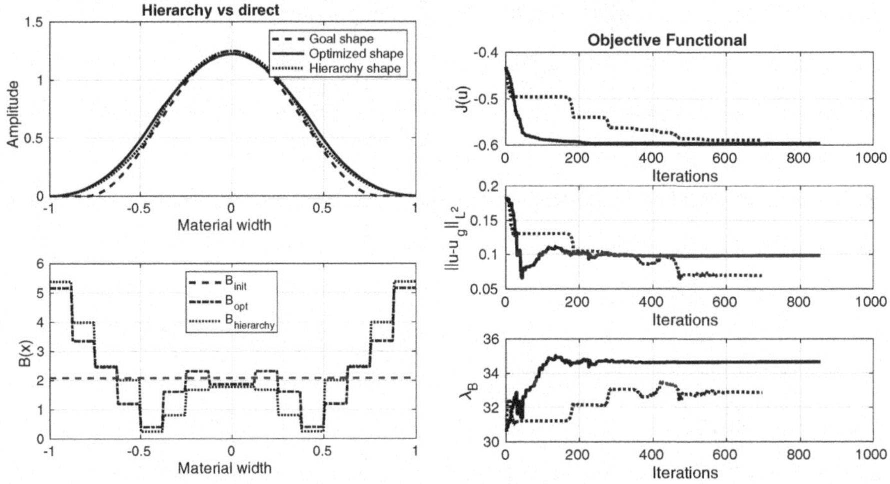

Fig. 4 Comparison of the results of the hierarchical and the standard approach. Both simulations use the Pareto-factor $\gamma = 0.98$

3.1 A Hierarchical Approach

The considered optimization problem allows easily for an hierarchical approach. As hierarchy we use an increasing number of segments for B. Here the stages are 4, 8 and 16 segments for B in the domain $\Omega = [-\ell, \ell]$. Figure 4 depicts the comparison between the hierarchical and the direct approach (cf. Fig. 1). This comparison shows that the considered optimal control problem easily reaches different local minima.

4 Outlook

Evidently, the next step is the coupling to the micro-structure, not being restricted to textiles. To this end a parametrized design space on a periodicity cell is needed. While we are interested in periodicity cells arising in textile structures, it is of course not limited to this specific use. For general examples we refer to [7] where several toy problems with different micro-structures and even analytic results for macroscopic properties are shown. The problem discussed in this work mostly relies on the effective bending stiffness. For instance, a very simple model for a textile is the regular grid of orthogonal beams, see Fig. 5. This is discussed within [7] and yields as bending rigidities $B_\alpha = \frac{Eb_\alpha h^3}{12t_\alpha}$ depending on Elastic modulus E and geometric parameters h, b_α and t_α (see Fig. 5) for the respective direction α (Fig. 6).

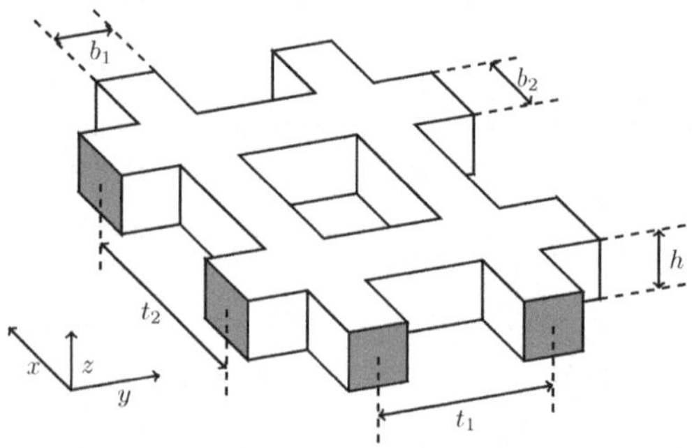

Fig. 5 Exemplary grid-like micro-structure consisting of beams. According to section 7.2.3 within [7]

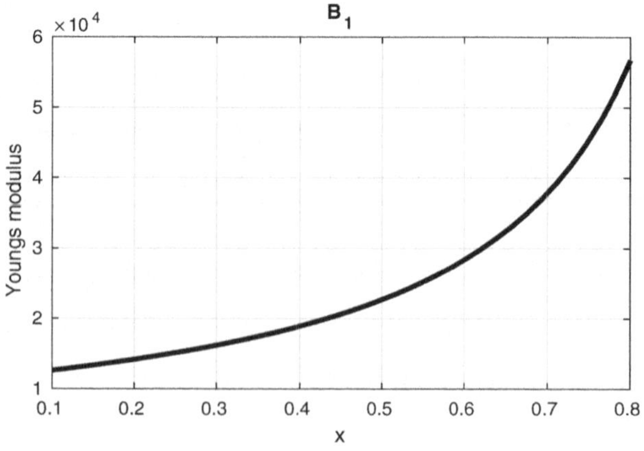

Fig. 6 Development of B_1 for the parameters $b_1 = b_2 = 0.2$, $t_1 = 2 - 2x$ and $t_2 = 4$ with $x \in [0.1, 0.8]$ capturing a broad range of bending rigidities

References

1. Berdichevsky, V., Variational Principles of Continuum Mechanics: I. Fundamentals. Springer, Berlin (2009)
2. Bourne, D.P., Conti, S., Müller, S.: Energy bounds for a compressed elastic film on a substrate. J. Nonlinear Sci. **27**, 453–494 (2017)
3. Cerda, E., Mahadevan, L.: Geometry and physics of wrinkling. Phys. Rev. Lett. **90**(7), 074302 (2003)
4. Ciarlet, P.G.: A justification of the von Kármán equations. Arch. Ration. Mech. Anal. **73**, 349–389 (1980)
5. Friesecke, G., James, R.D., Müller, S.: A hierarchy of plate models derived from nonlinear elasticity by gamma-convergence. Arch. Ration. Mech. Anal. **180**, 183–236 (2006)
6. Puntel, E., Deseri, L., Fried, E.: Wrinkling of a stretched thin sheet. J. Elast. **105**, 137–170 (2011)
7. Ventsel, E., Krauthammer, T.: Thin Plates and Shells: Theory: Analysis, and Applications. CRC Press, CRC Press (2001)

Optimal Control Simulations of Lateral and Tip Pinch Grasping

Uday Phutane, Michael Roller, and Sigrid Leyendecker

Abstract Grasping is a complex human movement. During grasping, when the hand closes around the object, the multibody system changes from a kinematic tree structure to a closed loop contact problem. To better understand work-related disorders or optimize execution of activities of daily life, an optimal control simulation to perform grasping is useful. We simulate the grasping action with a three-dimensional rigid multibody model composed of two fingers actuated by joint torques. The grasping movement is composed of a reaching phase (no contacts) and a grasping phase (closed contacts). The contact constraints are imposed first through distances between the fingers and the object surfaces and then through spherical joints. Thus, the dynamics of grasping is described by a hybrid dynamical system with a given switching sequence and unknown switching times. To determine a favourable trajectory for grasping action, we solve an optimal control problem (ocp). The ocp is solved using the direct transcription method DMOCC, leading to a structure preserving approximation of the continuous problem. An objective involving either the contact polygon centroid or the contol torques is minimized subject to discrete Euler-Lagrange equations, boundary conditions and path constraints. The dynamics of the object to grasp along with Coulomb friction is also taken into account.

U. Phutane (✉) · S. Leyendecker
Chair of Applied Dynamics, Friedrich-Alexander-Universität Erlangen-Nürnberg, Erlangen, Germany
e-mail: uday.phutane@fau.de; sigrid.leyendecker@fau.de

M. Roller
Fraunhofer ITWM, Kaiserslautern, Germany
e-mail: michael.roller@itwm.fraunhofer.de

© Springer Nature Switzerland AG 2019

I. Faragó et al. (eds.), *Progress in Industrial Mathematics at ECMI 2018*,
Mathematics in Industry 30, https://doi.org/10.1007/978-3-030-27550-1_64

1 Introduction

Humans have developed the skills and dexterity over years to perform grasping very naturally. However, to replicate it in simulation and practice with human precision is challenging task given the highly sophisticated coordination between the fingers. Although first treated as a purely kinematic problem, the dynamics concerning grasping simulations, such as the weight of the object and the magnitude of grasping forces are crucial. Using the paradigm of discrete mechanics and optimal control for constrained systems (DMOCC), see [3], we simulate a two-finger multibody system to predict motions for precision grasps subject to two objective functions.

2 Hand Model

We consider a two-finger model, as shown in Fig. 1, composed of the thumb and index finger, hereafter both are referred to as only 'finger', with the wrist and forearm.

The multibody system is modelled with $\boldsymbol{q}(t) \in \mathbb{R}^{108}$ time-dependent redundant coordinates. With $\boldsymbol{g}_{int} \in \mathbb{R}^{54}$ internal constraints and a combination of revolute, cardan, nino and fixed joints set up through external constraints $\boldsymbol{g}_{ext} \in \mathbb{R}^{41}$, see [4, 5], the model is reduced to thirteen degrees of freedom $\boldsymbol{u}(t)$. The model is actuated using joint torques $\boldsymbol{\tau}(t) \in \mathbb{R}^{13}$ to give the redundant forces $\boldsymbol{f}(t) \in \mathbb{R}^{108}$. The dynamics of the object to be grasped, with configuration $\boldsymbol{q}^O(t) \in \mathbb{R}^{12}$ is included in the model as well. The discrete Euler-Lagrange (DEL) equations of motion are derived through a discrete variational principle. This gives a symplectic time stepping scheme with structure preserving properties. We employ the discrete null-space method for both systems through null-space matrices $\boldsymbol{P}(\boldsymbol{q})$ and $\boldsymbol{P}^O(\boldsymbol{q}^O)$, and a discrete reparameterisation $\boldsymbol{F}_d(\boldsymbol{u}, \boldsymbol{q})$ for the hand to reduce the system size, see [4].

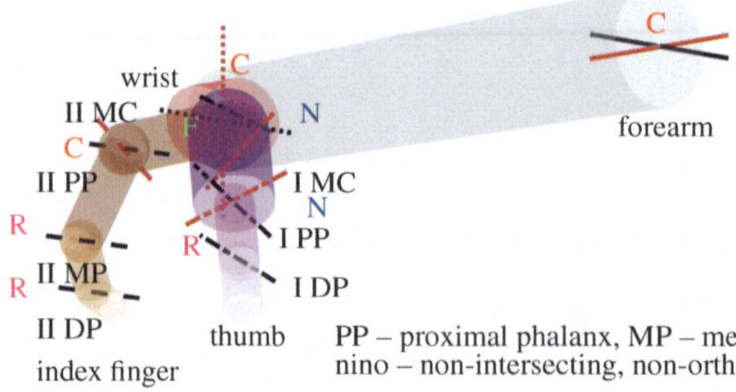

Joint type	
cardan	C
nino	N
revolute	R
fixed	F

PP – proximal phalanx, MP – medial phalanx, DP – distal phalanx
nino – non-intersecting, non-orthogonal

Fig. 1 The two finger model with the joint axes for different joints. The adjoining table shows the joint types. The thumb and index finger are denoted with roman numerals I and II

3 Contact Model

The contact is modelled through holonomic functions $\boldsymbol{g}_c(\boldsymbol{q}, \boldsymbol{q}^O) = \mathbf{0}$ with contact forces obtained through constraint Jacobians $\boldsymbol{G}_c$ and Lagrange multipliers $\boldsymbol{\lambda}_c(t)$. To close contacts at the end of the reaching phase, we use gap functions $\boldsymbol{g}_{c1} \in \mathbb{R}^{n_c}$. This is defined for n_c contact points, denoted by $\boldsymbol{\varrho}$, fixed on the finger digit surfaces. The contact points are not defined on the object surface(s). However, we apply inequality constraints $\boldsymbol{h}_{\varrho}$ to limit the contact points within the object dimensions. For the grasping phase, we use spherical joint constraints $\boldsymbol{g}_{c2} \in \mathbb{R}^{3 \cdot n_c}$ to maintain zero relative translational displacements. We also constrain the normal contact forces $\boldsymbol{h}_{normal}$ to press on the object and to lie in the friction cone $\boldsymbol{h}_{fric}$, using Coulomb's static friction law.

4 Optimal Control Problem for Grasping

Here we solve an optimal control problem (ocp) to determine the optimal trajectory, controls and contact forces for a grasping motion with regard to a certain objective. The ocp is formulated using a direct transcription method to transform it into a constrained optimization problem. We define a discrete objective function

$$J_d\left(\boldsymbol{u}_d, \boldsymbol{\tau}_d, \boldsymbol{q}_d^O, \boldsymbol{\lambda}_{c,d}, T_k, T_m\right) = \sum_{n=0}^{N-1} B_d\left(\boldsymbol{u}_n, \boldsymbol{u}_{n+1}, \boldsymbol{\tau}_n, \boldsymbol{q}_n^O, \boldsymbol{q}_{n+1}^O, \boldsymbol{\lambda}_{c,n}, T_k, T_m\right) \tag{1}$$

as a sum of a cost functional B_d, which has to be minimized. As side constraints, the DEL equations of motion, initial and final configuration and momentum conditions, and discrete path constraints have to be fulfilled.

As the grasping action is composed of two stages with different dynamics, the optimal control problem is composed of two phases, as shown in Fig. 2, see [2]. Here, we define a fixed number of time nodes N_k and N_m and unknown durations T_k and T_m for the reaching and grasping phases, respectively. For the reaching phase, we solve the dynamics of the hand and the object independently through the DEL equations for the two systems. After closing the contact at node N_k using the gap functions, as introduced in Sect. 3, the two systems are coupled through the contact constraints. Additionally, we have constraints on the contact forces as described in Sect. 3. Also, we have discrete path constraints such as joint angle limits as inequality constraints $\boldsymbol{h}_{limits}$. Finally, we define the initial configuration, initial and final momentum conditions for the complete system, and final configuration for the object.

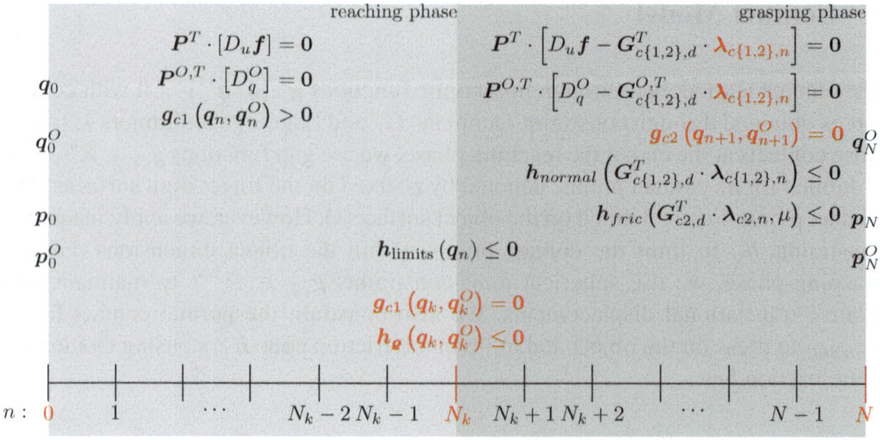

Fig. 2 Ocp setup with two phases. The placeholders $D_u f = D_1 L_d (q_n, q_{n-1}) + D_2 L_d (F_d (u_{n+1}, q_n), q_n) + f^+_{n-1} + f^-_n$ and $D^O_q = D_1 L_d (q^O_n, q^O_{n-1}) + D_2 L_d (q^O_{n+1}, q^O_n)$, see [3], define the DEL. The other terms are described in Sects. 2, 3 and 4

4.1 Objective Functions

The evolution of configurations, control torques and contact forces resulting from an ocp constitute a minimum of the objective function. In this work, two objective functions are chosen from two perspectives, namely a kinematic perspective which involves the contact points on the object $J_{1,d}$ and a biomechanical perspective in terms of control torques $J_{2,d}$.

Grasp contact polygon centroid $(J_{1,d})$ The objective here is to minimize the distance between the object center of mass φ^O and the contact polygon centroid ϱ_{cen}, see [6]. This maximizes the spread of the contact points around the object, thereby ensuring a better distribution of the contact forces.

$$J_{1,d}\left(u_d, q^O_d\right) = \min \frac{1}{2} ||\varrho_{cen} - \varphi^O||^2, \quad \text{where } \varrho_{cen} = \frac{1}{n_c} \sum_{i=1}^{n_c} \varrho_i \tag{2}$$

Rate of change of control torques $(J_{2,d})$ The objective here is to ensure a smooth movement of the fingers by minimizing changes in the control torques.

$$J_{2,d}(\tau_d, T_k, T_m) = \min \frac{1}{2} \sum_{n=0}^{k+m-2} (t_{n+1} - t_n) \left(\frac{\tau_{n+1} - \tau_n}{t_{n+1} - t_n} \right)^2 \tag{3}$$

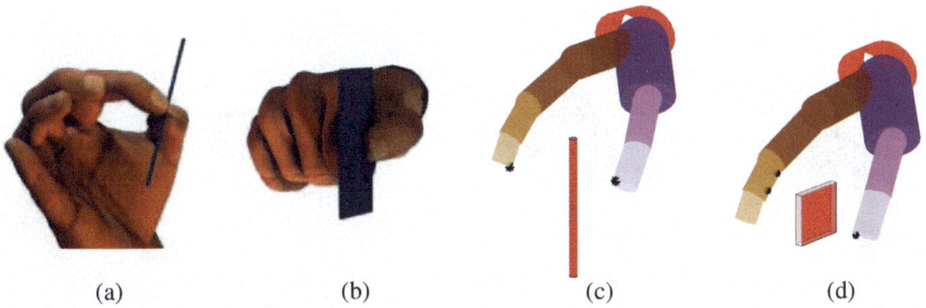

Fig. 3 The tip and lateral pinch postures as taken from [1] and initial configurations for the simulation with contact points (∗) defined on the fingers for the corresponding grasps

4.2 *Two Finger Grasp Taxonomy*

We simulate two (out of three possible) two-finger precision grasps as per grasp types defined in [1]. Being precision grasps, the hand holds objects with small dimensions. The tip pinch grasp, see Fig. 3a, holds thin cylindrical objects such as a toothpick or a candle. We simulate this grasp with two contact points, see Fig. 3c. The lateral pinch grasp, see Fig. 3b, holds thin objects with flat faces such as a key or a credit card. This is simulated with three contact points, see Fig. 3d.

5 Results

The simulations are performed as rest-to-rest actions with a fixed initial configuration for the object. For tip pinch, we lift a thin cylinder to a particular height. In the lateral pinch, we grasp a key, move it to a predefined location, and then rotate through a small angle. We show snapshots of the resulting grasping actions for tip pinch with $J_{1,d}$ and lateral pinch with $J_{2,d}$ in Figs. 4 and 5 respectively. For the same

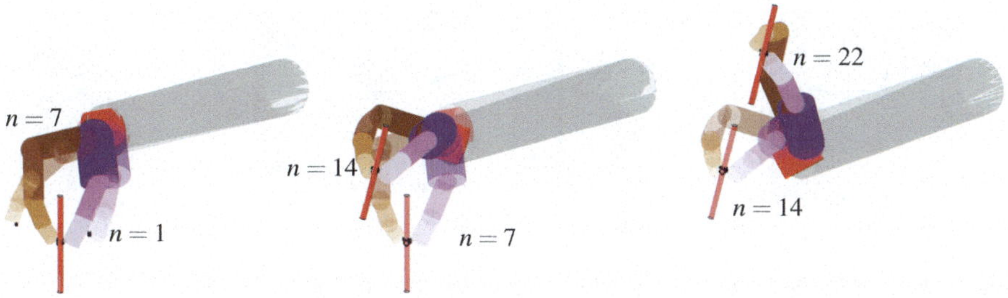

Fig. 4 The tip pinch configuration snapshots at time nodes $n = 1, 7, 14, 22$ for $J_{1,d}$ for a lifting motion. The contact is closed at $n = 7$ with points on the cylinder surface as close as possible to its center of mass

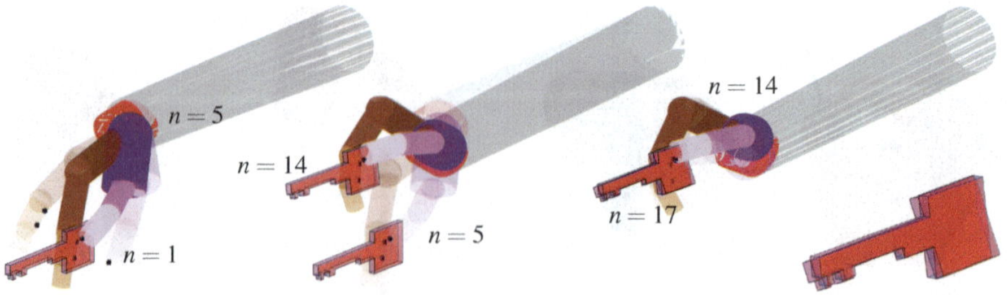

Fig. 5 The lateral pinch configuration snapshots at time nodes $n = 1, 5, 14, 17$ for $J_{2,d}$ for a lift and turn motion. The contact is closed at $n = 5$. Here, the square shaped key-head is defined as the area for the grasp. The image on the lower right shows the turning of the key through a small angle

Table 1 Phase durations for tip and lateral pinch grasps

Grasp	J_d	$T_k(s)$	$T_m(s)$
Tip	$J_{1,d}$	0.025	0.055
	$J_{2,d}$	0.067	0.079
Lateral	$J_{1,d}$	0.022	0.064
	$J_{2,d}$	0.12	0.35

cost functions, the observations are common for the different grasps, suggesting a higher influence of the objective rather than the action performed.

The reaching and grasping phase durations are noted to be higher for the $J_{2,d}$ than $J_{1,d}$, as noted in Table 1. This may have a physical interpretation to have less physically intensive motions by performing them slowly. The influence of the objective functions on the control torques is as expected, as shown in Fig. 6 for the tip pinch simulation with both cost functions. The torque evolution profiles are smoother for $J_{2,d}$ than $J_{1,d}$. This also has an additional affect on the magnitude of control torques, which are higher for $J_{1,d}$ than $J_{2,d}$, see Fig. 6, left. Contrarily, the grasping forces are of the same order for both cost functions as shown in Fig. 6, right, which shows that the contact force is dependent on object weight and static friction conditions.

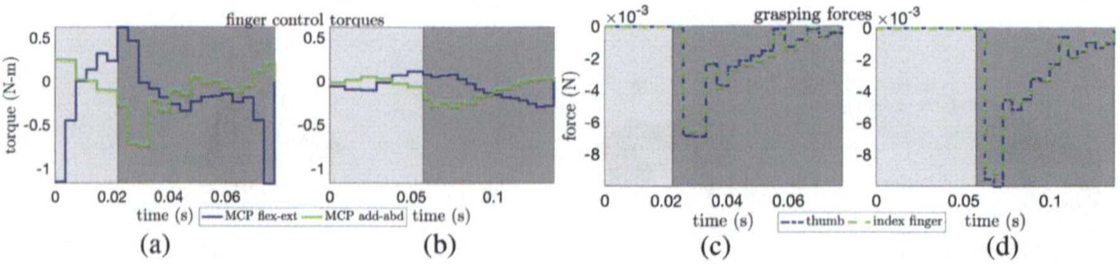

Fig. 6 Tip pinch: the evolution of control torques for the MCP joint motions, namely flexion-extension and adduction-abduction, are shown for $J_{1,d}$ (**a**) and $J_{2,d}$ (**b**). The torque evolution is smoother and of lesser order for $J_{2,d}$, rather $J_{1,d}$. The evolution of grasping forces for the fingers are shown for $J_{1,d}$ (**c**) and $J_{2,d}$ (**d**). For both objective functions, the grasping force is found to be higher after the initial contact, while stabilising over the grasping phase

Acknowledgements This work is supported by Fraunhofer Internal Programs under Grant No. MAVO 828424.

References

1. Feix, T., Romero, J., Schmiedmayer, H., Dollar, A.M., Kragic, D.: The GRASP taxonomy of human grasp types. IEEE Trans. Human-Machine Syst. **46**(1), 66–77 (2016)
2. Koch, M., Leyendecker, S.: Structure preserving simulation of monopedal jumping. Arch. Mech. Eng. **LX**, 2127–146 (2013)
3. Leyendecker, S., Ober-Blöbaum, S., Marsden, J., Ortiz, M.: Discrete mechanics and optimal control for constrained systems. Optimal Control Appl. Methods **31**, 505–528 (2010)
4. Maas, R., Leyendecker, S.: Biomechanical optimal control of human arm motion. J. Multibody Dyn. **227**(4), 375–389 (2013)
5. Phutane, U., Roller, M., Björkenstam, S., Linn, J., Leyendecker, S.: Kinematic validation of a human thumb model. In: ECCOMAS Conference Multibody Dynamics, Prague, June 2017 (2017)
6. Roa, M.A., Suárez, R.: Grasp quality measures: review and performance. Auton. Robot. **38**, 65–88 (2014)

Using Composite Finite Elements for Shape Optimization with a Stochastic Objective Functional

Matthias Bolten and Camilla Hahn

Abstract Shape optimization is an important tool to increase the reliability of mechanical components. The use of stochastic objective functionals is beneficial as the failure mechanism is usually described using stochastic models. Furthermore, stochastic objective functionals are smoother than, e.g., maxima of point stresses. Here, we consider a stochastic objective functional originating from modeling the failure of ceramic. Ceramic is a material frequently used in industry because of its favorable properties. We follow the approach above by minimizing the component's probability of failure under a given tensile load. Since the fundamental work of Weibull, the probabilistic description of the strength of ceramics is standard and has been widely applied. The resulting failure probability is used as objective function in PDE constrained shape optimization. Often the constraining PDE is discretized using finite elements, thus needing mesh morphing or re-meshing in every step of the optimization. This can be expensive and it can introduce noise. Instead, we propose to use composite finite elements for discretization. Using the Lagrangian formalism, the shape gradient via the adjoint equation is calculated at low computational cost.

1 Shape Optimization for Ceramic Components

In shape optimization, the common approach to optimize a mechanical component in order to improve its reliability is to consider the stress of the material and minimize its maximum value on the component. E.g., the von Mises yield criterion is often used. While it is a widely used objective it has several disadvantages, such as the non-differentiability of the maximum function. This motivates to consider, in the case of ceramic material, the components survival probability as the objective functional, see [6], which is differentiable as shown in [1]. It is introduced shortly in the following.

M. Bolten · C. Hahn (✉)
Bergische Universität Wuppertal, Wuppertal, Germany
e-mail: bolten@math.uni-wuppertal.de; hahn@math.uni-wuppertal.de

© Springer Nature Switzerland AG 2019
I. Faragó et al. (eds.), *Progress in Industrial Mathematics at ECMI 2018*,
Mathematics in Industry 30, https://doi.org/10.1007/978-3-030-27550-1_65

1.1 Problem Description

Let $\Omega \subseteq \mathbb{R}^d$, $d = 2, 3$, be a domain with Lipschitz boundary $\partial\Omega$. Assume that the domain Ω represents the ceramic component in its initial, force free state. Furthermore, assume that the boundary $\partial\Omega$ can be divided into three different parts $\partial\Omega = \overline{\partial\Omega_D} \cup \overline{\partial\Omega_{N_{fixed}}} \cup \overline{\partial\Omega_{N_{free}}}$. Here, $\partial\Omega_D$ is the part of the boundary where Dirichlet-boundary conditions hold and therefore it is supposed to be clamped, e.g., to a wall. $\partial\Omega_{N_{fixed}}$ is the Neumann-boundary part where surface forces may act. Finally, $\partial\Omega_{N_{free}}$ is the part of the boundary which can be modified and has zero-Neumann-boundary conditions. Forces may act on the object with the shape given by Ω. The volume force is represented by a function $f \in L^2(\Omega, \mathbb{R}^d)$, the surface force by a function $g \in L^2(\partial\Omega_N, \mathbb{R}^d)$.

As ceramic is a linear elastic material, the linear elasticity PDE,

$$B(u, v) = L(v) \, \forall v \in H_0^1(\Omega, \mathbb{R}^d), \tag{1}$$

$$B(u, v) := \int_\Omega \sigma(u) : \varepsilon(v) \, dx, \quad L(v) := \int_\Omega f \cdot v \, dx + \int_{\partial\Omega_N} g \cdot v \, dA, \tag{2}$$

must hold, where σ and ε describe the stress and the strain tensor, respectively. For existence and uniqueness see for instance [2].

1.2 The Objective Functional

The survival probability of a ceramic component is given by

$$J(\Omega, Du) := \frac{\Gamma(\frac{d}{2})}{2\pi^{\frac{d}{2}}} \int_\Omega \int_{S^{d-1}} \left(\frac{(n \cdot \sigma(u)n)^+}{\sigma_0} \right)^m dn dx, \tag{3}$$

where m is the Weibull modulus and σ_0 is a positive constant. For further understanding see [1].

2 Implementation

In [1] it is shown that (3) is differentiable. Therefore it is a reasonable next step to use it as an objective functional in a gradient-based shape optimization procedure. As a first step we consider a toy problem of a simple test object in $d = 2$ whose behavior during the optimization process is well understood. The material parameters E and ν are chosen to those of Aluminum oxide (Al_2O_3) ceramics and the Weibull module is chosen to be $m = 2$, which is smaller then the realistic value but leads to tractable numerics.

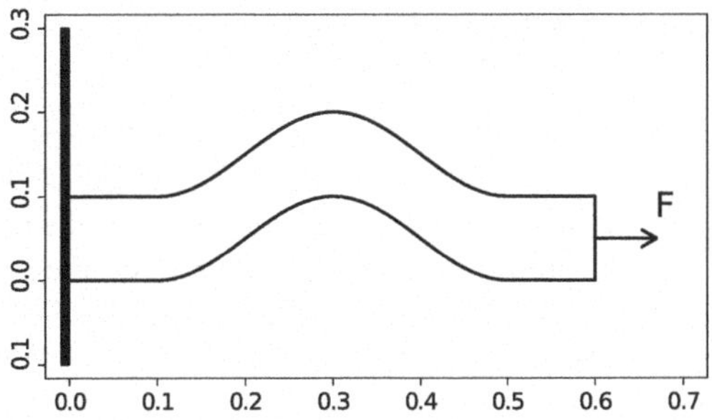

Fig. 1 Visualization of the toy problem

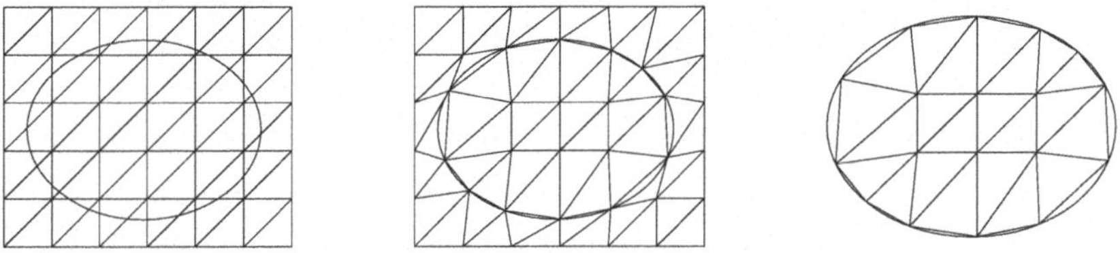

Fig. 2 Composite finite elements

As a test object we use a deformed rod of length 0.6 m and height 0.1 m, visualized in Fig. 1. It is clamped on the left part of the boundary, where the Dirichlet-boundary condition is imposed. The force is applied on the right Neumann-boundary. The expected bahavior of the gradient would be a smoothing of the inner bow.

The visualization in Fig. 1 is given without discretization. For the actual implementation we use composite finite elements, which are introduced in the following subsection.

2.1 Composite Finite Elements

Composite finite elements are a special type of finite elements introduced in [3–5].

In contrast to most approaches, we consider a supposed infinite mesh and then adapt this mesh to the shape of the component (Fig. 2). The idea is to assume a completely regular mesh and to superimpose the boundary of the considered component on that mesh. The nearest nodes are then adapted to the boundary. Thus only the elements forming the boundary are deformed, all other elements stay completely regular. The calculations are done only on the mesh inside the shape. This approach has several advantages: If the feasible region is described by the regular grid, the same grid can be used for each step of the optimization, yielding lower cost of calculation. At the same time problems of mesh morphing techniques, such as the degeneration of the elements, are avoided.

3 Calculation of the Shape Derivative

The linear elasticity equation and the objective functional (3) are discretized via composite finite elements with standard Lagrange interpolation. We use n-point Gauss quadrature for calculating the integrals. That gives us the discretized forms of the stiffness matrix $B(X)U$, the right hand side $F(X)$ and the objective functional $J(X, U(X))$.

Then, the shape gradient is of the form $\frac{dJ(X,U(X))}{dX} = \frac{\partial J(X,U(X))}{\partial X} + \frac{\partial J(X,U(X))}{\partial U}\frac{\partial U(X)}{\partial X}$. As the calculation of $\frac{\partial U(X)}{\partial X}$ is expensive, we use an adjoint approach. Hence, the set of equations

$$\begin{aligned} \frac{dJ(X,U(X))}{dX} &= \frac{\partial J(X,U)}{\partial X} + \Lambda^T \left[\frac{\partial F(X)}{\partial X} - \frac{\partial B(X)}{\partial X} U \right], \\ B^T(X)\Lambda &= \frac{\partial J(X,U)}{\partial U}, \\ B(X)U(X) &= F(X) \end{aligned} \tag{4}$$

give the discretized shape derivative, where Λ is the adjoint state.

Using a hand-written sparse solver, we obtain the shape gradient visualized in Fig. 3a. To demonstrate the universality of the chosen composite finite element discretization we placed the shape at an angle rather. In fact, the approach is independent from the location of the component to optimize and the location of the different types of boundaries. Regarding the result we observe that the gradient tends to smooth the inner bow, but also to blow up the volume. The former is what we want to see, the latter could for example be treated with bi-objective optimization techniques.

If we update the shape only slightly compared to the mesh width h, for example via $\Omega_t = \Omega_0 + \alpha p$, by choosing the search direction p to be the gradient and

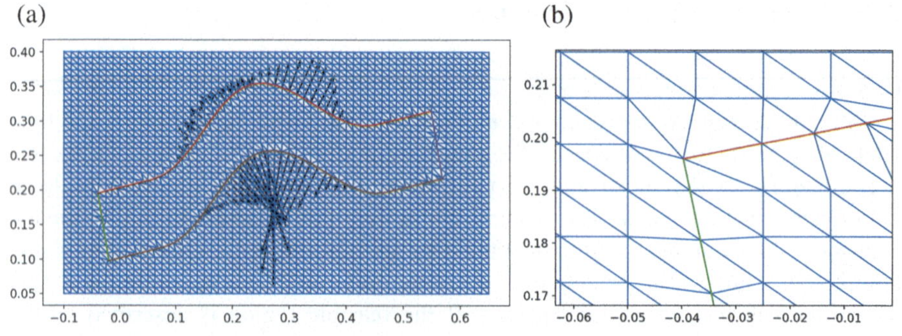

Fig. 3 Visualization of the shape gradient with CFE. (**a**) Test object. (**b**) Detail of (**a**)

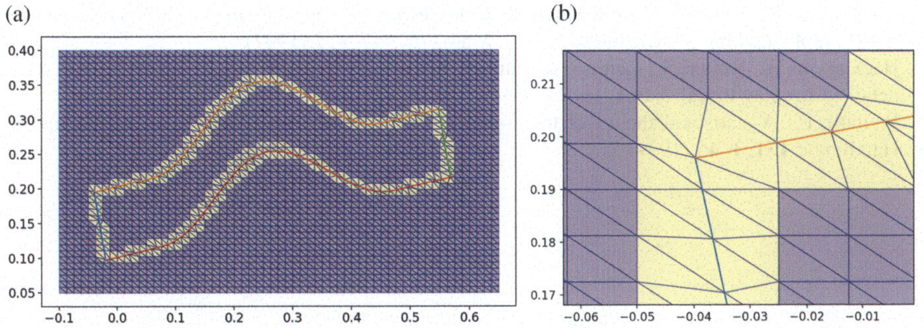

Fig. 4 Visualization of one iteration. (**a**) Test object. (**b**) Detail of (**a**)

a corresponding small step size α, we only have to recalculate the entries in the governing PDE corresponding to the elements highlighted in Fig. 4, thanks to the use of composite finite elements. As in this case only the boundary layer needs updating the time needed to do this is proportional to the surface volume of the optimized object. Larger step widths result in the inclusion or exclusion of more layers of cells, which can be handled with the chosen approach, as well.

4 Conclusion

We have seen that the functional describing the probability of failure of a ceramic component can be used as a meaningful objective in shape optimization at least in a toy problem. For the implementation of the problem we proposed to use composite finite elements. The numerical examples show that this discretization yields reasonable shape gradients and the computational demand needed to adapt the mesh in between optimization steps is small. A detailed analysis of the meshes, the obtained accuracies, as well as a comparison with other approaches that can be used in shape optimization are currently being prepared.

Acknowledgements This work was supported by BMBF (German Federal Ministry for Education and Research) as part of the project GIVEN (05M18PXA).

References

1. Bolten, M., Gottschalk, H., Schmitz, S: Minimal failure probability for ceramic design via shape control. J. Optim. Theory Appl. **166**, 983–1001 (2013)
2. Braess, D.: Finite Elements - Theory, Fast Solvers, and Applications in Solid Mechanics. Cambridge University Press, Cambridge (1997)
3. Hackbusch, W., Sauter, S.: Adaptive composite finite elements for the solution of PDEs containing nonuniformely distributed micro-scales. Math. Model. **8**, 31–43 (1996)

4. Hackbusch, W., Sauter, S.: Composite finite elements for the approximation of PDEs on domains with complicated micro-structures. Numer. Math. **75**, 447–472 (1997)
5. Hackbusch, W., Sauter, S.: Composite finite elements for problems containing small geometric details. Comput. Visual. Sci. **1**, 15–25 (1997)
6. Weibull, E.: A statistical theory of the strength of materials. Ingeniörsvetenskapsakedemiens Handlingar **151**, 1–45 (1939)

Reinforcement Learning in Order to Control Biomechanical Models

Simon Gottschalk and Michael Burger

Abstract In this paper, we address the challenge of controlling a biomechanical model to fulfill a prespecified task. We discuss the suitability of the Reinforcement Learning formulation as an optimal control problem and point out advantages of the Reinforcement Learning method in the particular biomechanical context. We conclude our paper with a numerical investigation of the performance of the presented method.

1 Introduction

These days, techniques belonging to the research field of Artificial Intelligence (AI) are widely applied and used. Researchers increasingly understand the possibilities and advantages of those techniques for new types of tasks as well as for solving problems which are studied for years and solved by well known solution techniques so far. One particular technique belonging to the field of AI is Reinforcement Learning [14]. The underlying optimization goal is comparable to the goals of optimal control problems, which motivates the following discussion of the similarities and differences.

2 Reinforcement Learning and Classical Optimal Control Techniques

In the optimal control context, one is interested in finding the essentially bounded optimal control $u \in L^\infty := L^\infty([t_0, t_f], \mathbb{R}^{n_c})$ for a given dynamical system on the time interval $[t_0, t_f]$, where $n_c \in \mathbb{N}$ is the dimension of the control space. One application of particular interest is the control of a human model in such a way

S. Gottschalk (✉) · M. Burger
Fraunhofer ITWM, Kaiserslautern, Germany
e-mail: simon.gottschalk@itwm.fraunhofer.de; michael.burger@itwm.fraunhofer.de

© Springer Nature Switzerland AG 2019
I. Faragó et al. (eds.), *Progress in Industrial Mathematics at ECMI 2018*,
Mathematics in Industry 30, https://doi.org/10.1007/978-3-030-27550-1_66

that a specific movement is executed. To model this task, a control framework and a biomechanical model is needed. Typically, biomechanical systems which can be seen as multibody systems are of the following form:

$$\begin{aligned} \dot{q}(t) &= v(t), \\ M(q(t))\dot{v}(t) &= G(q(t), v(t), u(t)), \; q_0 = q(t_0), \; v_0 = v(t_0), \end{aligned} \tag{1}$$

where q_0 and v_0 are given initial values, M is the mass matrix and G is a function depending on the position (state) $q(t)$, the velocity $v(t)$, and the control $u(t)$. In the following, we assume that for a given control $u \in L^\infty$ and given initial values, a unique position-velocity pair $x := (q, v) \in W^{1,\infty} := W^{1,\infty}([t_0, t_f], \mathbb{R}^{n_s}) \times W^{1,\infty}([t_0, t_f], \mathbb{R}^{n_s})$ fulfilling the differential equation (1) exists in the space of absolutely continuous functions with the additional assumption that the function itself and its weak derivative have a finite supremum norm. By $n_s \in \mathbb{N}$, we denote the dimension of the state space. In case of an ordinary differential equation, the Lipschitz conditions are sufficient for a unique solution.

The biomechanical system (1) constitutes the constraint in our optimal control framework. The objective function $J : W^{1,\infty} \times L^\infty \to \mathbb{R}$ describes what optimal means. Overall, we assume to have the following optimal control problem (cf. [5]):

$$\min_{x,u} \; J(x, u) = \phi(x(t_f)) + \int_{t_0}^{t_f} \Phi(x(t), u(t))dt, \text{ subject to (1)}, \tag{2}$$

where $\Phi : \mathbb{R}^{2n_s} \times \mathbb{R}^{n_c} \to \mathbb{R}$ is a mapping from the state and control space into the real numbers, and defines the objective function of the optimization problem together with $\phi : \mathbb{R}^{2n_s} \to \mathbb{R}$ rating the end position. Note that from the control theory point of view, this formulation is an open-loop optimization since the control u_t does not directly depend on the current state x_t.

In general, classical optimal control techniques can be divided into two classes: On one hand first-optimize-then-discretize techniques, which, for instance, derive optimality conditions of the continuous optimization problem in order to solve them numerically afterwards, and, on the other hand, first-discretize-then-optimize techniques. The latter mentioned techniques discretize the problem statement before further numerical methods are applied.

Reinforcement Learning [14] is able to tackle discrete optimal control problems. The most important assumption is that a Markov decision process (S, A, P_u, r) (see e.g. [10]) is present with state space S (e.g. $\mathbb{R}^{2n_s}$), control space A (e.g. $\mathbb{R}^{n_c}$), probability $P_u(x, x') = P(x_{t+1} = x'|x_t = x, u_t = u)$ for the next state x_{t+1} given previous state x_t, and control u_t (also called *action*), and the reward function $r : S \times A \to \mathbb{R}$ rating the current situation. Be aware of the assumption that the upcoming state only depends on the current action and the previous state, but not on older states (Markov property). The biomechanical system (1) is now hidden in a probability distribution. This means instead of having e.g. a discretized, deterministic forward simulation $x_{t+1} = f(x_t, u_t)$ we could have that $x_{t+1} \sim \mathcal{N}(f(x_t, u_t), \Sigma)$ is Gaussian distributed for a given variance Σ describing model imperfections. However, RL is not restricted to normal distributions. The objective function in the classical problem statement (2) is now replaced by the reward and the overall goal is to maximize the expected total reward $\mathbb{E}\left[\sum_{(x_t,u_t)\in\tau} r(x_t, u_t)\right]$ of a trajectory $\tau = \{(x_0, u_0), (x_1, u_1), \ldots\}$, where r is given as $r(x_t, u_t) = -h\Phi(x_t, u_t) - \mathbb{1}_{t=t_f}\phi(x_{t_f})$ with time step size h.

Since the information about the biomechanical system are given by a distribution, it becomes clear that the controller should be able to react to distributions and inaccuracies what motivates a closed-loop (respectively feedback) controller, where the next control depends on the current state. The control construction is typically called *policy* and can either be searched directly, which is then called *policy based*, or indirectly by improving a so called value function, which is an estimation of the expected future reward. The latter is called *value-function based* and an example is Q-learning (see e.g. [15]).

RL algorithms can be divided into *model-based* and *model-free* approaches. While the former techniques train its own approximation of the model with the observed transition tuple (x_t, u_t, x_{t+1}) and then use this approximation to improve the controls, the model-free methods do not use any models, but only the transition tuple itself. Model-based approaches can be sample efficient but only work if a good model can be learned. A famous example for model-based RL is PILCO [4].

In the numerical results, we use a model-free, policy-based RL approach. The policy $\pi_\theta(u_t|x_t)$ is parameterized by introducing an artificial neural network (ANN), which gives back the mean of a Gaussian function, which is used to sample the next action. The parameters θ are the weights and the biases of the ANN. Thus, instead of optimizing with respect to $(N_T \times n_c)$-variables with N_T denoting the number of time steps, we search for optimal weights and biases. The usage of the ANN is only an ansatz which we use here since ANN can be handled very simple, because of their structure, and their capabilities of approximating complex functions. There are a bunch of model-free, policy-based techniques based on neural networks: Gradient-based methods like Reinforce [16], Trust Region Policy Optimization (TRPO) [11], Proximal Policy Optimization (PPO) [13] and many more. In the numerics in Sect. 3, we focus on TRPO and its related PPO. Both are based on the idea that one simulates the dynamical system with the current policy π_θ and updates this policy by solving the following optimization problem (for a suitable $\delta \in \mathbb{R}_{>0}$) afterwards:

$$
\begin{aligned}
\max_{\tilde{\theta}} \quad & \mathbb{E}_{x\sim\rho_{\pi_\theta}, u\sim\pi_\theta}\left[\frac{\pi_{\tilde{\theta}}(u|x)}{\pi_\theta(u|x)} A_{\pi_\theta}(x, u)\right] \\
s.t. \quad & \mathbb{E}_{x\sim\rho_{\pi_\theta}}\left[D_{KL}\left[\pi_\theta(\cdot, x), \pi_{\tilde{\theta}}(\cdot, x)\right]\right] \leq \delta
\end{aligned}
\tag{3}
$$

with $D_{KL}\left[\pi_\theta(\cdot, x), \pi_{\tilde{\theta}}(\cdot, x)\right] = \sum_u \pi_\theta(u|x) \log \frac{\pi_\theta(u|x)}{\pi_{\tilde{\theta}}(u|x)}$ known as Kullback-Leibler divergence [8, 9]. $A_{\pi_\theta}(x, u)$ describes the expected reward if we take the action u at state x reduced by the reward one expects if we apply an average action at state x.

Working principle of TRPO:

Step 1 Initialize the parameters θ of the policy π_θ.

Step 2 Simulate M-times the dynamical system in Eq. (1) with controls generated by the policy π_θ. Store the trajectories $\tau_k = \{x_0, u_0, r_0, x_1, u_1, r_1, \ldots\}, k \in \{0, 1, \ldots, M\}$.

Step 3 Solve the trust region problem in (3). Use the stored trajectories in order to estimate the occurring expected values.

Step 4 Set $\theta := \tilde{\theta}$ and go to Step 2 as long as the policy seems to improve significantly or the maximal number of executed iterations is reached.

There are already examples where RL is applied in order to control a system. For instance, academic examples can be found on OpenAi Gym [2], which was designed to help to develop and compare RL algorithms. Furthermore, in [7] the authors consider musculoskeletal models of a human with the task to walk as long as possible without falling.

3 Biomechanical Application and Numerical Results

In the following, we consider a simplified biomechanical model of a human arm. The model is a multibody system describing the bones as rigid bodies equipped with Hill's muscle model [6] in order to actuate the system. Our model consists of two rigid bodies and three muscles. The upper arm (mass m_1, length l_1) is attached at the origin as well as the forearm (mass m_2, length l_2) is attached at the upper arm by a revolute joint. A sketch of the model can be seen in Fig. 1. Thus, it is enough to consider an inverted double pendulum with muscles. The differential equations of the well known double pendulum can be derived by Lagrangian formalism. The posture of the arm can be described by two angles (α and β) and the resulting

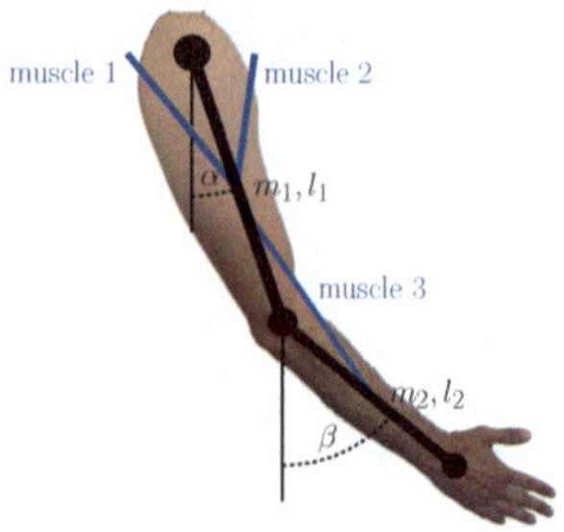

Fig. 1 Sketch of the arm model

Table 1 Parameters of the biomechanical system

Label	Definition	Label	Definition
m_1	1.924 kg	l_1	0.351 cm
m_2	1.502 kg	l_2	0.287 cm
$\tilde{\alpha}$	0 rad (respectively −0.3 rad)	$\tilde{\beta}$	0.5π rad
$r(\alpha, \beta)$	$-\sqrt{(\alpha-\tilde{\alpha})^2+(\beta-\tilde{\beta})^2+\eta}$	η	$\begin{cases} 100 & \text{task fulfilled} \\ 0 & \text{else} \end{cases}$

differential equation is an ordinary differential equation, which can be discretized by an explicit Euler discretization method.

Our task is moving the arm to a specific position and then to a next position. To be precise, we say that the task is fulfilled if, at some point, the upper arm is vertical ($\alpha = 0$) and the forearm ($\beta = 0.5\pi$) horizontal and if the hand is in the position $\alpha = -0.3$, $\beta = 0.5\pi$ at the end. In our case, we do not need to think too much about the transition between the two subtasks. We just define a not continuous reward function.

Now, we apply RL to the described problem statement by using the parameters which can be seen in Table 1. As implementation of the RL, we use a modified version of the proximal policy optimization algorithm of Coady [3]. It is based on the TRPO algorithm but instead of solving the constraint optimization problem in (3) directly, in the numerical results, we use an adaptive KL penalty coefficient as it is suggested in [13].The working environment is built in the programming language Python. For the neural network with three hidden layers, each with less than hundred neurons, and its learning updates, Tensorflow [1] is used. The advantage function in (3) is estimated by the generalized advantage estimator (see [12]).

Figure 2 shows the reward summed up over all trajectories in each iteration. This visualizes the learning process and shows how the policy becomes better. In Figs. 4 and 3, we consider one trajectory based on the learned policy at the end of the learning process. Figure 3 shows the activations of the muscles generating the path of the arm (Fig. 4). The intermediate state is reached at the dashed line whereas the dotted line indicates the completion of the final task.

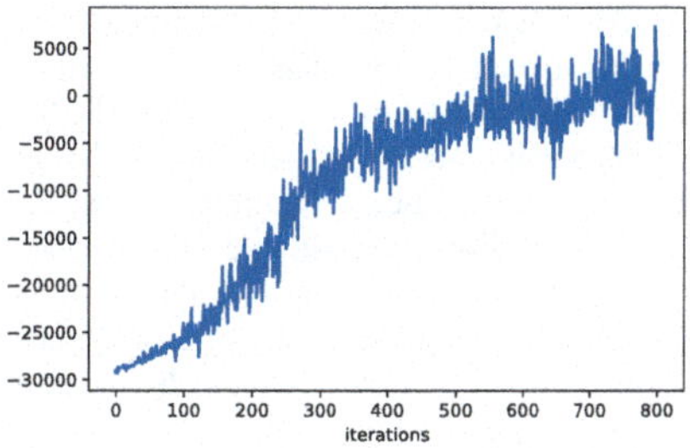

Fig. 2 Sum over all rewards in each iteration

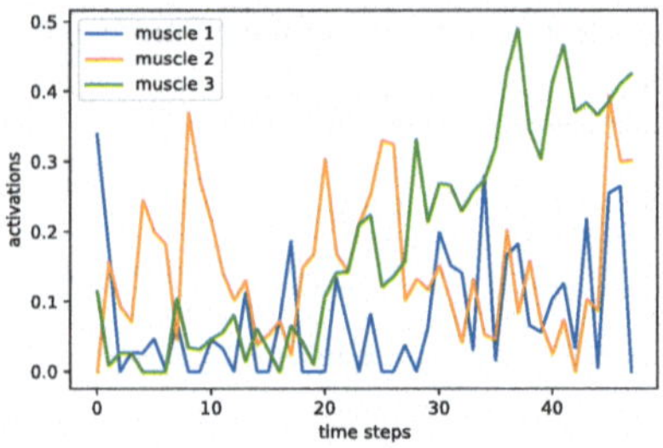

Fig. 3 Activations of the Hill's muscle models

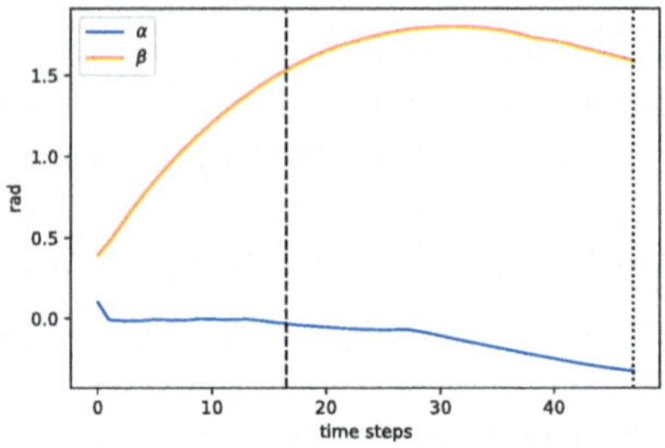

Fig. 4 State of one trajectory executed after the training

Conclusions We pointed out the connection between RL and classical optimal control problems. We have optimized a simplified biomechanical problem, which shows us how RL can be used in the optimal control context. We found a policy in order to move the model of a human arm as desired. Even the realization of two sequential tasks has been performed without thinking about the transition, which is a significant advantage. Further work includes the extension of biomechanical examples and a discussion of further advantages of RL techniques.

Acknowledgements The authors are grateful for the funding by the Federal Ministry of Education and Research of Germany (BMBF), project number 05M16UKD.

References

1. Abadi, M., Agarwal, A., Barham, P., Brevdo, E., Chen, Z., Citro, C., Corrado, G.S., Davis, A., Dean, J., Devin, M., Ghemawat, S., Goodfellow, I., Harp, A., Irving, G., Isard, M., Jia, Y., Jozefowicz, R., Kaiser, L., Kudlur, M., Levenberg, J., Mané, D., Monga, R., Moore, S., Murray, D., Olah, C., Schuster, M., Shlens, J., Steiner, B., Sutskever, I., Talwar, K., Tucker, P., Vanhoucke, V., Vasudevan, V., Viégas, F., Vinyals, O., Warden, P., Wattenberg, M., Wicke, M., Yu, Y., Zheng, X.: TensorFlow: large-scale machine learning on heterogeneous systems (2015). Software available from tensorflow.org
2. Brockman, G., Cheung, V., Pettersson, L., Schneider, J., Schulman, J., Tang, J., Zaremba, W.: Openai gym. CoRR (2016). abs/1606.01540
3. Coady, P.: Ai gym workout (2017). https://learningai.io/projects/2017/07/28/ai-gym-workout.html. Cited 26 Oct 2018

4. Deisenroth, M., Rasmussen, C.: PILCO: a model-based and data-efficient approach to policy search. In: Proceedings of the 28th International Conference on Machine Learning, ICML, pp. 465–472 (2011)
5. Gerdts, M.: Optimal Control of ODEs and DAEs. De Gruyter Textbook. De Gruyter, Berlin (2011)
6. Hill, A.V.: The heat of shortening and the dynamic constants of muscle. Proc. R. Soc. Lond. B Biol. Sci. **126**(843), 136–195 (1938)
7. Kidzinski, L., Mohanty, S.P., Ong, C.F., Hicks, J.L., Carroll, S.F., Levine, S., Salath, M., Delp, S.L.: Learning to Run Challenge: Synthesizing Physiologically Accurate Motion Using Deep Reinforcement Learning. CoRR (2018). abs/1804.00198
8. Kullback, S.: Information Theory and Statistics. Wiley, New York (1959)
9. Kullback, S., Leibler, R.A.: On Information and Sufficiency. Ann. Math. Stat. **22**(1), 79–86 (1951)
10. Puterman, M.L.: Markov Decision Processes: Discrete Stochastic Dynamic Programming, 1st edn. Wiley, New York (1994)
11. Schulman, J., Levine, S., Abbeel, P., Jordan, M.I., Moritz, P.: Trust Region Policy Optimization. In: ICML. Lille, France, pp. 1889–1897 (2015)
12. Schulman, J., Moritz, P., Levine, S., Jordan, M.I., Abbeel, P.: High-Dimensional Continuous Control Using Generalized Advantage Estimation. CoRR (2015). abs/1506.02438
13. Schulman, J., Wolski, F., Dhariwal, P., Radford, A., Klimov, O.: Proximal Policy Optimization Algorithms. CoRR (2017). abs/1707.06347
14. Sutton, R.S., Barto, A.G.: Introduction to Reinforcement Learning, 1st edn. MIT Press, Cambridge (1998)
15. Watkins, C.J.C.H., Dayan, P.: Q-learning. Machine Learning **8**(3), 279–292 (1992)
16. Williams, R.J.: Simple statistical gradient-following algorithms for connectionist reinforcement learning. Machine Learning **8**(3), 229–256 (1992)

Optimizing Majority Voting Based Systems Under a Resource Constraint for Multiclass Problems

Attila Tiba, András Hajdu, György Terdik, and Henrietta Tomán

Abstract Ensemble-based approaches are very effective in various fields in raising the accuracy of its individual members, when some voting rule is applied for aggregating the individual decisions. In this paper, we investigate how to find and characterize the ensembles having the highest accuracy if the total cost of the ensemble members is bounded. This question leads to Knapsack problem with non-linear and non-separable objective function in binary and multiclass classification if the majority voting is chosen for the aggregation. As the conventional solving methods cannot be applied for this task, a novel stochastic approach was introduced in the binary case where the energy function is discussed as the joint probability function of the member accuracy. We show some theoretical results with respect to the expected ensemble accuracy and its variance in the multiclass classification problem which can help us to solve the Knapsack problem.

1 Introduction

The ensemble creation is a rather popular and effective method in several problems to outperform the decision accuracy of individual approaches [5]. To aggregate the individual decisions of the members in the ensemble, the final decision is made by applying voting rule, such as the classic or weighted majority ones.

In a binary classification problem, each member of the ensemble makes true or false decision. It means that the classifier D_i with accuracy p_i ($0 \leq p_i \leq 1$, $i = 1, \ldots, n$) can be considered as Bernoulli distributed random variable η_i, where the probability of the correct classification by D_i is p_i. In this particular (Bernoulli distributed) case, the expected value of the i-th random variable η_i is p_i ($i = 1, \ldots, n$).

A. Tiba · A. Hajdu (✉) · G. Terdik · H. Tomán
Faculty of Informatics, University of Debrecen, Debrecen, Hungary
e-mail: tiba.attila@inf.unideb.hu; hajdu.andras@inf.unideb.hu; terdik.gyorgy@inf.unideb.hu; toman.henrietta@inf.unideb.hu

© Springer Nature Switzerland AG 2019
I. Faragó et al. (eds.), *Progress in Industrial Mathematics at ECMI 2018*, Mathematics in Industry 30, https://doi.org/10.1007/978-3-030-27550-1_67

In majority voting, that alternative is selected as the final decision which has majority in the ensemble (more than half of the n votes). In this case, the ensemble accuracy for $n \in \mathbb{N}$ independent binary classifiers [4] can be calculated as:

$$q_{binary} = \sum_{k=\lceil \frac{n}{2} \rceil}^{n} \Bigg(\sum_{\substack{I \subseteq \{1,\dots,n\} \\ |I|=k}} \prod_{i \in I} p_i \prod_{j \in \{1,\dots,n\} \setminus I} (1 - p_j) \Bigg). \quad (1)$$

In [2], the majority voting rule was extended to the spatial domain in a special object detection problem to find the optic disc (OD) in retinal images. The votes of the ensemble members (OD detectors) are given by single pixels as the centroid of the disc-like anatomical feature OD. The votes are required to fall inside a disc of a given diameter d_{OD} to vote together. To aggregate the outputs of individual OD detectors, the final decision is made by choosing the circle fulfilling the geometric constraint and containing the maximal number of the votes. To find the ensemble accuracy in this case, the term $p_{n,k}$ is introduced for the modified majority voting of the classifiers $D_1, \dots, D_n$: if k classifiers out of the n ones give a correct vote, then the good decision is made with probability $p_{n,k}$. By applying these notations, the ensemble accuracy (1) is transformed by the geometric restriction to the following formula:

$$q_{multi} = \sum_{k=0}^{n} p_{n,k} \Bigg(\sum_{\substack{I \subseteq \{1,\dots,n\} \\ |I|=k}} \prod_{i \in I} p_i \prod_{j \in \{1,\dots,n\} \setminus I} (1 - p_j) \Bigg). \quad (2)$$

For the given real numbers $p_{n,k}$ ($k = 0, 1, \dots, n$) in (2), we have that $0 \leq p_{n,0} \leq p_{n,1} \leq \dots \leq p_{n,n} \leq 1$.

In special case, we get back the classical majority voting scheme if the terms $p_{n,k}$ are chosen in the following way: $p_{n,k} = 1$, if $k > \lfloor n/2 \rfloor$, and $p_{n,k} = 0$, otherwise.

In the above spatial extension of the majority voting rule, the final decision is made by choosing from the candidates (circles) with respect to their cardinalities. The majority voting rule can be extended for a multiclass classification problem in a very similar way.

High accuracy for an ensemble system is a very important and natural requirement, mainly in clinical decision making. Besides the high accuracy, other performance parameters need to be discussed, as well. One of these parameters to be considered is the execution time. The ensemble creation is more resource demanding, because all the ensemble members have to be executed to make the final decision. In this paper, we solve the problem how to find the ensemble with the highest accuracy from the given possible ensemble members, with a constraint on the total execution time. These optimization problems, when the ensemble accuracy q_{binary} in (1) or q_{multi} in (2) is chosen as energy function, is very challenging, as both of them result in a non-linear, non-separable task. It means we cannot apply the classical solving methods, namely e.g. the dynamic programming, for finding the optimal solution. A Knapsack problem is formulated to handle the constraint for the total execution time. We give some theoretical results with respect to the multiclass classification problem which can help us to solve the Knapsack problem.

The rest of the paper is organized as follows. In Sect. 2, the proper formulation of the above optimization problem as Knapsack one is given. After discussing the multiclass classification problem in contrast with the binary one in Sect. 3, some theoretical and experimental results are enclosed for the multiclass classification problem in Sect. 4.

2 The Knapsack Problem with Total Time Constraint

As first step, the classic Knapsack problem is presented, then we formulate our ensemble creation issue and discuss why finding the solution is so difficult if the energy function of the Knapsack problem is selected as q_{multi} in (2).

To formulate the classic Knapsack problem, let n items be given, with value $v_1, \dots, v_n$ $(v_i \geq 0, i = 1, \dots, n)$ and weight $w_1, \dots, w_n$ $(w_i \geq 0, i = 1, \dots, n)$, respectively. Then let x_i $(x_i \in \{0, 1\}, i = 1, \dots, n)$ be the number of the i-th item to be packed. The maximal total weight of the knapsack is W $(W \geq 0)$. The aim is to find the maximal value of the target function $\sum\limits_{k=1}^{n} x_k v_k$ fulfilling the following conditions: $\sum\limits_{k=1}^{n} x_k w_k \leq W, \;\; x_k \in \{0, 1\} \;\; (k = 1, \dots, n)$.

With respect to the corresponding properties of the objective function coming from several different kinds of applications, many variations of the original Knapsack problem are considered: linear/non-linear, separable/non-separable, convex/non-convex objective functions with continuous/integer variables. Although some non-linear Knapsack problems are investigated in the literature, [1, 6], the vast majority of the works deal with Knapsack problems having linear or a separable convex non-linear objective function and linear constraint.

In the above presented ensemble creation motivated by the object detection problem, each possible ensemble member is an object detector. In Knapsack problem, the individual accuracy p_i of the i-th detector is considered as the value v_i, while the individual running time t_i is the weight w_i, where for the aggregation, a constrained majority voting is applied, that is, the ensemble accuracy q_{multi} given in (2) is the objective function. The problem is to find the most accurate ensemble with system accuracy q_T from these members with limited total execution time T:

$$q_T = \max_{\{i_1,\dots,i_s\}} \left\{ \sum_{k=0}^{s} p_{s,k} \Bigg(\sum_{\substack{I \subseteq \{i_1,\dots,i_s\} \\ |I|=k}} \prod_{i \in I} p_i \prod_{j \in \{i_1,\dots,i_s\} \setminus I} (1 - p_j) \Bigg) \right\} \tag{3}$$

with the following conditions:

$$\sum_{j=1}^{s} t_{i_j} \leq T, \quad \{i_1, \ldots, i_s\} \subseteq \{1, \ldots, n\} \quad (s = 1, \ldots, n). \tag{4}$$

The main challenge in solving this optimization problem is that the target function q_{multi} of the constrained majority voting is non-linear, non-separable. In general, Knapsack problems with these special kind of objective functions are investigated very rarely in the related papers, or only in that case when a strict restriction on their functional structure is given (e.g., the exponential type of target function is analyzed in [6]). That is, for a proper analysis we need some theoretical results for the optimization of the specific target function (2) within the Knapsack framework.

3 The Multiclass Classification Problem

In binary classification, the elements of a given set are classified into two classes (predicting which class each element belongs to). As first step, a Knapsack problem is investigated for ensemble creation with binary classifiers $D_1, D_2, \ldots, D_n$ as possible members of the ensemble, whose outputs are aggregated by applying the majority voting rule. It means that in this Knapsack problem, the objective function q_{binary} given in (1) is maximized when the total execution time of the selected members is bounded (see the condition in (4)).

In our proposed stochastic approach in [3], the selection of the items to the ensemble is based on the efficiency of the individual members. Instead of the usefulness values p_i/t_i considered in the classic greedy method, the system accuracy $q(p_i, t_i)$ of the ensemble containing maximal number of i-th items characterizes the efficiency of the i-th kind of item.

In our selection method, a discrete random variable depending on the efficiency values of the remaining items is applied in each step to determine the probability of choosing an item from the remaining set to add to the ensemble. This discrete random variable reflects that the more efficient the item is, the more probable it is selected to the ensemble in the next step.

To find and apply proper stopping criteria for this selection method, the behavior of the random variable q_{binary}, the joint distribution function based on the values p_i-s in (1) is investigated. Either the distribution of the values p_i is known, or it is fitted by Beta distribution, the knowledge on the behavior of the energy function q_{binary} (e.g. the expected ensemble accuracy, the probability to find more accurate ensembles) can be efficiently involved as a stopping rule in the stochastic search.

The multiclass classification can be interpreted in a similar way as the binary one, just in case the prediction of the class for each element where it belongs to is made for three or more classes [7]. We encounter similar problems to find the optimal solution q_T in (3) of multiclass Knapsack problem as in the binary case, but, besides the estimation of the behavior of the energy function q_{multi}, the terms $p_{n,k}$ need to be investigated, as well. It is reasonable to assume that the more classifiers out of the n ones give correct vote, the bigger probability $p_{n,k}$ for the good decision we get for the ensemble. Therefore, in the next section, the terms $p_{n,k}$ are considered as values of a function F such that $p_{n,k} = F\left(\frac{k}{n}\right)$, where $F(\cdot)$ is a cumulative distribution function on $[0, 1]$.

4 Stochastic Estimation of Ensemble Accuracy

We have the following theorem showing the behavior of the random variable q_{multi} (i.e. the expected ensemble accuracy and the variance), based on the random values of p_i-s.

Theorem 1 *Let $p \in [0, 1]$ be a random variable with $Ep = \mu$, $Var(p) = \sigma^2$, and p_i $(i = 1, 2, \ldots, n)$ are independent and identically distributed according to p. Furthermore let the energy function q_{multi} be defined by (2). Then for the expected ensemble accuracy $E(q_{multi})$ we have shown that*

$$E(q_{multi}) = \sum_{k=0}^{n} F\left(\frac{k}{n}\right)\binom{n}{k}\mu^k (1-\mu)^{n-k} . \tag{5}$$

Furthermore, if n is large then

$$\sum_{k=0}^{n} F\left(\frac{k}{n}\right)\binom{n}{k}\mu^k (1-\mu)^{n-k} \sim \int_0^1 F(y)\,\delta(\mu)\,dy = F(\mu) \tag{6}$$

where $\delta(\cdot)$ is the Dirac function.

In case of large n, we have the variance of the ensemble accuracy

$$0 \leq \text{Var}(q_{multi}) \leq F(\mu) - F^2(\mu) = F(\mu)(1 - F(\mu)) . \tag{7}$$

For practical issue, the following examples for the function F are important:

Arcsine law (distributed as Beta $(1/2, 1/2)$) with cumulative distribution function

$$F(y) = \frac{2}{\pi} \arcsin\left(\sqrt{y}\right), \quad y \in [0, 1], \tag{8}$$

and Generalized Arcsine law (distributed as Beta $(1 - \alpha, \alpha)$), as if the distribution of p is not known, then a Beta distribution is fitted to p.

From the results of the Theorem 1 with respect to the expected value and the variance of the ensemble accuracy, the decision in the multiclass case for relatively large n is considered to be Bernoulli variated with parameter $F(\mu)$.

While the binary classification problem is closely related to the results of the binomial distribution, then in the multiclass classification the multinomial coefficients are supposed to have very important role in finding a formula for the values of $p_{n,k}(d)$. As a first step, we simulated the multiclass classification problem for $d = 3$, $d = 4$ and $d = 5$ classes, by generating random numbers in [0, 1], to decide which class is chosen. From the results of the simulations, we get approximate values for the terms $p_{n,k}(d)$. In the next step, we give a closed formula for the values $p_{n,k}(d)$, as well.

Let the multinomial coefficients $b_{n,d}(x_1, x_2, \ldots, x_d)$ be given, $(x_i \geq 0, \sum x_i = n)$, $\underline{x} = (x_1, x_2, \ldots, x_d)$, and $\alpha_k(\underline{x})$ is defined as the card $\left(\underline{x} \middle| x_i = k\right) + 1$. Then for the terms $\mathfrak{p}_{n,k}(d)$ of accuracy in that case, we have the following formula,

$$\mathfrak{p}_{n,k}(d) = \frac{1}{d^{n-k}} \sum_{0 \leq \underline{x} \leq k} \frac{b_{n-k,d}(\underline{x})}{\alpha_k(\underline{x})}, \tag{9}$$

where $0 \leq \underline{x} \leq k := (x_i | 0 \leq x_i \leq k, i = 1, 2, \ldots, d)$.

Applying this formula, we get the same results for the values of $\mathfrak{p}_{n,k}(d)$ in case of $d = 3$, $d = 4$ and $d = 5$ classes as before with the simulations.

The closed formula for the values of $\mathfrak{p}_{n,k}(d)$ guarantee us that besides the experimental results (e.g. simulations), further theoretical investigation and characterization of the optimal solution of the Knapsack problem in multiclass classification can be achieved as our future plan.

Acknowledgements This work is supported in part by the project EFOP-3.6.2-16-2017-00015 supported by the European Union, co-financed by the European Social Fund.

References

1. Bretthauer, K.M., Shetty, B.: The nonlinear knapsack problem – algorithms and applications. Eur. J. Oper. Res. **138**, 459–472 (2002)
2. Hajdu, A., Hajdu, L., Jónás, A., Kovács, L., Tomán, H.: Generalizing the majority voting scheme to spatially constrained voting. IEEE Trans. Image Process. **22**, 4182–4194 (2013)
3. Hajdu, A., Tomán, H., Kovács, L., Hajdu, L.: Composing ensembles by a stochastic approach under execution time constraint. In: Proceedings of 23rd International Conference on Pattern Recognition (ICPR), pp. 222–227 (2016)
4. Kuncheva, L.: Combining Pattern Classifiers: Methods and Algorithms. Wiley, Hoboken (2004)
5. Lam, L., Suen, S.Y.: Application of majority voting to pattern recognition: an analysis of its behavior and performance. IEEE Trans. Syst. Man Cybern. **27**, 553–568 (1997)
6. Sharkey, T.C., Romeijn, H.E., Geunes, J.: A class of nonlinear nonseparable continuous knapsack and multiple-choice knapsack problems. Math. Program. **126**, 69–96 (2011)
7. Shiraishi, Y., Fukumizu, K.: Statistical approaches to combining binary classifiers for multi-class classification. Neurocomputing **74**, 680–688 (2011)

Part V
Processing and Using Measurements Data in Industrial Problems

Queues with Choice from a Symmetry Perspective

Juancho A. Collera

Abstract Recently, a deterministic queueing model where in customers are given the opportunity to choose between two queues was introduced. The information provided to the customers is not up-to-date but instead customers were given the queue length information some time units in the past. This time delay impacts the dynamical behavior of the queues and hence the decision-making process of the customers. We revisit this queues-with-choice model from a symmetry perspective. We show that the symmetry structure of the model can be used to classify the types and kinds of solutions that can occur. In particular, our results explain why only asynchronous periodic solutions and symmetric equilibrium solutions arise in such model, while synchronous periodic solutions and asymmetric equilibrium solutions do not occur. Our method can also be applied to study similar models with larger number of queues.

1 Introduction

Providing queue length information are common in the healthcare industry (e.g. emergency room waiting time), telecommunications systems (e.g. telephone call centers) and amusement parks (e.g. waiting times of various rides). Often times the information provided is not real-time or up-to-date but instead based on the information some time units in the past or based on moving average of the waiting times (see [11] and references therein). In [10], the deterministic model describing a single queue $\dot{z}(t) = a - bz(t)$ was given, where $z(t)$ represents the length of the queue and the parameters $a, b > 0$. Here, the rate of change with respect to time of a queue length is the difference between the arrival rate of the costumers and the rate at which customers are serviced. This was extended in [11] to a model where in the customers are given the opportunity to choose between two queues. Their model

J. A. Collera (✉)
University of the Philippines Baguio, Baguio City, Philippines
e-mail: jacollera@up.edu.ph

© Springer Nature Switzerland AG 2019
I. Faragó et al. (eds.), *Progress in Industrial Mathematics at ECMI 2018*, Mathematics in Industry 30, https://doi.org/10.1007/978-3-030-27550-1_68

is given by the following system of delay differential equations (DDEs)

$$\begin{cases} \dot{x}(t) &= a\dfrac{e^{-x(t-\tau)}}{e^{-x(t-\tau)} + e^{-y(t-\tau)}} - bx(t), \\ \\ \dot{y}(t) &= a\dfrac{e^{-y(t-\tau)}}{e^{-x(t-\tau)} + e^{-y(t-\tau)}} - by(t), \end{cases} \tag{1}$$

where $x(t)$ and $y(t)$ are the lengths of the first and second queues, respectively. The initial history functions $x(t) = \varphi_1(t) > 0$ and $y(t) = \varphi_2(t) > 0$ for $t \in [-\tau, 0]$ were used. The total arrival rate to both queues is equal to the constant rate a. Moreover, it is assumed that the arrival rates are based on delayed information. That is, the information given to the costumers is actually the queue length information τ time units in the past where the time delay parameter $\tau > 0$.

According to [11], if the same initial history functions were used in system (1), then the two queues are identical for all time and both converge to the equilibrium value. If different history functions were used, then the two queues are oscillating and are asynchronous. Moreover, if $\tau < \tau^*$, for some critical delay value τ^*, then the asynchronous queues both converge to the equilibrium value, while if $\tau > \tau^*$, the asynchronous queues are periodic. It is worth mentioning that no synchronous periodic solutions nor asymmetric equilibrium solutions were obtained in [11].

The rate at which customers are serviced, as well as how delay the provided information is might differ for each queue. However, in system (1), we see that the two queues are assumed to be *similar*, and thus exhibit some symmetry property. This special case organizes the dynamics for the asymmetric cases where corresponding parameters in each queue do not differ as much. Many applications, for example, in the physical sciences [1, 4, 6] and in the biological sciences [2, 3] are symmetric systems. In each of these examples, the symmetry properties of the model played a significant role in determining its dynamical behavior.

The goal of this research is to utilize the symmetry structure of model (1) in order to classify the types and kinds of solutions that this system can have. Using a technique from [5, 8], we classify the codimension-one bifurcations into regular and symmetry-breaking. The occurrence and non-occurrence of these bifurcations, as we vary the time delay parameter τ, allows us to determine the types and kinds of solutions that can only arise in model (1). These additional insights help us understand the effects of delayed information on the dynamical behavior of queues which are of great importance to both companies and their customers.

The rest of the paper is organized as follows. In the next section, we describe the symmetry of system (1) and its equilibria. We then use these symmetry properties in deriving our main results in Sect. 3. We end the paper with a summary and discussions of future directions of this research.

2 Symmetry Group and Symmetric Equilibrium

Consider the system of DDEs with a single discrete time delay $\tau > 0$ given by $\dot{X}(t) = f(X(t), X(t-\tau))$ where $X : \mathbb{R} \to \mathbb{R}^n$. If we let $\mathscr{C} = C([-\tau, 0], \mathbb{R}^n)$ be the space of continuous functions mapping the interval $[-\tau, 0]$ into $\mathbb{R}^n$, and $X_t \in \mathscr{C}$ means $X_t(\theta) = X(t+\theta)$ for $\theta \in [-\tau, 0]$, then the system of DDEs can be written as $\dot{X}(t) = F(X_t)$ where $F : \mathscr{C} \to \mathbb{R}^n$. The reader is referred to the text in [9] for more background on DDEs. We say that the system $\dot{X}(t) = F(X_t)$ is *G-equivariant* if there is a representation ρ of G such that for $(g, \phi) \in G \times \mathscr{C}$, we have $F(\rho(g)\phi) = \rho(g)F(\phi)$ where $\rho(g)\phi \in \mathscr{C}$ is given by $(\rho(g)\phi)(\theta) = \rho(g)\phi(\theta)$ for $\theta \in [-\tau, 0]$. This equivariance condition means that if $X(t)$ is a solution of the system, then so does $\rho(g)X(t)$. We also call G as a *symmetry group* of the system.

We now show that system (1) is $\mathbb{Z}_2$-equivariant. Let $\mathbb{Z}_2 = \langle\gamma\rangle$ and define the action of $\mathbb{Z}_2$ to the state variables as $\gamma \cdot [x(t), y(t)]' = [y(t), x(t)]'$. Writing the right-hand side of system (1) in the notation $F([x_t, y_t]')$, we see that $F(\gamma \cdot [x_t, y_t]') = F([y_t, x_t]') = \gamma \cdot F([x_t, y_t]')$. Therefore, system (1) has symmetry group $\mathbb{Z}_2$.

If we seek equilibrium solutions of system (1) that are fixed by $\mathbb{Z}_2$, that is $(x(t), y(t))$ with $\dot{x}(t) = 0$ and $\dot{y}(t) = 0$ and satisfying $\gamma \cdot [x(t), y(t)]' = [x(t), y(t)]'$, then we obtain $(x^*, y^*) := (a/2b, a/2b)$. For the rest of this paper, we call the equilibrium solution (x^*, y^*) as the *symmetric equilibrium* of system (1).

3 Local Stability Analysis of the Symmetric Equilibrium

The linearized system corresponding to system (1) around the symmetric equilibrium has characteristic equation $\det(\Delta(\lambda)) = 0$ with

$$\Delta(\lambda) = \begin{bmatrix} \lambda + b + (a/4)e^{-\lambda\tau} & -(a/4)e^{-\lambda\tau} \\ -(a/4)e^{-\lambda\tau} & \lambda + b + (a/4)e^{-\lambda\tau} \end{bmatrix}. \tag{2}$$

If we let $A := \lambda + b + \frac{1}{4}ae^{-\lambda\tau}$ and $B := -\frac{1}{4}ae^{-\lambda\tau}$ in Eq. (2), then the characteristic matrix $\Delta(\lambda)$ takes the form $L := \begin{bmatrix} A & B \\ B & A \end{bmatrix}$.

We now introduce a technique from [5, 8] which uses the symmetry structure of the system in order to classify steady-state and Hopf bifurcations into regular or symmetry-breaking. The action of the symmetry group $\mathbb{Z}_2$ on the physical space $\mathbb{R}^2$, yields the decomposition $\mathbb{R}^2 = \mathbb{T} \oplus \mathbb{A}$ where the isotypic components $\mathbb{T} := \{[v, v]',\ v \in \mathbb{R}\}$ and $\mathbb{A} := \{[-v, v]',\ v \in \mathbb{R}\}$ are orthogonal complement of each other, irreducible, and invariant under the symmetry group $\mathbb{Z}_2$. Moreover, the action of $\mathbb{Z}_2$ on the subspace $\mathbb{T}$ is by the trivial representation while the action of $\mathbb{Z}_2$ on the subspace $\mathbb{A}$ is by the alternating representation. Furthermore, observe that the action of L on the elements $v_0 \in \mathbb{T}$ and $v_1 \in \mathbb{A}$ are as follows: $Lv_0 = (A + B)[v, v]'$

and $Lv_1 = (A - B)[-v, v]'$. These imply that the characteristic roots of $L|_{\mathbb{T}}$ are those of $(A + B)$, while the characteristic roots of $L|_{\mathbb{A}}$ are those of $(A - B)$. Since $(A + B)$ corresponds to the action of L on the subspace $\mathbb{T}$ and the symmetry group $\mathbb{Z}_2$ acts trivially on $\mathbb{T}$, the critical characteristic roots from $(A + B)$ give rise to regular bifurcations. Meanwhile, because $(A - B)$ corresponds to the action of L on the subspace $\mathbb{A}$ and the symmetry group $\mathbb{Z}_2$ acts non-trivially on $\mathbb{A}$, the critical characteristic roots from $(A - B)$ give rise to symmetry-breaking bifurcations.

The isotypic decomposition also allows us to write $\det(\Delta(\lambda)) = (A+B)(A-B)$. Hence, the roots of the characteristic equation $\det(\Delta(\lambda)) = 0$ are the roots of the equations $(A + B) = 0$ and $(A - B) = 0$, that is, that of $\lambda + b = 0$ and

$$\lambda + b + (a/2)e^{-\lambda\tau} = 0. \tag{3}$$

At $\tau = 0$, the symmetric equilibrium is locally asymptotically stable (LAS) since both roots $\lambda = -b$ and $\lambda = -b - (a/2)$ of the characteristic equation are negative. That is, all roots of $\det(\Delta(\lambda)) = 0$ are in the open left-half plane when $\tau = 0$. We wanted to know if the symmetric equilibrium may switch stability for some $\tau > 0$, that is, if the roots of $\det(\Delta(\lambda)) = 0$ will cross the imaginary axis as τ is increased from zero. Observe that since $a, b > 0$, both equations $(A+B) = 0$ and $(A-B) = 0$ cannot have a zero root. Moreover, since $b > 0$, the equation $(A + B) = 0$ cannot have purely imaginary roots. We have the following results.

Theorem 1 *Steady-state bifurcations, both regular and symmetry-breaking, and regular or symmetry-preserving Hopf bifurcations will not occur in system (1).*

The non-occurrence of a symmetry-breaking steady-state bifurcation rules out asymmetric equilibrium solutions in system (1), while the non-occurrence of a regular Hopf bifurcation rules out synchronous periodic solutions in system (1).

Suppose now that the equation $(A - B) = 0$, given in Eq. (3), has a purely imaginary root $\lambda = i\omega$ with $\omega > 0$. Then, $i\omega + b + (a/2)e^{-i\omega\tau} = 0$. This gives $a\cos\omega\tau = -2b$ and $a\sin\omega\tau = 2\omega$, and hence $4\omega^2 = a^2 - 4b^2$. If $(a^2 - 4b^2) < 0$, then equation (3) cannot have purely imaginary roots. If $(a^2 - 4b^2) > 0$, then we obtain a positive value for ω given by $\omega_* := \sqrt{a^2 - 4b^2}/2$, and thus $\lambda = i\omega_*$ is a root of Eq. (3). Corresponding to the roots $\lambda = \pm i\omega_*$ of Eq. (3) is the sequence

$$\tau_n := \frac{1}{\omega_*}\left\{\cos^{-1}\left(-\frac{2b}{a}\right) + 2\pi n\right\} = \frac{\cos^{-1}\left(-\frac{2b}{a}\right) + 2\pi n}{\sqrt{a^2 - 4b^2}/2} \quad (n = 0, 1, 2, \ldots).$$

In view of the Hopf bifurcation theorem, we now show that the roots $\lambda = \pm i\omega_*$ of Eq. (3) that lie in the imaginary axis when $\tau = \tau_n$ move towards the right half-plane. That is, we need to show that $\frac{d}{d\tau}\,\mathrm{Re}(\lambda(\tau))|_{\tau=\tau_n} > 0$. Note that $\mathrm{sign}\left\{\frac{d}{d\tau}\mathrm{Re}(\lambda(\tau))\right\}_{\tau=\tau_n} = \mathrm{sign}\left\{\mathrm{Re}(d\lambda/d\tau)^{-1}\right\}_{\lambda=i\omega_*}$. So we first need to compute for $(d\lambda/d\tau)^{-1}$. Differentiating with respect to τ in Eq. (3) yields

$$\left(\frac{d\lambda}{d\tau}\right)^{-1} = \frac{1-(a/2)\tau e^{-\lambda\tau}}{(a/2)\lambda e^{-\lambda\tau}} = -\frac{1}{\lambda(\lambda+b)} - \frac{\tau}{\lambda}$$

since $(a/2)e^{-\lambda\tau} = -(\lambda+b)$ from Eq. (3). Consequently, we have

$$\text{sign}\left\{\frac{d}{d\tau}\text{Re}(\lambda(\tau))\right\}_{\tau=\tau_n} = \text{sign}\left\{\text{Re}\left(\frac{1}{-\lambda^2-b\lambda}\right)\right\}_{\lambda=i\omega_*} = \text{sign}\left\{\frac{1}{\omega_*^2+b^2}\right\}.$$

Therefore, $\frac{d}{d\tau}\,\text{Re}(\lambda(\tau))|_{\tau=\tau_n} > 0$. Taking $\tau^* := \min\{\tau_n \mid \tau_n > 0\}$, we have the following local stability results for the symmetric equilibrium.

Theorem 2 *If $(a^2-4b^2) < 0$, then the symmetric equilibrium (x^*, y^*) of system (1) is LAS for all $\tau > 0$. If $(a^2-4b^2) > 0$, then the symmetric equilibrium of system (1) is LAS for all $\tau \in (0, \tau^*)$ and is unstable for $\tau > \tau^*$. At $\tau = \tau^*$, system (1) undergoes a symmetry-breaking Hopf bifurcation at the symmetric equilibrium.*

We now illustrate our result in Theorem 2 using DDE-Biftool, which is a numerical continuation and bifurcation analysis tool for systems of DDEs [7].

Example 1 Consider system (1) with $a = 10$ and $b = 1$, so that $(x^*, y^*) = (5, 5)$ and $\tau^* = 0.361739$ approximately. We use the history functions $\varphi_1(t) = 4.50$ and $\varphi_2(t) = 4.00$. The left panel of Fig. 1 shows a branch of symmetric equilibria (horizontal line), that is LAS for $\tau < \tau^*$ (green) and is unstable for $\tau > \tau^*$ (magenta). The stability switch occurred at the Hopf bifurcation marked with asterisk where $\tau = \tau^*$. The branch of periodic solutions (green curve) that emerged from the Hopf bifurcation is stable. A profile plot of a periodic solution when $\tau = 0.4160$ is shown in the right panel of Fig. 1 where the two queues are periodic and asynchronous.

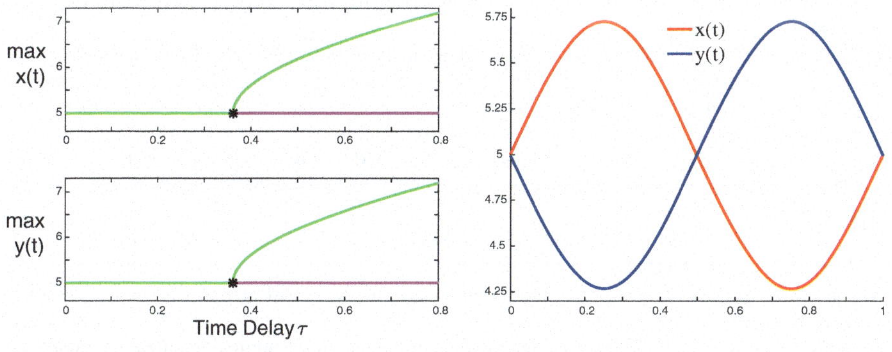

Fig. 1 (Left) Stability switch occurred at a Hopf bifurcation (HB) where a stable branch of periodic solutions emerges. (Right) Profile plot of a periodic solution from the branch that emerged from the symmetry-breaking HB showing that the two queues are periodic and asynchronous

4 Summary and Future Directions

We studied a deterministic queueing model with delayed information where in customers are given the opportunity to choose between two queues. By utilizing the symmetry structure of the system, we are able to classify the codimension-one bifurcations into regular and symmetry-breaking. This classification, in turn, allowed us to determine the types and kinds of solutions that can arise in our system and rule out those that will not occur. This additional insights in the dynamical behavior of queues based on the delayed information provided will help both companies and customers in making better decisions.

Our method can also be applied to study similar models with larger number of queues. Characteristic roots of higher multiplicity are common in symmetric systems and as a consequence, numerical bifurcation analysis of these systems is not as straightforward. However, our group-theoretic approach can be employed to identify codimension-one bifurcations correctly. This is the subject of an on-going research.

Acknowledgements This work was funded by the UP System Enhanced Creative Work and Research Grant (ECWRG 2018-1-001). The author also acknowledges the support of the UP Baguio through RLCs during the A.Y. 2018-2019.

References

1. Buono, P.-L., Collera, J.A.: Symmetry-breaking bifurcations in rings of delay-coupled semiconductor lasers. SIAM J. Appl. Dyn. Syst. **14**, 1868–1898 (2015)
2. Buono, P.-L., Eftimie, R.: Codimension-two bifurcations in animal aggregation models with symmetry. SIAM J. Appl. Dyn. Syst. **13**, 1542–1582 (2014)
3. Buono, P.-L., Palacios, A.: A mathematical model of motorneuron dynamics in the heartbeat of the leech. Phys. D **188**, 292–313 (2004)
4. Buono, P.-L., Chan, B.S., Palacios, A., In, V.: Dynamics and bifurcations in a Dn-symmetric Hamiltonian network. Application to coupled gyroscopes. Phys. D **290**, 8–23 (2015)
5. Collera, J.A.: Symmetry-breaking bifurcations in two mutually delay-coupled lasers. Phil. Sci. Tech. **8**, 17–21 (2015)
6. Collera, J.A.: Symmetry-breaking bifurcations in laser systems with all-to-all coupling. In: Bélair, J., Frigaard, I., Kunze, H., Makarov, R., Melnik, R., Spiteri, R. (eds.) Mathematical and Computational Approaches in Advancing Modern Science and Engineering, pp. 81–88. Springer, Cham (2016)
7. Engelborghs, K., Luzyanina, T., Samaey, G.: DDE-BIFTOOL v. 2.00: a Matlab package for bifurcation analysis of delay differential equations. Department of Computer Science, K. U. Leuven, Leuven (2001)
8. Golubitsky, M., Stewart, I., Schaeffer, D.G.: Singularities and Groups in Bifurcation Theory II. Springer, New York (1988)
9. Hale, J.K., Verduyn Lunel, S.M.: Introduction to Functional Differential Equations. Springer, New York (1993)
10. Pender, J., Rand, R.H., Wesson, E.: Delay-differential equations applied to queueing theory. In: Stépán, G., Csernák, G. (eds.) Proceedings of 9th European Nonlinear Dynamics Conference, ID 62. CongressLIne Ltd., Budapest (2017)
11. Pender, J., Rand, R.H., Wesson, E.: Queues with choice via delay differential equations. Int. J. Bifurcat. Chaos **27**, 1730016 (2017)

Finite Sample Confidence Region for EIV Systems Using Regression Model

Masoud Moravej Khorasani and Erik Weyer

Abstract Errors-In-Variables (EIV) models in which both input and output data are contaminated by noise have applications in signal processing. We propose a method for constructing non-asymptotic confidence regions for the parameters of EIV models. The method is based on the Leave-out Sign-dominant Correlation Regions (LSCR) principle which gives probabilistically guaranteed confidence region when the input is measured without noise. A regression model is utilized to extend LSCR to EIV systems. The newly established regression vector contains the past outputs and the estimated past inputs. It is shown that the corresponding prediction error has the desired properties such that it can be used to form correlation functions from which confidence regions can be constructed. For any finite number of data points it is proved that the region contains the true parameter with a user-chosen probability.

1 Introduction

The data generating system is

$$y(t) = \frac{B^0(q)}{A^0(q)} f(t) + e(t)$$
$$u(t) = f(t) + v(t) \tag{1}$$

where q^{-1} is the backward-shift operator and $f(t)$, $v(t)$, and $e(t)$ are the input, the input noise, and the output noise respectively. $B^0(q)$ and $A^0(q)$ are polynomials in the backward shift operator:

$$B^0(q) = b_1^0 q^{-1} + b_2^0 q^{-2} + \cdots + b_{n_b}^0 q^{-n_b}$$

M. Moravej Khorasani (✉) · E. Weyer
Department of Electrical and Electronic Engineering, The University of Melbourne, Parkville, VIC, Australia
e-mail: masoud.moravej@unimelb.edu.au; masoudm@unimelb.edu.au; ewey@unimelb.edu.au

© Springer Nature Switzerland AG 2019

I. Faragó et al. (eds.), *Progress in Industrial Mathematics at ECMI 2018*,
Mathematics in Industry 30, https://doi.org/10.1007/978-3-030-27550-1_69

$$A^0(q) = 1 + a_1^0 q^{-1} + a_2^0 q^{-2} + \cdots + a_{n_a}^0 q^{-n_a}.$$

Standard methods such as least squares are not able to identify EIV models without bias [6]. Several approaches to identification of EIV models under different assumptions and conditions have been proposed. An overview of methods for identification of EIV models can be found in [7] and [8].

A model without a statement about its quality is limited use and constructing confidence regions for the system parameters is the most common approach to assess the model quality. The corresponding regions are typically derived using asymptotic theory [4], but they may be unreliable with a finite number of data samples [3]. The Leave-out Sign-dominant Correlation Regions (LSCR) method [1, 2] on the other hand constructs confidence regions which are guaranteed to contain the true system parameters with a user-chosen probability for a finite number of data points. In this paper, we extend LSCR to EIV models.

This problem has previously been addressed in [5] under the condition that the output noise is Gaussian i.i.d. and all variances are known. The method presented in this paper find non-asymptotic confidence region when only a ratio between variances is known. Moreover, except having a Gaussian distribution there are no restrictive assumptions on the output noise e.g. it can be non-zero mean and correlated in time.

The paper is organized as follows. In Sect. 2, the data generating system and models are introduced. Section 3 develops LSCR for EIV system. The theoretical properties of the confidence regions are presented in Sect. 4. A simulation example is given in Sect. 5, followed by conclusions.

2 Preliminaries

We consider the data generating system (1). Let $\theta^{0\top} = [\theta_a^{0\top} \;\; \theta_b^{0\top}]$ be the vector of true system parameters, i.e. $\theta_a^0 = [a_1^0 \; a_2^0 \; \cdots \; a_{n_a}^0]^\top$ and $\theta_b^0 = [b_1^0 \; b_2^0 \; \cdots \; b_{n_b}^0]^\top$.

Assumption 1 *$\{f(t)\}$ and $\{v(t)\}$ are zero mean sequences of mutually independent and identically distributed (i.i.d.) Gaussian random variables with variances λ_f^2 and λ_v^2 respectively. $\{e(t)\}$ is a sequence of Gaussian random variables independent of $\{f(t)\}$ and $\{v(t)\}$.*

Assumption 2 *The degrees of the polynomials $B^0(q)$ and $A^0(q)$ and the ratio $\Gamma = \lambda_v^2/\lambda_f^2$ quantifying the accuracy of the observed input are known.*

Assumption 3 *The data generating system is strictly stable, and there are no pole/zero cancellations.*

Assumption 4 *The model parameters θ belong to closed set $\mathscr{D}$. For all values of $\theta \in \mathscr{D}$ the poles of $A(q)$ is strictly inside the unit circle.*

Existence of an independent sequence corresponding to the true parameter is key to LSCR algorithm. Knowing Γ (Assumption 2) allows us to compute output predictions such that the corresponding prediction errors are independent of the measured input. The predictions are given by

$$\hat{y}(t) = \phi(t)^\top \theta \tag{2}$$

where

$$\phi(t)^\top = \left[y(t-1) \quad \cdots \quad y(t-n_a) \quad \frac{1}{1+\Gamma} u(t-1) \quad \cdots \quad \frac{1}{1+\Gamma} u(t-n_b) \right] \tag{3}$$

and $\theta^\top = [a_1 \; a_2 \; \cdots \; a_{n_a} \; b_1 \; b_2 \; \cdots \; b_{n_b}]$ is the vector of model parameters. The dimension of parameter vector is $n_\theta = n_a + n_b$.

3 Construction of Confidence Region

The regression vector (3) contains delayed output and Least Minimum Mean Square Estimates (LMMSE) of true inputs given noisy inputs. This regression vector is used to form correlation functions in LSCR. The LSCR algorithm for EIV systems is as follows

For a given θ

1. Compute the prediction error: $\varepsilon(t, \theta) = y(t) - \phi^\top(t)\theta, \; t = 1, \cdots, N$
2. Form the vector $\xi(t) = [u(t-1) \; u(t-2) \; \cdots \; u(t-n_\theta)]^\top$ and compute

$$g_i(\theta) = \sum_{t=1}^{N} h_{i,t}\xi(t)\varepsilon(t, \theta), \; i = 0, 1, \cdots, M-1$$

where $h_{0,t} = 0$ for $t = 1, \cdots, N$ and the remaining $h_{i,t}$s are i.i.d. and with the following distribution:

$$h_{i,t} = \begin{cases} 0 & w.p. \; \frac{1}{2} \\ 1 & w.p. \; \frac{1}{2} \end{cases}$$

$g_i(\theta)$ is a n_θ-dimensional vector and its kth element is denoted by $g_i^k(\theta)$.

(continued)

3. Select an integer q in the interval $[1, (M+1)/2)$ and find the region of $\Theta_N^{(k)}$ such that at least q of the $g_i^k(\theta)$ functions are bigger than zero and at least q are smaller than zero. The confidence region is given by: $\Theta_N = \bigcap_{k=1}^{n_\theta} \Theta_N^{(k)}$

It can be show that $\{u(t-r)\}_{r=1}^{n_\theta}$ and $\varepsilon(t, \theta^0)$ are independent. As $g_i^k(\theta^0)$ is a sum of zero mean random variables with a symmetric distribution around zero, it is unlikely that the sequence $\{g_i^k(\theta^0)\}_{i=1}^{M-1}$ takes on either negative or positive values nearly all the times. Hence, the LSCR method excludes those values of θ for which there exist $k \in \{1, \cdots, n_a + n_b\}$ such that $g_i^k(\theta)$ no either negative or positive for most $i \in \{1, \cdots, M-1\}$.

4 Theoretical Results

In this section we prove two properties of the constructed confidence region. First, it is shown that the confidence region is probabilistically guaranteed. Then, we prove that any parameter different from the true system parameter θ^0 will eventually be excluded from the confidence region as the number of data points goes to infinity.

4.1 The Probability of the Confidence Region

The following theorems give the probability that the confidence region contains the true parameter. Note that θ^0 is deterministic while $\Theta_N^{(k)}$ and Θ_N are stochastic. The theorems hold true for finite number of data points.

Theorem 1 *Consider the system (1) and model (2), under Assumptions 1–2. The probability that θ^0 is in the confidence region $\Theta_N^{(k)}$ is*

$$\Pr\{\theta^0 \in \Theta_N^{(k)}\} = 1 - \frac{2q}{M}. \tag{4}$$

Sketch of Proof It can be shown that $\{u(t-r)\}_{r=1}^{n_\theta}$ and $\varepsilon(t, \theta^0)$ are uncorrelated and hence also independent since they are Gaussian. Let $\eta_t := u(t-r)\varepsilon(t, \theta^0)$. It has the same properties as the variable with the same name in the Proof of Theorem 1 in [2], and the proof follows along the same line as the proof of Theorem 1 in [2]. □

It is noteworthy that the parameters q and M are user-chosen.

Theorem 2 *Under the same assumptions as in Theorem 1*

$$\Pr\{\theta^0 \in \Theta_N\} \geq 1 - \frac{2q}{M} n_\theta. \tag{5}$$

Hence, to construct a confidence region with the minimum probability $p \in (0, 1)$, M and q should be chosen such that $\frac{q}{M} = \frac{1-p}{2n_\theta}$.

4.2 Asymptotic Convergence

Asymptotically, we have the following results.

Theorem 3 *Under Assumptions 1–4, for every fixed* $\theta \neq \theta^0$,

$$\Pr\{\exists \bar{N} | \theta \notin \Theta_N, \forall N > \bar{N}\} = 1 \tag{6}$$

Sketch of Proof It can be shown that $\frac{1}{N} g_i^k(\theta) \rightarrow \frac{1}{2}\mathbb{E}\{u(t-k)\varepsilon(t,\theta)\}$ for $k = 1, \cdots, n_\theta$ and $i = 1, \cdots, M-1$. The only parameter which satisfies the set of equations

$$\mathbb{E}\{\xi(t)\varepsilon(t,\theta)\} = 0$$

is the true parameter. From this it follows that any value different from the true value of the parameter will be excluded from the confidence region as the number of data points tends to infinity. □

5 Simulation

Here, the method is illustrated in a simulation example. The system is given by

$$\begin{aligned} y(t) &= \frac{b_1^0}{1 - a_1^0 q^{-1}} f(t) + e(t) \\ u(t) &= f(t) + v(t) \end{aligned} \tag{7}$$

where $b_1^0 = 1.0$, $a_1^0 = 0.5$, $\lambda_f^2 = 1$, $\lambda_v^2 = 0.2$, and $\lambda_e^2 = 0.3$.

The ratio Γ is 0.2 and the regression vector is

$$\phi(t) = [-y(t-1) \quad \frac{1}{1.2} u(t-1)].$$

To construct a 90% confidence region for θ^0, the parameters q and M in the LSCR algorithm are chosen as $q = 5,\ M = 200$.

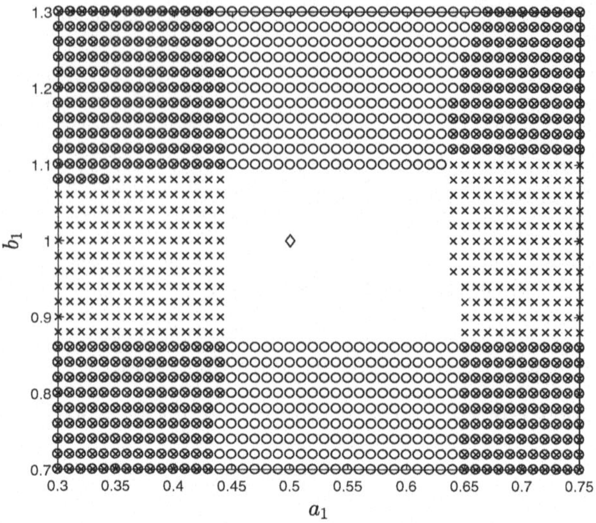

Fig. 1 90% confidence region obtained with 250 data points. $\times$ represents the regions where fewer than 5 of the $g_i^1(\theta)$ functions are smaller or larger than the zero function, and $\circ$ represents the same regions for $g_i^2(\theta)$. $\diamond$ is the true parameter $\theta^0 = [0.5 \;\; 1]^\top$

We compute $g_i^1(\theta) = \sum_{t=1}^{N} h_{i,t} u(t-1)\varepsilon(t,\theta)$ and $g_i^2(\theta) = \sum_{t=1}^{N} h_{i,t} u(t-2)\varepsilon(t,\theta)$ for $i = 0, 1, \cdots, M-1$. By excluding those values of θ for which fewer than $q = 5$ of the functions were smaller or bigger than the zero function, the confidence region shown in Fig. 1 was obtained. As proved in Theorem 2, the confidence region contains the true parameter with probability at least $1 - 2 \cdot \frac{q}{M} \cdot 2 = 0.9$.

6 Conclusion

An algorithm for construction of non-asymptotic confidence regions for EIV system has been developed. It assumes that the ratio between the variance of the noise on the input and the variance of the input is known and uses it to estimate the true input given the noisy input. Then a regression model is employed to compute prediction errors with desirable properties. The output noise is only assumed to be Gaussian and independent of the input and the noise on input. The confidence region contains the true parameters with a guaranteed user-chosen probability, and moreover, parameters different from true parameters will eventually be excluded from the confidence region.

References

1. Campi, M., Weyer, E.: Guaranteed non-asymptotic confidence regions in system identification. Automatica **41**, 1751–1764 (2005)
2. Campi, M., Weyer, E.: Non-asymptotic confidence sets for the parameters of linear transfer functions. IEEE Trans. Autom. Control. **55**, 2708–2720 (2010)

3. Garatti, S., Campi, M., Bittanti, S.: Assessing the quality of identified models through the asymptotic theory-when is the result reliable? Automatica **40**, 1319–1332 (2004)
4. Ljung, L.: System Identification-Theory for the User. Prentice Hall, Upper Saddle River (1999)
5. Moravej Khorasani, M., Weyer, E.: Non-asymptotic confidence regions for error-in-variables system. In: 18th IFAC Symposium on System Identification, pp. 2115–2120. IFAC, Stockholm (2018)
6. Söderström, T.: Identification of stochastic linear systems in presence of input noise. Automatica **17**, 713–725 (1981)
7. Söderström, T.: Errors-in-variables methods in system identification. Automatica **43**, 939–958 (2007)
8. Söderström, T.: Errors-in-variables methods in system identification. Springer, Cham (2018)

Moment Matching Based Model Order Reduction for Quadratic-Bilinear Systems

Nadine Stahl, Björn Liljegren-Sailer, and Nicole Marheineke

Abstract For model order reduction of quadratic-bilinear systems a moment matching approach has been recently proposed where univariate frequency responses are constructed by means of the associated transform onto the multivariate transfer functions. This approach comes with the obvious advantage of only one-dimensional interpolation frequencies to be considered, but suffers from the arising large size of the involved equation systems and the high computational demands that make the approach impractical for most applications. In this paper, by exploiting the problem-underlying sparse tensor structure, we propose a splitting algorithm that overcomes this curse of dimensionality. We demonstrate the performance of the extended univariate frequency approach and compare it with the well-established multimoment matching approach regarding accuracy, efficiency and need of memory.

1 Introduction

This paper deals with moment matching based model order reduction for single-input single-output quadratic-bilinear systems of the form

$$\dot{\mathbf{x}} = \mathbf{A}\,\mathbf{x} + \mathbf{H}\,\mathbf{x} \otimes \mathbf{x} + \mathbf{D}\,\mathbf{x}\,u + \mathbf{B}\,u, \qquad y = \mathbf{C}\,\mathbf{x} \tag{1}$$

with Kronecker product $\otimes$, constant coefficient matrices $\mathbf{A}, \mathbf{D} \in \mathbb{R}^{n,n}$, $\mathbf{H} \in \mathbb{R}^{n,n^2}$ and $\mathbf{B}, \mathbf{C}^T \in \mathbb{R}^n$, state vector $\mathbf{x}(t) \in \mathbb{R}^n$ as well as input $u(t) \in \mathbb{R}$ and output $y(t) \in \mathbb{R}$ at time t. To derive a reduced order model for $\mathbf{x}_r(t) \in \mathbb{R}^r$, $r \ll n$,

$$\dot{\mathbf{x}}_r = \mathbf{A}_r\,\mathbf{x}_r + \mathbf{H}_r\,\mathbf{x}_r \otimes \mathbf{x}_r + \mathbf{D}_r\,\mathbf{x}_r\,u + \mathbf{B}_r\,u, \qquad y_r = \mathbf{C}_r\,\mathbf{x}_r,$$

$$\mathbf{A}_r = \mathbf{V}^T\mathbf{A}\mathbf{V}, \quad \mathbf{H}_r = \mathbf{V}^T\mathbf{H}(\mathbf{V}\otimes\mathbf{V}), \quad \mathbf{D}_r = \mathbf{V}^T\mathbf{D}\mathbf{V}, \quad \mathbf{B}_r = \mathbf{V}^T\mathbf{B}, \quad \mathbf{C}_r = \mathbf{C}\mathbf{V}$$

N. Stahl (✉) · B. Liljegren-Sailer · N. Marheineke
Universität Trier, Trier, Germany
e-mail: nadine.stahl@uni-trier.de; bjoern.sailer@uni-trier.de; marheineke@uni-trier.de

© Springer Nature Switzerland AG 2019
I. Faragó et al. (eds.), *Progress in Industrial Mathematics at ECMI 2018*,
Mathematics in Industry 30, https://doi.org/10.1007/978-3-030-27550-1_70

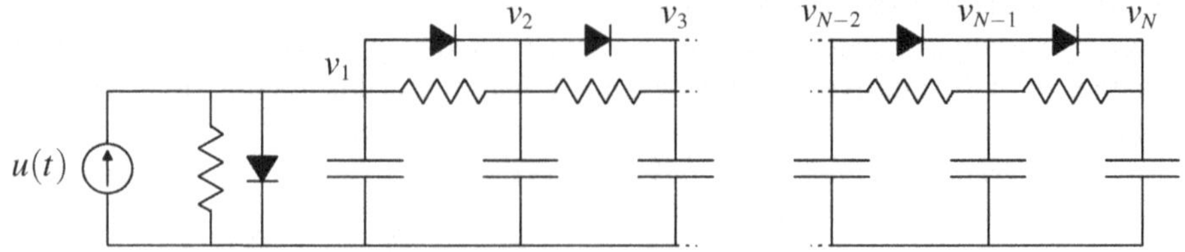

Fig. 1 Nonlinear RC-Ladder

capacitance, we end up with the following system ($2 \leq i \leq N-1$):

$$\dot{v}_1(t) = -2v_1(t) + v_2(t) - g(v_1(t)) - g(v_1(t) - v_2(t)) + u(t),$$
$$\dot{v}_i(t) = -2v_i(t) + v_{i-1}(t) + v_{i+1}(t) + g(v_{i-1}(t) - v_i(t)) - g(v_i(t) - v_{i+1}(t)),$$
$$\dot{v}_N(t) = -v_N(t) + v_{N-1}(t)g(v_{N-1}(t) - v_N(t)).$$

This nonlinear model can be transformed into an equivalent quadratic-bilinear system of size $n = 2N$ with first reformulating the upper system with the variables $x_1 = v_1, x_i = v_{i-1} - v_i$ for $i = 1, \ldots, N$ and then introducing the new variables $z_i = e^{40x_i} - 1$, cf. [4]. We obtain the following quadratic-bilinear system:

$$\begin{aligned}
\dot{x}_1 &= -x_1 - x_2 - z_1 - z_2 + u(t), \\
\dot{x}_2 &= -x_1 - 2x_2 + x_3 - z_1 - 2z_2 + z_3 + u(t), \\
\dot{x}_i &= x_{i-1} - 2x_i + x_{i+1} + z_{i-1} - 2z_i + z_{i+1}, && 3 \leq i \leq N-1, \\
\dot{x}_N &= x_{N-1} - 2x_N + z_{N-1} - 2z_N, \\
\dot{z}_i &= 40(z_i + 1)\dot{x}_i, && 1 \leq i \leq N.
\end{aligned}$$

As a numerical setup we use the input function $u(t) = 0.5(1 + \cos(0.2\pi t))$. The initial conditions are set to zero, i.e. $v_i(0) = 0$ and thus $z_i(0) = 0$, $i = 1, \ldots, N$.

Implementation was done in Matlab R2017b run on a Intel Xeon with 2.2 GHz on 88 Cores. The time integration of the ODE is performed via the implicit Euler method, and for solving the Lyapunov equation in the splitting algorithm we use the Matlab internal routine for full matrices and a sparse solver of [7], respectively.

Case 1 Choosing $N = 1000$, we compare the reduced models of size $r = 12$ that we obtain from the associated transform based moment matching without (AMOR) and with splitting (SAMOR) as well as from multimoment matching (MMOR). The relative error in the output over time with respect to the full order model is visualized in Fig. 2. As desired the associated versions yield the same results (overlapping of red dashed-dotted and yellow dotted lines) such that no additional errors are introduced by the numerical solution strategy. The approximation quality by (S)AMOR is slightly better than by MMOR. Concerning the computation times

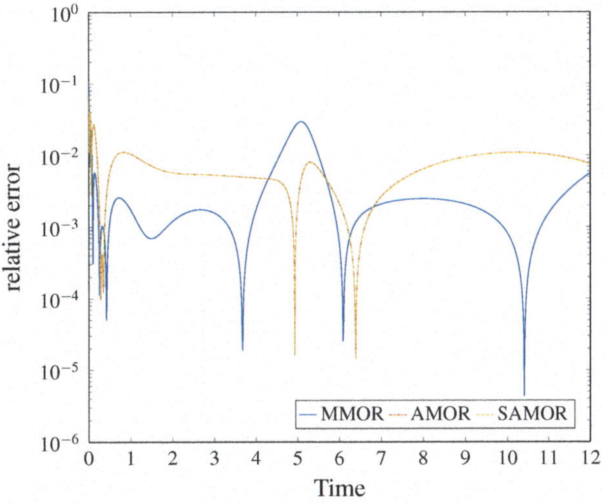

Fig. 2 Case 1: $N = 1000$, $r = 12$, $\sigma = 1$, MMOR: matching 3, 3, 3 moments for $\mathscr{H}_1$ and the bilinear and quadratic part of $\mathscr{H}_2$, (S)AMOR: matching 7 moments of $\mathring{\mathscr{H}}_1$ and 5 of $\mathring{\mathscr{H}}_2$

Table 1 Computational times, Case 1

Method	Offline phase [s]	Online phase [s]
Full nonlinear model	–	88
Full quadratic-bilinear model	–	1350
MMOR	0.3	3
AMOR	570	3
SAMOR	50	3

(cf. Table 1) the online phase is similar for all methods. But in the offline phase (i.e., computation of the projection matrix) MMOR is way faster than the associated versions. However, the proposed splitting algorithm yields a speed-up of a factor 10 in the univariate frequency approach (SAMOR vs. AMOR).

Case 2 To explore the performance of our proposed splitting algorithm in the univariate frequency approach, we use in addition to AMOR and SAMOR also a sparse Lyapunov solver (SSAMOR) for the Lyapunov equations arising in the splitting. Note that, in general, orthonormalization is needed while calculating the Krylov subspaces to keep the algorithm stable. This cannot be done for the sparse Lyapunov solver without introducing further approximation errors. Therefore, we restrict ourselves to cases when only a few moments are matched. The quality of the reduced model is then ensured by increasing the number of interpolation points. Here, using $\sigma \in \{1, 10^2, 10^4\}$, and matching 4 and 3 moments of $\mathring{\mathscr{H}}_1$ and $\mathring{\mathscr{H}}_2$, respectively, yields reduced models of size $r = 18$. While all algorithms provide the same approximation quality of the solution as desired (Fig. 3), the computational offline times for the different solution strategies strongly differ (cf. Table 2). SSAMOR gives an additional speedup of around 6. The improvement of the performance is even more pronounced when considering larger problems. Increasing the problem size from $N = 10^3$ to $N = 10^4$, we obtain a speedup of

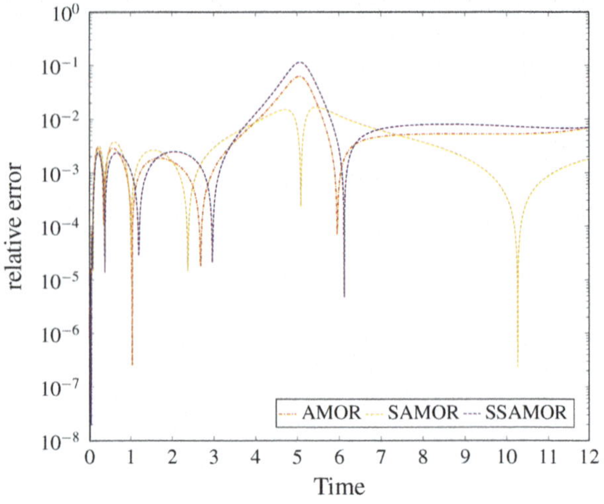

Fig. 3 Case 2: $N = 1000$, $r = 18$, $\sigma \in \{1, 10^2, 10^4\}$, (SS)AMOR: matching 4 moments of $\mathring{\mathscr{H}}_1$ and 3 of $\mathring{\mathscr{H}}_2$

Table 2 Computational times, Case 2, similar online phase for all strategies

Method	Offline phase [s] $N = 10^3$	Offline phase [s] $N = 10^4$
AMOR	1210	–
SAMOR	91	64,684
SSAMOR	15	226

more than 200, while the naive calculation in AMOR runs out of memory due to the large system size.

4 Discussion and Conclusion

For the associated transform based moment matching approach our proposed splitting algorithm yields a drastic reduction of the computational time by exploiting the inherent tensor structure, further improvements of the performance can be attained by using a sparse solver for the Lyapunov equations involved in the splitting. However, although in this approach only one-dimensional interpolation frequencies have to be considered, the computational effort to compute the projection matrix for the reduced model is much higher than for multimoment matching with its multivariate transfer functions due to the arising large systems. The approximation quality of the univariate frequency approach and the multimoment matching is comparably good.

As mentioned, the associated frequency representations here only characterize the response on the Dirac impulse. A slight generalization has been discussed in [8], but it turns out that a much more general class of inputs can be considered in a modified framework that we have recently developed in [5]. As deeper analysis reveals that gathering the associated moments in the reduction basis alone is not sufficient for the reduced system to match the respective moments, a second approximation condition naturally appears, for details we refer to [5].

The presented methods are certainly also relevant and applicable for more general classes of nonlinearities, since nonlinear systems can be transformed into quadratic-bilinear systems by introducing further variables. However, since this reformulation is not uniquely determined, the choice and the quality of the quadratization are problem-dependent.

Acknowledgements Supported by the German Federal Ministry for Economic Affairs and Energy, in the joint project: *MathEnergy – Mathematical Key Technologies for Evolving Energy Grids*.

References

1. Breiten, T.: Interpolatory methods for model reduction of large-scale dynamical systems. Ph.D Thesis, Max Planck Institute and Otto von Guericke University, Magdeburg (2013)
2. Brilliant, M.: Theory of the analysis of nonlinear systems. MIT RLE Technical Report 345 (1958)
3. Grimme, E.J.: Krylov projection methods for model reduction. Ph.D Thesis, ECE Department, University of Illinois, Urbana-Champaign (1997)
4. Gu, C.: Model order reduction of nonlinear dynamical systems. Ph.D. Thesis, Electrical Engineering and Computer Sciences, University of California, Berkeley (2011)
5. Liljegren-Sailer, B., Marheineke, N.: Input-tailored system-theoretic model order reduction for quadratic-bilinear systems. Preprint (2018). arXiv:1809.08979
6. Rugh, W.J.: Nonlinear System Theory. The Johns Hopkins University Press, Baltimore (1982)
7. Saak, J., Köhler, M., Benner, P.: M-M.E.S.S.-1.0.1 – the matrix equations sparse solvers library. (2016). https://doi.org/10.5281/zenodo.50575
8. Zhang, Y., Wong, N: Compact model order reduction of weakly nonlinear systems by associated transform. Int. J. Circuit Theory Appl. **44**, 1367–1384 (2016)

Data Science in Industry 4.0

Shirley Y. Coleman

Abstract Data science is piquing the interest of many large and small organisations and managers are asking universities for information and advice. Typically, the query is: I have many sensors and many measurements, what shall I do with all this data, and how can I get ready for Industry 4.0? The so-called fourth industrial revolution refers to automation and control based on data exchange in a digital environment where measurements are available on all aspects of production. Data science plays an intrinsic role in this scenario and is focused on understanding and using data. Data science requires a challenging mix of capability in data analytics and information technology, and business know-how. Statisticians need to work with computer scientists; data analytics includes machine learning and statistical analysis and these extract meaning from data in different ways. Moving towards increased use of data requires buy in from higher management and board members. Although serious progress involves a holistic approach, exemplars demonstrating potential value are also beneficial. This article considers the implications for mathematicians and statisticians of the growing industrial demands and discusses examples from ongoing research projects with industrial partners where data visualisation, multivariate statistical process control charts and funnel plots have made an important contribution.

1 Introduction

Data science is piquing the interest of many large and small organisations and managers are asking mathematicians and statisticians in universities for information and advice. Typically, the query is: I have many sensors and many measurements creating a virtual copy of the physical environment of my processes, what shall I do with all this data, and how can I get ready for Industry 4.0?

S. Y. Coleman (✉)
Industrial Statistics Research Unit School of Mathematics, Statistics and Physics, Newcastle University, Tyne, UK
e-mail: shirley.coleman@ncl.ac.uk

© Springer Nature Switzerland AG 2019

I. Faragó et al. (eds.), *Progress in Industrial Mathematics at ECMI 2018*, Mathematics in Industry 30, https://doi.org/10.1007/978-3-030-27550-1_71

The so-called fourth industrial revolution refers to automation and control based on data exchange in a digital environment where measurements are available on all aspects of production. Data science plays an intrinsic role in this scenario and is focused on understanding and using data.

Data science requires a challenging mix of capability in data analytics and information technology, and business know-how [3]. Mathematicians and statisticians need to work with computer scientists to exploit the growing interest in data science. Moving towards increased use of data requires buy in from higher management and board members so business domain knowledge is of vital importance to ensure that data science addresses business needs and enables evaluation of the benefits [2].

Case studies of sound data analytics carried out in different sectors [1] help us to show companies what data science can offer and what issues have to be considered to ensure value for money. This article gives examples from ongoing research projects with industrial partners where data visualisation, multivariate statistical analysis, process control charts and funnel plots have made an important contribution.

The next section gives background to Industry 4.0, Sect. 3 considers the role of mathematicians and statisticians in this new scenario. Section 4 gives two examples and the final section is the conclusion and offers a forward view.

2 Industry 4.0

Human civilisation has experienced many disruptive changes with far reaching consequences. Arguably the first big industrial revolution was in the mid-eighteenth century with the harnessing of water and steam power which transformed not only the way products were made but also the nature of society in manufacturing countries. This irreversible change was followed nearly a century later by mass production using assembly lines and electricity. Computers and automation revolutionised manufacture again in the late twentieth century giving rise to the third industrial revolution. The big change happening now and over the last decade is the domination of intelligent systems in which machines and control are connected by data flows independently interacting with each other. The interconnectivity facilitated by this disruptive technology is often referred to as the "Smart Factory" and is the basis of Industry 4.0.

Industry 4.0 leads to faster and more flexible production processes with greater efficiency of material supply and usage, and reduction of downtime. The connectivity between different parts of a company results in abundant data collected by sensors embedded in the manufacturing systems. The data creates a virtual copy of the physical world and is used in many diverse ways for operational and strategic purposes including monitoring quality, performance and the interaction between processes, failure detection and predictive modelling. The data can be viewed in reports and dashboards. The smart factory in Industry 4.0 is highly dependent on both technical connectivity and good communication between suppliers and manufacturers.

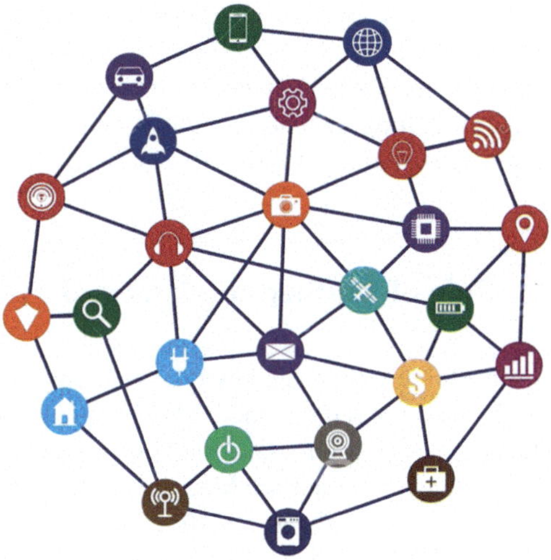

Fig. 1 Multiple data sources relating to diverse parts of a company

In the digital world, company personnel are faced with interconnection, not just in the factory but in all parts of the company. Figure 1 shows how all these data sources relate to each other. To make best use of all this data there needs to be integration and understanding between data providers and data users [5]. Vast data resources offer opportunities in terms of shedding light on operational performance and how it can be improved. There needs to be an interplay between strategic planners, data providers and data analysts to make sure that useful questions are asked and high quality data is available to address them.

Industry 4.0 represents another massive change to the way people live. Each industrial revolution has led to job losses and the necessity for people to adapt to new employment. This one may have the greatest impact so far. The Bank of England Chief Economist was recently reported in the Times newspaper [4] to say that there will be millions of job losses due to Industry 4.0.

> Andy Haldane said that the "fourth industrial revolution" would be on a much greater scale than those that played out in the eighteenth to twentieth centuries and would lead to widespread job losses and societal changes.

On the positive side, Industry 4.0 has led to creative advances in transportation with autonomous cars and automobile systems, medical monitoring, process control, robotics and avionics.

With all this Industry 4.0 data around, the key means of adding value is with data science. The essence of the many Venn diagrams about data science that started with one by Drew Conway in 2010, is that data science consists of:

- IT skills (hardware and software) to deal with data
- Business awareness of how data can help
- Data analytics: maths, statistics, machine learning, artificial intelligence and deep learning

Data analytics means more than the classical statistics around total quality management and in Six Sigma such as statistical process control and designed experiments. For Industry 4.0 we include all the advanced predictive modelling and meta-modelling techniques. These and the role of mathematicians and statisticians in the new Industry 4.0 world are discussed in the next section.

3 Role of Mathematicians and Statisticians

It is useful to review the key relevant components of data analytics, which makes up the third part of data science. Mathematical and statistical methods include summary statistics, correlation, data visualisation, pattern recognition, modelling, simulation and dimension reduction. Machine learning methods include decision trees, random forests, cluster analysis, t-SNE, neural networks and long short-term memory units.

Artificial Intelligence originated in simulating the working of the human mind. It has had varying academic respectability over the years partly because of differences of opinion as to the nature of the mind, the diversity of topics it embraces and the fear of unemployment due to the efforts of human beings becoming obsolete. It is currently in a resurgence of popularity partly because of the possibilities arising from advanced computing methods enabling so called deep learning.

Deep learning includes techniques such as Recurrent Neural Networks with multiple layers functioning like the 3D human brain. It is applied to tasks like speech and picture recognition. Rather than following ideas set by humans, deep learning allows the computer to learn by itself. The ImageNet challenge involves recognising 15 million images and network analysis with over 150 layers succeeded in making fewer than 5% errors in 2016 and beating human performance [8].

It is worth noting that many of the mathematical, statistical and machine learning techniques have been around for many years. What has changed and brought data analytics into its current prominence is the massive amount of data now being retained instead of just being noted and then overwritten. The realisation that profound insight can be gained from analysing this available data has led to a disruption in the order of professions with data scientist becoming a highly paid, glamorous and much sought after role. This sea change is encapsulated in the article in the New Yorker shown in Fig. 2. Previously business lauded the "Madison

Fig. 2 The sea change in professions

Avenue" New York creative advertising executives but now their influence is being overthrown by the mathematical analysts who can use data for behavioural profiling to segment customers and target advertising tailored precisely to individual tastes.

Data scientists are seen as useful and desirable. We now need to monitor the purity of the mathematics and statistics that form part of data science.

Many data analytical practices revolve around black box solutions in which data is fed into a black box, calculations are made and an answer comes out. Black box techniques mask the algorithms being used and just report an answer. There can be good reasons for encasing the methodology in a black box, for example if it is alarmingly complex or so that it cannot be interfered with. However, the downside of using a black box solution is that it alienates statistical thinkers and encourages a cavalier attitude to statistical detail and theoretical niceties that is frustrating and maybe damaging in the long term.

Black box algorithms may not be accessible. For example, in decision tree analysis, if two contender predictor variables are equal in their chi-square or f values with respect to the target variable, which does the algorithm choose? Is it possible to find out?

Black box data analytical solutions for prediction lack robustness against changes in influential variables. The solutions are often based on assemblies of models and predictions are averaged out using a range of methods, not all of which are appropriate for the type of data being input. It is not easy to find out which predictors have had the major influence in the models making up the prediction. One predictor may be much easier to collect than the other and have higher quality.

Black box users are not forced to check their data before analysis and there is often little emphasis on residual analysis so that variables with gross outliers are not detected. Users may miss data errors and opportunities for finding key subsets and obvious explanations for apparent patterns. But the black box user does not mind any of this provided their prediction is better than the one before and seems to work in the short term.

In summary, problems with black boxes include:

- Algorithms, tuning parameters, subtle effects and assemblies are not accessible
- Black box may work adequately for the short but not long term
- Robustness to change is not certain
- Skills to understand underlying methods are not nurtured

Business people, however, tend to side with black box approaches as they are easier to use and understand. Mathematicians and statisticians need to reclaim the field rather than let core black box services take over. They have very important roles to play which include constructing and validating models, checking the quality of input data, evaluating costs of obtaining variables and using proxies, and conducting sensitivity experiments.

Companies have a love-hate relationship with data science based on primordial fear and lust, they fear data but want the benefits of analysing it. Case studies are very important to show managers the sort of outcomes that are possible [6]; we need to encourage operational staff by emphasising the likely positive outcomes: less rework and waste, more time to do their job properly [7].

The Industrial Statistics Research Unit (ISRU) at Newcastle University, UK has been working with companies for many years and has extensive experience of what can be gained from applications of mathematics and statistics in business and industry. ISRU was set up by G.B. Wetherill and in the 1990s focussed on statistical process control and design of experiments in the process industries. The work incorporated Taguchi experiments and Six Sigma quality improvement with several important European Union contracts in the 2000s including pro-ENBIS which helped establish the European Network of Business and Industrial Statistics. A reorientation by ISRU to focus on data mining and the service sector followed, with ongoing work in utilities and healthcare. The most recent consultancy has been in big data analytics particularly in small to medium enterprises (SMEs), and the quantification of the impact of academic mathematics and statistics on society at large.

4 Examples

Current ISRU projects include a knowledge transfer partnership (KTP) funded by Innovate UK in the shipping sector [10]. This case study involves clear definition of the business issue which is to reduce the cost of fuel and the quantity of harmful emissions. Big data from engine sensors is relayed to a control centre and data analytics enables automatic shipping mode detection, evaluation of economic speed and calculation of emissions. Multivariate statistical analysis leads to the construction of control charts for whole journey fuel consumption through which the shipping company can understand the performance of its ships and the variation due to weather and tides.

Another KTP focuses on the insight that can be obtained from analysing vast quantities of data in the automotive aftersales market [9]. In this case study some of the business issues were identified by the customers of the SME business partner and some became apparent after applying data visualisation and exploratory statistical techniques to the vast integrated big data generated within this sector. One particular issue is the return rate of autoparts purchased by customers. Products with high return rates need to be investigated as they add cost to the business.

Millions of autoparts are purchased every day in quantities that vary considerably between different products. Evaluating this big data in tabular form is confusing. A funnel plot gives a convenient and effective tool for helping to prioritise which products to investigate. The funnel plot for return rates of air filters is shown in Fig. 3. There are many products with return rates outside the standard 95% control limits. Some points are due to type I errors and others apply to lower valued products which are not investigated. The circled points refer to more valuable products and are investigated; for example, it may be that their description in the catalogue is

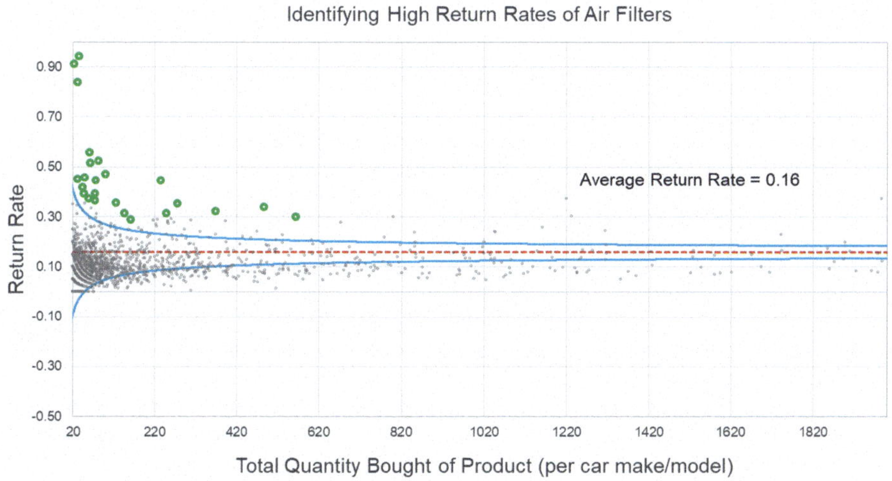

Fig. 3 Funnel plot of return rates for different autoparts

misleading. The type I error points are false alarms but do not detract from the benefits of using this control chart approach.

Analysing data using process control and predictive models, the data scientist is helping to improve the business and also potentially identifying new revenue streams.

5 Conclusions

Mathematics and statistics have everything to offer to data science and are vital to the successful exploitation of Industry 4.0. The teamwork implied by data science means that business domain experts interact with us to answer questions such as what, where, when and how to measure the key processes in a company and how to improve the business. In complex manufacturing these questions involve applications of pure science such as physics, chemistry, advanced materials and structural engineering where there are mechanistic models and well established laws. Mathematicians and statisticians are particularly suited to dealing with this part of the data science triangle providing input which is complementary to that of subject experts and computer scientists.

In some senses, we are in a golden era with great opportunities afforded by the expansion of data science before data analytics becomes so entrenched that there are fewer creative opportunities for developing bespoke solutions. Mathematicians and statisticians need to act now to engage with computer scientists and keep a firm presence in Industry 4.0 and the growing field of data science.

References

1. Ahlemeyer-Stubbe, A., Coleman, S.Y.: Monetising Data - How to Uplift Your Business. Wiley, Hoboken (2018)
2. Coleman, S.Y., Kenett, R.S.: The information quality framework for evaluating data science programs (2017). Available at SSRN: https://ssrn.com/abstract=2911557
3. Coleman, S.Y., Gob, R., Manco, G., Pievatolo, A., Tort-Martorell, X., Reis, M.: How can SMEs benefit from big data? Challenges and a path forward. J. Qual. Reliab. Eng. Int. (2016). http://onlinelibrary.wiley.com/doi/10.1002/qre.2008/full
4. Haldane, A.: https://www.thetimes.co.uk/article/artificial-intelligence-has-a-dark-side-andy-haldane-warns-6pbh2czw5 (2018). Accessed 27 Aug 2018
5. Kenett, R.S.: Quality assurance in the golden age of analytics. https://www.jmp.com/en_us/events/ondemand/analytically-speaking/quality-assurance-in-the-golden-age-of-analytics.html (2018). Accessed 27 Aug 2018
6. Mustafazade, F., Coleman, S., Bacardit, J.: Application of machine learning for decision support in social housing. In: Statistics and Data Science - New Developments for Business and Industrial Applications Conference in Turin, May (2018). http://www.sds2018.polito.it/
7. Pritchett, R.M., Coleman, S.Y., Campbell, J., Pabary, S.: Understanding the patient base: an introduction to data analytics in dental practice. Dent. Update **45**, 236–246 (2018)
8. Science Soft. The Great Expectations of the ImageNet Challenge. https://www.scnsoft.com/blog/imagenet-challenge-2017-expectations (2017). Accessed 1 Oct 2018
9. Smith, W., Coleman S., Bacardit, J.: Insight into aftermarket automotive sales, factory standards and predicting autopart replacement. In: European Network of Business and Industrial Statistics Conference in Nancy (2018). www.enbis.org
10. Zaman, I., Pazouki, K., Norman, R., Younessi, S., Coleman, S.Y.: Development of automatic mode detection system by implementing the statistical analysis of ship data to monitor the performance. Int. J. Maritime Eng. **159**(A3), A225–A235 (2017)

Sparse Multiple Data Assimilation with K-SVD for the History Matching of Reservoirs

Clement Etienam, Rossmary Villegas Velasquez, and Oliver Dorn

Abstract Calibrating subsurface reservoir models with historical well observations leads to a severely ill-posed inverse problem known as *history matching*. The recently proposed Ensemble Smoother with Multiple Data Assimilation (ES-MDA) method has proven to be a successful stochastic technique for solving this inverse problem, but its computational cost can be high in realistic scenarios and it remains challenging to incorporate certain non-Gaussian types of a-priori information into it. In this work we combine the ES-MDA method with Multiple-Point Statistics (MPS) and the K-SVD technique for building sparse dictionaries in order to obtain a novel sparsity-based history matching scheme that preserves non-Gaussian structural prior information and at the same time reduces computational cost. We present numerical experiments in 3D on a modified SPE10 benchmark reservoir model that demonstrate the performance of this new technique.

1 Introduction

The reconstruction of subsurface geological features from production data defines an inverse problem related to data assimilation, which has long been a challenge in the reservoir engineering community due to the small number of observations available [6]. Recently the Ensemble Smoother with Multiple Data Assimilation

C. Etienam
School of Chemical Engineering and Analytical Science, The University of Manchester, Manchester, UK
e-mail: clement.etienam@manchester.ac.uk

R. V. Velasquez
Mathematics and Natural Sciences, Prince Mohammad Bin Fahd University, Al Khobar, Kingdom of Saudi Arabia
e-mail: rvvelasquez@pmu.edu.sa

O. Dorn (✉)
School of Mathematics, The University of Manchester, Manchester, UK
e-mail: oliver.dorn@manchester.ac.uk

© Springer Nature Switzerland AG 2019
I. Faragó et al. (eds.), *Progress in Industrial Mathematics at ECMI 2018*, Mathematics in Industry 30, https://doi.org/10.1007/978-3-030-27550-1_72

(ES-MDA) method has become popular for this task [5]. However, the conventional ES-MDA framework fails to accurately capture non-Gaussian spatial distributions, for example in channelized reservoirs. In those cases, novel image processing techniques based on sparsity representations provide an interesting tool for enabling us to incorporate prior information in the assimilation task and thereby improve final results [6, 7]. In the work presented here we will reformulate this inverse problem as a sparse field recovery task which will then be solved, in contrast to previous work, by using a combination of ES-MDA and sparsity enhancing techniques, in particular K-SVD (an acronym for K-means and Singular Value Decomposition) and Orthogonal Matching Pursuit (OMP).

Our forward problem consists of a three-phase flow model of water, oil and gas fully derived using the combination of Darcy's law and continuity equations [3]. It consists of a system of coupled non-linear partial differential equations describing the evolution of the *dynamic* state variables inside the porous medium which change over time. These are in particular the water, oil and gas saturation levels S_γ, $\gamma \in \{w, o, g\}$, (where the subscripts $\{w, o, g\}$ refer to the corresponding three different phases water, oil and gas that are simultaneously present in the reservoir), and the associated pressure levels p_γ. For a more detailed description of the underlying fluid flow model we refer to [3, 8].

There are also *static* parameters involved in the description of the fluid flow problem that do not change over time. These are in particular the effective porosity φ and the absolute permeability K (all other static parameters of the fluid flow model are assumed known here) [3, 8]. Classically, those static parameters are the primary unknowns of the underlying inverse problem that need to be estimated from production data. Plugging those estimates into a simulator (e.g. [8]) will then (ideally) reproduce the correct dynamical state variables. Notice that those static parameters are related to the lithological structure of the reservoir of which some a-priori knowledge is usually available from independent investigations. Those are encoded in our training images used for generating the initial ensemble for the ES-MDA algorithm via Multiple-Point Statistics (MPS), as well as defining the dictionary conditioning our sparsity-based data assimilation procedure later on. Thereby, in our algorithm this information will be incorporated throughout the data assimilation algorithm to provide final reconstructions better satisfying this prior structural information. In this sense, our sparsity approach does not only speed up the reconstruction process, but also has a regularization effect on the final results.

2 Ensemble-Smoother Multiple Data Assimilation (ES-MDA)

ES-MDA is a Monte Carlo approach to the underlying data assimilation problem which was proposed in [5]. In ES-MDA, each given data set is assimilated multiple times as outlined in the following. The underlying statistical properties of the reservoir are represented by choosing an initial ensemble of size N_e of equi-probable parameter distributions. Let m_j denote the static petro-physical properties to be estimated, where j indicates the ensemble member ($j = 1, \ldots, N_e$), and let in particular m_j^f denote its current estimation in a given step of ES-MDA. Following the notation and the general approach outlined in [5], we denote the (perturbed) observed data by $d_{uc,j}$ and the predicted data, running our simulator on m_j^f, by d_j^f. Denote by N_a the total number of ES-MDA iteration steps taken [5]. The parameter update/assimilation step is then carried out N_a times as an iteration with update rule

$$m_j^a = m_j^f + \tilde{C}_{MD}^f \left(\tilde{C}_{DD}^f + \alpha C_D\right)^{-1} \times (d_{uc,j} - d_j^f) \tag{1}$$

for $j = 1, \ldots, N_e$. In (1), $\tilde{C}_{MD}^f$ is the cross-covariance matrix between the prior vector of model parameters, m_j^f, and the vector of predicted data, d_j^f. Furthermore, $\tilde{C}_{DD}^f$ is the auto-covariance matrix of predicted data, and C_D is the inflated data error covariance matrix. α is the data error covariance inflation factor at the given data assimilation step which is selected prior to the history matching loop for all iteration steps, see [5] for details on its choice and theoretical justifications.

We need an initial ensemble for starting the ES-MDA procedure. In this work, we use MPS [10] for the creation of this initial ensemble targeting a typical channelized model. Thereby the ensemble is conditioned on the information at the well locations and the corresponding analysis of statistical properties. Notice that, when introducing sparsity in this framework further below in this paper, we will then be able to replace some of the quantities that occur in (1) by the corresponding sparse representations, potentially leading to a significantly reduced computational cost in each iteration.

3 Dictionary Learning and OMP

A key factor of sparse coding is the identification of a basis (also often called dictionary or frame in this context) in which the field or signal under consideration permits a sparse representation [2, 7]. A classical approach is to use a general-purpose basis (dictionary) for this, for example involving wavelets or the Discrete Cosine Transform (DCT) [6]. The disadvantage of that choice is that the basis might not be optimal for the particular ensemble to be represented. Therefore, in this work we follow a different approach which employs a special *dictionary learning algorithm*, namely K-SVD, for determining a suitable basis [4]. To start with, we use MPS for generating a set of training signals y_i, $i = 1, \ldots, N_r$, each of them having length N_y (the dimension of the permeability field to be estimated), and arrange them as columns of a matrix Y of size $N_y \times N_r$. The K-SVD dictionary learning algorithm constructs now *simultaneously* a dictionary matrix D of size $N_y \times N_s$, whose columns consist of the "dictionary atoms", and an $N_s \times N_r$ matrix

X, whose columns consist of the representation vectors x_i, $i = 1, \ldots, N_r$, by the (joint) minimization task

$$\{X, D\} = \operatorname{argmin}_{\tilde{X}, \tilde{D}} \left\{ \left\| Y - \tilde{D}\tilde{X} \right\|_F^2 \right\} \text{ subject to } \forall i. \ \|\tilde{x}_i\|_0 \leq T_0 . \qquad (2)$$

In (2), the parameter T_0 indicates the sparsity level imposed on the algorithm and $\| \, \|_F$ denotes the Frobenius norm. Notice that, upon completion, (2) provides a sparse representation $y_i \approx Dx_i$, $i = 1, \ldots, N_r$. Practically, the K-SVD algorithm alternates between updates for D and updates for X for a given training set until optimality is reached. For more details of this algorithm, including a pseudo-code describing its individual stages, we refer to [1]. The creation of this over-complete dictionary is done off-line and only once before the history matching process starts. In our work, it is used to transform the permeability and porosity fields from a full spatial domain to a sparse domain and back.

Notice that the part of the K-SVD algorithm which finds an optimal sparse representation $\tilde{X}$ for a given training set Y, given the current iterate for the dictionary $\tilde{D}$, requires us to choose a specific sparsity promoting algorithms, for which we select here the well-known Orthogonal Matching Pursuit (OMP) algorithm [9]. We refer the reader to [1, 9] for further information on OMP. Notice that, in addition to its use inside this off-line K-SVD step, the OMP algorithm will also be used throughout our proposed ES-MDA algorithm whenever permeabilities and porosities need to be mapped from full to sparse representations.

The sparsity enhanced ES-MDA algorithm described above is summarized in the following Algorithm 1.

Algorithm 1 Sparsity-ensemble optimization method (SEOM)

procedure SEOM
  Generate N_r realizations of different permeability/porosity profiles using MPS. Learn dictionary D consisting of N_s atoms tailored to these realizations using K-SVD, Eq. (2).
  Independently select N_e initial ensemble realizations of permeability/porosity using again MPS for the channelized test case.
  Choose number of assimilation iterations N_a.
  for $k = 1$ to N_a **do**
    Progress the N_e realizations over time (using non-sparse representation,
    here with simulator ECLIPSE 100 [8])
    for $j = 1$ to N_e **do**
      Find sparse representation of all m_j using OMP with dictionary D.
      Carry out the ES-MDA analysis step (1) on sparse representation.
      Transform m_j back to the (non-sparse) spatial domain.
  STOP

4 Numerical Results

In order to demonstrate the performance of our SEOM algorithm, we run it on a modified version of the popular benchmark SPE10 model. Here we choose $N_r = 2000$, $N_s = 1500$ and $N_e = 100$. Typical values for N_a in ES-MDA and SEOM are $N_a \in \{2, 4, 6, 8\}$. The true model consists of five layers and can be seen in a 3D view on the top right side of Fig. 1. The individual five layers (permeability) can be seen in the first row on the left of the same figure. The second and third rows of Fig. 1 display two (of the $N_s = 1500$) members of the generated over-complete dictionary representing this reservoir. Row four of this figure shows the ensemble-mean starting profile used in the ES-MDA and SEOM algorithms for this test case. Row five displays the ensemble-mean result of standard ES-MDA (without sparsity), and row six shows the ensemble-mean result of the SEOM algorithm incorporating sparsity with respect to the learned dictionary. Visually, the SEOM estimate looks more 'channelized' compared to the

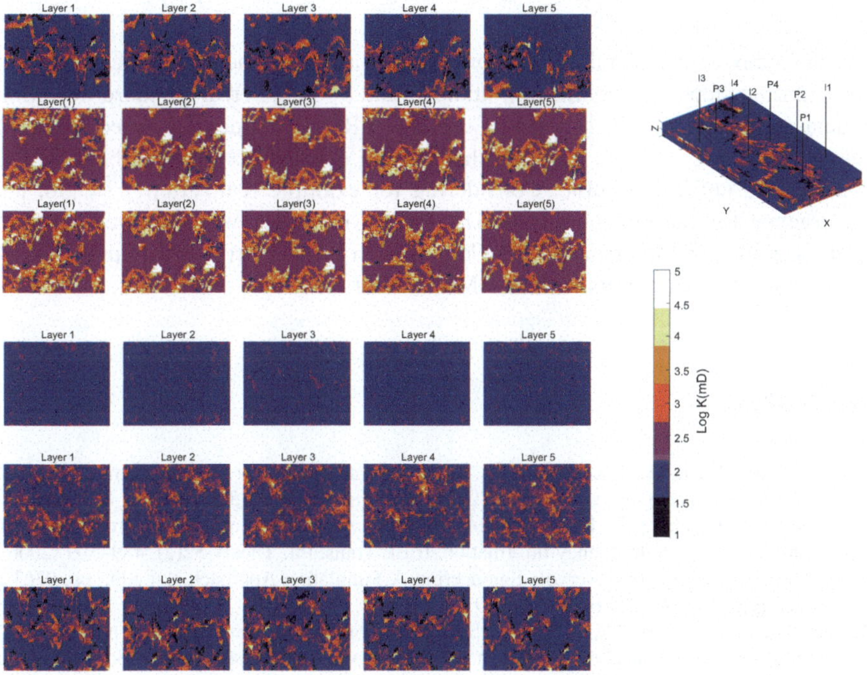

Fig. 1 On the left, all rows show from left to right layers 1–5 of the SPE10 reservoir as specified in the following. First row: true permeability profile; second and third rows: two (out of $N_s = 1500$) members of the over-complete dictionary; fourth row: ensemble-mean initial guess; fifth row: ensemble-mean final result of standard ES-MDA; sixth row: ensemble-mean final result of the proposed SEOM algorithm. On the right a 3D view of the reservoir is provided in the top image, where the bottom image shows the corresponding colour bar that applies to this figure

Table 1 RMSE values $RMSE(j)$

Realization j	Initial	Final (ES-MDA)	Final (SEOM)
13	49.33	24.21	4.32
56	43.83	12.8	1.34
92	56.87	24.58	6.44

ES-MDA reconstruction due to the incorporated a-priori information on the true model encoded in the dictionary. For evaluating further the performance of the techniques compared here, the estimated final porosity and permeability profiles are used for running the simulator from the initial time step to the final time step N_k. For the obtained data d for all 100 ensemble members we calculate the root-mean-square error (RMSE) for each ensemble member j, denoted by $RMSE(j)$, which is defined as

$$RMSE(j) = \left(\frac{1}{N_k N_d} \sum_{k=1}^{N_k} \sum_{\nu=1}^{N_d} \left(\frac{d^{\nu}_{uc,j}(k) - d^{f,\nu}_{j}(k)}{\sigma_\nu} \right)^2 \right)^{\frac{1}{2}}. \tag{3}$$

Here the index ν runs over the N_d observed well data components, k indicates the N_k physical time steps, j is realization number, and σ_ν is the error standard deviation of data type ν. Table 1 shows and compares RMSE values for three different realizations ($j \in \{13, 56, 92\}$) which we have chosen (somehow arbitrarily) to represent the entire ensemble. Considering for example realization $j = 56$, the convergence for the standard ES-MDA history matching was achieved in eight iterations with an RMS error of 12.8. Compared to that, the proposed method SEOM converged in four iterations with an RMS error of 1.34.

References

1. Aharon, M., Elad, M., Bruckstein, A.: K-SVD: an algorithm for designing over complete dictionaries for sparse representation. IEEE Trans. Signal Process. **54**(11), 4311–4322 (2006)
2. Candes, E.J., Romberg, J., Tao, T.: Robust uncertainty principles: exact signal reconstruction from high incomplete frequency information. IEEE Trans. Inf. Theory **52**(2), 489–509 (2006)
3. Crichlow, H.B.: Modern Reservoir Engineering-A Simulation Approach, 1st edn., pp. 07632. Prentice-Hall, Inc., Englewood Cliffs (1977)
4. Elsheikh, A., Wheeler, M., Hoteit, I.: Sparse calibration of subsurface flow models using nonlinear orthogonal matching pursuit and an iterative stochastic ensemble method. Adv. Water Resour. **56**(1), 14–26 (2013)
5. Emerick, A.A., Reynolds, A.C.: Ensemble smoother with multiple data assimilation. Comput. Geosci. **55**, 3–15 (2013)
6. Jafarpour, B.: Wavelet reconstruction of geologic facies from nonlinear dynamic flow measurements. IEEE Trans. Geosci. Remote Sens. **49**(5), 1520–1535 (2011)
7. Khaninezhad, M.M., Jafarpour, B., Li, L.: Sparse geologic dictionaries for subsurface flow model calibration: part 1, inversion formulation. Adv. Water Resour. **39**(1), 106–121 (2012)

8. Schlumberger GeoQuest: ECLIPSE 100 (Black Oil): Reference Manual and Technical Description, Houston. https://www.software.slb.com/products/eclipse (2014)
9. Tropp, J.A., Gilert, A.C.: Signal recovery from random measurements via orthogonal matching pursuit. IEEE Trans. Inf. Theory **53**(12), 4655–4666 (2007)
10. Wu, J., Boucher, A., Zhang, T.: A SGeMS code for pattern simulation of continuous and categorical variables: FILTERSIM. Comput. Geosci. **34**(12): 1863–1876 (2008)

Hypercomplex Fourier Transforms in the Analysis of Multidimensional Linear Time-Invariant Systems

Łukasz Błaszczyk

Abstract The aim of this paper is to further investigate the properties of octonion Fourier transform (OFT) of real-valued functions of three variables and its potential applications in signal and system processing. This is a continuation of the work started by Hahn and Snopek, in which they studied the octonion Fourier transform definition and its applications in the analysis of the hypercomplex analytic signals. First, the octonion algebra and the new quadruple-complex numbers algebra are introduced. Then, the OFT definition is recalled, together with some basic properties, proved in some earlier work. The main part of the article is devoted to new properties of the OFT, that allow us to use the OFT in the analysis of multidimensional signals and LTI systems, i.e. derivation and convolution of real-valued signals.

1 Introduction

The classical signal theory deals with $\mathbb{R}$- or $\mathbb{C}$-valued functions and their $\mathbb{C}$-valued spectra. However, in some practical applications, signals tend to be represented by hypercomplex algebras [4]. Hypercomplex Fourier transforms deserve special attention in this considerations. Quaternion Fourier transform (QFT) allows us to analyze two dimensions of the sampling grid independently, while the complex transform mixes those two dimensions. It enables us to use the Fourier transform in the analysis of some 2-D linear time-invariant (LTI) systems described by some linear partial differential equations (PDEs) [3].

In [2] we presented some preliminary results concerning the octonion Fourier transform (OFT). We showed that the OFT is well defined for $\mathbb{R}$-valued functions and proved some basic properties of the OFT, analogous to the properties of the classical FT and QFT. Our research follows previous results of Hahn and Snopek [6].

Ł. Błaszczyk (✉)
Faculty of Mathematics and Information Science, Warsaw University of Technology, Warszawa, Poland
e-mail: L.Blaszczyk@mini.pw.edu.pl

© Springer Nature Switzerland AG 2019

I. Faragó et al. (eds.), *Progress in Industrial Mathematics at ECMI 2018*,
Mathematics in Industry 30, https://doi.org/10.1007/978-3-030-27550-1_73

It should be noted that octonion signal processing have already found practical applications [5, 7], including image splicing detection [9] and neural networks [8].

In this paper, we introduce the most recent results, associating OFT (introduced in Sect. 3) with 3-D LTI systems of linear PDEs with constant coefficients. Properties of the OFT in context of signal-domain operations such as derivation and convolution of $\mathbb{R}$-valued functions are stated in Sect. 4. There are known results for QFT (see [3]), but they use the notion of other hypercomplex algebra, i.e. double-complex numbers. Results presented here require defining other higher-order hypercomplex structure, i.e. quadruple-complex numbers defined in Sect. 2. This hypercomplex generalization of the Fourier transformation provides an excellent tool for the analysis of 3-D LTI systems which is presented in Sect. 5. The paper is concluded in Sect. 6 with short discussion of those results.

2 Algebras of Octonions and Quadruple-Complex Numbers

Octonions ($\mathbb{O}$) are an example of Cayley-Dickson hypercomplex algebra [2, 6]. Its elements are of the form

$$o = x_0 + x_1\,\mathbf{e}_1 + x_2\,\mathbf{e}_2 + x_3\,\mathbf{e}_3 + x_4\,\mathbf{e}_4 + x_5\,\mathbf{e}_5 + x_6\,\mathbf{e}_6 + x_7\,\mathbf{e}_7, \quad x_0,\ x_1,\ \ldots,\ x_7 \in \mathbb{R},$$

where $\mathbf{e}_1,\ \mathbf{e}_2,\ \ldots,\ \mathbf{e}_7$ are seven imaginary units satisfying appropriate multiplication rules (presented in Table 1). Octonions form a *non-associative*, *non-commutative* (but alternative) composition and division algebra $\mathbb{O}$ of order 8 over the field of real numbers $\mathbb{R}$. Octonion algebra is endowed with the standard norm

$$|o| = \sqrt{o \cdot o^*} = \sqrt{x_0^2 + x_1^2 + \ldots + x_7^2},$$

where $o^* = x_0 - x_1\mathbf{e}_1 - \ldots - x_7\mathbf{e}_7$ is the octonion conjugate of o.

We define the octonion exponential function in a classical way—as the infinite sum $e^o := \sum_{k=0}^{\infty} \frac{o^k}{k!}$. Due to the fact, that octonion multiplication is non-

Table 1 Multiplication rules in octonion algebra

$\cdot$	1	$\mathbf{e}_1$	$\mathbf{e}_2$	$\mathbf{e}_3$	$\mathbf{e}_4$	$\mathbf{e}_5$	$\mathbf{e}_6$	$\mathbf{e}_7$
1	1	$\mathbf{e}_1$	$\mathbf{e}_2$	$\mathbf{e}_3$	$\mathbf{e}_4$	$\mathbf{e}_5$	$\mathbf{e}_6$	$\mathbf{e}_7$
$\mathbf{e}_1$	$\mathbf{e}_1$	-1	$\mathbf{e}_3$	$-\mathbf{e}_2$	$\mathbf{e}_5$	$-\mathbf{e}_4$	$-\mathbf{e}_7$	$\mathbf{e}_6$
$\mathbf{e}_2$	$\mathbf{e}_2$	$-\mathbf{e}_3$	-1	$\mathbf{e}_1$	$\mathbf{e}_6$	$\mathbf{e}_7$	$-\mathbf{e}_4$	$-\mathbf{e}_5$
$\mathbf{e}_3$	$\mathbf{e}_3$	$\mathbf{e}_2$	$-\mathbf{e}_1$	-1	$\mathbf{e}_7$	$-\mathbf{e}_6$	$\mathbf{e}_5$	$-\mathbf{e}_4$
$\mathbf{e}_4$	$\mathbf{e}_4$	$-\mathbf{e}_5$	$-\mathbf{e}_6$	$-\mathbf{e}_7$	-1	$\mathbf{e}_1$	$\mathbf{e}_2$	$\mathbf{e}_3$
$\mathbf{e}_5$	$\mathbf{e}_5$	$\mathbf{e}_4$	$-\mathbf{e}_7$	$\mathbf{e}_6$	$-\mathbf{e}_1$	-1	$-\mathbf{e}_3$	$\mathbf{e}_2$
$\mathbf{e}_6$	$\mathbf{e}_6$	$\mathbf{e}_7$	$\mathbf{e}_4$	$-\mathbf{e}_5$	$-\mathbf{e}_2$	$\mathbf{e}_3$	-1	$-\mathbf{e}_1$
$\mathbf{e}_7$	$\mathbf{e}_7$	$-\mathbf{e}_6$	$\mathbf{e}_5$	$\mathbf{e}_4$	$-\mathbf{e}_3$	$-\mathbf{e}_2$	$\mathbf{e}_1$	-1

commutative, for any $o_1, o_2 \in \mathbb{O}$ we have $e^{o_1+o_2} = e^{o_1} \cdot e^{o_2}$ if and only if $o_1 \cdot o_2 = o_2 \cdot o_1$.

Due to non-associativity and non-commutativity of octonion multiplication, many formulas concerning the Fourier transforms are quite complicated (see Sect. 4). To improve that, inspired by Ell [3], we introduce *the algebra of quadruple-complex numbers* $\mathbb{F}$, which elements can be written as

$$p = \underbrace{(p_0 + p_1\mathbf{e}_1)}_{=s_0\in\mathbb{C}} + \underbrace{(p_2 + p_3\mathbf{e}_1)}_{=s_1\in\mathbb{C}}\mathbf{e}_2 + \underbrace{(p_4 + p_5\mathbf{e}_1)}_{=s_2\in\mathbb{C}}\mathbf{e}_4 + \underbrace{(p_6 + p_7\mathbf{e}_1)}_{=s_3\in\mathbb{C}}\mathbf{e}_2\mathbf{e}_4.$$

Therefore, the algebra $\mathbb{F}$ consists of quadruples $(s_0, s_1, s_2, s_3) \in \mathbb{C}^4$ of complex numbers. Multiplication $\odot$ in $\mathbb{F}$ is given by the formula

$$\begin{aligned}(s_0, s_1, s_2, s_3) \odot (t_0, t_1, t_2, t_3) = (&s_0t_0 - s_1t_1 - s_2t_2 + s_3t_3, \quad s_0t_1 + s_1t_0 - s_2t_3 - s_3t_2,\\ &s_0t_2 + s_2t_0 - s_1t_3 - s_3t_1, \quad s_0t_3 + s_3t_0 + s_1t_2 + s_2t_1),\end{aligned}$$

where $(s_0, s_1, s_2, s_3), (t_0, t_1, t_2, t_3) \in \mathbb{F}$. It is easy to check that multiplication $\odot$ is associative and commutative, but not all nonzero elements of $\mathbb{F}$ are invertible with respect to $\odot$, e.g. $(1, 0, 0, 1) = 1 + \mathbf{e}_6 \in \mathbb{F}$ doesn't have an $\odot$-inverse.

3 Octonion Fourier Transform

Let $u : \mathbb{R}^3 \to \mathbb{R}$. The *octonion Fourier transform* (OFT) of u is defined by

$$U(\mathbf{f}) = \int_{\mathbb{R}^3} u(\mathbf{x}) \cdot e^{-2\pi\mathbf{e}_1 f_1 x_1} \cdot e^{-2\pi\mathbf{e}_2 f_2 x_2} \cdot e^{-2\pi\mathbf{e}_4 f_3 x_3}\, \mathbf{dx},$$

where $\mathbf{x} = (x_1, x_2, x_3)$, $\mathbf{f} = (f_1, f_2, f_3)$ and multiplication is done from left to right. Choice and order of imaginary units in the exponents is not accidental (see [2, 6]). Conditions of existence (and invertibility) are the same as for the classical (complex) Fourier transform. Let us recall the result from [2], where the inverse OFT formula was proved.

Theorem 1 *Let $u : \mathbb{R}^3 \to \mathbb{R}$ be a continuous and square-integrable. Then*

$$u(\mathbf{x}) = \int_{\mathbb{R}^3} U(\mathbf{f}) \cdot e^{2\pi\mathbf{e}_4 f_3 x_3} \cdot e^{2\pi\mathbf{e}_2 f_2 x_2} \cdot e^{2\pi\mathbf{e}_1 f_1 x_1}\, \mathbf{df},$$

where multiplication is done from left to right.

In fact, the abovementioned theorem holds for the general case of $\mathbb{O}$-valued functions (see [1]), but in this paper we will consider only the $\mathbb{R}$-valued functions.

In [2] we derived basic properties of the OFT, analogous to the properties of the classical Fourier transform. Let us recall some of those results.

Let U be the OFT of the $\mathbb{R}$-valued function u and let $\alpha_i(o) = -\mathbf{e}_i \cdot (o \cdot \mathbf{e}_i)$, where $\circ$ is standard function composition. We have the following octonion analogue of *Hermitian symmetry*:

$$\begin{aligned} U(-f_1, \quad f_2, \quad f_3) &= (\alpha_6 \circ \alpha_4 \circ \alpha_2)(U(f_1, f_2, f_3)),\\ U(\ f_1, -f_2, \quad f_3) &= (\alpha_5 \circ \alpha_4 \circ \alpha_1)(U(f_1, f_2, f_3)),\\ U(\ f_1, \quad f_2, -f_3) &= (\alpha_3 \circ \alpha_2 \circ \alpha_1)(U(f_1, f_2, f_3)). \end{aligned}$$

Moreover, if U^α, U^β and U^γ denote the OFTs of functions $u(x_1 - \alpha, x_2, x_3)$, $u(x_1, x_2 - \beta, x_3)$ and $u(x_1, x_2, x_3 - \gamma)$, respectively, then

$$\begin{aligned} U^\alpha(f_1, f_2, f_3) &= \cos(2\pi f_1\alpha)\, U(f_1, f_2, f_3) - \sin(2\pi f_1\alpha)\, U(f_1, -f_2, -f_3) \cdot \mathbf{e}_1,\\ U^\beta(f_1, f_2, f_3) &= \cos(2\pi f_2\beta)\, U(f_1, f_2, f_3) - \sin(2\pi f_2\beta)\, U(f_1, \quad f_2, -f_3) \cdot \mathbf{e}_2,\\ U^\gamma(f_1, f_2, f_3) &= \cos(2\pi f_3\gamma)\, U(f_1, f_2, f_3) - \sin(2\pi f_3\gamma)\, U(f_1, \quad f_2, \quad f_3) \cdot \mathbf{e}_4, \end{aligned}$$

which is the octonion version of *shift theorem*. We also have the *Plancherel* and *Rayleigh theorems*:

$$\int_{\mathbb{R}^3} u(\mathbf{x}) \cdot v^*(\mathbf{x})\, \mathbf{dx} = \int_{\mathbb{R}^3} U(\mathbf{f}) \cdot V^*(\mathbf{f})\, \mathbf{df}, \quad \Rightarrow \quad \int_{\mathbb{R}^3} |u(\mathbf{x})|^2\, \mathbf{dx} = \int_{\mathbb{R}^3} |U(\mathbf{f})|^2\, \mathbf{df},$$

where V is the OFT of the $\mathbb{R}$-valued function v. The above-presented theorems form the basis of the octonion signal theory and are the starting point for further research.

4 Recent Results

We will now present properties that are a key element in the analysis of multidimensional LTI systems described by a system of PDEs. In theorems stated below, we will denote the OFTs of the $\mathbb{R}$-valued functions u and v by U and V, respectively.

Theorem 2 (OFTs of Partial Derivatives) *Let U^{x_1}, U^{x_2} and U^{x_3} denote the OFTs of $\frac{\partial u}{\partial x_1}$, $\frac{\partial u}{\partial x_2}$ and $\frac{\partial u}{\partial x_3}$, respectively. Then*

$$\begin{aligned} U^{x_1}(f_1, f_2, f_3) &= U(f_1, -f_2, -f_3) \cdot (2\pi f_1 \mathbf{e}_1) = U(f_1, f_2, f_3) \odot (2\pi f_1 \mathbf{e}_1),\\ U^{x_2}(f_1, f_2, f_3) &= U(f_1, \quad f_2, -f_3) \cdot (2\pi f_2 \mathbf{e}_2) = U(f_1, f_2, f_3) \odot (2\pi f_2 \mathbf{e}_2),\\ U^{x_3}(f_1, f_2, f_3) &= U(f_1, \quad f_2, \quad f_3) \cdot (2\pi f_3 \mathbf{e}_4) = U(f_1, f_2, f_3) \odot (2\pi f_3 \mathbf{e}_4). \end{aligned}$$

Proof of this result follows from straightforward calculations and we leave details to the reader. It is worth noting, however, that the idea of this proof is to express the OFT of the derivative of u as a sum of components of different parity, i.e.

$$U^{x_\ell} = U^{x_\ell}_{eee} - U^{x_\ell}_{oee}\mathbf{e}_1 - U^{x_\ell}_{eoe}\mathbf{e}_2 + U^{x_\ell}_{ooe}\mathbf{e}_3 - U^{x_\ell}_{eeo}\mathbf{e}_4 + U^{x_\ell}_{oeo}\mathbf{e}_5 + U^{x_\ell}_{eoo}\mathbf{e}_6 - U^{x_\ell}_{ooo}\mathbf{e}_7, \tag{1}$$

where

$$U^{x_\ell}_{ijk}(\mathbf{f}) = \int_{\mathbb{R}^3} \frac{\partial u}{\partial x_\ell} \cdot F_i(2\pi f_1 x_1) \cdot F_j(2\pi f_2 x_2) \cdot F_k(2\pi f_3 x_3)\, \mathrm{d}\mathbf{x} \tag{2}$$

and $F_i(y) = \cos(y)$ if $i = e$, and $F_i(y) = \sin(y)$ if $i = o$ [2, 6]. The claim of the theorem follows from the integration by parts. Notice that treating octonions as elements of $\mathbb{F}$ and using the multiplication $\odot$, we get the same formulas as in classical theory.

The next result concerns function convolution. The convolution-multiplication duality is one of the key properties used in the frequency analysis of LTI systems [3]. Recall that the convolution of $u, v\colon \mathbb{R}^3 \to \mathbb{R}$ is given by the formula

$$(u * v)(\mathbf{x}) = \int_{\mathbb{R}^3} u(\mathbf{y}) \cdot v(\mathbf{x} - \mathbf{y})\, \mathrm{d}\mathbf{y}.$$

Convolution of functions is commutative and associative while the multiplication of octonions is not, hence the octonion version of duality theorem will have to differ significantly from its classical equivalent.

Theorem 3 (Convolution-Multiplication Duality) *Let $\mathscr{F}_{\mathrm{OFT}}\{u * v\}$ denote the OFT of the convolution of u and v, i.e. $u * v$. Then*

$$\begin{aligned} \mathscr{F}_{\mathrm{OFT}}\{u * v\}(\mathbf{f}) = \;& V(\;f_1,\;\;f_2,\;\;f_3) \cdot (\;\;U_{eee}(\mathbf{f}) \;\;\; - U_{eeo}(\mathbf{f})\,\mathbf{e}_4) \\ &+ V(\;f_1, -f_2, -f_3) \cdot (-U_{oee}(\mathbf{f})\,\mathbf{e}_1 + U_{ooe}(\mathbf{f})\,\mathbf{e}_3) \\ &+ V(\;f_1,\;\;f_2, -f_3) \cdot (-U_{eoe}(\mathbf{f})\,\mathbf{e}_2 + U_{oeo}(\mathbf{f})\,\mathbf{e}_5) \\ &+ V(-f_1,\;\;f_2, -f_3) \cdot (\;\;U_{eoo}(\mathbf{f})\,\mathbf{e}_6 - U_{ooo}(\mathbf{f})\,\mathbf{e}_7), \end{aligned} \tag{3}$$

where

$$U = U_{eee} - U_{oee}\mathbf{e}_1 - U_{eoe}\mathbf{e}_2 + U_{ooe}\mathbf{e}_3 - U_{eeo}\mathbf{e}_4 + U_{oeo}\mathbf{e}_5 + U_{eoo}\mathbf{e}_6 - U_{ooo}\mathbf{e}_7$$

is a sum of eight terms with different parity w.r.t. x_1, x_2, and x_3, similar to (1)–(2).

As in the previous theorem, this result follow from expressing the OFT as a sum of components of different parity. For details of such formulation see [2, 6]. Similar formulas concerning quaternion Fourier transform can be found in literature [3].

Notice that, as in the OFT of derivatives theorem, using the notion of quadruple-complex numbers we can improve the abovementioned formulas.

Corollary 1 *Using the* $\mathbb{F}$*-multiplication we can rewrite formula* (3) *in simple form:*

$$\mathscr{F}_{\mathrm{OFT}}\{u * v\}(\mathbf{f}) = U(\mathbf{f}) \odot V(\mathbf{f}).$$

Theorem 3 and Corollary 1 enable us to define the octonion frequency response of a system as the OFT of impulse response. It is worth mentioning that the notion of multiplication in $\mathbb{F}$ can be used to reduce parallel, cascade and feedback connections of linear systems into simple algebraic equations, as in classical system theory.

5 Multidimensional Linear Time-Invariant Systems

It is a well-known fact that the Fourier transform converts differential equations into algebraic equations. While the advantages of this approach in the 1-D case are obvious, in the case of partial derivatives the classic approach has some limitations.

Consider a function $u\colon \mathbb{R}^3 \to \mathbb{R}$ that is even w.r.t. all variables (making both classical FT and OFT $\mathbb{R}$-valued functions). The classical Fourier transform of $u_{x_1x_2}$ is $-U(\mathbf{f}) \cdot (2\pi f_1)(2\pi f_2)$, which is a $\mathbb{R}$-valued function. Therefore, we loose the information that the function u was differentiated at all. On the other hand, the OFT of $u_{x_1x_2}$ is $U(f_1, -f_2, -f_3) \cdot (2\pi f_1)(2\pi f_2)\mathbf{e}_3$, which is $\mathbb{O}$-valued (purely imaginary). This information indicates that the function has been differentiated by x_1 and x_2.

As a direct consequence of Theorem 2, every linear PDE with constant coefficients (i.e. every 3-D LTI system of PDEs) can be reduced to algebraic equation (with respect to multiplication in $\mathbb{F}$). Consider the heat equation in 2-D, i.e.

$$u_t(t, x_1, x_2) = u_{x_1x_1}(t, x_1, x_2) + u_{x_2x_2}(t, x_1, x_2) + f(t, x_1, x_2),$$

where we get

$$\left((2\pi f_1)^2 + (2\pi f_2)^2 + (2\pi \tau)\mathbf{e}_1\right) \odot U(\tau, f_1, f_2) = F(\tau, f_1, f_2).$$

It is easy to show that $\left((2\pi f_1)^2 + (2\pi f_2)^2 + (2\pi \tau)\mathbf{e}_1\right)^{-1}$ exists if and only if $(\tau, f_1, f_2) \neq (0, 0, 0)$ and is equal to

$$\left((2\pi f_1)^2 + (2\pi f_2)^2 + (2\pi \tau)\mathbf{e}_1\right)^{-1} = \frac{(2\pi f_1)^2 + (2\pi f_2)^2 - (2\pi \tau)\mathbf{e}_1}{\left((2\pi f_1)^2 + (2\pi f_2)^2\right)^2 + (2\pi \tau)^2}.$$

Hence

$$U(\tau, f_1, f_2) = \frac{(2\pi f_1)^2 + (2\pi f_2)^2 - (2\pi\tau)\mathbf{e}_1}{\left((2\pi f_1)^2 + (2\pi f_2)^2\right)^2 + (2\pi\tau)^2} \odot F(\tau, f_1, f_2).$$

We have thus obtained a simple formula for the system's response to the given stimulation. What's more, it wouldn't be possible using multiplication in $\mathbb{O}$.

6 Final Remarks

Presented results further develop the foundation of octonion-based signal and system theory. At the moment we are left to find real-life applications of the discussed theory. The results published in recent articles suggest that this is feasible, e.g. in the field of multispectral image processing [5, 7, 9]. However, it would be necessary to focus on the implementation of numerical algorithms for this purpose. It seems that extending octonion-based signal theory to discrete-variable signals may also be achieved by methods used so far.

Acknowledgements The research conducted by the author was supported by National Science Centre (Poland) grant No. 2016/23/N/ST7/00131.

References

1. Błaszczyk, Ł.: Octonion spectrum of 3D octonion-valued signals – properties and possible applications. In: Proceedings of 2018 26th European Signal Processing Conference (EUSIPCO), pp. 509–513 (2018)
2. Błaszczyk, Ł., Snopek, K.M.: Octonion Fourier Transform of real-valued functions of three variables – selected properties and examples. Signal Process. **136**, 29–37 (2017)
3. Ell, T.A.: Quaternion-Fourier transforms for analysis of 2-dimensional linear time-invariant partial-differential systems. In: Proceedings of 32nd IEEE Conference on Decision and Controll, vols. 1–4, pp. 1830–1841 (1993)
4. Ell, T.A., Le Bihan, N., Sangwine S.J.: Quaternion Fourier Transforms for Signal and Image Processing. Wiley, Hoboken (2014)
5. Grigoryan, A.M., Agaian, S.S.: Quaternion and Octonion Color Image Processing with MATLAB. SPIE, Bellingham (2018)
6. Hahn, S.L., Snopek, K.M.: The unified theory of complex and hypercomplex analytic signals. Bull. Polish Acad. Sci. Tech. Sci. **59**(2), 167–181 (2011)
7. Lazendić, S., De Bie, H., Pižurica, A.: Octonion sparse representation for color and multispectral image processing. In: Proceedings of 2018 26th European Signal Processing Conference (EUSIPCO), pp. 608–612 (2018)
8. Popa, C.A.: Global exponential stability of octonion-valued neural networks with leakage delay and mixed delays. Neural Netw. **105**, 277–293 (2018)
9. Sheng, H., Shen, X., Lyu, Y., Shi, Z., Ma, S.: Image splicing detection based on Markov features in discrete octonion cosine transform domain. IET Image Process. **12**(10), 1815–1823 (2018)

Analog-to-Probability Conversion—Efficient Extraction of Information Based on Stochastic Signal Models

Christian Adam, Michael H. Teyfel, and Dietmar Schroeder

Abstract Analog-to-probability conversion is introduced as a new concept for efficient parameter extraction from analog signals that can be described by nonlinear models. The current state of information about these parameters is represented by a multivariate probability distribution. Only a digital-to-analog converter and a comparator are required as acquisition hardware. The introduced approach reduces the number of comparisons to be done by the hardware and therefore the total energy consumption. As a proof of concept the algorithm is implemented on a system-on-chip and compared to a nonlinear least squares approach.

1 Introduction

Analog-to-digital converters (ADCs) are used in many electronic systems to convert analog signals like sensor data into a digital representation. This is typically done by discretizing the periodically sampled analog signal. However, in an increasing number of applications, where Edge Computing concepts [1] or Near-Sensor Data Analytics [3] are employed, the exact waveform is not important, since only certain features of the signal are of interest. The Probabilistic Computing effort [4] even goes one step further by working with probability distributions of the features. Probability is integrated into sensing systems in order to take uncertainties of real world measurements into account. Analog-to-probability conversion (APC) supports these ideas already close to the sensor by generating and updating probability distributions and expectation values of signal features directly from the analog signal.

C. Adam (✉) · M. H. Teyfel · D. Schroeder
Institute of Nano and Medical Electronics, Hamburg University of Technology, Hamburg, Germany
e-mail: christian.adam@tuhh.de; michael.teyfel@gmx.de; d-schroeder@tuhh.de

© Springer Nature Switzerland AG 2019

I. Faragó et al. (eds.), *Progress in Industrial Mathematics at ECMI 2018*,
Mathematics in Industry 30, https://doi.org/10.1007/978-3-030-27550-1_74

2 Analog-to-Probability Conversion

In analog-to-probability conversion the analog input is assumed to be described by a known signal model $r(\mathbf{X}, t)$, which depends on the unknown parameters $\mathbf{X}$ and the time t. The parameters, which have to be determined, are modeled by a multivariate Gaussian distribution. This distribution is completely defined by the vector of expected values $\mathbf{M_X}$ and the covariance matrix Σ_X, which are successively updated in order to increase the information about the signal parameters.

Figure 1a shows a block diagram of the sampling hardware, which consists of a digital-to-analog converter (DAC), a comparator and a control unit that executes the APC algorithm. The DAC generates a threshold value r_0 that is compared to the input signal. For linear signal models, the threshold corresponds to a hyperplane dividing the space spanned by the model parameters into points (half-spaces) that coincide with the comparator output and such that do not. This hyperplane is described by

$$r_0 = r(\mathbf{X}, t_s) \;, \tag{1}$$

where t_s is the sampling time. The comparator output then indicates in which half-space the true parameters defining the input signal are located, and the probability distribution is set to zero in the other half-space, which decreases the variance. After renormalization and approximation of the remaining distribution as a Gaussian distribution again, this procedure can be repeated iteratively until the signal parameters are determined with the desired accuracy. In Fig. 1b, c this method is illustrated for a two-dimensional example.

The updated expected vector $\mathbf{M}_\mathbf{X}^+ = E[\mathbf{X}]$ and covariance matrix $\Sigma_X^+ = E[(\mathbf{X} - \mathbf{M_X})(\mathbf{X} - \mathbf{M_X})^T]$ are calculated by integrating the probability density function over the coinciding half-space limited by the hyperplane defined in (1).

One advantage of the analog-to-probability conversion is that only one comparison at the comparator is necessary for each sample. In contrast, a typical 10-bit successive approximation ADC (SAR ADC) needs 10 comparisons for each sample,

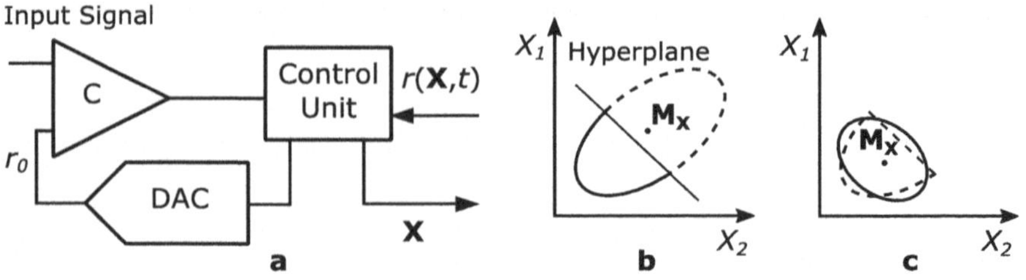

Fig. 1 Concept of the analog-to-probability conversion. (**a**) Block diagram of the sampling hardware. (**b**) Cutting the distribution along a hyperplane. (**c**) Renormalization to a Gaussian distribution

which additionally requires a sample and hold component to hold the varying input signal long enough for all comparisons to be done. Besides the reduction of comparisons, another advantage of the analog-to-probability conversion is that it provides a priori information about the next samples. Therefore the most efficient sampling time and comparison threshold can be chosen to maximize the expected reduction of the parameter variances in each iteration.

3 Non-linear Signal Model

In the case of nonlinear signal models like

$$r(\mathbf{X}, t) = X_1 + c\ \sin(X_2 t + X_3)\ , \tag{2}$$

the dividing hyperplane turns into a curved hypersurface. Therefore, the concept of Gaussian Mixture is introduced. It approximates the total probability distribution as a weighted sum $f(\mathbf{X}) = \sum_{i=1}^{L} w_i\ f_i(\mathbf{X})$ of several more dense Gaussian distributions $f_i(\mathbf{X})$, which are placed in a grid to cover the whole parameter space. For each of these Gaussian distributions the hypersurface can be approximated as a hyperplane defined by $r_0 = r(\mathbf{X_0}, t_s) + (\nabla r(\mathbf{X}, t_s)|_{\mathbf{X_0}})^T\ (\mathbf{X} - \mathbf{X_0})$ using a Taylor expansion about $\mathbf{X_0}$. Therefore the same method as described above for linear signal models can be applied to the mixture members. The expected vector and the variance of the true distribution are approximated from the sum of all Gaussian members. If the weight factor w_i of a Gaussian member falls below a certain threshold, the component is removed from the mixture.

4 Hardware Implementation

As a proof of concept a prototype of an ADC utilizing the introduced algorithm was successfully implemented using a Cypress PSoC 5LP system-on-chip [2]. A sinusoidal input signal described by the model in (2) with parameters $c = 0.3\,\text{V}$, $X_1 = 1.1\,\text{V}$ and $X_2 = 1.257\,\text{s}^{-1}$ was used. X_3 is irrelevant for the hardware system as it represents only the random phase shift between the input oscillation and the clock on the PSoC 5LP. The sampling rate was set to 2.2 Hz.

As a comparison, a more classical approach for parameter extraction, i.e. the nonlinear least squares method, was also implemented on the PSoC 5LP. A 10-bit SAR ADC was used for continuously sampling the input signal for the same time as the execution time of the full APC. Afterwards the nonlinear least squares method was applied to extract the signal parameters from the observed data. The same signal model and sampling rate were used for both approaches.

Table 1 shows the results of the comparison. It can be seen that compared to the nonlinear least squares approach the analog-to-probability conversion saves

Table 1 Comparison of performance between analog-to-probability conversion and nonlinear least squares approach (each value is averaged over ten repetitions of the same experiment)

	APC	Least squares	Least squares (Inef. $\mathbf{X^0}$)
Samples	36	45	45
Comparisons	36	450	450
X_1 (Relative error)	1.127 (2.46%)	1.109 (0.83%)	1.108 (0.74%)
X_2 (Relative error)	1.254 (0.18%)	1.344 (6.98%)	1.888 (50.34%)

20% of the required samples and 92% of the comparisons at the comparator. As the energy consumption of the acquisition hardware is directly proportional to the number of comparisons, this shows the high potential for energy saving. The performance of the nonlinear least squares method is highly dependent on the initial guess of the parameter vector $\mathbf{X^0}$. For a favorable choice it can be seen that the accuracy of both methods is approximately equal. However, for an unfavorable choice of $\mathbf{X^0}$ the performance of the least squares method is significantly reduced. Moreover, analog-to-probability conversion continuously provides approximations for the signal parameters with increasing accuracy, while the least squares method first needs to sample a set of measurements before it can be executed.

5 Conclusion

In this paper analog-to-probability conversion is introduced as an efficient method for parameter extraction from signals that can be described by nonlinear signal models. Compared to other approaches a simplified sampling hardware is sufficient, less samples are required and for each sample only one comparison has to be done. The signal model provides a priori information about the next samples, which can be used to select the most efficient sampling points and corresponding comparator thresholds. For many applications the energy consumption can be further reduced by implementing a distributed system where the sampling hardware is separated from the processing unit. In areas like biomedical implants this significantly reduces the energy consumption of the implant, while a more powerful processor can be used on a device outside the body.

References

1. Basu, S., Duch, L., Peón-Quirós, M., Atienza, D., Ansaloni, G., Pozzi, L.: Heterogeneous and inexact: maximizing power efficiency of edge computing sensors for health monitoring applications. In: International Symposium on Circuits and Systems (2018)
2. Cypress Semiconductor: CY8CKIT-059 PSoC 5LP prototyping kit with onboard programmer and debugger (2017). http://www.cypress.com/CY8CKIT-059

3. Das, S., Martin, K.J.M., Coussy, P., Rossi, D.: A heterogeneous cluster with reconfigurable accelerator for energy efficient near-sensor data analytics. In: International Symposium on Circuits and Systems (2018)
4. Moore, S.K.: Intel starts R&D effort in probabilistic computing for AI. In: IEEE Spectrum Automaton Blog (2018)

Clustering Algorithm Exploring Road Geometry in a Video-Based Driver Assistant System

Norbert Bogya, Róbert Fazekas, Judit Nagy-György, and Zsolt Vizi

Abstract In this paper we present two algorithms for an advanced driver assistance system to investigate road geometry. The proposed solutions can handle both simple and complex scenarios, e.g. construction zones. Our input data consists of segments and polygonal paths, whose clustering gives a proper input for a lane model. The presented methods use thresholding and spectral clustering approaches.

1 Motivation

An autonomous car is a vehicle that is able to sense environmental data and to navigate based on them without any human action. Nowadays, the development of self-driving cars and driver assistance systems belongs to the most dynamic industrial projects. Advanced driver assistance systems (ADAS) are electronic devices which help the drivers during the processing of driving. One of the most important task is to manage the data produced by sensors. In this research we considered a stereo video camera as a sensor of perception and we implemented two algorithms to provide a clustering of the available lane segments. Most of the solutions in the industry assume simple environments, e.g. highway, for a functionality like adaptive cruise control (ACC), but these concepts fail, if the scene has a complexity. In this work we provide algorithms, which are robust in more complicated situations, e.g. in construction zones. As we know, there are no state-of-the-art solution for these situations. The algorithms in use are very sensitive for the quality and quantity of input data, and most of the well-know algorithms are hybrid algorithms for multiple sensors.

N. Bogya (✉) · J. Nagy-György
University of Szeged, Bolyai Institute, Szeged, Hungary
e-mail: nbogya@math.u-szeged.hu; ngyj@math.u-szeged.hu

R. Fazekas · Z. Vizi
Robert Bosch Kft., Budapest, Hungary
e-mail: Robert.Fazekas@hu.bosch.com; zsvizi@math.u-szeged.hu

© Springer Nature Switzerland AG 2019

I. Faragó et al. (eds.), *Progress in Industrial Mathematics at ECMI 2018*,
Mathematics in Industry 30, https://doi.org/10.1007/978-3-030-27550-1_75

A stereo-camera includes two separate cameras, taking pictures of the same focus used to model the environment in 3-dimension. At first, the pictures go through a process that recognises the road markings and it makes a segmentation of them. This method is not considered in this paper and we use only a little part of the data given by the segmentation. We use only on the geometrical representation of the lane segment and ignore e.g. colour of the lane marking, which would improve the consistency of the lane model. Additionally, we have to care about the complexity of the provided algorithms, because these are implemented in an embedded system with a small computing capacity. In this environment, it is complicated to apply neural networks, mostly model-based algorithms are considered.

2 Basic Concepts

Our input is a set of polygonal paths, whose vertices are the points of the road markings given by the pre-segmentation method mentioned earlier in Sect. 1. Two points are connected with a straight segment if the pre-segmentation procedure marks them as points of a same road marking. In this work, we consider a polygonal path as one object and investigate a relationship between these objects.

Our goal is to classify which polygonal path belongs to which road marking, thus this can be interpreted as a clustering problem. A key concept in all clustering algorithm to define a similarity function, which measures the "conceptual distance" of the sample data. If H is the set of data points, similarity function f is an $H \times H \to \mathbb{R}^+$ symmetric map. We use a normalised similarity function in order to interpret $f(h_1, h_2)$ as the probability of h_1 and h_2 belong to the same cluster.

In [6], the authors investigate a very similar question raised from image processing, namely a finite set of objects is given, and what can be said about the global connection of the objects, if we have information about all connection of each pairs of objects. Their global concept is a clustering into two clusters (foreground and background). They use the concept of Gestalt psychology, that have several principles about what kind of segments can be thought as a continuation of each other. Following their idea, our similarity function is

$$f(i, j) = \exp\left(-\frac{d_{i,j}^2}{d_0^2} - \frac{2 - \cos(2\alpha_i) - \cos(2\alpha_j)}{\alpha_0} - \frac{g_{i,j}}{g_0}\right), \tag{1}$$

if i and j denotes different segments, otherwise $f(i, j)$ is zero. The distance $d_{i,j}$ is measured between the midpoints of the segments i and j; and $g_{i,j}$ denotes the distance between these segments. The angles α_i and α_j can be understood from Fig. 1. The constants d_0, α_0, g_0 are real parameters.

Notice that, the first term of the exponent is zero if the midpoints of the segments coincide. Moreover, the second term is zero if and only if the segments are parts of

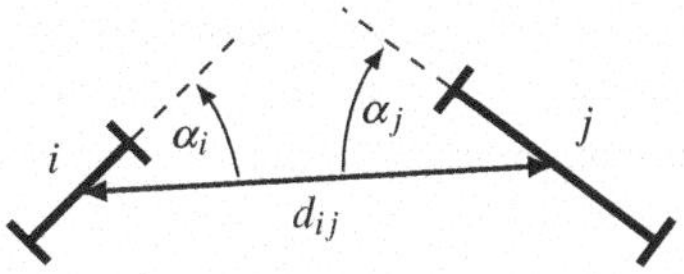

Fig. 1 Parameters for similarity of segments

the same straight line. Finally, the last term is zero if and only if the segments have a common endpoint. This similarity function takes its maximum, when the segments are the same, but with a good choice of the free parameters we can reach an almost maximum in that case, when the segments are the continuation of each other. This improvement can be obtained by using $\widetilde{d}_{i,j} = \left|\frac{\ell_i+\ell_j}{2} - d_{i,j}\right|$ (where ℓ_k denotes the length of segment k) instead of $d_{i,j}$ in the formula (1). Theoretically, this second version of similarity function should produce better results, but in practice, there are no significant difference between the two similarity functions. Therefore, we will present our results using (1) without any modification.

Instead of similarity of individual line segments we define similarity of provided polygonal paths. Investigating the inputs, we found that the connected polygonal paths can be contained in a very narrow but much longer rectangle. Hence, polygonal paths can be substituted by a segment between their first and last point.

From the similarity function, we can build a similarity matrix $W \in \mathbb{R}^{n\times n}$, where the rows and columns represent the polygonal paths and $W_{i,j} = f(i, j)$. If we consider this matrix as an edge-weight matrix of a graph G, the vertices of the defined graph are the polygonal paths and there is an edge between them if their similarity is positive.

In this approach, the original problem is converted to the clustering of the vertices of G and our goal now is to construct an efficient algorithm for edge-weighted graph clustering.

3 Algorithm #1

In [3], authors consider a thresholding method for clustering the vertices of the edge-weighted graph, i.e. in the first step the weight matrix W is transformed to a 0-1 matrix A:

$$A_{i,j} = \begin{cases} 1 & \text{if } W_{i,j} > \varepsilon, \\ 0 & \text{otherwise.} \end{cases}$$

The matrix A is symmetric, so it can be considered as an adjacency matrix of a graph, whose connected components give the clusters. We used depth-first search for finding these components.

The performance of the algorithm highly depends on the choice of threshold ε. In [3], the authors suggest an adaptive threshold $\varepsilon = \mu - \sigma + h$, where μ, σ and

h are the average, standard deviation and entropy of the nonzero elements of W, respectively. Our observations yield a larger threshold (thus stricter truncation of the elements), namely $\varepsilon = \mu + \sigma - h$, and this was implemented in our algorithm.

4 Algorithm #2

Our second algorithm is based on the so-called spectral clustering presented in [5]. Applying this method, the following question can be answered: how can we make vectors from the vertices of a graph, and embed them into $\mathbb{R}^k$ saving as much information as possible about the structure of the original graph [2, 4]. Let $S = \{s_1, \ldots, s_n\}$ be a set of d-dimensional points in the Euclidean space, but it can be used in arbitrary feature space with an appropriate measure of distance. The input is the edge-weight matrix W (which is calculated with the similarity function (1)), and the steps are the followings.

1. Let $D \in \mathbb{R}^{n \times n}$ be a diagonal matrix, such that $D_{i,i}$ is the sum of elements in the ith row of W.
2. We compute the $x_1, x_2, \ldots, x_k$ eigenvectors of matrix $D^{-1/2}WD^{-1/2}$ corresponding to the k largest eigenvalues. For eigenvalues with multiplicities greater than one, we choose the eigenvectors to be orthogonal to each other. We collect the vectors (as columns) in the matrix $X = [x_1, \ldots, x_k] \in \mathbb{R}^{n \times k}$.
3. We normalise the rows of X to obtain $Y \in \mathbb{R}^{n \times k}$.
4. We cluster the row vectors of Y, that are unit vectors.
5. The points $s_i, s_j \in S$ are declared to belong to the same cluster, if the corresponding row vectors of Y are declared into the same cluster.

In the second step, the matrix $M = D^{-1/2}WD^{-1/2}$ is used just for make the computations more simpler. If N is the symmetric normalised Laplacian of G, then $N = I - M$. The output of the algorithm is the same using either N or M. Intuitively, this step means an embedding of G into $\mathbb{R}^k$ according to the similarity matrix W. Then we project the vertices of this graph onto the unit sphere and make clusters from these points. More details about the mathematics behind the algorithm can be found in [1] or [7].

5 Comparison

In this chapter we present some results produced by our implemented MATLAB programs.

In the first row, on the left side of Fig. 2, the results of Algo #1 can be observed. The two most important lanes can be recognised easily: the green one is the original road marking, the red one is a detour. On the right side of this row, the result of the

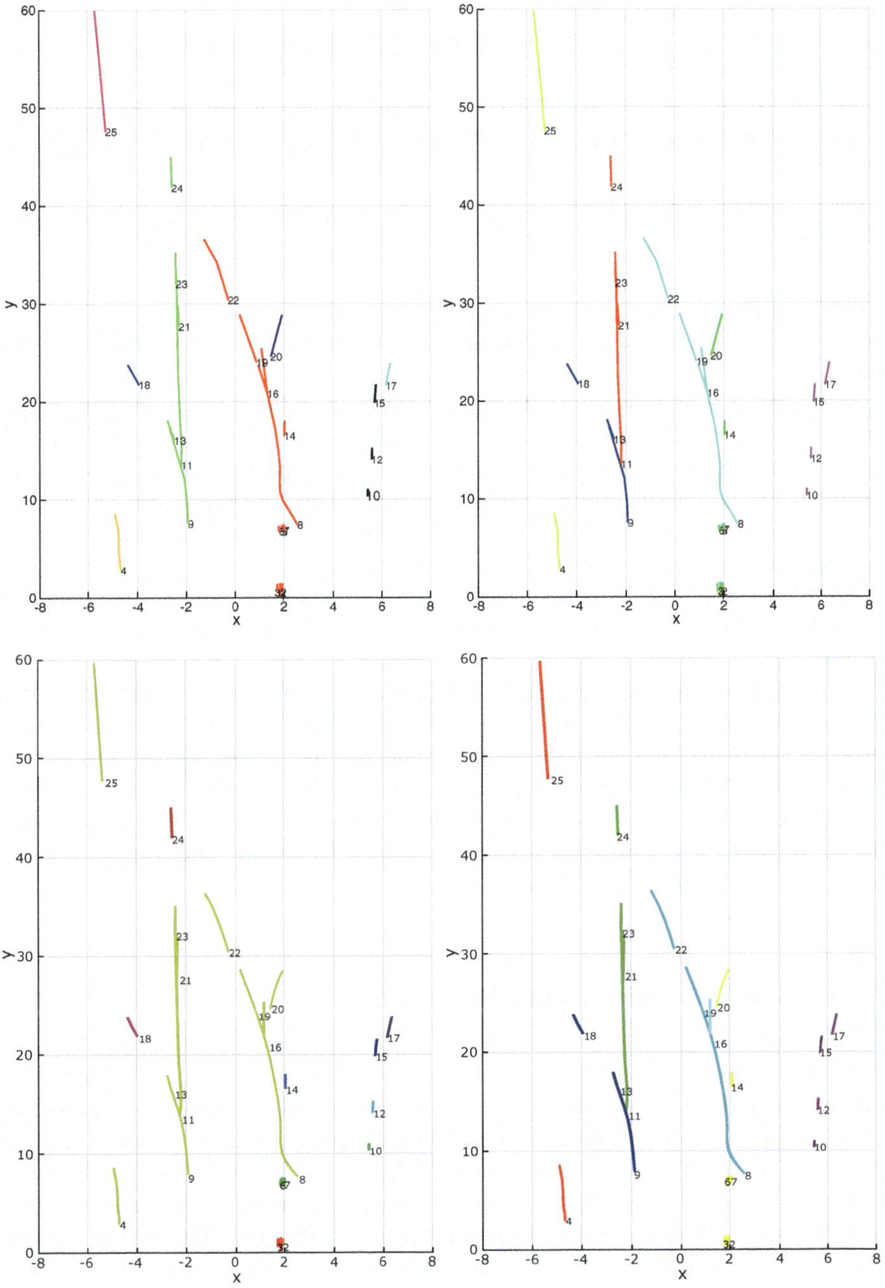

Fig. 2 First row: result of Algo #1 and Algo #2, respectively. Second row: result of algorithm in [3] and [6], respectively

Table 1 Run time of algorithms

	Drive #1	Drive #2	Drive #3	Drive #4	Drive #5	Drive #6	Drive #7	Drive #8
Algo #1	0.0179	0.0306	0.0379	0.0400	0.0440	0.0643	0.0666	0.0654
Algo #2 for 5 cl.	0.0273	0.0373	0.0440	0.0472	0.0509	0.0711	0.0717	0.0729
Algo #2 for 8 cl.	0.0285	0.0374	0.0476	0.0461	0.0559	0.0723	0.0761	0.0744

Table 2 Comparison of algorithms

Algo #1 won	Tie	Algo #2 won
27%	40%	33%

spectral clustering is presented, and the main lanes are almost the same as on the left.

As it can be seen, on the left side of the second row in Fig. 2 the method suggested in [3] with our threshold has a poor performance, since most of the line segments goes into the same cluster (the original threshold produces an even worse result). On the bottom right plot of Fig. 2 the result of the algorithm proposed in [6] for case of six clusters is shown. We can notice that this method could be parametrised to provide the same grouping as our algorithms.

In Table 1 we give some results about run times. Each drive consists of a sequence of pictures. The corresponding run times in the table are the average of these pictures.

In Table 2 we summarise a manual validation of our precesses. Tie means that the outputs are almost the same (99%) and they show the lanes correctly or they produce bad result, and it is hard to decide which is the worse. Algo #2 lost several times because it made connection between segments that are clearly not related to each other, just there was a pressure due to the predefined number of clusters. These numbers were chosen separately, manually by the pre-checking of images. Typically, they were between 3 and 6.

6 Conclusion and Further Research

As a conclusion, running the algorithms on several test cases, we cannot state, that one of the algorithms should be preferred over the other. With the thresholding method, a lot of information can be lost, and the theoretical investigation of spectral clustering suggests that it should be better, because more information can be saved and used. However, Algo #2 works with a predefined number of clusters, it may happen, that such polygonal paths can be clustered into the same group, that are probably not the parts of the same lane. For example, the segments 4 and 25 in Fig. 2 are calculated into the same cluster, but more likely, there is no connection between them, their distance is more than 35 m. Because of the predefined number of clusters, each path is put into one of the clusters, so cluster validation can be reasonable.

Another idea is to use machine learning methods. Both of the algorithms works with changeable parameters that can be tuned via training using good collections of input data.

Algo #1 can be implemented easier, and the clusters can be computed with depth-first search in contrast to the numeric eigenvalue problem of spectral method. And, of course, its running time is also less. The number of clusters are computed automatically and dynamically, we do not need any prediction about it.

Algo #2 uses k-means method, which needs an initialisation of centres (typically random) and pre-defined number of clusters. One idea to resolve both needs is combining the two algorithms: Algo #1 is run and centres of provided clusters initialises the k-mean method and we choose the number of clusters given by Algo #1.

More sophisticated evaluation of the algorithms and the latter idea will be investigated in the future.

Acknowledgements This research was supported by the project EFOP-3.6.2-16-2017-00015.

References

1. Chung, F.R.: Spectral graph theory. CBMS Regional Conference Series in Mathematics, vol. 92 (1997)
2. Hall, K.M.: An r-dimensional quadratic placement algorithm. Manag. Sci. Theory Ser. **17**(3), 219–229 (1970)
3. Kelly, A.R., Hancock, E.R.: Grouping-line segments using eigenclustering. In: Proceedings of the British Machine Vision Conference 2000, pp. 1–10 (2000)
4. Koren, Y.: On spectral graph drawing. In: Proceedings of the 9th Annual International Conference on Computing and Combinatorics, COCOON'03, pp. 496–508 (2003)
5. Ng, A.Y., Jordan, M.I., Weiss, Y.: On spectral clustering: analysis and an algorithm. In: Proceedings of the 14th International Conference on Neural Information Processing Systems: Natural and Synthetic, NIPS'01, pp. 849–856 (2002)
6. Perona, P., Freeman, W.: A factorization approach to grouping. In: Computer Vision ECCV'98. Lecture Notes in Computer Science, vol. 1406 (1998)
7. Shi, J., Malik, J.: Normalized cuts and image segmentation. IEEE Trans. Pattern Anal. Mach. Intel. **22**(8), 888–905 (2000)

Uncertainty Quantification for Real-Time Operation of Electromagnetic Actuators

František Mach and Karel Pospíšil

Abstract Model-based approach to fault detection of the linear electromagnetic actuators is proposed and validated in the framework of the electromagnetic actuator in the bistable valve operation. The forward uncertainty propagation for the nonlinear mathematical model is performed to determine the probability of faultless operation under aleatoric and epistemic uncertainties. A basic technique is then proposed to make a decision on occurred fault.

1 Technical Problem Formulation

Fault detection (FD), isolation, and recovery, are of huge importance in complex cyber-physical systems [4], especially for critical points of the system, such as high-performance electromechanical actuators. The uncertainties of the external influences can strongly reduce the safety and reliability of actuator operation, which brings many limitations for the cutting-edge applications. The major goal of the presented research is to develop a model-based and also sensorless fault detection approach for linear electromagnetic actuators in high-speed and fail-safe applications.

The proposed technique was designed originally for a bistable electromagnetic valve in a coaxial design, but it can be utilized for an arbitrary linear electromagnetic actuator, for example [1, 7]. Figure 1 shows a simplified arrangement of the considered valve. Its main part is represented by an axially symmetric segmented magnetic circuit supplemented with permanent magnets (PM) and a hollow cylin-

F. Mach (✉)
Regional Innovation Centre for Electrical Engineering, University of West Bohemia, Pilsen, Czech Republic
e-mail: fmach@rice.zcu.cz; fmach@kte.zcu.cz

K. Pospíšil
Department of Theory of Electrical Engineering, University of West Bohemia, Pilsen, Czech Republic
e-mail: pospisik@kte.zcu.cz

© Springer Nature Switzerland AG 2019
I. Faragó et al. (eds.), *Progress in Industrial Mathematics at ECMI 2018*, Mathematics in Industry 30, https://doi.org/10.1007/978-3-030-27550-1_76

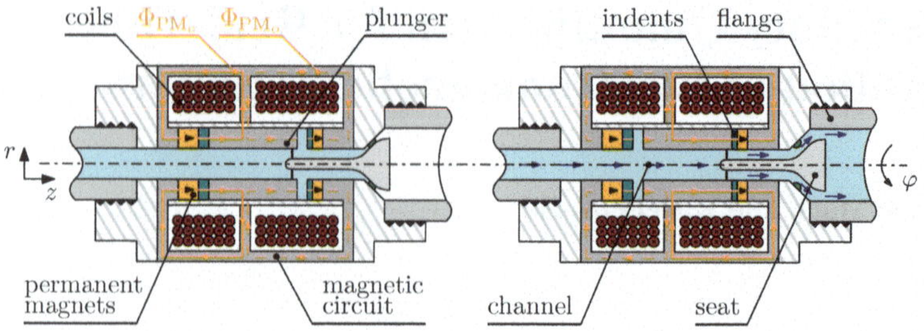

Fig. 1 A simplified arrangement of the bistable electromagnetic valve in a coaxial design. The left part shows a fully closed valve, and the right part shows a fully opened valve. Both positions are stable and secured by permanent magnets [6]

drical movable plunger. Two solenoidal coils are placed inside the magnetic circuit, isolated by a non-magnetic gasket from the channel in the centre of the valve. A ferromagnetic plunger is placed inside the valve channel, and its movement has just one degree of freedom (along the z-axis). The conical valve body is connected to the movable plunger by a thin shaft ([6] discusses the valve in more details).

The fluid flow in the channel is controlled by swapping of the movable plunger position (*on/off* mode). This is ensured by excitation of the field coils. An illustrative faultless and fault operations of the valve are shown in Fig. 2. Whereas presented FD and its recovery were found by direct measurement of the fluid flow by a flow meter, the proposed magnetic-flux FD is sensorless and takes advantage of the measurement of the induced voltage in the field coils.

The valve works in four modes, *closed*, *opening*, *open* and *closing*. The bistable modes (closed and opened) are secured by the cylindrical permanent magnets placed in the magnetic circuit. Both stable modes are shown in Fig. 1. Transition modes

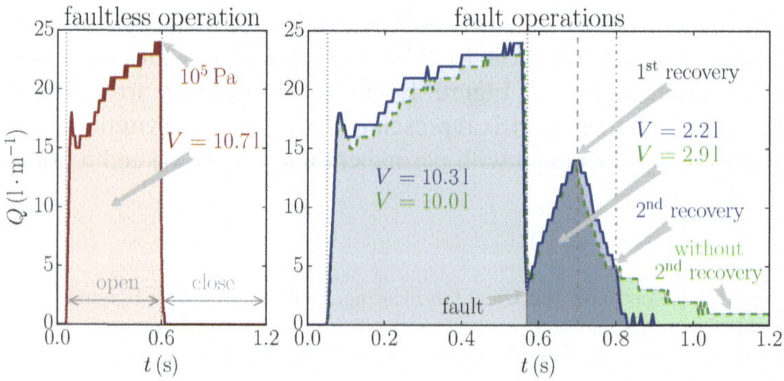

Fig. 2 Flow rates Q measured on laboratory prototype [6] of the valve. The left part shows the faultless operation of the valve and right part shows the operation when the fault was produced by shock pressure in the experimental circuit (V denotes the volume of liquid passing through the valve)

(opening and closing) are initiated by excitation of the appropriate field coil by DC current (the coils of the experimental prototype are designed for current $I = 2.5$ A). In the case of opening mode, magnetic flux Φ_E produced by the current i_c in the opening coil boosts magnetic flux Φ_0, the plunger is actuated by the magnetic field and starts moving to the open position. The closing mode is then driven by the magnetic flux Φ_E produced by the current i_c in the closing coil. Permanent magnets are not identical from the magnetic viewpoint ($\Phi_{PM_o} \neq \Phi_{PM_c}$).

2 Mathematical Model

The distribution of the magnetic field in the system for any position of the plunger is described by the equation for magnetic vector potential $\mathbf{A}$ in the form

$$\text{curl}\,\left(\mu(|\mathbf{B}|)^{-1}(\text{curl}\,\mathbf{A} + \mathbf{B}_r)\right) - \gamma\mathbf{v} \times \text{curl}\,\mathbf{A} + \gamma\frac{\partial\mathbf{A}}{\partial t} = \mathbf{J}_{\text{ext}}\,, \tag{1}$$

where μ denotes the magnetic permeability, which is a nonlinear function of the magnetic flux density $\mathbf{B} = \text{curl}\,\mathbf{A}$, symbol $\mathbf{B}_r$ stands for the remanent flux density, $\mathbf{v}$ stands for the plunger velocity, γ stands for the specific conductivity of the magnetic circuit and $\mathbf{J}_{\text{ext}}$ denotes the density of the field current in the field coils ($\mathbf{J}_{\text{ext}}(i_c, i_o)$).

Analysis of the magnetic field described by (1) represents a critical issue for fast simulation by the control algorithms. In the proposed technique, the magnetostatic problem ($\gamma = 0$) can be solved and magnetic flux Φ coupled with the field coils can be calculated only for the initial and final positions of the plunger in the form

$$\Phi = \iint_S \mathbf{B}\,\text{d}\mathbf{S}\,, \tag{2}$$

where $\mathbf{S}$ is the oriented internal cross-section of the field coil. The mathematical model can be solved as a 2D axisymmetric problem in the case of our prototype. The solution area is bounded by an artificial boundary characterized by the Dirichlet boundary condition in the form $\text{A}_z = 0$. Figure 3 shows the results of the numerical solution for the closing operation of the discussed valve.

3 Fault Detection Technique

The basic idea of the proposed FD technique is to observe changes of induced voltage u_{i_c} or u_{i_o} in the non-excited coil due to changes of plunger position δ and also due to time dependence of the currents i_c and i_o. Measurement of the voltage u_i can also be performed on the excited coil with consideration of the source voltage U_c or U_o, which is the DC component of the measured signal.

The technique compares differences in magnetic flux $\Delta\Phi = \Phi_E - \Phi_0$, where Φ_0 is magnetic flux coupled with the appropriate field coil at the beginning of the

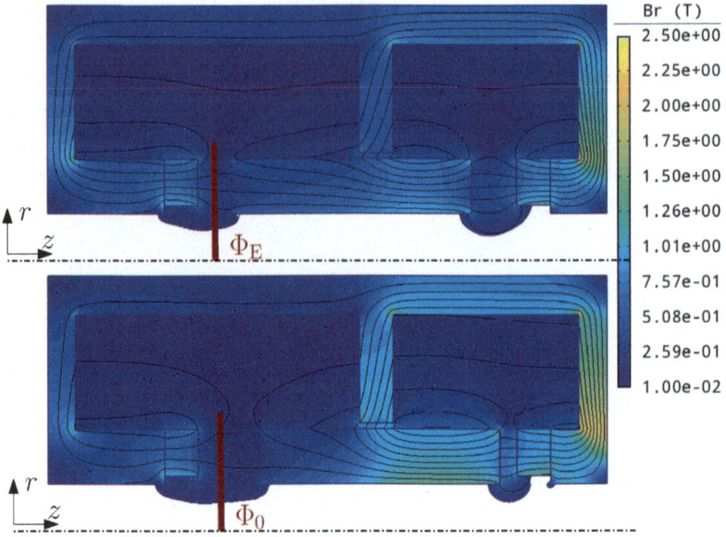

Fig. 3 Distribution of magnetic flux density **B** for two modes of the valve. The left part shows fully opened valve (plunger position $\delta = 5\,\text{mm}$) with non-excited coils, and the right part depicts the end of the closing operation (plunger position $\delta = 0\,\text{mm}$, closing coil is supplied by the current i_c). Magnetic flux $\triangle\Phi$ is calculated from the distribution of the magnetic flux density **B** in the internal area of the opening coil

operation, and Φ_E is magnetic flux after finishing of the transition mode. Flux $\Phi_0 \approx \Phi_{PM_c}$ or $\Phi_0 \approx \Phi_{PM_o}$ as long as $i_c = i_o = 0$ (see Fig. 1).

In the case of an FD, magnetic flux $\triangle\Phi$ can be calculated from the time integration of measured u_{i_c} or u_{i_o} during particular operations with thresholds determined from the calculated probability density function PDF($\triangle\Phi$). Whereas magnetic flux $\triangle\Phi$ is measured indirectly, PDF($\triangle\Phi$) can be obtained by the analysis of the forward uncertainty propagation using finite element analysis (FEA) of the model.

Whereas FD can be carried out on the basis of precalculated thresholds, fault isolation has to be performed by the solution of the inverse problem for observed $\triangle\Phi$ or by the advanced analysis of the measured voltage u_i with detailed knowledge of possible faults of the valve.

4 Experimental Validation

The uncertainty of the final position[1] of the plunger δ, which is caused mainly by the uncertainty of the fluid pressure and also by intents and gaskets plasticity and degradation, has the major impact on the probability of valve operation faults.

[1]The final position of the plunger is reached at the end of the operation. In the case of the opening mode, the final position is equal to $\delta = 5\,\text{mm}$.

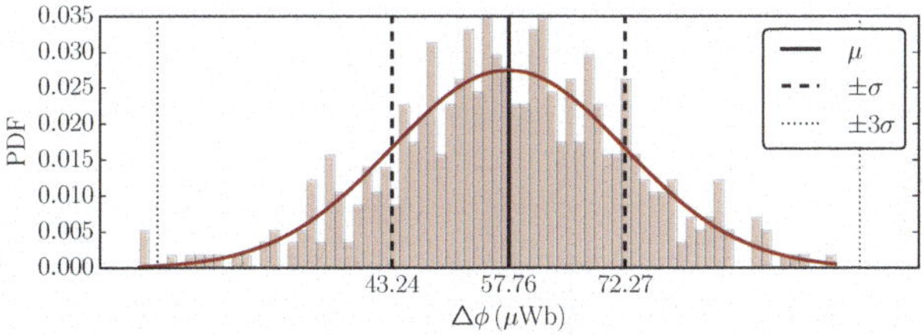

Fig. 4 Results of the forward uncertainty propagation defined by the PDF of the magnetic flux $\Delta\Phi$. Results are obtained for $\mathcal{N}_{U_o}(95\,\text{V}, 5\,\text{V})$ V and $\mathcal{N}_\delta(4.5\,\text{mm}, \sigma_\delta = 0.5\,\text{mm})$

Furthermore, the uncertainties of the external source voltages U_c and U_o are also non-negligible. PDFs for both uncertainties were experimentally estimated on the prototype and response of the model was determined by uncertainty propagation.

The analysis of uncertainty propagation in the nonlinear magnetic model (see Sect. 2) of the valve was performed using the latin hypercube sampling (LHS) with multidimensional uniformity [2, 3]. The parameter space for the final plunger position δ and the source voltage U_o of the opening coil was uniformly sampled by LHS. Inverse transform sampling was then used to the conversion of samples to normal distribution $\mathcal{N}_{U_o}(\mu_{U_o}, \sigma_{U_o})$ and normal distribution of the final plunger position $\mathcal{N}_\delta(\mu_\delta, \sigma_\delta)$. Normal distribution was used based on experimental investigation, and Agros2D [5] was used for the numerical solution of the model (1) by FEA.

The response of the model was characterized by the PDF of the magnetic flux $\Delta\Phi$. From the results can be concluded, that dominant influence on the model response has the final position of the plunger δ. While distribution $\mathcal{N}_{U_o}$ with variance $\sigma_{U_o} = 5\,\%\,\mu_{U_o}$ causes response of $\Delta\Phi$ equal to distribution $\mathcal{N}_{\Delta\Phi}$ with variance $\sigma_{\Delta\Phi} \approx 8\,\%\,\mu_{\Delta\Phi}$, distribution $\mathcal{N}_\delta$ with variance $\sigma_\delta = 10\,\%\,\mu_\delta$, causes response of $\Delta\Phi$ equal to distribution $\mathcal{N}_{\Delta\Phi}$ with variance $\sigma_{\Delta\Phi} \approx 24\,\%\,\mu_{\Delta\Phi}$.

Finally, Fig. 4 shows the response of the numerical model to uncertainty for $n = 500$ samples from two-dimensional parameter space for both parameters, voltage U_o and final plunger position δ ($\mathcal{N}_{U_o}(95\,\text{V}, 5\,\text{V})$, $\mathcal{N}_\delta(4.5\,\text{mm}, 0.5\,\text{mm})$). Distribution of PDF($\Delta\Phi$) is characterized by $\mu_{\Delta\Phi} = 57.76\,\mu\text{Wb}$ and variance $\sigma_{\Delta\Phi} \approx 25\,\%\,\mu_{\Delta\Phi}$. The faultless opening operation should occur in the thresholds $\Delta\Phi \in [14.2, 101.3]\,\mu\text{Wb}$.

Experimental validation of the technique was performed based on the thresholds obtained by the forward uncertainty propagation (see Fig. 4), and the results of several experiments are concluded in Fig. 5. Since the uncertainty of the final plunger position is difficult to simulate experimentally, validation was performed for several different voltages U_o. Insufficient voltage U_o of the opening coil excites small current i_o in the opening coil and the plunger does not reach the final position $\delta = 4.5$ mm. In this case, the plunger is returned to the closed position because of $\Phi_{\text{PM}_o} < \Phi_{\text{PM}_c}$ (plunger just jerks). On the basis of the measurement, the nominal opening voltage is equal to $U_{\text{oN}} = 97$ V and voltages $U_o <= 95$ V do not produce sufficient current i_o (difference with designed $U_{\text{oN}} = 95$ V is inflicted mainly by differences in the material parameters, especially H_c of the PM).

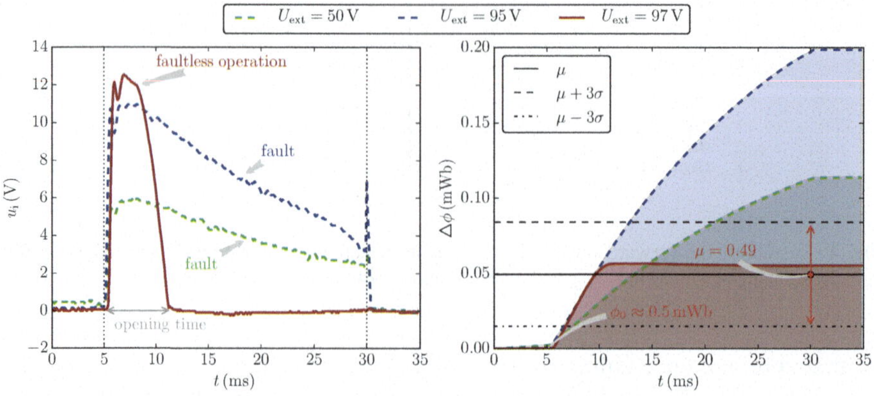

Fig. 5 Induced voltage u_{i_c} measured for faultless ($U_o = 97$ V) and fault operations ($U_o \leq 95$ V) (left part) and comparison of $\triangle\Phi$ obtained by integration of measured u_{i_c} with results of model-based uncertainty quantification depicted by thresholds (right part)

From the measurement of the voltage u_{i_c}, it is evident a huge difference between faultless and fault operations, if source voltage U_o is close to the nominal value. This difference is much expressed in the magnetic flux $\triangle\Phi$, where magnetic flux $\triangle\Phi = 55.56\,\mu$Wb for source voltage $U_o = 97$ V and $\triangle\Phi = 195.73\,\mu$Wb for voltage $U_o = 95$ V. Possible problems may be caused if the source voltage U_o is close to its nominal value ($U_o << U_{oN}$), where the voltage u_{i_c} is different, but $\triangle\Phi$ is close to the faultless operation. In this case, the additional criterion for $u_i \approx 0$ at the end of the operation process has to be added to the control algorithm.

5 Conclusion

Sensorless fault detection technique was proposed and validated on the laboratory prototype of the electromagnetic actuator in the valve operation. The technique enables to reduce the computational effort of the uncertainty propagation while maintaining the complexity of the mathematical model. Further research and development in the domain will be aimed at model-based fault isolation and recovery.

Acknowledgements This research has been supported by the Ministry of Education, Youth and Sports of the Czech Republic under the project LO1607 and by the University of West Bohemia under the project SGS-2018-043.

References

1. Ando, R., Koizumi, M., Ishikawa, T.: Development of a simulation method for dynamic characteristics of fuel injector. IEEE Trans. Magn. **37**(5), 3715–3718 (2001)
2. Davey, K.R.: Latin hypercube sampling and pattern search in magnetic field optimization problems. IEEE Trans. Magn. **44**(6), 974–977 (2008)
3. Davey, K.R.: Latin hypercube sampling with multidimensional uniformity. J. Stat. Plan. Inf. **142**(3), 763–772 (2012)
4. Guasp, M.R., Daviu, J.A., Capolino, G.A.: Advances in electrical machine, power electronic, and drive condition monitoring and fault detection: state of the art. IEEE Trans. Ind. Electr. **62**(3), 1746–1759 (2015)
5. Karban, P., Mach, F., Kůs, P.: Numerical solution of coupled problems using code Agros2D. Computing **95**(1), 381–408 (2013)
6. Mach, F., Kurfiřt, M., Doležel, I.: Bistable fully electromagnetic valve for high-speed and fail-safe operations. IEEE Trans. Ind. Electr. **66**(1), 349–357 (2019)
7. Mercorelli P.: A two-stage sliding-mode high-gain observer to reduce uncertainties and disturbances effects for sensorless control in automotive applications. IEEE Trans. Ind. Electr. **62**(9), 5929–5940 (2015)

Non-asymptotic Confidence Regions for Regularized Linear Regression Estimates

Balázs Csanád Csáji

Abstract Building *confidence regions* for regression models is of high importance, for example, they can be used for uncertainty quantification and are also fundamental for robust optimization. In practice, these regions are often computed from the asymptotic distributions, which however only lead to heuristic confidence sets. Sign-Perturbed Sums (SPS) is a resampling method which can construct *exact*, *non-asymptotic*, *distribution-free* confidence regions under very mild statistical assumptions. In its standard form, the SPS regions are built around the least-squares estimate of linear regression problems, and have favorable properties, such as they are star convex, strongly consistent, and have efficient ellipsoidal outer-approximations. In this paper, we extend the SPS method to *regularized* estimates, particularly, we present variants of SPS for ridge regression, LASSO and elastic net regularization.

1 Introduction

Estimating models based on noisy measurements is a fundamental problem for many scientific, engineering and economic applications. A very important issue in practice is to quantify the uncertainty of the obtained models. This is often done by building confidence regions for the models. While these regions are frequently built using the limiting distribution of the used point-estimate [6], such regions are not guaranteed for finite samples, and can only be seen as heuristics. It is of high importance to construct confidence regions with non-asymptotic guarantees, using minimal statistical assumptions. Resampling methods, such as bootstrap and Monte Carlo approaches, typically use some regularity of the noise to build such regions.

Sign-Perturbed Sums (SPS) is a recently developed resampling method with favorable properties. SPS can construct *exact*, distribution-free confidence regions

B. Cs. Csáji (✉)
EPIC Centre of Excellence, MTA SZTAKI: Institute for Computer Science and Control, Hungarian Academy of Sciences, Budapest, Hungary
e-mail: balazs.csaji@sztaki.mta.hu

© Springer Nature Switzerland AG 2019
I. Faragó et al. (eds.), *Progress in Industrial Mathematics at ECMI 2018*, Mathematics in Industry 30, https://doi.org/10.1007/978-3-030-27550-1_77

for finite samples [3, 7]. Its standard form constructs (star convex, strongly consistent) confidence sets around the least-squares estimate of linear regression problems.

Regularization is an important tool in regression which can help, for example, to handle ill-posed and ill-conditioned problems, reduce over-fitting, enforce sparsity, and in general to control the shape and smoothness of the regression function. The paper extends SPS to various regularized linear regression problems, particularly, to ridge regression (Tikhonov regularization), LASSO and elastic net regularization.

2 Preliminaries: Asymptotic Confidence Ellipsoids

We start by recalling the standard "textbook" approach to build (asymptotic) confidence ellipsoids around the least-squares estimate of linear regression problems.

Assume we are given a data sample, $\mathcal{D}_n \doteq \{(\varphi_1, y_1), \dots, (\varphi_n, y_n)\}$, with

$$y_t \;\doteq\; \varphi_t^{\mathrm{T}} \theta^* + \varepsilon_t, \qquad \text{for} \qquad t = 1, \dots, n \tag{1}$$

where y_t is the *output*, φ_t is the *input* or *regressor* and ε_t is the (non-observable) *noise* for measurement t. We aim at estimating the (constant) "true" parameter, $\theta^* \in \mathbb{R}^d$. We assume that $\{\varphi_t\} \subset \mathbb{R}^d$ are *deterministic* and the noise $\{\varepsilon_t\}$ is an *independent* sequence of random variables, each having a *symmetric* distribution about zero, that is the distribution of ε_t is the same as that of $-\varepsilon_t$. Finally, for simplicity, we assume that the *regressor matrix*, $\Phi \doteq [\,\varphi_1, \dots, \varphi_n\,]^{\mathrm{T}}$, is skinny ($n > d$) and full rank.

One of the standard estimators is the well-known *least-squares* (LS) method

$$\hat{\theta}_n \;\doteq\; \operatorname*{arg\,min}_{\theta \in \mathbb{R}^d} V(\theta \mid \mathcal{D}_n) \;=\; \operatorname*{arg\,min}_{\theta \in \mathbb{R}^d} \frac{1}{2} \| \, y - \Phi\theta \, \|_2^2, \tag{2}$$

where $y \doteq [y_1, \dots, y_n]^{\mathrm{T}}$; $\hat{\theta}_n$, can be obtained from the *normal equation*, that is

$$\nabla_\theta V(\hat{\theta}_n \mid \mathcal{D}_n) \;=\; \Phi^{\mathrm{T}} \Phi \, \hat{\theta}_n - \Phi^{\mathrm{T}} y \;=\; 0, \tag{3}$$

which has a unique analytical solution, famously given by $\hat{\theta}_n = (\Phi^{\mathrm{T}}\Phi)^{-1}(\Phi^{\mathrm{T}} y)$.

A crucial question is that how can we *quantify the uncertainty* of the so obtained estimate? This question can be answered, e.g., by constructing *confidence regions* around the point-estimate. More precisely, given a confidence probability $p \in (0, 1)$, we aim at finding a region, $\widehat{\Theta}_{\mathcal{D}_n, p}$ around $\hat{\theta}_n$, such that $\mathbb{P}\big(\theta^* \in \widehat{\Theta}_{\mathcal{D}_n, p}\big) \geq p$.

The standard method to build such regions is to use the *asymptotic distribution* of the estimate [6]. It is known that the (scaled) error of LS is asymptotically Gaussian,

$$\sqrt{n}\,(\hat{\theta}_n - \theta^*) \xrightarrow{\;d\;} \mathcal{N}(0, \sigma^2 R^{-1}), \qquad \text{as} \qquad n \to \infty, \tag{4}$$

where $\mathcal{N}(\mu, \Sigma)$ is the (multivariate) Gaussian distribution with mean μ and covariance Σ. This property holds under various conditions, e.g., if the regressors are bounded, there exits a positive definite matrix R as the limit of matrices $R_n \doteq \frac{1}{n}\Phi_n^{\mathrm{T}}\Phi_n$, and $\{\varepsilon_t\}$ are i.i.d. as well as $\mathbb{E}[\varepsilon_t] = 0$ and $\mathbb{E}[\varepsilon_t^2] = \sigma^2$, with $0 < \sigma^2 < \infty$.

Using the limiting distribution, a (heuristic) *confidence ellipsoid* can be built by

$$\widetilde{\Theta}_{n,p} \doteq \left\{ \theta \in \mathbb{R}^d : (\theta - \hat{\theta}_n)^{\mathrm{T}} R_n (\theta - \hat{\theta}_n) \leq \frac{q\,\hat{\sigma}_n^2}{n} \right\}, \tag{5}$$

where $p = F_{\chi^2(d)}(q)$, with $F_{\chi^2(d)}$ being the CDF of the χ^2 distribution with d degrees of freedom; and $\hat{\sigma}_n^2$ is an (unbiased) estimate of the noise variance, that is

$$\hat{\sigma}_n^2 \doteq \frac{1}{n-d} \sum_{t=1}^{n} (y_t - \varphi_t^{\mathrm{T}} \hat{\theta}_n)^2. \tag{6}$$

Then, we *approximately* have $\mathbb{P}(\theta^* \in \widetilde{\Theta}_{n,p}) \approx p$ (and, obviously, $\hat{\theta}_n \in \widetilde{\Theta}_{n,p}$).

However, the confidence regions constructed using the asymptotic distribution are *not guaranteed* for finite samples, and are typically *imprecise* if the sample size is small. Another drawback of the asymptotic approach is that it presupposes the *existence* of a limiting distribution, which cannot be guaranteed in certain cases.

3 Sign-Perturbed Sums: Non-asymptotic Confidence Regions

Now, we overview the *Sign-Perturbed Sums* (SPS) method [3, 7] that can construct *exact, non-asymptotic, distribution-free* confidence regions around the LS estimate.

As first glance, SPS can be seen as a *hypothesis testing* method. It tests the null hypothesis $\theta = \theta^*$, against the alternative hypothesis $\theta \neq \theta^*$. SPS is based on the idea that if $\theta = \theta^*$, then (1) we can compute the exact realization of the noise vector, $\varepsilon = (\varepsilon_1, \dots, \varepsilon_n)^{\mathrm{T}}$, by "inverting" the system, and (2) using some *regularity* of the noise (e.g., symmetry), alternative noise realizations can be generated, leading to alternative samples and estimates, which behave "similarly" (in the statistical sense) to the original ones. On the other hand, if $\theta \neq \theta^*$, then the residuals will be biased and the alternative samples and estimates based on them will behave statistically differently than the original ones. SPS applies a rank-test to decide whether the perturbed objects are similar to the original ones. Unlike other resampling based approaches, SPS avoids actually constructing the alternative samples and fitting surrogate models to them, as it directly perturbs the *gradient* of the objective function.

The principal building blocks of SPS are the following *evaluation functions*,

$$Z_i(\theta) \doteq \| \Psi^{1/2}\Phi^{\mathrm{T}} G_i\big(y - \Phi\theta\big) \|_2^2, \tag{7}$$

for $i \in \{0, 1, \ldots, m-1\}$, where $\Psi = (\Phi^{\mathrm{T}}\Phi)^{-1}$, $m > 0$ is a user-chosen integer, $G_0 \doteq I$, the identity matrix, and for $i \neq 0$, $G_i \doteq \mathrm{diag}(\alpha_{i,1}, \ldots, \alpha_{i,n})$; $\{\alpha_{i,j}\}$ are i.i.d. Rademacher variables[1]; and $\mathrm{diag}(\cdot)$ builds a diagonal matrix from its argument.

Notice that, apart from an (optional) linear transformation, $\Psi^{1/2}$, whose role is to make a covariance correction, $Z_0(\theta)$ is basically the norm of the (negative) *gradient* of the least-squares objective. The difference between $Z_0(\theta)$ and $Z_i(\theta)$, $i \neq 0$, is that in latter functions the signs of the residuals $(y - \Phi\theta)$ are perturbed in the gradient.

In case $\theta = \theta^*$, the residuals are the true noises, $y - \Phi\theta^* = \varepsilon$, and we know from the symmetry assumption that for all i, ε and $G_i\,\varepsilon$ have the same distribution, where G_i is a diagonal matrix containing random signs as defined above. Then,

$$Z_0(\theta^*) = \| \Psi^{1/2}\Phi^{\mathrm{T}}\varepsilon \|_2^2 \stackrel{d}{=} \| \Psi^{1/2}\Phi^{\mathrm{T}} G_i\,\varepsilon \|_2^2 = Z_i(\theta^*), \tag{8}$$

for $i = 1, \ldots, m-1$, where "$\stackrel{d}{=}$" denotes equality in distribution. Nevertheless, variables $\{Z_i(\theta^*)\}$ are of course not independent. On the other hand, it can be proved [3] that they are *conditionally i.i.d.*, conditioned on the σ-algebra generated by $\{|\varepsilon_t|\}$. Consequently, they are also *exchangeable* and hence each ordering[2] of them, $Z_{i_0}(\theta^*) \prec \cdots \prec Z_{i_{m-1}}(\theta^*)$, has the same probability, namely, $1/m!$.

If however, $\theta \neq \theta^*$, then this exchangeability argument does not hold, moreover, $Z_0(\theta)$ will eventually dominate $\{Z_i(\theta)\}_{i\neq 0}$ with high probability as $\|\theta - \theta^*\| \to \infty$.

To make these ideas more precise, let us define the *normalized rank* of $Z_0(\theta)$ as

$$\mathscr{R}(\theta) \doteq \frac{1}{m}\left[1 + \sum_{i=1}^{m-1} \mathbb{I}\left(Z_0(\theta) \prec Z_i(\theta)\right)\right], \tag{9}$$

where $\mathbb{I}(\cdot)$ is an indicator (its value is 1 if its argument is true and 0 otherwise).

Assume that the target confidence probability can be written as $p = 1 - q/m$ where $0 < q < m$ are user-chosen integers. Then, SPS accepts the null hypothesis, $\theta = \theta^*$, if $\mathscr{R}(\theta) \leq p$, and rejects it if $\mathscr{R}(\theta) > p$. As m and q are free-parameters, they are under our control, hence any (rational) probability can be achieved.

---

[1]Random variables which take values $+1$ and -1 with probability $1/2$ each.

[2]Relation "$\prec$" is a total order which we get from "$<$" by random tie-breaking, see [3].

Based on these observations, the *SPS confidence regions* can be defined as

$$\widehat{\Theta}_{n,p} \doteq \Big\{ \theta \in \mathbb{R}^d : \mathscr{R}(\theta) \leq p \Big\}. \tag{10}$$

It can be proved [3] that these regions have *exact* confidence $\mathbb{P}\big(\theta^* \in \widehat{\Theta}_{n,p}\big) = p$. Note that the exact confidence of the regions is guaranteed for *finite samples* despite no knowledge about the particular noise distributions is assumed, moreover, each noise term may have a different distribution with arbitrarily large variance.

There are several important properties of SPS confidence regions [3, 7]. For example, (1) they are *star convex* with the LS estimate as a star center; (2) they are uniformly *strongly consistent*; (3) they have asymptotically the *same size and shape* as the classical confidence ellipsoids; finally (4) they have *ellipsoidal outer approximation* that can be efficiently computed via semidefinite programming problems.

SPS has several generalizations, for example, it can be extended to general stochastic linear (dynamical) systems, even if they are operating in closed-loop [2], and to various non-linear dynamical systems, such as GARCH models [1].

Finally, we note that working with symmetric noises is not crucial for SPS as the theory can be extended to other noise distributions, as long as we know a group of transformations that leave the (joint) distribution of the noises unchanged. For example, one can assume that the noises are *exchangeable* and use random *permutation* matrices as $\{G_i\}$, see [5]. We refer to these generalized variants as (G)SPS.

4 Non-asymptotic Confidence Sets for Regularized Estimates

In this section we are going to extend the theory of (G)SPS, in order to construct non-asymptotic, distribution-free confidence regions around regularized estimates.

First, we consider *ridge regression* (RR) which has the objective function

$$V_{\mathrm{R}}(\theta) \doteq \frac{1}{2} \| y - \Phi\theta \|_2^2 + \frac{\lambda}{2} \| \theta \|_2^2, \tag{11}$$

for a $\lambda \geq 0$ hyper-parameter. It is well-known that RR can be reformulated as LS,

$$\widetilde{\Phi} = \begin{bmatrix} \Phi \\ \sqrt{\lambda}\, I \end{bmatrix}, \qquad \text{and} \qquad \widetilde{y} = \begin{bmatrix} y \\ 0 \end{bmatrix}, \tag{12}$$

where I is the identity matrix, after which we have $V_{\mathrm{R}}(\theta) = 1/2 \,\| \widetilde{y} - \widetilde{\Phi}\theta \|^2$.

Then, one might be tempted to apply standard SPS to the obtained (ordinary) LS formulation. However, we should proceed with caution, as the new problem has some auxiliary output terms, the zero part of $\widetilde{y}$, to which there are no real noise terms in the original problem. Therefore, the last d terms of the residual vector, $\widetilde{y} - \widetilde{\Phi}\theta$, should not be perturbed, as the distributional invariance was only assumed for the original noise vector. Consequently, the $\{G_i\}$ matrices should be extended by

$$\widetilde{G}_i \doteq \begin{bmatrix} G_i & 0 \\ 0 & I \end{bmatrix}, \tag{13}$$

for $i = 1, \ldots, m-1$. Then, using an analogue of (7) to the new LS system with $\{\widetilde{G}_i\}$ perturbations, we arrive at the *(G)SPS evaluation function for ridge regression*,

$$Z_i(\theta) \doteq \| \Psi_{\mathrm{R}}^{1/2} \Big[\Phi^{\mathrm{T}} G_i \, (y - \Phi\theta) - \lambda\theta \Big] \|_2^2, \tag{14}$$

where $\Psi_{\mathrm{R}} = (\Phi^{\mathrm{T}}\Phi + \lambda\, I)^{-1}(\Phi^{\mathrm{T}}\Phi)(\Phi^{\mathrm{T}}\Phi + \lambda\, I)^{-1}$ is a correction term from the covariance of RR. Based on this evaluation function, *exact* confidence regions can be built around the RR estimate, using the same steps as we had for standard SPS.

Now, let us consider *LASSO* (least absolute shrinkage and selection operator) which applies L1 regularization to enforce *sparsity*. It has the objective function

$$V_{\mathrm{L}}(\theta) \doteq \frac{1}{2} \| y - \Phi\theta \|_2^2 + \lambda \| \theta \|_1, \tag{15}$$

for $\lambda \geq 0$. This objective is no more quadratic and it cannot be traced back to LS. However, the underlying idea of SPS, i.e., to perturb the residuals in the (negative) gradient of the objective, can still be applied. A (sub-) gradient[3] of (15) is

$$\nabla_\theta V_{\mathrm{L}}(\theta) = \Phi^{\mathrm{T}}\Phi\,\theta - \Phi^{\mathrm{T}} y + \lambda\, \mathrm{sign}(\theta), \tag{16}$$

where the sign function is understood component-wise.

Then, we can proceed in the same way as before and perturb the residuals in (16) with $\{G_i\}$, leading to the *(G)SPS evaluation function for LASSO*,

$$Z_i(\theta) \doteq \| \Psi_{\mathrm{L}}^{1/2} \Big[\Phi^{\mathrm{T}} G_i \, (y - \Phi\theta) - \lambda\, \mathrm{sign}(\theta) \Big] \|_2^2, \tag{17}$$

where Ψ_{L} is an (optional) correction term, e.g., using the (asymptotic) results of [4], we may use $\Psi_{\mathrm{L}} = (\Phi^{\mathrm{T}}\Phi)^{-1}$. The correction matrix can be interpreted as the square-

[3]For our purposes, one of the subgradients is sufficient, thus we do not treat $\nabla_\theta V$ set-valued.

root of the (estimated) covariance of LASSO (modulo the variance of the noise, as multiplying each Z_i with the same positive scalar does not affect their ordering).

The last method that we discuss is the *elastic net* regularization with objective

$$V_{\mathrm{E}}(\theta) \doteq \frac{1}{2} \| y - \Phi\theta \|_2^2 + \lambda_1 \| \theta \|_1 + \frac{\lambda_2}{2} \| \theta \|_2^2, \tag{18}$$

were $\lambda_1, \lambda_2 \geq 0$ are hyper-parameters. As the objective is the combination of the ridge regression and LASSO objectives, it can be handled using similar ideas. That is we can compute a subgradient of the objective and perturb the residuals based on the transformations $\{G_i\}$ which leave the (joint) distribution of the true noise terms invariant. Then, the *(G)SPS evaluation function for elastic net regularization* is

$$Z_i(\theta) \doteq \| \Psi_{\mathrm{E}}^{1/2} \Big[\Phi^{\mathrm{T}} G_i \, (y - \Phi\theta) - \lambda_1 \, \mathrm{sign}(\theta) - \lambda_2 \, \theta \Big] \|_2^2, \tag{19}$$

where Ψ_{E} can again be an (optional) covariance estimate for the elastic net solution.

The exact confidence of the constructed regions easily follows from the related results for SPS. We leave the investigation of their other properties for further work.

Acknowledgements This research was partially supported by the National Research, Development and Innovation Office (NKFIH), grant numbers ED_18-2-2018-0006 and KH_17 125698, and by the János Bolyai Research Fellowship of the Hung. Academy of Sciences, BO/00217/16/6.

References

1. Csáji, B.Cs.: Score permutation based finite sample inference for generalized autoregressive conditional heteroskedasticity (GARCH) models. In: 19th International Conference on Artificial Intelligence and Statistics (AISTATS), Cadiz, Spain, pp. 296–304 (2016)
2. Csáji, B.Cs., Weyer, E.: Closed-loop applicability of the sign-perturbed sums method. In: 54th IEEE Conference on Decision and Control, Osaka, Japan, pp. 1441–1446 (2015)
3. Csáji, B.Cs., Campi, M.C., Weyer, E.: Sign-perturbed sums: a new system identification approach for constructing exact non-asymptotic confidence regions in linear regression models. IEEE Trans. Signal Process. **63**(1), 169–181 (2015)
4. Knight, K., Fu, W.: Asymptotics for LASSO-type estimators. Ann. Stat. **5**, 1356–1378 (2000)
5. Kolumbán, S.: System identification in highly non-informative environment. PhD thesis, Budapest University of Technology and Economics, Hungary, and Vrije Universiteit Brussels, Belgium (2016)
6. Ljung, L.: System Identification: Theory for the User, 2nd edn. Prentice-Hall, Upper Saddle River (1999)
7. Weyer, E., Campi, M.C., Csáji, B.Cs.: Asymptotic properties of SPS confidence regions. Automatica **82**, 287–294 (2017)

Detecting Periodicity in Digital Images by the LLL Algorithm

Lajos Hajdu, Balázs Harangi, Attila Tiba, and András Hajdu

Abstract In this paper we provide an algorithm to decide (or, to help the decision about) whether some repeatedly occurring pattern in a digital image can be considered to have periodical nature or not. Our approach extracts specific image components and represent them by single pixels. To decide upon the gridness nature of the resulting point set we use lattice theory and the LLL algorithm to fit lattices to the point set, and an efficient lattice point counting method of Barvinok. With this work we complete some of our corresponding former results, where the fitting of the lattice ignored possible holes inside the point set. Namely, now after some appropriate transformations we consider the convex hull of the point set which way we can detect and punish such fitted lattice points that fall in holes of the original point set, or equivalently image pattern. As a practical demonstration of our method we present how it can be applied to recognize segmentation errors of atypical/typical pigmented networks in skin lesion images.

1 Introduction

Patter analysis is a traditional task in digital image processing. In various fields the regularity/irregularity of the pattern directly relates to the underlying problem, so a proper decision on this phenomenon is the essence of the solution. To address the recognition of pattern regularity primarily some kind of periodicity check can be performed based on e.g. auto-correlation like in [8]. As an own contribution to this field, we have proposed a procedure based on the LLL algorithm [9] to find best approximating grids to an input point set. The theory is worked out for point sets, which can be composed e.g. via the extraction of dominant digital image

---

L. Hajdu
Institute of Mathematics, University of Debrecen, Debrecen, Hungary
e-mail: hajdul@science.unideb.hu

B. Harangi · A. Tiba · A. Hajdu (✉)
Faculty of Informatics, University of Debrecen, Debrecen, Hungary
e-mail: harangi.balazs@inf.unideb.hu; tiba.attila@inf.unideb.hu; hajdu.andras@inf.unideb.hu

© Springer Nature Switzerland AG 2019
I. Faragó et al. (eds.), *Progress in Industrial Mathematics at ECMI 2018*,
Mathematics in Industry 30, https://doi.org/10.1007/978-3-030-27550-1_78

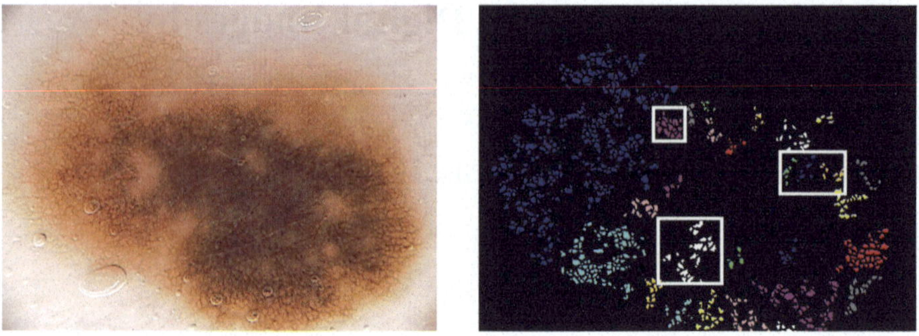

Fig. 1 Segmentation results for a skin lesion image to extract pigment network; original image (left) and some false segmentation results (right) causing holes in the pattern are marked with white rectangle

components and the representation of them with single points e.g. in terms of their centroids. However, segmentation errors may occur during this extraction steps causing holes in the pattern. Such a scenario can be observed in Fig. 1, where our intention is to extract pigment networks from skin lesion images and classify them as typical (regular pattern) or atypical (irregular one). In our former approach [5] we have required only that an approximating grid point should fall in a close environment of each point in the input set. However, this error measurement ignores possible holes in the input pattern since does not punish the reversed cases, when there are no base points close to the approximating grid points. Thus, to resolve this issue now we complete our former error measurement with a complementary check that the number of the approximating grid points should be close to that of the cardinality of the input point set. The proper extra condition can be formulated by counting lattice points in the convex hull of the original point set.

The structure of the paper is the following: first, we formulate the problem precisely and give a mathematical framework for it using lattice theory. Then, we propose an efficient method based on the LLL algorithm (see [9] and [4, 5, 11]) and lattice point counting in convex domains (see [1, 2]) to produce a descriptive value measuring that 'how much' the occurrence of the investigated pattern can be considered to be periodic. Finally, to demonstrate the practical applicability of our approach, we explain how to detect possible holes in extracted pigment networks to suppress segmentation errors.

2 Periodicity and Lattices

Suppose that we observe the occurrence of a certain pattern on a digital image repeatedly, and we wonder whether this occurrence can be regarded as periodic, or not. Such questions appear in several problems of image processing.

As a first step in building our model, we assume that we have to deal with a finite number of points on the plain. This can be achieved by standard discretization techniques, e.g. after considering the centroids of the occurring copies of the pattern.

Now let H be a subset of $\mathbb{R}^2$. Following the standard terminology, we say that H is periodic if there exist linearly independent vectors $\underline{u}, \underline{v} \in \mathbb{R}^2$ such that for any $\underline{h}$ in H, the vectors $\underline{h} \pm \underline{u}$ and $\underline{h} \pm \underline{v}$ are in H as well. Let

$$\Lambda = \{x\underline{u} + y\underline{v} : x, y \in \mathbb{Z}\}$$

be the lattice generated by $\underline{u}, \underline{v}$ in $\mathbb{R}^2$. Then the periodicity of H implies that $H = H + \Lambda$. (Indeed, $H \subseteq H + \Lambda$ is obvious, while $H + \Lambda \subseteq H$ follows directly from the definition, by noting that $h + xu + yv$ $(x, y \in \mathbb{Z})$ can be obtained from h by adding $|x|$ times u or $-u$, and $|y|$ times v or $-v$.) Clearly, any periodic set in $\mathbb{R}^2$ has to be infinite. However, we obviously need to consider the periodic property of *finite* subsets of $\mathbb{R}^2$. For this, first observe that if H is a countably infinite subset of $\mathbb{R}^2$ and H is periodic, then H is a shifted lattice, that is, H is of the shape $H = \underline{o} + \Lambda =: \Lambda'$ for some $\underline{o} \in \mathbb{R}^2$ and lattice Λ in $\mathbb{R}^2$. Let now T be a finite subset of Λ' with Λ' as above. Adopting the discrete convexity notion of Kim [6] (see also [7]), we say that T is convex if $T^c \cap \Lambda' = T$, where T^c is the convex hull of T in $\mathbb{R}^2$. Altogether, this is the property we use for the definition of periodicity. That is, a finite subset T of $\mathbb{R}^2$ is called periodic if there exists a shifted lattice Λ' in $\mathbb{R}^2$ such that T is a convex subset of Λ'. To measure the periodicity of finite subsets T of shifted lattices, we introduce the following function:

$$\text{per}(T) = \min_{\Lambda'} \frac{|T|}{|T^c \cap \Lambda'|},$$

where $|S|$ denotes the number of elements of a set S and the minimum is taken over all shifted lattices Λ' containing T. (Clearly, this minimum exists.) In this way, T is periodic if and only if $\text{per}(T) = 1$. (Indeed, it is clear that if T is periodic then $\text{per}(T) = 1$. On the other hand, if $\text{per}(T) = 1$ then there exists a Λ' as above, containing T, such that $|T| = |T^c \cap \Lambda'|$. As $T \subset T^c \cap \Lambda'$, the equality $|T| = |T^c \cap \Lambda'|$ shows that in fact $T = T^c \cap \Lambda'$, so T is a convex subset of Λ'.)

Then we can measure (decide about) the periodicity of an arbitrary finite subset $S = \{s_1, \ldots, s_k\} \subset \mathbb{R}^2$ of cardinality k in the following way.

Step 1 Following the method of Hajdu et al. [4] (based upon the LLL algorithm, see [9]) we can find a 'well approximating' shifted lattice Λ' for S. Namely, let the error of the approximation be calculated as

$$E_{approx} := \frac{\sqrt{\sum_{s \in S} |s - \Lambda'|^2}}{\Delta} \left(\frac{\text{diam } S}{\Delta}\right)^{\frac{2}{k-3}},$$

where diam S is the diameter of the point set S, and Δ is the square root of the lattice determinant of Λ'. If E_{approx} is 'too large' (see the paper of Tiba et al. [11]

for experiments; c.f. also [4] and [5]), then we can immediately say that the pattern in question does not appear periodically, and the forthcoming steps are superfluous.

Step 2 Write S' for the points of $\Lambda' = \underline{o} + \Lambda$ (with the previous notation) corresponding to the (approximated) points of S, and let $\underline{u}, \underline{v}$ be a basis of Λ. (They can be obtained by the already mentioned method of Hajdu et al. [4].) Put

$$A := \{(x, y) \in \mathbb{Z}^2 : \underline{o} + x\underline{u} + y\underline{v} \in S'\}.$$

Observe that $\text{per}(S') = \text{per}(A)$.

Step 3 Find a sublattice L of $\mathbb{Z}^2$ of largest index containing A. (Typically, L will be $\mathbb{Z}^2$ itself.) For this, observe that $L = \sum_{(x,y)\in A} (x, y)\mathbb{Z}$. Thus it is standard to find a basis $\underline{p}, \underline{q}$ of L; see e.g. p. 73 of Cohen's book [3], where an algorithm based upon the Hermite normal form of integer matrices is given. Then, transform A as follows:

$$B := \{(x, y) \in \mathbb{Z}^2 : x\underline{p} + y\underline{q} \in A\}.$$

In this way we have

$$\text{per}(S') = \text{per}(B) = \frac{|B|}{|B^c \cap \mathbb{Z}^2|}.$$

Step 4 Note that $|B| = |S'|$, so this number can be calculated easily. The number $|B^c \cap \mathbb{Z}^2|$ can be obtained in a very efficient way, based upon Barvinok's algorithm [2]. We use the Maple 15 [10] implementation of Baldoni et al. [1]. So, altogether we have an efficient way to calculate $\text{per}(S')$.

Step 5 We can combine the error of approximation E_{approx} obtained in Step 1 and the measure $\text{per}(S')$, e.g. say using a threshold, for deciding about the periodicity of S.

We illustrate our method by a simple example.

Example 1 Let our starting set of points be given by

$$S = \{(0, 0), (3.218875824, 3.891820298), (4.007333185, 4.510859506),$$
$$(4.795790546, 5.129898714), (6.405228458, 7.075808863),$$
$$(8.014666370, 9.021719012)\}.$$

In Step 1, we get $S' = S$ together with $\underline{u} = (1.609437912, 1.945910149)$, $\underline{v} = (4.007333185, 4.510859506)$. Further, we get $E_{approx} = 0$.

In Step 2, we obtain

$$A = \{(0, 0), (0, -2), (-1, 0), (-2, 2), (-2, 1), (-2, 0)\},$$

where the elements in A are the coefficients of the elements of S' in the basis $\underline{u}$, $\underline{v}$, in the given order. (In this case we have $\underline{o} = (0, 0)$.)

In Step 3, we see that $L = \mathbb{Z}^2$ and $\underline{p} = (1, 0)$, $\underline{q} = (0, 1)$. Thus

$$B = A = \{(0, 0), (0, -2), (-1, 0), (-2, 2), (-2, 1), (-2, 0)\}.$$

This follows from the fact that the gcd of the 2×2 subdeterminants of the matrix

$$\begin{pmatrix} 0 & 0 & -1 & -2 & -2 & -2 \\ 0 & -2 & 0 & 2 & 1 & 0 \end{pmatrix}$$

(composed of the entries of the elements of A) is 1.

In Step 4, using the Maple code of Baldoni, Berline and Vergne we obtain that $|B^c| = 9$. (In fact, in this simple case B^c is a 2×2 square, with vertices $(-2, 0)$, $(0, -2)$, $(0, 0)$, $(-2, 2)$.) Hence we get

$$\mathrm{per}(S') = \frac{|B|}{|B^c|} = \frac{6}{9} = \frac{2}{3}.$$

In Step 5, based upon $E_{approx} = 0$ and $\mathrm{per}(S') = 2/3$, depending on the actual application we are dealing with, we can decide whether we consider S to be periodic or not.

3 Application to Pigment Network Segmentation

As we have presented in the introduction, checking the periodicity (gridness) of a point set was motivated by the regularity analysis of pigment networks in skin lesion images. In this task we exploit the method introduced in the paper to check whether the extraction of the components on a pigment network was successful or not. The latter case generally occurs when our detector algorithm misses some components causing holes in the extracted pattern. With the proper details are given in [5], the extraction of the pigment cells can be summarized as follows:

- the input color image is converted to grayscale,
- for each pixel, the intensity profiles of lines passing through the given pixel are considered,
- second order derivative of the Gaussian filters are matched to the profiles,
- large filter response values are considered for pigment hole candidates,
- a hysteresis thresholding technique is applied to this response map for the final network components.

As it can be seen above, the number of extracted components can be increased with a corresponding threshold. Consequently, when a low periodicity score is found for an extracted pigment network, we lower the threshold to eliminate some holes

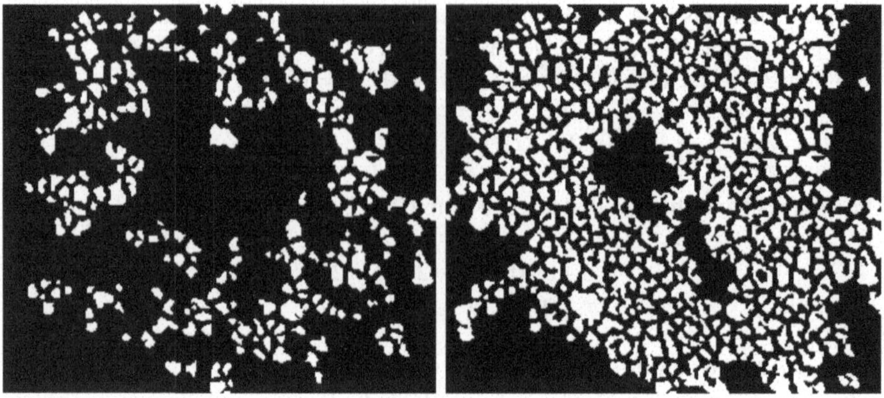

Fig. 2 Segmented pigment network in a skin lesion image with a low periodicity score (left) and the result of re-segmentation with a higher periodicity score (right)

in the pattern. As a demonstrative example, Fig. 2 depicts such a scenario, when an extracted network with a low periodicity score (0.14) has been improved to a higher score (0.91), since re-segmentation with a lower threshold has found more network components. As simple technical issues note that the point set is generated as the centroids of the binary components, and using low threshold in our segmentation method in an unjustified way is risky, since it can lead to over-segmentation.

Acknowledgements Research was supported in part by the project EFOP-3.6.2-16-2017-00015 supported by the European Union and the State of Hungary, co-financed by the European Social Fund, and by the NKFIH grants K115479 and K128088.

References

1. Baldoni, V., Berline, N., Vergne, M.: Summing a polynomial function over integral points of a polygon. User's guide (2009). arXiv:0905.1820 [cs.CG]
2. Barvinok, A.I.: A polynomial time algorithm for counting integral points in polyhedra when the dimension is fixed. In: 34th Annual Symposium of Foundations of Computer Science, pp. 566–572. IEEE, Piscataway (Nov 1993)
3. Cohen, H.: A Course in Computational Number Theory, Third, corrected printing. Springer, Berlin (1996)
4. Hajdu, A., Hajdu, L., Tijdeman, R.: Finding well approximating lattices for a finite set of points. Math. Comput. **88**, 369–387 (2019)
5. Hajdu, A., Harangi, B., Besenczi, R., Lázár, I., Emri, G., Hajdu, L., Tijdeman, R.: Measuring regularity of network patterns by grid approximations using the LLL algorithm. In: 23rd International Conference on Pattern Recognition (ICPR 2016), Cancun, Mexico, pp. 1525–1530 (2016)
6. Kim, C.E.: On the cellular convexity of complexes. IEEE Trans. Pattern Anal. Mach. Intel. **PAMI-3**(6), 617–625 (1981)

7. Kim, C.E., Rosenfeld, A.: Digital straight lines and convexity of digital regions. IEEE Trans. Pattern Anal. Mach. Intel. **PAMI-4**(2), 149–153 (1982)
8. Krupic, J., Burgess, N., OḰeefe, J.: Neural representations of location composed of spatially periodic bands. Science **337**(6096), 853–857 (2012)
9. Lenstra, A.K., Lenstra Jr., H.W., Lovász, L.: Factoring polynomials with rational coefficients. Math. Ann. **261**, 515–534 (1982)
10. Maple User Manual and Maple Programming Guide. Maplesoft, a division of Waterloo Maple Inc., Toronto (2011–2015)
11. Tiba, A., Harangi, B., Hajdu, A.: Efficient texture regularity estimation for second order statistical descriptors. In: 10th International Symposium on Image and Signal Processing and Analysis (ISPA), 2017, pp. 90–94. https://doi.org/10.1109/ISPA.2017.8073575

Fusion Markov Random Field Image Segmentation for a Time Series of Remote Sensed Images

Tamas Sziranyi, Andras Kriston, Andras Majdik, and Laszlo Tizedes

Abstract Change detection on images of very different time instants from remote sensing databases and up-to-date satellite born or UAV born imaging is an emerging technology platform today. Since outdoor sceneries, principally observation of natural reserves, agricultural meadows and forest areas, are changing in illumination, coloring, textures and shadows time-by-time, and the resolution and geometrical properties of the imaging conditions may be also diverse, robust and semantic level algorithms should be developed for the comparison of images of the same or similar places in very different times. Earlier, a new method, fusion Markov Random Field (fMRF) method has been introduced which applied unsupervised or partly supervised clustering on a fused image series by using cross-layer similarity measure, followed by a multi-layer Markov Random Field segmentation. This paper shows the effective parametrization of the fusion MRF segmentation method for the analysis of agricultural areas of fine details and difficult subclasses.

1 Introduction

As we have more and more remote sensing platforms for scanning the terrestrial surface, like satellite and airborne imaging, UAV based surveillance; and we have very different modalities as multi-band images, Lidar or Radar, the task to use them together in some fusion methodology and to find labelled changes among the very different scans needs new mathematical solutions. This paper presents a most recent methodology, where the different modalities of different time-instances can be fused to proceed a Markov Random Field (MRF) segmentation; the resulted label-map of fused MRF (fMRF [1]) is then used as master label-map to train the single layers for a forthcoming MRF segmentation procedure, where the single-layer MRF labeling can be compared to find labelled changes.

T. Sziranyi (✉) · A. Kriston · A. Majdik · L. Tizedes
MTA SZTAKI, Budapest, Hungary
e-mail: sziranyi@sztaki.mta.hu; kriston@sztaki.mta.hu; majdik@sztaki.mta.hu; tizedes@sztaki.mta.hu

© Springer Nature Switzerland AG 2019
I. Faragó et al. (eds.), *Progress in Industrial Mathematics at ECMI 2018*, Mathematics in Industry 30, https://doi.org/10.1007/978-3-030-27550-1_79

The multi-band remote imaging methods, including UAV scanning from 20 m to 100 m altitudes, help us to collect data for agriculture/environmental protection analysis. When applying an appropriate energy optimization algorithm we can exploit labelled maps and we can track the changes through time and modalities. In precision farm management biomass monitoring is crucial. As shown in the study presented in [2] red, green and blue imaging obtained from UAV found to be a good alternative to other sensors used for precision agriculture. Dynamic monitoring of agricultural terraces in China was efficiently performed by the application of UAVs, [3]. The use of UAVs play a more and more significant role in monitoring natural hazards due to their cost efficiency and versatility, [4].

The main challenge is that the images of the series are very different in lighting, color and micro-structure. The framework of fMRF makes it possible to merge different data-structures, even for semi-supervised parameter-setting, like in [5] for Lidar ground-truths.

In our tests we used 3DR Solo[1] UAV for data capturing, equipped with an RGB and a four channel (Green-550 nm, Red-660 nm, Red edge -735 nm, Near infrared -790 nm) multispectral camera system (i.e. Parrot Sequoia) over the farming site in the vicinity of Biatorbagy town, Budapest metropolitan area, Hungary. In the course of the multi-session mapping we captured more than 2100 narrow-band 1.2 mega pixel images covering 8 hectare (ha) of fruit and vineyard. Next, geo-referenced orthomosaic photos were computed corresponding to every multispectral band captured at different time instances using the Pix4D[2] software tools.

2 Fusion MRF

Fusion MRF (fMRF) has been introduced in [1] for remote sensing change detection. Different multilayer MRF based algorithms have been compared for change detection in [7], where fMRF proved to be most effective in some change detection tasks, as Fig. 1 shows. This fusion based method has been also used in a recent project, where aerial Lidar and satellite born images have been fused in a new MRF based solution to find small wetland areas [5].

In a series of N layers of remote sensing images, let $\overline{x}_s^{L_i}$ denote the feature vector at pixel s of layer L_i, $i = 1, 2, \ldots, N$. This feature vector might contain color, texture/micro-structural features, cross layer similarity measures, or mixture of these. Set $X = \{\overline{x}_s | s \in S\}$ marks the global image data. An example of a feature vector would be

$$\overline{x}_s^{L_i} = [\overline{x}_{C(s)}^{L_i}, \overline{x}_{M(s)}^{L_i}]^T \tag{1}$$

[1]3DR Solo: https://3dr.com/solo-drone/.

[2]Pix4d photogrammetry software: https://pix4d.com.

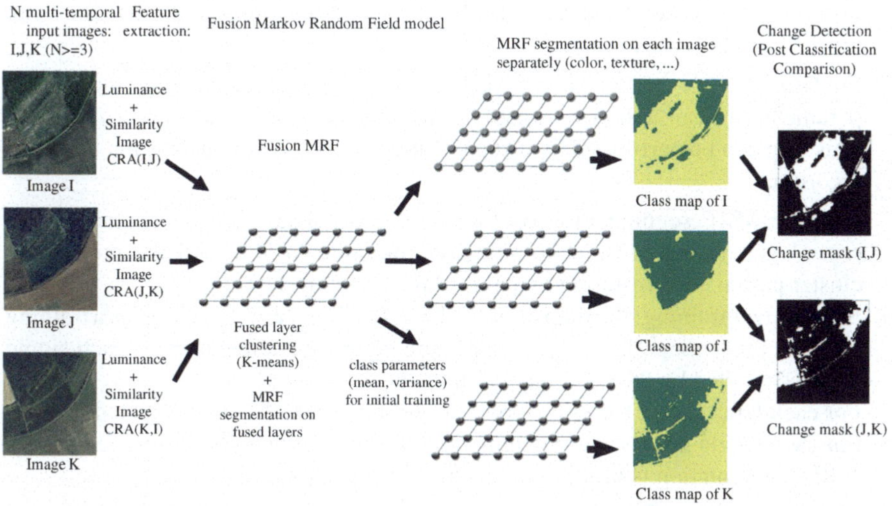

Fig. 1 The model of fusion MRF in case of three different time instants of scanning (Luminance—CRA [6] feature-set) [1, 7]

where $\overline{x}_{C(s)}^{L_i}$ contains the pixel's color values, and $\overline{x}_{M(s)}^{L_i}$ is the cross layer similarity measures between the image and other two or more images in the series. The cross layer similarity measure might be correlation, mutual information, or CRA. In this study Cluster Reward Algorithm (CRA) was used [6], defined between image pairs, calculated using the joint histogram of the two images and the marginal histograms, see more in [1].

The multiple layers of remote sensed image time series are characterized by the stack $\overline{x}_s^{L_{i_1 \ldots i_n}}$ of these vectors for a reasonable set of them, $n \leq N$:

$$\overline{x}_s^{L_{i_1 \ldots i_n}} = \{\overline{x}_s^{L_{i_1}}, \overline{x}_s^{L_{i_2}}, \ldots \overline{x}_s^{L_{i_n}}\} \tag{2}$$

2.1 *Fusion-MRF: Multi-Layer Segmentation and Change Detection*

For MRF segmentation, more details can be found in [8–11]. Once feature vectors are generated, the six steps of the algorithm proposed here are applied, as it is introduced in [1]. This segmentation and change detection procedure contains different levels of MRF optimization in the following main steps:

1. Selecting and registering the image layers; In case of professional data suppliers orthonormed and geographically registered images are given; no further registration is needed. In our method no color-constancy or any shape/color semantic

information is needed; the color of the corresponding areas and the texture can differ strongly layer-by-layer.

2. Finding clusters in the set of vectors $(\overline{x}_s^{L_{i_1 \dots i_n}})$ and calculating the cluster parameters (mean and covariance) for the fusion based "*multi-layer clusters*". This step can be performed by using unsupervised methods such as the K-means algorithm.
3. Running MRF segmentation on the fused layer data $(\overline{x}_s^{L_{i_1 \dots i_n}})$ containing the cross-layer measures (refer to similarity measure in [1]), and the multi-layer cluster parameters, resulting in a multi-layer labeling $\Omega_{L_{i_1 \dots i_n}}$;
4. Single-layer training: the map of multi-layer labeling $\Omega_{L_{i_1 \dots i_n}}$ is used as a training map for each image layer L_i: cluster parameters are calculated for each single layer controlled by the label map of multi-layer clusters.
5. For each label $k \in \Lambda$ the corresponding subspace of $(\overline{x}_s^{ML_n})$ is collected;
6. For each single layer L_i (containing only its color and maybe texture features) a MRF segmentation is processed, resulting in a labeling: Ω_{L_i};
7. The consecutive image layers $(\dots, (i-1), (i), \dots)$ are compared to find the changes among the different label maps to get the change map.

The above fusion MRF model and its processing can be seen in Fig. 1, [1, 7]. In our application a graph cut based α-expansion algorithm was used for energy minimization of MRF, with the adherent implementation of [10].

3 Segmentation of UAV Based Remote Sensed Image Time Series

3.1 Datasets and Features

During the measurement series, data was collected from April, June and July. The vegetation—thus the difference between images—is expected to change quite much between April and June and less between June and July. The vegetation intensity, health of plants, soil composition, presence of water or building can be analyzed by the near infrared band of the image. Since living plants with more chlorophyll reflect more near-infrared energy they appear darker on the images which makes their classification more accurate. Non-supervised segmentation is performed to see the evolution of vegetation from one month to the other and to observe locations where changes occur.

In this study the red, green and red-edge bands were used to create the feature vectors for the fMRF model and two different feature-sets and their combination were tested:

- In the *Multi-band feature-set* a twelve element vector was created for each pixel of the three image stack: red–green–red-edge values, Normalized Difference Vegetation Index (NDVI); NDVI = (red-edge − red)/(red-edge + red); It has advantages when microstructure is not so characteristic, but color spectra have discriminative power.

- In the *microstructure* based description (*Luminance-CRA*) the intensity values are used together with the mutual microstructure measure, defined by Cluster Reward Algorithm (CRA) [6]. It has advantages in case of higher resolution where texture can characterize the fine details.
- A composition of Multi-band and Luminance-CRA has high dimensionality; however, including all the possible features in one feature-vector has a higher complexity at an increased noise level as well.

In the case of both levels of processing steps (fused layer clustering and MRF; single layer MRF) the above detailed feature vectors were applied. This feature setting is powerful when there are image layers with color variation of nearly obvious clusters.

3.2 *Evaluation of the Time-Series Images*

Figure 2 shows the original input images from three consecutive months, then the fused master fMRF images originated from the inputs for different CRA window

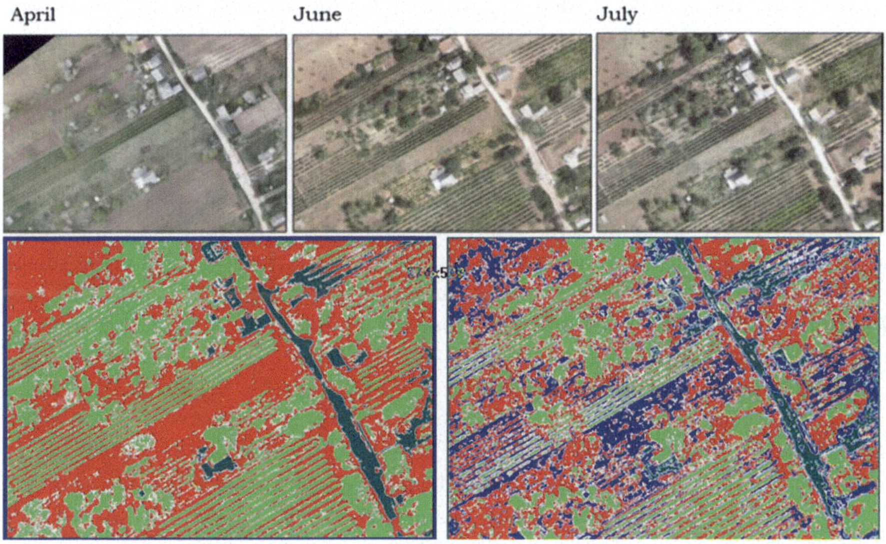

Fig. 2 Original images used for creating fused layer by K-means clustering and fMRF segmentation based on four color channels and CRA fine-texture statistical distance: above is the series of three input images of different time instants, while below the fused master-maps: left with CRA windows of 15, right with CRA windows of 7. Clusters are: meadows, vinegard, trees, road, house

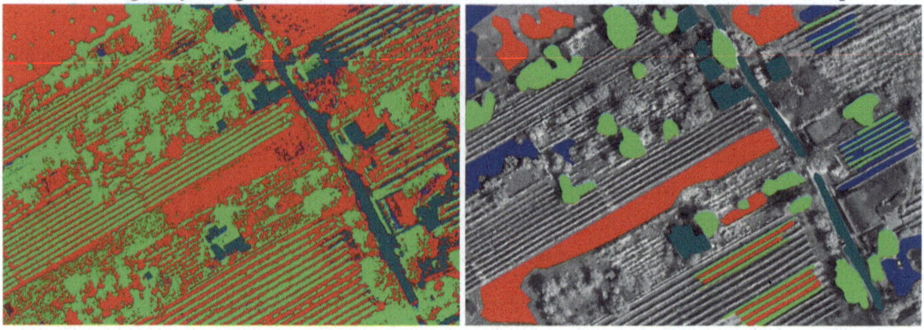

Fig. 3 Single layer mapping; left: segmented single layer of July, trained by the label-map of fMRF master (CRA width: 15); right: GoundTruth label-map for comparison

sizes. We see that the level of details can be tuned by the CRA window, characteristic for the microstructure similarity.

Based on the master fMRF label-maps, segmentation for each single layer can be proceeded, getting unified labelling to all the layers. Figure 3 shows a segmentation result on the left, with a GroundTruth label-map on the right. The label-mapping Recall error evaluation can be found in Table 1.

Besides unsupervised segmentation, supervised segmentation using predefined class masks was also performed. Figure 4 shows the selected training mask that were used for supervised segmentation. Class parameters were calculated from the pixels covered by masked regions followed by MRF segmentation.

Results of supervised segmentation (additional to unsupervised k-means segmentation) on fused and single images are given in Fig. 5. Calculated recall values (using GroundTruth reference label-map) are presented in Table 3.

Recall values show, given by Table 1 that in the case of applying *four channels multi-band set + CRA* detection of Low Intensity Vegetation areas was more accurate compared to the other feature-sets. However, colored labeled result image shows that Bare Soil class disappeared, merged into Low Intensity Vegetation class. *Precision* values were uniformly very good for building/roads cluster in all three cases, see Table 2, and low for bare soil.

Table 1 Recall values for the unsupervised segmentation results, for the image taken in July

Class	4-channel multi-band set	Luminance + CRA statistical measure	4-channel multi-band set + CRA
High intensity vegetation	0.84	0.52	0.44
Low intensity vegetation	0.26	0.62	0.73
Buildings/roads	0.82	0.8	0.81
Bare soil	0.67	0.16	0.02

Column names indicate which layers was used to create the master image

Fig. 4 Training masks used for supervised segmentation

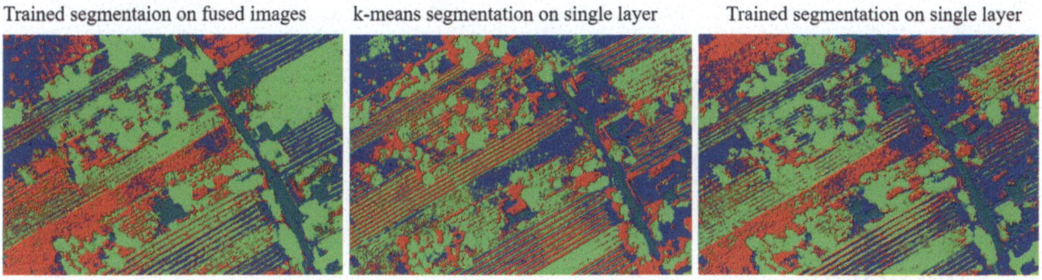

Fig. 5 Training label maps were used for training the four classes. Results on fused (April, June, July), and single layer (July) are presented, furthermore a simple k-means segmentation was performed on single layer (July). Recall values are presented in Table 3

Table 2 Precision values for the unsupervised segmentation results, for the image taken in July

Class	4-channel multi-band set	Luminance + CRA statistical measure	4-channel multi-band set + CRA
High intensity vegetation	0.74	0.7	0.45
Low intensity vegetation	0.6	0.48	0.51
Buildings/roads	0.78	0.77	0.75
Bare soil	0.35	0.19	0.34

Column names indicate which layers was used to create the master image

The best results were achieved by trained fMRF segmentation, given by Table 3. However, training areas were selected on the July image, recall values are lower compared to the fused layer results, except the Bare Soil class. Similarly to unsupervised segmentation, precision in Table 4 was uniformly high for building/roads cluster and lower for bare soil, but still higher than in the case of unsupervised segmentation.

Table 3 Recall values for the supervised segmentation (with training masks) on fused and single layer (July) results, and for simple K-means segmentation on single layer (July) using only spectral channels and calculated NDVI map

Class	Supervised fMRF on multi-layer	K-means on single layer	Supervised MRF on single layer
High intensity vegetation	0.88	0.78	0.79
Low intensity vegetation	0.68	0.15	0.66
Buildings/roads	0.95	0.63	0.88
Bare soil	0.67	0.61	0.7

Table 4 Precision values for the supervised segmentation (with training masks) on fused and single layer (July) results, and for simple K-means segmentation on single layer (July) using only spectral channels and calculated NDVI map

Class	Supervised fMRF on multi-layer	K-means on single layer	Supervised MRF on single layer
High intensity vegetation	0.82	0.63	0.82
Low intensity vegetation	0.83	0.3	0.74
Buildings/roads	0.89	0.91	0.84
Bare soil	0.57	0.33	0.54

4 Conclusion

Unsupervised clustering on multilayer image dataset combined with MRF segmentation is a powerful tool to segment multispectral images into relevant classes. When MRF segmentation is used for multi-band+CRA images we may get proper results, however low texture classes, like bare soil, is likely to merged into another similar class. Additional CRA layer might help to segment classes with similar spectral properties. Supervised clustering could achieve even more accurate segmentation, but one has to give accurate training areas. In case of severe combination of subclasses due to heavy differences in lighting and color content, the fused MRF segmentation is better to use fine-structure information, as the CRA.

Acknowledgements The research reported in this paper was supported by Hungarian Scientific Research Fund (No. NKFIH OTKA K-120499 and KH-126513).

References

1. Sziranyi, T., Shadaydeh, M.: Segmentation of remote sensing images using similarity-measure-based fusion-mrf model. IEEE Geosci. Remote Sens. Lett. **11**, 1544–1548 (2014)
2. Ballesteros, R., Ortega, J.F., Hernandez, D., Moreno, M.A.: Onion biomass monitoring using uav-based rgb imaging. Precis. Agric. **19**(5), 840–857 (2018)

3. Wei, Z., Han, Y., Li, M., Yang, K., Yang, Y., Luo, Y., Ong, S.H.: A small UAV based multi-temporal image registration for dynamic agricultural terrace monitoring. Remote Sens. **9**, 904 (2017)
4. Giordan, D., Manconib, A., Remondinoc, F., Nexd, F.: Use of unmanned aerial vehicles in monitoring application and management of natural hazards. Geomat. Nat. Haz. Risk **8**, 1–4 (2017)
5. Shadaydeh, M., Zlinszky, A., Manno-Kovacs, A., Sziranyi, T.: Wetland mapping by fusion of airborne laser scanning and multi-temporal multispectral satellite imagery. Int. J. Remote Sens. **38**, 7422–7440 (2017)
6. Inglada, J., Giros, A.: On the possibility of automatic multisensor image registration. IEEE Trans. Geosci. Remote Sens. **42**, 2104–2120 (2004)
7. Benedek, C., Shadaydeh, M., Kato, Z., Sziranyi, T., Zerubia, J.: Multilayer Markov random field models for change detection in optical remote sensing images. ISPRS J. Photogrammetry Remote Sens. **107**, 22–37 (2015). Multitemporal remote sensing data analysis
8. Benedek, C., Sziranyi, T.: Change detection in optical aerial images by a multilayer conditional mixed Markov model. IEEE Trans. Geosci. Remote Sens. **47**, 3416–3430 (2009)
9. Geman, S., Geman, D.: Stochastic relaxation, Gibbs distributions and the Bayesian restoration of images. IEEE Trans. Pattern Anal. Mach. Intell. **PAMI-6**, 721–741 (1984)
10. Szeliski, R., Zabih, R., Scharstein, D., Veksler, O., Kolmogorov, V., Agarwala, A., Tappen, M., Rother, C.: A comparative study of energy minimization methods for Markov Random Fields. In: 9th European Conference on Computer Vision, vol. 2, pp. 16–29 (2006)
11. Kato, Z., Zerubia, J., Berthod, M.: Unsupervised parallel image classification using Markovian models. Pattern Recogn. **32**, 591–604 (1999)

Stochastic Order Relations in a Gambling-Type Environment

Sándor Guzmics

Abstract In this work we examine some stochastic ordering relations, namely the increasing convex order and the Lorenz order, between random variables which arise from a simple lottery setting as well as the relation between their natural continuous variants. We will provide stochastic ordering results for the continuized random variables.

1 Introduction

The notion of stochastic dominance has been established originally for comparing the riskiness of possible scenarios in financial and insurance mathematics. Later these concepts have been also applied in other probabilistic environments, for instance in gamblings. We examine the structure of a lottery type gambling by introducing advanced indicators that stem from the distribution of the random variables which are naturally associated with the corresponding game. Our work is motivated by a standard 90/5 type lottery setting. The outcomes are described by five-tuples, and we consider the ordered sample, and investigate the ordered differences between the elements of the ordered sample with respect to the increasing convex order and the Lorenz order. We illustrate our computations by a data set obtained from Hungarian lottery history from 1957 to 2018. It consists of 3217 five-tuples drawn from the set $\{1, \ldots, 90\}$, and it is available under the link https://bet.szerencsejatek.hu/cmsfiles/otos.html. In addition we will examine a natural continuization of the above setting, which possesses nicer mathematical properties than the original discrete one. In particular, the ordered differences in the discrete setting are not ordered in the Lorenz order, in contrast to the continuous setting. Finally we present some ideas for possible extensions.

S. Guzmics (✉)
University of Vienna, Department of Statistics and Operations Research, Vienna, Austria
e-mail: sandor.guzmics@univie.ac.at

© Springer Nature Switzerland AG 2019
I. Faragó et al. (eds.), *Progress in Industrial Mathematics at ECMI 2018*,
Mathematics in Industry 30, https://doi.org/10.1007/978-3-030-27550-1_80

2 The Discrete Setting

Let us consider a standard lottery setting, where five numbers are drawn form the fundamental set $H = \{1, \ldots, 90\}$ without replacement. In accordance with this, the players have to fill in a lottery coupon by crossing five numbers from the set H. It is well known, that the number of scores follows a hypergeometric distribution and it is also obvious, that if we denote the result of one draw by the set-valued random variable $X = \{X_1, X_2, X_3, X_4, X_5\}$, then X is uniformly distributed on the 5-element subsets of H, i.e., on the set $\mathscr{H} = \{h \subseteq H : |h| = 5\}$ and $\mathbb{P}(X = h) = 1/\binom{90}{5} = 1/43949268 \simeq 2.2754 \times 10^{-8}$ for all $h \in \mathscr{H}$. (Since in the following it will play an important role that H consists of equally placed numbers, we will sometimes refer to such a discrete set as a *grid*.)

Let us introduce the usual notation X_j^* for the ordered sample, i.e., in our case X_j^* is j-th smallest out of the five drawn numbers ($j = 1, \ldots, 5$). Due to the current setting $X_1^* < X_2^* < X_3^* < X_4^* < X_5^*$ holds with probability 1. It is easy to see that the probability distribution function of X_j^* ($j = 1, \ldots, 5$) is

$$\mathbb{P}(X_j^* = k) = \frac{\binom{k-1}{j-1} \cdot \binom{90-k}{5-j}}{\binom{90}{5}} \quad \text{for } k = j, \ldots, 85 + j, \tag{1}$$

and its expectation is

$$\mathbb{E}(X_j^*) \simeq 15.1667 \cdot j \,. \tag{2}$$

In order to obtain a better insight into the structure of this lottery, we focus on the differences

$$D_j = X_{j+1}^* - X_j^* \quad (j = 1, \ldots, 4) \tag{3}$$

between the neighbouring elements in the ordered sample $(X_1^*, X_2^*, X_3^*, X_4^*, X_5^*)$, and we will investigate D_j^* $(j = 1, \ldots, 4)$, i.e., the ordered sample of the random variables D_j. Notice that each inequality of the general relation $D_1^* \leq D_2^* \leq D_3^* \leq D_4^*$ can also hold with equality (with positive probability). First we study the range of the sample, which coincides with the sum of the differences defined in (3).

The Range of the Sample

$$Z := \max_{1 \leq j \leq 5} \{X_j\} - \min_{1 \leq j \leq 5} \{X_j\} = X_5^* - X_1^* = \sum_{j=1}^{4} D_j = \sum_{j=1}^{4} D_j^*. \tag{4}$$

It is easy to see that $\mathbb{P}(Z = k) = c \cdot (90 - k) \cdot (k - 1) \cdot (k - 2) \cdot (k - 3)$ for $k = 4, 5, \ldots, 89$, where $c = \frac{20}{90 \cdot 89 \cdot 88 \cdot 87 \cdot 86} \simeq 3.7923 \times 10^{-9}$. The expected value is $\mathbb{E}(Z) \simeq 60.6667$, while the mode of Z is 68.

The Distribution of D_j^* $(j = 1, \ldots, 4)$
We provide the distribution of D_j^* $(j = 1, \ldots, 4)$ via an implicit, combinatorial consideration, namely in terms of certain partitioning problems. The derivation of an explicit form will only be possible for D_1^*. As an introductory step, for each $j = 1, \ldots, 4$ we give N_j, the largest possible value of D_j^*, which are respectively $N_1 = 22$, $N_2 = 29$, $N_3 = 43$, $N_4 = 86$.

Let us introduce the notion of a *gap*, the number of integers lying between two neighbouring drawn numbers in some realization. As a distinction, we will call *initial gap* the number of integers smaller than X_1^*. In notation

$$l_0 := X_1^* - 1, \quad l_j := X_{j+1}^* - X_j^* - 1 \;\; (j = 1, \ldots, 4). \tag{5}$$

It is clear that the sequence of gaps l_0, l_1, l_2, l_3, l_4 uniquely determines $X_1^*, \ldots, X_5^*$. Since a difference of size k between neighbouring drawn numbers corresponds to a gap of size $k-1$, it is worth to introduce the following index sets:

$$I_1 := \{1 \le i \le 4 \mid l_i \ge k - 1\}, \quad I_2 := \{1 \le i \le 4 \mid l_i > k - 1\}. \tag{6}$$

Using I_1 and I_2, the distribution of D_j^* can be written as

$$\mathbb{P}(D_j^* = k) = \frac{\#\{(l_0, l_1, l_2, l_3, l_4) \mid \text{Condition 1., 2., 3. hold}\}}{\binom{90}{5}} \quad \text{for } k = 1, \ldots, N_j, \tag{7}$$

where

Condition 1. $l_0 + l_1 + l_2 + l_3 + l_4 \le 85$,
Condition 2. $|I_1| \ge 5 - j$,
Condition 3. $|I_2| \le 4 - j$.

Figure 1 shows the probability distributions of D_1^*, D_2^*, $D_3^*.D_4^*$ along with their realizations in the data that we have described in Sect. 1. Note that for visualizing purposes we display the probability distributions with continuous curves, but meanwhile we have to keep their discreteness in mind.

Explicit Formula for the Probability Distribution Function of D_1^*
We do not attempt to solve the combinatorial problems given in (7), but by another combinatorial consideration we get $\mathbb{P}(D_1^* \ge k) = \binom{94-4k}{5} / \binom{90}{5}$, which implies $\mathbb{P}(D_1^* = k) = \left(\binom{94-4k}{5} - \binom{90-4k}{5}\right) / \binom{90}{5}$ $(k = 1, \ldots, 22)$. The expected value is $\mathbb{E}(D_1^*) \simeq 4.1844$, and since $\mathbb{P}(D_1^* = k)$ is decreasing in k, the mode of D_1^* is 1.

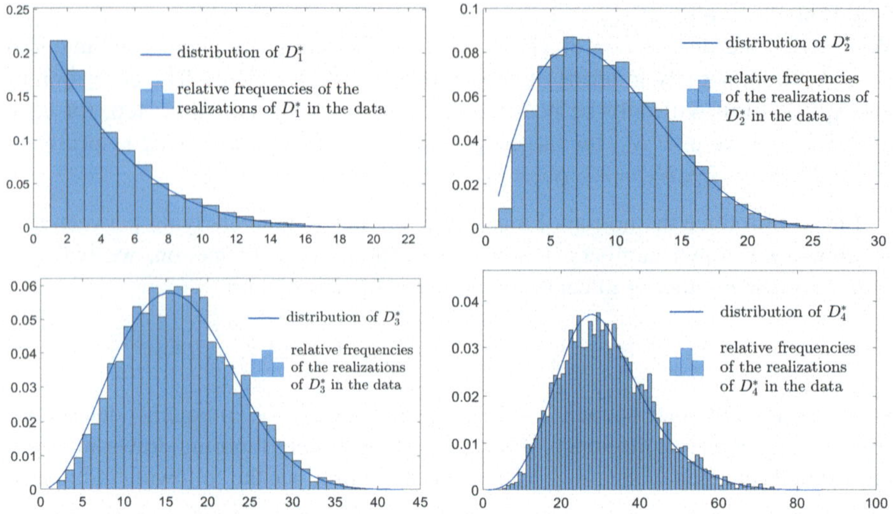

Fig. 1 The pdfs of the ordered differences along with their realizations in the data. For the sake of convenience the pdfs are visualized by continuous curves

Numerical Evaluations of the Probability Distribution Function of D_2^*, D_3^*, D_4^* We succeeded in determining the pdfs $\mathbb{P}(D_j^* = k)$, $(j = 1, 2, 3)$ numerically (look at also Fig. 1), and we computed the expected value and the mode of the distributions: $\mathbb{E}(D_2^*) = 9.0528$ and its mode is 7, $\mathbb{E}(D_3^*) \simeq 16.3838$ and its mode is 15, $\mathbb{E}(D_4^*) \simeq 31.0457$, and its mode is 28.

What would be a natural continuous analogue of the lottery setting described above? We define such a continuous analogy of the discrete setting, where the expectations of $X_1^*, \ldots, X_5^*$ and $D_1^*, \ldots, D_4^*$ nearly coincide with those of the discrete setting. In order to obtain this, we suggest the following continuous model.

3 The Continuous Setting

Let $\{X_1, X_2, X_3, X_4, X_5\}$ be the sample drawn from the discrete grid $\{1, \ldots, 90\}$. Then

$$Y_j := X_j + U_j \qquad j = 1, \ldots, 5,$$

where $U_j \sim UNI[-0.5, 0.5]$ are independent of X_j and of each other. It is obvious that $Y_j \sim UNI[0.5, 90.5]$, furthermore the construction has the favourable property that three or more Y_i cannot fall very close to each other, so an important feature of the discrete grid is preserved.

They also fulfil our previously described aim, that is, their expected values nearly coincide with the expected values of the corresponding discrete variables, as the following table shows. For sake of simplicity we will use the notation $D_1^* \leq \ldots \leq D_4^*$ for both the discrete and the continuous setting and we will always make it clear which variant is actually meant.

Discr. setting[a]	$\mathbb{E}(X_1^*)$	$\mathbb{E}(X_2^*)$	$\mathbb{E}(X_3^*)$	$\mathbb{E}(X_4^*)$	$\mathbb{E}(X_5^*)$	$\mathbb{E}(D_1^*)$	$\mathbb{E}(D_2^*)$	$\mathbb{E}(D_3^*)$	$\mathbb{E}(D_4^*)$
	15.1667	30.3333	45.5000	60.6667	75.8333	4.1844	9.0528	16.3838	31.0457
Cont. setting[b]	$\mathbb{E}(Y_1^*)$	$\mathbb{E}(Y_2^*)$	$\mathbb{E}(Y_3^*)$	$\mathbb{E}(Y_4^*)$	$\mathbb{E}(Y_5^*)$	$\mathbb{E}(D_1^*)$	$\mathbb{E}(D_2^*)$	$\mathbb{E}(D_3^*)$	$\mathbb{E}(D_4^*)$
	15.1612	30.3281	45.4965	60.6652	75.8320	4.1606	9.0610	16.3918	31.0573

[a] The values are exact and they are displayed up to four decimal place accuracy
[b] Values based on a sample of 10 Million drawn from the distribution $(Y_1, \ldots, Y_5)$

4 Stochastic Order Relations in the Lorenz Order and in the Increasing Convex Order

It is worth to examine whether some stochastic order relation holds between the ordered differences $D_1^*, \ldots, D_4^*$. Here we will consider the increasing convex order and the Lorenz order. For their definitions we refer to Shaked and Shantikumar [5, 6], Denuit et al. [1], Scarsini [4], Lorenz [3], and Kämpke and Radermacher [2].

Investigations in the Lorenz Order
We found that the ordered differences in the discrete setting are *not* ordered in the Lorenz order, while in the continuous case they are, i.e., $D_1^* \preceq_L D_2^* \preceq_L D_3^* \preceq_L D_4^*$. Figure 2 depicts the Lorenz curves. Looking at Fig. 2a one might conjecture an order relation for the discrete case, too, but by examining the lower tails of the distributions carefully (the left part of the Lorenz curves), the opposite can be concluded.

Proposition 1

1. *In the continuous setting* $D_i^* \preceq_L D_j^*$ *for* $1 \leq i < j \leq 4$.
2. *In the discrete setting* $D_i^* \npreceq_L D_j^*$ *for* $i \neq j$.

Sketch of the Proof We have to examine the pointwise orderedness of the Lorenz curves. We omit the details but provide a graphical justification (Fig. 2).

Investigations in the Increasing Convex Order
Proposition 2 *In both settings* $D_i^* \preceq_{ICVX} D_j^*$ *for* $1 \leq i < j \leq 4$.

Sketch of the Proof According to Denuit et al. [1] Proposition 3.4.6 we have to examine whether $\mathbb{E}((D_i^* - t)^+) \leq \mathbb{E}((D_j^* - t)^+)$ holds for all $t \in \mathbb{R}$ when $i < j$.

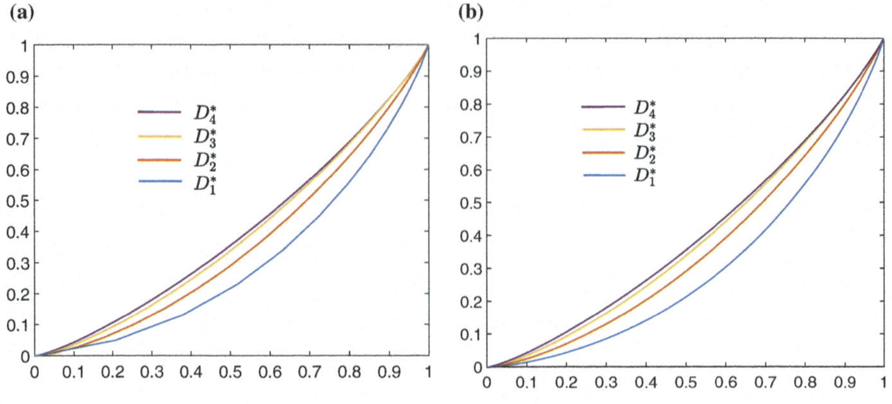

Fig. 2 Lorenz curves of D_j^* $(j = 1, \ldots, 4)$ in the discrete and continuous settings. (**a**) Discrete setting. (**b**) Continuous setting

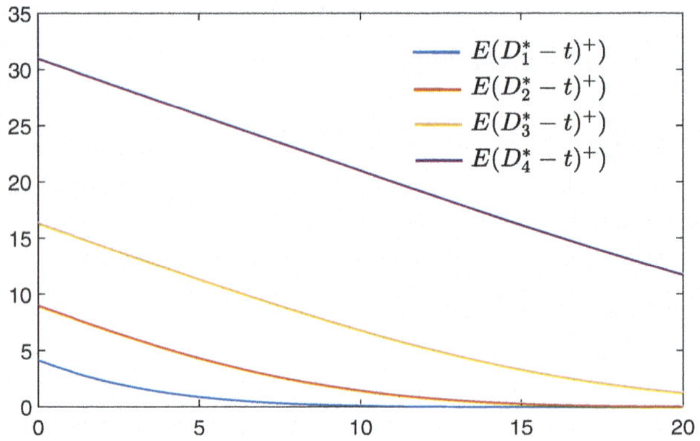

Fig. 3 The expressions $\mathbb{E}(D_i^* - t)^+$ $(i = 1, \ldots, 4)$ are plotted as functions of t to illustrate the stochastic order relations in the increasing convex order: $D_i^* \preceq_{ICVX} D_j^*$ for $1 \le i < j \le 4$

In the discrete setting it is enough the examine this relation for $t \in \{1, \ldots, 43\}$. In the continuous setting Fig. 3 confirms the statement.

5 Summary

We have seen that the application of stochastic order relations can lead to a better understanding in some settings which inherently possess a stochastic nature, such as a lottery game. The perspective is wider, since some extensions of the discussed tools (e.g., the multivariate Lorenz dominance) might enable us to investigate stochastic dominance in multivariate settings. That can be the topic of possible future research.

References

1. Denuit, M., Dhaene, J., Goovaerts, M., Kaas, R.: Actuarial Theory for Dependent Risks: Measures, Orders and Models. Wiley, Hoboken (2005)
2. Kämpke, T., Radermacher, F.J.: Lorenz curves and partial orders. In: Income Modeling and Balancing. Lecture Notes in Economics and Mathematical Systems, vol. 679. Springer, Cham (2015)
3. Lorenz, M.O.: Methods of measuring the concentration of wealth. Publ. Am. Stat. Assoc. **9**(New Series, No. 70), 209–219 (1905). https://doi.org/10.2307/2276207
4. Scarsini, M.: Multivariate convex orderings, dependence, and stochastic equality. J. Appl. Probab. **35**, 93–103 (1998)
5. Shaked, M., Shanthikumar, J.G.: Stochastic Orders and Their Applications. Academic, Boston (1994)
6. Shaked, M., Shanthikumar, J.G.: Stochastic Orders. Springer, New York (2007)

Damage Detection in Thin Plates via Time-Harmonic Infrared Thermography

Manuel Pena and María-Luisa Rapún

Abstract Non-destructive damage detection has become a very active research topic recently. This paper is devoted to the processing of time-harmonic thermograms (color images of one side of the sample to be inspected, obtained by a thermal camera) for structural health monitoring of thin plates. Our approach is based on the evaluation of an indicator function, the so-called topological derivative, which will identify the regions inside the plate where damage is located.

1 Statement of the Problem

Infrared thermography has become a powerful tool for non-destructive testing in a wide range of applications, ranging from medical imaging, to building and material diagnosis. In this work we aim at finding small defects inside metallic plates by processing time-harmonic thermograms, which are obtained after heating the plate to be inspected by a time-harmonic excitation from one lamp at the same side of the sample where the thermogram is taken, see Fig. 1.

The plate $\mathscr{R} \subset \mathbb{R}^d$ (where $d = 2$ or 3) is assumed to be surrounded by air at room temperature, T_{air}, with whom it exchanges heat by radiation and natural convection. The convection coefficient h is assumed to be constant and the surface of the plate is modeled as a gray body with absorptance α and emissivity ε. The lamp is modeled as a point source which radiates in a time-harmonic manner with an amplitude I and frequency ω. The defects conform a region $\mathscr{D} \subset \mathscr{R}$. For simplicity, we assume the

M. Pena (✉) · M.-L. Rapún
Universidad Politécnica de Madrid, Madrid, Spain
e-mail: manuel.pena@upm.es; marialuisa.rapun@upm.es

© Springer Nature Switzerland AG 2019

I. Faragó et al. (eds.), *Progress in Industrial Mathematics at ECMI 2018*,
Mathematics in Industry 30, https://doi.org/10.1007/978-3-030-27550-1_81

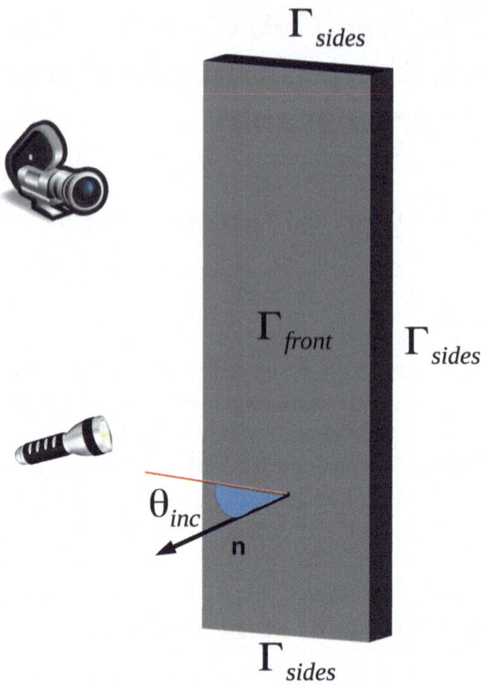

Fig. 1 Layout of the experiments: the plate $\mathscr{R}$ is defined as a bounded box region of $\mathbb{R}^2$ or $\mathbb{R}^3$, with one of its dimensions much smaller than the remaining ones. The lamp and the thermal camera are located at the same side of the plate. The illuminated side is denoted as Γ_{front}, and the opposite side is Γ_{back}. The remaining sides, much smaller in area, are denoted as Γ_{sides}. The angle between the incoming light rays and the normal $\mathbf{n}$ is denoted as θ_{inc}

thermal conductivity κ, the density ρ and the specific heat capacity c to be piecewise constant functions, i.e.:

$$\kappa(\mathbf{x}) = \begin{cases} \kappa_{\mathrm{e}} & \mathbf{x} \in \mathscr{R} \setminus \overline{\mathscr{D}} \\ \kappa_{\mathrm{i}} & \mathbf{x} \in \mathscr{D} \end{cases}, \quad \rho(\mathbf{x}) = \begin{cases} \rho_{\mathrm{e}} & \mathbf{x} \in \mathscr{R} \setminus \overline{\mathscr{D}} \\ \rho_{\mathrm{i}} & \mathbf{x} \in \mathscr{D} \end{cases}, \quad c(\mathbf{x}) = \begin{cases} c_{\mathrm{e}} & \mathbf{x} \in \mathscr{R} \setminus \overline{\mathscr{D}} \\ c_{\mathrm{i}} & \mathbf{x} \in \mathscr{D} \end{cases}. \tag{1}$$

Then, the complex amplitude $T(\mathbf{x})$ of the time-harmonic temperature distribution $\mathscr{T}(\mathbf{x}, t) = \widetilde{T}(\mathbf{x}) + \mathrm{Re}\left(T(\mathbf{x})e^{-\imath\omega t}\right)$ (where $\tilde{T}(\mathbf{x})$ is a steady mean value) satisfies the set of equations:

$$\begin{cases} \mathrm{div}\,(\kappa_{\mathrm{e}}\nabla T) + i\omega\rho_{\mathrm{e}}c_{\mathrm{e}}T = 0 & \text{in } \mathscr{R} \setminus \overline{\mathscr{D}} \\ \mathrm{div}\,(\kappa_{\mathrm{i}}\nabla T) + i\omega\rho_{\mathrm{i}}c_{\mathrm{i}}T = 0 & \text{in } \mathscr{D} \\ T^{+} - T^{-} = 0 & \text{on } \partial\mathscr{D} \\ \kappa_{\mathrm{e}}\partial_{\mathbf{n}}T^{+} - \kappa_{\mathrm{i}}\partial_{\mathbf{n}}T^{-} = 0 & \text{on } \partial\mathscr{D} \\ \kappa_{\mathrm{e}}\partial_{\mathbf{n}}T = 0 & \text{on } \Gamma_{\mathrm{sides}} \\ \kappa_{\mathrm{e}}\partial_{\mathbf{n}}T + AT = -\alpha\, q_{\mathrm{s}} & \text{on } \Gamma_{\mathrm{front}} \\ \kappa_{\mathrm{e}}\partial_{\mathbf{n}}T + AT = 0 & \text{on } \Gamma_{\mathrm{back}} \end{cases} \tag{2}$$

where $A = h + 4\varepsilon\sigma T_{\text{air}}^3$ is a constant that takes into account both the effects of convective terms and a linearization around T_{air} of the radiative terms. The heat $q_s(\mathbf{x})$ coming from a lamp located at a point $\mathbf{s}$ is modeled by

$$q_s(\mathbf{x}) = \frac{I}{2\pi}\frac{\cos\theta_{\text{inc}}(\mathbf{x})}{|\mathbf{x}-\mathbf{s}|} \quad \text{or} \quad \frac{I}{4\pi}\frac{\cos\theta_{\text{inc}}(\mathbf{x})}{|\mathbf{x}-\mathbf{s}|^2} \tag{3}$$

for the two-dimensional and three-dimensional cases respectively.

Given a thermogram T_{front}, that is, a measurement of the temperature distribution along Γ_{front} for a given experiment, we would like to obtain the domain $\mathscr{D}_{\text{app}}$ such that $T_{\mathscr{D}_{\text{app}}}(\mathbf{x}) = T_{\text{front}}(\mathbf{x})$, for all $\mathbf{x} \in \Gamma_{\text{front}}$, where $T_{\mathscr{D}_{\text{app}}}$ stands for the solution of the set of equations (2) setting $\mathscr{D} = \mathscr{D}_{\text{app}}$. Since experimental errors of very different nature are expected, we will consider a less demanding formulation and seek for $\mathscr{D}_{\text{app}}$ such that the functional

$$\mathscr{J}\left(\mathscr{R}\setminus\overline{\mathscr{D}_{\text{app}}}\right) = \int_{\Gamma_{\text{front}}}\left|T_{\mathscr{D}_{\text{app}}}(\mathbf{x}) - T_{\text{front}}(\mathbf{x})\right|^2 d\ell \tag{4}$$

attains a global minimum. This will be done by computing its topological derivative, defined in the next section.

2 Topological Derivative

The topological derivative (TD in the sequel) of the shape functional $\mathscr{J}$ at a point $\mathbf{x}$ measures the sensitivity of such functional to locating an infinitesimal ball $\mathrm{B}_\epsilon(\mathbf{x})$ of radius $\epsilon > 0$ at $\mathbf{x}$, providing the asymptotic expansion (see [6]):

$$\mathscr{J}\left(\mathscr{R}\setminus\overline{\mathrm{B}}_\epsilon(\mathbf{x})\right) = \mathscr{J}(\mathscr{R}) + \mathrm{D_T}(\mathbf{x})\, f(\epsilon) + o(f(\epsilon)) \quad \text{as } \epsilon \to 0^+. \tag{5}$$

where $f(\epsilon)$ is a positive increasing function chosen such that the expansion (5) holds. In our case, we can take $f(\epsilon)$ to be the measure of $\mathrm{B}_\epsilon(\mathbf{x})$. In view of expansion (5), the points where $\mathrm{D_T}$ attains large negative values are the most effective in minimizing the functional (4), and therefore, our guess of $\mathscr{D}_{\text{app}}$ will be defined as [1, 5]:

$$\mathscr{D}_{\text{app}} := \left\{\mathbf{x} \in \mathscr{R};\ \mathrm{D_T}(\mathbf{x}) < \lambda \min_{\mathbf{y}\in\mathscr{R}} \mathrm{D_T}(\mathbf{y})\right\}, \tag{6}$$

where $0 < \lambda < 1$ is a parameter that can be tuned.

Formula (5) is not practical from the numerical point of view. Adapting the results in [1, 2, 4], we obtain a closed-form formula for the TD: for all $\mathbf{x} \in \mathscr{R}$,

$$\mathrm{D_T}(\mathbf{x}) = \Re\mathfrak{e}\left(\frac{d\kappa_e(\kappa_e - \kappa_i)}{(d-1)\kappa_e + \kappa_i}\nabla T^0(\mathbf{x})\cdot\overline{\nabla V^0(\mathbf{x})} - i\omega(\rho_e c_e - \rho_i c_i)\, T^0(\mathbf{x})\,\overline{V^0(\mathbf{x})}\right), \tag{7}$$

where T^0 is solution to the direct problem

$$\begin{cases} \operatorname{div}\left(\kappa_e \nabla T^0\right) + i\omega\rho_e c_e T^0 = 0 & \text{in } \mathscr{R} \\ \kappa_e \partial_{\mathbf{n}} T^0 = 0 & \text{on } \Gamma_{\text{sides}} \\ \kappa_e \partial_{\mathbf{n}} T^0 + A T^0 = -\alpha\, q_{\mathbf{s}} & \text{on } \Gamma_{\text{front}} \\ \kappa_e \partial_{\mathbf{n}} T^0 + A T^0 = 0 & \text{on } \Gamma_{\text{back}} \end{cases} \tag{8}$$

and V^0 is solution to its associated adjoint problem

$$\begin{cases} \operatorname{div}\left(\kappa_e \nabla V^0\right) - i\omega\rho_e c_e V^0 = 0 & \text{in } \mathscr{R} \\ \kappa_e \partial_{\mathbf{n}} V^0 = 0 & \text{on } \Gamma_{\text{sides}} \\ \kappa_e \partial_{\mathbf{n}} V^0 + A V^0 = T_{\text{front}} - T^0 & \text{on } \Gamma_{\text{front}} \\ \kappa_e \partial_{\mathbf{n}} V^0 + A V^0 = 0 & \text{on } \Gamma_{\text{back}} \end{cases}. \tag{9}$$

As can be seen, for the computation of the TD no a priori information is needed about the number or size of defects, as both T_0 and V_0 are defined on the plate without any defect. The adjoint problem compares the thermogram expected at a healthy plate T^0 with the measured thermogram T_{front}.

In general, we will have several experiments with the lamp at a number N_{lamps} of different positions $\mathbf{s}_i$ and a number N_{freq} of different frequencies ω_j. In that case, we replace the functional (4) by a functional of the form

$$\mathscr{J}\left(\mathscr{R} \setminus \overline{\mathscr{D}_{\text{app}}}\right) = \sum_{i=1}^{N_{\text{lamps}}} \sum_{j=1}^{N_{\text{freq}}} p_{ij} \int_{\Gamma_{\text{front}}} \left| T^{(i,j)}_{\mathscr{D}_{\text{app}}}(\mathbf{x}) - T^{(i,j)}_{\text{front}}(\mathbf{x}) \right|^2 d\ell, \tag{10}$$

where the superscripts stand for the different configurations (namely $T^{(i,j)}$ corresponds to the temperature associated with the i-th position of the lamp and the j-th frequency), and $p_{ij} > 0$ are weights that can be tuned. By linearity, the TD of (10) is nothing but the linear combination of each individual derivative. The weights p_{ij} are defined in terms of the inverse of the largest negative value of each individual TD, as done in [3].

3 Numerical Experiments

In this section we present a couple of numerical experiments. The following parameters model an aluminum plate with air defects:

- $\kappa_e = 200$ W/(m · K), $\rho_e = 2700$ Kg/m^3 and $c_e = 900$ J/(Kg · K)
- $\kappa_i = 0.025$ W/(m · K), $\rho_i = 1$ Kg/m^3 and $c_i = 1000$ J/(Kg · K)
- $\alpha = 0.4$, $\varepsilon = 0.08$, $T_{\text{air}} = 290$ K and $h = 15$ W/(m^2 · K)

The lamp will have an amplitude $I = 6000$ W and will be located at a distance of 0.15 m from the plate.

Given that the thermograms are no actual measures but simulated data, a gaussian random error is added to them to simulate noisy experimental data, see [5] for further details about the generation of such error.

First, we present a two dimensional example, where the plate is the box $[0, \ell_x] \times [0, \ell_y]$ with $\ell_x = 0.01$ m and $\ell_y = 1$ m, and contains three different defects: an elliptical hole located at $(0.5\ell_x, 0.25\ell_y)$ and semi-axis of $0.1\ell_x$ and $0.3\ell_y$, a circular hole located at $(0.6\ell_x, 0.6\ell_y)$ with radius $0.125\ell_x$, and a circular hole located at $(0.4\ell_x, 0.8\ell_y)$ with radius $0.1\ell_x$.

In Fig. 2 we represent the TD for two different data sets. The true defects have been superimposed in white. The plate is distorted in the non-zoomed drawing for a better visualization. The TD is normalized in such a way that its largest negative value is equal to -1 for an easier comparison between both experiments. It can be seen that the TD accurately pinpoints the position, size and number of defects (regions in blue) even for a relatively high noise level. However it is unable to provide the correct depths, since the largest negative values are always attained in regions close to Γ_{front}. The biggest region corresponds to the elliptical hole, which is bigger in size. Reconstructions correlate not only with size but also with depth. We observe that the smaller and less deep circular hole is better identified than the

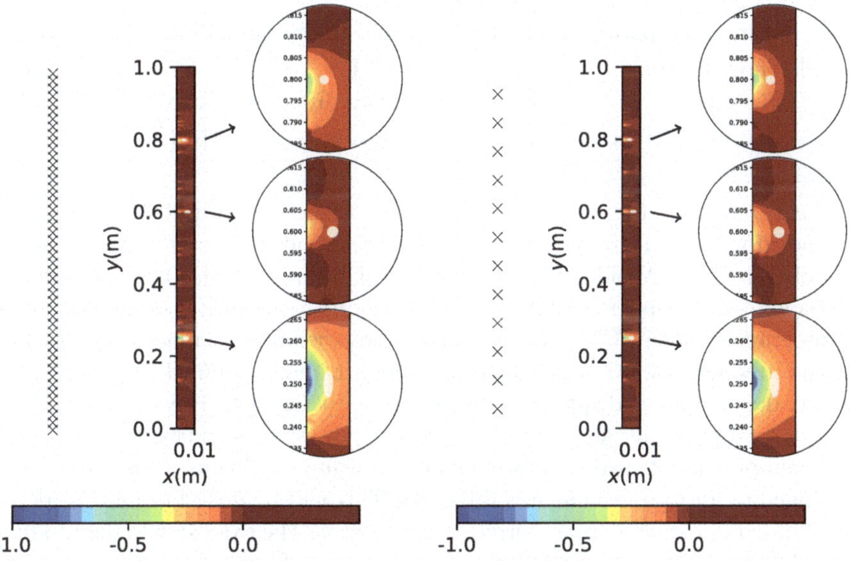

Fig. 2 Left: TD for $N_{\text{lamps}} = 48$ different lamp positions marked as $\times$ at frequency of 1 Hz. Right: TD for $N_{\text{lamps}} = 12$ lamp positions at $N_{\text{freq}} = 4$ linearly spaced frequencies between 0.8 and 2 Hz. The level of noise in the thermograms for both experiments is 20%

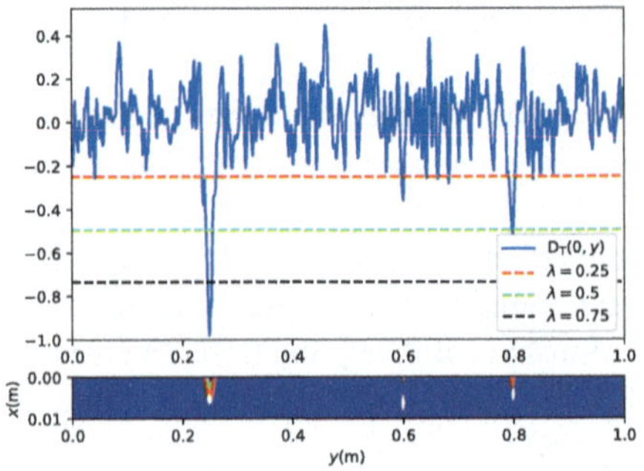

Fig. 3 TD at Γ_{front} and reconstructed defects for several values of λ

remaining one, which is bigger but is farther located from Γ_{front}. When comparing both figures, we see that for a fixed number of thermograms, reconstructions are sharper when thermograms correspond to both several locations of the lamp and several excitation frequencies.

In Fig. 3 we represent the TD at Γ_{front} to better visualize the sharpness of the minima. The sharper the minimum the less dependent is the reconstruction on the parameter λ in (6). In the same figure we also represent the (rotated) plate where the three true holes are in white and the reconstructed holes corresponding to different values of λ are shown in different color regions.

To illustrate the performance of the method in the three dimensional case, we consider now the plate $[-\frac{\ell_x}{2}, \frac{\ell_x}{2}] \times [-\frac{\ell_y}{2}, \frac{\ell_y}{2}] \times [-\frac{\ell_z}{2}, \frac{\ell_z}{2}]$ with $\ell_x = 0.5$ m, $\ell_y = 0.01$ m and $\ell_z = 1$ m, which contains two different defects: a spherical hole located at $(0, 0, 0)$ with radius of $0.3\ell_y$, and a box hole defined by $[\frac{\ell_x - 3\ell_y}{4}, \frac{\ell_x + 3\ell_y}{4}] \times [-\frac{\ell_y + 3\ell_y}{6}, \frac{-\ell_y + 3\ell_y}{6}] \times [\frac{\ell_z - 6\ell_y}{4}, \frac{\ell_z + 6\ell_y}{4}]$.

The TD at Γ_{front} for two different data sets is shown in Fig. 4. In the first one, four positions and six frequencies are combined, and the thermograms contain a 5% relative error. We identify the position of the rectangular box, however we can barely see the spherical defect. For the second experiment, we combine noisy thermograms with a 10% level of noise, corresponding to nine lamp positions and six frequencies. Although thermograms are more polluted, we can clearly identify the position and approximate size of the two defects. However, the method has again problems in detecting the correct depth. More sophisticated (and much more computational costly) iterative methods using the TD as a first step can be developed to try to overcome this difficulty. This will be done in future work. We have limited our study to time-harmonic excitations. The extension to the full time-dependent heat equation could also overcome this problem, and will be considered in future.

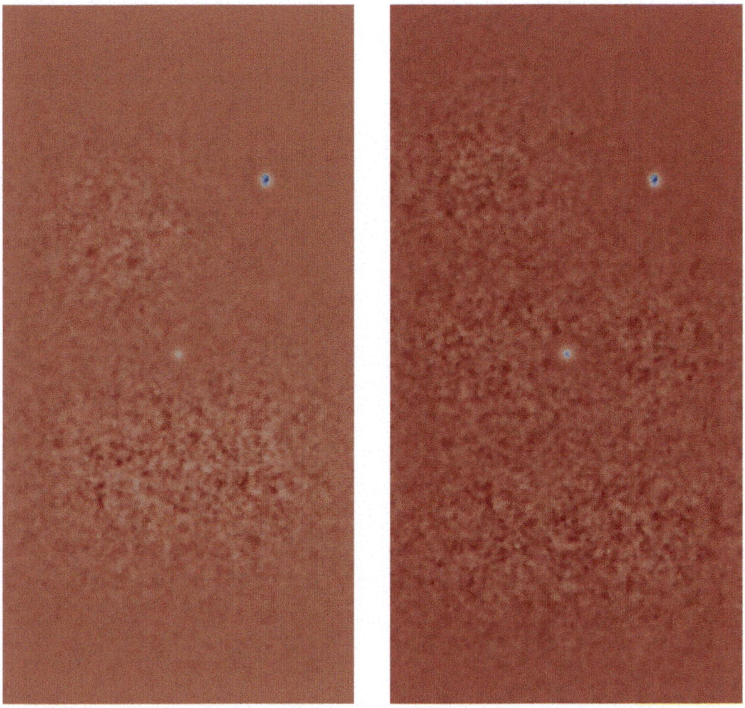

Fig. 4 Left: TD for $N_{\text{lamps}} = 4$ lamp positions and $N_{\text{freq}} = 6$ linearly spaced frequencies between 0.8 and 2 Hz. Right: TD for $N_{\text{lamps}} = 9$ lamp positions and $N_{\text{freq}} = 6$ linearly spaced frequencies between 0.8 and 2 Hz

Acknowledgements The authors are supported by the Spanish Ministry of Economy and Competitiveness under the research projects MTM2014-56948-C2-1-P and TRA2016-75075-R.

References

1. Carpio, A., Rapún, M.L.: Solving inhomogeneous inverse problems by topological derivative methods. Inv. Probl. **24**, art. 045014 (2008)
2. Carpio, A., Rapún, M.L.: Hybrid topological derivative and gradient-based methods for electrical impedance tomography. Inv. Probl. **28**, art. 095010 (2012)
3. Funes, J.F., Perales, J.M., Rapún, M.-L., Vega, J.M.: Defect detection from multifrequency limited data via topological sensitivity. J. Math. Imaging Vis. **55**, 19–35 (2016)
4. Guzina, B.B., Bonnet, M.: Small-inclusion asymptotic of misfit functionals for inverse problems in acoustics. Inv. Probl. **22**(5), 1761–1786 (2006)
5. Pena, M., Rapún, M.L.: Detecting damage in thin plates by processing infrared thermographic data with topological derivatives. Adv. Math. Phys. **4**, 1–18 (2019)
6. Sokolowski, J., Zochowski, A.: On the topological derivative in shape optimization. SIAM J. Control Optim. **38**, 1251–1272 (1999)

Author Index

© Springer Nature Switzerland AG 2019

I. Faragó et al. (eds.), *Progress in Industrial Mathematics at ECMI 2018*,
Mathematics in Industry 30, https://doi.org/10.1007/978-3-030-27550-1

The manufacturer's authorised representative in the EU is Springer Nature Customer Service Centre GmbH, Europaplatz 3, 69115 Heidelberg, Germany. If you have any concerns regarding our products, please contact ProductSafety@springernature.com

Printed and bound by CPI Group (UK) Ltd, Croydon, CR0 4YY
05/10/2026
02228123-0018